Chalcidoidea of Turkey (Insecta: Hymenoptera)

Chalcidoidea of Turkey (Insecta: Hymenoptera)

By

Elif Sakaltaş Arıyak

Cambridge
Scholars
Publishing

Chalcidoidea of Turkey (Insecta: Hymenoptera)

By Elif Sakaltaş Arıyak

This book first published 2024

Cambridge Scholars Publishing

Lady Stephenson Library, Newcastle upon Tyne, NE6 2PA, UK

British Library Cataloguing in Publication Data
A catalogue record for this book is available from the British Library

ISBN (10): 1-5275-5245-4
ISBN (13): 978-1-5275-5245-6

This book is dedicated to valuable researchers who have conducted, and will continue to conduct, studies on Chalcidoidea in Turkey.

TABLE OF CONTENTS

ACKNOWLEDGEMENTS

I would like to express my endless thanks and appreciation to Prof. Dr. Mikdat Doğanlar, the pioneer of Chalcidoidea research in Turkey. His valuable guidance, support, knowledge, skills, and experiences have been very helpful in the creation of this book. I would also like to extend my heartfelt gratitude to my dear husband, Dr. Göker Arıyak, whose presence, support and suggestions have been invaluable.

I would also like to express my gratitude to my dear family and my beloved daughter Elif Eda for their motivational and financial supports throughout my studies.

Lastly, I would like to thank Dr. Salvador Vitanza for allowing me to use the tour photo of the family Tetrastichinae on the cover of this work, and Ramazan Saygılı for allowing me to use the map of Turkey retrieved from http://cografyaharita.com/

INTRODUCTION

CHALCIDOIDEA OF TURKEY (INSECTA: HYMENOPTERA)

Turkey has been a small continent in terms of biological diversity because of the change in climatic and geographical characteristics at short intervals. It has three biogeographic regions: Europe-Siberia, Mediterranean and Iran-Turan, and along with their transition zones. Moreover, Turkey serves as a vital bridge connecting Asia and Europe. This unique position grants Turkey access to many resources of vital importance in terms of food security. Turkey also carries the responsibility of conserving and utilizing them wisely for the well-being of future generations. In this regard, Turkey can be classified as one of the most fortunate countries in the world (Republic of Turkey Ministry of Forestry and Agriculture, Biodiversity database, 2021).

Turkey is the home to many living species thanks to its geographical diversity. One of them is the Chalcidoidea superfamily, which is in the Hymenoptera order and has an important place in biological control. In this regard, there are faunistic and systematic studies about the Chalcidoidea superfamily in Turkey (Avcı, 2000; 2003; 2009; Çalışkan et al., 2016; Çalışkan et al., 2018; Çam, 2011; 2012; Çıkman and Doğanlar 2006; Çıkman, 2012; Dzhaposhvili, 2005; Doğanlar, 1984; 1985a; 1985b; 1987; 1990; 1993a; 1993b; 2003; 2006; 2007; 2011; 2011a; 2011b; 2012; 2013; 2014; 2015; 2016; 2016a; 2016b; 2016c; 2016d; 2016f; 2006g; 2017a; 2017b; 2017c; 2017d; 2018a, 2018b, 2018c, 2019; 2020; Doğanlar and Çam, 1991; Doğanlar and Mendel, 2007; Doğanlar and Triapitsyn, 2007; Doğanlar and Doğanlar, 2008a; 2008b; 2009; 2010; Doğanlar et al. 2009a; 2009b; Doğanlar and Hassan, 2010; Doğanlar et al., 2010; Doğanlar and Yiğit, 2011; Doğanlar et al., 2011a; 2011b; Doğanlar and Doğanlar, 2012; 2013; 2013a; 2013b; 2014a; 2014b; Doğanlar et al., 2013; Doğanlar and Aslan, 2018; Doğanlar and Zengin, 2018; Doğanlar et al., 2018a; Doğanlar and Yiğit, 2019; Doğanlar et al., 2020; Doğanlar and Laz, 2022; 2023; Gençer and Doğanlar 1995; Gençer, 2004; 2009a; 2009b;

Güler and Kodon 2010; İpekdal et al., 2014; Japoshvili et al., 2002; Japoshvili and Karaca, 2002; Japoshvili and Khan, 2004; Japoshvili, 2007; Japoshvili and Çelik, 2010; Japoshvili, 2012; Japoshvili and Noyes, 2005; Kaçar and Doğanlar, 2020; Karaca and Uygun, 1990; Karaca et al., 2003; Kaydan and Japoshvili, 2010; Mete and Mergen, 2017; Öncüer, 1991; Özbek et al., 1999; Özbek and Çoruh, 2010; Özgen et al., 2018; Polat and Tozlu, 2010; Sakaltaş and Gençer, 2005; Sakaltaş Arıyak and Tüzün, 2014; 2016; Stojanova et al., 2012; Ulgenturk and Toros, 2001; Yıldırım and Çalmaşur, 2002; Comrade, 1995; Zerova and Seryogina, 2003).

There are more than 26,000 identified species of Chalcidoidea worldwide, and with unidentified species, this number is estimated at more than 100,000. Current studies suggested that there may be more than 500,000 species (Noyes, 2019; Ghahari et al., 2021). Many species of Chalcidoidea have bright, metallic, and eye-catching colors. Their body length ranges between 0.3-17 mm. Their small size can make them extremely difficult to collect and investigate. Therefore, they have received comparatively little attention from taxonomists. If it is not for their importance as parasitoids of insect pests, the group would receive even less attention. Although their general morphology is in the form of the features described in Hymenoptera, they can be easily distinguished from other Hymenoptera groups with some of their distinct features. These features are usually seen in the antenna, wing and lateral region of the mesosome. Elbowed antennae and the presence of multiporius plate sensilla (longitudinal sensilla) with free apical ends in the antennae of both sexes (Gibson, 1986; Başıbüyük and Quicke, 1999) make them easily distinguished from other small Hymenoptera. Another common feature that distinguishes Chalcids from other Hymenoptera is the presence of a triangular scleritis, the prepectus, lateral to the mesosoma between the tegula and the pronotum. However, in most Chalcids the prepectus is very small, so it would be difficult to distinguish by considering the prepectus. In Chalcidoids, the wing is well developed, its veins are reduced, it is reduced to a single longitudinal vein anteriorly, and the cellular structure in the wing branches only at the end of the longitudinal vein to form the postmarginal and stigmal vein. Most of the time there is no postmarginal vein. Sometimes, the stigmal vein may also become smaller. The first abdominal segment in the metasoma is described as the propodeum. The second abdominal and the next 7 or 8 abdominal segments are called the gasters. The majority of Chalcidoids have a pair of sensory cerci in the 8th metasomal tergum.

Chalcidoids are crucial both as biological control agents of agricultural pests and as pests themselves as well. Therefore, more than 800 different species have been associated with biocontrol programs in one way or another and many Chalcidoid families have been utilized successfully in this regard. However, two families, namely Aphelinidae and Encyrtidae, have been proven extremely successful in the biological control of insect pests. Generally, the beneficial effects of Chalcidoids emerge only when the injudicious use of pesticides leads to their eradication and a consequent explosion of the pest population. Among the Aphelinidae, the genus Encarsia is one of the most important parasitoid groups exploited for biological control. Besides, many species have proven their importance in the biological control of whiteflies (Aleyrodidae) and armored-scale insects (Diaspididae). Despite these beneficial effects, more than 80 species are also known to be pests of agriculture, and most of these species belong to the families Eurytomidae and Torymidae (Noyes, 2019).

The economic importance, interesting biological features, and host relationships of Chalcidoids have made Chalcidoid biosystematics a valuable and important field of study. Most species are parasitoid, and their hosts are plant pests. Some of these are Coleoptera, Lepidoptera, Diptera, Hemiptera, Homoptera, Hymenoptera, Neuroptera, Odanata, Orthoptera, Psocoptera, Siphonaptera, Strepsiptera, Thysonoptera, as well as spiders' egg sacs, ticks, mites, and nematodes (Boucek, 1988; LaSalle, 1999; Noyes, 2019).

Prof. Dr. Mikdat Doğanlar, who conducted the most comprehensive studies on Chalcidoidea in Turkey, donated his collection consisting of over 12.000 Chalcidoidea specimens to Ankara-Plant Protection Central Research Institute Agricultural Fauna and Microflora department. This collection is an indispensable comparison resource for researchers working on Chalcidoidea systematics, plant protection and biological diversity.

The Chalcidoidea superfamily has been recently divided into 19 separate families with more than 90 subfamilies recognized. Currently, the largest family is the Eulophidae with nearly 4,500 included species and followed closely by the Encyrtidae and Pteromalidae (Noyes, 2019). In this study, the families, genera, and species belonging to the Chalcidoidea fauna of Turkey were based on the Universal Chalcidoidea Database and the studies conducted in Turkey. Some information about the tours (publications with Turkey records, hosts, associates, parasitoids, and general distributions) is also presented in this book. Accordingly, it was determined that 19 families, 51 subfamilies, 269 genus and 1024 species (see Table 1) of Chalcidoidea were distributed in Turkey. The highest number of species belonged to the families such as Eulophidae, Pteromalidae, and Encyrtidae (Noyes, 2019). In addition, the species names of Chalcidoidea found in Turkey are presented in the appendix section of the book. The aim of this study is to reveal the place of Chalcidoidea, an agriculturally important parasitoid group, in the fauna of Turkey and to help researchers with similar purposes.

Table 1: Number of families, subfamilies, genus, and species of Chalcodoidea identified and registered in Turkey according to Universal Chalcidoidea Database (Noyes, 2019)

Family	**Genus**	**Species**
Agaonidae		
Agaoninae	1	1
Blastophaginae	1	1
Total	**2**	**2**
Aphelinidae		
Aphelininae	3	21
Calesinae	1	1
Coccophaginae	5	24
Eretmocerinae	1	3
Total	**10**	**49**
Azotidae	**1**	**3**
Chalcididae		
Chalcidinae	1	10
Dirhininae	1	1
Haltichellinae	11	28
Total	**13**	**39**
Encyrtidae		
Encyrtinae	45	118
Tetracneminae	14	46
Total	**59**	**164**
Eucharitidae		
Eucharitinae	2	6
Total	**2**	**6**
Eulophidae		
Entedoninae	6	81
Entiinae	2	4
Eulophinae	22	76
Opheliminae	1	1
Tetrastichinae	17	83
Total	**48**	**245**
Eupelmidae		
Calosotinae	2	3

Eupelminae	2	14
Total	**4**	**17**
Eurytomidae		
Eurytominae	8	96
Rileyinae	1	2
Total	**9**	**98**
Leucospidae	**1**	**6**
Megastigmidae	**3**	**20**
Mymaridae	**9**	**19**
Ormyridae	**1**	**12**
Perilampidae		
Chrysolampinae	4	17
Perilampinae	1	2
Total	**5**	**19**
Pteromalidae		
Asaphinae	1	2
Ceinae	1	1
Cerocephalinae	1	1
Cleonyminae	1	1
Colotrechninae	1	9
Diparinae	1	1
Epichrysomallinae	1	1
Eunotinae	2	2
Miscogastrinae	5	21
Ormocerinae	2	2
Otitesellinae	1	1
Pireninae	2	6
Pteromalinae	46	124
Spalangiinae	1	4
Sycoryctinae	1	3
Total	**67**	**179**
Signiphoridae	**2**	**3**
Tetracampidae		
Mongolocampinae	1	1
Platynocheilinae	1	2
Tetracampinae	2	5
Total	**4**	**8**
Torymidae		
Chalcimerinae	1	1
Erimerinae	7	49
Glyphomerinae	1	2
Monodontomerinae	2	6
Podagrioninae: Podagrionini	2	3
Toryminae: Torymini	2	22
Toryminae: Torymoidini	4	28
Total	**19**	**111**

Trichogrammatidae		
Oligositinae: Chaetostrichini	**6**	**8**
Oligositinae: Oligositini	**2**	**5**
Trichogrammatinae: Trichogrammatini	2	11
Total	**10**	**24**
CHALCIDOIDEA TOTAL	**269 Genus**	**1024 species**

References

Avcı, Mustafa. 2000. "Investigations on structure of egg-batches, parasitism and egg layinh habits of *Thaumetopoea pityocampa* (Den. & Schiff.) (Lepidoptera: Thaumetopoeidae) in various regions of Turkey". *Turkish Journal of Entomology*, 24(3):167-178.

—. 2003. "Parasitism of egg-batches of the cedar processionary moth *Traumatocampa ispartaensis* in Turkey". *Phytoparasitica*, 31(2):118-123.

—. 2009. "Parasitoid complex and new host plants of the gypsy moth, *Lymantria dispar* L. in the Lakes District, Turkey". *Journal of Animal and Veterinary Advances*, 8(7):1402-1405.

Başıbüyük, Hasan Hüseyin, and Quicke, Donald L. J., 1999. "Gross morphology of multiporous plate sensilla in the Hymenoptera (Insecta)". *Zoologica Scripta*, 28 (1-2): 51-67.

Boucek, Zdenek. 1988. *Australasian Chalcidoidea (Hymenoptera) A biosystematic revision of genera of fourteen families, with a reclassification of species*. CAB International, Wallingfort, Oxon, UK.

Çalışkan Keçe, Asime Filiz, and Kahya, Doğancan, and Hayat, Mohammad, and Ulusoy, Mehmet Rifat. 2018. "A new record of a parasitoid (Hymenoptera: Encyrtidae) of an invasive meaybug *Phenacoccus solenopsis* Tinsley (Hemiptera: Pseudococcidae) from Turkey". *Türkiye Biyolojik Mücadele Dergisi*, 9(1):31-37.

Çalışkan, Asime Filiz, and Hayat, Mohammad, and Ulusoy, Mehmet Rifat, and Kaydan, Mehmet Bora. 2016. "Parasitoids (Hymenoptera: Encyrtidae) of an invasive mealybug *Phenacoccus solenopsis* Tinsley (Hemiptera: Pseudococcidae) in Turkey". *Turkish Journal of Entomology*, 40(2):133-148.

Çam, Halit. 2011. "A checklist of the Eurytomidae (Hymenoptera, Chalcidoidea) species of Turkey". *Zootaxa*, 3113:53-64.

—. 2012. "Updated checklist of the Eurytomidae (Hymenoptera, Chalcidoidea) species of Turkey", *Archives of Biological Sciences*, 64 (2), 667-674.

Çikman, Emine, and Doğanlar, Mikdat. 2006. "Parasitoids of natural populations of *Phytomyza orobanchia* (Kaltenbach, 1864) (Diptera:

Agromyzidae) in southeastern Anatolia", *Journal of Applied Science Research*, 2 (6), 327-330.

Çikman, Emine. 2012. "Parasitoids of the leafminers (Diptera: Agromyzidae) from Elazig Province, Turkey". *African Journal of Agricultural Research*, 7(12):1937-1943.

Doğanlar, Mikdat, and Aslan, M. Murat. 2018. "Two new species of Torymidae (Hymenoptera: Chalcidoidea) reared from flower heads of *Echinops orientalis* Trautvetter in south-eastern Turkey". *Zoology in the Middle East*, 64(3):238.243.

Doğanlar, Mikdat, and Çam, Halit. 1991. "The species of Eurytoma Ill. with two-three strong setae on hind tibiae from Türkiye, and description of a new species from Tokat, Türkiye (Hymenoptera, Eurytomidae)", *Turkish Journal of Entomology*, 5 (3), 143-151.

Doğanlar, Mikdat, and Doğanlar, Oğuzhan. & Boyadzhiev, S. Boyadziev. 2013. "Hypopygia of the sparetus species group of *Entedon* Dalman (Hymenoptera: Eulophidae), with descriptions of two new species". *Munis Entomology and Zoology*, 8(2):831-851.

Doğanlar, Mikdat, and Doğanlar, Oğuzhan. 2010. "Review of the species of *Gugolzia* Delucchi and Steffan (Hymenoptera: Pteromalidae) in Europe and Turkey, with descriptions of new species". Turkish Journal of Zoology, 34(1):23-34.

—. 2012. "Palearctic species of *Chrysomalla* Förster, 1859 (Hymenoptera: Perilampidae: Chrysolampinae), with descriptions of new species from Turkey". *Turkish Journal of Entomology*, 36(4):519-532.

—. 2013. "Systematics of *Urfacus* Doğanlar, 2003 (Hymenoptera, Eulophidae, Ceranisinae), with descriptions of new species in Turkey". *Journal of Biological Sciences*, 7(1):76-90.

—. 2013a. "Hypopygia of the hercyna species group of *Entedon* Dalman (Hymenoptera: Eulophidae), with descriptions of new species". *Munis Entomology and Zoology*, 8 (2):812-824.

—. 2013b. "Revision of the cyanellus species group of *Entedon* Dalman (Hymenoptera: Eulophidae), with descriptions of new species". *Munis Entomology and Zoology*, 8 (2):532-548.

—. 2014a. "A review of some species of *Epomphale* Girajult (Hymenoptera: Eulophidae: Ceranisinae) from several parts of the world, with description of new species". *Munis Entomology and Zoology*, 9(2):727-755.

—. 2014b. "Review of Chrysolampinae Dalla Torre 1898 (Hymenoptera: Perilampidae) with descriptions of new species from Turkey". *Entomofauna*, 35(4):49-84.

Doğanlar, Mikdat, and Gözüaçık, Celalettin, and Subaşı, Yüksel. 2018. "A New Species of *Gugolzia* Delucchi & Steffan (Hym., Pteromalidae) from Iğdir, Turkey, Reared from fruit of Sophora alopecuroides L. (Fabaceae)". *Munis Entomology and Zoology*. Vol. 13, No: 2; 374-377.

Doğanlar, Mikdat, and Gumovsky, Alex, and Doğanlar, Oğuzhan. 2009b. "A new species of the menes species group of the genus *Ceranisus* (Hymenoptera: Eulophidae) from Turkey". *Turkish Journal of Entomology*, 33(4):265-278.

Doğanlar, Mikdat, and Hassan, Errol. 2010. "Review of Australian species of *Megastigmus* (Hymenoptera: Torymidae) associated with Eucalyptus, with descriptions of new species". *Australian Journal of Basic and Applied Sciences*, 4(10): 5059-5120.

Doğanlar, Mikdat, and Lasalle, J., Sertkaya, E. & Doğanlar, O. 2009a. "A new species of *Quadrastichus* Girault, 1913 (Hymenoptera: Eulophidae: Tetrastichinae) from Hatay, Turkey", Turkish Journal of Zoology, 33 (3), 309-314.

Doğanlar, Mikdat, and Laz, Bülent. 2022. "Notes on The Parasitoid Community of Bethylidae and Chalcidoidea (Hymenoptera) reared from *Gundelia tournefortii* L. (Asterales: Asteraceae) in Kahramanmaraş, Turkey". *Munis Entomology and Zoology*. 17 (2): 1104-1111.

Doğanlar, Mikdat, and Laz, Bülent. 2023. "*Chalcedectus balachowskyi* Steffan, 1968 (Hymenoptera: Pteromalidae), Collected from bark of *Robinia pseudoacacia* (Umbraculifera), in Kahramanmaraş, Turkey". *Munis Entomology and Zoology*. 18 (1):167-169.

Doğanlar, Mikdat, and Mendel, Zvi. 2007. "Note: First record of the eucalyptus gall wasp *Ophelimus maskelli* and its parasitoid, *Closterocerus chamaeleon*, in Turkey", Phytoparasitica, 35 (4), 333-335.

Doğanlar, Mikdat, and Sertkaya, Erdal, and & Skuhravá, Marcela. 2011a. "Pest status of olive leaf gall midge *Dasineura oleae* (Angelini, 1831), description of *Lasioptera oleicola* Skuhravá sp. new (Diptera: Cecidomyiidae) and effectiveness of parasitoids on their populations in Hatay, Turkey". *Turkish Journal of Entomology*, 35(2):265-284.

Doğanlar, Mikdat and Triapitsyn, Serguei. 2007. "Review of Ceranisus (Hymenoptera: Eulophidae) of Turkey, with description of a new species". European Journal of Entomology, 104(1):105-110.

Doğanlar, Mikdat, Vatansever Sakin, G. and Laz, B. 2020. "Four new species of *Zeytinus* Doğanlar, 2011 (Hymenoptera: Eulophidae: Tetrastichinae) from Turkey and a redescription of the type species, *Zeytinus hatayensis* Doğanlar, 2011". *Türk. Biyo. Mücadele Derg.* 11(1):103-117.

Doğanlar, Mikdat, and Yiğit, Abdurrahman. 2011. "Parasitoids complex of the tomato leaf miner, *Tuta absoluta* (Meyrick 1917), (Lepidoptera: Gelechiidae) in Hatay Turkey", *Kahramanmaraş Sütçü İmam Üniversitesi Doğa Bilimleri Dergisi*, 14(4), sy.28-37.

Doğanlar, Mikdat, and Yiğit, Talip. 2019. "A new species of *Gastrancistrus* Westwood (Hymenoptera: Pteromalidae) reared from apricot flower midge, *Contarina pruniflorum* Coutin & Rambier (Diptera: Cecidomyiidae) in Malatya province, Turkey". *Munis Entomology and Zoology*, 14 (1): 24-27.

Doğanlar, Mikdat, and Zengin, Erdal. 2018. "A new species of *Megastigmus* (Hymenoptera: Torymidae: Megastigminae) from Usak, Turkey". *Munis Entomology and Zoology*, 13(1):288-291.

Doğanlar, Mikdat, Gumovsky, Alex, and Doğanlar, Oğuzhan. 2011b. "A review of *Ceranisus* (Hymenoptera: Eulophidae) of Ukraine, with description of two new species". *Turkish Journal of Entomology*, 35(2):215-229.

Doğanlar, Mikdat. 1984. "Notes on Chalcidoidea of Turkey. I. Chalcididae, Eurytomidae, Torymidae, Ormyridae, Perilampidae, Eucharitidae", *Türkiye Bitki Koruma Dergisi*, 8 :151-158.

—. 1985a. "Notes on Chalcidoidea of Turkey. III. Encyrtidae, Tetracampidae, Aphelinidae, Eulophidae and Elasmidae". *Türkiye Bitki Koruma Dergisi*, 9(2):91-103.

—. 1985b. "A new cotton pest of Turkey from Igdir Plain, *Pexicopia malvella* (Hb.) (Lepidoptera; Gelechiidae) and its parasites". *Türkiye Bitki Koruma Dergisi*, 9(3):165-172.

—. 1987. "Erzurum ve çevresindeki elma ve armut ağaçlarında bulunan yaprak bükenler ve benzer şekilde beslenen diğer Lepidopter'ler ile bunların parazitleri üzerinde araştırmalar", *Doğa Tübitak Tarım ve Orman Dergisi*, 11 (1), 86-93.

—. 1990. "A new species of *Hockeria* Walker (Hymenoptera, Chalcididae), a pupal parasite of Heliothis viriplaca Hufn. (Lepidoptera, Noctuidae) on lentil (Lens esculenta) and chickpea (Cicer sp.) in southestern Anatolia". *Turkish Journal of Entomology*, 14(3):149-153.

—. 1993a. "Notes on *Baryscapus* Förster, 1856, with description of four new species (Hymenoptera, Eulophidae, Tetrastichinae). Entomofauna", *Zeitschrift für Entomologie*, 14(23):381-389.

—. 1993b. "Notes on the species of *Neotrichoporoides* Girault, 1913, with descriptions of some new species from Ghana and Turkey (Hymenoptera, Eulophidae, Tetrastichinae)". *Entomofauna, Zeitschrift für Entomologie*, 14(23):369-380.

—. 2003. "A new genus and a new species of Entedontinae (Hymenoptera, Eulophidae) from southeastern Anatolia, Turkey". *Turkish Journal of Zoology*, 27(3):181-185.

—. 2006. "Systematic studies on some species of *Halticoptera* Spinola, 1811 (Hymenoptera: Chalcidoidea, Pteromalidae), with descriptions of new species". *Journal of Applied Sciences Research*, 2(3):168-183.

—. 2007. "A new species of *Muscidifurax* Girault & Sanders, 1910 (Hymenoptera: Pteromalidae) from Adana province, Turkey". *Turkish Journal of Entomology*, 31(4):243-252.

—. 2011a. "Review of Palearctic and Australian species of *Bootanomyia* Girault 1915 (Hymenoptera: Torymidae: Megastigminae), with descriptions of new species", *Turkish Journal of Zoology*, 35(2): 123-157.

—. 2011b. "Parastioid complex of the olive leaf gall midges, *Dasineura oleae* (Angelini 1831) and *Lasioptera oleicola* Skuhravá, 2011 (Diptera: Cecidomyiidae) in Hatay Turkey, with descriptions of new genus and species from Tetrastichinae (Hymenoptera: Eulophidae)". *Turkish Journal of Entomology*, 35(2):245-264.

—. 2012. "Occurrence of fig wasps (Hymenoptera: Chalcidoidea) in *Ficus carica* and *F. microcarpa* in Hatay, Turkey", *Turkish Journal of Zoology*, 36(5):721-724.

—. 2013. "A peculiar new species of *Entedon* Dalman, 1820 (Hymenoptera: Eulophidae) from Sanliurfa Province, Turkey". *Munis Entomology and Zoology*, 8(2):654-658.

—. 2014. "First reports on the natural enemy fauna of the chestnut gall wasp, *Dryocosmus kuriphilus* Yasumatsu (Hymenoptera: Cynipidae) in Yalova, Turkey". *Türkiye Biyolojik Mucadele Dergisi*, 5(1):67-74.

—. 2015. "Diagnosis of *Megastigmus spp*. (Hymenoptera: Torymidae) reared from galls of *Leptocybe invasa* Fisher & Lasalle, 2004 (Hymenoptera: Eulophidae) on *Eucalyptus spp*. (Myrtaceae), with description of a new species from South Africa". *Entomofauna*, 36(43):561-580.

—. 2016. "*Ditropinotus* Crawford, 1907 (Hymenoptera: Torymidae: Microdontomerini) species from Turkey, with descriptions of new species". *Munis Entomology and Zoology*, 11(2):269-277.

—. 2016a. "Species of *Idiomacromerus* Crawford in Turkey (Hymenoptera: Torymidae: Microdontomerini)". *Munis Entomology and Zoology*, 11(2):312-354.

—. 2016b. "Species of *Pseudotorymus* Masi in Turkey (Hymenoptera: Torymidae: Microdontomerini)". *Munis Entomology and Zoology*, 11(2):393- 432.

—. 2016c. "Turkish species of *Microdontomerus* Crawford, 1907 with descriptions of new species (Hymenoptera: Toryidae: Microdontomerini)", *Munis Entomology and Zoology*, 11(2):363-381.

—. 2016d. "Species of *Microdontomerus* (Crawford, 1907) and *Eridontomerus* (Crawford, 1907) (Hymenoptera: Torymidae: Microdontomerini) from Turkey, with descriptions of a new species". *Entomofauna*, 37(33):505-520.

—. 2016e. "The morphology of hypopygia and its importance in taxonomy of the genera of Torymidae (Hymenoptera), with review of the genera and species of Turkey". *Entomofauna Zeitschrift Für Entomologie*, Band 37, Heft 34: 521-544.

—. 2016f. "*Pseuderimerus* Gahan, 1919 (Hymenoptera: Torymidae: Microdontomerini) species from Turkey, with Descriptions of New Species". *Munis Entomology and Zoology*. Vol. 11, No. 1.

—. 2016g. "The morphology of hypopygia and its importance in taxonomy of the genera of Torymidae (Hymenoptera), with review of the genera and species of Turkey". *Entomofauna Zeitschrift Für Entomologie*, Band 37, Heft 34: 521-544.

—. 2017a. "Turkish species of *Torymus* Dalman, 1820 (Hymenoptera: Toryidae [sic]: Torymini), with descriptions of new species". *Munis Entomology and Zoology*, 12(1):5-22.

—. 2017b. "Species of *Spalangia* Latreille, 1805 (Hymenoptera: Pteromalidae, Spalangiinae) from Turkey, with descriptions of new species". *Entomofauna Zeitschrift Für Entomologie*, Band 38, Heft 21: 477-492.

—. 2017c. "Turkish Species of *Torymus* Dalman, 1820 (Hymenoptera: Toryidae: Torymini), with Descriptions of New Species". *Munis Entomology and Zoology.* Vol. 12, No:1.

—. 2017d. "Studies on the Remnants of the Parasitoid Larvae of *Dryocosmus kuriphilus* Yasumatsu (Hymenoptera: Cynipidae), in the galls collected in two places in Italy for determining parasitism level and type of parasitoids". *Munis Entomology and Zoology*. Vol. 12, No. 2: 399-407.

—. 2018a. "A New Species of *Lampoterma* (Hymenoptera: Pteromalidae) from Erzurum, Turkey". *Munis Entomology and Zoology*. Vol. 13, No. 2. 386-394.

—. 2018b. "World species of *Arthrolytus* (Hymenoptera: Pteromalidae, Pteromalinae), with description of new species". *Linzer biol. Beitr.* 50 (1): 255-289.

—. 2018c. "Species of *Macroglenes* Westwood, 1832 (Hymenoptera: Pteromalidae, Spalangiinae) from Turkey, with description of a new species". *Entomofauna*. 39/1, Heft 14: 317-324.

—. 2019. "Species of *Gugolzia* Delucchi & Steffan (Hymenoptera: Pteromalidae) from Turkey, with descriptions of two new species". *Munis Entomology and Zoology*. Vol. 14, No. 1:1-7.

—. 2020. "The species of the Compressus-Group, Graham, 1969 of *Gastrancistrus* Westwood, 1833 (Hymenoptera: Pteromalidae) from Turkey with Description of Three New Species". *Munis Entomology and Zoology*. Vol. 15, No. 2:336-347.

Doğanlar, Oğuzhan, and Doğanlar, Mikdat, and Frary, Anne. 2010. "Preliminary phylogeny of thrips parasitoids of Turkey based on some morphological scales and 28S D2 rDNA, with description of a new species". *Turkish Journal of Zoology*, 34(3):279-289.

Doğanlar, Oğuzhan, and Doğanlar, Mikdat. 2008a. "A new species of *Idarnotorymus* Masi, 1916 (Hymenoptera: Torymidae) from Turkey". *Turkish Journal of Zoology*, 32:85-90.

—. 2008b. "First record of the eucalyptus seed gall wasp, *Quadrastichodella nova* Girault, 1922, (Eulophidae: Tetrastichinae) from Turkey". *Turkish Journal of Zoology*, 32(4):457-459.

—. 2009. "*Podagrionella* Girault, 1913 (Hymenoptera: Torymidae: Podagrionidae) species from Turkey with descriptions of new species". *Turkish Journal of Zoology*, 33(2):181-186.

Dzhaposhvili, G.O. 2005. "A new species of encyrtids (Hymenoptera, Chalcidoidea, Encyrtidae) from Turkey", *Zoologicheskiĭ Zhurnal*, 84(4):524-527.

Gençer, Lütfiye, and Doğanlar, Mikdat. 1995. "Sivas-Merkez'de Tetrastichinae (Hymenoptera: Eulophidae) Üzerine Faunistik Bir Araştırma", *C.Ü. Fen- Edebiyat Fak. Fen Bilimleri Dergisi*, 18: 32-42.

Gençer, Lütfiye. 2004. "Contribution to the knowledge of the family Eulophidae (Hymenoptera: Chalcidoidea) from Turkey, with some new records", *Phytoparasitica*, 32, pages 264–273.

—. 2009a. "A new species of Anaprostocetus Graham 1987 (Hymenoptera: Eulophidae) from Turkey". *Entomologica Fennica*, 20 (4) Helsinki: Entomological Society of Finland, 225-227.

—. 2009b. "Contribution to the Knowledge of the Chalcid Parasitoid Complex (Hymenoptera: Chalcidoidea) of Agromyzid Leafminers (Diptera: Agromyzidae) from Turkey, with New Hosts and Records". *Journal of Plant Protection Research*. No. 49(2). 158-161.

Ghahari, Hassan, and Gibson, Gary A.P., and Viggiani, Gennaro. 2021. "Chalcidoidea of Iran (Insecta: Hymenoptera)", https://www.cabi.org/bookshop/book/9781789248463/ (accessed 23.05.2021).

Gibson, Gary A. P., 1986. "Evidence for monophyly and relationships of Chalcidoidea, Mymaridae and Mymarommatidae (Hym: Terebrantes)". *Canadian Entomology*, 118: 205-240.

Güler, Yasemin, and Kodan, Münevver. 2010. "A new Record of Ectoparasitoid *Melittobia acasta* (Walker, 1839) (Hymenoptera: Eulophidae) from Turkey", *Mellıfera*, HARUM, 10(1): 34.

İpekdal, Kahraman, and Senan Coşkuncu, Kıymet, and Aytar, Fatih, and Doğanlar, Mikdat. 2014. "Chestnut Gall Wasp *Dryocosmus kuriphilus* Yasumatsu (Hymenoptera: Cynipidae): An update for its situation on the world and in Turkey and its management". *Türkiye Entomoloji Bülteni*, 4 (4): 241-257.

Japoshvili, George. 2012. "New data on encyrtid (Hymenoptera: Chalcidoidea: Encyrtidae) parasitoids of Coccoids (Hemiptera: Coccoidea) from Turkey, with description of five new species". *Entomologica Fennica*, 23 (2): 72-82.

Kaçar, Gülay, and Doğanlar, Mikdat. 2020. "Two new species of *Conomorium* Masi, 1924 (Hymenoptera: Pteromalidae) and additions of some new records of the Turkish species". *Munis Entomology and Zoology*, 15 (2): 711-719.

LaSalle, J., 1999. *Ekonomik Önemi Olan Hymenopterler*. Biyoloji ve Taksonomi Kursu Notları. C. Ü. Fen-Edebiyat Fak. Biyoloji Bölümü, Sivas.

Mete, Özlem, and Mergen, Orhan. 2017. "The community components associated with two common rose gall wasps (Hymenoptera: Cynipidae: Diplolepidini) in Turkey". *Turkish Journal of Zoology*, 41:696-701.

Noyes, John S. 2019. *Universal Chalcidoidea Database*. Available at: https://www.nhm.ac.uk/our-science/data/chalcidoids/ (accessed January 2022).

Öncüer, Cezmi. 1991. *A catalogue of the parasites and predators of insect pests of Turkey*. pp.163.

Özbek, Hikmet, and Çoruh, Saliha. 2010. "Egg parasitoids of *Malacosoma neustria* (Linnaeus, 1758) (Lepidoptera: Lasiocampidae) in Erzurum province of Turkey". *Turkish Journal of Entomology*, 34(4):551-560.

Özbek, Hikmet, and Güçlü, Şaban, and Tozlu, Göksel. 1999. "Biology and natural enemies of *Diplolepis mayri* Schld. (Hymenoptera: Cynipidae), a pest of *Rosa canina* L. in Erzurum Province". *Turkish Journal of Entomology*, 23(1):39-50.

Özgen, İnanç, and Doğanlar, Mikdat, and Özdemir, Mustafa, and Kaplan, Mehmet. 2018. "The additional record in Turkey: A larva-pupa parasitoid of orchard ermine moth *Yponomeuta padellus* Linnaeus, 1758 (Lepidoptera: Yponomeutidae): *Baryscapus evonymellae* Bouche, 1834 (Hymenoptera: Eulophidae)". *International Journal of Fauna and Biological Studies*, 5(1): 145-147.

Polat, Alper, and Tozlu, Göksel. 2010. "Investigations on the brief biology, host plants and parasitoids of *Archips rosana* (Linnaeus, 1758) (Lepidoptera: Tortricidae) in Erzurum". *Turkish Journal of Entomology*, 34(4):529-542.

Protasov, Alex, and Doğanlar, Mikdat, La Salle, John, and Mendel, Zvi. 2008. "Occurrence of two local Megastigmus species parasitic on the Eucalyptus gallwasp *Leptocybe invasa* in Israel and Turkey". *Phytoparasitica*, 36(5):449-459.

Sakaltaş Arıyak, Elif, and Tüzün, Ayla. 2014. "Contribution to the knowledge of the *Aprostocetus* Westwood, 1833 (Hymenoptera: Tetrastichinae) from Kırıkkale, Çankırı, Turkey, with some new records", *Turkish Journal of Entomology*, 38 (4): 415-426.

—. 2016. "Contribution to the knowledge of the Tetrastichinae (Hymenoptera: Eulophidae) from Kırıkkale, Çankırı, Turkey, with some new records", *Turkish Journal of Zoology*. 40 (5): 806-813.

Sakaltaş, Elif, and Gençer, Lütfiye. 2005. "Contribution to the knowledge of the Tetrastichinae (Hymenoptera: Eulophidae) from Ankara, Turkey, with Some New Records". *Acta Phytopathologica et Entomologica Hungarica*, 40 (3-4): 383-390.

Stojanova, Anelia, and Civelek, Hasan Sungur, and Yörük, Berna, and Sari, Sabahat, and Atahan, Tuğçe. 2012. "Checklists of Turkish Eurytomidae Walker, 1832 and Torymidae Walker, 1833 (Hymenoptera: Chalcidoidea)". *Turkish Journal of Entomology*, 36(1):69-82.

T.C. Orman ve Tarım Bakanlığı, Biyoçeşitlilik Veri Tabanı, http://www.nuhungemisi.gov.tr/Library/TurkiyeBiyocesitlilik (accessed 27.05.2021).

Yefremova, Zoya A. 2002. "Catalogue of the Eulophidae (Hymenoptera: Chalcidoidea) of Russia". *Linzer Biologische Beiträge*, 34 (1) 563-618.

Yefremova, Zoya A., and Civelek, Hasan Sungur, and Boyadzhiev, Peter, and Dursun, Oktay, and Eskin, Ata. 2010. "Contributions to the Turkish Eulophidae (Hymenoptera, Chalcidoidea) with new records". *Turkish Journal of Entomology*, 34 (4): 447-463.

Yıldırım, Erol, and Çalmaşur, Önder, and Madl, Michael. 2002. "Leucospidae of Turkey (Hymenoptera, Chalcidoidea)", *Linzer biol. Beitr*. 34/ 2, 1185-1189.

Yoldaş, Zeynep. 1995. "Investigation on the effectiveness of *Encarsia formosa* (Gahan) (Hymenoptera, Aphelinidae) in the biological control of *Bemisia tabaci* (Genn.) (Homoptera, Aleyrodidae) in cucumbers Türkiye". *Entomoloji Dergisi*, 19(2): 95-100.

Zerova, Marina, and Seryogina, Liudmila, and Çam, Halit. 2003. "New species of Eurytomidae (Hymenoptera: Chalcidoidea) from Turkey". *Zool. Med. Leiden*, 77 (14), 259-266.

CHAPTER ONE

FAMILY: AGAONIDAE WALKER, 1846

The Agaonidae is a medium-sized family found in both tropical and subtropical regions. They are generally only collected on figs. The family is characterized by the head lacking a horseshoe-shaped occipital carina (which is typical of Torymidae), and their bodies are generally smooth or have only very faint sculpturing. They also have at least one of the following characteristics: mid-femur being distinctly thinner than either fore or hind legs, mandibles in females are modified into rasp-like structures, and female metasoma is turned into terminal 'tail' composed of either ovipositor alone or ovipositor and terminal tergites. In addition, the stigmal vein of the forewing is often thin and at a right angle to the wing margin. Species in this family are often strongly sexually dimorphic with males being wingless and having reduced appendages, making them hardly recognizable as wasps.

All agaonids are associated with figs (*Ficus*), with some acting as pollinators (Agaoninae) and others as phytophagous or possibly parasitoids. All species are associated with the fig inflorescences (syconia). The relationship between figs and their Agaonine pollinators is an obligate mutualism because *Ficus* species can only be pollinated by the right species of fig wasps, and fig wasps are unable to produce progeny outside a narrow spectrum of fig species. The biology of non-pollinator species of Agaonidae is poorly known; however, it is believed that most of these species, which have generally been considered as parasitoids, are phytophagous. The larvae feed on the fig ovaries or may be seed-eaters or gall-formers within the fig syconia (Stringer et al., 2012).

While there are currently 76 genera and 757 species in six subfamilies in the world, there are two species belonging to two subfamilies in Turkey (Noyes, 2019).

Subfamily Agaoninae (Walker, 1846)
Genus *Eupristina* (Saunders, 1882)
Eupristina verticillata (Waterston, 1921)
Publications with Turkey records: Doğanlar (2012), Noyes (2019).
Associates: *Odontofroggatia sp., Walkerella kurandensis* (Pteromalidae), *Ficus macrocarpa, Ficus microcarpa, Ficus retusa, Ficus aurea* (Moraceae) (Noyes, 2019).
Hosts: *Ficus aurea, Ficus microcarpa* (Moraceae) (Noyes, 2019).
Parasitoids: *Odontofroggatia galili, Philotrypesis okinavensis, Philotrypesis taiwanensis, Sycoryctes moneres, Sycoscapter gajimaru* (Hymenoptera; Agaonidae) (Noyes, 2019).
Distribution: Afrotropical, Australia, Bermuda, Brazil (Sao Paulo), Canary Islands, El Salvador, Hawaii, Honduras, India, Indonesia, Italy, Japan, Madeira, Malaysia, Mexico, People's Republic of China, Hong Kong, Philippines, Puerto Rico, Solomon Islands, Taiwan, Tunisia, Turkiye, United States of America (Noyes, 2019).

Subfamily Blastophaginae (Kirchner, 1867)
Genus *Blastophaga* (Gravenhorst, 1829)
Blastophaga psenes (Linnaeus, 1758)
Publications with Turkey records: Doğanlar (2012), Noyes (2019).
Associates: *Philotrypesis caricae* (Hymenoptera: Agaonidae), *Fusarium sp., Fusarium moniliforme, Fusarium verticillioides* (Fungus), *Crematogaster scutllaris* (Formicidae), *Schistonchus caprifici* (Nematoda: Aphelenchoididae) (Noyes, 2019).
Hosts: *Ficus carica, Ficus palmata* (Moraceae), *Philotrypesis caricae* (Hymenoptera: Agaonidae) (Noyes, 2019).
Parasitoids: *Philotrypesis caricae* (Hymenoptera: Agaonidae) (Noyes, 2019).
Distribution: Afghanistan, Algeria, Armenia, Australia (New South Wales), Canary Islands, Caucasus, Croatia, Eritrea, Ethiopia, Europe, France, India, Iran, Israel, Italy, Nearctic, Nepal, North Africa, Pakistan, South Africa, Turkey, Ukraine, United States of America (Arizona, California, Texas), USSR, Central Asia, Yemen (Noyes, 2019).

References

Doğanlar, Mikdat. (2012) "Occurrence of fig wasps (Hymenoptera: Chalcidoidea) in *Ficus carica* and *F. microcarcapa* in Hatay, Turkey". *Turkish Journal of Zoology*, 36(5):721-724.

Noyes, John S. 2019. *Universal Chalcidoidea Database*. Available at: https://www.nhm.ac.uk/our-science/data/chalcidoids/ (accessed January 2022).

Stringer, Danielle N., and Mantel, Sarah, and Jennings, John T., and Austin, Andrew D. 2012. *Family AGAONIDAE*, Australian Centre for Evolutionary Biology and Biodiversity, and the School of Earth and Environmental Science. The University of Adelaide, Australia. (Available from: https://www.researchgate.net/publication/264784089_Family_AGAONIDAE)

CHAPTER TWO

FAMILY: APHELINIDAE THOMSON, 1876

The Aphelinidae are a medium-sized family of Chalcids that are often overlooked due to their extremely small size (0.5–1.5 mm). They are characterised by the antenna having four or less funicular segments, the tarsi having 4–5 segments, the fore tibial spur being large, curved and bifid, the metasoma being broadly joined to the thorax (not the petiolate), and the notauli being complete, straight, deep, and widely separated posteriorly (generally by more than the length of the scutellum).

Aphelinids are commonly reared from a large variety of homopteran hosts. Most are primary ecto- or endoparasitoids of Hemiptera: Sternorrhyncha (especially Coccoidea, Aleyrodoidea, and Aphidoidea). However, others attack the eggs of these groups or of Orthoptera (grasshoppers and crickets), and still others are obligatory or facultative hyperparasitoids. Several of these species are of particular interest because males and females of the same species may develop in very different ways. In well-studied species, the females are always primary endoparasitoids. However, the males may develop as primary ectoparasitoids on the same host, as hyperparasitoids on their own species, as facultative hyperparasitoids on their own or other species, or as obligate hyperparasitoids on other species. There are numerous examples of aphelinids being used successfully in classical biological control, and this family boasts more such successes than any other group of entomophagous insects. The majority of aphelinids belong to the subfamilies Aphelininae (antenna with six or less segments and linea calva present) and Coccophaginae (antenna with 7–9 segments and linea calva absent) (Stringer et al., 2012).

The family currently includes 33 genera and 1168 species classified in seven subfamilies. There are 49 species in four subfamilies in Turkey (Noyes, 2019).

Subfamily Aphelininae (Thomson, 1876)
Genus *Aphelinus* (Dalman, 1820)
Aphelinus asychis (Walker, 1839)
Publications with Turkey records: Noyes (2019), Öncüer (1991).
Associates: *Paecilomyces fumosoroseus* (Fungus), *Aphidius ervi* (Braconidae), *Aphidencyrtus annulipes, Aphidencyrtus aphidivorus* (Encyrtidae), *Asaphes suspensus, Pachyneuron aphidis* (Pteromalidae), *Helianthus sp., Helianthus annuus, Sonchus sp., Sonchus oleraceus* (Asteraceae), *Heliotropium europaeum* (Boraginaceae), *Brassica napus, Brassica oleracea, Raphanus sativus* (Brassicaceae), *Cucumis sativus* (Cucurbitaceae), *Medicago sativa, Phaseolus radiatus, Trifolium pratense, Vicia faba, Wisteria sp.* (Fabaceae), *Paecilomyces fumosoroseus* (Fungus), *Ocimum sp.* (Lamiaceae), *Gossypium hirsutum* (Malvaceae), *Picea pungens* (Pinaceae), *Hordeum vulgare, Saccharum officinarum, Sorghum sp., Triticum aestivum* (Poaceae), *Emex australia, Rumex obtusifolius* (Polygonaceae), *Digitalis sp.* (Scrophulariaceae), *Capsicum annuum, Lycopersicon esculentum, Solanum tuberosum* (Solanaceae) (Noyes, 2019).
Hosts: *Phytomyza atricornis* (Agromyzidae), *Acyrthosiphon sp., Acyrthosiphon kondoi, Acyrthosiphon pisum,* Aphis sp., *Aphis chloris, Aphis citricola, Aphis craccivora, Aphis gossypii, Aphis helianthi, Aphis nerii, Aulacorthum solani, Brachycaudus persicae, Brachycolus korotnewi, Brachycolus noxius, Brevicoryne sp., Brevicoryne brassicae, Chaetosiphon fragaefolii, Dactynotus helianthicola, Diuraphis sp., Diuraphis noxia, Diuraphis tritici, Elatobium abietinum, Holcaphis tritici, Hyperomyzus lactucae, Lipaphis sp., Lipaphis erysimi, Macrosiphoniella sp., Macrosiphum sp., Macrosiphum avenae, Macrosiphum convolvuli, Macrosiphum euphorbiae, Megoura viciae, Metopolophium dirhodum, Metopolophium festucae, Myzaphis sp., Myzaphis rosarum, Myzodes persicae, Myzus sp., Myzus convolvuli, Myzus ornatus, Myzus persicae, Nasonovia ribisnigri, Neomyzus sp., Rhopalosiphum maidis, Rhopalosiphum padi, Rungsia maydis, Schizaphis sp., Schizaphis graminum, Sipha flava, Sipha glyceriae, Sitobion avenae, Therioaphis maculata, Therioaphis trifolii, Toxoptera citricidus* (Hemiptera: Aphididae) (Noyes, 2019).
Distribution: Afghanistan, Angola, Argentina, Australia, Azerbaijan, Belgium, Brazil, Canada, Canary Islands, Chile, Colombia, Croatia, Czech Republic, Egypt, Europe, Finland, France, Georgia, Germany, Greece, Hawaii, Hungary, India, Iran, Iraq, Israel, Italy, Japan, Kazakhstan, Mexico, Morocco, Nearctic, Nepal, Netherlands, Norway, Pakistan, Palaearctic, People's Republic of China, Portugal, Russia, Slovakia, South Africa, Spain, Sweden, Transcaucasus, Tselinograd Obl., Turkiye, Ukraine, United Kingdom, United States of America, USSR, Central Asia (Noyes, 2019).

Aphelinus chaonia (Walker, 1839)
Publications with Turkey records: Noyes (2019), Öncüer (1991).
Associates: *Yucca sp.* (Agavaceae), *Cirsium arvense, Taraxacum dens-leonis* (Asteraceae), *Sambucus nigra* (Caprifoliaceae), *Gypsophila struthium* (Caryophyllaceae), *Euonymus europaea* (Celastraceae), *Beta vulgaris* (Chenopodiaceae), *Retama sphaerocarpa, Vicia faba* (Fabaceae), *Philadelphus coronarius* (Hydrangeaceae), *Rosa sp.* (Rosaceae), *Citrus sp.* (Rutaceae), *Tamarix canariensis* (Tamaricaceae) (Noyes, 2019).
Hosts: *Aphis sp., Aphis craccivora, Aphis euonymi, Aphis fabae, Aphis gossypii, Aphis sambuci. Aphis spiraecola, Aphis taraxacicola, Brachycaudus helichrysi, Doralis sp., Dysaphis sp., Phorodon sp., Phorodon crataegaria, Rhopalosiphum padi, Semiaphis atriplicis, Toxoptera sp., Toxoptera aurantii* (Hemiptera: Aphididae) (Noyes, 2019).
Distribution: Austria, Azerbaijan, Belgium, Brazil, Canary Islands, Chile, Croatia, Czech Republic, Czechoslovakia, Egypt, Europe, France, Georgia, Germany, Hungary, Lithuania, Madeira, Montenegro, Nepal, Netherlands, Norway, Pakistan, Palaearctic, People's Republic of China, Poland, Portugal, Russia, Serbia, Slovakia, Spain, Sweden, Turkiye, Ukraine, England, United States of America, USSR (Noyes, 2019).

Aphelinus mali (Haldeman, 1851)
Publications with Turkey records: Noyes (2019), Öncüer (1991).
Associates: *Quadraspidiotus perniciosus* ((Hemiptera: Diaspididae), *Trioxys communis* (Hymenoptera: Braconidae), *Cirsium heterophyllum* (Asteraceae), *Gossypium barbadense* (Malvaceae), *Panicum sp.* (Poaceae), *Malus sp., Malus communis, Malus pumila, Pyrus communis, Pyrus malus, Sorbus aucuparia* (Rosaceae) (Noyes, 2019).
Hosts: *Anuraphis schwartzi, Aphis sp., Aphis gossypii, Aphis monardae, Aphis pomi, Aphis spiraecola, Aphis tavaresi, Aphis travaresi, Brevicoryne brassicae, Ceratovacuna lanigera, Colopha eragrostidis, Colopha graminis, Dactynotus floridae, Eriosoma sp., Eriosoma americanum, Eriosoma crataegi, Eriosoma lanigerum, Eriosoma pyricola, Macrosiphum sp., Macrosiphum floridae, Macrosiphum rosae, Mordwilkoia vagabunda, Myzus lythri, Myzus persicae, Prociphilus fraxinifolii, Schizaphis graminum, Schizoneura americanum, Tetraneura graminis, Toxoptera aurantii* (Hemiptera: Aphididae), *Aspidiotus destructor, Aspidiotus hederae, Lepidosaphes ulmi* (Hemiptera: Diaspididae), *Pseudococcus sp.* (Hemiptera: Pseudococcidae), *Coleophora malivorella* (Lepidoptera: Coleophoridae), *Crocidolomia binotalis* (Lepidoptera: Pyralidae) (Noyes, 2019).

Distribution: Argentina, Australia, Austria, Azerbaijan, Belgium, Bolivia, Brazil, Bulgaria, Canada, Canary Islands, Caucasus, Chile, Colombia, Costa Rica, Croatia, Cyprus, Czech Republic, Czechoslovakia, Ecuador, Egypt, Europe, France, Georgia, Germany, India, Indonesia, Iraq, Israel, Italy, Japan, Korea, Lebanon, Malta, Mexico, Moldova, Morocco, Nearctic, Neotropical, Netherlands, New Zealand, North Africa, Norway, Pakistan, Paraguay, People's Republic of China, Peru, Philippines, Poland, Portugal, Puerto Rico, Romania, Russia, Saudi Arabia, Senegal, Slovakia, South Africa, Spain, Sweden, Switzerland, Tadzhikistan, Trinidad & Tobago, Turkey, Ukraine, United Kingdom, England, United States of America, Uruguay, USSR, Uzbekistan, Venezuela, Yugoslavia (pre-1991), Zambia, Zimbabwe (Noyes, 2019).

Aphelinus semiflavus (Howard, 1908)
Publications with Turkey records: Noyes (2019).
Associates: *Brassica chinensis, Brassica oleracea* (Brassicaceae), *Medicago sativa, Melilotus alba* (Fabaceae), *Aquilegia sp.* (Ranunculaceae), *Solanum pseudocapsicum, Solanum tuberosum* (Solanaceae) (Noyes, 2019).
Hosts: *Acyrthosiphon cornelli, Acyrthosiphon kondoi, Acyrthosiphon pisum, Anuraphis viburnicola, Aphididae unspecified sp., Aphis gossypii, Aphis maidis, Aphis rumicis, Brevicoryne brassicae, Capitophorus rosarum, Chaetosiphon fragaefolii, Chaetosiphon thomasi, Chaitophorus viminalis, Hydaphis erysimi, Lipaphis erysimi, Lipaphis pseudobrassicae, Macrosiphoniella sanborni, Macrosiphum avenae, Macrosiphum cornelli, Macrosiphum euphorbiae, Macrosiphum granarium, Macrosiphum pisi, Myzaphis rosarum, Myzodes persicae, Myzus sp., Myzus circumflexus, Myzus persicae, Neocerusaphis viburnicola, Neomyzus circumflexus, Pterocallidium maculatum, Rhopalomyzus ascalonicus, Rhopalomyzus poae, Rhopalosiphoninus staphyleae, Rhopalosiphum maidis, Rhopalosiphum pseudobrassicae, Schizaphis graminum, Therioaphis maculata, Therioaphis riehmi, Therioaphis trifolii, Toxoptera aurantii* (Hemiptera: Aphididae), *Heliothrips haemorrhoidalis* (Thysanoptera: Thripidae) (Noyes, 2019).
Parasitoids: *Aphidencyrtus aphidivorus* (Hymenoptera: Encyrtidae), *Asaphes vulgaris* (Hymenoptera: Pteromalidae) (Noyes, 2019).
Distribution: Argentina, Brazil, Canada, France, Germany, Hawaii, India, Iran, Israel, Italy, Korea, Mexico, Nearctic, Spain, Turkey, United States of America, USSR (Noyes, 2019).

Aphelinus varipes (Foerster, 1841)
Publications with Turkey records: Elmali and Toros (1994), Noyes (2019), Öncüer (1991).
Associates: *Lipara sp.* (Diptera: Chloropidae), *Colocasia esculentum* (Araceae), *Artemisia sp., Chrysanthemum coronarium, Helianthus sp.* (Asteraceae), *Brassica campestris, Brassica oleracea, Descurainia sophia, Raphanus sativus* (Brassicaceae), *Cucumis sativus, Cucurbita sp.* (Cucurbitaceae), *Glycine max* (Fabaceae), *Arundo donax, Phragmites australis, Sorghum sp., Sorghum bicolor, Triticum aestivum, Zea mays* (Poaceae), *Solanum melongena, Solanum tuberosum* (Solanaceae), *Cayratia japonica* (Vitaceae) (Noyes, 2019).
Hosts: Aleyrodidae (Hemiptera), *Acyrthosiphon pisum, Aphis sp., Aphis fabae, Aphis gossypii, Aphis helianthi, Brachycolus noxius, Diuraphis muehlei, Diuraphis noxia, Hysteroneura setariae, Melanaphis donacis, Metopolophium dirhodum, Myzus persicae, Rhopalosiphum insertum, Rhopalosiphum maidis, Rhopalosiphum padi, Schizaphis graminum, Sipha sp., Sitobion avenae, Toxoptera graminum* (Hemiptera: Aphididae), *Pseudococcus sp.* (Hemiptera: Pseudococcidae), *Asaphes vulgaris* (Hymenoptera: Pteromalidae) (Noyes, 2019).
Parasitoids: *Marietta graminicola, Marietta picta* (Hymenoptera: Aphelinidae), *Aphidencyrtus aphidivorus* (Hymenoptera: Encyrtidae), *Pachyneuron aphidis* (Hymenoptera: Pteromalidae) (Noyes, 2019).

Distribution: Australia, Azores, Belgium, Canary Islands, Chile, Croatia, Czech Republic, Egypt, Europe, France Georgia, Germany, Hungary, Israel, Italy, Japan, Kazakhstan, Madeira, Mexico, Morocco, Nearctic, Nepal, Netherlands, Norway, Pakistan, Palaearctic, Paraguay, Portugal, Russia, Serbia, Slovakia, South Africa, Spain, Sweden, Transcaucasus, Tselinograd Obl., Turkey, Ukraine, United Kingdom, England, United States of America, USSR (Noyes, 2019).

Genus *Aphytis* (Howard, 1900)
Aphytis abnormis (Howard, 1881)
Publications with Turkey records: Noyes (2019), Öncüer (1991).
Associates: *Salix sp., Salix caprea* (Salicaceae), *Vitis vinifera* (Vitaceae) (Noyes, 2019).
Hosts: *Coccus hesperidum* (Hemiptera: Coccidae), *Chionaspis salicis, Dynaspidiotus abietis, Epidiaspis leperii, Lepidosaphes sp., Lepidosaphes ulmi, Leucaspis pini, Leucaspis signoreti, Mytilaspis sp., Targionia vitis* (Hemiptera: Diaspididae) (Noyes, 2019).

Distribution: France, Greece, Hungary, Nearctic, Spain, Turkey, United States of America (Noyes, 2019).

Aphytis aonidiae (Mercet, 1911)
Publications with Turkey records: Japoshvili and Karaca (2002), Noyes (2019), Öncüer (1991).
Associates: *Pistacia terebinthus* (Anacardiaceae), *Buxus sempervirens* (Buxaceae), *Quercus sp., Quercus sessiliflora* (Fagaceae), *Laurus sp., Laurus nobilis* (Lauraceae), *Ficus sp.* (Moraceae), *Olea europaea* (Oleaceae), *Rhamnus alaternus* (Rhamnaceae), *Amygdalus sp., Chaenomeles japonica, Crataegus pyracantha, Cydonia oblonga, Prunus domestica, Prunus dulcis, Prunus persica, Pyracantha hybrida, Rosa sp.* (Rosaceae), *Citrus sp., Citrus aurantium, Citrus sinensis, Citrus unshiu* (Rutaceae), *Populus sp.* (Salicaceae), *Solanum tuberosum* (Solanaceae), *Ampelopsis sp.* (Vitaceae) (Noyes, 2019).
Hosts: *Asterolecanium arabidis* (Hemiptera: Asterolecaniidae), *Eulecanium corni* (Hemiptera: Coccidae), *Aonidia lauri, Aonidiella aurantii, Aonidiella citrina, Aonidiella orientalis, Aspidaspis densiflorae, Aspidiotus hederae, Aspidiotus nerii, Aulacaspis rosae, Carulaspis carueli, Chrysomphalus dictyospermi, Comstockaspis perniciosa, Diaspidiotus caucasicus, Diaspidiotus perniciosus, Diaspidiotus prunorum, Diaspidiotus pyri, Dynaspidiotus britannicus, Hemiberlesia lataniae, Insulaspis juniperi, Lepidosaphes sp., Lepidosaphes ulmi, Leucaspis loewi, Leucaspis pusilla, Nuculaspis abietis, Parlatoria pergandii,Quadraspidiotus juglansregiae, Quadraspidiotus ostreaeformis, Quadraspidiotus pelagijae, Quadraspidiotus perniciosus, Quadraspidiotus zonatus, Unaspis yanonensis* (Hemiptera: Diaspididae) (Noyes, 2019).
Distribution: Argentina, Armenia, Caucasus, Chile, Cyprus, Czech Republic, Egypt, Europe, Georgia, Greece, Hungary, Iran, Israel, Italy, Japan, Mexico, Moldova, Montenegro, Romania, Slovakia, Spain, Turkey, Ukraine, United Kingdom, England, United States of America, Uruguay, USSR, Yugoslavia (pre-1991) (Noyes, 2019).

Aphytis chilensis (Howard, 1900)
Publications with Turkey records: Noyes (2019), Öncüer (1991).
Associates: *Wolbachia sp.* (Wolbachiaceae: Bacteria), *Mangifera indica, Pistacia sp.* (Anacardiaceae), *Nerium oleander* (Apocynaceae), *Hedera helix* (Araliaceae), *Cereus sp.* (Cactaceae), *Ficus nitida* (Moraceae), *Olea europaea* (Oleaceae), *Picea sp., Pinus sp., Pinus radiata* (Pinaceae), *Citrus sp.* (Rutaceae) (Noyes, 2019).

Hosts: *Coccus hesperidum* (Hemiptera: Coccidae), *Aonidia lauri, Aonidiella aurantii, Aonidiella citrina, Aspidiotus sp., Aspidiotus camelliae, Aspidiotus hederae, Aspidiotus hedericola, Aspidiotus nerii, Carulaspis carueli, Chionaspis berlesii, Chionaspis heterophyllae, Chionaspis pinifoliae, Chrysomphalus sp., Chrysomphalus dictyospermi, Diaspis sp., Diaspis echinocacti, Diaspis zamiae, Furchadaspis zamiae, Hemiberlesia lataniae, Hemiberlesia palmae, Hemiberlesia rapax, Lepidosaphes japonica, Parlatoria oleae, Parlatoria pergandii, Parlatoria pittospori, Phenacaspis pinifoliae, Pseudaulacaspis pentagona, Quadraspidiotus perniciosus, Rolaspis sp., Rolaspis incisa* (Hemiptera: Diaspididae) (Noyes, 2019).
Distribution: Algeria, Argentina, Australia, Caucasus, Chile, Cyprus, Egypt, Europe, France, Germany, Greece, Italy, Lebanon, Mexico, Middle East, Nearctic, Neotropical, New Zealand, North Africa, Peru, South Africa, Spain, Tunisia, Turkey, United States of America, USSR (Noyes, 2019).

Aphytis chrysomphali (Mercet, 1912)
Publications with Turkey records: Erler and Tunç (2001), Noyes (2019), Öncüer (1991).
Associates: *Wolbachia sp.* (Bacteria: Wolbachiaceae), *Aphytis melinus* (Hymenoptera: Aphelinidae), *Mangifera indica* (Anacardiaceae), *Annona sp.* (Annonaceae), *Nerium oleander* (Apocynaceae), *Hedera helix* (Araliaceae), *Cocos nucifera* (Arecaceae), *Opuntia sp.* (Cactaceae), *Elaeagnus sp.* (Elaeagnaceae), *Acacia sp., Robinia pseudoacacia* (Fabaceae), *Persea americana* (Lauraceae), *Ficus sp., Ficus nitida* (Moraceae), *Musa sp.* (Musaceae), *Ligustrum sp., Olea sp., Olea europaea* (Oleaceae), *Plantago sp.* (Plantaginaceae), *Crataegus sp., Rosa sp.* (Rosaceae), *Citrus sp., Citrus deliciosa, Citrus limon, Citrus sinensis, Citrus unshiu* (Rutaceae), *Populus sp.* (Salicaceae), *Solanum tuberosum* (Solanaceae) (Noyes, 2019).
Hosts: *Chloropulvinaria psidii* (Hemiptera: Coccidae), *Acutaspis umbonifera, Aonidiella sp., Aonidiella aurantii, Aonidiella citrina, Aonidiella orientalis, Aonidiella taxus, Aspidiotus camelliae, Aspidiotus cyanophylli, Aspidiotus destructor, Aspidiotus hederae, Aspidiotus lataniae, Aspidiotus nerii, Aspidiotus orientalis, Aspidiotus palmae, Aulacaspis rosae, Carulaspis minima, Chionaspis minor, Chrysomphalus sp., Chrysomphalus aonidum, Chrysomphalus dictyospermi, Chrysomphalus ficus, Chrysomphalus personatus, Chrysomphalus pinnulifer, Chrysomphalus rossi, Diaspidiotus spurcatus, Diaspis bromeliae, Diaspis echinocacti, Dynaspidiotus britannicus, Epidiaspis leperii, Fiorinia fioriniae, Fulaspis*

sp., Fulaspis mazoensis, Hemiberlesia cyanophylli, Hemiberlesia lataniae, Hemiberlesia rapax, Ischnaspis longirostris, Lepidosaphes alba, Lepidosaphes auriculata, Lepidosaphes beckii, Lepidosaphes gloverii, Lepidosaphes japonica, Lepidosaphes newsteadi, Lepidosaphes tapleyi, Lepidosaphes vermiculus, Lindingaspis floridana, Melanaspis inopinata, Mycetaspis personata, Mytilococcus beckii, Parlatoria oleae, Parlatoria ziziphi, Phenacaspis syringae, Pinnaspis strachani, Pinnaspis temporaria, Pseudaonidia sp., Pseudaonidia trilobitiformis, Pseudaulacaspis pentagona, Pseudischnaspis linearis, Pseudoparlatoria giffardi, Selenaspidius articulatus, Temnaspidiotus destructor, Unachionaspis bambusae (Hemiptera: Diaspididae), *Protargionia larreae* (Hymenoptera: Diprionidae) (Noyes, 2019).

Parasitoids: *Marietta leopardina* (Hymenoptera: Aphelinidae) (Noyes, 2019).

Distribution: Afghanistan, Algeria, Argentina, Australasian, Australia, Belgium, Bermuda, Brazil, Caribbean, Caucasus, Chile, Costa Rica, Cuba, Cyprus, Dominican Republic, Egypt, El Salvador, Europe, Fiji, France, French Polynesia, Georgia, Germany, Greece, Guam, Guyana, Haiti, Hawaii, Hungary, India, Indonesia, Iran, Israel, Italy, Japan, Lebanon, Malaysia, Mauritius, Mexico, Morocco, Nearctic, New Caledonia, North Africa, Palearctic, Panama, People's Republic of China, Peru, Philippines, Puerto Rico, Seychelles, Somalia, South Africa, Spain, Surinam, Swaziland, Taiwan, Tanzania, Trinidad & Tobago, Tunisia, Turkey, Ukraine, United States of America, Uruguay, USSR (Eastern USSR, European),Yugoslavia (pre-1991), Zimbabwe (Noyes, 2019).

Aphytis diaspidis (Howard, 1881)

Publications with Turkey records: Noyes (2019), Öncüer (1991).

Associates: *Wolbachia sp.* (Bacteria: Wolbachiaceae), *Nerium oleander* (Apocynaceae), *Ilex cornuta* (Aquifoliaceae), *Ficus sp.* (Moraceae), *Olea sp., Olea europaea, Osmanthus fragrans* (Oleaceae), *Picea sp., Pinus sp., Pinus mugo, Pinus nigra, Pinus resinosa, Pinus sylvestris* (Pinaceae), *Malus pumila, Prunus armeniaca, Prunus mume, Prunus persica, Pyrus communis, Rosa sp.* (Rosaceae), *Citrus sp.* (Rutaceae), *Populus sp.* (Salicaceae), *Solanum auriculatum* (Solanaceae), *Camellia japonica* (Theaceae) (Noyes, 2019).

Hosts: *Aleyrodes ricini, Dialeurodes citri* (Hemiptera: Aleyrodidae), *Coccus hesperidum* (Hemiptera: Coccidae), *Abgrallaspis oxycoccus, Acutaspis agavis, Acutaspis perseae, Aonidia mesembryanthemi, Aonidiella aurantii, Aonidiella citrina, Aonidomytilus albus, Aonidomytilus espinosai,Aspidiotus hederae, Aspidiotus lataniae,*

Aspidiotus nerii, Aspidiotus perniciosus, Aulacaspis cinnamomi, Aulacaspis rosae, Chionaspis sp., Chionaspis furfura, Chionaspis pinifoliae, Chionaspis salicis, Chionaspis salicisnigrae, Chrysomphalus agavis, Chrysomphalus aonidum, Chrysomphalus bifaciculatus, Chrysomphalus dictyospermi, Chrysomphalus ficus, Comstockiella sabalis, Diaspis bromeliae, Diaspis calyptroides, Diaspis echinocacti, Diaspis pentagona, Duplachionaspis sansevieriae, Entaspidiotus lounsburyi, Fiorinia theae, Hemiberlesia sp., Hemiberlesia lataniae, Hemiberlesia rapax, Hemichionaspis minor, Howardia biclavis, Insulaspis gloverii, Lepidosaphes sp., Lepidosaphes alba, Lepidosaphes beckii, Lepidosaphes camelliae, Lepidosaphes tubulorum, Lepidosaphes ulmi, Melanaspis sitreana, Mytilococcus beckii, Parlatoria oleae, Parlatoria pergandii, Parlatoria pittospori, Phenacaspis pinifoliae, Pinnaspis aspidistrae, Pinnaspis buxi, Pinnaspis strachani, Pinnaspis temporaria, Pseudaonidia duplex, Pseudaulacaspis pentagona, Quadraspidiotus juglansregiae, Quadraspidiotus perniciosus, Quadraspidiotus schneideri, Selenaspidius articulatus (Hemiptera: Diaspididae) (Noyes, 2019).
Parasitoids: *Marietta carnesi* (Hymenoptera: Aphelinidae) (Noyes, 2019).
Distribution: Argentina, Australia, Austria, Bermuda, Brazil, Canada, Cape Verde Islands, Chile, Cyprus, Dominican Republic, Egypt, Eritrea, Ethiopia, France, Greece, Grenada, Haiti, Hawaii, India, Iran, Israel, Italy, Jamaica, Japan, Korea, South, Lebanon, Madagascar, Mauritius, Mexico, Netherlands, New Zealand, North Africa, Pakistan, People's Republic of China, Peru, Poland, Réunion, Rwanda, South Africa, Spain, Sri Lanka, Switzerland, Trinidad & Tobago, Turkey, United Kingdom, United States of America, Uruguay (Noyes, 2019).

Aphytis hispanicus (Mercet, 1912)
Publications with Turkey records: Noyes (2019), Öncüer (1991).
Associates: *Aphytis comperei* (Hymenoptera: Aphelinidae), *Ficus nitida* (Moraceae), *Olea europaea* (Oleaceae), *Malus pumila* (Rosaceae), *Citrus sp., Citrus aurantium, Citrus sinensis* (Rutaceae) (Noyes, 2019).
Hosts: *Coccus hesperidum* (Hemiptera: Coccidae), *Acutaspis scutiformis, Aonidiella sp., Aonidiella aurantii, Aspidiotus hederae, Aspidiotus nerii, Aulacaspis tubercularis, Carulaspis carueli, Chrysomphalus dictyospermi, Comstockaspis perniciosa, Cornuaspis beckii, Diaspidiotus hungaricus, Diaspidiotus perniciosus, Hemiberlesia sp., Hemiberlesia lataniae, Insulaspis pallida, Lepidosaphes beckii, Lepidosaphes ficus, Lepidosaphes pallida, Lopholeucaspis japonica, Mytilaspis sp., Mytilaspis conchiformis, Parlatoria cinerea, Parlatoria crotonis, Parlatoria oleae, Parlatoria*

pergandii, Quadraspidiotus ostreaeformis, Quadraspidiotus perniciosus (Hemiptera: Diaspididae) (Noyes, 2019).
Distribution: Algeria, Argentina, Barbados, Brazil, Caucasus, Chile, Cook Islands, Czech Republic, Egypt, Europe, France, Georgia, Haiti, Hungary, India, Israel, Italy, Jamaica, Lebanon, Mexico, Morocco, Myanmar, Nearctic, Neotropical, People's Republic of China, South Africa, Spain, Sweden, Taiwan, Transcaucasus, Trinidad & Tobago, Turkey, United States of America (Noyes, 2019).

Aphytis lepidosaphes (Compere, 1955)
Publications with Turkey records: Noyes (2019), Öncüer (1991).
Associates: *Citrus sp., Citrus deliciosa, Citrus sinensis* (Rutaceae) (Noyes, 2019).
Hosts: *Aonidiella aurantii, Chrysomphalus aonidum, Cornuaspis beckii, Insulaspis gloverii, Lepidosaphes sp., Lepidosaphes beckii, Lepidosaphes gloverii, Mytilococcus beckii* (Hemiptera: Diaspididae) (Noyes, 2019).
Parasitoids: *Marietta javensis* (Hymenoptera: Aphelinidae), *Thysanus flavopalliatus* (Hymenoptera: Signiphoridae) (Noyes, 2019).
Distribution: Argentina, Australia, Brazil, Chile, Costa Rica, Cyprus, Ecuador, Egypt, El Salvador, Fiji, France, Greece, Guadeloupe, Hawaii, India, Israel, Italy, Jamaica, Lebanon, Mexico, Myanmar (Burma), Neotropical, New Caledonia, Pakistan, People's Republic of China, Peru, Philippines, Puerto Rico, South Africa, Spain, Taiwan, Thailand, Trinidad & Tobago, Turkey, United States of America (Noyes, 2019).

Aphytis libanicus (Traboulsi, 1969)
Publications with Turkey records: Noyes (2019), Japoshvili and Karaca (2002).
Associates: *Ceratonia siliqua* (Fabaceae), *Olea sp., Olea europaea* (Oleaceae), *Ulmus sp.* (Ulmaceae) (Noyes, 2019).
Hosts: *Dynaspidiotus britannicus, Lepidosaphes conchiformis, Lepidosaphes malicola, Leucaspis riccae, Mytilaspis conchiformis, Parlatoria oleae* (Hemiptera: Diaspididae) (Noyes, 2019).
Distribution: Egypt, Iran, Israel, Italy, Lebanon, Turkey (Noyes, 2019).

Aphytis lingnanensis (Compere, 1955)
Publications with Turkey records: Noyes (2019).
Associates: *Wolbachia sp.* (Bacteria: Wolbachiaceae), *Aphytis melinus* (Hymenoptera: Aphelinidae), *Yucca sp.* (Agavaceae), *Mangifera indica* (Anacardiaceae), *Nerium oleander* (Apocynaceae), *Cocos nucifera, Metroxylon sagu* (Arecaceae), *Opuntia sp.* (Cactaceae), *Carica papaya*

(Caricaceae), *Ipomoea batatas* (Convolvulaceae), *Citrullus vulgaris, Cucurbita moschata* (Cucurbitaceae), *Cycas revoluta* (Cycadaceae), *Acacia sp.* (Fabaceae), *Quercus virginiana* (Fagaceae), *Laurus nobilis, Persea americana* (Lauraceae), *Azadirachta indica* (Meliaceae), *Artocarpus altilis*, *Ficus sp.*(Moraceae), *Musa sp.* (Musaceae), *Pinus sp.* (Pinaceae), *Saccharum officinarum* (Poaceae), *Rosa sp.* (Rosaceae), *Citrus sp., Citrus limon, Citrus medica, Citrus sinensis* (Rutaceae), *Aspidiotus hederae, Parlatoria pergandii* (Hemiptera: Diaspididae) (Noyes, 2019).

Hosts: *Abgrallaspis cyanophylli, Aonidia lauri, Aonidiella sp., Aonidiella aurantii, Aonidiella citrina, Aonidiella orientalis, Aonidomytilus albus, Aspidiotus sp., Aspidiotus aurantii, Aspidiotus destructor, Aspidiotus hederae, Aspidiotus nerii, Aulacaspis tegalensis, Aulacaspis tubercularis, Chrysomphalus aonidum, Chrysomphalus dictyospermi, Chrysomphalus ficus, Cornuaspis beckii, Diaspis echinocacti, Hemiberlesia lataniae, Hemiberlesia palmae, Hemiberlesia rapax, Insulaspis gloverii, Insulaspis pallidula, Lepidosaphes beckii, Lepidosaphes gloverii, Lepidosaphes pineti, Lindingaspis fusca, Mytilaspis sp., Parlatoria ziziphi, Pinnaspis sp., Pinnaspis strachani, Pseudaonidia trilobitiformis, Pseudaulacaspis cockerelli, Pseudaulacaspis pentagona, Selenaspidius articulatus, Temnaspidiotus destructor, Unaspis citri, Unaspis lingnanensis, Unaspis yanonensis* (Hemiptera: Diaspididae) (Noyes, 2019).

Distribution: Argentina, Australia, Bermuda, Bolivia, Brazil, Caribbean (including West Indies), Chile, Costa Rica, Cuba, Cyprus, Egypt, El Salvador, Europe, Fiji, India, Indonesia, Israel, Italy, Jamaica, Japan, Malaysia, Mauritius, Mexico, Morocco, Nearctic, Neotropical, New Caledonia, Pakistan, People's Republic of China, Peru, Philippines, Puerto Rico, Solomon Islands, South Africa, Spain, Taiwan, Thailand, Trinidad & Tobago, Turkey, United States of America, Uruguay (Noyes, 2019).

Aphytis maculicornis (Masi, 1911)

Publications with Turkey records: Erler and Tunç (2001), Japoshvili and Karaca (2002), Noyes (2019), Öncüer (1991).

Associates: *Coccophagoides utilis* (Hymenoptera: Aphelinidae), *Alternanthera sp.* (Amaranthaceae), *Opuntia sp.* (Cactaceae), *Ligustrum vulgare, Olea sp., Olea europaea* (Oleaceae), *Rhamnus pallasii* (Rhamnaceae), *Malus communis, Malus pumila, Prunus persica, Pyrus communis, Rosa sp.* (Rosaceae), *Citrus sp.* (Rutaceae), *Vitis vinifera* (Vitaceae) (Noyes, 2019).

Hosts: *Aonidiella aurantii, Aspidiotus cyanophylli, Aspidiotus nerii, Chionaspis salicis, Chrysomphalus dictyospermi, Chrysomphalus ficus, Diaspis echinocacti, Epidiaspis leperii, Hemiberlesia cyanophylli,*

Hemiberlesia lataniae, Lepidosaphes beckii, Lepidosaphes conchiformis, Lepidosaphes gloverii, Lepidosaphes ulmi, Parlatoreopsis longispinus, Parlatoria sp., Parlatoria camelliae, Parlatoria oleae, Parlatoria pergandii, Parlatoria theae, Quadraspidiotus juglansregiae (Diaspididae: Hemiptera) (Noyes, 2019).
Distribution: Afghanistan, Algeria, Argentina, Armenia, Azerbaijan, Egypt, Europe, France, Georgia, Greece, India, Iran, Iraq, Israel, Italy, Mexico, Morocco, North Africa, Pakistan, People's Republic of China, Russia, South Africa, Spain, Sweden, Tadzhikistan, Transcaucasus, Turkey, Ukraine, United States of America, USSR (Noyes, 2019).

Aphytis melinus (DeBach, 1959)
Publications with Turkey records: Erler and Tunç (2001), Karaca (1998), Karaca et al. (2001), Noyes (2019), Öncüer (1991), Sengonca et al. (1998).
Associates: *Ixeuticus sp.* (Araneae: Desidae), *Aphytis lingnanensis* (Hemiptera: Aphidoidea), *Aphytis chrysomphali, Encarsia perniciosi* (Hymenoptera: Aphelinidae), *Camponotus nylanderi, Formica aerata, Iridomyrmex humilis, Linepithema humile, Solenopsis xyloni, Tapinoma erraticum* (Hymenoptera: Formicidae), *Agave sp.* (Agavaceae), *Mangifera indica, Pistacia sp.* (Anacardiaceae), *Nerium oleander* (Apocynaceae), *Hedera helix* (Araliaceae), *Opuntia sp.* (Cactaceae), *Carica papaya* (Caricaceae), *Citrullus vulgaris, Cucurbita maxima, Curcubita maxima* (Cucurbitaceae), *Acacia saligna* (Fabaceae), *Moraea sp.* (Iridaceae), *Laurus nobilis, Persea americana* (Lauraceae), *Magnolia virginiana* (Magnoliaceae), *Azadirachta indica* (Meliaceae), *Ligustrum sp., Olea europaea* (Oleaceae), *Malus pumila, Prunus persica, Rosa sp.* (Rosaceae), *Citrus sp., Citrus clementina, Citrus grandis, Citrus limettioides, Citrus limon, Citrus limonia, Citrus maxima, Citrus paradisi, Citrus reticulata, Citrus sinensis, Citrus unshiu* (Rutaceae), *Salix sp.* (Salicaceae), *Solanum tuberosum* (Solanaceae) (Noyes, 2019).
Hosts: *Abgrallaspis cyanophylli, Aonidiella sp., Aonidiella aurantii, Aonidiella citrina, Aonidiella orientalis, Aspidiotus aurantii, Aspidiotus citrina, Aspidiotus destructor, Aspidiotus hederae, Aspidiotus nerii, Aspidiotus orientalis, Chrysomphalus dictyospermi, Diaspidiotus perniciosus, Diaspidis sp., Diaspis bromeliae, Diaspis echinocacti, Hemiberlesia lataniae, Lindingaspis ferrisi, Octaspidiotus tamarindi, Parlatoria oleae, Parlatoria pergandii, Quadraspidiotus juglansregiae, Quadraspidiotus perniciosus, Temnaspidiotus destructor*(Hemiptera: Diaspididae).

Distribution: Argentina, Australia, Chile, Cyprus, France, Georgia, Greece, India, Israel, Italy, Mexico, Morocco, Nearctic, Netherlands, Pakistan, Paraguay, People's Republic of China, Peru, South Africa, Spain, Turkey, United States of America, Uruguay, Yugoslavia (pre-1991) (Noyes, 2019).

Aphytis mytilaspidis (Le Baron, 1870)
Publications with Turkey records: Erler and Tunç (2001), Erol and Yasar (1999), Japoshvili and Karaca (2002), Noyes (2019), Öncüer (1991).
Associates: *Acer rubrum* (Aceraceae), *Yucca sp.* (Agavaceae), *Alternanthera sp.* (Amaranthaceae), *Mangifera indica, Pistacia terebinthus* (Anacardiaceae), *Nerium oleander* (Apocynaceae), *Hedera helix* (Araliaceae), *Opuntia sp., Opuntia littoralis, Opuntia vulgaris, Opuntia zacuapanensis* (Cactaceae), *Celastrus scaudens* (Celastraceae), *Cupressus sp., Cupressus sempervirens, Juniperus sp., Juniperus chinensis, Juniperus foetidissima, Thuja sp., Thuja horizontalis* (Cupressaceae), *Hippophae rhamnoides* (Elaeagnaceae), *Caragana sp., Ceratonia siliqua,Genista sp.* (Fabaceae), *Quercus sp., Quercus robur* (Fagaceae), *Ribes odoratum* (Grossulariaceae), *Juglans sp.* (Juglandaceae), *Laurus nobilis* (Lauraceae), *Ficus sp., Ficus carica* (Moraceae), *Psidium guajava* (Myrtaceae), *Jasminum sp., Ligustrum sp., Ligustrum japonicum, Ligustrum vulgare, Olea europaea, Syringa vulgaris* (Oleaceae), *Picea sp., Pinus sp., Pinus eldarica, Pinus mugo, Pinus nigra, Pinus resinosa, Pinus sylvestris* (Pinaceae), *Platanus sp., Platanus orientalis* (Platanaceae), *Phragmites communis* (Poaceae), *Rhamnus pallasii* (Rhamnaceae), *Crataegus sp., Heteromeles sp., Malus pumila, Prunus sp., Prunus armeniaca, Prunus domestica, Prunus persica, Pyrus communis, Rosa sp.* (Rosaceae), *Citrus sp., Citrus sinensis, Ptelea trifoliata* (Rutaceae), *Populus sp., Populus alba, Populus hybrida, Populus transcaucasica, Salix sp., Salix caprea. Salix nigra* (Salicaceae), *Tilia sp.* (Tiliaceae), *Ampelopsis sp., Vitis vinifera* (Vitaceae) (Noyes, 2019).
Hosts: *Asterolecanium pustulans, Asterolecanium variolosum* (Hemiptera: Asterolecaniidae), *Chloropulvinaria psidii, Pulvinaria psidii* (Hemiptera: Coccidae), *Acutaspis agavis, Aonidia lauri, Aonidiella aurantii, Aonidiella citrina, Aonidiella orientalis, Aspidiotus britannicus, Aspidiotus cyanophylli, Aspidiotus hederae, Aspidiotus hedericola, Aspidiotus nerii, Aspidiotus ostreaeformis, Aspidiotus perniciosus, Aspidiotus pyri, Aulacaspis rosae, Aulacaspis rosarum, Aulacaspis tegalensis, Aulacaspis tubercularis, Carulaspis carueli, Carulaspis juniperi, Carulaspis minima, Carulaspis visci, Chionaspis sp., Chionaspis berlesii, Chionaspis furfura,*

Chionaspis heterophyllae, Chionaspis micropori, Chionaspis pinifoliae, Chionaspis salicis, Chionaspis salicisnigrae, Chionaspis stantophris, Chionaspis striata, Chrysomphalus dictyospermi, Comstockiella sabalis, Cupressaspis juniperi, Diaspidiotus sp., Diaspidiotus caucasicus, Diaspidiotus lataniae, Diaspidiotus ostreaeformis, Diaspidiotus perniciosus, Diaspidiotus spurcatus, Diaspis echinocacti, Diaspis visci, Duplachionaspis graminis, Dynaspidiotus abietis, Dynaspidiotus britannicus, Epidiaspis leperii, Hemiberlesia cyanophylli, Hemiberlesia lataniae, Insulaspis juniperi, Lepidosaphes sp., Lepidosaphes beckii, Lepidosaphes conchiformis, Lepidosaphes ficus, Lepidosaphes gloverii, Lepidosaphes malicola, Lepidosaphes newsteadi, Lepidosaphes pinnaeformis, Lepidosaphes populi, Lepidosaphes ulmi, Leucaspis loewi, Leucaspis pini, Leucaspis pusilla, Leucaspis riccae, Lineaspis striata, Mytilaspis conchiformis, Mytilococcus beckii, Mytilococcus ulmi, Parlatoria sp., Parlatoria blanchardii, Parlatoria oleae, Phenacaspis pinifoliae, Pinnaspis aspidistrae, Pinnaspis strachani, Pseudaulacaspis pentagona, Quadraspidiotus gigas, Quadraspidiotus juglansregiae, Quadraspidiotus marani, Quadraspidiotus ostreaeformis, Quadraspidiotus perniciosus, Quadraspidiotus pyri, Quadraspidiotus schneideri, Quadraspidiotus zonatus, Salicicola kermanensis, Unaspis permutans (Hemiptera: Diaspididae) (Noyes, 2019).
Distribution: Algeria, Argentina, Azerbaijan, Belgium, Bermuda, Bulgaria, Canada, Caucasus, Chile, Croatia, Cyprus, Czech Republic, Czechoslovakia, Egypt, Europe, France, Georgia, Germany, Greece, Hungary, India, Indonesia, Iran, Iraq, Israel, Italy, Japan, Kazakhstan, Lebanon, Macedonia, Mauritania, Mauritius, Mexico, Moldova, Montenegro, Morocco, Nearctic, Netherlands, New Zealand, North Africa, People's Republic of China, Poland, Romania, Russia, Saudi Arabia, Serbia, Slovakia, Slovenia, South Africa, Spain, Sri Lanka, Sweden, Switzerland, Taiwan, Transcaucasus, Tselinograd Obl. Turkey, Ukraine, United Kingdom, England, United States of America, USSR, Yugoslavia (pre-1991) (Noyes, 2019).

Aphytis neuter (Yasnosh and Myartseva, 1971)
Publications with Turkey records: Noyes (2019).
Associates: *Populus sp.* (Salicaceae) (Noyes, 2019).
Hosts: *Quadraspidiotus slavonicus* (Hemiptera: Diaspididae) (Noyes, 2019).
Distribution: Tadzhikistan, Turkey, Turkmenistan, USSR, Uzbekistan (Noyes, 2019).

Aphytis proclia (Walker, 1839)
Publications with Turkey records: Japoshvili and Karaca (2002), Noyes (2019), Öncüer (1991).
Associates: *Epidiaspis leperii* (Hemiptera: Diaspididae), *Acer rubrum* (Aceraceae), *Actinidia chinensis, Actinidia deliciosa* (Actinidiaceae), *Betula sp., Betula platyphylla, Corylus avellana* (Betulaceae), *Euonymus sp.* (Celastraceae), *Cucurbita sp.* (Cucurbitaceae), *Quercus robur* (Fagaceae), *Aesculus hippocastanum* (Hippocastanaceae), *Juglans sp.* (Juglandaceae), *Magnolia pumila* (Magnoliaceae), *Morus sp.* (Moraceae), *Syringa vulgaris* (Oleaceae), *Pandanus sp.* (Pandanaceae), *Picea sp., Pinus sp.* (Pinaceae), *Ziziphus sativa* (Rhamnaceae), *Malus pumila, Prunus sp., Prunus persica, Pyrus americana, Pyrus communis* (Rosaceae), *Citrus sp., Citrus sinensis* (Rutaceae), *Populus hybrida, Populus transcaucasica, Salix sp.* (Salicaceae), *Tilia sp.* (Tiliaceae),*Ulmus americana* (Ulmaceae) (Noyes, 2019).
Hosts: *Dialeurodes citri* (Hemiptera: Aleyrodidae), *Coccus hesperidum* (Coccidae), *Aonidiella aurantii, Aonidiella citrina, Aspidiotus sp., Aspidiotus destructor, Aspidiotus hederae, Aspidiotus nerii, Aulacaspis rosae, Carulaspis juniperi, Carulaspis visci, Chionaspis americana, Chionaspis furfura, Chionaspis heterophyllae, Chionaspis pinifoliae, Chionaspis salicis, Chionaspis salicisnigrae, Chrysomphalus aonidum, Chrysomphalus dictyospermi, Chrysomphalus ficus, Diaspidiotus sp., Diaspidiotus ancylus, Diaspidiotus caucasicus, Diaspidiotus gigas, Diaspidiotus perniciosus, Diaspidiotus prunorum, Diaspidiotus pyri, Diaspidiotus spurcatus, Diaspidiotus uvae, Diaspis bromeliae, Diaspis echinocacti, Dynaspidiotus abietis, Dynaspidiotus britannicus, Epidiaspis leperii, Furchadaspis zamiae, Hemiberlesia lataniae, Hemiberlesia rapax, Kuwanaspis hikosani, Lepidosaphes sp., Lepidosaphes beckii, Lepidosaphes gloverii, Lepidosaphes machili, Lepidosaphes malicola, Lepidosaphes pinnaeformis, Lepidosaphes salicina, Lepidosaphes tubulorum, Lepidosaphes ulmi, Melanaspis tenebricosa, Mytilaspis conchiformis, Nuculaspis californica, Parlatoria oleae, Parlatoria pergandii, Parlatoria ziziphi, Phenacaspis pinifoliae, Pinnaspis aspidistrae, Pinnaspis strachani, Pseudaonidia duplex, Pseudaulacaspis pentagona, Quadraspidiotus forbesi, Quadraspidiotus gigas, Quadraspidiotus juglansregiae, Quadraspidiotus ostreaeformis, Quadraspidiotus perniciosus, Quadraspidiotus pyri, Quadraspidiotus zonatus, Tecaspis prunorum, Unaspis citri, Unaspis euonymi* (Diaspididae), *Trifolium pratense* (Fabaceae), *Phacelia sp.* (Hydrophyllaceae) (Noyes, 2019).
Parasitoids: *Azotus perspeciosus, Prospaltella perniciosi* (Hymenoptera: Aphelinidae) (Noyes, 2019).

Distribution: Algeria, Argentina, Austria, Azerbaijan, Bermuda, Bulgaria, Canada, Caucasus, Chile, Croatia, Cyprus, Czech Republic, Czechoslovakia, El Salvador, Europe, France, Georgia, Germany, Grenada, Hawaii, Hungary, India, Iran, Italy, Japan, Kazakhstan, Korea, , Macedonia, Mexico, Moldova, Morocco, Myanmar (Burma), Nearctic, Neotropical, Netherlands, North Africa, Pakistan, People's Republic of China, Poland, Portugal, Russia, Serbia, Slovakia, Slovenia, South Africa, Spain, Sweden, Switzerland, Taiwan, Tselinograd Obl., Turkey, United Kingdom, England, United States of America, USSR, Yugoslavia (pre-1991) (Noyes, 2019).

Genus *Marietta* (Motschulsky, 1863)
Marietta leopardina (Motschulsky, 1863)
Publications with Turkey records: Noyes (2019), Öncüer (1974; 1991).
Associates: *Mangifera indica* (Anacardiaceae), *Boscia foetida* (Capparidaceae), *Manihot esculenta* (Euphorbiaceae), *Hibiscus rosa-sinensis* (Malvaceae), *Ficus sp., Ficus retusa* (Moraceae), *Psidium guajava* (Myrtaceae), *Olea sp., Olea europaea* (Oleaceae), *Rosa sp.* (Rosaceae), *Citrus sp., Citrus grandis, Citrus paradisi, Citrus sinensis, Murraya paniculate* (Rutaceae), *Santalum album* (Santalaceae), *Tamarix sp.* (Tamaricaceae) (Noyes, 2019).
Hosts: *Artemisaspis abdita, Psectrosema manii* (Diptera: Cecidomyiidae), *Leucopis nigricornis* (Chamaemyiidae), *Asterodiaspis quercicola, Asterolecanium bambusae, Asterolecanium coffeae, Asterolecanium pustulans, Asterolecanium quercicola* (Hemiptera: Asterolecaniidae), *Cerococcus sp., Cerococcus hibisci, Cerococcus indicus* (Cerococcidae), *Amblerdiella sp.* (Cicadellidae), *Ceroplastes sp., Ceroplastes floridensis, Ceroplastes pseudoceriferus, Ceroplastes rusci, Ceroplastodes sp., Ceroplastodes cajani, Chloropulvinaria sp., Chloropulvinaria polygonate, Chloropulvinaria psidii, Coccus hesperidum, Coccus longulus, Coccus pseudomagnoliarum, Coccus viridis, Gascardia brevicauda, Gascardia sinoiae, Inglisia bivalvata, Lecanium sp., Lecanium coffeae, Lecanium viride, Parasaissetia nigra, Protopulvinaria mangiferae, Pulvinaria floccifera, Pulvinaria maxima, Pulvinaria mesembryanthemi, Pulvinaria psidii, Saissetia sp., Saissetia coffeae, Saissetia nigra, Saissetia oleae (Coccidae), Andaspis leucophloeae, Aonidiella aurantia, Aonidiella citrina, Aonidiella orientalis, Aspidiotus orientalis, Chionaspis sp., Chrysomphalus aonidum, Chrysomphalus dictyospermi, Chrysomphalus ficus, Contigaspis sp., Duplachionaspis sp., Hemiberlesia lataniae, Hemiberlesia rapax, Lepidosaphes beckii, Lepidosaphes ulmi, Melanaspis glomerata, Octaspidiotus tamarindi, Parlatoria oleae, Parlatoria ziziphin,*

Pseudaulacaspis pentagona, Quadraspidiotus perniciosus, Rolaspis whitehilli, Selenaspidius articulates, Separaspis capensis, Umbaspis regularis (Diaspididae), *Kerria lacca, Laccifer lacca, Paratachardina lobata, Tachardia sp., Tachardia lobata, Tachardina actinella, Tachardina affluens* (Kerriidae), *Lecanodiaspis tarsalis* (Lecanodiaspididae), *Crypticerya bicolor* (Margarodidae), *Ferrisia virgata, Nipaecoccus vastator, Nipaecoccus viridis, Phenacoccus manihoti, Planococcoides robustus, Planococcus citri, Planococcus lilacinus, Rastrococcus iceryoides, Rastrococcus invadens* (Pseudococcidae), *Diaphorina citri* (Psyllidae), *Trioza erytreae* (Triozidae), *Leucospis nigricornis* (Hymenoptera: Leucospidae), *Lymantria dispar, Porthetria dispar* (Lepidoptera: Lymantriidae) (Noyes, 2019).

Parasitoid hosts: *Aneristus ceroplastae, Aphytis africanus, Aphytis chrysomphali, Aphytis holoxanthus, Aphytis lepidosaphes, Coccophagus sp., Coccophagus ceroplastae, Coccophagus lycimnia, Promuscidea unfasciativentris* (Hymenoptera: Aphelinidae), *Anagyrus sp., Anagyrus mirzai, Comperiella bifasciata, Diaphorencyrtus diaphorinae, Epidinocarsis lopezi, Gyranusoidea mirzai, Gyranusoidea tebygi, Habrolepis aspidioti, Habrolepis exitiosa, Habrolepis fanari, Habrolepis rouxi, Leptomastix dactylopii, Metaphycus sp., Microterys flavus, Microterys newcombi* (Encyrtidae), *Anysis alcocki, Anysis saissetiae, Cephaleta brunniventris, Pachyneuron leucopiscida* (Pteromalidae), *Chartocerus sp.* (Signiphoridae) (Noyes, 2019).

Distribution: Afrotropical, Bangladesh, Cameroon, Congo, Democratic Republic of Congo (Zaire), Egypt, Gambia, Guinea-Bissau, India, Indonesia, Israel, Japan, Kenya, Libya, Madagascar, Malaysia, Mauritius, Morocco, Namibia, Nearctic, Nigeria, Oriental, Pakistan, Paraguay, Philippines, Saudi Arabia, Singapore, South Africa, Sri Lanka, Swaziland, Taiwan, Togo, Turkey, United States of America, West Africa (Noyes, 2019).

Marietta picta (Andre, 1878)

Publications with Turkey records: Kaydan and Japoshvili (2010), Noyes (2019), Öncüer (1991).

Associates: *Pistacia vera* (Anacardiaceae), *Robinia pseudoacacia* (Fabaceae), *Morus sp.* (Moraceae), *Jasminum fructicans* (Oleaceae), *Festuca circummediterranea* (Poaceae), *Malus communis, Prunus sp., Prunus cerasifera* (Rosaceae), *Rubus sp.* (Rubiaceae), *Salix sp.* (Salicaceae), *Tamarix sp., Tamarix hohenackeri* (Tamaricaceae), *Scythia aetnensis, Stotzia ephedrae* (Hemiptera: Coccidae), *Trabutina crassispinosa* (Pseudococcidae) (Noyes, 2019).

Hosts: *Aclerda subterranean* (Hemiptera: Aclerdidae), *Aleurothrixus floccosus* (Aleyrodidae), *Sipha maydis* (Aphididae), *Planchonia arabidis* (Asterolecaniidae), *Acantholecanium haloxyloni, Ceroplastes mimosa, Chloropulvinaria aurantia, Coccus hesperidum, Didesmococcus unifasciatus, Eriopeltis araxis, Eriopeltis festucae, Eriopeltis lichtensteinii, Eulecanium sp., Eulecanium mali, Filippia asperagi, Filippia folicularis, Lecanium prunastri, Lichtensia viburni, Parthenolecanium corni, Sphaerolecanium prunastri, Stotzia ephedrae* (Coccidae), *Aulacaspis rosae, Chionaspis salicis, Chrysomphalus aonidum, Lepidosaphes ulmi, Parlatoria oleae, Quadraspidiotus slavonicus* (Diaspididae), *Acanthococcus sp., Eriococcus sp., Eriococcus greeni, Eriococcus obscurus, Rhizococcus sp.* (Eriococcidae), *Kermes sp.* (Kermesidae), *Antonina phragmitis, Antonina purpurea, Dysmicoccus multivorus, Nipaecoccus viridis, Planococcus citri, Planococcus ficus, Pseudococcus sp., Pseudococcus citri, Pseudococcus comstocki, Pseudococcus ficus, Spilococcus nanae, Trabutina crassispinosa* (Pseudococcidae), *Agonoscena succincta, Psylla pyri* (Psyllidae), *Trioza centranthi, Trioza urticae* (Triozidae) (Noyes, 2019).
Parasitoid hosts: *Aphelinus toxopteraphidis* (Hymenoptera: Aphelinidae), *Aphidius sp.* (Braconidae), *Anagyrus bohemani, Anagyrus diversicornis, Anagyrus pseudococci, Aphidencyrtus sp., Cerapterocerus mirabilis, Discodes aeneus, Habrolepis pascuorum, Metaphycus insidiosus, Paraphaenodiscus subterraneus, Phaenodiscus aeneus, Prochiloneurus bolivari, Psyllaephagus triozae* (Encyrtidae), *Tetrastichus obscuratus* (Eulophidae), *Pachyneuron sp.* (Pteromalidae), *Chartocerus conjugalis* (Signiphoridae) (Noyes, 2019).
Distribution: Armenia, Canada, Czech Republic, Czechoslovakia, Egypt, England, Europe, France, Georgia, Germany, Greece, Hungary, India, Iran, Italy, Kazakhstan, Korea, Mexico, Moldova, Palaearctic, People's Republic of China, Peru, Romania, Russia, Serbia, Slovakia, Spain, Sweden, Transcaucasus, Tselinograd Obl., Turkey, Turkmenistan, Ukraine, USSR, Yugoslavia (Federal Republic) (Noyes, 2019).

Subfamily Calesinae (Mercet, 1929)
Genus *Cales* (Howard, 1907)
Cales noacki (Howard, 1907)
Publications with Turkey records: Noyes (2019), Ulusoy et al. (2003).
Associates: *Beauveria bassiana* (Fungus), *Eretmocerus sp.* (Hymenoptera: Aphelinidae), *Iridomyrmex humilis* (Hymenoptera: Formicidae), *Trichogramma sp., Trichogramma evanescens* (Hymenoptera: Trichogrammatidae), *Rhus cotinus* (Anacardiaceae),

Plumeria rubra (Apocynaceae), *Ilex perado* (Aquifoliaceae), *Syagrus romanzoffiana* (Arecaceae), *Catalpa longissima* (Bignoniaceae), *Blechnum spicans* (Blechnaceae), *Viburnum tinus* (Caprifoliaceae), *Persea americana* (Lauraceae), *Swietenia sp.* (Meliaceae), *Eugenia uniflora, Psidium guajava* (Myrtaceae), *Prunus myrtifolia* (Rosaceae), *Coffea arabica* (Rubiaceae), *Citrus sp., Citrus aurantium, Citrus limon, Citrus reticulata, Citrus sinensis* (Rutaceae) (Noyes, 2019).

Hosts: *Aleurocanthus woglumi, Aleurothrixus sp., Aleurothrixus floccosus Aleurothrixus porteri, Aleurotrachelus sp., Aleurotrachelus atratus, Aleurotrachelus jelinekii, Aleurotrachelus rhamnicola, Aleurotuba jelinekii, Aleurotulus nephrelepidis, Aleyrodes lonicerae, Bemisia afer, Crenidorsum aroidephagus, Parabemisia myricae, Paraleyrodes sp., Tetraleurodes sp., Tetraleurodes perseae, Trialeurodes vaporariorum* (Hemiptera: Aleyrodidae), *Lepidosaphes sp., Pseudaulacaspis pentagona* (Hemiptera: Diaspididae), *Orthezia sp.* (Ortheziidae), *Phalera bucephala* (Lepidoptera: Notodontidae) (Noyes, 2019).

Distribution: Argentina, Azores, Brazil, Canary Islands, Chile, El Salvador, Europe, France, Greece, Haiti, Israel, Italy, Madeira, Malta, Mexico, Morocco, Nearctic, New World, Palaearctic, Peru, Réunion, Spain, Tunisia, Turkey, United States of America, Uruguay (Noyes, 2019).

Subfamily Coccophaginae (Foerster, 1878)

Genus *Coccobius* (Ratzeburg, 1852)

Coccobius annulicornis (Ratzeburg, 1852)

Publications with Turkey records: Japoshvili and Karaca (2002; 2010), Noyes (2019), Öncüer (1991), Özgen et al. (2009).

Associates: *Pistacia vera* (Anacardiaceae), *Hippophae rhamnoides* (Elaeagnaceae), *Ficus sp.* (Moraceae), *Malus communis, Malus pumila, Pyrus malus* (Rosaceae), *Salix sp., Salix caprea* (Salicaceae), *Ulmus sp.* (Ulmaceae) (Noyes, 2019).

Parasitoid hosts: *Comperiella bifasciata* (Hymenoptera: Encyrtidae) (Noyes, 2019).

Hosts: *Coccus sp.* (Hemiptera: Coccidae), *Chionaspis salicis, Contigaspis kochiae, Diaspidiotus gigas, Diaspidiotus jaapi, Diaspidiotus prunorum, Diaspidiotus slavonicus, Diaspidiotus transcaspiensis, Diaspis zamiae, Furchadaspis zamiae, Lepidosaphes sp., Lepidosaphes beckii, Lepidosaphes conchiformis, Lepidosaphes ficus, Lepidosaphes malicola, Lepidosaphes mesasiatica, Lepidosaphes pistaciae, Lepidosaphes ulmi, Lineaspis junipericola Lineaspis striata, Mytilaspis conchiformis, Parlatoria oleae, Quadraspidiotus gigas, Quadraspidiotus jaapi, Quadraspidiotus perniciosus, Quadraspidiotus slavonicus, Tecaspis*

asiatica, Tecaspis prunorum, Zepidaphes mesasiatica (Diaspididae), *Eriococcus spurius, Gossyparia spuria* (Eriococcidae) (Noyes, 2019).
Distribution: Azerbaijan, Caucasus, Croatia, Czech Republic, Egypt, Europe, France, Georgia, Germany, Hungary, Iran, Italy, Lebanon, Moldova, Montenegro, Nearctic, Palaearctic, People's Republic of China, Poland, Romania, Slovakia, Spain, Turkey, Ukraine, United Kingdom, United States of America, USSR, Uzbekistan (Noyes, 2019).

Genus Coccophagus (Westwood, 1833)
Coccophagus differens (Yasnosh, 1966)
Publications with Turkey records: Noyes (2019), Ulgenturk (2001).
Associates: *Prunus cerasifera* (Rosaceae) (Noyes, 2019).
Hosts: *Didesmococcus unifasciatus, Eulecanium ficiphilum, Palaeolecanium bituberculatum, Sphaerolecanium prunastri* (Hemiptera: Coccidae) (Noyes, 2019).
Distribution: Georgia, Moldova, Transcaucasus, Turkey, Ukraine (Noyes, 2019).

Coccophagus insidiator (Dalman, 1826)
Publications with Turkey records: Noyes (2019), Ulgenturk (2001), Ulgenturk and Toros (2001).
Associates: *Betula sp., Betula verrucosa* (Betulaceae), *Viburnum sp.* (Caprifoliaceae), *Rhododendron dahuricum* (Ericaceae), *Caragana sp.* (Fabaceae), *Luzula sp.* (Juncaceae), *Olea europaea* (Oleaceae), *Picea abies* (Pinaceae), *Rosa sp.* (Rosaceae), *Populus sp.* (Salicaceae), *Ulmus sp.* (Ulmaceae) (Noyes, 2019).
Hosts: *Eriopeltis sp., Eulecanium sp., Eulecanium bituberculatum, Eulecanium caraganae, Eulecanium ciliatum, Eulecanium corni, Eulecanium douglasi, Eulecanium franconicum, Filippia folicularis, Filippia oleae, Filippia viburni, Lichtensia chilianthi, Lichtensia viburni, Palaeolecanium bituberculatum, Parthenolecanium corni, Physokermes hemicryphus, Physokermes piceae, Physokermes sugonjaevi, Pulvinaria vitis* (Hemiptera: Coccidae); *Chionaspis salicis* (Diaspididae); *Eriococcus henryi, Eriococcus spurius, Gossyparia spuria* (Eriococcidae) (Noyes, 2019).
Distribution: Canada, Europe, Finland, France, Germany, Greece, Hungary, Italy, Kazakhstan, Mongolia, People's Republic of China, Poland, Russia, Sweden, Tselinograd Obl., Turkey, United Kingdom, United States of America, USSR, European, Siberia (Noyes, 2019).

Coccophagus kvavadze (Japoshvili and Karaca, 2002)
Publications with Turkey records: Japoshvili and Karaca (2002), Noyes (2019).
Associates: Graminaceae (Poaceae) (Noyes, 2019).
Hosts: *Eriopeltis sp.* (Hemiptera: Coccidae) (Noyes, 2019).
Distribution: Turkey (Noyes, 2019).

Coccophagus lycimnia (Walker, 1839)
Publications with Turkey records: Japoshvili and Karaca (2002), Noyes (2019), Öncüer (1974; 1991), Ozsemerci and Aksit (2003), Sengonca et al. (1998), Ulgenturk (2001), Ulgenturk et al. (2001).
Associates: *Acer sp., Acer campestre* (Aceraceae), *Agave sp.* (Agavaceae), *Pistacia sp., Pistacia mutica* (Anacardiaceae), *Nerium oleander* (Apocynaceae), *Asplenium nidus* (Aspleniaceae), *Baccharis pilularis* (Asteraceae), *Corylus avellana* (Betulaceae), *Viburnum lantana* (Caprifoliaceae), *Euonymus europaea* (Celastraceae), *Cornus sanguinea* (Cornaceae), *Cupressus sp., Juniperus sp., Thuja sp., Thuja occidentalis* (Cupressaceae), *Carex brizoides* (Cyperaceae), *Diospyros virginiana* (Ebenaceae), *Arctostaphylos sp., Vaccinium myrtillus* (Ericaceae), *Stillengia sp.* (Euphorbiaceae), *Bauhinia variegata, Robinia pseudoacacia* (Fabaceae), *Quercus sp.* (Fagaceae), *Ephedra procera* (Gnetaceae), *Gossypium sp.* (Malvaceae), *Melia azedarach* (Meliaceae), *Ficus sp., Ficus carica, Maclura pomifera, Morus sp.* (Moraceae), *Psidium guajava* (Myrtaceae), *Fraxinus sp., Fraxinus excelsior, Jasminum fructicans, Olea europaea* (Oleaceae), *Abies cephalonica, Pinus radiata, Pinus taeda* (Pinaceae), *Platanus occidentalis* (Platanaceae), *Agropyron sp., Agropyron intermedium* (Poaceae), *Rhamnus cathartica, Rhamnus pallasii* (Rhamnaceae), *Adenostoma sp, Armeniaca vulgaris, Cerasus japonica, Crataegus sp., Heteromeles sp., Prunus sp., Prunus armeniaca, Prunus cerasifera, Prunus domestica, Prunus padus, Prunus persica, Prunus salicina, Prunus spinosa* (Rosaceae), *Citrus sp., Citrus limon, Citrus sinensis, Citrus unshiu, Schefflera actinophylla* (Rutaceae), *Salix sp.* (Salicaceae), *Taxus sp., Taxus baccata* (Taxaceae), *Camellia sp.* (Theaceae), *Tilia sp.* (Tiliaceae), *Celtis caucasica, Ulmus scabra* (Ulmaceae), *Parthenocissus quinquefolia, Vitis vinifera* (Vitaceae) (Noyes, 2019).
Hosts: *Asterodiaspis variolosa, Planchonia arabidis* (Hemiptera: Asterolecaniidae), *Anapulvinaria pistaciae, Ceroplastes ceriferus, Ceroplastes floridensis, Ceroplastes japonicus, Ceroplastes rusci, Ceroplastes sinensis, Chloropulvinaria aurantii, Chloropulvinaria floccifera, Chloropulvinaria psidii, Coccus sp., Coccus hesperidum,*

Coccus pseudomagnoliarum, Coccus viridis, Didesmococcus koreanus, Didesmococcus megriensis, Eriopeltis agropyri, Eriopeltis festucae, Eriopeltis lichtensteinii, Eriopeltis luzulae, Eucalymnatus tessellatus, Eulecanium sp., Eulecanium arion, Eulecanium bituberculatum, Eulecanium cerasifax, Eulecanium corni, Eulecanium coryli, Eulecanium crudum, Eulecanium ficiphilum, Eulecanium fletcheri, Eulecanium graniforme, Eulecanium kunoense, Eulecanium mali, Eulecanium nigrofasciatum, Eulecanium persicae, Eulecanium pomeranicum, Eulecanium pruinosum, Eulecanium pulchrum, Eulecanium rufulum, Eulecanium spiraeae, Eulecanium tesselatus, Eulecanium turanicum, Eulecanium unifasciatum, Eupulvinaria hydrangeae, Eupulvinaria peregrina, Filippia folicularis, Filippia viburni, Lecanium sp., Lecanium corni, Lecanium coryli, Lecanium fletcheri, Lecanium hesperidum, Lecanium kunoensis, Lecanium nigrofasciatum, Lecanium persicae, Lecanium pomeranicum, Lecanium pruinosum, Lecanium quercifex, Lecanium tiliae, Lichtensia viburni, Luzulaspis bisetosa, Luzulaspis frontalis, Luzulaspis luzulae, Mesolecanium nigrofasciatum, Nemolecanium graniformis, Neopulvinaria imeretina, Neopulvinaria innumerabilis, Palaeolecanium bituberculatum, Parasaissetia nigra, Parthenolecanium corni, Parthenolecanium fletcheri, Parthenolecanium persicae, Parthenolecanium pomeranicum, Parthenolecanium quercifex, Parthenolecanium rufulum, Philephedra lutea, Philephedra tuberculosa, Physokermes insignicola, Physokermes piceae, Psilococcus ruber, Pulvinaria sp., Pulvinaria acericola, Pulvinaria amygdali, Pulvinaria aurantii, Pulvinaria betulae, Pulvinaria bigeloviae, Pulvinaria elongata, Pulvinaria floccifera, Pulvinaria horii, Pulvinaria innumerabilis, Pulvinaria mesembryanthemi, Pulvinaria phaiae, Pulvinaria polygonata, Pulvinaria pruni, Pulvinaria psidii, Pulvinaria regalis, Pulvinaria vitis, Rhodococcus turanicus, Saissetia sp., Saissetia coffeae, Saissetia hemisphaerica, Saissetia miranda, Saissetia neglecta, Saissetia nigra, Saissetia oleae, Sphaerolecanium prunastri, Stotzia maxima, Toumeyella liriodendri, Toumeyella pini, Toumeyella pinicola, Vittacoccus longicornis (Coccidae), *Aonidiella aurantii, Aonidiella citrina, Aspidiotus hederae, Aspidiotus perniciosus, Chrysomphalus dictyospermi, Lepidosaphes beckii, Lepidosaphes japonica, Lepidosaphes ulmi, Parlatoria oleae, Parlatoria pergandii, Pinnaspis strachani, Quadraspidiotus perniciosus, Unaspis euonymi* (Diaspididae), *Acanthococcus aceris, Eriococcus aceris, Eriococcus buxi, Eriococcus obscurus, Gossyparia spuria, Rhizococcus obscurus* (Eriococcidae), *Kermes sp., Kermococcus quercus* (Kermesidae), *Icerya purchasi* (Margarodidae), *Coccura ussuriensis, Phenacoccus acericola,*

Planococcus citri (Pseudococcidae), *Psylla pyri* (Psyllidae), *Stictococcus gowdeyi* (Stictococcidae) (Noyes, 2019).
Parasitoid hosts: *Coccophagus lycimnia* (Hymenoptera: Aphelinidae), *Blastothrix longipennis, Microterys flavus* (Encyrtidae) (Noyes, 2019).
Parasitoids: *Marietta javensis, Marietta leopardina, Marietta mexicana* (Aphelinidae), *Aprostocetus minutus, Tetrastichus minutus* (Eulophidae), *Pachyneuron sp., Pachyneuron solitarium* (Pteromalidae) (Noyes, 2019).
Distribution: Argentina, Armenia, Australia, Azerbaijan, Belgium, Bermuda, Brazil, Bulgaria, Canada, Caucasus, Chile, Colombia, Croatia, Cuba, Cyprus, Czech Republic, Czechoslovakia, Egypt, El Salvador, Europe, France, Georgia, Germany, Greece, Hawaii, Hungary, India, Indonesia, Israel, Italy, Jamaica, Japan, Kazakhstan, Mexico, Moldova, Montenegro, Morocco, Nearctic, Neotropical, Netherlands, Oriental, People's Republic of China, Peru, Poland, Portugal, Puerto Rico, Romania, Russia, Serbia, Slovakia, South Africa, Spain, St Vincent & Grenadines, Sweden, Switzerland, Taiwan, Trinidad & Tobago, Tselinograd Obl., Turkey, Uganda, Ukraine, England, United States of America, USSR, Uzbekistan, Yugoslavia (Federal Republic) (Noyes, 2019).

Coccophagus palaeolecanii (Yasnosh, 1957)
Publications with Turkey records: Erol and Yasar (1999), Noyes (2019), Öncüer (1991), Ulgenturk (2001).
Associates: *Hedera helix* (Araliaceae), *Ephedra procera* (Gnetaceae), *Ligustrum vulgare, Olea europaea* (Oleaceae), *Crataegus sp., Malus pumila* (Rosaceae) (Noyes, 2019).
Hosts: *Eulecanium bituberculatum, Filippia folicularis, Filippia viburni, Lichtensia viburni, Neopulvinaria, Palaeolecanium bituberculatum, Sphaerolecanium prunastri, Stotzia maxima* (Hemiptera: Coccidae) (Noyes, 2019).
Parasitoids: *Pachyneuron solitarium* (Hymenoptera: Pteromalidae) (Noyes, 2019).
Distribution: Caucasus, Georgia, Germany, Moldova, Romania, Transcaucasus, Turkey, USSR (Noyes, 2019).

Coccophagus piceae (Erdos, 1956)
Publications with Turkey records: Noyes (2019), Öncüer (1991), Ulgenturk (2001).
Associates: *Betula sp.* (Betulaceae) (Noyes, 2019).
Hosts: *Pulvinaria sp., Pulvinaria betulae, Pulvinaria vitis* (Hemiptera: Coccidae) (Noyes, 2019).

Distribution: Caucasus, Czech Republic, Hungary, Poland, Russia, Slovakia, Sweden, Turkey (Noyes, 2019).

Coccophagus proximus (Yasnosh, 1966)
Publications with Turkey records: Noyes (2019), Ulgenturk (2001).
Associates: *Armeniaca vulgaris, Prunus sp., Prunus cerasifera* (Rosaceae) (Noyes, 2019).
Hosts: *Sphaerolecanium prunastri* (Hemiptera: Coccidae) (Noyes, 2019).
Distribution: Azerbaijan, Georgia, Moldova, Romania, Turkey (Noyes, 2019).

Coccophagus scutellaris (Dalman, 1826)
Publications with Turkey records: Japoshvili and Karaca (2002), Noyes (2019), Öncüer (1974; 1991).
Associates: *Pistacia mutica* (Anacardiaceae), *Nerium sp.* (Apocynaceae), *Baccharis sp., Baccharis pilularis* (Asteraceae), *Cucurbita sp.* (Cucurbitaceae), *Ficus carica* (Moraceae), *Psidium guajava* (Myrtaceae), *Olea sp., Olea europaea* (Oleaceae), *Bambusa sp., Saccharum officinarum* (Poaceae), *Prunus armeniaca, Prunus persica* (Rosaceae), *Rubus sp.* (Rubiaceae), *Citrus sp., Citrus deliciosa, Citrus limon, Citrus nobilis, Citrus unshiu* (Rutaceae), *Populus hybrida, Populus transcaucasica* (Salicaceae), *Parthenocissus quinquefolia, Vitis vinifera* (Vitaceae) (Noyes, 2019).
Hosts: *Saccharum officinarum* (Poaceae), *Alichtensia orientalis, Anapulvinaria pistaciae, Ceroplastes sp., Ceroplastes floridensis, Ceroplastes rusci, Chloropulvinaria aurantia, Chloropulvinaria floccifera, Coccus sp., Coccus elongatus, Coccus hesperidum, Coccus longulus, Coccus mangiferae, Coccus pseudomagnoliarum, Coccus viridis, Eriopeltis araxis, Eulecanium sp., Eulecanium bituberculatum, Eulecanium corni, Eulecanium coryli, Eulecanium fletcheri, Eulecanium persicae, Filippia folicularis, Filippia viburni, Lecanium sp., Lecanium corni, Lecanium coryli, Lecanium hesperidum, Lecanium persicae, Lecanium prunastri, Neopulvinaria imeretina, Parasaissetia nigra, Parthenolecanium corni, Physokermes piceae, Pulvinaria sp., Pulvinaria aurantia, Pulvinaria betulae, Pulvinaria floccifera, Pulvinaria mesembryanthemi, Pulvinaria populi, Pulvinaria psidii, Pulvinaria regalis, Pulvinaria tenuivalvata, Pulvinaria vitis, Saissetia sp., Saissetia coffeae, Saissetia hemisphaerica, Saissetia nigra, Saissetia oleae, Sphaerolecanium prunastri* (Hemiptera: Coccidae), *Aonidiella aurantia, Aonidiella citrina, Aspidiotus hederae, Hemiberlesia rapax* (Diaspididae),

Icerya purchase (Margarodidae), *Ferrisiana oleae* (Pseudococcidae) (Noyes, 2019).
Parasitoid hosts: *Coccophagus scutellaris* (Hymenoptera: Aphelinidae) (Noyes, 2019).
Distribution: Afrotropical, Argentina, Australia, Bermuda, Brazil, Canada, Canary Islands, Caucasus, Czechoslovakia, Dominican Republic, Egypt, Europe, France, Georgia, Germany, Greece, Hungary, Iran, Israel, Italy, Japan, Lebanon, Mexico, Morocco, Nearctic, Neotropical, North Africa, People's Republic of China, Poland, Portugal, Puerto Rico, Russia, South Africa, Spain, Sweden, Switzerland, Turkey, Ukraine, United States of America, Uruguay, USSR, Venezuela, Yugoslavia (pre1991) (Noyes, 2019).

Coccophagus spartanus (Japoshvili and Karaca, 2002)
Publications with Turkey records: Japoshvili and Karaca (2002a; 2002b), Noyes (2019).
Associates: *Armeniaca vulgaris, Prunus sp., Prunus divaricate* (Rosaceae) (Noyes, 2019).
Host: *Sphaerolecanium prunastri* (Hemiptera: Coccidae) (Noyes, 2019).
Distribution: Turkey (Noyes, 2019).

Genus *Diaspiniphagus* (Silvestri, 1927)
Diaspiniphagus moeris (Walker, 1839)
Publications with Turkey records: Erler and Tunç (2001), Japoshvili and Karaca (2002b), Noyes (2019).
Associates: *Corylus avellana* (Betulaceae), *Elaeagnus angustifolia* (Elaeagnaceae), *Olea sp.* (Oleaceae), *Cedrus sp., Pinus eldarica* (Pinaceae), *Amygdalus sp.* (Rosaceae), *Populus sp., Populus hybrida, Populus transcaucasica* (Salicaceae), *Vitis vinifera* (Vitaceae) (Noyes, 2019).
Hosts: *Asterolecanium minus* (Hemiptera: Asterolecaniidae), *Acanthomytilus cedricola, Chionaspis lepineyi, Chionaspis salicis, Diaspidiotus elaeagni, Diaspidiotus gigas, Diaspidiotus hungaricus, Diaspidiotus ostreaeformis, Diaspidiotus prunorum, Diaspidiotus spurcatus, Diaspidiotus viticola, Diaspis boisduvalii, Lepidosaphes ulmi, Leucaspis loewi, Leucaspis pusilla, Mytilaspis conchiformis, Nuculaspis abietis, Parlatoria oleae, Quadraspidiotus gigas, Quadraspidiotus ostreaeformis, Quadraspidiotus perniciosus, Quadraspidiotus slavonicus, Quadraspidiotus zonatus, Salicicola kermanensis, Targionia vitis, Tecaspis prunorum, Unaspis euonymi, Zepidaphes mesasiatica* (Diaspididae) *Coccura ussuriensis* (Pseudococcidae) (Noyes, 2019).

Parasitoid hosts: *Azotus celsus, Coccophagoides similis* (Hymenoptera: Aphelinidae) (Noyes, 2019).
Parasitoids: *Azotus celsus* (Hymenoptera: Aphelinidae) (Noyes, 2019).
Distribution: Austria, Azerbaijan, Bulgaria, Caucasus, Czech Republic, Egypt, England, Europe, France, Georgia, Germany, Hungary, Italy, Nearctic, Poland, Russia, Spain, Sweden, Switzerland, Turkey, United States of America, USSR, Uzbekistan (Noyes, 2019).

Genus *Encarsia* (Foerster, 1878)
Encarsia aspidioticola (Mercet, 1929)
Publications with Turkey records: Noyes (2019).
Associates: *Hedera helix* (Araliaceae), *Ruscus aculeatus, Ruscus racemosus* (Liliaceae) (Noyes, 2019).
Hosts: *Bemisia tabaci* (Hemiptera: Aleyrodidae), *Aspidiotus abietis, Dynaspidiotus abietis, Dynaspidiotus Britannicus, Nuculaspis abietis* (Diaspididae) (Noyes, 2019).
Parasitoid hosts: *Encarsia aspidioticola* (Hymenoptera: Aphelinidae) (Noyes, 2019).
Distribution: Caucasus, Georgia, Hungary, Italy, Spain, Switzerland, Turkey, USSR (Noyes, 2019).

Encarsia berlesei (Howard, 1906)
Publications with Turkey records: Altay et al. (1972), Erkilic and Uygun (1995), Noyes (2019), Öncüer (1991), Sengonca et al. (1998).
Associates: *Pteroptrix orientalis* (Aphelinidae), *Actinidia chinensis, Actinidia deliciosa* (Actinidiaceae), *Catalpa bignonioides* (Bignoniaceae), *Sophora japonica* (Fabaceae), *Quercus palustris* (Fagaceae), *Carya illinoinensis* (Juglandaceae), *Morus sp.* (Moraceae), *Passiflora edulis* (Passifloraceae), *Pinus sylvestris* (Pinaceae), *Amygdalus communis, Malus pumila, Prunus domestica, Prunus persica* (Rosaceae), *Citrus aurantium* (Rutaceae), *Camellia sinensis* (Theaceae) (Noyes, 2019).
Hosts: *Aulacaspis cinnamomic, Chrysomphalus dictyospermi, Chrysomphalus obscurus, Diaspis pentagona, Melanaspis obscura, Nuculaspis abietis, Parlatoria pergandii, Pinnaspis minor, Pinnaspis strachani, Pinnaspis temporaria, Pseudaulacaspis pentagona* (Hemiptera: Diaspididae) (Noyes, 2019).
Distribution: Afghanistan, Argentina, Austria, Bermuda, Bolivia, Brazil, Caucasus, Chile, Colombia, Croatia, Cuba, Dominican Republic, Egypt, Europe, France, Germany, Greece, Hungary, Iran, Italy, Japan, Korea, Montenegro, Nearctic, Neotropical, People's Republic of China, Peru, Puerto Rico, Russia, Slovenia, Spain, Sri Lanka, Switzerland, Trinidad &

Tobago, Turkey, United States of America, Uruguay, USSR, Western Samoa, Yugoslavia (pre-1991) (Noyes, 2019).

Encarsia citrina (Craw, 1891)

Publications with Turkey records: Bozan and Yildirim (1992), Japoshvili and Karaca (2002), Noyes (2019), Öncüer (1991), Soydanbay (1977), Tuncyürek and Öncüer (1974), Tuncyürek-Soydanbay and Erkin (1979; 1981).

Associates: *Actinidia chinensis* (Actinidiaceae), *Yucca sp.* (Agavaceae), *Alternanthera sp.* (Amaranthaceae), *Mangifera indica* (Anacardiaceae), *Hedera sp., Hedera helix* (Araliaceae), *Cocos nucifera* (Arecaceae), *Corylus avellana* (Betulaceae), *Myrtillocactus geometrizans, Opuntia sp.* (Cactaceae), *Carica papaya* (Caricaceae), *Euonymus sp., Euonymus fortune, Euonymus japonica* (Celastraceae), *Cucurbita sp.* (Cucurbitaceae), *Cupressus sp., Juniperus foetidissima, Thuja orientalis* (Cupressaceae), *Hevea brasiliensis* (Euphorbiaceae), *Pithecellobium sp.* (Fabaceae), *Actinidia chinensis, Beilschmiedia taraire, Laurus sp., Laurus nobilis, Persea americana* (Lauraceae), *Ruscus hypoglossum* (Liliaceae), *Hibiscus sp., Hibiscus roseus* (Malvaceae), *Ficus sp.* (Moraceae), *Olea europaea* (Oleaceae), *Picea sp., Pinus sp., Pinus eldarica, Pinus nigra, Tsuga sp., Tsuga canadensis* (Pinaceae), *Pittosporum sp.* (Pittosporaceae), *Plantago sp.* (Plantaginaceae), *Podocarpus sp.* (Podocarpaceae), *Rhamnus pallasii* (Rhamnaceae), *Amygdalus sp., Malus pumila, Prunus sp., Prunus persica, Pyrus communis, Rosa sp.* (Rosaceae), *Rubus idaeus* (Rubiaceae), *Amyris madrensis, Citrus sp., Citrus medica, Citrus paradisi, Citrus reticulata, Citrus sinensis, Citrus unshiu* (Rutaceae), *Santalum album* (Santalaceae), *Cupaniopsis sp.* (Sapindaceae), *Camellia sinensis* (Theaceae) (Noyes, 2019).

Hosts: *Aleurocanthus sp., Aleurotrachelus micheliae* (Hemiptera: Aleyrodidae), *Asterolecanium pustulans* (Asterolecaniidae), *Ceroplastes floridensis, Coccus hesperidum, Saissetia oleae* (Coccidae), *Fagisuga triloba* (Conchaspididae), *Abgrallaspis cyanophylli, Aonidia lauri, Aonidiella sp., Aonidiella aurantia, Aonidiella citrina, Aonidiella lauri, Aonidiella orientalis, Aonidiella taxus, Aonidiella tinerfensis, Aspidiotus sp., Aspidiotus beilschmiediae, Aspidiotus Britannicus, Aspidiotus camelliae, Aspidiotus cryptomeriae, Aspidiotus cyanophylli, Aspidiotus destructor, Aspidiotus hederae, Aspidiotus hedericola, Aspidiotus nerii, Aspidiotus palmae, Aspidiotus perniciosus, Aspiditous nerii, Aulacaspis rosae, Aulacaspis rosarum, Aulacaspis tubercularis, Borchseniaspis palmae, Carulaspis carueli, Carulaspis juniperi, Carulaspis minima, Carulaspis visci, Chionaspis sp., Chionaspis bambusae, Chionaspis* *berlesii, Chionaspis euonymi, Chionaspis graminis, Chionaspis heterophyllae, Chionaspis lepineyi, Chionaspis pinifoliae, Chionaspis salicis, Chionaspis striata, Chrysomphalus sp., Chrysomphalus agavis, Chrysomphalus aonidum, Chrysomphalus dictyospermi, Chrysomphalus ficus, Chrysomphalus pinnulifer, Chrysomphalus rossi, Chrysomphalus tenebricosus, Comstockiella sabalis, Cornuaspis beckii, Cornuaspis gloverii, Diaspidiotus alni, Diaspidiotus ancylus, Diaspidiotus gigas, Diaspidiotus hungaricus, Diaspidiotus ostreaeformis, Diaspidiotus perniciosus, Diaspidiotus prunorum, Diaspidiotus pyri, Diaspidiotus spurcatus, Diaspidiotus uvae, Diaspidiotus wuenni, Diaspidiotus zonatus, Diaspis boisduvalii, Diaspis bromeliae, Diaspis calyptroides, Diaspis echinocacti, Diaspis visci, Duplachionaspis berlesei, Duplachionaspis graminis, Dynaspidiotus britannicus, Dynaspidiotus californica, Fiorinia externa, Fiorinia fioriniae, Fiorinia stricta, Furchadaspis zamiae, Genaparlatoria pseudaspidiotus, Gonaspidiotus minimus, Hemiberlesia cyanophylli, Hemiberlesia lataniae, Hemiberlesia minima, Hemiberlesia palmae, Hemiberlesia pitysophila, Hemiberlesia rapax, Insulaspis gloverii, Insulaspis pallidula, Insulaspis tapleyi, Kuwanaspis bambusae, Kuwanaspis hikosani, Kuwanaspis howardi, Kuwanaspis pseudoleucaspis, Lepidosaphes sp., Lepidosaphes alba, Lepidosaphes auriculata, Lepidosaphes beckii, Lepidosaphes gloverii, Lepidosaphes japonica, Lepidosaphes machili, Lepidosaphes pallida, Lepidosaphes pineti, Lepidosaphes pinnaeformis, Lepidosaphes tapleyi, Lepidosaphes ulmi, Lepidosaphes vermiculus, Leucaspis sp., Leucaspis loewi, Leucaspis pini, Leucaspis pusilla, Leucaspis signoreti, Lindingaspis floridana, Lindingaspis rossi, Lineaspis striata, Lopholeucaspis japonica, Melanaspis sitreana, Melanaspis tenebricosa, Mytilaspis conchiformis, Mytilococcus beckii, Nuculaspis californica, Nuculaspis tsugae, Odonaspis secreta, Parlatoria oleae, Parlatoria pergandii, Parlatoria pittospori, Parlatoria ziziphin, Phenacaspis eugeniae, Phenacaspis pinifoliae, Pinnaspis aspidistrae, Pinnaspis buxi, Pinnaspis strachani, Pinnaspis temporaria, Pseudaonidia duplex, Pseudaonidia paeoniae, Pseudaulacaspis pentagona, Pseudoparlatoria giffardi, Quadraspidiotus gigas, Quadraspidiotus juglansregiae, Quadraspidiotus ostreaeformis, Quadraspidiotus pelagijae, Quadraspidiotus perniciosus, Quadraspidiotus pyri, Quadraspidiotus zonatus, Selenaspidius articulates, Silvestraspis sp., Temnaspidiotus destructor, Tsugaspidiotus tsugae, Unachionaspis bambusae, Unaspis citri, Unaspis euonymi, Unaspis yanonensis* (Diaspididae), *Gossyparia spuria* (Eriococcidae) (Noyes, 2019).

Parasitoid hosts: *Archenomus bicolor* (Hymenoptera: Aphelinidae), *Adelencyrtus aulacaspidis, Apterencyrtus microphagus* (Encyrtidae) (Noyes, 2019).
Parasitoids: *Marietta carnesi* (Aphelinidae) (Noyes, 2019).
Distribution: Afrotropical, Algeria, Argentina, Australia, Azerbaijan, Barbados, Belgium, Bermuda, Bolivia, Brazil, Bulgaria, Canada, Caucasus, Chile, Colombia, Cook Islands, Cuba, Czechoslovakia, Dominican Republic, Egypt, El Salvador, Europe, Fiji, France, French Polynesia, Georgia, Germany, Ghana, Greece, Grenada, Guam, Guyana, Haiti, Hawaii, Hungary, India, Indonesia, Iran, Italy, Jamaica, Japan, Korea, Macedonia, Malaysia, Mauritius, Mexico, Moldova, Morocco, Nearctic, Neotropical, Netherlands, New Zealand, North Africa, Oriental, People's Republic of China, Peru, Philippines, Portugal, Puerto Rico, Russia, Sao Tomé and Principe, Saudi Arabia, Serbia, Slovakia, South Africa, Spain, Sri Lanka, Sweden, Switzerland, Taiwan, Transcaucasus, Trinidad & Tobago, Tunisia, Turkey, Ukraine England, United States of America, USSR, Wallis and Futuna Islands, Yugoslavia (pre-1991), Zimbabwe (Noyes, 2019).

Encarsia formosa (Gahan, 1924)
Publications with Turkey records: Noyes (2019), Öncüer et al. (1994), Yasarakinci and Hincal (1997), Yoldas (1995), Yoldas et al. (1999).
Associates: *Amaranthus spinosus* (Amaranthaceae), *Xanthosoma sp.* (Araceae), *Bidens pilosa, Chrysanthemum sp., Conyza sp., Emilia sonchifolia, Erigeron sp., Gerbera sp., Gerbera jamesoni, Lactuca serriola, Sonchus sp., Sonchus oleraceus, Tagetes sp., Tagetes erecta, Xanthium occidentale* (Asteraceae), *Betula pendula* (Betulaceae), *Brassica sp., Brassica oleracea* (Brassicaceae), *Ipomoea sp., Ipomoea batatas* (Convolvulaceae), *Citrullus lanatus, Cucumis sp., Cucumis melo, Cucumis sativus, Cucurbita sp., Cucurbita maxima, Cucurbita moschata, Cucurbita pepo* (Cucurbitaceae), *Vaccinium corymbosum* (Ericaceae), *Euphorbia sp., Euphorbia peplus, Euphorbia pulcherrima, Manihot esculenta, Poinsettia sp., Poinsettia pulcherrima, Ricinus communis* (Euphorbiaceae), *Glycine max, Phaseolus sp., Phaseolus vulgaris, Sophora secundiflora, Vicia faba* (Fabaceae), *Acremonium strictum, Oidium lycopersium (Fungus), Geranium sp., Pelargonium sp.* (Geraniaceae), *Salvia sp.* (Lamiaceae), *Alstroemeria sp.* (Liliaceae), *Cuphea carthaginensis* (Lythraceae), *Alcea rosea, Anoda cristata, Gossypium sp., Gossypium barbadense, Gossypium hirsutum, Hibiscus sp., Hibiscus rosa-sinensis, Malvastrum coromandelianum* (Malvaceae), *Azadirachta indica* (Meliaceae), *Fuchsia sp.* (Onagraceae), (Poaceae), *Fragaria sp.* (Rosaceae), *Bouvardia sp.*

(Rubiaceae), *Citrus sp., Citrus sinensis, Ruta sp., Ruta officinalis* (Rutaceae), *Calceolaria andina* (Scrophulariaceae), *Atropa belladonna, Capsicum sp., Capsicum annuum, Datura sp., Lycopersicon esculentum, Nicotiana sp., Nicotiana glauca, Nicotiana tabacum, Solanum melongena, Solanum tuberosum (Solanaceae), Corchorus olitorius, Tilia sp.* (Tiliaceae), *Urtica sp.* (Urticaceae), *Lantana sp., Lantana camara, Verbena sp., Verbena bonariensis* (Verbenaceae), *Bemisia tabaci, Trialeurodes vaporariorum* (Hemiptera: Aleyrodidae) (Noyes, 2019).

Hosts: *Ageratum sp.* (Asteraceae), *Aleuroglandulus malangae, Aleuroglandulus subtilis, Aleurolobus malangae, Aleurothrixus floccosus, Aleurotrachelus trachoides, Aleyrodes sp., Aleyrodes lonicerae, Aleyrodes proletella, Aleyrodes singularis, Aleyrodes spiraeoides, Aleyrodidae unspecified, Asterochiton vaporariorum, Bemisia sp., Bemisia argentifolii, Bemisia tabaci, Dialeurodes chittendeni, Dialeurodes citri, Lipaleyrodes atriplex, Lipaleyrodes euphorbiae, Massilieurodes chittendeni, Tetraleurodes sp., Tetraleurodes mori, Trialeurodes sp., Trialeurodes abutiloneus, Trialeurodes ricini, Trialeurodes vaporariorum, Trialeurodes variablis* (Hemiptera: Aleyrodidae) (Noyes, 2019).

Parasitoid hosts: *Encarsia formosa* (Hymenoptera: Aphelinidae) (Noyes, 2019).

Parasitoids: *Encarsia sp., Encarsia pergandiella, Encarsia tricolor* (Hymenoptera: Aphelinidae) *Syrphophagus sp.* (Encyrtidae), *Aschersonia aleyrodis, Beauveria bassiana, Paecilomyces fumosoroseus* (Fungus), *Bacillus thuringiensis* (Bacteria: Bacillaceae), *Wolbachia sp.* (Wolbachiaceae), *Bemisia tabaci, Trialeurodes vaporariorum* (Hemiptera: Aleyrodidae), *Macrolophus caliginosus, Nesidiocaris tenuis* (Miridae), *Encarsia pergandiella, Encarsia tricolor, Eretmocerus eremicus, Eretmocerus mundus* (Hymenoptera: Aphelinidae) (Noyes, 2019).

Distribution: Algeria, Argentina, Australia, Austria, Azores, Barbados, Belarus, Belgium, Bermuda, Brazil, Bulgaria, Canada, Canary Islands, Caribbean, Caucasus, Channel Islands (British Is), Colombia, Costa Rica, Czech Republic, Czechoslovakia, Denmark, Dominican Republic, Egypt, Europe, Fiji, Finland, France, French Polynesia, Georgia, Germany, Greece, Hawaii, Hungary, Iran, Israel, Italy, Japan, Jordan, Korea, Korea, Madeira, Malta, Mexico, Moldova, Nearctic, Netherlands, New Zealand, Norway, Pakistan, People's Republic of China, Poland, Portugal, Puerto Rico, Réunion, Romania, Russia, Serbia, Slovakia, Slovenia, Scotland, South Africa, Spain, Sweden, Switzerland, Thailand, Tonga, Trinidad & Tobago, Turkey, Ukraine, England, United States of America, USSR, Uzbekistan, Yugoslavia (pre1991) (Noyes, 2019).

Encarsia inaron (Walker, 1839)
Publications with Turkey records: Noyes (2019), Ulusoy and Ulgenturk (2003).
Associates: *Hedera helix* (Araliaceae), *Conyza sp., Lactuca sp., Lactuca serriola, Sonchus sp.* (Asteraceae), *Corylus avellana* (Betulaceae), *Brassica oleracea* (Brassicaceae), *Ipomoea sp.* (Convolvulaceae), *Cucumis sativus* (Cucurbitaceae), *Azalea sp., Rhododendron sp.* (Ericaceae), *Mercurialis annua, Ricinus communis* (Euphorbiaceae), *Robinia sp.* (Fabaceae), *Punica granatum* (Lythraceae), *Gossypium sp., Gossypium hirsutum* (Malvaceae), *Fraxinus sp., Fraxinus uhdei, Fraxinus velutina, Phillyrea sp.* (Oleaceae), *Chelidonium majus* (Papaveraceae), *Picea excelsa* (Pinaceae), *Thalictrum minus* (Ranunculaceae), *Crataegus sp., Pyrus sp., Pyrus calleryana, Pyrus communis, Rosa beggeriana* (Rosaceae), *Citrus sp., Citrus aurantifolia* (Rutaceae), *Lycopersicon esculentum, Solanum sp., Solanum nigrum* (Solanaceae), *Ulmus campestris, Ulmus carpinifolia* (Ulmaceae), *Urtica dioica* (Urticaceae), *Lantana camara* (Verbenaceae), *Vitis sp.* (Vitaceae) (Noyes, 2019).
Hosts: *Acaudaleyrodes citri, Acaudaleyrodes rachipora, Aleurocanthus woglumi, Aleurothrixus floccosus, Aleyrodes sp., Aleyrodes brassicae, Aleyrodes elevates, Aleyrodes lonicerae, Aleyrodes proletella, Aleyrodes singularis, Aleyrodes spiraeae, Asterobemisia avellanae, Asterobemisia carpini, Asterobemisia paveli, Bemisia sp., Bemisia tabaci, Bulgarialeurodes cotesii, Pealius azalea, Pealius madeirensis, Pealius quercus, Siphoninus sp., Siphoninus granati, Siphoninus immaculatus, Siphoninus phillyreae, Tetraleurodes hederae, Trialeurodes hederae, Trialeurodes ricini, Trialeurodes vaporariorum* (Hemiptera: Aleyrodidae), *Coccus hesperidum, Eulecanium coryli, Pulvinaria vitis* (Coccidae), *Lepidosaphes ulmi* (Diaspididae), *Spodoptera littoralis* (Lepidoptera: Noctuidae), *Ephestia kuehniella, Ostrinia nubilalis* (Pyralidae), *Cydia pomonella, Eupoecilia ambiguella* (Tortricidae), *Plutella xylostella* (Yponomeutidae) (Noyes, 2019).
Parasitoids: *Encarsia tricolor* (Hymenoptera: Aphelinidae) (Noyes, 2019).
Distribution: Afghanistan, Austria, Azerbaijan, Brazil, Bulgaria, Canary Islands, Caucasus, Egypt, England, Europe, Finland, France, Georgia, Germany, Greece, Hungary, India, Iran, Israel, Italy, Jordan, Kazakhstan, Lebanon, Madeira, Mexico, Montenegro, Morocco, Nearctic, New Zealand, North Africa, Pakistan, Palaearctic, People's Republic of China, Russia, Serbia, South Africa, Spain, Sweden, Switzerland, Syria, Taiwan, Thailand, Tselinograd Obl., Turkey, Turkmenistan, Ukraine, United States of America, USSR, Uzbekistan, Venezuela (Noyes, 2019).

Encarsia lahorensis (Howard, 1911)
Publications with Turkey records: Noyes (2019), Ozkan et al. (1996), Ulusoy et al. (1996), Uygun et al. (1996), Uygun et al. (1997), Yoldas and Öncüer (1992).
Associates: *Diospyros kaki, Diospyros virginiana* (Ebenaceae), *Ligustrum sp.* (Oleaceae), *Citrus sp., Citrus bigaradia, Citrus limon, Citrus reticulata* (Rutaceae), *Corchorus trilocularis* (Tiliaceae) (Noyes, 2019).
Hosts: *Aleurocanthus spiniferus, Aleurodicus disperses, Aleyrodes citri, Aleyrodes elevates, Aleyrodes ricini, Dialeurodes citri, Dialeurodes citrifolii, Dialeurodes kirkaldyi, Tuberaleyrodes machili* (Hemiptera: Aleyrodidae) (Noyes, 2019).
Parasitoid hosts: *Encarsia lahorensis, Prospaltella lahorensis* (Hymenoptera: Aphelinidae) (Noyes, 2019).
Distribution: Egypt, Europe, France, Georgia, Greece, India, Iran, Israel, Italy, Mexico, Pakistan, People's Republic of China, Russia, Turkey, United States of America, USSR, Uzbekistan (Noyes, 2019).

Encarsia leucaspidis (Mercet, 1912)
Publications with Turkey records: Noyes (2019), Japoshvili and Karaca (2002).
Associates: *Pinus sp., Pinus eldarica, Pinus nigra, Pinus sylvestris* (Pinaceae) (Noyes, 2019).
Hosts: *Anamaspis loewi, Chionaspis pinifoliae, Insulaspis newsteadi, Lepidosaphes conchiformis, Lepidosaphes japonica, Lepidosaphes newsteadi, Leucaspis sp., Leucaspis loewi, Leucaspis pini, Leucaspis pusilla, Leucaspis signoreti, Nuculaspis abietis* (Hemiptera: Diaspididae) (Noyes, 2019).
Parasitoids: *Azotus atomon* (Hymenoptera: Aphelinidae) (Noyes, 2019).
Distribution: Caucasus, Czech Republic, Europe, France, Georgia, Germany, Hungary, Italy, North Africa, Palaearctic, Poland, Romania, Spain, Sweden, Switzerland, Turkey, Ukraine, USSR, Yugoslavia (pre1991) (Noyes, 2019).

Encarsia lounsburyi (Berlese and Paoli, 1916)
Publications with Turkey records: Noyes (2019), Öncüer (1991).
Associates: *Mangifera indica* (Anacardiaceae), *Hedera helix* (Araliaceae), *Aleurites sp.* (Euphorbiaceae), *Citrus sp., Citrus latifolia* (Rutaceae), *Camellia japonica* (Theaceae) (Noyes, 2019).
Hosts: *Abgrallaspis cyanophylli, Acutaspis perseae, Aonidia lauri, Aonidiella sp., Aonidiella aurantia, Aonidiella citrina, Aonidiella taxus, Aspidiotus sp., Aspidiotus destructor, Aspidiotus hederae, Aspidiotus nerii,*

Carulaspis juniperi, Carulaspis minima, Carulaspis visci, Chionaspis stantophris, Chrysomphalus sp., Chrysomphalus aonidum, Chrysomphalus dictyospermi, Chrysomphalus ficus, Chrysomphalus personatus, Chrysomphalus pinnulifer, Cornuaspis beckii, Cupidiaspis sp., Diaspis boisduvalii, Diaspis echinocacti, Fiorinia fioriniae, Hemiberlesia sp., Hemiberlesia cyanophylli, Hemiberlesia lataniae, Hemiberlesia rapax, Insulaspis pallidula, Lepidosaphes beckii, Lepidosaphes gloverii, Lepidosaphes newsteadi, Lepidosaphes pinnaeformis, Lepidosaphes ulmi, Leucaspis loewi, Lineaspis striata, Mytilococcus beckii, Parlatoria oleae, Parlatoria pergandii, Parlatoria proteus, Parlatoria ziziphin, Pinnaspis aspidistrae, Pinnaspis strachani, Pseudaulacaspis pentagona, Selenaspidius articulates, Unaspis citri, Unaspis yanonensis (Hemiptera: Diaspididae)(Noyes, 2019).

Distribution: Argentina, Australia, Bermuda, Brazil, Canada, Chile, Cuba, Dominican Republic, Egypt, France, Haiti, Indonesia, Iran, Italy, Madeira, Malawi, Mexico, Morocco, Mozambique, Nearctic, New Caledonia, North Africa, People's Republic of China, Peru, Philippines, Puerto Rico, Sao Tomé and Principe, Senegal, South Africa, Switzerland, Taiwan, Trinidad & Tobago, Turkey, United States of America, Venezuela (Noyes, 2019).

Encarsia perniciosi (Tower, 1913)

Publications with Turkey records: Bozan and Yildirim (1992), Noyes (2019), Öncüer (1991).

Associates: *Ceroplastes rubens* (Hemiptera: Coccidae), *Aphytis melinus* (Hymenoptera: Aphelinidae), *Mangifera indica* (Anacardiaceae), *Betula sp., Betula platyphylla* (Betulaceae), *Citrullus lanatus, Cucumis melo, Cucurbita sp., Cucurbita maxima* (Cucurbitaceae), *Robinia viscosa* (Fabaceae), *Ribes grossularia* (Grossulariaceae), *Morus sp.* (Moraceae), *Malus sp., Malus pumila, Prunus sp., Prunus dulcis, Prunus persica* (Rosaceae), *Citrus sp., Citrus grandis, Citrus limettioides, Citrus paradisi, Citrus sinensis, Murraya exotica* (Rutaceae), *Populus sp., Populus nigra* (Salicaceae)(Noyes, 2019).

Parasitoid hosts: *Aphytis proclia, Prospaltella perniciosi* (Hymenoptera: Aphelinidae) (Noyes, 2019).

Parasitoids: *Marietta carnesi* (Hymenoptera: Aphelinidae) (Noyes, 2019).

Hosts: *Aonidiella sp., Aonidiella aurantia, Aonidiella citri, Aspidiotus perniciosus, Comstockaspis perniciosa, Diaspidiotus gigas, Diaspidiotus perniciosus, Insulaspis pallidula, Lepidosaphes pallida, Lepidosaphes ulmi, Parlatoria acalcarata, Pseudaulacaspis pentagona, Quadraspidiotus*

gigas, Quadraspidiotus perniciosus, Unaspis citri (Hemiptera: Diaspididae) (Noyes, 2019).
Distribution: Argentina, Australia, Austria, Azerbaijan, Bermuda, Brazil, Bulgaria, Canada, Caucasus, Chile, Croatia, Czech Republic, Czechoslovakia, Egypt, Europe, France, Georgia, Germany, Greece, Hungary, India, Iran, Israel, Italy, Korea, Mexico, Moldova, Nearctic, Pakistan, People's Republic of China, Portugal, Russia, Serbia, Slovakia, Slovenia, South Africa, Spain, Switzerland, Taiwan, Turkey, United States of America, Uruguay, USSR, Yugoslavia (pre-1991) (Noyes, 2019).

Encarsia sophia (Girault and Dodd, 1915)
Publications with Turkey records: Noyes (2019), Ulusoy (1999).
Associates: *Delphastus pusillus* (Coleoptera: Coccinellidae), *Bemisia afer, Bemisia tabaci, Trialeurodes vaporariorum* (Hemiptera: Aleyrodidae), *Eretmocerus eremicus* (Hymenoptera: Aphelinidae), *Chrysoperla carnea* (Neuroptera: Chrysopidae), *Ilex sp.* (Aquifoliaceae), *Emilia sonchifolia, Helianthus annuus, Lactuca sp., Lactuca serriola, Sonchus oleraceus, Xanthium sp., Xanthium occidentale* (Asteraceae), *Brassica sp., Brassica oleracea* (Brassicaceae), *Opuntia sp.* (Cactaceae), *Carica papaya* (Caricaceae), *Ipomoea batatas* (Convolvulaceae), *Cucumis melo, Cucumis sativus, Cucurbita moschata* (Cucurbitaceae), *Chamaesyce hirta, Chamaesyce hyssopifolia, Euphorbia sp., Euphorbia pulcherrima, Manihot esculenta, Poinsettia sp., Poinsettia pulcherrima, Ricinus communis* (Euphorbiaceae), *Arachis hypogaea, Cajanus cajan, Dalbergia sissoo, Desmodium tortuosum, Glycine max, Medicago sativa, Phaseolus vulgaris, Prosopis juliflora* (Fabaceae), *Pelargonium sp.* (Geraniaceae), *Gossypium sp., Gossypium hirsutum, Hibiscus sp., Hibiscus mutabilis, Hibiscus rosa-sinensis* (Malvaceae), *Azadirachta indica* (Meliaceae), *Morus sp., Morus alba* (Moraceae), *Eucalyptus sp. Eucalyptus grandifolia, Eucalyptus leucoxylon* (Myrtaceae), *Sesamum indicum* (Pedaliaceae), *Oryza sativa, Saccharum officinarum* (Poaceae), *Grevillea decurrens, Hakea undulata* (Proteaceae), *Gardenia sp. Mussaenda sp.* (Rubiaceae), *Citrus sp., Citrus limon, Citrus sinensis, Ruta sp.* (Rutaceae), *Capsicum annuum, Lycopersicon esculentum, Nicotiana glauca, Solanum melongena, Solanum tuberosum* (Solanaceae), *Lantana sp., Lantana camara* (Verbenaceae) (Noyes, 2019).
Hosts: *Acaudaleyrodes rachipora, Aleurocybotus indicus, Aleurodicus disperses, Aleuroduplidens eucalyptifolia, Aleurolobus sp., Aleurolobus niloticus, Aleurothrixus floccosus, Aleyrodes hibisci, Aleyrodes proletella, Aleyrodes singularis, Aleyrodes vaporariorum, Aleyrodicinae unspecified sp., Aleyrodidae unspecified sp., Asterobemisia carpini, Asterochiton sp.,*

Asterochiton sonchi, Bemisia sp., Bemisia afer, Bemisia argentifolii, Bemisia hibisci, Bemisia myricae, Bemisia tabaci, Bemisia tuberculate, Chitonaleyrodes sp., Dialeurodes citri, Dialeuropora decempuncta, Parabemisia myricae, Pealius hibisci, Pealius longispinus, Pealius mori, Singhius hibisci, Trialeurodes sp., Trialeurodes ricini, Trialeurodes vaporariorum, Trialeurodes variablis, Vasdavidius indicus, Xenaleyrodes sp. (Hemiptera: Aleyrodidae), *Aphis sacchari* (Aphididae), *Diaspis sp.* (Diaspididae), *Diaphorina citri* (Psyllidae) (Noyes, 2019).

Parasitoid hosts: *Encarsia sophia, Encarsia transvena, Eretmocerus eremicus* (Hymenoptera: Aphelinidae) (Noyes, 2019).

Parasitoids: *Encarsia sophia, Encarsia transvena* (Hymenoptera: Aphelinidae) (Noyes, 2019).

Distribution: Afghanistan, Afrotropical, Algeria, Argentina, Australia, Burundi, Canary Islands, Cape Verde Islands, Caribbean (including West Indies), Dominican Republic, Egypt, Ethiopia, French Polynesia, Guadeloupe, Hawaii, Honduras, India, Indonesia, Iran, Israel, Italy, Ivory Coast, Japan, Kenya, Malawi, Malaysia, Martinique, Mexico, Morocco, Niger, North Africa, Pakistan, People's Republic of China, Philippines, Russia, Sierra Leone, Somalia, Spain, Sri Lanka, Taiwan, Thailand, Tunisia, Turkey, United States of America, Zimbabwe (Noyes, 2019).

Encarsia tricolor (Foerster, 1878)

Publications with Turkey records: Noyes (2019), Ulusoy (1999).

Associates: *Hedera helix* (Araliaceae), *Bidens pilosa*, *Lactuca serriola, Tolips laciniata* (Asteraceae), *Brassica oleracea* (Brassicaceae), *Canarina canariensis* (Campanulaceae), *Viburnum sp.* (Caprifoliaceae), *Ipomoea batatas* (Convolvulaceae), *Cucumis sativus, Sechium edule* (Cucurbitaceae), *Rhododendron sp*., (Ericaceae), *Euphorbia helioscopica, Euphorbia pulcherrima, Mercurialis annua, Poinsettia pulcherrima* (Euphorbiaceae), *Phaseolus vulgaris* (Fabaceae), *Quercus ilex* (Fagaceae), *Salvia glutinosa*, *Salvia incendiae* (Lamiaceae), *Ficus carica* (Moraceae), *Eugenia uniflora* (Myrtaceae), *Lysimachia nemorum* (Primulaceae), *Rubia peregrina, Rubus sp., Rubus fruticosus* (Rubiaceae), *Ruta officinalis* (Rutaceae), *Lycopersicon esculentum, Nicotiana glauca, Solanum muricatum* (Solanaceae)(Noyes, 2019).

Parasitoid hosts: *Encarsia formosa, Encarsia inaron, Encarsia lutea, Encarsia partenopea, Encarsia pergandiella, Encarsia tricolor* (Hymenoptera: Aphelinidae), *Euderomphale chelidonii* (Eulophidae) (Noyes, 2019).

Hosts: *Aleurotrachelus sp., Aleurotuba jelinekii, Aleyrodes sp., Aleyrodes brassicae, Aleyrodes elevates, Aleyrodes lonicerae, Aleyrodes philadelphi,*

Aleyrodes proletella, Aleyrodes singularis, Bemisia sp., Bemisia tabaci, Dialeurodes citri, Pealius sp., Pealius azalea, Siphoninus phillyreae, Tetraleurodes hederae, Trialeurodes vaporariorum (Hemiptera: Aleyrodidae) (Noyes, 2019).
Distribution: Azores, Belgium, Canary Islands, Caucasus, Czech Republic, England, Europe, France, Germany, Greece, Hungary, Italy, Madeira, Netherlands, Palaearctic, Portugal, Russia, Serbia, Spain, Sweden, Turkey, Ukraine, USSR (Noyes, 2019).

Genus *Pteroptrix* (Westwood, 1833)
Pteroptrix bicolor (Howard, 1898)
Publications with Turkey records: Japoshvili and Karaca (2002), Noyes (2019).
Associates: *Liquidambar styraciflua* (Hamamelidaceae), *Crataegus oxyacantha, Pyrus communis* (Rosaceae), *Populus sp., Populus hybrida, Populus transcaucasica* (Salicaceae) (Noyes, 2019).
Hosts: *Aspidiotus sp., Aulacaspis sp., Aulacaspis rosae, Carulaspis juniperi, Chionaspis sp., Chionaspis salicis, Diaspidiotus sp., Diaspidiotus alni, Diaspidiotus bavaricus, Diaspidiotus caucasicus, Diaspidiotus gigas, Diaspidiotus hungaricus, Diaspidiotus ostreaeformis, Diaspidiotus perniciosus, Diaspidiotus prunorum, Diaspidiotus pyri, Diaspidiotus spurcatus, Epidiaspis leperii, Lepidosaphes ulmi, Parlatoria crotonis, Phenacaspis vitis, Pseudaulacaspis sp., Pseudaulacaspis pentagona, Quadraspidiotus sp., Quadraspidiotus gigas, Quadraspidiotus ostreaeformis, Quadraspidiotus perniciosus, Quadraspidiotus pyri, Quadraspidiotus schneideri, Quadraspidiotus zonatus, Salicicola kermanensis, Syngenaspis parlatoriae, Targionia vitis, Tecaspis asiatica* (Hemiptera: Diaspididae) (Noyes, 2019).
Distribution: Argentina, Austria, Azerbaijan, Caucasus, Czech Republic, Egypt, England, Europe, France, Georgia, Germany, Hungary, Indonesia, Iran, Italy, Japan, Kazakhstan, Nearctic, North Africa, People's Republic of China, Russia, Slovakia, Spain, Sri Lanka, Switzerland, Tselinograd Obl., Tunisia, Turkey, Ukraine, United States of America, USSR (Noyes, 2019).

Pteroptrix lauri (Mercet, 1911)
Publications with Turkey records: Japoshvili and Karaca (2002), Noyes (2019).
Associates: *Platanus sp.* (Platanaceae), *Amygdalus sp.* (Rosaceae), *Populus sp., Populus hybrida, Populus nigra, Populus transcaucasica*

(Salicaceae), *Diaspidiotus caucasicus* (Hemiptera: Diaspididae) (Noyes, 2019).

Hosts: *Aonidiella lauri, Chionaspis salicis, Diaspidiotus caucasicus, Diaspidiotus gigas, Diaspidiotus ostreaeformis, Diaspidiotus perniciosus, Diaspidiotus prunorum, Diaspidiotus pyri, Diaspidiotus spurcatus, Fiorinia fioriniae, Hemiberlesia pitysophila, Lepidosaphes malicola, Lepidosaphes ulmi, Mytilaspis conchiformis, Odonaspis secreta, Parlatoria oleae, Quadraspidiotus armeniacus, Quadraspidiotus caucasicus, Quadraspidiotus gigas, Quadraspidiotus ostreaeformis, Quadraspidiotus perniciosus, Quadraspidiotus pyri. Salicicola kermanensis, Tecaspis prunorum* (Hemiptera: Diaspididae) (Noyes, 2019).

Distribution: Azerbaijan, Bulgaria, Caucasus, Czechoslovakia, Georgia, Italy, Moldova, North Africa, People's Republic of China, Poland, Russia, Slovakia, Spain, Turkey, USSR, Yugoslavia (pre1991) (Noyes, 2019).

Subfamily Eretmocerinae (Shafee and Khan, 1978)

Genus *Eretmocerus* (Haldeman, 1850)

Eretmocerus debachi (Rose and Rosen, 1992)

Publications with Turkey records: Noyes (2019), Sengonca et al. (1993; 1995; 1998), Ulusoy (1996), Uygun et al. (1990), Uygun et al. (1994; 1996; 1997).

Associates: *Punica granatum* (Lythraceae), *Morus sp., Morus alba* (Moraceae), *Prunus persica* (Rosaceae), *Citrus sp., Citrus aurantifolia, Citrus aurantium, Citrus bigaradia, Citrus limon, Citrus paradisi, Citrus sinensis, Poncirus trifoliatus* (Rutaceae), *Salix sp.* (Salicaceae), *Vitis vinifera* (Vitaceae) (Noyes, 2019).

Hosts: *Parabemisia myricae* (Hemiptera: Aleyrodidae) (Noyes, 2019).

Distribution: France, Iran, Israel, Italy, Spain, Tunisia, Turkey, United States of America (Noyes, 2019).

Eretmocerus diversiciliatus (Silvestri, 1915)

Publications with Turkey records: Noyes (2019), Öncüer (1991).

Associates: *Amaranthus retroflexus* (Amaranthaceae), *Lawsonia alba, Punica granatum* (Lythraceae), *Hibiscus esculentus* (Malvaceae), *Citrus sp.* (Rutaceae), *Lantana camara* (Verbenaceae) (Noyes, 2019).

Hosts: *Acaudaleyrodes citri, Aleurocanthus hansfordi, Aleuroclava similis, Bemisia gossypiperda, Bemisia tabaci, Siphoninus phillyreae* (Hemiptera: Aleyrodidae) (Noyes, 2019).

Distribution: Egypt, Iran, Israel, Nigeria, Sudan, Turkey (Noyes, 2019).

Eretmocerus mundus (Mercet, 1931)
Publications with Turkey records: Noyes (2019), Öncüer (1991).
Associates: *Cucumis sp.* (Cucurbitaceae), *Wolbachia sp.* (Bacteria: Wolbachiaceae), *Encarsia formosa, Encarsia lutea, Encarsia pergandiella, Eretmocerus queenslandensis* (Hymenoptera: Aphelinidae), *Gerbera sp., Sonchus sp., Sonchus oleraceus* (Asteraceae), *Brassica sp., Brassica oleracea* (Brassicaceae), *Cleome viscosa* (Capparidaceae), *Clethra arborea* (Clethraceae), *Ipomoea sp., Ipomoea batatas* (Convolvulaceae), *Cucumis melo, Cucumis sativus* (Cucurbitaceae), *Euphorbia sp., Euphorbia bertheloti, Euphorbia pulcherrima, Manihot esculenta, Phyllanthus amarus, Poinsettia sp., Poinsettia pulcherrima* (Euphorbiaceae), *Glycine max, Medicago sativa, Phaseolus vulgaris* (Fabaceae), *Fagus sylvatica* (Fagaceae), *Ajuga gorganica, Salvia officinalis* (Lamiaceae), *Lawsonia alba, Punica granatum* (Lythraceae), *Abelmoschus esculentum, Gossypium sp., Gossypium barbadense, Gossypium hirsutum, Hibiscus sp., Hibiscus esculentus, Hibiscus rosa-sinensis* (Malvaceae), *Jasminum sambac* (Oleaceae), *Crataegus microphylla, Rosa sp.* (Rosaceae), *Citrus sp.* (Rutaceae), *Salvadora persica* (Salvadoraceae), *Capsicum sp., Capsicum annuum, Lycopersicon esculentum, Nicotiana glauca, Solanum melongena, Solanum muricatum* (Solanaceae), *Lantana sp., Lantana camara, Lantana indica* (Verbenaceae) (Noyes, 2019).
Hosts: *Acaudaleyrodes citri, Aleuroplatus cadabae, Aleyrodes sp., Aleyrodes lonicerae, Aleyrodes proletella, Asterobemisia avellanae, Asterobemisia carpini, Bemisia sp., Bemisia afer, Bemisia argentifolii, Bemisia ovata, Bemisia tabaci, Dialeurodes kirkaldyi, Neomaskellia bergii, Rosanovia hulthemiae, Siphoninus phillyreae, Trialeurodes ricini, Trialeurodes vaporariorum* (Hemiptera: Aleyrodidae) (Noyes, 2019).
Parasitoids: *Encarsia pergandiella* (Hymenoptera: Aphelinidae) (Noyes, 2019).
Distribution: Australasian, Australia, Bangladesh, Canary Islands, Cyprus, Czech Republic, Egypt, Europe, France, Georgia, Germany, India, Iran, Israel, Italy, Jordan, Kenya, Madeira, Malawi, Mali, Malta, Mexico, Micronesia, Moldova, Nauru, Netherlands, Pakistan, Palaearctic, People's Republic of China, Russia, Spain, Sudan, Syria, Taiwan, Tunisia, Turkey, Turkmenistan, Ukraine, United Kingdom, United States of America, USSR (Noyes, 2019).

References

Bozan, İbrahim, and Yıldırım, A. Faik. 1992. "Investigations on the population fluctuations of the species of scales (*Aonidiella auranii*

Mask, *A. citrina* Coq. and *Chrysomphalus dictyospermi* Morg.) causing damage to citrus trees in the eastern Black Sea region of Turkey". *Zirai Mücadele Arastirma Yıllığı,* Turkey (Plant Protection Research Annual), No 22-23:64-67.

Elmali, Meryem, and Toros, S. 1994. "Investigations on the natural enemies of aphids damaging wheat in Konya Province". *Proceedings of the Third Turkish National Congress of Biological Control*, Izmir, 25-28 January, 1994, pp.13-28 ISBN 975-483-267-6.

Erkiliç, Lerzan, and Uygun, Nedim. 1995. "Distribution, population fluctuations and natural enemies of the white peach scale, *Pseudaulacaspis pentagona* (Targioni Tozzetti) (Homoptera: Diaspididae) in the east mediterranean region of Turkey". *Israel Journal of Entomology*, 29:191-198.

Erler, Fedai, and Tunç, İrfan. 2001. "A survey (1992-1996) of natural enemies of Diaspididae species in Antalya, Turkey". *Phytoparasitica*, 29(4):299-305.

Erol, Tülin, and Yaşar, Bülent. 1999. "Investigations on natural enemies, some biological characteristics and population fluctuations of *Lepidosaphes ulmi* (L.) (Homoptera, Diaspididae) and *Palaeolecanium bituberculatum* (Targ. and Tozz.) (Homoptera, Coccidae) harmful to apple trees in Van province". *Turkish Journal of Agriculture and Forestry*, 23(2):151-164.

Japoshvili, George, and Karaca, İsmail. 2002a. "Two new subspecies of the genus *Coccophagus* (Hymenoptera: Aphelinidae) from Turkey". *Bulletin of the Georgian Academy of Sciences*, 166(1):139-143.

—. 2002b. "*Coccid* (Homoptera: Coccoidea) species of Isparta Province, and their parasitoids from Turkey and Georgia". *Turkish Journal of Zoology*, 26(4):371-376.

—. 2010. "The transcaucasian species of *Coccobius* Ratzeburg 1852 (Chalcidoidea: Aphelinidae), with the description of three new species from Georgia". *Journal of Hymenoptera Research*, 19(1):121-127.

Karaca, İsmail, and Şenal, Derya, and Çölkesen, Tülin, and Özgökçe, Mehmet Salih. 2001. "Observations on the oleander scale, *Aspidiotus nerii* Bouche (Hemiptera: Diaspididae) and its natural enemies on blueleaf wattle in Adana Province, Turkey". Entomologica, Bari, 33:407-412.

Karaca, İsmail, and Uygun, Nedim, and Elekçioğlu, Naime Zülal, and Şenal, Derya. 2001. "Population development of *Aonidiella aurantii* (Maskell) and *Parlatoria pergandii* Comstock (Homoptera: Diaspididae) in Cukurova region of Turkey". *Bollettino di Zoologia Agraria e Bachicoltura*, 33(3):313-317.

Karaca, İsmail. 1998. "Parasitization efficacy of *Aphytis melinus* DeBach (Hymenoptera: Aphelinidae) as affected by host size and size distribution of *Aonidiella aurantii* (Maskell) (Homoptera: Diaspididae) in a lemon orchard". *Turkish Journal of Entomology*, 22(2):101-108.

Kaydan, Mehmet Bora, and Japoshvili, George. 2010. "The aphelinid and encyrtid parasitoids (Hymenoptera: Chalcidoidea) of Pseudococcidae (Hemiptera: Coccoidea) in the Van Lake basin of Turkey". *Turkish Journal of Entomology*, 34(4):465-476.

Noyes, John S. 2019. Universal Chalcidoidea Database. Available at: https://www.nhm.ac.uk/our-science/ data/chalcidoids/ (accessed January 2022).

Öncüer, Cezmi, and Yoldaş, Zeynep, and Madanlar, Nilgün, and Gül, A. 1994. "Biological control practices against insect pests of greenhouse vegetables in the Aegean region". *Proceedings of the Third Turkish National Congress of Biological Control* (25-28 January, İzmir-Turkey), 395-407.

Öncüer, Cezmi. 1974. "Ege bölgesinde turunçgil bahçelerinde zararli *Coccus* (Homoptera: Coccidae) türlerinin taninmasi yayilisi ve dogal düsmanlari üzerinde araştırmalar". *Bitki Koruma Bülteni*, Supplement, 1:1-59.

—. 1991. "A catalogue of the parasites and predators of insect pests of Turkey". pp.163.

Özgen, İnanç, and Karsavuran, Yusuf, and Japoshvili, George. 2009. "The first record of *Coccobius testaceus* (Masi, 1909) (Hymenoptera: Aphelinidae) on *Lepidosaphes pistaciae* (Archangelskaya) (Homoptera: Diaspididae) in Turkey". *Munis Entomology and Zoology*, 4(2):470-471.

Özkan, Ahmet, and Türkyilmaz, Naci, and Çiftci, Kemal, and Kaplan, Mehmet. 1996. "Researches on the mass production of citrus whitefly, *Dialeurodes citri* (Ashm.) (Homoptera: Aleyrodidae) and its parasitoid, *Encarsia lahorensis* (How.) (Hymenoptera: Aphelinidae) and attempts for its colonization in nature". *Bitki Koruma Bülteni*, 36(1-2):39-53.

Özsemerci, Fatma, and Aksit, Tülin. 2003. "Investigations on some biological characteristics and populations fluctuation of *Ceroplastes rusci* L. (Homoptera: Coccidae) harmful to fig trees in Aydin province". *Turkish Journal of Entomology*, 27(1):13-25.

Şengonca, Çetin, and Uygun, Nedim, and Kersting, Ulrich, and Ulusoy, Mehmet Rifat. 1993. "Successful colonization of *Eretmocerus debachi* (Hymenoptera: Aphelinidae) in the easter Mediterranean citrus region of Turkey". *Entomophaga*, 38(3):383-390.

—. 1995. "Efficiency of *Eretmocerus debachi* Rose and Rosen (Hym., Aphelinidae) in controlling the whitefly Parabemisia myricae (Kuwana) (Hom., Aleyrodidae)". *Anzeiger für Schädlingskunde, Pflanzen- und Umweltschutz*, 68(6):123-127.

—. 1996. "Successful biological control of *Parabemisia myricae* (Kuwana) by *Eretmocerus debachi* Rose et Rosen in the east Mediterranean citrus-region of Turkey". XX International Congress of Entomology, Firenze, Italy, 25-31 August, 1996, Abstract No: 20-164.

—. 1998. "Population dynamics of *Parabemisia myricae* (Kuwana) (Homoptera: Aleyrodidae) and its parasitoid *Eretmocerus debachi* Rose and Rosen (Hymenoptera, Aphelinidae) on non-citrus host plants". *Zeitschrift für Pflanzenkrankheiten und Pflanzenschutz*, 105(2):149-156.

Şengonca, Çetin, Uygun, Nedim, and Karaca, İsmail. 1998. "Primary studies on the parasitoid fauna of Coccoidea in cultivated and non-cultivated areas in the east Mediterranean region of Turkey". *Anzeiger für Schädlingskunde, Pflanzenschutz und Umweltschutz*, 71(7):128-131.

Soydanbay, Mine. 1977. "The effect of aphelinid parasites on the population of citrus scale insects in the Aegean region". *Fruits*, Paris, 32(6):428-431.

Stringer, Danielle N., and Mantel, Sarah, and Jennings, John T., and Austin, Andrew D. 2012. "Family Aphelinidae", *Australian Centre for Evolutionary Biology and Biodiversity, and the School of Earth and Environmental Science*. The University of Adelaide, Australia (Available from: https://www.researchgate.net/publication/262008500_Family_APHELINIDAE).

Tunçyürek, Mine, and Öncüer, Cezmi. 1974. "Studies on aphelinid parasites and their hosts, citrus diaspine scale insects, in citrus orchards in the Aegean region". Bulletin. Section Regionale Ouest Palaearctique, Organisation Internationale de Lutte Biologique, No 3:95-108.

Tunçyürek-Soydanbay, Mine, and Erkin, Enis. 1979. "Investigations on the citrus armoured scale insects (Diaspididae) and on their parasite *Aspidiotiphagus citrinus* Craw in western Anatolia". *Bitki Koruma Bülteni*, 19(4):218-236.

—. 1981. "Species-distribution of citrus armored scales amaging citrus trees in the Aegean region, with the influence of parasitic activity on population fluctuations". *Bitki Koruma Bülteni*, 21(4):173-196.

Ülgentürk, Selma, and Kaydan, Mehmet Bora, and Zeki, Cevdet, and Toros, Seval. 2001. "*Sphaerolecanium prunastri* Boyer de Fonscolombe (Hemiptera: Coccidae): distribution, host plants and natural enemies in the Turkish Lake District". *Bolletino Zoologia Agraria di Bachicoltura Ser.* II, 33:3, 357-363.

Ülgentürk, Selma, and Toros, Seval. 2001. "Natural enemies of the oak scale insect, *Eulecanium ciliatum* (Douglas) (Hemiptera: Coccidae) in Turkey". *Entomologica*, Bari 33:219-224.

Ülgentürk, Selma. 2001. "Parasitoids and predators of Coccidae (Homoptera: Coccoidea) species on ornamental plants in Ankara, Turkey". *Acta Phytopathologica et Entomologica Hungarica*, 36(3/4):369-375.

Ulusoy, Mehmet Rifat, and Ülgentürk, Selma. 2003. "The natural enemies of whiteflies (Hemiptera: Aleyrodidae) in southern Anatolia". *Zoology in the Middle East*, 28:119-124.

Ulusoy, Mehmet Rifat, and Uygun, Nedim, and Kersting, Ulrich, and Karaca, İsmail, and Satar, Serdar. 1996. "Present status of citrus whiteflies (Homoptera: Aleyrodidae) in Turkey and their control". *Zeitschrift für Pflanzenkrankheiten und Pflanzenschutz*, 103(4):397-402.

Ulusoy, Mehmet Rifat, and Vatansever, Gonca, and Erkilic, Lerzan, and Uygun, Nedim. 2003. "Studies on *Aleurothrixus floccosus* (Maskell) (Homoptera, Aleyrodidae) and its parasitoid, Cales noacki Howard (Hymenoptera, Aphelinidae) in the east Mediterranean region of Turkey". *Anzeiger für Schädlingskunde*, 76(6):163-169.

Ulusoy, Mehmet Rifat. 1999. "Aphelinidae species (Hymenoptera: Chalcidoidea) obtained from the species of Aleyrodidae (Homoptera) in the Mediterranean region of Turkey". *Turkish Journal of Entomology*, 23(4):251-258.

Uygun, Nedim, and Ohnesorge, Bernhard, and Ulusoy, Mehmet Rifat. 1990. "Two species of whiteflies on citrus in eastern mediterranean *Parabemisia myricae* Kuwana and *Dialeurodes citri* Ashmead: morphology, biology, host plants and control in southern Turkey". Journal of Applied Entomology, 110(5):471-482.

Uygun, Nedim, and Ulusoy, Mehmet Rifat, and Karaca, İsmail, and Kersting, Ulrich. 1996. "Aproaches [sic] on biological control of *Dialeurodes citri* (Ashmead) in Turkey". (Abstract 22-050), *Proceedings XX International Congress of Entomology*, Firenze, Italy, (August 25-31).

Uygun, Nedim, and Ulusoy, Mehmet Rifat, and Karaca, İsmail, and Kersting, Ulrich. 1997. "Approaches to biological control of *Dialeurodes citri* (Ashmead) in Turkey". *Bulletin*. Section Regionale Ouest Palaearctique, Organisation Internationale de Lutte Biologique, 20(7):52-62.

Uygun, Nedim, and Ulusoy, Mehmet Rifat, and Şekeroglu, Erdal, and Ohnesorge, Bernhard, and Gozel, Uğur. 1994. "Interactions between two introduced species of whiteflies in the Mediterranean area of Turkey: *Dialeurodes citri* (Ashmead) and *Parabemisia myricae* (Kuwana) (Hom., Aleyrodidae)". *Journal of Applied Entomology*, 118(4-5):365-369.

Yaşarakıncı, Nilgün, and Hıncal, Pervin. 1997. "Determining the pests and beneficial species and their population densities on tomato, cucumber, pepper and lettuce greenhouses in İzmir". *Bitki Koruma Bülteni*, 37(1/2):79-89.

Yoldaş, Zeynep, and Madanlar, Nilgün, and Gül, A. & Onogur, E. 1999. "Investigations on integrated control practices in vegetable glasshouses in Izmir". Acta Horticulturae, Wageningen, No 491:453-460.

Yoldaş, Zeynep, and Öncüer, Cezmi. 1992. "An investigation on the colonization of *Encarsia lahorensis* (How.), an introduced parasitoid of the citrus whitefly, *Dialeurodes citri* (Ashm.), in Izmir". *Proceedings of the Second Turkish National Congess of Entomology* (28-31 January, 1992), pp.79-87.

Yoldaş, Zeynep. 1995. "Investigation on the effectiveness of *Encarsia formosa* (Gahan) (Hymenoptera, Aphelinidae) in the biological control of *Bemisia tabaci* (Genn.) (Homoptera, Aleyrodidae) in cucumbers". *Turkish Journal of Entomology*, 19(2): 95-100.

CHAPTER THREE

FAMILY: AZOTIDAE NIKOLSKAYA AND YASNOSH, 1966

Azotidae was first proposed as the subfamily Azotinae of Aphelinidae by Nikolskaya and Yasnosh (1966), but it was raised to family status by Heraty et al. (2013). These tiny wasps are normally hyperparasitoids and are associated with the Aleyrodidae and Coccoidea as well as the eggs of a variety of other insects. Overall, there are over 90 species of *Ablerus* (Ghahari et al., 2021), and three species in Turkey (Noyes, 2019).

Genus *Ablerus* (Howard, 1894)
Ablerus atomon (Walker, 1847)
Publications with Turkey records: Japoshvili and Karaca (2002), Noyes (2019).
Associates: *Cupressus sp.* (Cupressaceae), *Pinus sp., Pinus eldarica* (Pinaceae), Poaceae, *Amygdalus sp.* (Rosaceae), *Citrus sp.* (Rutaceae), *Populus hybrida, Populus transcaucasica* (Salicaceae) (Noyes, 2019).
Hosts: *Aspidiotus ostreaeformis, Aspidiotus pyri, Chionaspis stantophris, Diaspidiotus caucasicus, Diaspidiotus gigas, Diaspidiotus ostreaeformis, Diaspidiotus perniciosus, Diaspidiotus prunorum, Diaspidiotus pyri, Diaspidiotus uvae, Duplachionaspis natalensis, Lepidosaphes ulmi, Leucaspis sp., Leucaspis loewi, Leucaspis pini, Leucaspis pusilla, Leucaspis signoreti, Mytilaspis conchiformis, Nuculaspis abietis, Pseudaulacaspis pentagona, Quadraspidiotus sp., Quadraspidiotus gigas, Quadraspidiotus ostreaeformis* (Hemiptera: Diaspididae), *Dendrolimus pini* (Lepidoptera: Lasiocampidae) (Noyes, 2019).
Parasitoid hosts: *Prospaltella leucaspidis* (Hymenoptera: Aphelinidae) (Noyes, 2019).
Distribution: Argentina, Australia, Austria, Azerbaijan, Caucasus, Czech Republic, Egypt, Europe, France, Georgia, Germany, Hungary, Iran, Italy, Nearctic, Palearctic, Poland, Russia, Slovakia, Spain, Sweden, Turkey, United States of America, USSR (Noyes, 2019).

Ablerus celsus (Walker, 1839)
Publications with Turkey records: Japoshvili and Karaca (2002), Noyes (2019), Öncüer (1991).
Associates: *Corylus avellana* (Betulaceae), *Juglans sp.* (Juglandaceae), *Pinus halepensis* (Pinaceae), *Platanus sp.* (Platanaceae), *Malus communis, Malus pumila* (Rosaceae), *Populus hybrida, Populus transcaucasica* (Salicaceae) (Noyes, 2019).
Parasitoid hosts: *Azotus celsus, Coccophagoides similis* (Hymenoptera: Aphelinidae), *Mayrencyrtus merceti* (Encyrtidae) (Noyes, 2019).
Hosts: *Chionaspis salicis, Diaspidiotus gigas, Diaspidiotus ostreaeformis, Diaspidiotus perniciosus, Diaspidiotus prunorum, Diaspidiotus slavonicus, Diaspidiotus viticola, Lepidosaphes ulmi, Quadraspidiotus sp., Quadraspidiotus gigas, Quadraspidiotus ostreaeformis, Quadraspidiotus perniciosus, Quadraspidiotus zonatus, Salicicola kermanensis, Targionia vitis* (Hemiptera: Diaspididae), *Kuwania rubra* (Margarodidae), *Malacosoma neustria* (Lepidoptera: Lasiocampidae) (Noyes, 2019).
Distribution: Bulgaria, Caucasus, Croatia, Czech Republic, England, Europe, France, Georgia, Italy, Moldova, Montenegro, Spain, Turkey, Ukraine, USSR (Noyes, 2019).

Ablerus perspeciosus (Girault, 1916)
Publications with Turkey records: Erkilic and Uygun (1995), Erler and Tunç (2001), Noyes (2019).
Parasitoid hosts: *Aphytis diaspidis* (Hymenoptera: Aphelinidae) (Noyes, 2019).
Hosts: *Ceroplastes sp., Coccus hesperidum, Lecanium sp.* (Hemiptera: Coccidae), *Chrysomphalus ficus* (Diaspididae) (Noyes, 2019).
Distribution: Argentina, Egypt, France, India, Israel, Italy, Japan, Nearctic, People's Republic of China, Serbia, Thailand, Turkey, United States of America (Noyes, 2019).

References

Erkılıç, Lerzan, and Uygun, Nedim. 1995. "Distribution, population fluctuations and natural enemies of the white peach scale, *Pseudaulacaspis pentagona* (Targioni Tozzetti) (Homoptera: Diaspididae) in the east mediterranean region of Turkey". *Israel Journal of Entomology*, 29:191-198.

Erler, Fedai, and Tunç, İrfan. 2001. "A survey (1992-1996) of natural enemies of Diaspididae species in Antalya, Turkey". *Phytoparasitica*, 29(4):299-305.

Ghahari, Hassan, and Gibson, Gary, and Viggiani, Gennaro. 2021. "Chalcidoidea of Iran (Insecta: Hymenoptera)", https://www.cabi.org/bookshop/book/9781789248463/ (accessed 23.05.2021)

Japoshvili, Gerorge, and Karaca, İsmail. 2002. "Coccid (Homoptera: Coccoidea) species of Isparta Province, and their parasitoids from Turkey and Georgia". *Turkish Journal of Zoology*, 26 (4):371-376.

Noyes, John S. 2019. *Universal Chalcidoidea Database*. Available at: www.nhm.ac.uk/chalcidoids (accessed January 2022).

Öncüer, Cezmi. 1991. "A catologue of the parasites and predators of insect pests of Turkey". *Ege University Agricultural Faculty Publication 505*. İzmir. pp. 354.

CHAPTER FOUR

FAMILY: CHALCIDIDAE LATREILLE, 1817

The Chalcididae is a medium-sized family of generally larger body size (2–15 mm) compared to other members of the superfamily Chalcidoidea. They are characterized by having a swollen hind femur with a toothed ventral surface. They are typically black and non-metallic, but often have yellow or red patches on the legs. Most species are primary endoparasitoids of lepidopteran (moth and butterfly) and dipteran (fly) larvae and pupae, although they also attack Hymenoptera, Coleoptera (beetles), and Neuroptera (lacewings). The family also includes hyperparasitoids, ectoparasitoids and gregarious species. The majority of Chalcididae species are idiobiont pupal or larval parasitoids, although some are known to be koinobionts (Stringer et al., 2012).

The family currently includes 87 genera and 1464 species placed in 5 subfamilies, and Turkey is home to 39 species in three subfamilies (Noyes, 2019).

Subfamily Chalcidinae (Latreille, 1817)
Genus *Brachymeria* (Westwood, 1829)
Brachymeria femorata (Panzer, 1801)
Publications with Turkey records: Noyes (2019), Öncüer (1991).
Associates: *Mesophleps oxycedrella* (Lepidoptera: Gelechiidae), *Eryngium campestre* (Apiaceae), *Brassica sp., Brassica oleracea* (Brassicaceae), *Juniperus phoeniciae, Juniperus thurifera* (Cupressaceae), *Crataegus sp.* (Rosaceae) (Noyes, 2019).
Hosts: *Cleonus punctiventris* (Coleoptera: Curculionidae), *Hyphantria cunea* (Lepidoptera: Arctiidae), *Lasiocampa quercus* (Lasiocampidae), *Heliothis sp., Mamestra oleracea* (Noctuidae), *Melitaea didyma, Pyrameis cardui, Vanessa cardui, Vanessa urticae* (Nymphalidae), *Aporia crataegi, Pieris sp., Pieris brassicae, Pieris rapae, Pontia daplidice* (Pieridae), *Plutella maculipennis* (Yponomeutidae) (Noyes, 2019).

Parasitoid hosts: *Echinomyia grossa* (Diptera: Tachinidae) (Noyes, 2019).
Distribution: Bosnia Hercegovina, Bulgaria, Caucasus, Central Asia, Croatia, Cyprus, Czech Republic, Czechoslovakia, Egypt, England, Europe, France, Germany, Greece, Hungary, India, Indonesia, Iran, Iraq, Israel, Italy, Kazakhstan, Macedonia, Moldova, Mongolia, Pakistan, People's Republic of China, Philippines, Poland, Romania, Russia, Serbia, Slovakia, Siberia, Spain, Sweden, Switzerland, Tselinograd Obl., Turkey, Ukraine, Yugoslavia (Federal Republic), Yugoslavia (pre1991) (Noyes, 2019).

Brachymeria inermis (Fonscolombe, 1840)
Publications with Turkey records: Noyes (2019), Öncüer (1991).
Associates: *Onopordum sp.* (Asteraceae), *Juniperus thurifera* (Cupressaceae), *Tamarix canariensis* (Tamaricaceae) (Noyes, 2019).
Parasitoid hosts: *Bracon telengai, Rogas bicolor, Rogas rossicus* (Hymenoptera: Braconidae) (Noyes, 2019).
Hosts: *Cassida sp., Cassida nebulosa, Cassida nobilis, Cassida rubiginosa, Cassida viridis, Metriona thais* (Coleoptera: Chrysomelidae), *Zagulajevia tadzhikiella* (Lepidoptera: Coleophoridae), *Apocheima cinerarius* (Geometridae), *Vanessa sp.* (Nymphalidae), *Apotomis lutozana, Spilonota ocellana* (Tortricidae) (Noyes, 2019).
Distribution: Austria, Croatia, Czech Republic, Czechoslovakia, France, Greece, Hungary, India, Italy, Japan, Macedonia, Moldova, Montenegro, Romania, Serbia, Slovakia, Spain, Syria, Turkey, Turkmenistan (Noyes, 2019).

Brachymeria minuta (Linnaeus, 1767)
Publications with Turkey records: Noyes (2019), Öncüer (1991).
Associates: *Helix sp.* (Mollusca: Helicidae), *Arachis hypogaea, Glycine max* (Fabaceae), *Quercus suber* (Fagaceae), *Populus sp.* (Salicaceae) (Noyes, 2019).
Hosts: *Calliphora vomitoria, Lucilia cuprina* (Diptera: Calliphoridae), *Sarcophaga sp.* (Sarcophagidae), *Cimbex femorata* (Hymenoptera: Cimbicidae), *Diprion sp., Diprion polytomum* (Diprionidae), *Tyria jacobaeae* (Lepidoptera: Arctiidae), *Aproaerema modicella* (Gelechiidae), *Gegenes nostrodamus* (Hesperiidae), *Dendrolimus sibiricus, Dendrolimus spectabilis, Malacosoma neustria* (Lasiocampidae), *Dasychira fascelina, Euproctis phaeorrhoea, Euproctis similis, Leucoma salicis, Lymantria dispar, Ocneria dispar, Orgyia antiqua, Porthetria dispar, Stilpnotia salicis* (Lymantriidae), *Aporia crataegi, Pieris brassicae* (Pieridae), *Archips xylosteanus, Clysia ambiguella, Lobesia botrana, Polychrosis*

botrana, Tortrix viridana (Tortricidae), *Yponomeuta malinellus, Yponomeuta padellus* (Yponomeutidae) (Noyes, 2019).

Parasitoid hosts: *Agria affinis, Agria mamillata, Parasarcophaga brevicornis, Parasarcophaga dux, Pseudosarcophaga mamillata, Ravinia striata, Sarcophaga sp., Sarcophaga haemorrhoidalis, Sarcophaga harpax, Sarcophaga hirtipes, Sarcophaga melanura, Sarcophaga pseudoscoparia, Sarcophaga uliginosa* (Diptera: Sarcophagidae), *Bessa selecta, Echinomyia fera, Eurysthaea scutellaris, Exorista larvarum, Masicera zimini, Nemorilla floralis, Phebelia glauca, Sturmia scutellate* (Tachinidae) (Noyes, 2019).

Distribution: Australia, Bosnia Hercegovina, Bulgaria, Canary Islands, Caucasus, Central Asia, Croatia, Cyprus, Czech Republic, Czechoslovakia, Egypt, England, Europe, France, Germany, Hungary, India, Iran, Israel, Italy, Japan, Kazakhstan, Malaysia, Moldova, Montenegro, Netherlands, Papua New Guinea, Poland, Romania, Russia, Serbia, Siberia, Scotland, Slovakia, Slovenia, Spain, Sweden, Switzerland, Syria, Thailand, Tselinograd Obl., Turkey, Ukraine, Uzbekistan, Vietnam, Yugoslavia (pre-1991) (Noyes, 2019).

Brachymeria moerens (Ruschka, 1922)

Publications with Turkey records: Noyes (2019), Öncüer (1991).

Hosts: *Barbitistes fischeri* (Orthoptera: Tettigoniidae) (Noyes, 2019).

Distribution: Austria, Bosnia Hercegovina, Bulgaria, Caucasus, Czechoslovakia, France, Germany, Hungary, Italy, Macedonia, Moldova, Slovakia, Transcaucasus, Turkey, Ukraine, Yugoslavia (Federal Republic) (Noyes, 2019).

Brachymeria obtusata (Foerster, 1859)

Publications with Turkey records: Bayram et al. (1998), Noyes (2019), Öncüer (1991).

Associates: *Helix sp.* (Mollusca: Helicidae), *Rosmarinus officinalis* (Family: Lamiaceae) (Noyes, 2019).

Hosts: *Lucilia caesar* (Diptera: Calliphoridae), *Eristalis tenax* (Syrphidae), *Porthetria dispar* (Lepidoptera: Lymantriidae) (Noyes, 2019).

Parasitoid hosts: *Sarcophaga sp.* (Diptera: Sarcophagidae) (Noyes, 2019).

Distribution: Albania, Austria, Croatia, Cyprus, Czech Republic, Czechoslovakia, Egypt, England, Europe, France, Germany, Greece, Hungary, Iran, Iraq, Israel, Italy, Japan, Moldova, Netherlands, North Africa, Romania, Serbia, Slovakia, Slovenia, Spain, Sweden, Transcaucasus, Turkey, Turkmenistan, Ukraine, USSR, Yugoslavia (Federal Republic) (Noyes, 2019).

Brachymeria pseudorugosa (Masi, 1951)
Publications with Turkey records: Noyes (2019).
Associates: *Styrax officinalis* (Styracaceae) (Noyes, 2019).
Hosts: *Archips sp., Archips rosanus* (Lepidoptera: Tortricidae) (Noyes, 2019).
Distribution: Italy, Spain, Switzerland, Turkey (Noyes, 2019).

Brachymeria rugulosa (Foerster, 1859)
Publications with Turkey records: Noyes (2019), Öncüer (1991).
Associates: *Andricus argenteus* (Hymenoptera: Cynipidae), *Quercus sp., Quercus pubescens* (Fagaceae), *Punica granatum* (Lythraceae), *Larix decidua* (Pinaceae), *Citrus sp.* (Rutaceae), *Styrax officinalis* (Styracaceae) (Noyes, 2019).
Hosts: *Andricus kollari, Cynips kollari* (Hymenoptera: Cynipidae), *Depressaria depressella* (Lepidoptera: Oecophoridae), *Apomyelois ceratoniae, Ectomyelois ceratoniae* (Pyralidae), *Archips sp., Archips rosanus, Cacoecia rosanus, Grapholita funebrana, Gypsonoma minutana, Ptycholomoides aeriferanus, Tortrix viridana* (Tortricidae) (Noyes, 2019).
Distribution: Austria, Bulgaria, Croatia, Czech Republic, Czechoslovakia, France, Germany, Greece, Hungary, Iran, Italy, Macedonia, Moldova, Romania, Serbia, Slovakia, Slovenia, Spain, Turkey, Turkmenistan, Yugoslavia (Federal Republic) (Noyes, 2019).

Brachymeria secundaria (Ruschka, 1922)
Publications with Turkey records: Noyes (2019), Öncüer (1991).
Associates: *Populus sp.* (Salicaceae), *Tamarix canariensis* (Tamaricaceae) (Noyes, 2019).
Hosts: *Apanteles melitaearum* (Hymenoptera: Braconidae), *Agriopis leucophaearia* (Lepidoptera: Geometridae), *Euproctis chrysorrhoea, Euproctis flava, Lymantria dispar, Notolophus aurolimbatus, Nygmia phaeorrhoea, Ocneria dispar, Orgyia guadarramensis, Stilpnotia salicis* (Lymantriidae), *Helicoverpa armigera, Heliothis armigera* (Noctuidae), *Cerura vinula* (Notodontidae), *Euphydryas aurinia, Mellicta sp.* (Nymphalidae), *Pieris daplidice* (Pieridae), *Laspeyresia pomonella, Pandemis cerasana, Polychrosis botrana* (Tortricidae), *Plutella maculipennis* (Yponomeutidae) (Noyes, 2019).
Parasitoid hosts: *Apanteles sp., Apanteles liparidis, Apanteles ordinarius, Meteorus sp., Meteorus pulchricornis, Meteorus rubens, Meteorus versicolor, Rhogas sp., Rhogas drymoniae* (Hymenoptera: Braconidae), *Campoletis chlorideae* (Ichneumonidae) (Noyes, 2019).

Distribution: Austria, Bosnia Hercegovina, Bulgaria, Croatia, Czechoslovakia, Europe, France, Germany, Hungary, India, Italy, Japan, Kazakhstan, Lithuania, Moldova, Romania, Russia, Serbia, Slovakia, Spain, Transcaucasus, Tselinograd Obl., Turkey, Ukraine, USSR, Yugoslavia (Federal Republic), Yugoslavia (pre1991) (Noyes, 2019).

Brachymeria tibialis (Walker, 1834)
Publications with Turkey records: Avci (2009), Noyes (2019), Öncüer (1991), Polat and Tozlu (2010).
Associates: *Blepharipa pratensis* (Diptera: Tachinidae), *Andricus quercustozae, Cynipidae sp.* (Hymenoptera: Cynipidae), *Dibrachys cavus* (Pteromalidae), *Pistacia atlantica* (Anacardiaceae), *Eryngium campestre, Thapsia villosa* (Apiaceae), *Onopordum nervosum* (Asteraceae), *Arbutus sp.* (Ericaceae), *Robinia pseudoacacia* (Fabaceae), *Quercus sp., Quercus cerris, Quercus frainetto, Quercus persica, Quercus prinus, Quercus pubescens, Quercus robur, Quercus suber, Quercus trojana* (Fagaceae), *Liriodendron sp.* (Magnoliaceae), *Pinus radiata* (Pinaceae), *Zea mays* (Poaceae), *Malus pumila, Prunus sp.* (Rosaceae), *Citrus sp.* (Rutaceae), *Populus sp., Salix sp.* (Salicaceae), *Styrax officinalis* (Styracaceae), *Tamarix sp.* (Tamaricaceae), *Urtica sp.* (Urticaceae), *Vitis vinifera* (Vitaceae) (Noyes, 2019).
Hosts: *Resseliella piceae* (Diptera: Cecidomyiidae), *Compsilura sp., Exorista sp., Sturmia sp.* (Tachinidae), *Brachymeria intermedia* (Hymenoptera: Chalcididae), *Andricus quercustozae* (Cynipidae), *Diprion abieticola* (Diprionidae), *Holomelina lamae, Hyphantria cunea, Hypocrita jacobaeae, Tyria jacobaeae* (Lepidoptera: Arctiidae), *Anarsia lineatella* (Gelechiidae), *Thymelicus lineola* (Hesperiidae), *Dendrolimus pini, Malacosoma neustria* (Lasiocampidae), *Euproctis sp., Euproctis chrysorrhoea, Euproctis Melania, Euproctis phaeorrhoea, Hemerocampa leucostigma, Hypogymna morio, Leucoma salicis, Leucoma wiltshiroi, Lymantria sp., Lymantria dispar, Lymantria obfuscate, Lymantria wiltshirei, Lymantria xylina, Nygmia phaeorrhoea, Ocneria dispar, Ocneria joeniteris, Ocneria terebynthina, Ocnerogyia amanda, Parocneria terebynthina, Porthetria dispar, Stilpnotia salicis* (Lymantriidae), *Barathra brassicae, Mamestra oleracea, Mamestra susa, Polia oleracea, Sesamia cretica, Trichoplusia ni* (Noctuidae), *Heterocampa guttivitta* (Notodontidae), *Euphydryas phaeton, Limenitis archippus, Melitaea harrisii, Nymphalis milberti, Nymphalis xanthomelas, Vanessa atalanta, Vanessa polychloros, Vanessa urticae* (Nymphalidae), *Depressaria depressella* (Oecophoridae), *Iphiclides podalirius, Papilio glaucus, Papilio podalirius* (Papilionidae), *Aporia crataegi, Colias hyale,*

Pieris brassicae, Pieris daplidice, Pieris rapae (Pieridae), *Dioryctria abietella, Galleria mellonella, Pleuroptya ruralis, Psammotis hyalinalis* (Pyralidae), *Cricula trifenestrata* (Saturniidae), *Archips sp., Archips argyrospilus, Archips griseus, Archips pronubana, Archips rosanus, Archips xylosteanus, Argyroploce nubiferana, Barbara herrichiana, Cacoecia sp., Carpocapsa pomonella, Choristoneura fumiferana, Eucosma ocellana, Oenophthira pilleriana, Pandemis sp., Pandemis ribeana, Sparganothis pilleriana, Sphaeroeca ocellana, Spilonota ocellana, Tmetocera ocellana, Tortrix viridana* (Tortricidae), *Theresimima ampelophaga, Zygaena sp., Zygaena achilleae, Zygaena angelicae, Zygaena carniolica, Zygaena ephialtes, Zygaena esseni, Zygaena filipendulae, Zygaena haberhaueri, Zygaena laeta, Zygaena Lydia, Zygaena meliloti, Zygaena olivieri, Zygaena osterodensis, Zygaena Sarpedon, Zygaena transalpine, Zygaena trifolii* (Zygaenidae) (Noyes, 2019).

Parasitoid hosts: *Compsilura concinnata, Masicera sylvatica, Sturmia sp., Tachina larvarum, Tricholyga sorbillans* (Diptera: Tachinidae), *Rhogas sp., Rhogas pallidator* (Hymenoptera: Braconidae), *Brachymeria compsilurae* (Chalcididae), *Casinaria tenuiventris, Charops cantator, Diadegma sp.* (Ichneumonidae) (Noyes, 2019).

Parasitoids: *Entedon sp.* (Eulophidae), *Pteromalus obscurus* (Pteromalidae), *Monodontomerus aereus* (Torymidae) (Noyes, 2019).

Distribution: Albania, Algeria, Armenia, Austria, Azerbaijan, Bosnia Hercegovina, Bulgaria, Canada, Caucasus, Croatia, Cyprus, Czech Republic, Czechoslovakia, England, Europe, France, Germany, Greece, Hungary, India, Iran, Iraq, Israel, Italy, Kazakhstan, Lebanon, Macedonia, Moldova, Morocco, Nearctic, Netherlands, North Africa, Pakistan, Poland, Portugal, Romania, Russia, Serbia, Slovakia, Slovenia, Spain, Sweden, Switzerland, Syria, Tselinograd Obl., Tunisia, Turkey, Ukraine, United States of America, USSR, Uzbekistan, Yugoslavia (Federal Republic), Yugoslavia (pre-1991) (Noyes, 2019).

Brachymeria vitripennis (Foerster, 1859)

Publications with Turkey records: Noyes (2019).

Associates: *Silene vulgaris* (Caryophyllaceae) (Noyes, 2019).

Hosts: *Cassida sp., Cassida algirica, Cassida atrata, Cassida azurea, Cassida deflorate, Cassida murraea, Cassida nebulosa, Cassida nobilis, Cassida rubiginosa, Cassida vibex, Cassida viridis, Cassida vittate* (Coleoptera: Chrysomelidae), *Byctiscus populi* (Rhynchitidae) (Noyes, 2019).

Distribution: Bulgaria, Croatia, Czech Republic, Czechoslovakia, Germany, Greece, Hungary, Iran, Italy, Macedonia, Moldova, Romania, Serbia, Slovakia, Switzerland, Turkey, Yugoslavia (Federal Republic) (Noyes, 2019).

Subfamily Dirhininae (Ashmead, 1904)
Genus *Dirhinus* (Dalman, 1818)
Dirhinus hesperidum (Rossi, 1790)
Publications with Turkey records: Noyes (2019), Öncüer (1991).
Hosts: *Musca domestica* (Diptera: Muscidae), *Sarcophaga sp.* (Sarcophagidae) (Noyes, 2019).
Distribution: Austria, Bosnia Hercegovina, Croatia, Czechoslovakia, France, Hungary, Iran, Italy, Japan, Korea, Moldova, North Africa, Romania, Slovakia, Turkey, Ukraine (Noyes, 2019).

Subfamily Haltichellinae (Ashmead, 1904)
Genus *Anachalcis* (Steffan, 1951)
Anachalcis major (Boucek, 1952)
Publications with Turkey records: Noyes (2019).
Hosts: Unknown (Noyes, 2019).
Parasitoids: Unknown (Noyes, 2019).
Distribution: Turkey (Noyes, 2019).

Anachalcis rubra (Steffan, 1951)
Publications with Turkey records: Noyes (2019), Öncüer (1991).
Hosts: Unknown (Noyes, 2019).
Parasitoids: Unknown (Noyes, 2019).
Distribution: Bulgaria, Caucasus, Europe, Greece, Macedonia, Montenegro, Turkey (Noyes, 2019).

Genus *Aphasganophora* (Nikolskaya, 1952)
Aphasganophora armeniaca (Boucek, 1952)
Publications with Turkey records: Noyes (2019), Öncüer (1991).
Hosts: Unknown (Noyes, 2019).
Parasitoids: Unknown (Noyes, 2019).
Distribution: Armenia, Caucasus, Central Asia, Turkey (Noyes, 2019).

Genus *Belaspidia* (Masi, 1916)
Belaspidia nigra (Siebold, 1856)
Publications with Turkey records: Noyes (2019).
Hosts: *Apterona sp., Apterona crenulella, Apterona helix, Epichnopterix plumella, Psyche helix* (Lepidoptera: Psychidae) (Noyes, 2019).

Distribution: Austria, Azerbaijan, Belgium, Czech Republic, Czechoslovakia, France, Hungary, Iran, Moldova, Spain, Switzerland, Tunisia, Turkey (Noyes, 2019).

Belaspidia obscura (Masi, 1916)
Publications with Turkey records: Noyes (2019), Öncüer (1991).
Associates: *Suaeda vera* (Chenopodiaceae) (Noyes, 2019).
Hosts: *Coleophora hemerobiella* (Lepidoptera: Coleophoridae), *Apterona sp., Apterona crenulella, Cochliotheca crenulella, Rebelia kruegeri* (Psychidae) (Noyes, 2019).
Distribution: Austria, Belgium, Caucasus, Central Asia, Croatia, Czech Republic, Czechoslovakia, Europe, France, Germany, Greece, Hungary, Iran, Italy, Kazakhstan, Macedonia, Moldova, Morocco, Serbia, Slovakia, Spain, Syria, Tadzhikistan, Transcaucasus, Tselinograd Obl., Turkey, Turkmenistan, Ukraine, Yugoslavia (Federal Republic) (Noyes, 2019).

Genus *Euchalcis* (Dufour, 1861)
Euchalcis magna (Boucek, 1952)
Publications with Turkey records: Noyes (2019), Öncüer (1991).
Hosts: *Zygaena sp.* (Lepidoptera: Zygaenidae) (Noyes, 2019).
Distribution: Central Asia, Croatia, Cyprus, Czech Republic, Czechoslovakia, Europe, Germany, Hungary, Israel, Kazakhstan, Moldova, Romania, Russia, Serbia, Slovakia, Spain, Swaziland, Switzerland, Tselinograd Obl., Turkey, Ukraine, Yugoslavia (Federal Republic), Yugoslavia (pre-1991) (Noyes, 2019).

Euchalcis miegii (Dufour, 1861)
Publications with Turkey records: Noyes (2019), Öncüer (1991).
Associates: *Onopordum nervosum* (Asteraceae), *Retama sphaerocarpa* (Fabaceae) (Noyes, 2019).
Distribution: Albania, Algeria, Armenia, Europe, Greece, Israel, Italy, Jordan, Kirgizia, Spain, Tunisia, Turkey (Noyes, 2019).
Euchalcis unicolor (Lucas, 1871)
Publications with Turkey records: Noyes (2019), Öncüer (1991).
Hosts: *Psychidae sp.* (Lepidoptera: Psychidae) (Noyes, 2019).
Distribution: Algeria, Armenia, Caucasus, Europe, Israel, Spain, Turkey (Noyes, 2019).

Genus *Haltichella* (Spinola, 1811)
Haltichella rufipes (Olivier, 1791)
Publications with Turkey records: Noyes (2019), Öncüer (1991).

Hosts: *Ptilinus pectinicornis* (Coleoptera: Anobiidae), *Lagria hirta* (Tenebrionidae), *Grapholita molesta, Polychrosis botrana* (Lepidoptera: Tortricidae) (Noyes, 2019).
Distribution: Austria, Caucasus, Croatia, Czech Republic, Czechoslovakia, Democratic Republic of Congo (Zaire), England, Europe, France, Germany, Hungary, Iran, Italy, Kazakhstan, Moldova, Netherlands, Norway, Romania, Russia, Serbia, Slovakia, Spain, Sweden, Tselinograd Obl., Turkey, USSR, Yugoslavia (Federal Republic) (Noyes, 2019).

Genus *Hockeria* (Walker, 1834)
Hockeria apani (Doğanlar, 1990)
Publications with Turkey records: Doğanlar (1990), Noyes (2019).
Associates: *Cicer arietinum, Lens esculenta* (Fabaceae) (Noyes, 2019).
Hosts: *Heliothis viriplaca* (Lepidoptera: Noctuidae) (Noyes, 2019).
Distribution: Turkey (Noyes, 2019).

Hockeria bifasciata (Walker, 1834)
Publications with Turkey records: Noyes (2019), Öncüer (1991).
Associates: *Astragalus sp.* (Fabaceae), *Olea europaea* (Oleaceae) (Noyes, 2019).
Hosts: *Eublemma scitula* (Lepidoptera: Noctuidae), *Fumea nudella, Luffia lapidella, Rebelia nudella* (Psychidae), *Grapholita molesta* (Tortricidae), *Prays oleellus* (Yponomeutidae) (Noyes, 2019).
Distribution: Algeria, Austria, Canary Islands, Croatia, Czech Republic, Czechoslovakia, Europe, France, Greece, Hungary, India, Italy, Japan, Lebanon, Montenegro, Slovakia, Slovenia, Spain, Sweden, Turkey, Yugoslavia (Federal Republic) (Noyes, 2019).

Hockeria brachygaster (Boucek, 1956)
Publications with Turkey records: Noyes (2019), Öncüer (1991).
Hosts: *Casama innotata* (Lepidoptera: Lymantriidae) (Noyes, 2019).
Associates: *Acacia karroo* (Fabaceae) (Noyes, 2019).
Distribution: India, Israe, Libya, Turkey (Noyes, 2019).

Hockeria karatasensis (Boucek, 1952)
Publications with Turkey records: Noyes (2019), Öncüer (1991).
Associates: Unknown (Noyes, 2019).
Hosts: Unknown (Noyes, 2019).
Distribution: Turkey (Noyes, 2019).

Hockeria mengenillarum (Silvestri, 1943)
Publications with Turkey records: Noyes (2019), Öncüer (1991).
Hosts: *Mengenilla quaesita* (Strepsiptera: Mengenillidae) (Noyes, 2019).
Parasitoid hosts: *Mengenilla laevigata, Mengenilla nigritula, Mengenilla parvula, Mengenilla quaesita, Mengenilla spinulosa, Mengenilla subnigrescens* (Strepsiptera: Mengenillidae) (Noyes, 2019).
Distribution: Italy, Turkey (Noyes, 2019).

Hockeria unicolor (Walker, 1834)
Publications with Turkey records: Noyes (2019), Öncüer (1991).
Associates: *Eryngium campestre, Thapsia villosa* (Apiaceae), *Artemisia herba-alba, Onopordum nervosum* (Asteraceae), *Lithodora fruticose* (Boraginaceae), *Gypsophila struthium* (Caryophyllaceae), *Atriplex halimus, Salsola vermiculata, Suaeda vera* (Chenopodiaceae), *Ononis tridentata, Retama sphaerocarpa* (Fabaceae), *Rosmarinus officinalis* (Lamiaceae), *Olea europaea* (Oleaceae), *Pinus halepensis* (Pinaceae), *Rhamnus lycioides* (Rhamnaceae), *Malus sp.* (Rosaceae), *Citrus sp.* (Rutaceae), *Tamarix canariensis* (Tamaricaceae), *Vitis vinifera* (Vitaceae) (Noyes, 2019).
Hosts: *Lixus scabricollis* (Coleoptera: Curculionidae), *Exoteleia dodecella* (Lepidoptera: Gelechiidae), *Gymnoscelis pumilata* (Geometridae), *Phyllonorycter robiniella* (Gracillariidae), *Vulcaniella pomposella* (Momphidae), *Spodoptera litura* (Noctuidae), *Apterona helicoidella, Cochliotheca crenulella* (Psychidae), *Evetria sp., Laspeyresia sp., Laspeyresia leplastriana, Polychrosis sp., Rhyacionia buoliana* (Tortricidae), *Prays oleae, Yponomeuta malinellus* (Yponomeutidae), *Theresimima ampelophaga, Zygaena sp.* (Zygaenidae) (Noyes, 2019).
Distribution: Algeria, Austria, Bosnia Hercegovina, Bulgaria, Canary Islands, Croatia, Czech Republic, Czechoslovakia, Europe, France, Greece, Hungary, Iran, Italy, Kazakhstan, Moldova, Morocco, Netherlands, Poland, Romania, Russia, Senegal, Serbia, Slovakia, Slovenia, Spain, Sweden, Transcaucasus, Turkey, Ukraine, USSR (Noyes, 2019).

Genus *Kriechbaumerella* (Dalla Torre, 1897)
Kriechbaumerella mansues (Nikolskaya, 1952)
Publications with Turkey records: Noyes (2019), Öncüer (1991).
Associates: Unknown (Noyes, 2019).
Hosts: Unknown (Noyes, 2019).
Distribution: Croatia, Cyprus, Iraq, Israel, Transcaucasus, Turkey, Ukraine, USSR (Noyes, 2019).

Genus *Lasiochalcidia* (Masi, 1929)
Lasiochalcidia dargelasii (Latreille, 1805)
Publications with Turkey records: Noyes (2019), Öncüer (1991).
Hosts: *Distolen sp., Megistopus sp.* (Neuroptera: Myrmeleontidae) (Noyes, 2019).
Distribution: Algeria, Austria, Croatia, Czech Republic, Czechoslovakia, Europe, France, Hungary, India, Iran, Italy, Moldova, North Africa, Serbia, Slovakia, Transcaucasus, Turkey, USSR, Yugoslavia (Federal Republic) (Noyes, 2019).

Genus *Neochalcis* (Kirby, 1883)
Neochalcis fertoni (Kieffer, 1899)
Publications with Turkey records: Noyes (2019), Öncüer (1991).
Associates: *Anthidium lituratum* (Apidae), *Thapsia garganica* (Apiaceae), *Rubus sp.* (Rubiaceae) (Noyes, 2019).
Hosts: *Anthidium lituratum, Paranthidiellum lituratum* (Hymenoptera: Apidae), *Leptochilus gallicus, Odynerus gallicus* (Vespidae) (Noyes, 2019).
Distribution: Algeria, Croatia, Cyprus, Czech Republic, Czechoslovakia, England, Europe, France, Hungary, Iran, Italy, Kazakhstan, Macedonia, Moldova, Morocco, North Africa, Russia, Slovakia, Spain, Transcaucasus, Tselinograd Obl., Tunisia, Turkey, USSR (Noyes, 2019).

Neochalcis hippotoides (Masi, 1916)
Publications with Turkey records: Noyes (2019), Öncüer (1991).
Associates: Unknown (Noyes, 2019).
Hosts: Unknown (Noyes, 2019).
Distribution: Hungary, Israel, Italy, Spain, Turkey (Noyes, 2019).

Genus *Proconura* (Dodd, 1915)
Proconura nigripes (Fonscolombe, 1832)
Publications with Turkey records: Noyes (2019), Öncüer (1991).
Associates: *Eryngium campestre* (Apiaceae), *Gypsophila struthium* (Caryophyllaceae), *Retama sphaerocarpa* (Fabaceae), *Teucrium capitatum* (Lamiaceae) (Noyes, 2019).
Hosts: *Recurvaria sp., Recurvaria pistaciicola, Schneidereria pistaciicola* (Lepidoptera: Gelechiidae), *Ephestia cautella* (Pyralidae), *Acrolepia sp., Acrolepia assectella* (Yponomeutidae) (Noyes, 2019).
Distribution: Albania, Algeria, Canary Islands, Croatia, Czech Republic, Czechoslovakia, Europe, France, Hungary, Iran, Italy, Kazakhstan, Libya, Macedonia, Moldova, Montenegro, North Africa, Slovakia, Slovenia,

Spain, Transcaucasus, Tselinograd Obl., Tunisia, Turkey, Ukraine, USSR (Noyes, 2019).

Genus *Psilochalcis* (Kieffer, 1905)
Psilochalcis benoisti (Steffan, 1948)
Publications with Turkey records: Noyes (2019), Öncüer (1991).
Associates: *Thapsia villosa* (Apiaceae), *Artemisia herba-alba* (Asteraceae) (Noyes, 2019).
Distribution: Algeria, Cyprus, Czech Republic, Czechoslovakia, Europe, France, Germany, Hungary, Iran, Italy, Kazakhstan, Morocco, Slovakia, Spain, Transcaucasus, Tselinograd Obl., Turkey, Ukraine, USSR (Noyes, 2019).

Psilochalcis immaculata (Rossi, 1792)
Publications with Turkey records: Noyes (2019), Öncüer (1991).
Associates: Unknown (Noyes, 2019).
Hosts: Unknown (Noyes, 2019).
Distribution: Albania, Caucasus, Croatia, Cyprus, Europe, France, Greece, Israel, Italy, Kazakhstan, Macedonia, Tselinograd Obl., Turkey (Noyes, 2019).

Psilochalcis ligustica (Masi, 1929)
Publications with Turkey records: Noyes (2019), Öncüer (1991).
Associates: Unknown (Noyes, 2019).
Hosts: Unknown (Noyes, 2019).
Distribution: Caucasus, Croatia, Cyprus, Czech Republic, Czechoslovakia, Europe, Greece, Hungary, India, Iran, Italy, Romania, Slovakia, Syria, Tunisia, Turkey, USSR (Noyes, 2019).

Psilochalcis mirabilis (Boucek, 1952)
Publications with Turkey records: Noyes (2019), Öncüer (1991).
Associates: *Arachis hypogaea* (Fabaceae) (Noyes, 2019).
Hosts: *Stegasta bosqueella* (Lepidoptera: Gelechiidae), *Elasmopalpus lignosellus* (Pyralidae) (Noyes, 2019).
Distribution: Czechoslovakia, Europe, Hungary, Kazakhstan, Slovakia, Tselinograd Obl., Turkey, United States of America, USSR (Noyes, 2019).

Psilochalcis nitida (Boucek, 1952)
Publications with Turkey records: Noyes (2019), Öncüer (1991).
Associates: Unknown (Noyes, 2019).
Hosts: Unknown (Noyes, 2019).

Distribution: Turkey (Noyes, 2019).

Psilochalcis rufitarsis (Illiger, 1807)
Publications with Turkey records: Noyes (2019), Öncüer (1991).
Associates: *Eryngium campestre* (Apiaceae), *Juniperus thurifera* (Cupressaceae) (Noyes, 2019).
Hosts: *Myelois sp., Myelois cincticornis, Myelois cinctipalpella* (Lepidoptera: Pyralidae) (Noyes, 2019).
Distribution: Croatia, Cyprus, Europe, France, Hungary, Iran, Italy, Kazakhstan, Moldova, Slovakia, Spain, Sweden, Syria, Transcaucasus, Tselinograd Obl., Turkey, Ukraine, USSR (Noyes, 2019).

Psilochalcis subaenea (Masi, 1929)
Publications with Turkey records: Noyes (2019), Öncüer (1991).
Hosts: *Caryedon acaciae* (Coleoptera: Bruchidae) (Noyes, 2019).
Distribution: Czechoslovakia, Europe, France, Hungary, Iran, Italy, Kazakhstan, Slovakia, Spain, Transcaucasus, Tselinograd Obl., Turkey, Ukraine, United Kingdom (Noyes, 2019).

Psilochalcis subarmata (Foerster, 1855)
Publications with Turkey records: Noyes (2019), Öncüer (1991).
Associates: *Atriplex halimus* (Chenopodiaceae), *Retama sphaerocarpa* (Fabaceae), *Reseda lutea* (Resedaceae), *Tamarix canariensis* (Tamaricaceae) (Noyes, 2019).
Distribution: Canary Islands, Czech Republic, England, Europe, France, Germany, Greece, Hungary, India, Iran, Kazakhstan, Moldova, Slovakia, Spain, Sweden, Tselinograd Obl., Turkey, Ukraine, USSR (Noyes, 2019).

References

Avcı, Mustafa. 2009. "Parasitoid complex and new host plants of the gypsy moth, *Lymantria dispar* L. in the Lakes District, Turkey". *Journal of Animal and Veterinary Advances*, 8(7):1402-1405.

Bayram, Şerife, and Ülgentürk, Selma, and Toros, Seval. 1998. "Research on the insects causing galls on dog rose (*Rosa spp.*) and their parasitoids in Ankara province". *Turkish Journal of Entomology*, 22(4):259-268.

Doğanlar, Mikdat. 1990. "A new species of *Hockeria* Walker (Hymenoptera, Chalcididae), a pupal parasite of *Heliothis viriplaca* Hufn. (Lepidoptera, Noctuidae) on lentil (Lens esculenta) and chickpea

(*Cicer sp.*) in southestern Anatolia". *Turkish Journal of Entomology*, 14(3):149-153.

Noyes, John S. 2019. *Universal Chalcidoidea Database*. Available at: www.nhm. ac.uk/chalcidoids (accessed January 2022).

Öncüer, Cezmi. 1991. "A catologue of the parasites and predators of insect pests of Turkey". *Ege University Agricultural Faculty Publication 505*, İzmir. 354 pp.

Polat, Alper, and Tozlu, Göksel. 2010. "Investigations on the brief biology, host plants and parasitoids of Archips rosana (Linnaeus, 1758) (Lepidoptera: Tortricidae) in Erzurum". *Turkish Journal of Entomology*, 34(4):529-542.

Stringer, Danielle N., and Mantel, Sarah, and Jennings, John T., and Austin, Andrew D. 2012. "Family Chalcididae". *Australian Centre for Evolutionary Biology and Biodiversity, and the School of Earth and Environmental Science*. The University of Adelaide, Australia (Available from: https://www.researchgate.net/publication/ 262008500_Family_CHALCIDIDAE).

Yaşarakıncı, Nilgün, and Kornoşor, Serpil. 1990. "Güneydoğu Anadolu Bölgesi'nde Nohut ve Mercimekte Zarar Yapan *Heliothis viriplaca* (Huff.) (Lepidoptera: Noctuidae)'nın Doğal Düşmanları ve Parazitoidlerinin Etkinlikleri Üzerinde Araştırmalar". *Türkiye II. Biyolojik Mücadele Kongresi*, 26-29 Eylül, Ankara, 83-91.

CHAPTER FIVE

FAMILY: ENCYRTIDAE WALKER, 1837

The Encyrtidae is a large family of Chalcidoidea with over 4500 described species worldwide (Noyes, 2019). Although they are small in size (0.5–4 mm) and taxonomically challenging, the family encompasses a variety of bizarre and unusual forms. They are characterized by having a large and convex mesopleuron that is longer than its height and lacks a median groove or furrow, an enlarged spur of the mid tibia, the mid coxa inserted anterior to the midline of the mesopleuron, and the cerci typically positioned along the metasoma, some distance from the apex. Many small encyrtids can be mistaken for Aphelinids; however, in Encyrtids, the axillae meet medially while in aphelinids, they are widely separated.

Many Encyrtids are primary parasitoids of various scale insects (Hemiptera: Coccoidea), with the Tetracneminae being restricted to this host group. The other subfamily, Encyrtinae, has a very broad host range, including hosts from many insect orders, as well as mites, ticks (Acarina) and spiders (Araneae). Encyrtids are all endoparasitoids, and some are polyembryonic, meaning a single egg divides to form two or more identical embryos, often in hundreds, within larvae of Lepidoptera (moths and butterflies).

After the Aphelinids, the Encyrtidae are the second most important family of Hymenoptera in biological control with numerous successful examples of their use in controlling insect pests (Stringer et al., 2012).

The family currently includes 460 genera and 3735 species placed in two subfamilies Encyrtinae and Tetracneminae. In Turkey, there are two subfamilies, 59 genera, and 164 species (Noyes, 2019).

Subfamily Encyrtinae (Walker, 1837)
Genus *Acerophagus* (Smith, 1880)
Acerophagus malinus (Gahan, 1946)
Publications with Turkey records: Kaydan and Japoshvili (2010), Noyes (2019).
Associates: *Morus sp., Morus alba* (Moraceae), *Malus pumila, Prunus sp., Pyrus communis* (Rosaceae), *Rubus sp.* (Rubiaceae), *Citrus sp.* (Rutaceae), *Populus sp.* (Salicaceae) (Noyes, 2019).
Hosts: Hemiptera: Aphididae, *Pulvinaria vitis* (Coccidae), *Kermococcus quercus* (Kermesidae), *Coccura comari, Coccura suwakoensis, Coccura ussuriensis, Dysmicoccus brevipes, Pseudococcus comstocki, Pseudococcus cryptus* (Pseudococcidae) (Noyes, 2019).
Distribution: Armenia, Azerbaijan, Canada, Georgia, Germany, Hawaii, Israel, Japan, Kazakhstan, Kirgizia, Korea, Korea (South), Moldova, People's Republic of China, Russia, Tadzhikistan, Tselinograd Obl., Turkey, Turkmenistan, Ukraine, United States of America, USSR, Uzbekistan (Noyes, 2019).

Genus *Achalcerinys* (Girault, 1915)
Achalcerinys lindus (Mercet, 1921)
Publications with Turkey records: Japoshvili and Noyes (2005), Noyes (2019), Trjapitzin and Doğanlar (1997).
Associates: *Phragmites australis* (Poaceae) (Noyes, 2019).
Hosts: *Cynodon dactylon* (Poaceae), *Acanthococcus obscurus, Eriococcus obscurus, Rhizococcus obscurus* (Hemiptera: Eriococcidae), *Antonina phragmitis, Chaetococcus phragmitis, Pseudococcus copiosus, Trionymus copiosus* (Pseudococcidae) (Noyes, 2019).
Parasitoid hosts: *Asitus phragmitis, Platyrhopus meridionalis* (Hymenoptera: Encyrtidae) (Noyes, 2019).
Distribution: Armenia, Azerbaijan, Bulgaria, Croatia, Czech Republic, Czechoslovakia, Ethiopia, Europe, France, Hungary, India, Italy, Moldova, Netherlands, Palaearctic, Romania, Saudi Arabia, Slovakia, Spain, Tadzhikistan, Transcaucasus, Turkey, Turkmenistan, Ukraine, Uzbekistan, Yugoslavia (pre1991) (Noyes, 2019).

Genus *Adelencyrtus* (Ashmead, 1900)
Adelencyrtus aulacaspidis (Brethes, 1914)
Publications with Turkey records: Japoshvili and Noyes (2005), Noyes (2019), Öncüer (1991).

Associates: *Castanea crenata* (Fagaceae), *Myrica rubra* (Myricaceae), *Rosa canina* (Rosaceae), *Rubus sp., Rubus idaeus* (Rubiaceae) (Noyes, 2019).
Hosts: *Lecanopsis nevesi* (Hemiptera: Coccidae), *Aulacaspis difficilis, Aulacaspis rosae, Chionaspis salicis, Dynaspidiotus britannicus, Lepidosaphes cupressi, Pseudaulacaspis pentagona, Quadraspidiotus macroporanus, Unaspis yanonensis* (Diaspididae) (Noyes, 2019).
Parasitoids: *Aspidiotiphagus citrinus* (Hymenoptera: Aphelinidae) (Noyes, 2019).
Distribution: Argentina, Australia, Azerbaijan, Bulgaria, Chile, Croatia, Czech Republic, Czechoslovakia, England, Europe, France, Georgia, Germany, Hungary, Italy, Japan, Kazakhstan, Madagascar, Moldova, Nearctic, Neotropical, New Zealand, Norway, People's Republic of China, Russia, Slovenia, Spain, Switzerland, Tselinograd Obl., Turkey, Ukraine, United States of America, USSR, Yugoslavia (pre-1991) (Noyes, 2019).

Adelencyrtus intersectus (Fonscolombe, 1832)
Publications with Turkey records: Japoshvili and Karaca (2002), Noyes (2019), Öncüer (1991).
Associates: *Pistacia vera* (Anacardiaceae), *Corylus avellana* (Betulaceae), *Juniperus bermudana* (Cupressaceae), *Hippophae rhamnoides* (Elaeagnaceae), *Castanea crenata, Quercus sp.* (Fagaceae), *Abies alba, Pinus nigra* (Pinaceae), *Malus sp., Malus pumila* (Rosaceae), *Populus sp., Populus gracilis, Populus tremula, Salix sp., Salix caprea* (Salicaceae), *Paulownia sp.* (Scrophulariaceae), *Camellia sinensis* (Theaceae)(Noyes, 2019).
Hosts: *Aclerda subterranea* (Hemiptera: Aclerdidae), *Asterolecanium sp., Asterolecanium minus, Asterolecanium quercicola* (Asterolecaniidae), *Ceroplastes rubens, Eulecanium rufulum, Sphaerolecanium prunastri* (Coccidae), *Aspidiotus perniciosus, Carulaspis minima, Diaspidiotus bavaricus, Diaspidiotus ostreaeformis, Diaspidiotus perniciosus, Diaspidiotus prunorum, Diaspidiotus pyri, Diaspidiotus spurcatus, Lepidosaphes sp., Lepidosaphes malicola, Lepidosaphes newsteadi, Lepidosaphes populi, Lepidosaphes tubulorum, Lepidosaphes ulmi, Leucaspis pusilla, Mytilococcus ulmi, Pseudaonidia duplex, Pseudaonidia paeoniae, Pseudaulacaspis pentagona, Quadraspidiotus gigas, Quadraspidiotus macroporanus, Quadraspidiotus ostreaeformis, Quadraspidiotus pernici*osus (Diaspididae) (Noyes, 2019).
Distribution: Armenia, Azerbaijan, Belarus, Bermuda, Bulgaria, Canada, Canary Islands, Croatia, Czech Republic, Czechoslovakia, England, Europe, Finland, France, Georgia, Germany, Hawaii, Hungary, India, Iran,

Italy, Japan, Madeira, Moldova, Nearctic, Netherlands, New Zealand, Norway, People's Republic of China, Poland, Romania, Russia, Slovakia, Spain, Sweden, Switzerland, Transcaucasus, Tunisia, Turkey, Ukraine, United States of America, USSR, Yugoslavia (pre-1991) (Noyes, 2019).

Genus *Ageniaspis* (Dahlbom, 1857)
Ageniaspis fuscicollis (Dalman, 1820)
Publications with Turkey records: Noyes (2019), Öncüer (1991), Trjapitzin and Doğanlar (1997).
Associates: *Dacus oleae* (Diptera: Tephritidae), *Euonymus sp., Euonymus europaea, Euonymus verrucosa* (Celastraceae), *Olea europaea* (Oleaceae), *Crataegus oxyacantha, Malus sp., Malus pumila, Prunus sp., Prunus armeniaca, Prunus domestica, Prunus mahaleb, Prunus padus, Prunus spinosa, Pyrus sp.* (Rosaceae), *Citrus limon* (Rutaceae), *Populus sp., Salix sp., Salix alba* (Salicaceae) (Noyes, 2019).
Hosts: *Lecanium caryae, Lecanium corni* (Hemiptera: Coccidae), *Phyllocnistis citrella* (Lepidoptera: Gracillariidae), *Autographa sp.* (Noctuidae), *Anchylopera plumbata, Cydia pomonella* (Tortricidae), *Acrolepia assectella, Acrolepiopsis assectella, Argyresthia ephippella, Dyscedestis farinatella, Ocnerostoma sp., Ocnerostoma piniariella, Prays citri, Prays oleae, Prays oleellus, Yponomeuta sp., Yponomeuta cagnagellus, Yponomeuta cognatellus, Yponomeuta evonymellus, Yponomeuta irrorellus, Yponomeuta mahalabellus, Yponomeuta malinellus, Yponomeuta orientalis, Yponomeuta padellus, Yponomeuta plumbellus, Yponomeuta rorellus, Zelleria oleastrella* (Yponomeutidae) (Noyes, 2019).
Distribution: Argentina, Armenia, Azerbaijan, Bosnia Hercegovina, Bulgaria, Canada, Caucasus, Central Asia, Croatia, Czechoslovakia, England, Europe, Finland, France, Georgia, Germany, Greece, Hungary, Iran, Ireland, Israel, Italy, Kazakhstan, Kirgizia, Korea, Lebanon, Lithuania, Moldova, Mongolia, Morocco, Nearctic, Netherlands, Norway, People's Republic of China, Poland, Portugal, Romania, Russia, Serbia, Siberia, Spain, Sweden, Switzerland, Transcaucasus, Tselinograd Obl., Tunisia, Turkey, Turkmenistan, Ukraine, United States of America, Uzbekistan, Yugoslavia (pre-1991) (Noyes, 2019).

Genus *Anicetus* (Howard, 1896)
Anicetus italicus (Masi, 1917)
Publications with Turkey records: Japoshvili and Noyes (2005), Noyes (2019), Öncüer (1991), Trjapitzin and Doğanlar (1997).

Associates: *Dalbergia sissoo* (Fabaceae), *Ficus sp., Ficus carica* (Moraceae), *Ulmus sp.* (Ulmaceae) (Noyes, 2019).
Hosts: *Ceroplastes sp., Ceroplastes rusci, Coccus elongatus* (Hemiptera: Coccidae), *Gossyparia spuria* (Eriococcidae) (Noyes, 2019).
Distribution: Albania, Algeria, Cyprus, Egypt, France, Greece, India, Iran, Israel, Italy, Montenegro, Turkey, Yugoslavia (pre-1991), Zimbabwe (Noyes, 2019).

Genus *Aphycoides* (Mercet, 1921)
Aphycoides clavellatus (Dalman, 1820)
Publications with Turkey records: Noyes (2019), Trjapitzin and Doğanlar (1997), Ulgenturk (2001).
Associates: *Abies excelsa, Picea sp., Picea abies, Picea pungens* (Pinaceae) (Noyes, 2019).
Hosts: *Eulecanium graniforme, Lecanium piceae, Physokermes sp., Physokermes abietis, Physokermes fasciatus, Physokermes hemicryphus, Physokermes jezoensis, Physokermes piceae, Physokermes sugonjaevi* (Hemiptera: Coccidae) (Noyes, 2019).
Distribution: Austria, Bulgaria, Canada, Czech Republic, Czechoslovakia, Denmark, England, Estonia, Europe, Finland, France, Georgia, Germany, Hungary, Japan, Kazakhstan, Kirgizia, Moldova, Mongolia, Montenegro, Norway, Poland, Romania, Russia, Slovakia, Siberia, Sweden, Tselinograd Obl., Turkey, Ukraine, United States of America, Yugoslavia (pre-1991) (Noyes, 2019).

Genus *Aphycus* (Mayr, 1876)
Aphycus moravicus (Hoffer, 1952)
Publications with Turkey records: Japoshvili (2012), Noyes (2019).
Hosts: Unknown (Noyes, 2019).
Distribution: Czech Republic, Czechoslovakia, Hungary, Moldova, Norway, Russia, Slovakia, Turkey, Ukraine (Noyes, 2019).

Aphycus primus (Mercet, 1917)
Publications with Turkey records: Japoshvili and Noyes (2005b), Noyes (2019).
Associates: Unknown (Noyes, 2019).
Hosts: *Chionaspis pinifoliae* (Hemiptera: Diaspididae) (Noyes, 2019).
Distribution: Montenegro, Russia, Spain, Turkey (Noyes, 2019).

Genus *Blastothrix* (Mayr, 1876)
Blastothrix erythrostetha (Walker, 1847)
Publications with Turkey records: Japoshvili and Noyes (2005), Noyes (2019).
Associates: *Quercus sp., Quercus glauca, Quercus mongolica, Quercus robur* (Fagaceae) (Noyes, 2019).
Hosts: *Ceroplastes mimosa, Coccus hesperidum, Parthenolecanium corni, Sphaerolecanium prunastri* (Hemiptera: Coccidae), *Kermes sp., Kermes bacciformis, Kermes corticalis, Kermes ilicis, Kermes miyasakii, Kermes nakagawae, Kermes nawae, Kermes quercus, Kermes roboris, Kermococcus sp., Kermococcus corticalis, Kermococcus quercus, Kermococcus roboris* (Kermesidae) (Noyes, 2019).
Distribution: Armenia, Austria, Bulgaria, Egypt, England, Europe, France, Georgia, Germany, Hungary. Israel, Italy, Japan, Moldova, Poland, Romania, Russia, Spain, Transcaucasus, Turkey, Ukraine, USSR (Noyes, 2019).

Blastothrix gurselae (Japoshvili and Karaca, 2004)
Publications with Turkey records: Japoshvili (2012), Japoshvili and Celik (2010), Japoshvili and Noyes (2005), Japoshvili and Karaca, (2004), Noyes (2019).
Associates: *Quercus coccifera* (Fagaceae) (Noyes, 2019).
Hosts: *Kermes palestinensis* (Hemiptera: Kermesidae) (Noyes, 2019).
Distribution: Spain, Turkey (Noyes, 2019).

Blastothrix hungarica (Erdos, 1959)
Publications with Turkey records: Japoshvili and Noyes (2005), Noyes (2019), Ulgenturk (2001).
Associates: *Juniperus sp.* (Cupressaceae), *Ficus carica, Morus sp.* (Moraceae), *Prunus sp., Spiraea sp.* (Rosaceae), *Taxus sp.* (Taxaceae), *Vitis vinifera* (Vitaceae) (Noyes, 2019).
Hosts: *Eulecanium sp., Eulecanium rugulosum, Parthenolecanium persicae, Rhodococcus spiraeae* (Hemiptera: Coccidae) (Noyes, 2019).
Distribution: Afghanistan, Azerbaijan, Bulgaria, Caucasus, Georgia, Germany, Hungary, Iran, Japan, Poland, Transcaucasus, Turkey, Ukraine, USSR, Yugoslavia (pre1991) (Noyes, 2019).

Blastothrix longipennis (Howard, 1881)
Publications with Turkey records: Japoshvili and Hansen (2013), Noyes (2019), Öncüer (1991), Ulgenturk (2001).
Associates: *Corylus sp., Corylus avellana* (Betulaceae), *Cornus mas, Thelycrania australis* (Cornaceae), *Cupressus sp., Thuja sp.* (Cupressaceae), *Acacia sp., Cercis siliquastrum, Cytisus scoparius, Pisum sativum, Robinia pseudoacacia* (Fabaceae), *Quercus sp., Quercus robur* (Fagaceae), *Juglans regia* (Juglandaceae), *Fraxinus sp.* (Oleaceae), *Crataegus sp., Malus sp., Padus racemose, Prunus sp., Prunus armeniaca, Prunus cerasifera, Prunus domestica* (Rosaceae), *Ulmus foliacea* (Ulmaceae), *Vitis vinifera* (Vitaceae) (Noyes, 2019).
Hosts: *Brevicoryne brassicae* (Hemiptera: Aphididae), *Eulecanium cerasorum, Eulecanium corni, Eulecanium coryli, Eulecanium fletcheri, Eulecanium tiliae, Lecanium sp., Lecanium cerasorum, Lecanium corni, Lecanium fletcheri, Lecanium pruinosum, Lecanium quercifex, Lecanium tarsale, Lecanium tiliae, Parthenolecanium sp., Parthenolecanium corni, Parthenolecanium fletcheri, Parthenolecanium pomeranicum, Parthenolecanium pruinosum, Parthenolecanium quercifex, Parthenolecanium rufulum, Pulvinaria regalis, Sphaerolecanium prunastri* (Coccidae), *Gossyparia spuria* (Eriococcidae), *Kermes pubescens* (Kermesidae), *Evagora milleri* (Lepidoptera: Gelechiidae) (Noyes, 2019).
Parasitoids: *Coccophagus lycimnia* (Hymenoptera: Aphelinidae), *Aprostocetus minutus, Tetrastichus minutus* (Eulophidae), *Pachyneuron sp., Pachyneuron solitarium* (Pteromalidae) (Noyes, 2019).
Distribution: Armenia, Azerbaijan, Bulgaria, Canada, Caucasus, Costa Rica, Croatia, Czech Republic, Czechoslovakia, England, Europe, Finland, France, Georgia, Germany, Hungary, Italy, Japan, Kazakhstan, Kirgizia, Moldova, Montenegro, Nearctic, North Africa, Norway, People's Republic of China, Poland, Romania, Russia, Slovakia, Slovenia, Spain, Sweden, Transcaucasus, Tselinograd Obl., Turkey, Ukraine, United States of America, USSR, Uzbekistan, Yugoslavia (Federal Republic) (Noyes, 2019).

Blastothrix sericea (Dalman, 1820)
Publications with Turkey records: Japoshvili and Noyes (2005), Noyes (2019), Öncüer (1991).
Associates: *Pistacia vera* (Anacardiaceae), *Corylus avellana* (Betulaceae), *Diospyros virginiana* (Ebenaceae), *Acacia sp., Robinia pseudoacacia* (Fabaceae), *Quercus sp.* (Fagaceae), *Ficus sp., Ficus carica* (Moraceae), *Fraxinus sp.* (Oleaceae), *Pseudotsuga sp.* (Pinaceae), *Amygdalus communis, Cerasus japonica, Malus pumila, Prunus sp., Prunus*

domestica, Prunus spinosa, Rosa sp., Spiraea hypericifolia (Rosaceae), *Ulmus americana* (Ulmaceae), *Vitis vinifera* (Vitaceae) (Noyes, 2019).
Hosts: *Baccha fascipennis* (Diptera: Syrphidae), *Anapulvinaria pistaciae, Coccidae unspecified sp., Coccus persicae, Ericerus pela, Eulecanium sp., Eulecanium ciliatum, Eulecanium corni, Eulecanium coryli, Eulecanium douglasi, Eulecanium giganteum, Eulecanium graniforme, Eulecanium persicae, Eulecanium pruinosum, Eulecanium rugulosum, Eulecanium slavum, Eulecanium spiraeae, Eulecanium tiliae, Eulecanium turanicum, Lecanium sp., Lecanium corni, Lecanium coryli, Lecanium fletcheri, Lecanium nigrofasciatum, Lecanium tiliae, Parthenolecanium sp., Parthenolecanium corni, Parthenolecanium rufulum, Physokermes sp., Pulvinaria betulae, Pulvinaria vitis, Rhodococcus spiraeae, Sphaerolecanium prunastri* (Hemiptera: Coccidae), *Aulacaspis rosae* (Diaspididae), *Kermes sp., Kermes pubescens, Kermococcus quercus* (Kermesidae), *Phenacoccus acericola* (Pseudococcidae) (Noyes, 2019).
Distribution: Canada, Croatia, Czechoslovakia, England, Europe, France, Georgia, Germany, Greece, Hungary, India, Iran, Ireland, Italy, Nearctic, Pakistan, People's Republic of China, Poland, Romania, Russia, Serbia, Spain, Sweden, Turkey, Ukraine, United States of America, USSR, Yugoslavia (Federal Republic) (Noyes, 2019).

Blastothrix turanica (Sugonjaev, 1964)
Publications with Turkey records: Japoshvili and Noyes (2005), Noyes (2019), Trjapitzin and Doğanlar (1997).
Associates: *Rhamnus pallasii* (Rhamnaceae) (Noyes, 2019).
Hosts: *Rhodococcus turanicus* (Hemiptera: Coccidae) (Noyes, 2019).
Distribution: Afghanistan, Armenia, Azerbaijan, Georgia, Iran, Kazakhstan, Pakistan, Russia, Tselinograd Obl., Turkey, Uzbekistan (Noyes, 2019).

Genus *Bothriothorax* (Ratzeburg, 1844)
Bothriothorax clavicornis (Dalman, 1820)
Publications with Turkey records: Japoshvili and Noyes (2005), Noyes (2019), Trjapitzin and Doğanlar (1997).
Associates: *Triticum sp.* (Poaceae) (Noyes, 2019).
Hosts: *Episyrphus balteatus, Neocnemodon fulvimanus, Scaeva sp., Syrphus sp., Syrphus balteatus, Syrphus ribesii, Syrphus vitripennis* (Diptera: Syrphidae) (Noyes, 2019).
Distribution: Armenia, Belarus, Czech Republic, Czechoslovakia, Denmark, Europe, Finland, Germany, Hungary, Ireland, Italy, Lithuania, Moldova, Mongolia, Montenegro, Netherlands, Norway, People's

Republic of China, Poland, Romania, Russia, Slovakia, Sweden, Switzerland, Turkey, United Kingdom, Uzbekistan, Yugoslavia (pre-1991) (Noyes, 2019).

Bothriothorax serratellus (Dalman, 1820)
Publications with Turkey records: Japoshvili and Noyes (2005), Noyes (2019), Trjapitzin and Doğanlar (1997).
Hosts: *Heringia heringi, Syrphus albostriatus* (Diptera: Syrphidae) (Noyes, 2019).
Distribution: Armenia, Czech Republic, Czechoslovakia, Denmark, Europe, Finland, Iran, Mongolia, Montenegro, Netherlands, Norway, Russia, Slovakia, Sweden, Switzerland, Tadzhikistan, Turkey, United Kingdom, Yugoslavia (pre-1991) (Noyes, 2019).

Genus *Cerapterocerus* (Westwood, 1833)
Cerapterocerus mirabilis (Westwood, 1833)
Publications with Turkey records: Japoshvili (2012), Japoshvili and Karaca (2002), Noyes (2019), Öncüer (1991), Trjapitzin and Doğanlar (1997), Ulgenturk (2001), Ulgenturk et al. (2001).
Associates: *Artemisia sp.* (Asteraceae), *Cupressus sp., Thuja sp.* (Cupressaceae), *Quercu coccifera* (Fagaceae), *Olea europaea* (Oleaceae), *Brachypodium ramosum, Festuca circummediterranea* (Poaceae), *Atraphaxis sp.* (Polygonaceae), *Amygdalus sp., Armeniaca vulgaris, Prunus sp., Prunus cerasifera, Prunus domestica, Prunus salicina, Prunus spinosa* (Rosaceae), *Rubus sp.* (Rubiaceae), *Citrus sp., Citrus deliciosa* (Rutaceae), *Tamarix canariensis* (Tamaricaceae) (Noyes, 2019).
Hosts: *Aclerda subterranean* (Hemiptera: Aclerdidae), *Acantholecanium haloxyloni, Ceroplastes sp., Ceroplastes elytropappi, Coccidae unspecified sp., Coccus hesperidum, Coccus pseudomagnoliarum, Didesmococcus unifasciatus, Eriopeltis sp., Eriopeltis agropyri, Eriopeltis araxis, Eriopeltis festucae, Eriopeltis lichtensteinii, Eriopeltis sachalinensis, Eulecanium sp., Eulecanium corni, Eulecanium coryli, Eulecanium prunastri, Filippia viburni, Gascardia rustica, Lecanium corni, Lecanium pruinosum, Lecanium prunastri, Luzulaspis grandis, Parthenolecanium sp., Parthenolecanium corni, Parthenolecanium coryli, Rhizopulvinaria sp., Scythia aetnensis, Scythia stipae, Sphaerolecanium sp., Sphaerolecanium prunastri, Stotzia maxima* (Coccidae), *Aonidiella orientalis* (Diaspididae), *Kermes palestinensis* (Kermesidae), *Antonina phragmitis, Antonina purpurea, Coccura circumscripta, Coccura comari, Mediococcus circumscriptus* (Pseudococcidae) (Noyes, 2019).

Parasitoid hosts: *Anagyrus diversicornis, Discodes sp., Discodes aeneus, Discodes coccophagus, Encyrtus masii, Metaphycus punctipes, Paraphaenodiscus subterraneus* (Hymenoptera: Encyrtidae) (Noyes, 2019).
Parasitoids: *Marietta picta* (Aphelinidae), *Discodes coccophagus* (Encyrtidae) (Noyes, 2019).
Distribution: Afrotropical, Armenia, Austria, Azerbaijan, Bosnia Hercegovina, Bulgaria, Canary Islands, Caucasus, Croatia, Czech Republic, Czechoslovakia, Denmark, England, Europe, Finland, France, Georgia, Germany, Greece, Hungary, India, Iran, Israel, Italy, Japan, Kazakhstan, Lithuania, Mexico, Moldova, Mongolia, Montenegro, Netherlands, Poland, Romania, Russia, Serbia, Slovakia, South Africa, Spain, Sweden, Switzerland, Tadzhikistan, Tselinograd Obl., Turkey, Turkmenistan, Ukraine, USSR, Uzbekistan, Yugoslavia (Federal Republic) (Noyes, 2019).

Genus *Cerchysiella* (Girault, 1914)
Cerchysiella planiscutellum (Mercet, 1921)
Publications with Turkey records: Noyes (2019), Trjapitzin and Doğanlar (1997).
Associates: *Brassica sp.* (Brassicaceae) (Noyes, 2019).
Hosts: *Meligethes aeneus* (Coleoptera: Nitidulidae) (Noyes, 2019).
Distrubution: Afghanistan, Andorra, Armenia, Azerbaijan, Bulgaria, Czech Republic, Czechoslovakia, England, Europe, France, Georgia, Hungary, Kazakhstan, Macedonia, Montenegro, Netherlands, Norway, Portugal, Russia, Serbia, Spain, Sweden, Tadzhikistan, Tselinograd Obl., Turkey, Turkmenistan, Ukraine, USSR, Yugoslavia (pre-1991) (Noyes, 2019).

Genus *Cerchysius* (Westwood, 1832)
Cerchysius subplanus (Dalman, 1820)
Publications with Turkey records: Japoshvili and Noyes (2005), Noyes (2019), Trjapitzin and Doğanlar (1997).
Associates: *Balanococcus singularis* (Hemiptera: Pseudococcidae), *Quercus sp.* (Fagaceae), *Arundo donax, Festuca ovina* (Poaceae) (Noyes, 2019).
Hosts: *Chamaemyia juncorum, Leucopomyia silesiaca, Lipoleucopis praecox* (Diptera: Chamaemyiidae), *Greenisca brachypodii* (Hemiptera: Eriococcidae), *Adelosoma phragmitidis* (Pseudococcidae) (Noyes, 2019).
Distrubution: Afghanistan, Armenia, Austria, Azerbaijan, Canary Islands, Croatia, Czech Republic, Europe, Georgia, Germany, Hungary, India,

Korea, Latvia, Madeira, Moldova, Mongolia, Netherlands, Norway, People's Republic of China, Portugal, Romania, Russia, Slovakia, Sweden, Thailand, Transcaucasus, Turkey, United Kingdom, USSR (Noyes, 2019).

Genus *Cheiloneurus* (Westwood, 1833)
Cheiloneurus claviger (Thomson, 1876)
Publications with Turkey records: Japoshvili (2012), Japoshvili and Celik (2010), Noyes (2019), Öncüer (1991), Trjapitzin and Doğanlar (1997).
Associates: *Ilex sp.* (Aquifoliaceae), *Artemisia sp.* (Asteraceae), *Corylus sp.* (Betulaceae), *Cornus mas* (Cornaceae), *Quercus sp., Quercus ilex* (Fagaceae), *Ephedra procera* (Gnetaceae), *Hibiscus sp.* (Malvaceae), *Ficus carica, Morus sp.* (Moraceae), *Fraxinus sp., Olea europaea* (Oleaceae), *Picea abies* (Pinaceae) *Prunus sp., Prunus cerasifera, Prunus scoparia, Spiraea sp.* (Rosaceae), *Citrus nobilis, Citrus unshiu* (Rutaceae), *Populus sp., Populus hybrida, Populus transcaucasica* (Salicaceae), *Vitis vinifera* (Vitaceae) (Noyes, 2019).
Hosts: *Acanthopulvinaria orientalis, Ceroplastes ceriferus, Ceroplastes japonicus, Chloropulvinaria aurantia, Coccus hesperidum, Didesmococcus unifasciatus, Ericerus pela, Eulecanium ciliatum, Eulecanium corni, Eulecanium giganteum, Eulecanium kunoense, Eulecanium kuwanai, Eulecanium quercifex, Eulecanium rugulosum, Eulecanium tiliae, Filippia folicularis, Filippia viburni, Parthenolecanium corni, Parthenolecanium persicae, Parthenolecanium quercifex, Parthenolecanium rufulum, Physokermes fasciatus, Physokermes hemicryphus, Pulvinaria sp., Pulvinaria aurantia, Pulvinaria betulae, Pulvinaria idesiae, Pulvinaria populi, Pulvinaria vitis, Rhizopulvinaria sp., Rhodococcus spiraeae, Rhodococcus turanicus, Saissetia oleae, Sphaerolecanium prunastri, Stotzia maxima, Takahashia japonica* (Hemiptera: Coccidae), *Eriococcus brachypodii, Greenisca brachypodii, Neoacanthococcus tamaricicol* (Eriococcidae), *Kermes miyasakii, Kermes vermilio* (Kermesidae), *Maconellicoccus hirsutus, Nesticoccus sinensis, Nipaecoccus filamentosus, Phenacoccus aceris, Phenacoccus mespili, Planococcus citri* (Pseudococcidae) (Noyes, 2019).
Parasitoid hosts: *Coccophagus aterrimus* (Hymenoptera: Aphelinidae), *Blastothrix hungarica, Blastothrix longipennis, Blastothrix scenographica, Blastothrix turanica, Metaphycus insidiosus, Microterys sp., Microterys cuprinus, Microterys intermedius, Microterys praedator, Microterys sylvius* (Encyrtidae) (Noyes, 2019).

Distribution: Armenia, Austria, Azerbaijan, Bulgaria, Croatia, Czech Republic, Czechoslovakia, Egypt, England, Europe, France, Georgia, Germany, Greece, Hungary, Iran, Israel, Italy, Japan, Kazakhstan, Moldova, Montenegro, Netherlands, Norway, Palearctic, People's Republic of China, Romania, Russia, Serbia, Slovakia, Spain, Sweden, Tadzhikistan, Transcaucasus,Tselinograd Obl., Turkey, Turkmenistan, Ukraine, United Kingdom, USSR, Uzbekistan, Yugoslavia (Federal Republic), Yugoslavia (pre-1991) (Noyes, 2019).

Cheiloneurus elegans (Dalman, 1820)
Publications with Turkey records: Japoshvili (2012), Japoshvili and Celik (2010), Noyes (2019).
Associates: *Pistacia mutica* (Anacardiaceae), *Carduus pycnocephalus* (Asteraceae), *Manihot esculenta* (Euphorbiaceae), *Brachypodium ramosum, Hordeum leporinum* (Poaceae), *Reseda lutea* (Resedaceae) (Noyes, 2019).
Hosts: *Mayetiola destructor, Phytophaga destructor* (Diptera: Cecidomyiidae), *Aclerda subterranean* (Hemiptera: Aclerdidae), *Anapulvinaria pistaciae, Eulecanium franconicum, Physokermes piceae, Pulvinaria vitis* (Coccidae), *Kermes sp.* (Kermesidae), *Antonina purpurea, Phenacoccus hordei, Phenacoccus manihoti, Trionymus aberrans* (Pseudococcidae) (Noyes, 2019).
Parasitoid hosts: *Epidinocarsis lopezi* (Hymenoptera: Encyrtidae), *Platygaster zosine* (Platygastridae) (Noyes, 2019).
Distribution: Argentina, Armenia, Austria, Azerbaijan, Bulgaria, Canada, Croatia, Czechoslovakia, Denmark, England, Europe, Finland, France, Georgia, Germany, Hungary, India, Israel, Italy, Kazakhstan, Lithuania, Macedonia, Mexico, Moldova, Mongolia, Nearctic, Netherlands, Nigeria, Palearctic, Romania, Russia, Saudi Arabia, Serbia, Spain, Sweden, Switzerland, Tadzhikistan, Transcaucasus, Tselinograd Obl., Turkey, Turkmenistan, Ukraine, United States of America, Uzbekistan, Yugoslavia (pre-1991) (Noyes, 2019).

Cheiloneurus paralia (Walker, 1837)
Publications with Turkey records: Japoshvili (2012), Japoshvili and Celik (2010), Japoshvili and Noyes (2005), Noyes (2019), Trjapitzin and Doğanlar (1997).
Associates: *Marietta javensis* (Aphelinidae), *Pachyneuron concolor* (Pteromalidae), *Lactuca serriola* (Asteraceae), *Cupressus sp., Thuja sp.* (Cupressaceae), *Carex sp.* (Cyperaceae), *Calluna vulgaris, Erica tetralix* (Ericaceae), *Quercus robur* (Fagaceae), *Ephedra procera* (Gnetaceae),

Thymus sp. (Lamiaceae), *Olea europaea* (Oleaceae), *Picea abies* (Pinaceae), *Agropyron repens, Arrhenatherum elatius, Brachypodium sylvaticum, Festuca sp., Lolium sp., Lolium italicum* (Poaceae), *Rosa sp.* (Rosaceae), *Citrus sp.* (Rutaceae) (Noyes, 2019).

Hosts: *Asterolecanium arabidis* (Hemiptera: Asterolecaniidae), *Cerococcus sp.* (Cerococcidae), *Ceroplastes floridensis, Coccus hesperidum, Eriopeltis araxis, Eriopeltis festucae, Eulecanium sp., Eulecanium ciliatum, Eulecanium corni, Eulecanium franconicum, Eulecanium persicae, Eulecanium rufulum, Eulecanium turanicum, Filippia sp., Lecanium coryli, Luzulaspis bisetosa, Parthenolecanium corni, Parthenolecanium rufulum, Physokermes hemicryphus, Physokermes piceae, Pulvinaria vitis, Rhodococcus turanicus, Saissetia oleae, Sphaerolecanium prunastri, Stotzia maxima* (Coccidae), *Acanthococcus desertus, Acanthococcus greeni, Eriococcus sp., Eriococcus desertus, Eriococcus devoniensis, Eriococcus greeni, Eriococcus insignis, Greenisca brachypodii, Rhizococcus sp., Rhizococcus devoniensis* (Eriococcidae), *Kermococcus sp.* (Kermesidae), *Antonina purpurea, Peliococcus kimmericus* (Pseudococcidae) (Noyes, 2019).

Parasitoid hosts: *Aenigmaphycus paluster, Aphycoides clavellatus, Aphycoides speciosus, Blastothrix Britannica, Blastothrix trjaptzini, Encyrtus sp., Microterys flavus, Microterys lunatus, Microterys nietneri, Microterys sylvius* (Hymenoptera: Encyrtidae) (Noyes, 2019).

Distribution: Afghanistan, Armenia, Austria, Azerbaijan, Bulgaria, Croatia, Czech Republic, Czechoslovakia, Denmark, England, Europe, France, Georgia, Germany, Hungary, Iran, Israel, Italy, Kazakhstan, Lithuania, Mongolia, Netherlands, North Africa, Norway, Palearctic, Poland, Romania, Russia, Serbia, Slovakia, Spain, Sweden, Transcaucasus, Tselinograd Obl., Turkey, Turkmenistan, Ukraine, USSR, Yugoslavia (Federal Republic), Yugoslavia (pre-1991) (Noyes, 2019).

Cheiloneurus quercus (Mayr, 1876)

Publications with Turkey records: Japoshvili and Noyes (2005), Japoshvili and Karaca (2004), Noyes (2019).

Associates: *Quercus sp.* (Fagaceae) (Noyes, 2019).

Hosts: *Eulecanium sp., Pulvinaria vitis* (Hemiptera: Coccidae), *Kermes miyasakii, Kermes nakagawae, Kermes quercus, Kermococcus sp., Kermococcus miyasakii, Kermococcus nakagawae* (Kermesidae), *Coccura ussuriensis, Phenacoccus polyphagus* (Pseudococcidae) (Noyes, 2019).

Distribution: Austria, Czech Republic, Czechoslovakia, Hungary, Italy, Japan, People's Republic of China, Russia, Turkey, USSR (Noyes, 2019).

Genus *Choreia* (Westwood, 1833)
Choreia maculata (Hoffer, 1954)
Publications with Turkey records: Japoshvili and Noyes (2005), Noyes (2019).
Associates: *Arrhenatherum elatius* (Poaceae), *Rubus sp.* (Rubiaceae) (Noyes, 2019).
Hosts: *Eriopeltis sp., Eriopeltis araxis, Eriopeltis festucae, Eulecanium festucae, Scythia stipae* (Hemiptera: Coccidae) (Noyes, 2019).
Distribution: Armenia, Azerbaijan, Bulgaria, Czechoslovakia, Georgia, Hungary, Iran, Kazakhstan, Moldova, Russia, Slovakia, Transcaucasus, Tselinograd Obl., Turkey, Turkmenistan, Ukraine, USSR, Uzbekistan (Noyes, 2019).

Genus *Coelopencyrtus* (Timberlake, 1919)
Coelopencyrtus arenarius (Erdos, 1957)
Publications with Turkey records: Noyes (2019).
Hosts: *Heriades sp., Hylaeus sp., Hylaeus communis, Hylaeus gibbus, Hylaeus nigritus, Prosopis sp., Prosopis communis, Prosopis nigrita* (Hymenoptera: Apidae) (Noyes, 2019).
Distribution: Czech Republic, Czechoslovakia, England, Finland, Germany, Hungary, Moldova, Norway, Russia, Serbia, Slovakia, Sweden, Turkey, Ukraine, USSR (Noyes, 2019).

Coelopencyrtus callidii (Jansson, 1957)
Publications with Turkey records: Japoshvili and Noyes (2005), Noyes (2019), Trjapitzin and Doğanlar (1997).
Associates: Coleoptera: Cerambycidae, Scolytidae, Hymenoptera: Apidae (Noyes, 2019).
Hosts: *Callidium aeneum* (Coleoptera: Cerambycidae) *Hylaeus communis, Hylaeus gibbus, Megachile rotundata, Prosopis communis* (Hymenoptera: Apidae), *Trypoxylon scutatum* (Sphecidae) (Noyes, 2019).
Distribution: Denmark, England, Finland, Iran, Kazakhstan, Moldova, Mongolia, Norway, Russia, Sweden, Tselinograd Obl., Turkey, Ukraine (Noyes, 2019).

Genus *Comperiella* (Howard, 1906)
Comperiella bifasciata (Howard, 1906)
Publications with Turkey records: Japoshvili and Noyes (2005), Karaca and Uygun (1990), Noyes (2019), Öncüer (1991), Sengonca et al. (1998), Uygun and Sekeroglu (1987).

Associates: *Marietta javensis, Prospaltella perniciosi* (Aphelinidae), *Iridomyrmex humilis, Linepithema humile, Solenopsis xyloni* (Formicidae), *Fourcroya gigantea* (Agavaceae), *Mangifera indica* (Anacardiaceae), *Nerium oleander* (Apocynaceae), *Cocos nucifera, Metroxylon sagu* (Arecaceae), *Euonymus japonica* (Celastraceae), *Cycas revoluta* (Cycadaceae), *Liquidambar formosana* (Hamamelidaceae), *Aspidistra elatior, Aspidistra lurida* (Liliaceae), *Ficus sp.* (Moraceae), *Eugenia jambolana, Eugenia vaccinifolia, Psidium guajava* (Myrtaceae), *Passiflora edulis* (Passifloraceae), *Pinus kochiana* (Pinaceae), *Podocarpus sp., Podocarpus macrophylla* (Podocarpaceae), *Ziziphus jujuba* (Rhamnaceae), *Prunus laurocerasus, Rosa sp.* (Rosaceae), *Gardenia sp.* (Rubiaceae), *Citrus sp., Citrus limon* (Rutaceae), *Formica aerate* (Hymenoptera: Formicidae) (Noyes, 2019).

Hosts: *Saissetia coffeae* (Hemiptera: Coccidae), *Aonidiella aurantia, Aonidiella citrina, Aonidiella eremocitri, Aonidiella inornate, Aonidiella orientalis, Aonidiella taxus, Aspidiotus sp., Aspidiotus camelliae, Aspidiotus cryptomeriae, Aspidiotus destructor, Aspidiotus nerii, Aspidiotus orientalis, Chrysomphalus sp., Chrysomphalus aonidum, Chrysomphalus aurantia, Chrysomphalus bifaciculatus, Chrysomphalus dictyospermi, Chrysomphalus diversicolor, Chrysomphalus ficus, Clavaspis sp., Diaspidiotus gigas, Diaspis echinocacti, Dynaspidiotus abietis, Hemiberlesia lataniae, Hemiberlesia rapax, Lindingaspis fusca, Morganella longispina, Nuculaspis abietis, Parlatoria pergandii, Pseudaonidia trilobitiformis, Pseudaulacaspis sp., Pseudaulacaspis pentagona, Quadraspidiotus gigas, Quadraspidiotus perniciosus, Temnaspidiotus destructor, Unaspis yanonensis* (Diaspididae) (Noyes, 2019).

Parasitoids: *Coccobius debachi, Marietta carnesi, Marietta exitiosa, Marietta javensis, Marietta leopardine, Physcus sp., Physcus testaceus* (Hymenoptera: Aphelinidae), *Signiphora flavopalliata, Thysanus flavopalliatus* (Signiphoridae) (Noyes, 2019).

Distribution: Afrotropical, Algeria, Argentina, Australia, Azerbaijan, Bermuda, Chile, Cyprus, Czech Republic, Czechoslovakia, Egypt, Fiji, France, Georgia, Greece, Guam, Haiti, Hawaii, Hungary, India, Indonesia, Iran, Israel, Italy, Japan, Madagascar, Mauritius, Mexico, Micronesia, Moldova, Montenegro, Morocco, Nearctic, New Caledonia, New Zealand, Oriental, Pakistan, People's Republic of China, Philippines, Russia, South Africa, Spain, Sri Lanka, Swaziland, Taiwan, Turkey, Ukraine, United States of America, USSR, Yugoslavia (pre-1991), Zambia, Zimbabwe (Noyes, 2019).

Genus *Copidosoma* (Ratzeburg, 1844)
Copidosoma agrotis (Fonscolombe, 1832)
Publications with Turkey records: Noyes (2019), Öncüer (1991), Trjapitzin and Doğanlar (1997).
Associates: Poaceae (Noyes, 2019).
Hosts: *Agrotis sp., Agrotis segetum, Euxoa obelisca, Euxoa oberthuri, Mamestra brassicae* (Lepidoptera: Noctuidae), *Pyrameis cardui, Vanessa cardui* (Nymphalidae) (Noyes, 2019).
Distribution: Andorra, Armenia, Austria, Croatia, Czech Republic, Czechoslovakia, Denmark, Europe, Finland, France, Georgia, Germany, Hungary, Italy, Kazakhstan, Mongolia, Norway, People's Republic of China, Russia, Slovakia, Spain, Sweden, Switzerland, Tselinograd Obl., Turkey, Turkmenistan, United Kingdom, USSR, Yugoslavia (Federal Republic), Yugoslavia (pre-1991) (Noyes, 2019).

Copidosoma aithyia (Walker, 1837)
Publications with Turkey records: Japoshvili et al. (2013), Noyes (2019), Zeki et al. (1999).
Associates: *Pastinaca sativa, Pimpinella anisum* (Apiaceae), *Centaurea nigra, Centaurea scabiosa, Serratula tinctoria* (Asteraceae) (Noyes, 2019).
Hosts: *Euxoa obelisca, Euxoa temera* (Lepidoptera: Noctuidae), *Agonopterix sp., Agonopterix alpigena, Agonopterix arenella, Agonopterix heracliana, Depressaria sp., Depressaria alpigena, Depressaria alpigenella, Depressaria fruticosella, Depressaria heracliana, Depressaria pastinacella* (Oecophoridae), *Mecyna lutealis* (Pyralidae) (Noyes, 2019).
Parasitoids: *Baryscapus endemus* (Hymenoptera: Eulophidae) (Noyes, 2019).
Distribution: Austria, Bulgaria, Czech Republic, Czechoslovakia, Denmark, France, Germany, Hungary, Norway, Poland, Russia, Slovakia, Spain, Sweden, Switzerland, Turkey, United Kingdom, USSR (Noyes, 2019).

Copidosoma boucheanum (Ratzeburg, 1844)
Publications with Turkey records: Japoshvili and Noyes (2005), Noyes (2019), Öncüer (1991).
Associates: *Populus sp., Populus deltoides, Populus euramericana, Populus nigra, Populus tremula* (Salicaceae) (Noyes, 2019).
Hosts: *Pemphigus bursarius* (Hemiptera: Aphididae), *Anacampsis populella, Gelechia sp., Gelechia nigra, Gelechia pinguinella, Gelechia turpella* (Lepidoptera: Gelechiidae), *Euproctis chrysorrhoea, Leucoma*

salicis, Stilpnotia salicis (Lymantriidae), *Archips oporanus, Archips podanus* (Tortricidae), *Yponomeuta evonymellus, Yponomeuta malinellus, Ypsolopha vitelli, Ypsolophus vitellus* (Yponomeutidae) (Noyes, 2019).
Distribution: Armenia, Belarus, Bulgaria, Czech Republic, Czechoslovakia, England, Europe, France, Germany, Hungary, Iran, Kazakhstan, Lithuania, Mongolia, Netherlands, Norway, People's Republic of China, Russia, Slovakia, Spain, Sweden, Switzerland, Tselinograd Obl., Turkey, USSR, Yugoslavia (Federal Republic) (Noyes, 2019).

Copidosoma breviclava (Hoffer, 1970)
Publications with Turkey records: Japoshvili and Noyes (2005), Noyes (2019).
Associates: Unknown (Noyes, 2019).
Hosts: Unknown (Noyes, 2019).
Distribution: Austria, Bulgaria, Czech Republic, Czechoslovakia, Georgia, Greece, Hungary, Italy, Macedonia, Moldova, Slovakia, Spain, Turkey, Ukraine, Yugoslavia (Federal Republic), Yugoslavia (pre-1991) (Noyes, 2019).

Copidosoma floridanum (Ashmead, 1900)
Publications with Turkey records: Japoshvili and Noyes (2005), Noyes (2019), Trjapitzin and Doğanlar (1997).
Associates: Coleoptera: Scolytidae, *Glyptapanteles pallipes, Microplitis demolitor* (Hymenoptera: Braconidae), *Apium graveolens* (Apiaceae), *Helianthus annuus* (Asteraceae), *Brassica sp., Brassica oleracea* (Brassicaceae), *Cajanus cajan, Glycine max, Medicago sativa* (Fabaceae), *Gossypium hirsutum* (Malvaceae), *Sesamum indicum* (Pedaliaceae), *Oryza sativa, Saccharum officinarum* (Poaceae), *Fagopyrum esculentum* (Polygonaceae), *Lycopersicon esculentum, Solanum tuberosum* (Solanaceae) (Noyes, 2019).
Hosts: (Hemiptera: Aphididae), *Empicoris variolosus* (Reduviidae: Lepidoptera), *Calpodes ethlius, Panoquina ethlius* (Hesperiidae), *Acharia bonaerensis, Sibine bonaerensis* (Limacodidae), *Acanthoplusia aganta, Acronicta intermedia, Agrapha agnata, Alabama argillacea, Amyna octo, Anomis sabulifera, Argyrogramma signatum, Autographa gamma, Autographa signata, Cerapteryx graminis, Chrysodeixis sp., Chrysodeixis acuta, Chrysodeixis argentifera, Chrysodeixis chalcites, Chrysodeixis eriosoma, Cosmophila sabulifera, Hadena luteago, Helicoverpa armigera, Nebrarctia obliqua, Phytometra sp., Phytometra confuse, Phytometra gamma, Plusia sp., Plusia agnata, Plusia chalcites, Plusia confuse, Plusia*

gamma, Plusia intermedia, Plusia nu, Plusia oo, Plusia signata, Polychrysia moneta, Prodenia litura, Pseudoplusia includens, Rachiplusia nu, Spodoptera litura, Thysanoplusia intermixta, Thysanoplusia orichalcea, Trichoplusia ni (Noctuidae), *Ptilodon capucina* (Notodontidae), *Cydia molesta, Grapholita molesta* (Tortricidae) (Noyes, 2019).
Distribution: Afrotropical, Antilles, Argentina, Australia, Azerbaijan, Bangladesh, Barbados, Belarus, Belgium, Belize, Bermuda, Brazil, Bulgaria, Canada, Canary Islands, Cape Verde Islands, Central America, Cuba, Czech Republic, Ecuador, Europe, France, Georgia, Germany, Greece, Guyana, Hawaii, Hungary, India, Indonesia, Iran, Italy, Ivory Coast, Jamaica, Japan, Korea, Malaysia, Mexico, Netherlands, New Zealand, Norway, People's Republic of China , Portugal, Russia, Saudi Arabia, Senegal, Serbia, Slovakia, Spain, Sri Lanka, Sweden, Thailand, Transcaucasus, Trinidad & Tobago, Turkey, United Kingdom, United States of America, Uruguay, Venezuela, Yugoslavia (Federal Republic), Yugoslavia (pre-1991) (Noyes, 2019).

Copidosoma hispanicum (Mercet, 1921)
Publications with Turkey records: Noyes (2019).
Associates: Unknown (Noyes, 2019).
Hosts: Unknown (Noyes, 2019).
Distribution: Andorra, France, Italy, Portugal, Spain, Turkiye (Noyes, 2019).

Copidosoma primulum (Mercet, 1921)
Publications with Turkey records: Japoshvili et al. (2013), Noyes (2019).
Associates: *Sorghum sp.* (Poaceae) (Noyes, 2019).
Hosts: *Chloridea obsolete, Helicoverpa armigera, Heliothis armigera, Heliothis peltigera, Heliothis zea, Raghuva albipunctella* (Lepidoptera: Noctuidae) (Noyes, 2019).
Distribution: Czechoslovakia, Hungary, Norway, People's Republic of China, Slovakia, Spain, Sudan, Turkey, Turkmenistan, Uzbekistan (Noyes, 2019).

Copidosoma truncatellum (Dalman, 1820)
Publications with Turkey records: Japoshvili et al. (2013), Japoshvili and Noyes (2005), Noyes (2019), Öncüer (1991).
Associates: *Voria ruralis* (Diptera: Tachinidae), *Microplitis brassicae* (Hymenoptera: Braconidae), *Hyposoter exiguae* (Ichneumonidae), *Ferula*

communis, Thapsia gigantica (Apiaceae), *Helianthus sp.* (Asteraceae), *Brassica sp., Brassica oleracea* (Brassicaceae), *Arachis hypogaea, Glycine max, Medicago sativa* (Fabaceae), *Gossypium sp.* (Malvaceae), *Olea europaea* (Oleaceae), *Passiflora edulis* (Passifloraceae), *Zea mays* (Poaceae), *Lycopersicon esculentum, Nicotiana glauca* (Solanaceae), *Phloeotribus scarabaeoides* (Coleoptera: Scolytidae), *Coleophora pruniella* (Lepidoptera: Coleophoridae) Zeuzera pyrina (Cossidae), Eupithecia centaureata (Geometridae), Hepialus humuli (Hepialidae), *Actinotia polyodont, Agrotis sp., Agrotis diatrapezium, Agrotis malefida, Alabama argillacea, Anomis erosa, Apamea monoglypha, Apamea sublustris, Autographa sp., Autographa brassicae, Autographa californica, Autographa chalcites, Autographa gamma, Autographa ou, Autoplusia egena, Catocala electa, Chorizagrotis adumbrate, Chrysodeixis chalcites, Chrysodeixis includens, Euchalcia modestoides, Euxoa sp., Euxoa lidia, Euxoa obelisca, Euxoa temera, Feltia malefida, Hadena luteago, Lamprotes c-aureum, Mamestra brassicae, Mocis latipes, Noctuidae unspecified sp., Phytometra gamma, Plusia sp., Plusia californica, Plusia c-aureum, Plusia egena, Plusia epigaea, Plusia festucae, Plusia gamma, Plusia modesta, Plusia moneta, Plusia ni, Plusia nu, Plusia oo, Polychrysia moneta, Prodenia sp., Prodenia ornithogalli, Pseudoplusia includens, Rachiplusia nu, Rachiplusia ou, Spodoptera sp., Spodoptera ornithogalli, Syngrapha epigaea, Trachea monoglypha, Trichoplusia ni, Xestia ditrapezium* (Noctuidae), *Notodonta ziczac* (Notodontidae), *Agonopterix heracliana, Agonopterix nervosa, Agonopterix nodiflorella, Agonopterix thapsiella, Depressaria heracliana, Depressaria nervosa* (Oecophoridae), *Parnassius apollo* (Papilionidae) (Noyes, 2019).

Distribution: Andorra, Argentina, Austria, Brazil, Canada, Cape Verde Islands, Caribbean , Chile, Colombia, Cuba, Czech Republic, Czechoslovakia, Denmark, Egypt, Europe, Finland, France, Georgia, Germany, Greece, Hawaii, Hungary, Italy, Macedonia, Mexico, Montenegro, Nearctic, Netherlands, Norway, People's Republic of China, Poland, Puerto Rico, Romania, Russia, Serbia, Slovakia, Slovenia, Spain, Sweden, Switzerland, Turkey, United Kingdom, United States of America, Uruguay, USSR, Venezuela, Yugoslavia (Federal Republic) (Noyes, 2019).

Copidosoma varicorne (Nees, 1834)

Publications with Turkey records: Noyes (2019).

Associates: *Acer tarica, Acer tatarica* (Aceraceae), *Elaeagnus sp.* (Elaeagnaceae), *Acacia ataxacantha, Ammodendron connolyi, Caragana arborescens, Dalbergia sissoo, Medicago sp., Ulex europaeus* (Fabaceae),

Malus domestica, Prunus armeniaca, Prunus domestica, Prunus persica (Rosaceae) (Noyes, 2019).
Hosts: *Eumenes coarctatus* (Hymenoptera: Vespidae), *Anarsia sp., Anarsia eleagnella, Anarsia ephippias, Anarsia lineatella, Anarsia sagmatica, Anarsia spartiella, Compsolechia anisogramma, Dichomeris eridantis, Gelechiidae unspecified sp.* (Lepidoptera: Gelechiidae), *Haritalodes derogate, Sylepta derogate* (Pyralidae), Acleris hippophaeana, Carpocapsa pomonella, Cydia funebrana, Cydia molesta, Cydia pomonella, Eucosma sp., *Grapholita funebrana, Grapholita molesta, Gypsonoma minutana, Laspeyresia funebrana, Lobesia incultana, Polychrosis incultana, Tortrix viridana* (Tortricidae), *Yponomeuta sp.* (Yponomeutidae) (Noyes, 2019).
Distribution: Afghanistan, Afrotropical, Armenia, Azerbaijan, Bulgaria, Croatia, Czech Republic, Czechoslovakia, England, Europe, France, Germany, Hungary, India, Iran, Israel, Italy, Ivory Coast, Kazakhstan, Lebanon, Moldova, Montenegro, Pakistan, People's Republic of China, Poland, Romania, Russia, Senegal, South Africa, Spain, Transcaucasus, Tselinograd Obl., Turkey, Turkmenistan, Ukraine, USSR, Yugoslavia (Federal Republic), Yugoslavia (pre-1991), Zimbabwe (Noyes, 2019).

Genus *Cowperia* (Girault, 1919)
Cowperia areolata (Walker, 1872)
Publications with Turkey records: Japoshvili and Noyes (2005), Noyes (2019), Öncüer (1991), Trjapitzin and Doğanlar (1997).
Hosts: *Cryptolaemus montrouzieri, Scymnus apetzi* (Coleoptera: Coccinellidae), *Pseudococcus sp.* (Hemiptera: Pseudococcidae) (Noyes, 2019).
Distribution: Armenia, Bulgaria, France, Georgia, Hungary, Italy, Montenegro, Spain, Tadzhikistan, Turkey, Ukraine (Noyes, 2019).

Genus *Discodes* (Foerster, 1856)
Discodes aeneus (Dalman, 1820)
Publications with Turkey records: Noyes (2019), Öncüer (1991).
Associates: *Hedera sp., Hedera helix* (Araliaceae), *Betula nana* (Betulaceae), *Calluna vulgaris* (Ericaceae), *Festuca circummediterranea* (Poaceae), *Cerasus japonica, Prunus sp., Prunus domestica, Prunus salicina* (Rosaceae) (Noyes, 2019).
Hosts: *Asterolecanium sp., Asterolecanium fimbriatum* (Hemiptera: Asterolecaniidae), *Eriopeltis sp., Eulecanium sp., Eulecanium caraganae, Eulecanium corni, Eulecanium douglasi, Eulecanium franconicum, Eulecanium persicae, Eulecanium spiraeae, Lecanium corni, Lecanium*

coryli, Lecanium prunastri, Parthenolecanium corni, Parthenolecanium persicae, Pulvinaria betulae, Saissetia sp., Saissetia coffeae, Scythia aetnensis, Sphaerolecanium prunastri (Coccidae), *Aulacaspis rosae* (Diaspididae), *Planococcus citri* (Pseudococcidae) (Noyes, 2019).
Parasitoids: *Marietta picta* (Hymenoptera: Aphelinidae) (Noyes, 2019).
Distribution: Croatia, Czech Republic, Czechoslovakia, England, Europe, France, Georgia, Germany, Hungary, Israel, Italy, Moldova, Montenegro, Netherlands, Norway, Poland, Romania, Russia, Serbia, Slovakia, Spain, Sweden, Switzerland, Turkey, Ukraine, United States of America, USSR (Noyes, 2019).

Discodes coccophagus (Ratzeburg, 1848)
Publications with Turkey records: Japoshvili and Karaca (2002), Japoshvili et al. (2002), Karaca et al. (2003), Noyes (2019), Öncüer (1991), Trjapitzin and Doğanlar (1997), Ulgenturk (2001), Ulgenturk et al. (2001).
Associates: *Armeniaca vulgaris, Persica vulgaris, Prunus sp., Prunus cerasifera, Prunus divaricate, Prunus domestica, Prunus scoparia, Prunus spinosa* (Rosaceae) (Noyes, 2019).
Hosts: *Sphaerolecanium prunastri* (Hemiptera: Coccidae) (Noyes, 2019).
Parasitoid hosts: *Cerapterocerus mirabilis* (Hymenoptera: Encyrtidae) (Noyes, 2019).
Distribution: Armenia, Azerbaijan, Belarus, Bulgaria, Caucasus, Czechoslovakia, Europe, France, Georgia, Germany, Greece, Hungary, Iran, Israel, Italy, Kirgizia, Moldova, Poland, Romania, Russia, Sweden, Transcaucasus, Turkey, Turkmenistan, Ukraine, Yugoslavia (Federal Republic), Yugoslavia (pre-1991) (Noyes, 2019).

Discodes obscuriclavus (Myartseva, 1981)
Publications with Turkey records: Japoshvili and Noyes (2005), Noyes (2019).
Associates: *Halocnemum strobilaceum, Haloxylon persicum, Kalidium capsicum, Salsola sp., Salsola richleri* (Chenopodiaceae) (Noyes, 2019).
Hosts: *Acanthopulvinaria orientalis* (Hemiptera: Coccidae) (Noyes, 2019).
Distribution: Iran, Turkey, Turkmenistan, Uzbekistan (Noyes, 2019).

Discodes trjapitzini (Herthevtzian, 1979)
Publications with Turkey records: Japoshvili and Noyes (2005), Noyes (2019).
Associates: *Ulmus sp.* (Ulmaceae) (Noyes, 2019).

Distribution: Armenia, Austria, Azerbaijan, Bulgaria, Czech Republic, Czechoslovakia, France, Georgia, Moldova, Russia, Slovakia, Turkey, Ukraine (Noyes, 2019).

Genus *Echthroplexiella* (Mercet, 1921)
Echthroplexiella aeneiventris (Mercet, 1921)
Publications with Turkey records: Japoshvili and Noyes (2005), Noyes (2019).
Hosts: *Scythia stipae* (Hemiptera: Coccidae) (Noyes, 2019).
Associates: *Rubus sp.* (Rubiaceae) (Noyes, 2019).
Distribution: Armenia, Azerbaijan, Czech Republic, Czechoslovakia, Georgia, Greece, Hungary, Iran, Mongolia, Russia, Slovakia, Spain, Turkey, Turkmenistan, Ukraine (Noyes, 2019).

Genus *Encyrtus* (Latreille, 1809)
Encyrtus aurantii (Geoffroy, 1785)
Publications with Turkey records: Japoshvili and Karaca (2002), Japoshvili and Noyes (2005), Noyes (2019), Öncüer (1974; 1991), Trjapitzin and Doğanlar (1997).
Associates: *Mangifera indica* (Anacardiaceae), *Nerium oleander, Tabernaemontana divaricate* (Apocynaceae), *Hedera sp., Schefflera sp.* (Araliaceae), *Calotropis gigantea* (Asclepiadaceae), *Lucilia sp.* (Asteraceae), *Casuarina equisetifolia* (Casuarinaceae), *Convolvulus arvensis* (Convolvulaceae), *Coccinia cordifolia* (Cucurbitaceae), *Diospyros sp.* (Ebenaceae), *Elaeagnus sp.* (Elaeagnaceae), *Croton bonplandianus, Ricinus communis* (Euphorbiaceae), *Acacia sp., Erythrina sp., Pongamia pinnata, Robinia sp., Tephrosia vogelii* (Fabaceae), *Gladiolus sp.* (Iridaceae), *Cinnamomum camphora, Laurus nobilis* (Lauraceae), *Gossypium sp.* (Malvaceae), *Azadirachta indica* (Meliaceae), *Ficus sp., Ficus carica, Morus sp., Morus alba* (Moraceae), *Psidium guajava* (Myrtaceae), *Olea europaea* (Oleaceae), *Amygdalus sp., Crataegus sp., Malus sp.* (Rosaceae), *Plectronia parviflora* (Rubiaceae), *Citrus sp., Citrus aurantifolia, Murraya paniculate* (Rutaceae), *Santalum album* (Santalaceae), *Dimocarpus longan* (Sapindaceae), *Solanum tuberosum* (Solanaceae), *Camellia sinensis* (Theaceae), *Ulmus sp.* (Ulmaceae), *Lantana camara* (Verbenaceae)(Noyes, 2019).
Hosts: Hemiptera: Aleyrodidae, *Ceroplastes actiniformis, Ceroplastes rusci, Ceroplastes sinensis, Chloropulvinaria sp., Chloropulvinaria psidii, Coccus sp. Coccus hesperidum, Coccus pseudomagnoliarum, Coccus viridis, Eucalymnatus tessellatus, Eulecanium sp., Eulecanium corni, Eulecanium coryli, Eulecanium tiliae, Lecanium sp., Lecanium corni,*

Lecanium fletcheri, Lecanium hesperidum, Lecanium nigrum, Parasaissetia nigra, Pulvinaria sp., Pulvinaria camelicola, Saissetia sp., Saissetia coffeae, Saissetia hemisphaerica, Saissetia nigra, Saissetia oleae, Saissetia privigna, Sphaerolecanium prunastri (Coccidae), *Diaspis boisduvalii* (Diaspididae), *Eriococcus buxi* (Eriococcidae), *Planococcus sp., Planococcus citri, Pseudococcidae indet. sp., Pseudococcus adonidum, Pseudococcus longispinus, Pseudococcus njalensis, Rastrococcus sp.* (Pseudococcidae) (Noyes, 2019).

Distribution: Afrotropical, Algeria, Argentina, Armenia, Australia, Austria, Azerbaijan, Bangladesh, Belgium, Bermuda, Brazil, Cameroon, Canada, Cape Verde Islands, Caucasus, Chagos Archipelago, Colombia, Costa Rica, Croatia, Cyprus, Czech Republic, Dominica, Ecuador, England, Finland, France, French Polynesia, Gabon, Georgia, Germany, Ghana, Greece, Grenada, Hawaii, India, Indonesia, Iran, Israel, Italy, Jamaica, Japan, Lebanon, Madagascar, Malawi, Malaysia, Mauritius, Mexico, Montenegro, Montserrat, Morocco, Mozambique, Nearctic, Netherlands, New Caledonia, New Zealand, Pakistan, Panama, People's Republic of China, Peru, Philippines, Poland, Puerto Rico, Russia, Senegal, Serbia, Seychelles, Singapore, South Africa, Spain, Sri Lanka, St Christopher (Kitts) & Nevis, St Vincent & Grenadines, Sweden, Tadzhikistan, Thailand,, Transcaucasus, Trinidad & Tobago, Turkey, Turkmenistan, Uganda, Ukraine, United States of America, USSR, Uzbekistan, Vanuatu, Venezuela, Vietnam, Yugoslavia (pre-1991), Zimbabwe (Noyes, 2019).

Encyrtus infelix (Embleton, 1902)

Publications with Turkey records: Noyes (2019), Öncüer (1991).

Associates: *Aphelandra squarrosa* (Acanthaceae), *Nerium oleander* (Apocynaceae), *Wedelia sp.* (Asteraceae), *Cycas sp.* (Cycadaceae), *Ceratiola ericoides* (Empetraceae), *Miconia calvescens* (Melastomataceae), *Ficus carica* (Moraceae), *Coffea sp.* (Rubiaceae), *Citrus reticulata* (Rutaceae), *Ceroplastes sp.* (Hemiptera: Coccidae) (Noyes, 2019).

Hosts: *Ceroplastes destructor, Ceroplastes rusci, Coccus sp., Coccus hesperidum, Gascardia madagascariensis, Lecanium hemisphaericum, Protopulvinaria pyriformis, Pulvinaria innumerabilis, Pulvinaria urbicola, Saissetia sp., Saissetia coffeae, Saissetia hemisphaerica, Saissetia nigra, Saissetia oleae* (Hemiptera: Coccidae) (Noyes, 2019).

Distribution: Afrotropical, Algeria, Australia, Austria, Bermuda, Brazil, Canary Islands, Cape Verde Islands, Colombia, Costa Rica, Cuba, Dominica, Dominican Republic, Ecuador, Egypt, Fiji, French Polynesia,

Germany, Hawaii, India, Israel, Jamaica, Kenya, Madagascar, Madeira, Malaysia, Mexico, Morocco, Nearctic, Netherlands, New Caledonia, New Zealand, Peru, Philippines, Puerto Rico, Russia, Seychelles, Sri Lanka, St Vincent & Grenadines, Switzerland, Trinidad & Tobago, Turkey, United Kingdom, United States of America, Venezuela, Vietnam, Western Samoa, Zimbabwe (Noyes, 2019).

Encyrtus infidus (Rossi, 1790)

Publications with Turkey records: Japoshvili and Noyes (2005), Noyes (2019), Öncüer (1991), Trjapitzin and Doğanlar (1997), Ulgenturk and Toros (2001).

Associates: *Pistacia vera* (Anacardiaceae), *Caragana arborescens* (Fabaceae), *Quercus ilex* (Fagaceae), *Ficus carica* (Moraceae), *Pinus eldarica* (Pinaceae), *Amygdalus sp., Prunus sp.* (Rosaceae), *Tilia platyphyllos* (Tiliaceae), *Ulmus sp.* (Ulmaceae) (Noyes, 2019).

Hosts: *Eulecanium sp., Eulecanium caraganae, Eulecanium ciliatum, Eulecanium coryli, Eulecanium douglasi, Eulecanium franconicum, Eulecanium kunoense, Eulecanium rugulosum, Eulecanium tiliae, Eulecanium transcaucasicum, Lecanium caryae, Lecanium coryli, Lecanium kunoensis, Lecanium zebrinum, Sphaerolecanium prunastri* (Hemiptera: Coccidae), *Kermes sp., Kermes vermilio, Kermococcus sp.* (Kermesidae)(Noyes, 2019).

Distribution: Armenia, Azerbaijan, Canada, Croatia, Czech Republic, England, Europe, France, Georgia, Germany, Greece, Iran, Italy, Japan, Kazakhstan, Kirgizia, Korea, Moldova, Mongolia, Morocco, Nearctic, Netherlands, Norway, Palearctic, People's Republic of China, Poland, Russia, Slovakia, Spain, Sweden, Transcaucasus, Tselinograd Obl., Turkey, Turkmenistan, United States of America, USSR, Uzbekistan (Noyes, 2019).

Genus *Eugahania* (Mercet, 1926)

Eugahania fumipennis (Ratzeburg, 1852)

Publications with Turkey records: Japoshvili and Noyes (2005), Noyes (2019), Öncüer (1991), Trjapitzin and Doğanlar (1997).

Hosts: *Macropsis vicina* (Hemiptera: Cicadellidae) (Noyes, 2019).

Distribution: Armenia, Austria, Bosnia Hercegovina, Croatia, Czechoslovakia, Europe, Finland, Germany, Hungary, Italy, Japan, Kazakhstan, Kirgizia, Macedonia, Moldova, Mongolia, Norway, People's Republic of China, Romania, Russia, Slovakia, Spain, Sweden, Tadzhikistan, Tselinograd Obl., Turkey, Turkmenistan, Ukraine, USSR, Yugoslavia (pre-1991) (Noyes, 2019).

Genus *Ginsiana* (Erdos and Novicky, 1955)

Ginsiana carpetana (Mercet, 1921)

Publications with Turkey records: Japoshvili and Noyes (2005), Noyes (2019), Trjapitzin and Doğanlar (1997).

Associates: Unknown (Noyes, 2019).

Hosts: Unknown (Noyes, 2019).

Distribution: Armenia, Azerbaijan, Bulgaria, Czech Republic, Czechoslovakia, Denmark, England, Finland, France, Georgia, Hungary, Israel, Lithuania, Mongolia, Netherlands, Norway, Poland, Russia, Spain, Sweden, Tadzhikistan, Transcaucasus, Turkey, Turkmenistan, Ukraine (Noyes, 2019).

Genus *Habrolepis* (Foerster, 1856)

Habrolepis dalmanni (Westwood, 1837)

Publications with Turkey records: Japoshvili and Noyes (2005), Noyes (2019).

Associates: *Pistacia vera* (Anacardiaceae), *Quercu coccifera*, *Quercus sp., Quercus pedunculata, Quercus robur, Quercus sessilis* (Fagaceae), *Fraxinus sp.* (Oleaceae), *Prunus dulcis* (Rosaceae), *Coffea sp.* (Rubiaceae) (Noyes, 2019).

Hosts: *Asterodiaspis sp., Asterodiaspis quercicola, Asterodiaspis variolosa, Asterolecanium sp., Asterolecanium ilicicola, Asterolecanium minus, Asterolecanium quercicola, Asterolecanium variolosum* (Hemiptera: Asterolecaniidae), *Didesmococcus sp.* (Coccidae), *Lepidosaphes ulmi, Melanaspis inopinata, Targionia vitis* (Diaspididae), *Pseudococcus sp.* (Pseudococcidae), *Leucoptera sp.* (Lepidoptera: Lyonetiidae) (Noyes, 2019).

Distribution: Afrotropical, Algeria, Armenia, Australia, Austria, Azerbaijan, Bulgaria, Chile, Croatia, Czech Republic, Czechoslovakia, Denmark, England, Europe, France, Georgia, Germany, Hungary, Iran, Israel, Italy, Moldova, Nearctic, Netherlands, New Zealand, North Africa, Poland, Romania, Russia, Slovakia, South Africa, Spain, Sweden, Turkey, Uganda, Ukraine, United States of America, USSR, Yugoslavia (pre-1991) (Noyes, 2019).

Habrolepis italicus (Delucchi, 1965)

Publications with Turkey records: Japoshvili and Noyes (2005), Noyes (2019).

Hosts: *Asterodiaspis sp., Asterodiaspis ilicicola, Asterodiaspis variolosa, Asterolecanium sp.* (Hemiptera: Asterolecaniidae) (Noyes, 2019).

Distribution: Algeria, Bulgaria, Italy, Montenegro, North Africa, Norway, Turkey, Ukraine, Yugoslavia (pre-1991) (Noyes, 2019).

Genus *Homalotylus* (Mayr, 1876)
Homalotylus flaminius (Dalman, 1820)
Publications with Turkey records: Japoshvili and Noyes (2005), Noyes (2019), Öncüer (1991).
Associates: *Parlatoria pergandii, Quadraspidiotus gigas, Unaspis yanonensis* (Hemiptera: Diaspididae), *Drosicha stebbengii, Icerya Pilosa, Icerya purchase* (Margarodidae), *Nipaecoccus sp., Nipaecoccus viridis, Peliococcus kimmericus, Phenacoccus manihoti* (Pseudococcidae), *Lactuca serriola* (Asteraceae), *Euonymus japonica* (Celastraceae), *Chenopodium sp.* (Chenopodiaceae), *Manihot esculenta* (Euphorbiaceae), *Acacia erioloba* (Fabaceae), *Saccharum officinarum* (Poaceae), *Citrus aurantium* (Rutaceae), *Salvadora persica* (Salvadoraceae), *Solanum sp., Solanum gilo* (Solanaceae), *Tamarix canariensis* (Tamaricaceae), *Peganum harmala* (Zygophyllaceae) (Noyes, 2019).
Hosts: *Galeruca calmariensis* (Coleoptera: Chrysomelidae), *Adalia bipunctata, Adalia deficiens, Adonia variegate, Anatis ocellata, Brumoides suturalis, Brumus suturalis, Cheilomenes sp., Cheilomenes lunata, Cheilomenes propinqua, Chilocorus sp., Chilocorus bipustulatus, Chilocorus kuwanae, Chilocorus renipustulatus, Coccinella algerica, Coccinella arcuate, Coccinella bruckii, Coccinella kingi, Coccinella repanda, Coccinella septempunctata, Coleomegilla maculate, Cycloneda limbifer, Cycloneda sanguinea, Epilachna chrysomelina, Eriopis connexa, Exochomus sp., Exochomus concavus, Exochomus flavipes, Exochomus flaviventris, Exochomus nigromaculatus, Exochomus nigromarginatus, Exochomus quadripustulatus, Harmonia conglobate, Harmonia quatuordecimpunctata, Henosepilachna bifasciata, Hippodamia convergens, Hippodamia tibialis, Hippodamia tredecimpunctata, Hyperaspis campestris, Hyperaspis festiva, Hyperaspis guttulate, Hyperaspis senegalensis, Lioadala flavomaculata, Menochilus sexmaculatus, Neomysia oblongoguttata, Nephus bipunctatus, Orcus chalibaeus, Orcus chalybaeus, Orcus janthinus, Orcus laferti, Orcus nummularis, Pharoscymnus numidicus, Pharoscymnus ovoideus, Platynaspis sp., Rodolia sp., Rodolia cardinalis, Rodolia fumida, Rodolia guerini, Rodolia guerni, Scymnus sp., Scymnus interruptus, Scymnus nigrinus, Scymnus ornatus, Scymnus soudanensis, Scymnus subvillosus, Scymnus suturalis, Scymnus syriacus, Thea vigintiduopunctata, Verania frenata* (Coccinellidae), *Eulecanium corni, Saissetia oleae* (Hemiptera: Coccidae), *Nipaecoccus viridis, Planococcus citri, Pseudococcus citri* (Pseudococcidae) (Noyes, 2019).
Parasitoids: *Cheiloneurus cyanonotus, Prochiloneurus aegyptiacus* (Hymenoptera: Encyrtidae), *Lygocerus sp.* (Megaspilidae), *Chartocerus sp.* (Signiphoridae) (Noyes, 2019).

Distribution: Afghanistan, Afrotropical, Argentina, Australia, Austria, Bahamas, Barbados, Brazil, Bulgaria, Cameroon, Colombia, Congo, Croatia, Cyprus, Czech Republic, Democratic Republic of Congo, Egypt, England, Europe, France, Gambia, Georgia, Germany, Ghana, Greece, Guinea-Bissau, Guyana, Hungary, India, Indonesia, Iran, Israel, Italy, Japan, Jordan, Kazakhstan, Korea, Madagascar, Mali, Mexico, Moldova, Mongolia, Montenegro, Morocco, Namibia, Nearctic, Nigeria, North Africa, Norway, Palearctic, People's Republic of China, Peru, Puerto Rico, Romania, Russia, Saudi Arabia, Senegal, Serbia, Slovakia, South Africa, Spain, Sweden, Tadzhikistan, Togo, Tselinograd Obl., Tunisia, Turkey, Turkmenistan, United States of America, Uruguay, USSR, Uzbekistan, Venezuela, Vietnam, West Africa (Noyes, 2019).

Homalotylus nigricornis (Mercet, 1921)
Publications with Turkey records: Noyes (2019), Öncüer (1991).
Associates: *Aphis gossypii* (Hemiptera: Aphididae), *Artemisia sp.* (Asteraceae), *Malus communis* (Rosaceae).
Hosts: *Scymnus sp., Scymnus subvillosus, Scymnus suturalis* (Coleoptera: Coccinellidae) (Noyes, 2019).
Distribution: Armenia, Azerbaijan, Bulgaria, Canary Islands, Czechoslovakia, Hungary, Iran, Italy, Montenegro, Spain, Tadzhikistan, Turkey, Turkmenistan, USSR, Uzbekistan, Yugoslavia (pre-1991) (Noyes, 2019).

Homalotylus quaylei (Timberlake, 1919)
Publications with Turkey records: Japoshvili and Noyes (2005), Noyes (2019), Öncüer (1991).
Associates: *Ferrisia virgata, Nipaecoccus viridis, Phenacoccus manihoti, Planococcus citri* (Hemiptera: Pseudococcidae), *Phoenix dactylifera* (Arecaceae), *Juniperus sp.* (Cupressaceae), *Manihot esculenta* (Euphorbiaceae), *Malus communis* (Rosaceae), *Citrus sp.* (Rutaceae) (Noyes, 2019).
Hosts: Coccinellidae, *Hyperaspis marmottani, Hyperaspis polita, Nephus sp., Nephus bipunctatus, Nephus includens, Nephus phenacoccophagus, Nephus quattuormaculatus, Nephus regularis, Nephus vetustus, Pharoscymnus setulosus, Pharoscymnus varius, Scymnus sp., Scymnus fenestratus, Scymnus includens, Scymnus quadrimaculatus, Scymnus subvillosus, Scymnus suturalis, Sidis sp.* (Coleoptera: Coccinellidae), *Eulecanium persicae* (Hemiptera: Coccidae), *Nipaecoccus viridis, Phenacoccus herreni, Planococcus citri, Planococcus ficus, Planococcus vovae* (Pseudococcidae) (Noyes, 2019).

Parasitoids: *Prochiloneurus aegyptiacus* (Hymenoptera: Encyrtidae) (Noyes, 2019).
Distribution: Armenia, Azerbaijan, Brazil, Bulgaria, Canary Islands, Egypt, Gabon, Georgia, India, Iran, Israel, Italy, Jordan, Mauritania, Spain, Turkey, Turkmenistan, USSR, Uzbekistan, Venezuela (Noyes, 2019).

Genus *Hoplopsis* (De Stefani, 1889)
Hoplopsis minuta (Fabricius, 1793)
Publications with Turkey records: Japoshvili and Noyes (2005), Noyes (2019), Trjapitzin and Doğanlar (1997).
Associates: *Agropyron sp.* (Poaceae) (Noyes, 2019).
Hosts: *Eulecanium corni, Lecanopsis festucae* (Hemiptera: Coccidae) (Noyes, 2019).
Distribution: Armenia, Azerbaijan, Bulgaria, Croatia, Czechoslovakia, Europe, France, Hungary, Iran, Italy, Kazakhstan, Moldova, Montenegro, Netherlands, Romania, Russia, Slovakia, Spain, Sweden, Transcaucasus, Tselinograd Obl., Turkey, Turkmenistan, Ukraine, Yugoslavia (pre-1991) (Noyes, 2019).

Genus *Isodromus* (Howard, 1887)
Isodromus flaviscutum (Hoffer and Trjapitzin, 1978)
Publications with Turkey records: Japoshvili and Noyes (2005), Noyes (2019), Trjapitzin and Doğanlar (1997).
Hosts: *Chrysopa phyllochroma* (Neuroptera: Chrysopidae) (Noyes, 2019).
Distribution: Armenia, Azerbaijan, Bulgaria, Canary Islands, Croatia, Cyprus, Czechoslovakia, France, Georgia, Greece, Hungary, Kazakhstan, Moldova, Russia, Spain, Transcaucasus, Tselinograd Obl., Turkey, Turkmenistan, Ukraine, Uzbekistan, Yugoslavia (pre-1991) (Noyes, 2019).

Isodromus niger (Ashmead, 1900)
Publications with Turkey records: Japoshvili and Noyes (2005), Noyes (2019), Öncüer (1991).
Hosts: *Chilocorus similis* (Coleoptera: Coccinellidae), *Porthetria dispar* (Lepidoptera: Lymantriidae), *Chrysopa sp., Chrysopa californica, Chrysopa carnea, Chrysopa majuscule, Chrysopa oculate, Chrysopa septempunctata* (Neuroptera: Chrysopidae), *Sympherobius angustus* (Sympherobiidae) (Noyes, 2019).
Distribution: Georgia, Hungary, Japan, Kazakhstan, Moldova, Nearctic, People's Republic of China, Romania, Tadzhikistan, Taiwan, Tselinograd Obl., Turkey, Ukraine, United States of America, Uzbekistan (Noyes, 2019).

Genus *Lamennaisia* (Girault, 1922)
Lamennaisia ambigua (Nees, 1834)
Publications with Turkey records: Japoshvili and Noyes (2005), Noyes (2019), Trjapitzin and Doğanlar (1997).
Associates: *Olea europaea* (Oleaceae), *Pinus sp.* (Pinaceae), *Prunus sp.* (Rosaceae).
Hosts: *Bruchus brachialis* (Coleoptera: Bruchidae), *Melanophthalma sp.* (Lathrididae), *Icerya purchase* (Hemiptera: Margarodidae) (Noyes, 2019).
Distribution: Afghanistan, Armenia, Austria, Azerbaijan, Bulgaria, Canary Islands, Croatia, Czech Republic, Czechoslovakia, Denmark, England, Europe, France, Georgia, Germany, Hungary, India, Indonesia, Israel, Italy, Kazakhstan, Lithuania, Mexico, Moldova, Mongolia, Montenegro, Nearctic, Netherlands, New Zealand, People's Republic of China, Romania, Russia, Saudi Arabia, Spain, Sweden, Tadzhikistan, Thailand, Transcaucasus, Tselinograd Obl., Turkey, Turkmenistan, Ukraine, United States of America, USSR, Yugoslavia (pre-1991) (Noyes, 2019).

Genus *Mahencyrtus* (Masi, 1917)
Mahencyrtus comara (Walker, 1837)
Publications with Turkey records: Japoshvili (2012), Japoshvili and Celik (2010), Noyes (2019).
Associates: *Agropyron repens, Agrostis vulgaris, Festuca ovina, Holcus mollis, Phragmites australis* (Poaceae), *Rosa canina* (Rosaceae) (Noyes, 2019).
Hosts: *Pulvinaria vitis* (Hemiptera: Coccidae), *Antonina phragmitis, Balanococcus singularis, Chaetococcus phragmitis, Heterococcus nudus, Pseudococcus copiosus, Trionymus copiosus, Trionymus dactylis, Trionymus perrisii* (Pseudococcidae) (Noyes, 2019).
Distribution: Armenia, Canary Islands, Croatia, Czech Republic, Czechoslovakia, Denmark, England, Europe, Finland, Georgia, Germany, Hungary, Ireland, Italy, Kazakhstan, Lithuania, Madeira, Moldova, Mongolia, Netherlands, Norway, People's Republic of China, Poland, Romania, Russia, Slovakia, Spain, Sweden, Transcaucasus, Tselinograd Obl., Turkey, Turkmenistan, Ukraine, Yugoslavia (pre1991) (Noyes, 2019).

Genus *Mayrencyrtus* (Hincks, 1944)
Mayrencyrtus imandes (Walker, 1837)
Publications with Turkey records: Japoshvili (2012), Noyes (2019).
Associates: Unknown (Noyes, 2019).
Hosts: Unknown (Noyes, 2019).

Distribution: Austria, Czech Republic, Czechoslovakia, Denmark, England, Europe, Germany, Hungary, Moldova, Norway, Romania, Sweden, Turkey, Ukraine (Noyes, 2019).

Genus *Mayridia* (Mercet, 1921)
Mayridia formosula (Mercet, 1921)
Publications with Turkey records: Japoshvili and Noyes (2005), Noyes (2019).
Associates: *Agropyron sp., Lygeum sp., Triticum sp.* (Poaceae) (Noyes, 2019).
Hosts: *Trionymus aberrans* (Hemiptera: Pseudococcidae) (Noyes, 2019).
Distribution: Armenia, Azerbaijan, Bulgaria, Croatia, Czech Republic, Czechoslovakia, Europe, Greece, Hungary, Italy, Kazakhstan, Moldova, Mongolia, Portugal, Romania, Russia, Slovakia, Spain, Tselinograd Obl., Turkey, Turkmenistan, Ukraine, Uzbekistan, Yugoslavia (pre-1991) (Noyes, 2019).

Mayridia merceti (Trjapitzin, 1969)
Publications with Turkey records: Japoshvili and Noyes (2005), Noyes (2019), Trjapitzin and Doğanlar (1997).
Associates: *Phragmites australis, Triticum sp.* (Poaceae) (Noyes, 2019).
Hosts: *Neotrionymus monstatus, Trionymus aberrans* (Hemiptera: Pseudococcidae) (Noyes, 2019).
Distribution: Armenia, Germany, Hungary, Mongolia, Romania, Spain, Turkey, Turkmenistan (Noyes, 2019).

Mayridia procera (Mercet, 1921)
Publications with Turkey records: Japoshvili (2012), Noyes (2019).
Associates: *Quercus sp.* (Fagaceae) (Noyes, 2019).
Distribution: Bulgaria, Croatia, England, Europe, Georgia, Germany, Greece, Hungary, India, Israel, Kazakhstan, Moldova, Romania, Russia, Slovakia, Spain, Tselinograd Obl., Turkey, Ukraine, United States of America, Yugoslavia (Federal Republic), Yugoslavia (pre-1991) (Noyes, 2019).

Mayridia pulchra (Mercet, 1921)
Publications with Turkey records: Japoshvili and Noyes (2005), Noyes (2019).
Associates: *Artemisia herba-alba* (Asteraceae), *Stachys sp.* (Lamiaceae), *Phragmites sp., Phragmites communis* (Poaceae) (Noyes, 2019).
Hosts: *Trionymus multivorus* (Hemiptera: Pseudococcidae) (Noyes, 2019).

Distribution: Armenia, Azerbaijan, Bulgaria, Croatia, Europe, Greece, Hungary, India, Iran, Kazakhstan, Moldova, Portugal, Romania, Russia, Saudi Arabia, Spain, Sweden, Tadzhikistan, Tselinograd Obl., Turkey, Turkmenistan, Ukraine (Noyes, 2019).

Genus *Metanotalia* (Mercet, 1921)
Metanotalia maderensis (Walker, 1872)
Publications with Turkey records: Japoshvili and Noyes (2005), Noyes (2019).
Associates: *Sideritis sp.* (Lamiaceae), *Avena sp., Brachypodium sp., Brachypodium ramosum* (Poaceae) (Noyes, 2019).
Hosts: *Antonina sp., Antonina purpurea, Phenacoccus madeirensis, Trionymus multivorus* (Hemiptera: Pseudococcidae) (Noyes, 2019).
Distribution: Algeria, Bulgaria, Canary Islands, France, Israel, Italy, Madeira, New Zealand, Russia, Spain, Turkey, Ukraine, United States of America, Yugoslavia (pre-1991) (Noyes, 2019).

Genus *Metaphycus* (Mercet, 1917)
Metaphycus asterolecanii (Mercet, 1923)
Publications with Turkey records: Japoshvili and Noyes (2005), Noyes (2019).
Associates: *Quercus sp., Quercus ilex, Quercus petraea, Quercus robur, Quercus sessilis* (Fagaceae) (Noyes, 2019).
Hosts: *Asterodiaspis sp., Asterodiaspis ilicicola, Asterodiaspis quercicola, Asterodiaspis variolosa, Asterolecanium sp., Asterolecanium ilicicola, Asterolecanium minus, Asterolecanium quercicola, Asterolecanium variolosum* (Hemiptera: Asterolecaniidae), *Saissetia oleae* (Coccidae) (Noyes, 2019).
Distribution: Algeria, Andorra, Armenia, Bulgaria, Czech Republic, Czechoslovakia, Denmark, England, Europe, Finland, France, Georgia, Germany, Greece, Hungary, Italy, Netherlands, North Africa, Poland, Russia, Saudi Arabia, Spain, Turkey, USSR (Noyes, 2019).

Metaphycus delos (Guerrieri and Noyes, 2000)
Publications with Turkey records: Noyes (2019).
Associates: Unknown (Noyes, 2019).
Hosts: Unknown (Noyes, 2019).
Distribution: France, Germany, Greece, Spain, Turkey (Noyes, 2019).

Metaphycus discolor (De Santis, 1970)
Publications with Turkey records: Japoshvili (2012), Noyes (2019).

Associates: *Casearia silvestris* (Flacourtiaceae) (Noyes, 2019).
Hosts: *Ceroplastes sp.* (Hemiptera: Coccidae) (Noyes, 2019).
Distribution: Brazil, Turkey (Noyes, 2019).

Metaphycus dispar (Mercet, 1925)
Publications with Turkey records: Japoshvili and Karaca (2002), Japoshvili and Noyes (2005), Noyes (2019), Öncüer (1991), Ozsemerci and Aksit (2003), Ulgenturk (2001), Ulgenturk and Toros (2001).
Associates: *Corylus avellana* (Betulaceae), *Cercis siliquastrum* (Fabaceae), *Ficus carica, Morus sp.* (Moraceae), *Crataegus oxyacantha, Prunus domestica* (Rosaceae) (Noyes, 2019).
Hosts: *Ceroplastes rusci, Didesmococcus koreanus, Ericerus pela, Eulecanium sp., Eulecanium ciliatum, Eulecanium douglasi, Eulecanium kunoense, Eulecanium rugulosum, Eulecanium secretum, Eulecanium tiliae, Lecanium kunoensis, Parthenolecanium corni, Parthenolecanium persicae, Pulvinaria sp., Pulvinaria vitis, Rhodococcus turanicus, Sphaerolecanium prunastri* (Hemiptera: Coccidae) (Noyes, 2019).
Distribution: Armenia, Azerbaijan, Canary Islands, Czech Republic, England, Finland, France, Georgia, Greece, Hungary, Italy, Japan, Kazakhstan, Kirgizia, Madeira, Montenegro, People's Republic of China, Russia, Serbia, Slovakia, Spain, Tadzhikistan, Transcaucasus, Tselinograd Obl., Turkey, Turkmenistan, United States of America, USSR, Uzbekistan, Yugoslavia (pre-1991) (Noyes, 2019).

Metaphycus flavus (Howard, 1881)
Publications with Turkey records: Japoshvili and Karaca (2002), Japoshvili and Noyes (2005), Noyes (2019), Öncüer (1974; 1991), Sengonca et al. (1998), Trjapitzin and Doğanlar (1997).
Associates: *Inglisia sp.* (Hemiptera: Coccidae), *Mangifera indica, Pistacia palaestina, Rhus coriaria* (Anacardiaceae), *Nerium oleander* (Apocynaceae), *Hedera sp.* (Araliaceae), *Bixa Orellana* (Bixaceae), *Carica papaya* (Caricaceae), *Codiaeum variegatum* (Euphorbiaceae), *Hibiscus sp.* (Malvaceae), *Ficus carica* (Moraceae), *Psidium guajava* (Myrtaceae), *Olea europaea* (Oleaceae), *Saccharum officinarum* (Poaceae), *Prunus spinosa* (Rosaceae), *Coffea arabica* (Rubiaceae), *Citrus sp*. (Rutaceae) (Noyes, 2019).
Hosts: *Cerococcus parahybensis* (Hemiptera: Cerococcidae), *Ceroplastes sp., Ceroplastes cirripediformis, Ceroplastes floridensis, Chloropulvinaria psidii, Coccus sp., Coccus capparidis, Coccus deltae, Coccus hesperidum, Coccus mangiferae, Coccus perlatus, Coccus pseudomagnoliarum, Coccus viridis, Eulecanium corni, Eulecanium kunoense, Eulecanium*

perinflatum, Eulecanium persicae, Filippia folicularis, Filippia oleae, Lecanium sp., Lecanium deltae, Lecanium perinflatum, Lecanium persicae, Mesolecanium deltae, Milviscutulus mangiferae, Parthenolecanium corni, Parthenolecanium persicae, Philephedra lutea, Philephedra tuberculosa, Protopulvinaria pyriformis, Pulvinaria sp., Pulvinaria acericola, Pulvinaria convexa, Pulvinaria elongate, Pulvinaria flavescens, Pulvinaria floccifera, Pulvinaria iceryi, Pulvinaria maxima, Pulvinaria mesembryanthemi, Pulvinaria minuta, Pulvinaria platensis, Pulvinaria psidii, Pulvinaria pyriformis, Pulvinariella mesembryanthemi, Saissetia coffeae, Saissetia nigra, Saissetia oleae, Stictolecanium sp.,Toumeyella liriodendri (Coccidae), *Chrysomphalus aonidum, Chrysomphalus dictyospermi, Chrysomphalus ficus, Chrysomphalus pinnulifer, Lepidosaphes beckii, Lepidosaphes gloverii, Mytilococcus beckii (Diaspididae), Eriococcus joergenseni* (Eriococcidae), *Kerria sp., Tachardia sp.* (Kerriidae) (Noyes, 2019).
Parasitoid hosts: *Cheiloneurus pulvinariae* (Hymenoptera: Encyrtidae) (Noyes, 2019).
Distribution: Albania, Antilles, Argentina, Australia, Azerbaijan, Barbados, Bermuda, Brazil, Canary Islands, Chile, Costa Rica, Cuba, Cyprus, Czech Republic, Egypt, France, Georgia, Greece, Grenada, Hawaii, Honduras, India, Iran, Israel, Italy, Jamaica, Lebanon, Madeira, Mexico, Montenegro, Morocco, Nearctic, Paraguay, Peru, Portugal, Puerto Rico, Russia, Spain, Trinidad & Tobago, Tunisia, Turkey, United States of America, Uruguay (Noyes, 2019).

Metaphycus garmon (Guerrieri and Noyes, 2000)
Publications with Turkey records: Noyes (2019).
Associates: Unknown (Noyes, 2019).
Hosts: Unknown (Noyes, 2019).
Distribution: France, Greece, Italy, People's Republic of China, Spain, Turkey (Noyes, 2019).

Metaphycus gennaroi (Guerrieri and Noyes, 2000)
Publications with Turkey records: Noyes (2019).
Associates: *Quercus ilex* (Fagaceae) (Noyes, 2019).
Hosts: *Kermes sp., Kermes vermilio* (Hemiptera: Kermesidae) (Noyes, 2019).
Distribution: Algeria, Croatia, Greece, Italy, Montenegro, Turkey (Noyes, 2019).

Metaphycus helvolus (Compere, 1926)
Publications with Turkey records: Japoshvili and Noyes (2005), Noyes (2019), Öncüer (1991).
Associates: *Metaphycus sp., Metaphycus lounsburyi* (Hymenoptera: Encyrtidae), *Aphelandra squarrosa* (Acanthaceae), *Acer sp.* (Aceraceae), *Mangifera indica* (Anacardiaceae), *Nerium oleander* (Apocynaceae), *Baccharis pilularis* (Asteraceae), *Persea americana* (Lauraceae), *Myoporum sp.* (Myoporaceae), *Melaleuca hypericifolia, Psidium guajava* (Myrtaceae), *Olea sp., Olea europaea* (Oleaceae), *Citrus sp., Citrus sinensis* (Rutaceae), *Solanum tuberosum* (Solanaceae)(Noyes, 2019).
Hosts: *Ceroplastes sp., Ceroplastes destructor, Ceroplastes helichrysi, Ceroplastes pseudoceriferus, Ceroplastes sinensis, Chloropulvinaria polygonate, Chloropulvinaria psidii, Chloropulvinaria urbicola, Coccus aethiopicus, Coccus hesperidum, Coccus proteae, Coccus pseudomagnoliarum, Coccus viridis, Eucalymnatus tessellatus, Eulecanium corni, Eulecanium persicae, Lecanium corni, Lecanium persicae, Marsipococcus proteae, Parasaissetia sp., Parasaissetia litorea, Parasaissetia nigra, Parthenolecanium corni, Parthenolecanium persicae, Protopulvinaria pyriformis, Pulvinaria sp., Pulvinaria aethiopica, Pulvinaria mesembryanthemi, Pulvinaria psidii, Pulvinaria urbicola, Pulvinariella mesembryanthemi, Saissetia sp., Saissetia coffeae, Saissetia hemisphaerica, Saissetia nigra, Saissetia nigrella, Saissetia oleae, Saissetia somereni* (Hemiptera: Coccidae), *Aonidiella aurantia, Aspidiotus sp., Parlatoria oleae* (Diaspididae), *Rastrococcus mangiferae* (Pseudococcidae)(Noyes, 2019).
Distribution: Afrotropical, Argentina, Australia, Bangladesh, Bermuda, Channel Islands, Chile, Cuba, Cyprus, Ecuador, Egypt, England, Eritrea, France, Greece, Guam, Hawaii, Honduras, India, Iran, Israel, Italy, Mexico, Morocco, Namibia, Netherlands, Pakistan, Peru, South Africa, Spain, Sweden, Switzerland, Tunisia, Turkey, United States of America, USSR (Noyes, 2019).

Metaphycus ibericus (Mercet, 1921)
Publications with Turkey records: Noyes (2019).
Associates: *Ulex sp.* (Fabaceae), *Quercus rubra* (Fagaceae) (Noyes, 2019).
Hosts: *Aspidiotus trabuti, Chrysomphalus dictyospermi, Diaspidiotus ceconii, Hemiberlesia trabuti* (Hemiptera: Diaspididae) (Noyes, 2019).
Distribution: Hungary, Montenegro, Morocco, Spain, Turkey (Noyes, 2019).

Metaphycus insidiosus (Mercet, 1921)
Publications with Turkey records: Japoshvili and Karaca (2002), Japoshvili and Noyes (2005), Noyes (2019), Öncüer (1991).
Associates: *Corylus avellana* (Betulaceae), *Cornus mas, Cornus sanguinea (Cornaceae), Cupressus sp., Juniperus sp., Thuja sp., Thuja occidentalis* (Cupressaceae), *Acacia sp., Robinia pseudoacacia, Sarothamnus sp.* (Fabaceae), *Quercus borealis, Quercus pubescens, Quercus robur* (Fagaceae), *Ribes sp.* (Grossulariaceae), *Fraxinus sp., Fraxinus excelsior* (Oleaceae), *Platanus occidentalis* (Platanaceae), *Crataegus sp., Crataegus monogyna, Prunus sp., Prunus cerasifera, Prunus domestica, Prunus padus* (Rosaceae), *Rubus fruticosus* (Rubiaceae), *Populus gracilis, Populus hybrida, Populus transcaucasica, Salix sp.* (Salicaceae), *Taxus sp., Taxus baccata* (Taxaceae), *Vitis vinifera* (Vitaceae), *Pulvinaria vitis* (Hemiptera: Coccidae) (Noyes, 2019).
Hosts: *Eulecanium sp., Eulecanium corni, Eulecanium coryli, Eulecanium crudum, Eulecanium pomeranicum, Eulecanium rufulum, Eulecanium taxi, Eulecanium tiliae, Parthenolecanium sp., Parthenolecanium corni, Parthenolecanium fletcheri, Parthenolecanium persicae, Parthenolecanium pomeranicum, Parthenolecanium rufulum, Pulvinaria sp., Pulvinaria betulae, Pulvinaria populi, Pulvinaria vitis, Sphaerolecanium prunastri* (Hemiptera: Coccidae) (Noyes, 2019).
Parasitoids: *Marietta picta* (Hymenoptera: Aphelinidae), *Pachyneuron solitarium* (Pteromalidae) (Noyes, 2019).
Distribution: Andorra, Armenia, Austria, Azerbaijan, Bulgaria, Canary Islands, Caucasus, Czech Republic, Czechoslovakia, Denmark, Europe, Finland, France, Georgia, Germany, Greece, Hungary, Italy, Kazakhstan, Moldova, Norway, People's Republic of China, Poland, Romania, Russia, Slovakia, Spain, Sweden, Switzerland, Transcaucasus, Tselinograd Obl., Turkey, United Kingdom, United States of America, USSR (Noyes, 2019).

Metaphycus lounsburyi (Howard, 1898)
Publications with Turkey records: Japoshvili (2012), Japoshvili and Celik (2010), Noyes (2019).
Associates: *Metaphycus helvolus, Metaphycus swirskii* (Encyrtidae), *Linepithema humile* (Formicidae), *Scutellista cyanea* (Pteromalidae), *Nerium oleander* (Apocynaceae), *Calotropis gigantea* (Asclepiadaceae), *Baccharis pilularis* (Asteraceae), *Cucurbita moschata* (Cucurbitaceae), *Ficus carica* (Moraceae), *Olea europaea* (Oleaceae), *Coffea sp.* (Rubiaceae), *Citrus sp.* (Rutaceae), *Solanum nigrum, Solanum tuberosum* (Solanaceae) (Noyes, 2019).

Hosts: *Asterolecanium coffeae* (Hemiptera: Asterolecaniidae), *Ceroplastes floridensis, Ceroplastes mimosa, Coccus sp., Coccus capparidis, Coccus hesperidum, Coccus pseudomagnoliarum, Eulecanium tiliae, Lichtensia viburni, Pulvinaria psidii, Saissetia sp., Saissetia coffeae, Saissetia oleae, Waxiella mimosa* (Coccidae), *Parlatoria oleae* (Diaspididae) (Noyes, 2019).
Parasitoid hosts: *Cheiloneurus inimicus* (Hymenoptera: Encyrtidae), *Marietta mexicana* (Aphelinidae), *Ammonoencyrtus californicus, Cheiloneurus inimicus, Cheiloneurus lineascapus, Cheiloneurus noxius, Quaylea whittieri* (Encyrtidae), *Aprostocetus minutus, Tetrastichus minutus* (Eulophidae), *Eupelmus inyoensis* (Eupelmidae), *Pachyneuron concolor* (Pteromalidae), *Chartocerus niger* (Signiphoridae) (Noyes, 2019).
Distribution: Afrotropical, Albania, Argentina, Australia, Bermuda, Chile, Cyprus, Egypt, Eritrea, Ethiopia, France, Greece, Guam, Hawaii, Iran, Israel, Italy, Kenya, Libya, Mexico, Morocco, Nearctic, Netherlands, New Zealand, Peru, Portugal, South Africa, Spain, Tanzania, Tunisia, Turkey, United States of America, Uruguay (Noyes, 2019).

Metaphycus melanostomatus (Timberlake, 1916)
Publications with Turkey records: Japoshvili and Noyes (2005), Noyes (2019).
Associates: *Alnus sp., Corylus avellana* (Betulaceae), *Fagus sp., Quercus robur* (Fagaceae), *Crataegus oxyacantha* (Rosaceae), *Tilia sp.* (Tiliaceae) (Noyes, 2019).
Hosts: *Coccus pseudomagnoliarum, Eulecanium franconicum, Eulecanium mali, Eulecanium rugulosum, Eulecanium tiliae, Lecanium sp., Lecanium corni, Lecanium coryli, Lecanium fuscum, Pulvinaria regalis, Pulvinaria vitis* (Hemiptera: Coccidae) (Noyes, 2019).
Distribution: Armenia, Azerbaijan, Belarus, Belgium, Bulgaria, Croatia, Czech Republic, Denmark, Europe, France, Georgia, Germany, Greece, Iran, Ireland, Italy, Japan, Kazakhstan, Moldova, Netherlands, Russia, Spain, Sweden, Tadzhikistan, Transcaucasus, Tselinograd Obl., Turkey, Turkmenistan, United Kingdom, United States of America, Uzbekistan, Yugoslavia (pre1991) (Noyes, 2019).

Metaphycus petitus (Walker, 1851)
Publications with Turkey records: Japoshvili (2012), Japoshvili and Celik (2010), Noyes (2019).
Associates: *Calluna sp.* (Ericaceae) (Noyes, 2019).
Hosts: *Eriococcus insignis* (Hemiptera: Eriococcidae) (Noyes, 2019).

Distribution: Bulgaria, Czech Republic, Czechoslovakia, Denmark, Finland, Georgia, Greece, Ireland, Italy, Moldova, Norway, Slovakia, Spain, Turkey, United Kingdom (Noyes, 2019).

Metaphycus piceus (Hoffer, 1954)
Publications with Turkey records: Noyes (2019).
Associates: Unknown (Noyes, 2019).
Hosts: Unknown (Noyes, 2019).
Distribution: Bulgaria, Czech Republic, Czechoslovakia, England, Moldova, Romania, Serbia, Slovakia, Turkey, Yugoslavia (Federal Republic) (Noyes, 2019).

Metaphycus silvestrii (Sugonjaev, 1970)
Publications with Turkey records: Japoshvili and Karaca (2002), Japoshvili and Noyes (2005), Noyes (2019), Öncüer (1991), Trjapitzin and Doğanlar (1997), Ulgenturk (2001).
Associates: *Terminalia sp.* (Combretaceae), *Quercus sp.* (Fagaceae), *Armeniaca vulgaris, Prunus sp., Prunus americana, Prunus cerasifera, Prunus domestica, Prunus spinosa* (Rosaceae) (Noyes, 2019).
Hosts: *Eulecanium corni, Parthenolecanium corni, Sphaerolecanium prunastri* (Hemiptera: Coccidae) (Noyes, 2019).
Distribution: Armenia, Azerbaijan, Belarus, Bulgaria, Croatia, Georgia, Germany, Hungary, Iran, Italy, Moldova, Poland, Romania, Russia, Transcaucasus, Turkey, Ukraine, Yugoslavia (Federal Republic), Yugoslavia (pre1991) (Noyes, 2019).

Metaphycus stagnarum (Hoffer, 1954)
Publications with Turkey records: Japoshvili and Noyes (2005), Noyes (2019).
Associates: *Alnus sp., Betula sp., Corylus avellana* (Betulaceae), *Buxus sp.* (Buxaceae), *Robinia pseudoacacia* (Fabaceae), *Ribes sanguineum* (Grossulariaceae), *Picea abies* (Pinaceae), *Crataegus oxyacantha, Cydonia sp.* (Rosaceae), *Vitis vinifera* (Vitaceae) (Noyes, 2019).
Hosts: *Eulecanium corni, Parthenolecanium corni, Physokermes piceae, Pulvinaria sp., Pulvinaria floccifera, Pulvinaria hydrangea, Pulvinaria vitis* (Hemiptera: Coccidae), *Planococcus ficus* (Pseudococcidae) (Noyes, 2019).
Distribution: Belgium, Czech Republic, Czechoslovakia, Denmark, Europe, Finland, Germany, Iran, Isle of Man, Italy, Kazakhstan, Netherlands, Norway, Poland, Romania, Russia, Slovakia, Tselinograd Obl., Turkey, United Kingdom, USSR (Noyes, 2019).

Metaphycus stanleyi (Compere, 1940)
Publications with Turkey records: Japoshvili (2012), Noyes (2019).
Associates: *Hedera helix* (Araliaceae), *Codiaeum variegatum* (Euphorbiaceae), *Persea americana* (Lauraceae), *Ficus retusa* (Moraceae), *Olea chrysophylla* (Oleaceae), *Coffea sp.* (Rubiaceae), *Citrus sp.* (Rutaceae), *Solanum tuberosum* (Solanaceae) (Noyes, 2019).
Hosts: *Ceroplastes brevicauda, Chloropulvinaria psidii, Coccus sp., Coccus alpinus, Coccus celatus, Coccus elongatus, Coccus hesperidum, Coccus pseudomagnoliarum, Coccus subhemisphaericus, Coccus viridis, Eucalymnatus tessellatus, Gascardia sp., Gascardia brevicauda, Lichtensia chilianthi, Parasaissetia sp., Parasaissetia nigra, Protopulvinaria pyriformis, Pulvinaria mesembryanthemi, Pulvinaria psidii, Pulvinariella mesembryanthemi, Saissetia sp., Saissetia coffeae, Saissetia nigra, Saissetia nigrella, Saissetia oleae, Saissetia somereni, Saissetia subhemisphaerica* (Hemiptera: Coccidae) (Noyes, 2019).
Parasitoids: *Cheiloneurus noxius* (Hymenoptera: Encyrtidae) (Noyes, 2019).
Distribution: Afrotropical, Argentina, Bermuda, Canada, Canary Islands, Chile, Costa Rica, Cuba, Eritrea, Ethiopia, Greece, Guam, Hawaii, Honduras, Iran, Israel, Italy, Kenya, Mexico, Namibia, Nearctic, Papua New Guinea, Peru, South Africa, Spain, St Helena, Turkey, Uganda, United States of America (Noyes, 2019).

Metaphycus swirskii (Annecke and Mynhardt, 1979)
Publications with Turkey records: Japoshvili (2012), Japoshvili and Celik (2010), Noyes (2019).
Associates: *Coccus sp., Saissetia sp., Icerya pattersoni* (Margarodidae), *Metaphycus bartletti* (Hymenoptera: Encyrtidae), *Ficus sp.* (Moraceae), *Coffea sp.* (Rubiaceae), *Citrus sp.* (Rutaceae), *Solanum tuberosum* (Solanaceae) (Noyes, 2019).
Hosts: *Chloropulvinaria floccifera, Coccus capparidis, Coccus hesperidum, Coccus viridis, Protopulvinaria sp., Protopulvinaria pyriformis, Pulvinaria floccifera, Saissetia sp., Saissetia coffeae, Saissetia oleae, Saissetia somereni* (Hemiptera: Coccidae) (Noyes, 2019).
Distribution: Afrotropical, France, Greece, Israel, Italy, Kenya, South Africa, Turkey (Noyes, 2019).

Metaphycus zebratus (Mercet, 1917)
Publications with Turkey records: Japoshvili (2012), Japoshvili and Celik (2010), Noyes (2019).

Associates: *Hedera sp., Hedera helix* (Araliaceae), *Cupressus sp., Juniperus thurifera, Thuja sp.* (Cupressaceae), *Dalbergia sissoo, Robinia sp., Robinia pseudoacacia* (Fabaceae), *Quercus sp., Quercus petraea* (Fagaceae), *Ficus sp.* (Moraceae), *Jasminum fructicans, Olea europaea* (Oleaceae), *Pinus halepensis* (Pinaceae), *Brachypodium pinnatum, Elymus repens, Festuca sp.* (Poaceae), *Cerasus japonica, Prunus cerasifera, Prunus domestica, Rosa damascena* (Rosaceae), *Rubus sp.* (Rubiaceae), *Citrus sp.* (Rutaceae) (Noyes, 2019).
Hosts: *Planchonia sp., Planchonia arabidis* (Hemiptera: Asterolecaniidae), *Ceroplastes floridensis, Ceroplastodes cajani, Drepanococcus cajani, Eriopeltis sp., Eriopeltis agropyri, Eriopeltis araxis, Eriopeltis festucae, Eriopeltis lichtensteinii, Eriopeltis stammeri, Eriopeltis zolotarevae, Eulecanium corni, Eulecanium persicae, Eulecanium pomeranicum, Eulecanium rufulum, Lecanopsis formicarum, Luzulaspis luzulae, Parafairmairia sp., Parthenolecanium sp., Parthenolecanium corni, Parthenolecanium persicae, Parthenolecanium pomeranicum, Parthenolecanium rufulum, Pulvinaria vitis, Rhodococcus bulgariensis, Rhodococcus perornatus, Saissetia oleae* (Coccidae), *Aonidiella orientalis* (Diaspididae), *Acanthococcus roboris, Eriococcus roboris* (Eriococcidae), *Nipaecoccus sp., Phenacoccus aceris, Trionymus perrisii* (Pseudococcidae) (Noyes, 2019).
Distribution: Azerbaijan, Belgium, Bulgaria, Croatia, Czech Republic, Czechoslovakia, Denmark, Egypt, Europe, Finland, France, Georgia, Germany, Greece, Hungary, India, Iran, Israel, Italy, Kazakhstan, Lithuania, Moldova, Mongolia, Montenegro, Netherlands, Norway, Palearctic, Poland, Portugal, Romania, Russia, Serbia, Slovakia, Slovenia, Spain, Tselinograd Obl., Turkey, Turkmenistan, Ukraine, United Kingdom, United States of America, USSR, Yugoslavia (Federal Republic), Yugoslavia (pre-1991) (Noyes, 2019).
Genus *Microterys* (Thomson, 1876)
Microterys annulatus (Erdos, 1957)
Publications with Turkey records: Japoshvili and Noyes (2005), Noyes (2019).
Associates: Unknown (Noyes, 2019).
Hosts: Unknown (Noyes, 2019).
Distribution: Czechoslovakia, England, Hungary, Slovakia, Sweden, Turkey (Noyes, 2019).

Microterys bellae (Trjapitzin, 1968)
Publications with Turkey records: Japoshvili and Karaca (2002), Japoshvili and Noyes (2005), Noyes (2019).

Associates: *Rosa sp., Rosa acicularis, Rosa damascene* (Rosaceae) (Noyes, 2019).
Hosts: *Eulecanium caraganae, Eulecanium sachalinense, Eulecanium secretum, Rhodococcus bulgariensis, Rhodococcus perornatus, Rhodococcus rosaeluteae, Rhodococcus spiraeae* (Hemiptera: Coccidae) (Noyes, 2019).
Distribution: Iran, Kazakhstan, Kirgizia, Norway, People's Republic of China, Russia, Tselinograd Obl., Turkey (Noyes, 2019).

Microterys cneus (Trjapitzin and Sugonjaev, 1976)
Publications with Turkey records: Japoshvili and Noyes (2005), Noyes (2019).
Associates: *Betula sp., Betula nana, Betula verrucose* (Betulaceae), *Cydonia sp.* (Rosaceae) (Noyes, 2019).
Hosts: *Pulvinaria betulae* (Hemiptera: Coccidae), *Planococcus ficus* (Pseudococcidae) (Noyes, 2019).
Distribution: Iran, Russia, Turkey (Noyes, 2019).

Microterys darevskii (Trjapitzin, 1968)
Publications with Turkey records: Japoshvili (2012), Japoshvili and Noyes (2005), Noyes (2019).
Associates: *Medicago sp.* (Fabaceae), *Ephedra sp.* (Gnetaceae), (Poaceae), *Galium sp.* (Rubiaceae) (Noyes, 2019).
Hosts: *Stotzia maxima* (Hemiptera: Coccidae), *Puto sp.* (Pseudococcidae) (Noyes, 2019).
Distribution: Armenia, Azerbaijan, Turkey, Turkmenistan (Noyes, 2019).

Microterys golcukus (Japoshvili, 2012)
Publications with Turkey records: Japoshvili (2012), Noyes (2019).
Associates: Unknown (Noyes, 2019).
Hosts: Unknown (Noyes, 2019).
Distribution: Turkey (Noyes, 2019).

Microterys hortulanus Erdos, 1956
Publications with Turkey records: Japoshvili and Karaca (2002), Noyes (2019), Trjapitzin and Doğanlar (1997), Ulgenturk et al. (2001), Ulgenturk and Toros (2001).
Associates: *Amygdalus sp., Armeniaca vulgaris, Persica vulgaris, Prunus sp., Prunus cerasifera, Prunus dulcis, Prunus spinosa* (Rosaceae) (Noyes, 2019).

Hosts: *Didesmococcus unifasciatus, Eulecanium ciliatum, Eulecanium coryli, Eulecanium prunastri, Sphaerolecanium prunastri* (Hemiptera: Coccidae) (Noyes, 2019).
Distribution: Armenia, Azerbaijan, Bulgaria, Caucasus, Czech Republic, Czechoslovakia, Europe, Georgia, Germany, Greece, Hungary, Iran, Israel, Italy, Kirgizia, Moldova, Poland, Romania, Russia, Slovakia, Spain, Turkey, Ukraine, United Kingdom, USSR, Uzbekistan, Yugoslavia (Federal Republic) (Noyes, 2019).

Microterys lunatus (Dalman, 1820)
Publications with Turkey records: Japoshvili and Karaca (2002), Noyes (2019), Öncüer (1991).
Associates: *Quercus ilex* (Fagaceae), *Olea europaea* (Oleaceae), *Picea sp., Picea abies, Picea orientalis* (Pinaceae), *Cydonia oblonga, Malus sp., Prunus domestica, Pyrus malus* (Rosaceae), *Citrus nobilis, Citrus unshiu* (Rutaceae) (Noyes, 2019).
Hosts: *Ceroplastes japonicus, Eulecanium coryli, Filippia folicularis, Lecanium corni, Lecanium prunastri, Parthenolecanium corni, Parthenolecanium coryli, Physokermes sp., Physokermes hemicryphus, Physokermes piceae, Physokermes sugonjaevi, Pulvinaria vitis, Saissetia oleae, Sphaerolecanium prunastri* (Hemiptera: Coccidae), *Kermococcus roboris, Nidularia pulvinate* (Kermesidae), *Ferrisiana oleae* (Pseudococcidae) (Noyes, 2019).
Distribution: Austria, Azerbaijan, Bulgaria, Croatia, Czech Republic, Czechoslovakia, Denmark, England, Estonia, Europe, Finland, France, Georgia, Germany, Greece, Hungary, Iran, Italy, Kazakhstan, Lithuania, Mongolia, Nearctic, North Africa, Norway, Poland, Romania, Russia, Serbia, Slovakia, Sweden, Switzerland, Tselinograd Obl., Turkey, Ukraine, United States of America, USSR, Yugoslavia (pre1991) (Noyes, 2019).

Microterys masii (Silvestri, 1919)
Publications with Turkey records: Japoshvili and Noyes (2005), Noyes (2019), Öncüer (1991), Trjapitzin and Doğanlar (1997).
Associates: *Hedera helix* (Araliaceae), *Viburnum tinus* (Caprifoliaceae), *Olea europaea* (Oleaceae) (Noyes, 2019).
Hosts: *Coccus hesperidum, Eulecanium corni, Eulecanium coryli, Eulecanium tiliae, Euphilippia olivine, Filippia folicularis, Filippia oleae, Filippia viburni, Lichtensia viburni, Parthenolecanium sp., Parthenolecanium corni, Philippia oleae, Pulvinaria floccifera, Sphaerolecanium prunastri* (Hemiptera: Coccidae), *Eriococcus buxi* (Eriococcidae), *Ferrisiana oleae* (Pseudococcidae) (Noyes, 2019).

Distribution: Algeria, Croatia, Cyprus, England, Europe, France, Germany, Greece, Iran, Israel, Italy, Montenegro, North Africa, Spain, Tunisia, Turkey, Ukraine, USSR, Yugoslavia (pre-1991) (Noyes, 2019).

Microterys nietneri (Motschulsky, 1859)
Publications with Turkey records: Noyes (2019), Öncüer (1991).
Associates: *Crossandra infundibuliformis* (Acanthaceae), *Mangifera indica* (Anacardiaceae), *Hedera sp.* (Araliaceae), *Carica papaya, Carica sinensis* (Caricaceae), *Citrullus vulgaris* (Cucurbitaceae), *Erythrina lithosperma* (Fabaceae), *Laurus sp.* (Lauraceae), *Psidium guajava* (Myrtaceae), *Olea europaea* (Oleaceae), *Prunus sp.* (Rosaceae), *Coffea sp.* (Rubiaceae), *Citrus sp., Citrus reticulata, Citrus sinensis* (Rutaceae), *Thea sinensis* (Theaceae) (Noyes, 2019).
Hosts: *Ceroplastes sp., Ceroplastes cirripediformis, Ceroplastes floridensis, Ceroplastes japonicus, Ceroplastes rubens, Chloropulvinaria polygonate, Chloropulvinaria psidii, Coccus sp., Coccus hesperidum, Coccus longulus, Coccus mangiferae, Coccus piperis, Coccus pseudomagnoliarum, Coccus viridis, Eucalymnatus tessellatus, Gascardia sp., Gascardia destructor, Lecanium sp., Lecanium corni, Lecanium fletcheri, Lecanium piperis, Maacoccus piperis, Parasaissetia nigra, Parasaissetia oleae, Parthenolecanium sp., Parthenolecanium corni, Protopulvinaria mangiferae, Protopulvinaria pyriformis, Pulvinaria sp., Pulvinaria floccifera, Pulvinaria iceryi, Pulvinaria mesembryanthemi, Pulvinaria peregrina, Pulvinaria polygonate, Pulvinaria psidii, Pulvinaria regalis, Pulvinaria vitis, Pulvinariella mesembryanthemi, Saissetia sp., Saissetia coffeae, Saissetia hemisphaerica, Saissetia miranda, Saissetia nigra, Saissetia oleae, Sphaerolecanium prunastri* (Hemiptera: Coccidae), *Hemichionaspis theae, Pinnaspis theae* (Diaspididae), *Rastrococcus iceryoides* (Pseudococcidae) (Noyes, 2019).
Parasitoids: *Marietta exitiosa, Marietta javensis, Marietta leopardine* (Hymenoptera: Aphelinidae), *Cheiloneurus noxius, Cheiloneurus paralia, Eusemion longipennis* (Encyrtidae), *Tetrastichus sp., Tetrastichus ceroplastophilus* (Eulophidae), *Moranila californica, Pachyneuron concolor* (Pteromalidae) (Noyes, 2019).
Distribution: Afrotropical, Argentina, Australasian, Australia, Azerbaijan, Bangladesh, Belgium, Bermuda, Bulgaria, Canada, Canary Islands, Caucasus, Croatia, Cuba, Cyprus, Egypt, Fiji, France, Georgia, Germany, Hawaii, Hungary, India, Iran, Israel, Italy, Japan, Lebanon, Malaysia, Mexico, Montenegro, Nearctic, Netherlands, New Zealand, Pakistan, Palearctic, Papua New Guinea, Peoples', Republic of China, Peru, South

Africa, Spain, Sri Lanka, Transcaucasus, Turkey, United States of America, USSR, Yugoslavia (pre-1991) (Noyes, 2019).

Microterys sylvius (Dalman, 1820)
Publications with Turkey records: Noyes (2019), Trjapitzin and Doğanlar (1997).
Associates: *Corylus manshurica* (Betulaceae), *Cupressus sp., Juniperus sp., Thuja sp.* (Cupressaceae), *Diospyros virginiana* (Ebenaceae), *Quercus sp.* (Fagaceae), *Ephedra procera* (Gnetaceae), *Liquidambar styraciflua* (Hamamelidaceae), *Ficus sp., Morus sp.* (Moraceae), *Prunus sp., Prunus dulcis, Spiraea sp., Spiraea hypericifolia* (Rosaceae), *Taxus sp.* (Taxaceae), *Vitis vinifera* (Vitaceae) (Noyes, 2019).
Hosts: *Didesmococcus unifasciatus, Eulecanium sp., Eulecanium corni, Eulecanium coryli, Eulecanium douglasi, Eulecanium ficiphilum, Eulecanium nocivum, Eulecanium persicae, Eulecanium rufulum, Eulecanium sericeum, Eulecanium tiliae, Lecanium corni, Lecanium coryli, Parthenolecanium corni, Parthenolecanium persicae, Parthenolecanium rufulum, Physokermes jezoensis, Pulvinaria vitis, Rhodococcus bulgariensis, Rhodococcus perornatus, Rhodococcus spiraeae, Rhodococcus turanicus, Sphaerolecanium prunastri, Stotzia maxima* (Hemiptera: Coccidae) (Noyes, 2019).
Parasitoids: *Pachyneuron solitarium* (Hymenoptera: Pteromalidae) (Noyes, 2019).
Distribution: Armenia, Austria, Azerbaijan, Belarus, Bulgaria, Canada, Caucasus, Czech Republic, Czechoslovakia, Denmark, England, Europe, Finland, Georgia, Germany, Hungary, Iran, Italy, Japan, Lebanon, Moldova, Nearctic, Poland, Romania, Russia, Serbia, Slovakia, Spain, Sweden, Switzerland, Turkey, Turkmenistan, Ukraine, United States of America, USSR, Yugoslavia (Federal Republic) (Noyes, 2019).

Microterys tricoloricornis (De Stefani, 1886)
Publications with Turkey records: Japoshvili and Karaca (2002), Japoshvili and Noyes (2005), Noyes (2019), Öncüer (1974; 1991).
Associates: *Hedera sp.* (Araliaceae), *Corylus avellana* (Betulaceae), *Laurocerasus officinalis* (Rosaceae), *Citrus sp., Citrus unshiu* (Rutaceae), *Celtis caucasica* (Ulmaceae) (Noyes, 2019).
Hosts: *Coccus hesperidum, Coccus pseudomagnoliarum, Eulecanium ciliatum, Eulecanium diminutum, Pulvinaria mesembryanthemi, Saissetia oleae* (Hemiptera: Coccidae) (Noyes, 2019).

Distribution: Georgia, Greece, Israel, Italy, Mexico, Palearctic, Russia, Spain, Transcaucasus, Turkey, Turkmenistan, United States of America (Noyes, 2019).

Genus *Parasauleia* (Hoffer, 1968)
Parasauleia trjapitzini (Hoffer, 1968)
Publications with Turkey records: Japoshvili and Karaca (2002), Japoshvili and Noyes (2005), Noyes (2019).
Associates: *Pinus sp.* (Pinaceae) (Noyes, 2019).
Hosts: *Anamaspis loewi, Lepidosaphes newsteadi, Leucaspis pini, Leucaspis pusilla, Leucaspis signoreti, Pinnaspis aspidistrae* (Hemiptera: Diaspididae) (Noyes, 2019).
Distribution: Czech Republic, Czechoslovakia, France, Turkey (Noyes, 2019).

Genus *Paratetracnemoidea* (Girault, 1915)
Paratetracnemoidea malenotti (Mercet, 1918)
Publications with Turkey records: Noyes (2019).
Associates: Unknown (Noyes, 2019).
Hosts: Unknown (Noyes, 2019).
Distribution: Armenia, Azerbaijan, Bulgaria, Croatia, Czech Republic, Czechoslovakia, France, Gambia, Greece, Hungary, India, Italy Moldova, Montenegro, Nigeria, People's Republic of China, Portugal, Russia, Slovakia, Spain, Sudan, Turkey, Turkmenistan, Yugoslavia (pre1991) (Noyes, 2019).

Genus *Pentacladocerus* (Erdos, 1963)
Pentacladocerus araraticus (Trjapitzin, 1962)
Publications with Turkey records: Japoshvili and Noyes (2005), Noyes (2019), Öncüer (1991), Trjapitzin and Doğanlar (1997).
Associates: *Ulmus sp.* (Ulmaceae) (Noyes, 2019).
Hosts: *Eriosoma sp.* (Hemiptera: Aphididae) (Noyes, 2019).
Distribution: Armenia, Turkey (Noyes, 2019).
Genus *Prochiloneurus* (Silvestri, 1915)
Prochiloneurus bolivari (Mercet, 1919)
Publications with Turkey records: Kaydan and Japoshvili (2010), Noyes (2019).
Associates: *Juniperus thurifera* (Cupressaceae), *Euphorbia sp., Manihot esculenta* (Euphorbiaceae), *Retama sphaerocarpa* (Fabaceae), *Morus sp.* (Moraceae), *Pinus halepensis* (Pinaceae), *Cynodon dactylon* (Poaceae), *Prunus dulcis* (Rosaceae), *Rubus sp.* (Rubiaceae), *Verbascum sp.* (Scrophulariaceae), *Vitis vinifera* (Vitaceae) (Noyes, 2019).

Hosts: *Rhizopulvinaria armeniaca* (Hemiptera: Coccidae), *Acanthococcus desertus, Eriococcus insignis, Neoacanthococcus tamaricicola* (Eriococcidae), *Atrococcus sp., Dysmicoccus multivorus, Ferrisia virgata, Heliococcus bohemicus, Heterococcopsis opertus, Maconellicoccus hirsutus, Naiacoccus serpentinus, Peliococcus calluneti, Peliococcus mesasiaticus, Peliococcus zillae, Phenacoccus sp., Phenacoccus arthrophyte, Phenacoccus madeirensis, Phenacoccus manihoti, Phenacoccus zillae, Planococcus sp., Planococcus citri, Planococcus ficus, Pseudococcus sp., Pseudococcus citri, Pseudococcus comstocki, Pseudococcus copiosus, Pseudococcus ficus, Puto pilosellae, Spinococcus calluneti, Trionymus copiosus, Trionymus multivorus* (Pseudococcidae) (Noyes, 2019).
Parasitoid hosts: *Anagyrus sp., Blepyrus insularis, Clausenia purpurea, Epidinocarsis lopezi, Leptomastix flava* (Hymenoptera: Encyrtidae) (Noyes, 2019).
Parasitoids: *Marietta picta* (Aphelinidae) (Noyes, 2019).
Distribution: Afghanistan, Afrotropical, Algeria, Armenia, Austria, Azerbaijan, Bulgaria, Congo, Croatia, Czech Republic, Czechoslovakia, Democratic Republic of Congo (Zaire), Denmark, Egypt, Europe, Finland, France, Georgia, Hungary, India, Iran, Israel, Italy, Japan, Lithuania, Moldova, Mongolia, Nigeria, Norway, Palearctic, Poland, Romania, Russia, Sao Tomé and Principe, Slovakia, South Africa, Spain, Sweden, Transcaucasus, Turkey, Turkmenistan, Ukraine, United Kingdom, USSR, Uzbekistan, Yugoslavia (pre-1991) (Noyes, 2019).

Prochiloneurus uyguni (Hayat, 2016)
Publications with Turkey records: Caliskan et al. (2016), Noyes (2019).
Associates: *Lantana camara* (Verbenaceae) (Noyes, 2019).
Hosts: *Phenacoccus solenopsis* (Hemiptera: Pseudococcidae) (Noyes, 2019).
Distribution: Turkey (Noyes, 2019).

Genus *Pseudococcobius* (Timberlake, 1916)
Pseudococcobius obenbergeri (Novickij, 1926)
Publications with Turkey records: Japoshvili (2012), Japoshvili and Celik (2010), Japoshvili and Noyes (2005), Noyes (2019).
Associates: *Achillea sp.* (Asteraceae), *Suaeda sp.* (Chenopodiaceae), *Festuca rubra* (Poaceae), *Rubus sp.* (Rubiaceae)
Hosts: *Eriococcus devoniensis* (Hemiptera: Eriococcidae), *Peliococcus sp., Pseudococcus multivorus, Trionymus multivorus, Trionymus perrisii* (Pseudococcidae) (Noyes, 2019).

Distribution: Armenia, Azerbaijan, Bulgaria, Czech Republic, Czechoslovakia, Denmark, England, Georgia, Hungary, Ireland, Moldova, Romania, Russia, Slovakia, Turkey, Ukraine (Noyes, 2019).

Genus *Psilophrys* (Mayr, 1876)
Psilophrys bella (Japoshvili and Noyes, 2006)
Publications with Turkey records: Noyes (2019).
Associates: *Quercu coccifera* (Fagaceae) (Noyes, 2019).
Hosts: Kermes palestinensis (Hemiptera: Kermesidae) (Noyes, 2019).
Distribution: Turkey (Noyes, 2019).

Psilophrys ghilarovi (Japoshvili, 2005)
Publications with Turkey records: Japoshvili and Noyes (2005), Noyes (2019).
Associates: *Quercu coccifera, Quercus sp.* (Fagaceae) (Noyes, 2019).
Hosts: *Kermes bacciformis, Kermes palestinensis* (Hemiptera: Kermesidae) (Noyes, 2019).
Distribution: Italy, Turkey (Noyes, 2019).

Psilophrys parvulus (Guerrieri and Viggiani, 1991)
Publications with Turkey records: Japoshvili et al. (2009), Noyes (2019).
Associates: *Quercus sp., Quercus cerris* (Fagaceae) (Noyes, 2019).
Hosts: *Kermes bacciformis* (Hemiptera: Kermesidae) (Noyes, 2019).
Distribution: Italy, Turkey (Noyes, 2019).

Psilophrys tenuicornis (Graham, 1969)
Publications with Turkey records: Japoshvili and Noyes (2005), Japoshvili and Karaca (2004), Noyes (2019).
Associates: *Quercu coccifera, Quercus sp.* (Fagaceae) (Noyes, 2019).
Hosts: *Eulecanium coryli, Physokermes piceae* (Hemiptera: Coccidae), *Kermes bacciformis, Kermes corticalis, Kermes ilicis, Kermes nawae, Kermes palestinensis, Kermes roboris, Kermococcus sp., Kermococcus corticalis, Kermococcus quercus, Kermococcus roboris* (Kermesidae) (Noyes, 2019).
Distribution: Armenia, Austria, Czech Republic, Czechoslovakia, Europe, France, Germany, Hungary, Italy, Moldova, People's Republic of China, Poland, Russia, Spain, Turkey, Ukraine (Noyes, 2019).

Genus *Psyllaephagus* (Ashmead, 1900)
Psyllaephagus caillardiae (Sugonjaev, 1968)

Publications with Turkey records: Japoshvili and Noyes (2005), Noyes (2019).
Associates: *Haloxylon aphyllum, Haloxylon persicum* (Chenopodiaceae) (Noyes, 2019).
Hosts: *Caillardia azurea, Caillardia notata, Caillardia robusta* (Hemiptera: Psyllidae) (Noyes, 2019).
Distribution: Kazakhstan, Mongolia, Tadzhikistan, Tselinograd Obl., Turkey, Turkmenistan, Uzbekistan (Noyes, 2019).

Psyllaephagus elaeagni (Trjapitzin, 1967)
Publications with Turkey records: Japoshvili and Noyes (2005), Noyes (2019).
Associates: *Elaeagnus sp., Elaeagnus angustifolia* (Elaeagnaceae), *Rhamnus sp., Rhamnus pallasii* (Rhamnaceae) (Noyes, 2019).
Hosts: *Cacopsylla rhamnicola, Psylla ramnicola* (Hemiptera: Psyllidae), *Trioza magnisetosa* (Triozidae) (Noyes, 2019).
Distribution: Armenia, Azerbaijan, Georgia, Kazakhstan, Turkey, Turkmenistan (Noyes, 2019).

Psyllaephagus euphyllurae (Masi, 1911)
Publications with Turkey records: Japoshvili and Noyes (2005), Noyes (2019), Öncüer (1991), Trjapitzin and Doğanlar (1997).
Associates: *Olea europaea* (Oleaceae) (Noyes, 2019).
Hosts: *Euphyllura olivine, Euphyllura phillyreae, Euphyllura straminea* (Hemiptera: Psyllidae) (Noyes, 2019).
Distribution: Egypt, Greece, Italy, Portugal, Spain, Tunisia, Turkey (Noyes, 2019).

Psyllaephagus pistaciae (Ferriere, 1961)
Publications with Turkey records: Japoshvili and Noyes (2005), Noyes (2019), Trjapitzin and Doğanlar (1997).
Associates: *Pistacia sp., Pistacia vera* (Anacardiaceae) (Noyes, 2019).
Hosts: *Agonoscena pistaciae, Agonoscena targionii* (Hemiptera: Psyllidae) (Noyes, 2019).
Distribution: Greece, Iran, Turkey (Noyes, 2019).

Psyllaephagus procerus (Mercet, 1921)
Publications with Turkey records: Noyes (2019), Ulgenturk and Toros (2001).
Associates: *Fraxinus sp.* (Oleaceae), *Pyrus communis* (Rosaceae) (Noyes, 2019).

Hosts: *Eulecanium ciliatum* (Hemiptera: Coccidae), *Cacopsylla sp., Cacopsylla pyri, Psyllopsis fraxini* (Psyllidae) (Noyes, 2019).
Distribution: Armenia, Azerbaijan, Europe, Georgia, Hungary, Italy, Moldova, Mongolia, Poland, Russia, Spain, Turkey, Ukraine (Noyes, 2019).

Psyllaephagus pulchellus (Mercet, 1921)
Publications with Turkey records: Japoshvili and Noyes (2005), Noyes (2019), Trjapitzin and Doğanlar (1997).
Associates: *Pistacia sp.* (Anacardiaceae), *Olea europaea* (Oleaceae) (Noyes, 2019).
Hosts: *Anapulvinaria pistaciae* (Hemiptera: Coccidae), *Euphyllura olivine* (Psyllidae) (Noyes, 2019).
Distribution: Croatia, Greece, Montenegro, Spain, Turkey, Yugoslavia (pre1991) (Noyes, 2019).

Psyllaephagus tokgaevi (Myartseva, 1979)
Publications with Turkey records: Japoshvili and Noyes (2005), Noyes (2019).
Associates: *Tamarix sp.* (Tamaricaceae) (Noyes, 2019).
Hosts: *Colposcenia kiritschenkoi, Crastina tamaricina* (Hemiptera: Psyllidae) (Noyes, 2019).
Distribution: Georgia, Kazakhstan, Turkey, Turkmenistan (Noyes, 2019).

Genus *Syrphophagus* (Ashmead, 1900)
Syrphophagus aeruginosus (Dalman, 1820)
Publications with Turkey records: Noyes (2019), Trjapitzin and Doğanlar (1997).
Associates: *Brassica oleracea* (Brassicaceae), *Triticum aestivum* (Poaceae), *Prunus persica, Pyrus communis, Rosa damascene* (Rosaceae), *Rubus idaeus* (Rubiaceae) (Noyes, 2019).
Hosts: *Episyrphus balteata, Eupeodes corollae, Melanostoma mellinum, Neocnemodon dreyfusiae, Neocnemodon fulvimanus, Paragus sp., Paragus bicolor, Paragus quadrifasciatus, Paragus tibialis, Platychirus discimanus, Sphaerophoria sp., Syrphidae unspecified sp., Syrphus sp., Syrphus auricollis, Syrphus balteatus, Syrphus bifasciatus, Syrphus corollae, Syrphus pyrastri, Syrphus ribesii, Syrphus torvus, Syrphus vitripennis* (Diptera: Syrphidae), *Aphididae unspecified sp., Aphis maidis, Brachycaudus amygdalinus, Brachycolus noxius, Rhopalosiphum padi, Toxoptera graminum* (Hemiptera: Aphididae), *Physokermes piceae* (Coccidae), *Cacopsylla pyri* (Psyllidae) (Noyes, 2019).

Distribution: Armenia, Austria, Azerbaijan, Bulgaria, Canary Islands, Croatia, Czech Republic, Europe, Finland, France, Georgia, Germany, Hungary, India, Iran, Israel, Macedonia, Madeira, Moldova, Mongolia, Montenegro, Morocco, Netherlands, Norway, People's Republic of China, Portugal, Romania, Russia, Serbia, Slovakia, Sweden, Switzerland, Turkey, Ukraine, United Kingdom, Yugoslavia (pre-1991) (Noyes, 2019).

Syrphophagus aphidivorus (Mayr, 1876)
Publications with Turkey records: Bulut and Gocmen (2000), Noyes (2019), Öncüer (1991), Trjapitzin and Doğanlar (1997).
Associates: *Aphelinus asychis* (Aphelinidae), *Lasius niger, Myrmica laevinodis* (Formicidae), *Acer platanoides* (Aceraceae), *Pistacia vera* (Anacardiaceae), *Carduus sp., Carthamus sp., Carthamus tinctorius, Cirsium sp., Cirsium arvense, Sonchus sp., Tanacetum vulgare* (Asteraceae), *Corylus avellana* (Betulaceae), *Brassica campestris, Brassica oleracea* (Brassicaceae), *Euonymus japonica* (Celastraceae), *Alhagi camlorum, Glycine max, Glycyrrhiza glabra, Leucaena leucocephala, Medicago sp., Medicago sativa, Phaseolus sp., Phaseolus vulgaris, Pisum sativum, Trifolium pratense, Vicia faba* (Fabaceae), *Quercus robur* (Fagaceae), *Philadelphus coronarius* (Hydrangeaceae), *Juglans sp.* (Juglandaceae), *Punica granatum* (Lythraceae), *Gossypium sp.* (Malvaceae), *Sorghum sp., Sorghum bicolor, Zea mays* (Poaceae), *Armeniaca vulgaris, Cerasus macrocarpa, Malus sp., Malus pumila, Prunus armeniaca, Prunus divaricate, Pyrus malus* (Rosaceae), *Populus sp., Salix sp.* (Salicaceae), *Capsicum sp., Nicotiana tabacum, Solanum nigrum* (Solanaceae) (Noyes, 2019).
Hosts: (Diptera: Agromyzidae), *Leucopis obscura* (Chamaemyiidae), *Trialeurodes vaporariorum* (Hemiptera: Aleyrodidae), *Acyrthosiphon cornelli, Acyrthosiphon gossypii, Acyrthosiphon pisum, Anuraphis bakeri, Anuraphis persicaeniger, Anuraphis schwartzi, Aphididae unspecified, Aphis craccivora, Aphis fabae, Aphis gossypii, Aphis laburni, Aphis maidis, Aphis medicaginis, Aphis medicagnis, Aphis pomi, Aphis punicae, Aphis rumicis, Aphis saliceti, Aspidaphis aquatica, Aulacorthum solani, Brachycaudus sp., Brachycaudus cardui, Brachycaudus prunicola, Brachycolus korotnewi, Brachycolus noxius, Brevicoryne brassicae, Cavariella aegopodii, Cavariella aquatica, Chaitophorus sp., Chromaphis juglandicola, Dactynotus compositae, Diuraphis noxia, Drepanosiphum platanoidis, Dysaphis reaumuri, Eriosoma lanigerum, Hyalopterus pruni, Hysteroneura setariae, Lipaphis erysimi, Macrosiphum avenae, Macrosiphum cornelli, Macrosiphum granarium, Macrosiphum liriodendri, Macrosiphum rosae, Myzocallis coryli, Myzus cerasi, Myzus*

persicae, Nearctaphis bakeri, Pemphigus populi, Periphyllus aceris, Pterochlorus persicae, Rhopalosiphonius solani, Rhopalosiphum insertum, Rhopalosiphum maidis, Rhopalosiphum padi, Roepkea marchali, Sanbornia juniperi, Schizaphis graminum, Therioaphis maculate, Therioaphis trifolii, Toxoptera aurantia, Toxoptera graminum, Tuberculoides annulatus, Uroleucon compositae, Uroleucon sonchi (Aphididae), *Agonoscena pistaciae, Agonoscena targionii, Heteropsylla cubana* (Psyllidae), *Praon volucre, Trioxys pallidus* (Hymenoptera: Braconidae), *Phaenoglyphis villosa* (Cynipidae) (Noyes, 2019).

Parasitoid hosts: *Sphaerophoria javana* (Diptera: Syrphidae), *Aphelinus sp., Aphelinus asychis, Aphelinus flavus, Aphelinus jucundus, Aphelinus maidis, Aphelinus mali, Aphelinus nigritus, Aphelinus sanborniae, Aphelinus semiflavus, Aphelinus toxopteraphidis* (Hymenoptera: Aphelinidae), *Arge scapularis* (Argidae), *Aphidius sp., Aphidius avenae, Aphidius cardui, Aphidius ervi, Aphidius medicaginis, Aphidius picipes, Aphidius smithi, Aphidius sonchi, Aphidius testaceipes, Aphidius uzbekistanicus, Binodoxys communis, Diaeretiella rapae, Diaretiella rapae, Ephedrus lacertosus, Ephedrus persicae, Ephedrus plagiator, Lysiphlebus cardui, Lysiphlebus dissolutus, Lysiphlebus fabarum, Monoctonus caricis, Pauesia antennata, Praon exsoletum, Praon palitans, Praon volucre, Trioxys indicus, Trioxys pallidus, Trioxys utilis* (Braconidae), *Psyllaephagus sp.* (Encyrtidae), *Alloxysta fractricornis* (Figitidae) (Noyes, 2019).

Parasitoids: *Dendrocerus carpenter* (Megaspilidae) (Noyes, 2019).

Distribution: Argentina, Armenia, Austria, Azerbaijan, Brazil, Bulgaria, Chile, Croatia, Cuba, Czech Republic, Egypt, France, Georgia, Germany, Greece, Hawaii, Hungary, India, Iran, Iraq, Italy, Jordan, Kenya, Mexico, Moldova, Mongolia, Montenegro, Nearctic, Netherlands, Pakistan, Palearctic, People's Republic of China, Peru, Poland, Portugal, Puerto Rico, Romania, Russia, Saudi Arabia, Serbia, Slovakia, Spain, Sudan, Sweden, Turkey, Turkmenistan, Ukraine, United Kingdom, United States of America, USSR, Yugoslavia (pre-1991) (Noyes, 2019).

Genus *Trechnites* (Thomson, 1876)

Trechnites flavipes (Mercet, 1921)

Publications with Turkey records: Japoshvili and Noyes (2005), Noyes (2019).

Associates: *Hammada leptoclada* (Chenopodiaceae), *Calluna vulgaris* (Ericaceae), *Halimodendron halodendron* (Fabaceae) (Noyes, 2019).

Hosts: (Hemiptera: Aphididae), *Eulecanium franconicum* (Coccidae), *Caillardia sp., Cyanophila sp.* (Psyllidae) (Noyes, 2019).

Distribution: Armenia, Azerbaijan, Czech Republic, Czechoslovakia, Denmark, France, Hungary, Italy, Kazakhstan, Mongolia, Norway,

Poland, Russia, Slovakia, Spain, Tselinograd Obl., Turkey, USSR (Noyes, 2019).

Trechnites insidiosus (Crawford, 1910)
Publications with Turkey records: Japoshvili and Noyes (2005), Noyes (2019).
Associates: *Crataegus oxyacantha, Pyrus sp., Pyrus communis* (Rosaceae) (Noyes, 2019).
Hosts: *Cacopsylla sp., Cacopsylla pyri, Cacopsylla pyricola, Psylla sp., Psylla pyri, Psylla pyricola, Psylla pyrisuga, Psylla vasiljevi* (Hemiptera: Psyllidae) (Noyes, 2019).
Parasitoids: *Aphidencyrtus mamitus, Syrphophagus mamitus* (Hymenoptera: Encyrtidae) (Noyes, 2019).
Distribution: Armenia, Austria, Azerbaijan, Belarus, Canada, Czech Republic, Czechoslovakia, England, Europe, France, Georgia, Germany, Greece, Hungary, Iran, Italy, Kirgizia, Moldova, Mongolia, Nearctic, Poland, Russia, Serbia, Slovakia, Spain, Tadzhikistan, Turkey, Turkmenistan, Ukraine, United States of America, Uzbekistan (Noyes, 2019).

Genus *Trichomasthus* (Thomson, 1876)
Trichomasthus albimanus (Thomson, 1876)
Publications with Turkey records: Japoshvili and Karaca (2002), Noyes (2019), Öncüer (1991), Trjapitzin and Doğanlar (1997).
Associates: *Hedera sp.* (Araliaceae), *Cupressus sp., Thuja sp.* (Cupressaceae), *Robinia sp.* (Fabaceae), *Jasminum fructicans* (Oleaceae), *Festuca sp.* (Poaceae), *Crataegus sp., Prunus sp., Prunus divaricate, Spiraea sp.* (Rosaceae), *Rubus sp.* (Rubiaceae), *Citrus sp., Citrus unshiu* (Rutaceae) (Noyes, 2019).
Hosts. *Planchonia arabidis* (Hemiptera: Asterolecaniidae), *Coccus hesperidum, Coccus pseudomagnoliarum, Eriopeltis sp., Eriopeltis araxis, Eriopeltis festucae, Eriopeltis lichtensteinii, Eulecanium sp., Eulecanium bituberculatum, Eulecanium corni, Eulecanium tiliae, Luzulaspis luzulae, Palaeolecanium sp., Palaeolecanium bituberculatum, Parthenolecanium corni, Parthenolecanium pomeranicum, Phyllostroma myrtillin, Pulvinaria regalis, Pulvinaria ribesiae, Pulvinariella mesembryanthemi, Rhodococcus spiraeae* (Coccidae), *Eriococcus greeni* (Eriococcidae) (Noyes, 2019).
Distribution: Armenia, Azerbaijan, Belarus, Bulgaria, Croatia, Czechoslovakia, Denmark, Europe, Finland, France, Georgia, Germany, Hungary, Italy, Lithuania, Moldova, Mongolia, Montenegro, Poland,

Portugal, Russia, Serbia, Slovenia, Spain, Sweden, Transcaucasus, Turkey, Ukraine, United States of America, USSR, Yugoslavia (Federal Republic), Yugoslavia (pre-1991) (Noyes, 2019).

Trichomasthus cyaneus (Dalman, 1820)
Publications with Turkey records: Noyes (2019).
Associates: *Brachypodium sylvaticum* (Poaceae), *Rubus sp.* (Rubiaceae) (Noyes, 2019).
Hosts: *Eriopeltis sp., Eriopeltis agropyri, Eriopeltis araxis, Eriopeltis festucae, Eriopeltis lichtensteinii, Eulecanium sp., Palaeolecanium sp., Parafairmairia gracilis, Parthenolecanium sp., Pulvinaria vitis* (Hemiptera: Coccidae), *Anophococcus insignis, Eriococcus pseudinsignis, Greenisca brachypodii* (Eriococcidae) (Noyes, 2019).
Distribution: Armenia, Austria, Azerbaijan, Bulgaria, Czech Republic, Czechoslovakia, England, Europe, Finland, France, Georgia, Germany, Hungary, Ireland, Italy, Russia, Serbia, Slovakia, Spain, Sweden, Transcaucasus, Turkey, Ukraine, Uzbekistan, Yugoslavia (pre-1991) (Noyes, 2019).

Genus *Zaomma* (Ashmead, 1900)
Zaomma lambinus (Walker, 1838)
Publications with Turkey records: Japoshvili (2012), Japoshvili and Celik (2010), Japoshvili and Karaca (2002), Noyes (2019), Öncüer (1991).
Associates: *Aphelinus mytilaspidis* (Aphelinidae), *Pistacia vera* (Anacardiaceae), *Nerium oleander* (Apocynaceae), *Cocos nucifera* (Arecaceae), *Hippophae rhamnoides* (Elaeagnaceae), *Quercus sp.* (Fagaceae), *Juglans regia* (Juglandaceae), *Fraxinus sp.* (Oleaceae), *Pinus kochiana* (Pinaceae), *Platanus sp.* (Platanaceae), *Malus pumila, Prunus sp., Prunus mume, Rosa sp., Rosa canina* (Rosaceae), *Rubus sp., Rubus idaeus* (Rubiaceae), *Citrus sp., Murraya paniculate* (Rutaceae), *Populus sp., Populus alba, Populus gracilis, Salix sp., Salix caprea* (Salicaceae), *Tamarix sp.* (Tamaricaceae), *Ulmus sp.* (Ulmaceae) (Noyes, 2019).
Hosts: *Leucopis vorax* (Diptera: Chamaemyiidae), *Asterodiaspis variolosa, Asterolecanium sp., Asterolecanium variolosum* (Hemiptera: Asterolecaniidae), *Ceroplastes rubens, Eulecanium nigrofasciatum, Lecanium nigrofasciatum, Mesolecanium nigrofasciatum, Phyllostroma myrtilli* (Coccidae), *Acutaspis paulista, Adiscodiaspis tamaricicola, Aonidia sp., Aspidiotus sp., Aspidiotus destructor, Aspidiotus hederae, Aspidiotus juglansregiae, Aspidiotus nerii, Aspidiotus perniciosus, Aspidiotus spurcatus, Aulacaspis sp., Aulacaspis difficilis, Aulacaspis rosae, Chionaspis alnus, Chionaspis micropori, Chionaspis salicis,*

Chrysomphalus sp., Chrysomphalus aonidum, Chrysomphalus ficus, Diaspidiotus sp., Diaspidiotus juglansregiae, Diaspidiotus marani, Diaspidiotus ostreaeformis, Diaspidiotus perniciosus, Diaspidiotus prunorum, Diaspidiotus pyri, Diaspidiotus zonatus, Diaspis rosae, Dynaspidiotus abietis, Lepidosaphes sp., Lepidosaphes conchiformis, Lepidosaphes malicola, Lepidosaphes mesasiatica, Lepidosaphes pistaciae, Lepidosaphes tubulorum, Lepidosaphes ulmi, Melanaspis palustris, Melanaspis paulista, Mytilaspis sp., Mytilaspis pomorum, Mytilococcus ulmi, Nuculaspis abietis, Phenacaspis alnus, Prodiaspis tamaricicola, Pseudaulacaspis pentagona, Quadraspidiotus sp., Quadraspidiotus juglansregiae, Quadraspidiotus ostreaeformis, Quadraspidiotus perniciosus, Quadraspidiotus pyri, Quadraspidiotus schneideri, Quadraspidiotus zonatus, Tecaspis asiatica (Diaspididae), *Acanthococcus aceris, Eriococcus spurius, Gossyparia spuria* (Eriococcidae) (Noyes, 2019).

Parasitoids: *Aspidiotiphagus citrinus* (Aphelinidae) (Noyes, 2019).

Parasitoid hosts: *Aphytis mytilaspidis, Coccobius testaceus, Encarsia berlesei, Physcus testaceus* (Hymenoptera: Aphelinidae) (Noyes, 2019).

Distribution: Afrotropical, Algeria, Argentina, Armenia, Austria, Azerbaijan, Brazil, Bulgaria, Canada, Croatia, Czech Republic, Czechoslovakia, Denmark, England, Finland, France, Georgia, Germany, Hawaii, Hungary, India, Indonesia, Iran, Ireland, Italy, Japan, Korea, Lebanon, Mauritius, Moldova, Mongolia, Montenegro, Nearctic, Netherlands, New Zealand, North Africa, Norway, People's Republic of China, Peru, Philippines, Poland, Romania, Russia, Serbia, Slovakia, Spain, Sweden, Switzerland, Tunisia, Turkey, Ukraine, United States of America, USSR, Uzbekistan, Yugoslavia (pre-1991) (Noyes, 2019).

Subfamily Tetracneminae (Howard, 1892)

Genus *Aenasius* (Walker, 1846)

Aenasius arizonensis (Girault, 1915)

Publications with Turkey records: Çaliskan Keçe et al. (2018), Noyes (2019).

Associates: *Gazania rigens, Helianthus annuus, Parthenium hysterophorus* (Asteraceae), *Lagenaria siceraria* (Cucurbitaceae), *Cyamopsis tetragonoloba* (Fabaceae), *Abelmoschus esculentum, Abutilon hirtum, Abutilon indicum, Gossypium hirsutum, Hibiscus rosa-sinensis* (Malvaceae), *Cynodon dactylon* (Poaceae), *Datura stramonium, Lycopersicon esculentum, Solanum melongena, Solanum nigrum* (Solanaceae), *Lantana camara* (Verbenaceae) (Noyes, 2019).

Hosts: *Phenacoccus solenopsis, Pseudococcus longispinus* (Hemiptera: Pseudococcidae) (Noyes, 2019).
Parasitoids: *Prochiloneurus stenopterus* (Hymenoptera: Encyrtidae) (Noyes, 2019).
Distribution: India, Pakistan, People's Republic of China, Turkey, United States of America (Noyes, 2019).

Genus *Anagyrus* (Howard, 1896)
Anagyrus alienus (Japoshvili, 2012)
Publications with Turkey records: Japoshvili (2012), Noyes (2019).
Associates: Unknown (Noyes, 2019).
Hosts: Unknown (Noyes, 2019).
Distribution: Turkey (Noyes, 2019).

Anagyrus aligarhensis (Agarwal and Alam, 1959)
Publications with Turkey records: Japoshvili (2012), Japoshvili and Celik (2010), Japoshvili and Noyes (2005), Trjapitzin and Doğanlar (1997).
Associates: *Kiritshenkella sacchari* (Hemiptera: Pseudococcidae), *Teucrium capitatum* (Lamiaceae), *Morus alba* (Moraceae), *Saccharum officinarum* (Poaceae), *Citrus sp.* (Rutaceae) (Noyes, 2019).
Hosts: *Cheilomenes sexmaculata* (Coleoptera: Coccinellidae), *Adelosoma phragmitidis, Antonina graminis, Antonina purpurea, Dysmicoccus multivorus, Kiritshenkella sacchari, Nipaecoccus viridis, Phenacoccus aceris, Phenacoccus arthrophyte, Phenacoccus polyphagus, Pseudococcus comstocki, Pseudococcus copiosus, Pseudococcus cryptus, Pseudococcus multivorus, Saccharicoccus sacchari, Trionymus copiosus, Trionymus multivorus* (Hemiptera: Pseudococcidae) (Noyes, 2019).
Distribution: Armenia, Azerbaijan, Canary Islands, Croatia, Czech Republic, Czechoslovakia, England, Europe, France, Georgia, Greece, Hungary, India, Indonesia, Iran, Israel, Italy, Madeira, Mexico, Moldova, Nepal, People's Republic of China, Portugal, Romania, Russia, Spain, Turkey, Turkmenistan, United States of America, USSR, Uzbekistan, Yugoslavia (pre-1991) (Noyes, 2019).

Anagyrus dactylopii (Howard, 1898)
Publications with Turkey records: Kaydan and Japoshvili (2010), Noyes (2019).
Associates: *Ceroplastes sp.* (Coccidae), *Icerya aegyptiaca* (Margarodidae), *Annona squamosa* (Annonaceae), *Casuarina sp.* (Casuarinaceae), *Euphorbia hirta* (Euphorbiaceae), *Acacia sp., Dalbergia*

sissoo, Leucaena glauca, Tamarindus indica, Tephrosia purpurea (Fabaceae), *Abelmoschus esculentum, Gossypium sp., Gossypium hirsutum, Hibiscus rosa-sinensis* (Malvaceae), *Artocarpus heterophyllus, Ficus sp., Morus alba, Morus indica* (Moraceae), *Pinus elliotti* (Pinaceae), *Ziziphus sp., Ziziphus jujuba, Ziziphus mauritiana* (Rhamnaceae), *Citrus sp., Citrus aurantifolia, Citrus maxima, Citrus medica* (Rutaceae), *Tamarix sp.* (Tamaricaceae), *Vitis vinifera* (Vitaceae) (Noyes, 2019).
Hosts: *Cerococcus sp.* (Hemiptera: Cerococcidae), *Drosicha stebbengii* (Margarodidae), *Dactylopius vastator, Ferrisia virgata, Maconellicoccus hirsutus, Nipaecoccus sp., Nipaecoccus filamentosus, Nipaecoccus vastator, Nipaecoccus viridis, Oracella acuta, Phenacoccus solenopsis, Planococcus citri, Pseudococcus sp., Pseudococcus citriculus, Pseudococcus comstocki, Pseudococcus cryptus, Pseudococcus filamentosus, Rastrococcus sp., Rastrococcus cappariae, Rastrococcus iceryoides, Saccharicoccus sacchari, Trabutina crassispinosa, Trionymus sacchari* (Pseudococcidae) (Noyes, 2019).
Distribution: Australia, Barbados, Fiji, Hawaii, India, Indonesia, Iran, Iraq, Japan, Oman, People's Republic of China, Peru, Philippines, Taiwan, Thailand, Turkey (Noyes, 2019).

Anagyrus descriptus (Japoshvili, 2012)
Publications with Turkey records: Japoshvili (2012), Noyes (2019).
Associates: Unknown (Noyes, 2019).
Hosts: Unknown (Noyes, 2019).
Distribution: Turkey (Noyes, 2019).

Anagyrus kilinceri (Japoshvili, 2010)
Publications with Turkey records: Kaydan and Japoshvili (2010), Noyes (2019).
Associates: *Cynodon dactylon* (Poaceae) (Noyes, 2019).
Hosts: *Heterococcopsis opertus* (Hemiptera: Pseudococcidae) (Noyes, 2019).
Distribution: Turkey (Noyes, 2019).

Anagyrus matritensis (Mercet, 1921)
Publications with Turkey records: Japoshvili and Noyes (2005), Noyes (2019).
Associates: *Artemisia sp., Lactuca serriola* (Asteraceae), *Perovskia abrotanoides* (Lamiaceae), *Pinus halepensis* (Pinaceae) (Noyes, 2019).
Hosts: *Filippia oleae, Saissetia oleae* (Hemiptera: Coccidae), *Acanthococcus sp., Acanthococcus turkmenicus* (Eriococcidae), *Ferrisiana*

oleae, Peliococcus kimmericus, Trionymus sp., Trionymus multivorus (Pseudococcidae) (Noyes, 2019).

Distribution: Armenia, Austria, Azerbaijan, Canary Islands, Croatia, Czechoslovakia, Europe, Germany, Hungary, Iran, Kazakhstan, Moldova, Mongolia, People's Republic of China, Romania, Russia, Spain, Tadzhikistan, Transcaucasus, Tselinograd Obl., Tunisia, Turkey, Turkmenistan, Ukraine, Uzbekistan, Yugoslavia (pre-1991) (Noyes, 2019).

Anagyrus pseudococci (Girault, 1915)

Publications with Turkey records: Kaydan and Japoshvili (2010), Noyes (2019), Öncüer (1991), Trjapitzin and Doğanlar (1997).

Associates: *Anagyrus sp.* (Encyrtidae), *Mangifera indica* (Anacardiaceae), *Nerium oleander* (Apocynaceae), *Hedera sp.* (Araliaceae), *Cupressus sp., Juniperus sp.* (Cupressaceae), *Cycas revoluta* (Cycadaceae), *Codiaeum variegatum, Manihot esculenta* (Euphorbiaceae), *Persea americana* (Lauraceae), *Ficus sp., Ficus carica, Morus sp.* (Moraceae), *Jasminum fructicans* (Oleaceae), *Pittosporum sp.* (Pittosporaceae), *Platanus orientalis* (Platanaceae), *Saccharum officinarum* (Poaceae), *Citrus sp., Citrus limon, Citrus paradisi, Citrus reticulata, Citrus sinensis* (Rutaceae), *Theobroma cacao* (Sterculiaceae), *Tamarix sp.* (Tamaricaceae), *Vitis sp., Vitis vinifera* (Vitaceae) (Noyes, 2019).

Hosts: *Planchonia arabidis* (Hemiptera: Asterolecaniidae), *Antonina graminis, Dysmicoccus brevipes, Maconellicoccus hirsutus, Nipaecoccus filamentosus, Nipaecoccus vastator, Nipaecoccus viridis, Peliococcus mesasiaticus, Phenacoccus sp., Phenacoccus aceris, Phenacoccus gossypii, Phenacoccus herreni, Phenacoccus hirsutus, Phenacoccus manihoti, Planococcoides njalensis, Planococcus sp., Planococcus citri, Planococcus ficus, Planococcus vovae, Pseudococcus sp., Pseudococcus adonidum, Pseudococcus affinis, Pseudococcus calceolariae, Pseudococcus citri, Pseudococcus citriculus, Pseudococcus comstocki, Pseudococcus cryptus, Pseudococcus filamentosus, Pseudococcus gahani, Pseudococcus longispinus, Pseudococcus maritimus, Pseudococcus njalensis, Pseudococcus vovae, Rastrococcus iceryoides, Saccharicoccus sacchari, Trabutina crassispinosa* (Pseudococcidae) (Noyes, 2019).

Parasitoids: *Marietta picta* (Hymenoptera: Aphelinidae), *Prochiloneurus aegyptiacus* (Encyrtidae), *Pachyneuron sp., Pachyneuron coccorum* (Pteromalidae) (Noyes, 2019).

Distribution: Afrotropical, Argentina, Armenia, Azerbaijan, Bermuda, Brazil, Canary Islands, Chile, Congo, Croatia, Cyprus, Democratic Republic of Congo, Egypt, England, Europe, France, Gambia, Georgia,

Ghana, Guyana, Hawaii, Hungary, India, Iran, Iraq, Israel, Italy, Kenya, Mexico, Moldova, Montenegro, Nearctic, Neotropical, Netherlands, Nigeria, Pakistan, People's Republic of China, Peru, Portugal, Russia, Rwanda, Saudi Arabia, Senegal, South Africa, Spain, Sweden, Tadzhikistan, Taiwan, Transcaucasus, Turkey, Turkmenistan, Ukraine, United States of America, Uruguay, USSR, Uzbekistan, Yugoslavia (pre-1991) (Noyes, 2019).

Anagyrus schmuttereri (Ferriere, 1955)
Publications with Turkey records: Japoshvili (2012), Japoshvili and Celik (2010), Noyes (2019).
Associates: *Picea sp.* (Pinaceae) (Noyes, 2019).
Hosts: *Quadraspidiotus ostreaeformis* (Hemiptera: Diaspididae), *Paroudablis piceae, Phenacoccus piceae* (Pseudococcidae) (Noyes, 2019).
Distribution: Germany, Russia, Sweden, Turkey (Noyes, 2019).

Anagyrus schoenherri (Westwood, 1837)
Publications with Turkey records: Japoshvili and Noyes (2005), Noyes (2019).
Associates: *Fraxinus sp.* (Oleaceae), *Crataegus oxyacantha, Malus pumila* (Rosaceae), *Vitis vinifera* (Vitaceae) (Noyes, 2019).
Hosts: *Eulecanium sp., Eulecanium coryli, Eulecanium douglasi, Eupulvinaria sp., Eupulvinaria horri, Lecanium coryli, Pulvinaria sp., Pulvinaria betulae, Pulvinaria vitis* (Hemiptera: Coccidae), *Eriococcus sp.* (Eriococcidae), *Phenacoccus sp., Phenacoccus aceris, Phenacoccus azalea, Phenacoccus Fraxinus, Phenacoccus mespili, Phenacoccus pergandei, Phenacoccus polyphagus, Planococcus citri, Pseudococcus sp.* (Pseudococcidae) (Noyes, 2019).
Distribution: Armenia, Azerbaijan, Belarus, Bulgaria, Caucasus, Czech Republic, Czechoslovakia, Denmark, England, Estonia, Europe, Finland, France, Germany, Hungary, Iran, Italy, Japan, Moldova, Mongolia, Montenegro, Netherlands, People's Republic of China, Poland, Romania, Russia, Slovakia, Sweden, Turkey, Ukraine, United States of America, USSR, Yugoslavia (pre-1991) (Noyes, 2019).

Anagyrus securicornis (Domenichini, 1953)
Publications with Turkey records: Japoshvili (2012), Noyes (2019).
Associates: *Artemisia sp.* (Asteraceae), *Parietaria officinalis* (Urticaceae) (Noyes, 2019).
Distribution: Austria, Bulgaria, Czech Republic, Czechoslovakia, Denmark, England, Europe, Finland, Germany, Hungary, Italy,

Kazakhstan, Moldova, Netherlands, Norway, Romania, Russia, Tselinograd Obl., Turkey, Ukraine, Wales (Noyes, 2019).

Anagyrus tamaricicola (Trjapitzin, 1968)
Publications with Turkey records: Kaydan and Japoshvili (2010), Noyes (2019).
Associates: *Tamarix sp., Tamarix hohenackeri, Tamarix ramosissima* (Tamaricaceae) (Noyes, 2019).
Hosts: *Trabutina crassispinosa, Trabutina mannipara* (Hemiptera: Pseudococcidae) (Noyes, 2019).
Distribution: Georgia, Russia, Turkey, Turkmenistan (Noyes, 2019).

Genus *Callipteroma* (Motschulsky, 1863)
Callipteroma sexguttata (Motschulsky, 1863)
Publications with Turkey records: Japoshvili and Noyes (2005), Noyes (2019).
Associates: *Juniperus thurifera* (Cupressaceae), *Salvia lavandulifolia* (Lamiaceae), *Cocculus trilobus* (Menispermaceae), *Pinus halepensis* (Pinaceae), *Digitaria pentzii, Paspalum dilatatum, Saccharum officinarum* (Poaceae) (Noyes, 2019).
Hosts: *Birendracoccus saccharifolii, Heliococcus summervillei* (Hemiptera: Pseudococcidae) (Noyes, 2019).
Distribution: Armenia, Australia, Azerbaijan, Bulgaria, Burkina Faso, Cameroon, Croatia, Czechoslovakia, Europe, Gambia, Georgia, Ghana, Greece, Hungary, India, Indonesia, Israel, Japan, Kazakhstan, Madagascar, Malawi, Malaysia, Moldova, Mongolia, Nigeria, Pakistan, Papua New Guinea, People's Republic of China, Portugal, Russia, Saudi Arabia, Singapore, Slovakia, South Africa, Spain, Sri Lanka, Thailand, Transcaucasus, Tselinograd Obl., Turkey, Turkmenistan, Ukraine, USSR, Yugoslavia (pre-1991), Zambia, Zimbabwe (Noyes, 2019).

Genus *Charitopus* (Foerster, 1856)
Charitopus bulentyasari (Japoshvili, 2012)
Publications with Turkey records: Japoshvili (2012), Noyes (2019).
Associates: Unknown (Noyes, 2019).
Hosts: Unknown (Noyes, 2019).
Distribution: Turkey (Noyes, 2019).

Charitopus desertus (Myartseva, 1981)
Publications with Turkey records: Kaydan and Japoshvili (2010), Noyes (2019).

Associates: Unknown (Noyes, 2019).
Hosts: Unknown (Noyes, 2019).
Distribution: Turkey, Turkmenistan, Uzbekistan (Noyes, 2019).

Charitopus eristoi (Japoshvili, 2010)
Publications with Turkey records: Kaydan and Japoshvili (2010), Noyes (2019).
Associates: *Cynodon dactylon* (Poaceae) (Noyes, 2019).
Hosts: *Heterococcopsis opertus* (Hemiptera: Pseudococcidae) (Noyes, 2019).
Distribution: Turkey (Noyes, 2019).

Charitopus fulviventris Foerster, 1860
Publications with Turkey records: Japoshvili (2012), Japoshvili and Noyes (2005), Noyes (2019).
Associates: *Artemisia herba-alba* (Asteraceae), *Lithodora fruticose* (Boraginaceae), *Koeleria glauca* (Poaceae) (Noyes, 2019).
Distribution: Algeria, Armenia, Azerbaijan, Canary Islands, Croatia, Czech Republic, Czechoslovakia, Europe, Finland, Germany, Hungary, India, Kazakhstan, Moldova, Mongolia, North Africa, Norway, Palearctic, People's Republic of China, Romania, Russia, Serbia, Slovakia, South Africa, Spain, Sweden, Tadzhikistan, Transcaucasus, Tselinograd Obl., Turkey, Turkmenistan, Ukraine, United Arab Emirates, United Kingdom, USSR (Noyes, 2019).

Charitopus ismailkaracai (Japoshvili, 2012)
Publications with Turkey records: Japoshvili (2012), Noyes (2019).
Associates: Unknown (Noyes, 2019).
Hosts: Unknown (Noyes, 2019).
Distribution: Turkey (Noyes, 2019).

Genus *Dicarnosis* (Mercet, 1921)
Dicarnosis sugonjaevi (Myartseva, 1977)
Publications with Turkey records: Japoshvili and Noyes (2005), Noyes (2019).
Associates: Unknown (Noyes, 2019).
Hosts: Unknown (Noyes, 2019).
Distribution: Kazakhstan, Mongolia, Tadzhikistan, Tselinograd Obl., Turkey, Turkmenistan (Noyes, 2019).

Dicarnosis superba (Mercet, 1921)
Publications with Turkey records: Japoshvili and Noyes (2005), Noyes (2019).
Associates: *Rosmarinus officinalis* (Lamiaceae), *Pinus halepensis* (Pinaceae) (Noyes, 2019).
Distribution: Algeria, Azerbaijan, Bulgaria, Canary Islands, Croatia, Cyprus, France, Greece, Italy, Kuwait, Spain, Turkey, Yugoslavia (pre-1991) (Noyes, 2019).

Genus *Dinocarsis* (Foerster, 1856)
Dinocarsis hemiptera (Dalman, 1820)
Publications with Turkey records: Japoshvili and Noyes (2005), Noyes (2019), Trjapitzin and Doğanlar (1997).
Associates: *Koeleria vallesiaca* (Poaceae) (Noyes, 2019).
Hosts: *Rhodania flava* (Hemiptera: Pseudococcidae) (Noyes, 2019).
Distribution: Armenia, Azerbaijan, Czech Republic, Czechoslovakia, Europe, France, Germany, Italy, Mongolia, Netherlands, People's Republic of China, Slovakia, Spain, Sweden, Turkey, Ukraine, United Kingdom (Noyes, 2019).

Genus *Dusmetia* (Mercet, 1921)
Dusmetia ceballosi (Mercet, 1921)
Publications with Turkey records: Japoshvili (2012), Japoshvili and Celik (2010), Noyes (2019).
Hosts: *Euripersia brevispina, Euripersia europea, Fonscolombia europaea* (Hemiptera: Pseudococcidae) (Noyes, 2019).
Distribution: Australia, Azerbaijan, Croatia, Czech Republic, Palearctic, Romania, Serbia, Slovakia, Slovenia, Spain, Transcaucasus, Turkey, United Kingdom, United States of America, USSR (Noyes, 2019).

Genus *Ericydnus* (Haliday, 1832)
Ericydnus aeneus (Nikolskaya, 1952)
Publications with Turkey records: Japoshvili and Noyes (2005), Noyes (2019).
Associates: *Lactuca sp.* (Asteraceae), *Juniperus thurifera* (Cupressaceae), *Quercu coccifera* (Fagaceae), *Rosmarinus officinalis, Sideritis sp.* (Lamiaceae) (Noyes, 2019).
Hosts: *Trionymus multivorus, Trionymus perrisii* (Hemiptera: Pseudococcidae) (Noyes, 2019).
Distribution: Azerbaijan, Europe, Norway, Portugal, Romania, Spain, Transcaucasus, Turkey, Uzbekistan (Noyes, 2019).

Ericydnus apterogenes (Mayr, 1876)
Publications with Turkey records: Japoshvili (2012), Noyes (2019).
Associates: Unknown (Noyes, 2019).
Hosts: Unknown (Noyes, 2019).
Distribution: Austria, Czech Republic, France, Georgia, Germany, Norway, Romania, Slovakia, Sweden, Turkey, United Kingdom (Noyes, 2019).

Ericydnus caudatus (Erdos, 1957)
Publications with Turkey records: Kaydan and Japoshvili (2010), Noyes (2019).
Associates: *Hordeum sp.* (Poaceae) (Noyes, 2019).
Hosts: *Trionymus aberrans* (Hemiptera: Pseudococcidae) (Noyes, 2019).
Distribution: Armenia, Czechoslovakia, Hungary, Moldova, Romania, Russia, Turkey (Noyes, 2019).

Ericydnus robustior (Mercet, 1921)
Publications with Turkey records: Japoshvili and Celik (2010), Japoshvili and Noyes (2005), Noyes (2019).
Associates: *Lactuca sp.* (Asteraceae), *Chenopodium album* (Chenopodiaceae), *Sideritis sp.* (Lamiaceae), *Cynodon dactylon* (Poaceae) (Noyes, 2019).
Hosts: *Dysmicoccus multivorus, Mirococcus inermis, Peliococcopsis priesneri, Trionymus multivorus, Trionymus perrisii* (Hemiptera: Pseudococcidae) (Noyes, 2019).
Distribution: Azerbaijan, Croatia, Czech Republic, Czechoslovakia, Europe, France, Georgia, Germany, Montenegro, Norway, Portugal, Spain, Sweden, Turkey, Turkmenistan, Uzbekistan, Yugoslavia (pre-1991) (Noyes, 2019).

Ericydnus sipylus (Walker, 1837)
Publications with Turkey records: Japoshvili (2012), Japoshvili and Celik (2010), Kaydan and Japoshvili (2010), Noyes (2019).
Associates: *Carduus pycnocephalus, Onopordum nervosum* (Asteraceae), *Lepidium subulatum* (Brassicaceae), *Ephedra nebrodensis* (Gnetaceae), *Salvia sp.* (Lamiaceae), *Hordeum leporinum* (Poaceae), *Tamarix canariensis* (Tamaricaceae), *Vitis vinifera* (Vitaceae) (Noyes, 2019).
Hosts: *Heliococcus bohemicus, Heterococcus nudus, Heterococcus pulverarius, Phenacoccus pumilis* (Hemiptera: Pseudococcidae) (Noyes, 2019).

Distribution: Austria, Azerbaijan, Canary Islands, Croatia, Czech Republic, England, Europe, France, Georgia, Germany, Hungary, Madeira, Mongolia, Romania, Russia, Spain, Sweden, Switzerland, Turkey, Turkmenistan, Ukraine, United States of America (Noyes, 2019).

Genus *Leptomastidea* (Mercet, 1916)
Leptomastidea abnormis (Girault, 1915)
Publications with Turkey records: Japoshvili and Noyes (2005), Noyes (2019), Öncüer (1991).
Associates: *Icerya pattersoni* (Hemiptera: Margarodidae), *Annona sp.* (Annonaceae), *Phoenix dactylifera* (Arecaceae), *Aristolochia sp.* (Aristolochiaceae), *Stephanotis sp.* (Asclepiadaceae), *Begonia semporflorens* (Begoniaceae), *Jacaranda sp.* (Bignoniaceae), *Maytenus sp.* (Celastraceae), *Juniperus phoeniciae, Juniperus thurifera* (Cupressaceae), *Codiaeum variegatum* (Euphorbiaceae), *Inga inga* (Fabaceae), *Ficus sp.* (Moraceae), *Psidium guajava* (Myrtaceae), *Olea europaea* (Oleaceae), *Passiflora quadrangularis* (Passifloraceae), *Saccharum officinarum* (Poaceae), *Coffea sp., Coffea arabica* (Rubiaceae), *Citrus sp., Citrus sinensis* (Rutaceae), *Vitis sp., Vitis vinifera* (Vitaceae) (Noyes, 2019).
Hosts: *Coccus viridis, Saissetia oleae* (Hemiptera: Coccidae), *Dysmicoccus brevipes, Dysmicoccus ryani, Ferrisia virgata, Ferrisiana virgata, Maconellicoccus hirsutus, Phenacoccus gossypii, Planococcoides njalensis, Planococcus sp., Planococcus citri, Planococcus ficus, Planococcus kenyae, Planococcus kraunhiae, Planococcus vitis, Pseudococcus sp., Pseudococcus calceolariae, Pseudococcus citri, Pseudococcus citriculus, Pseudococcus comstocki, Pseudococcus cryptus, Pseudococcus ficus, Pseudococcus gahani, Pseudococcus kenyae, Pseudococcus kraunhiae, Pseudococcus longispinus, Pseudococcus maritimus, Pseudococcus njalensis, Saccharicoccus sacchari, Trionymus sacchari* (Pseudococcidae) (Noyes, 2019).
Parasitoids: *Chartocerus niger* (Hymenoptera: Signiphoridae) (Noyes, 2019).
Distribution: Afrotropical, Algeria, Argentina, Australia, Azerbaijan, Belgium, Bermuda, Brazil, Canada, Canary Islands, Chile, Costa Rica, Cuba, Cyprus, Dominica, Dominican Republic, Egypt, England, Ethiopia, Europe, France, Georgia, Ghana, Greece, Hawaii, India, Iran, Israel, Italy, Japan, Kenya, Mexico, Morocco, Nearctic, Neotropical, Netherlands, New Zealand, Palearctic, People's Republic of China, Peru, Puerto Rico, Senegal, Sierra Leone, South Africa, Spain, Sweden, Tadzhikistan, Turkey, Turkmenistan, United States of America, USSR, Uzbekistan (Noyes, 2019).

Genus *Leptomastix* (Foerster, 1856)

Leptomastix algirica (Trjapitzin, 1989)

Publications with Turkey records: Japoshvili and Noyes (2005), Noyes (2019).

Associates: *Persea americana* (Lauraceae), *Citrus sp.* (Rutaceae) (Noyes, 2019).

Hosts: *Maconellicoccus hirsutus, Planococcus citri, Pseudococcus cryptus* (Hemiptera: Pseudococcidae) (Noyes, 2019).

Distribution: Algeria, Cyprus, Egypt, Greece, Israel, Turkey (Noyes, 2019).

Leptomastix dactylopii (Howard, 1885)

Publications with Turkey records: Japoshvili (2012), Japoshvili and Celik (2010), Japoshvili and Noyes (2005), Noyes (2019), Özkan et al. (2001), Süzer et al. (1992).

Associates: *Cryptolaemus montrouzieri* (Coleoptera: Coccinellidae), *Camponotus nylanderi, Linepithema humile, Tapinoma erraticum* (Hymenoptera: Formicidae), *Mangifera indica* (Anacardiaceae), *Annona sp., Annona muricata, Annona reticulata* (Annonaceae), *Nerium oleander* (Apocynaceae), *Phoenix dactylifera* (Arecaceae), *Stephanotis sp.* (Asclepiadaceae), *Combretum splendens* (Combretaceae), *Ipomoea batatas* (Convolvulaceae), *Cucurbita sp., Cucurbita moschata, Sechium edule* (Cucurbitaceae), *Dioscorea sp., Dioscorea cayennensis* (Dioscoreaceae), *Codiaeum variegatum* (Euphorbiaceae), *Acacia sp., Erythrina lysistemon* (Fabaceae), *Coleus blumei, Epimeredi indicus, Solenostemon scutellarioides* (Lamiaceae), *Persea americana* (Lauraceae), *Punica granatum* (Lythraceae), *Azadirachta indica* (Meliaceae), *Tambourissa sp.* (Monimiaceae), *Ficus sp., Ficus carica* (Moraceae), *Psidium sp., Psidium guajava* (Myrtaceae), Paeonia sp. (Paeoniaceae), Saccharum officinarum (Poaceae), Ziziphus mauritiana (Rhamnaceae), Coffea sp., *Coffea arabica, Coffea robusta* (Rubiaceae), *Citrus sp., Citrus aurantifolia, Citrus latifolia, Citrus limon, Citrus sinensis* (Rutaceae), *Manilkara zapota* (Sapotaceae), *Solanum tuberosum* (Solanaceae), *Theobroma cacao* (Sterculiaceae), *Vitis vinifera* (Vitaceae), *Planococcus citri* (Hemiptera: Pseudococcidae) (Noyes, 2019).

Hosts: *Aspidiella hartii* (Hemiptera: Diaspididae), *Birendracoccus saccharifolii, Dactylopius destructor, Delottococcus proteae, Delottococcus quaesitus, Dysmicoccus brevipes, Ferrisia virgata, Ferrisiana virgata, Formicococcus njalensis, Nipaecoccus vastator, Nipaecoccus viridis, Phenacoccus gossypii, Phenacoccus hargreavesi, Phenacoccus madeirensis, Phenacoccus saccharifolii, Phenacoccus*

solani, Planococcoides lamabokensis, Planococcoides njalensis, Planococcus sp., Planococcus aemulor, Planococcus citri, Planococcus ficus, Planococcus inamabilis, Planococcus kenyae, Planococcus kraunhiae, Planococcus lamabokensis, Planococcus lilacinus, Planococcus minor, Planococcus njalensis, Planococcus pacificus, Planococcus vitis, Planococcus vovae, Pseudococcus sp., Pseudococcus adonidum, Pseudococcus bukobensis, Pseudococcus calceolariae, Pseudococcus citri, Pseudococcus comstocki, Pseudococcus concavocerarii, Pseudococcus gahani, Pseudococcus hargreavesi, Pseudococcus inamabilis, Pseudococcus kenyae, Pseudococcus lilacinus, Pseudococcus longispinus, Pseudococcus maritimus, Pseudococcus njalensis, Pseudococcus occiduus, Pseudococcus proteae, Pseudococcus saccharifolii, Pseudococcus solani (Pseudococcidae) (Noyes, 2019).
Parasitoids: *Coccophagus sp., Coccophagus gurneyi, Marietta javensis, Marietta leopardine, Promuscidea unfasciativentris* (Hymenoptera: Aphelinidae), *Coccidoctonus trinidadensis, Gahaniella tertia, Prochiloneurus dactylopii, Prochiloneurus indicus, Prochiloneurus seini* (Encyrtidae), *Aprostocetus purpureus* (Eulophidae), *Pachyneuron coccorum* (Pteromalidae) (Noyes, 2019).
Distribution: Afrotropical, Antilles, Australia, Barbados, Belgium, Benin, Bermuda, Brazil, Cameroon, Canada, Cape Verde Islands, Caribbean, Chile, Costa Rica, Cuba, Cyprus, Dominican Republic, Egypt, England, Eritrea, Ethiopia, Europe, France, Gabon, Gambia, Georgia, Ghana, Greece, Grenada, Guadeloupe, Guyana, Hawaii, Honduras, India, Iran, Israel, Italy, Ivory Coast, Jamaica, Japan, Kenya, Libya, Madagascar, Malawi, Mauritius, Morocco, Nearctic, Netherlands, Nigeria, Pakistan, Peru, Philippines, Portugal, Puerto Rico, Saudi Arabia, Senegal, Seychelles, Sierra Leone, Somalia, South Africa, Spain, St Vincent & Grenadines, Sweden, Taiwan, Tanzania, Togo, Trinidad & Tobago, Turkey, Turkmenistan, Uganda, United States of America, USSR, Uzbekistan, Yugoslavia (pre-1991), Zambia, Zimbabwe (Noyes, 2019).

Leptomastix ephyra (Noyes and Hayat, 1994)
Publications with Turkey records: Japoshvili (2012), Noyes (2019).
Associates: Unknown (Noyes, 2019).
Hosts: Unknown (Noyes, 2019).
Distribution: India, Turkey (Noyes, 2019).

Leptomastix flava (Mercet, 1921)
Publications with Turkey records: Japoshvili and Noyes (2005), Noyes (2019) Trjapitzin and Doğanlar (1997).

Associates: *Pistacia vera* (Anacardiaceae), *Acacia sp., Prosopis sp.* (Fabaceae) (Noyes, 2019).
Parasitoids: *Prochiloneurus bolivari* (Hymenoptera: Encyrtidae) (Noyes, 2019).
Hosts: *Eulecanium rugulosum* (Hemiptera: Coccidae), *Dysmicoccus multivorus, Nipaecoccus filamentosus, Nipaecoccus vastator, Nipaecoccus viridis, Peliococcus kimmericus, Peliococcus mesasiaticus, Phenacoccus aceris, Planococcus sp., Planococcus citri, Planococcus ficus, Pseudococcus sp., Pseudococcus citriculus, Pseudococcus comstocki, Pseudococcus cryptus, Pseudococcus filamentosus, Pseudococcus multivorus, Trionymus multivorus* (Pseudococcidae) (Noyes, 2019).
Distribution: Afrotropical, Armenia, Azerbaijan, Bulgaria, Croatia, Egypt, Europe, Georgia, Iran, Israel, Moldova, Montenegro, Romania, Russia, Senegal, Spain, Transcaucasus, Turkey, Turkmenistan, Ukraine, United States of America, Yugoslavia (pre-1991) (Noyes, 2019).

Leptomastix tanasijtshuki (Sharkov, 1983)
Publications with Turkey records: Kaydan and Japoshvili (2010), Noyes (2019).
Associates: *Euphorbia sp*. (Euphorbiaceae) (Noyes, 2019).
Hosts: *Puto pilosellae* (Hemiptera: Pseudococcidae) (Noyes, 2019).
Distribution: Norway, Russia, Turkey (Noyes, 2019).

Genus *Monodiscodes* (Hoffer, 1953)
Monodiscodes dimorphus (Mercet, 1921)
Publications with Turkey records: Japoshvili and Noyes (2005), Noyes (2019), Öncüer (1991), Trjapitzin and Doğanlar (1997).
Associates: *Artemisia herba-alba* (Asteraceae) (Noyes, 2019).
Distribution: Armenia, Azerbaijan, France, Kazakhstan, Russia, Spain, Tselinograd Obl., Turkey, Turkmenistan, Uzbekistan (Noyes, 2019).

Monodiscodes intermedius (Mayr, 1876)
Publications with Turkey records: Noyes (2019), Trjapitzin and Doğanlar (1997).
Associates: *Lepidium subulatum* (Brassicaceae), *Herniaria sp.* (Caryophyllaceae), *Juniperus thurifera* (Cupressaceae), *Caragana grandiflora* (Fabaceae), *Elymus repens, Festuca sp.* (Poaceae) (Noyes, 2019).
Hosts: *Eriopeltis festucae, Rhizopulvinaria armeniaca* (Hemiptera: Coccidae), *Rhizococcus agropyri* (Eriococcidae) (Noyes, 2019).

Distribution: Armenia, Austria, Azerbaijan, Bulgaria, Canary Islands, Croatia, Czech Republic, Czechoslovakia, Europe, Georgia, Greece, Hungary, Iran, Italy, Kazakhstan, Macedonia, Moldova, Montenegro, Romania, Russia, Slovakia, Spain, Transcaucasus, Tselinograd Obl., Turkey, Turkmenistan, Ukraine, Uzbekistan, Yugoslavia (pre-1991) (Noyes, 2019).

Genus *Ooencyrtus* (Ashmead, 1900)
Ooencyrtus kuvanae (Howard, 1910)
Publications with Turkey records: Avci (2009), Japoshvili and Noyes (2005), Noyes (2019), Öncüer (1991), Trjapitzin and Doğanlar (1997).
Associates: *Casuarina sp.* (Casuarinaceae), *Robinia pseudoacacia, Vigna unguiculata* (Fabaceae), *Quercus sp., Quercus suber* (Fagaceae), *Pinus sp.* (Pinaceae), *Pyrus communis* (Rosaceae), *Populus sp., Salix sp.* (Salicaceae) (Noyes, 2019).
Hosts: *Anoplocnemis curvipes, Clavigralla tomentosicollis* (Hemiptera: Coreidae), *Lycorma deliculata* (Fulgoridae), *Dendrolimus spectabilis, Malacosoma neustria* (Lepidoptera: Lasiocampidae), *Dasychira pinicola, Hemerocampa leucostigma, Leucoma salicis, Lymantria dispar, Lymantria fumida, Lymantria monachal, Lymantria xylina, Nygmia phaeorrhoea, Orgyia sp., Orgyia antiqua, Orgyia gonostigma, Porthetria dispar, Stilpnotia salicis* (Lymantriidae), *Eriogyna pyretorum, Hemileuca oliviae, Saturnia pyretorum* (Saturniidae), *Chrysoperla carnea, Suarius fedtschenkoi* (Neuroptera: Chrysopidae) (Noyes, 2019).
Parasitoid hosts: *Apanteles melanoscelus, Cotesia melanoscela* (Hymenoptera: Braconidae), *Anastatus disparis* (Eupelmidae) (Noyes, 2019).
Parasitoids: *Tyndarichus navae* (Encyrtidae), *Anastatus disparis* (Eupelmidae) (Noyes, 2019).
Distribution: Afrotropical, Algeria, Austria, Bosnia Hercegovina, Bulgaria, Canada, Croatia, Czech Republic, Czechoslovakia, Europe, France, Georgia, Germany, Iran, Italy, Japan, Kazakhstan, Kirgizia, Korea, Macedonia, Moldova, Morocco, Nearctic, Nigeria, North Africa, People's Republic of China, Portugal, Romania, Russia, Serbia, Slovakia, Spain, Taiwan, Tselinograd Obl., Tunisia, Turkey, Ukraine, United States of America, USSR, Uzbekistan, Yugoslavia (pre-1991) (Noyes, 2019).

Ooencyrtus masii (Mercet, 1921)
Publications with Turkey records: Avci (2003), Noyes (2019).
Associates: *Pistacia sp.* (Anacardiaceae), *Quercus sp.* (Fagaceae), *Cedrus libani* (Pinaceae) (Noyes, 2019).

Hosts: *Malacosoma neustria* (Lepidoptera: Lasiocampidae), *Lymantria dispar, Ocneria terebynthina, Porthetria dispar* (Lymantriidae), *Thaumetopoea processionea, Traumatocampa ispartaensis* (Notodontidae) (Noyes, 2019).
Distribution: Bulgaria, Europe, Iran, Italy, Kazakhstan, Morocco, People's Republic of China, Spain, Tselinograd Obl., Turkey, Ukraine, USSR (Noyes, 2019).

Ooencyrtus neustriae (Mercet, 1925)
Publications with Turkey records: Noyes (2019), Özbek and Coruh (2010).
Associates: *Quercus ilex* (Fagaceae) (Noyes, 2019).
Hosts: *Malacosoma castrensis, Malacosoma neustria* (Lepidoptera: Lasiocampidae), *Dicranura vinula* (Notodontidae) (Noyes, 2019).
Distribution: France, Italy, Portugal, Romania, Spain, Turkey, Ukraine, USSR (Noyes, 2019).

Ooencyrtus pityocampae (Mercet, 1921)
Publications with Turkey records: Avci (2000; 2003), Japoshvili and Noyes (2005), Melan (1994), Noyes (2019), Öncüer (1991), Trjapitzin and Doğanlar (1997).
Associates: *Tetrastichus servadei* (Hymenoptera: Eulophidae), *Cedrus sp., Cedrus libani, Pinus sp., Pinus brutia, Pinus halepensis, Pinus insignis, Pinus nigra, Pinus pinaster, Pinus pinea, Pinus radiata* (Pinaceae), *Triticum aestivum* (Poaceae) (Noyes, 2019).
Hosts: *Enoplops cornutus, Gonocerus juniperi, Leptoglossus occidentalis* (Hemiptera: Coreidae), *Aela rostrata, Aelia sp., Aelia acuminata, Aelia cognata, Aelia germari, Carpocoris sp., Carpocoris fuscispinus, Carpocoris pudicus, Dolycoris sp., Dolycoris baccarum, Dolycoris numidicus, Eurydema sp., Eurydema oleracea, Eurydema ornatum, Eurydema ventral, Graphosoma lineatum, Graphosoma semipunctatum, Nezara viridula, Odontotarsus grammicus, Piezodorus lituratus, Rhaphigaster nebulosa, Solenostethium lynceum, Stenozygium coloratum* (Pentatomidae), *Pyrrhocoris apterus* (Pyrrhocoridae), *Rhinocoris iracundus* (Reduviidae), *Eurygaster sp., Eurygaster austriaca, Eurygaster integriceps, Eurygaster maura, Ventocoris nigellae* (Scutelleridae), *Bombyx mori* (Lepidoptera: Bombycidae), *Bombycopsis bufo, Dendrolimus pini, Eriogaster philippi, Pachypasa otus, Taragama repanda* (Lasiocampidae), *Phalera bucephaloides, Thaumetopoea bonjeani, Thaumetopoea pinivora, Thaumetopoea pityocampa, Thaumetopoea wilkinsoni, Traumatocampa ispartaensis, Traumatocampa*

pityocampa (Notodontidae), *Acherontia atropos, Agrius convolvuli, Amorpha populi, Hyloicus pinastri* (Sphingidae) (Noyes, 2019).
Parasitoids: *Baryscapus transversalis, Pediobius bruchicida* (Hymenoptera: Eulophidae) (Noyes, 2019).
Distribution: Albania, Algeria, Bosnia Hercegovina, Bulgaria, Croatia, Cyprus, Europe, France, Greece, Iran, Israel, Italy, Mexico, Morocco, Poland, Portugal, Spain, Tunisia, Turkey, Yugoslavia (pre-1991) (Noyes, 2019).

Ooencyrtus tardus (Ratzeburg, 1844)
Publications with Turkey records: Noyes (2019), Öncüer (1991).
Hosts: *Malacosoma neustria* (Lepidoptera: Lasiocampidae), *Ocneria dispar, Porthetria dispar* (Lymantriidae) (Noyes, 2019).
Distribution: Belarus, Bulgaria, Czechoslovakia, Europe, Georgia, Germany, Hungary, Kazakhstan, Lithuania, Moldova, Netherlands, North Africa, Poland, Romania, Russia, Tselinograd Obl., Turkey, Ukraine, USSR (Noyes, 2019).

Ooencyrtus telenomicida (Vassiliev, 1904)
Publications with Turkey records: Japoshvili and Noyes (2005), Noyes (2019), Öncüer (1991), Öncüer and Kivan (1995), Simsek *et al*. (1994), Trjapitzin and Doğanlar (1997).
Associates: *Rhopalomyia sp.* (Diptera: Cecidomyiidae), *Aphanurus scutellaris* (Hymenoptera: Scelionidae), *Pistacia vera* (Anacardiaceae), *Artemisia austriaca* (Asteraceae), *Capparis spinosa* (Capparidaceae), *Triticum aestivum* (Poaceae), *Crataegus monogyna, Prunus domestica, Prunus persica, Prunus spinosa* (Rosaceae) (Noyes, 2019).
Hosts: *Coreus marginatus, Coriomeris denticulatus, Enoplops cornutus, Gonocerus acuteangulatus, Gonocerus juniperi, Mesocerus marginatus* (Hemiptera: Coreidae), *Acrosternum sp., Aelia sp., Aelia acuminata, Aelia cognata, Aelia germari, Aelia rostrata, Aspongopus viduatus, Brachynema germarii, Carpocoris fuscispinus, Carpocoris pudicus, Dolycoris sp., Dolycoris baccarum, Dolycoris numidicus, Dolycoris penicillatus, Eurydema festivum, Eurydema ornatum, Graphosoma sp., Graphosoma lineatum, Graphosoma semipunctatum, Nezara viridula, Palomena sp., Palomena prasine, Solenostethium lynceum, Stenozygium coloratum* (Pentatomidae), *Pyrrhocoris apterus* (Pyrrhocoridae), *Rhinocoris iracundus* (Reduviidae), *Eurygaster sp., Eurygaster austriaca, Eurygaster integriceps, Eurygaster maura* (Scutelleridae), *Microphanurus eurydemae* (Hymenoptera: Scelionidae), *Taragama repanda* (Lepidoptera: Lasiocampidae), *Euproctis phaeorrhoea* (Lymantriidae), *Thaumetopoea*

pityocampa (Notodontidae), *Iphiclides podalirius* (Papilionidae), *Amorpha populi* (Sphingidae) (Noyes, 2019).
Parasitoid hosts: *Asolcus sp., Asolcus semistriatus, Telenomus sokolowi* (Hymenoptera: Scelionidae) (Noyes, 2019).
Distribution: Armenia, Azerbaijan, Bulgaria, Croatia, Czech Republic, Czechoslovakia, Egypt, Europe, France, Georgia, Germany, Greece, Hungary, India, Iran, Israel, Italy, Kazakhstan, Madeira, Montenegro, Morocco, Norway, Pakistan, People's Republic of China, Portugal, Romania, Russia, Serbia, Slovakia, Spain, Sweden, SyriaTselinograd Obl., Turkey, Turkmenistan, Ukraine, USSR, Uzbekistan, Yugoslavia (pre-1991) (Noyes, 2019).

Ooencyrtus zoeae (Guerrieri and Samra, 2018)
Publications with Turkey records: Noyes (2019).
Associates: *Capparis spinosa* (Capparidaceae) (Noyes, 2019).
Hosts: *Stenozygium coloratum* (Hemiptera: Pentatomidae), *Bombyx mori* (Lepidoptera: Bombycidae) (Noyes, 2019).
Distribution: Israel, Turkey (Noyes, 2019).

Genus *Rhopus* (Foerster, 1856)
Rhopus flavidus (Mercet, 1921)
Publications with Turkey records: Japoshvili (2012), Japoshvili and Celik (2010), Japoshvili and Noyes (2005), Kaydan and Japoshvili (2010), Noyes (2019).
Associates: *Juniperus thurifera* (Cupressaceae), *Cynodon dactylon* (Poaceae) (Noyes, 2019).
Hosts: *Pseudococcus sp.* (Hemiptera: Pseudococcidae) (Noyes, 2019).
Distribution: Czech Republic, Czechoslovakia, Hungary, Spain, Turkey (Noyes, 2019).

Rhopus trjapitzini (Myartseva, 1982)
Publications with Turkey records: Kaydan and Japoshvili (2010), Noyes (2019).
Hosts: *Phenacoccus tergrigorianae* (Hemiptera: Pseudococcidae) (Noyes, 2019).
Distribution: Georgia, Turkey, Turkmenistan (Noyes, 2019).

Genus *Tetracnemus* (Westwood, 1837)
Tetracnemus heydeni (Mayr, 1876)
Publications with Turkey records: Japoshvili and Noyes (2005), Noyes (2019).

Associates: *Caragana grandiflora* (Fabaceae), *Elytrigia intermedia* (Poaceae) (Noyes, 2019).
Hosts: *Rhizopulvinaria armeniaca* (Hemiptera: Coccidae) (Noyes, 2019).
Distribution: Austria, Azerbaijan, Czech Republic, Czechoslovakia, Europe, Georgia, Germany, Greece, Hungary, Kazakhstan, Moldova, Mongolia, Romania, Serbia, Slovakia, Spain, Tselinograd Obl., Turkey, Yugoslavia (pre-1991) (Noyes, 2019).

Tetracnemus terminassianae (Herthevtzian, 1978)
Publications with Turkey records: Japoshvili and Noyes (2005), Noyes (2019).
Associates: Unknown (Noyes, 2019).
Hosts: Unknown (Noyes, 2019).
Distribution: Armenia, Azerbaijan, France, Greece, Turkey (Noyes, 2019).

References

Avcı, Mustafa. 2000. "Investigations on structure of egg-batches, parasitism and egg layinh habits of Thaumetopoea pityocampa (Den. & Schiff.) (Lepidoptera: Thaumetopoeidae) in various regions of Turkey". *Turkish Journal of Entomology*, 24(3):167-178.

—. 2003. "Parasitism of egg-batches of the cedar processionary moth Traumatocampa ispartaensis in Turkey". *Phytoparasitica*, 31(2):118-123.

—. 2009. "Parasitoid complex and new host plants of the gypsy moth, Lymantria dispar L. in the Lakes District, Turkey". *Journal of Animal and Veterinary Advances*, 8(7):1402-1405.

Bulut, Ebrucan, and Göçmen, Hüseyin. 2000. "Pests and their natural enemies on greenhouse vegetables in Antalya". *Bulletin. Section Regionale Ouest Palaearctique, Organisation Internationale de Lutte Biologique*, 23(1):33-37.

Çalışkan Asime Filiz, and Hayat, Mohammad, and Ulusoy, Mehmet Rifat, and Kaydan, Mehmet Bora. 2016. "Parasitoids (Hymenoptera: Encyrtidae) of an invasive mealybug *Phenacoccus solenopsis* Tinsley (Hemiptera: Pseudococcidae) in Turkey". *Turkish Journal of Entomology*, 40(2):133-148.

Çalışkan Keçe, and Asime Filiz, and Kahya, Doğancan, and Hayat, Mohammad, and Ulusoy, Mehmet Rifat. 2018. "A new record of a parasitoid (Hymenoptera: Encyrtidae) of an invasive meaybug *Phenacoccus solenopsis* Tinsley (Hemiptera: Pseudococcidae) from Turkey". *Türkiye Biyolojik Mücadele Dergisi*, 9(1):31-37.

Japoshvili, George O., and Karaca, İsmail. 2004. "New records of encyrtid parasitoids of *Kermes palestiniensis* Balachowsky (Hemiptera: Kermesidae), with the description of a new species of *Blastothrix* Mayr (Hymenoptera: Encyrtidae) from Turkey". *Entomological News*, 114(4):187-191.

Japoshvili, George, and Çelik, H. 2010. "Fauna of Encyrtidae, parasitoids of coccids in Golcuk Natural Park". *Entomologia Hellenica*, 19(2), 132–136.

Japoshvili, George, and Karaca, İsmail, and Demirozer, Ozan. 2002. "The globose scale population regulation and their parasitoid complex in Isparta province (Turkey)". (Abstract 031) Abstracts, VIIth European Congress of Entomology, October 7-13-2002, Thessaloniki, Greece pp.93, Hellenic Entomological Society.

Japoshvili, George, and Karaca, İsmail. 2002. "Coccid (Homoptera: Coccoidea) species of Isparta Province, and their parasitoids from Turkey and Georgia". *Turkish Journal of Zoology*, 26(4):371-376.

Japoshvili, George, and Noyes, John S. 2005. "Checklist and new data on Encyrtidae of Transcaucasia and Turkey (Hymenoptera: Chalcidoidea)". *Zoosystematica Rossica*, 14(1):135-144.

Japoshvili, George. 2007. "New data on species of *Syrphophagus* (Hymeoptera: Encyrtidae) from Transcaucasia and Turkey". *Annals of the Entomological Society of America*, 100 (5):683-687.

—. 2012. "New data on encyrtid (Hymenoptera: Chalcidoidea: Encyrtidae) parasitoids of Coccoids (Hemiptera: Coccoidea) from Turkey, with description of five new species". *Entomologica Fennica*, 23(2):72-82.

Karaca, İsmail, and Japoshvili, George, and Demirozer, Ozan. 2003. "The chalcid parasitoid complex (Hymenoptera: Chalcidoidea) associated with the globose scale (*Sphaerolecanium prunasttri* Fonscolombe) (Hemiptera: Coccoidea) in Isparta Province, Turkey and some European countries". *Zeitschrift für Pflanzenkrankheiten und Pflanzenschutz*, 110(5):505-511.

Karaca, İsmail, and Uygun, Nedim. 1990. "Doğu Akdeniz Bolgesi turuncgillerinde zararli *Aonidiella aurantii* (Maskell) (Homoptera: Diaspididae)'nin dogal dusmanlari ve bunlarin degisik turuncgil tur ve cesitlerinde populasyon gelismesinin saptanmasi". *Turkiye Il. Biyolojik Mucadele Kongresi*, 26-29 Eylul 1990, Ankara, pp. 97-108.

Kaydan, Mehmet Bora, and Japoshvili, George. 2010. "The aphelinid and encyrtid parasitoids (Hymenoptera: Chalcidoidea) of Pseudococcidae (Hemiptera: Coccoidea) in the Van Lake basin of Turkey". *Turkish Journal of Entomology*, 34(4):465- 476.

Melan, Kadir. 1994. "Sunn pest species and their egg parasitoids in Thrace". Proceedings of the Third Turkish National Congress of Biological Control, Izmir, 25-28 January1994, pp.147-154.

Noyes, John S. 2019. *Universal Chalcidoidea Database*. Available at: www.nhm.ac.uk/chalcidoids (accessed January 2022).

Öncüer, Cezmi, 1974. "Ege bölgesinde turunçgil bahçelerinde zararli Coccus (Homoptera: Coccidae) türlerinin taninmasi yayilisi ve dogal düsmanlari üzerinde arastirmalar". *Bitki Koruma Bülteni*, Supplement 1:1-59.

Öncüer, Cezmi, and Kıvan, Müjgan. 1995. "Determination and distribution of *Eurygaster* Lap. species and the biology and natural enemies of *Eurygaster integriceps* Put. (Heteroptera: Scutelleridae) in Tekirdag and surroundings". Turkish Journal of Agriculture and Forestry, 19 (4): 223-230.

Öncüer, Cezmi. 1991. "A catologue of the parasites and predators of insect pests of Turkey". *Ege University Agricultural Faculty Publication 505.* İzmir. 354 pp.

Özbek, Hikmet, and Çoruh, Saliha. 2010. "Egg parasitoids of *Malacosoma neustria* (Linnaeus, 1758) (Lepidoptera: Lasiocampidae) in Erzurum province of Turkey". *Turkish Journal of Entomology*, 34(4): 551-560.

Özkan, Ahmet, and Gürol, Müzehher, and Uysal, Hüseyin, and Çelik, Göçeri, and Akteke, Ş. Ali, and Aytekin, Hüseyin, and Arslan, Mustafa, and Kaplan, Mehmet, and Dalka, Yasin, and Akyel, Erhan, and Tuncer, Hüseyin. 2001. "Integrated pest management in citrus orchards in Antalya (1995-1999)". *Bitki Koruma Bülteni*, 41(3/4):135-166.

Şengonca, Çetin, and Uygun, Nedim, and Karaca, İsmail. 1998. "Primary studies on the parasitoid fauna of Coccoidea in cultivated and non-cultivated areas in the east Mediterranean region of Turkey". *Anzeiger für Schädlingskunde, Pflanzenschutz und Umweltschutz*, 71(7):128-131.

Şimşek, Necip, and Güllü, Mustafa, and Yasarbaş, M. 1994. "Investigation on the sunn pest (*Eurygaster integriceps* Put.) natural enemies and their effectiveness in the mmediterranean region". *Proceedings of the Third Turkish National Congress of Biological Control*, Izmir, 25-28 January1994, pp. 155-164.

Stringer, Danielle N., and Mantel, Sarah, and Jennings, John T., and Austin, Andrew D. 2012. "Family Encyrtidae", Australian Centre for Evolutionary Biology and Biodiversity, and the School of Earth and Environmental Science. The University of Adelaide, Australia

(Available from: https://www.researchgate.net/publication/ 262008500 _Family_ENCYRTIDAE).

Süzer, Turgut, and Aytaş, Mehtap, and Yumruktepe, Rabia. 1992. "Chemical experiment on citrus white fly (*Dialeurodes citri* Ashmead), citrus red mite (*Panonychus citri* McGregor) and citrus rust mite (*Phyllocoptruta oleivora* Ashmead) in the Mediterranean region". *Zirai Mücadele Araştırma Yıllığı*, Turkey, No 22-23:61-63.

Trjapitzin, Vladimir A., and Doğanlar, Mikdat. 1997. "A review of encyrtids (Hymenoptera, Encyrtidae) of Turkey". Entomologicheskoe Obozrenie, 76 (1):213-222.

Ülgentürk, Selma, and Kaydan, Mehmet Bora, and Zeki, Cevdet, and Toros, Seval. 2001. "*Sphaerolecanium prunastri* Boyer de Fonscolombe (Hemiptera: Coccidae): distribution, host plants and natural enemies in the Turkish Lake district". *Bolletino Zoologia Agraria di Bachicoltura Ser*. II, 33:3, 357-363.

Ülgentürk, Selma, and Toros, Seval. 2001. "Natural enemies of the oak scale insect, *Eulecanium ciliatum* (Douglas) (Hemiptera: Coccidae) in Turkey". *Entomologica*, Bari 33:219-224.

Ülgentürk, Selma. 2001. "Parasitoids and predators of Coccidae (Homoptera: Coccoidea) species on ornamental plants in Ankara, Turkey". *Acta Phytopathologica et Entomologica Hungarica*, 36(3-4):369-375.

Uygun, Nedim, and Şekeroğlu, Erdal. 1987. "Potential use of imported natural enemies for biological control in Çukurova". *Türkiye I. Entomoloji Kongresi Bildirileri*, 13-16 Ekim 1987, Ege Universitesi, Bornova, Izmir, pp.553-562.

Zeki, Huriye, and Özdem, Ayşe, and Bozkurt, Vildan. 1999. "Natural enemies of larvae of *Depressaria cf. daucivorella* Rag. (Lepidoptera: Oecophoridae) on anise (*Pimpinella anisum* L.) and the parasitism rates on Burdur province of Turkey". Bitki Koruma Bülteni, 39(1/2):35-43.

CHAPTER SIX

FAMILY: EUCHARITIDAE WALKER, 1846

The Eucharitidae is a relatively small family that is infrequently collected. They appear to be most closely related to the Perilampidae family. Both families have a free-living, mobile first instar larva known as planidium. The family is characterized by the pronotum being small and invisible in dorsal view, the thorax being dorsally vaulted in lateral view, the metasoma being distinctly petiolate, and the scutellum often being armed with a pair of conspicuous backwards-projecting forked processes.

Eucharitids are all parasitoids of ants, except for the aberrant subfamily Echthrodapinae, which parasitizes twig-nesting bees. Their biology is rather unusual. Eggs are laid in very large numbers (up to 15,000) on vegetation. The first instar larvae are actively moving planidia. They attach themselves to passing ants and are carried back to the ants' nest. Once inside the nest, the planidia actively search out ant larvae upon which they feed. Males can often be found swarming around the openings of ant nests, waiting for the newly emerged females to leave the nest (Stringer *et al.*, 2012).

The family currently includes 55 genera and 423 species placed in three subfamilies, and in Turkey, there is one subfamily, two genera and six species (Noyes, 2019).

Subfamily Eucharitinae (Walker, 1846)
Genus *Eucharis* (Latreille, 1804)
Eucharis adscendens (Fabricius, 1787)
Publications with Turkey records: Noyes (2019), Öncüer (1991).
Associates: *Falcaria vulgaris, Peucedanum oreoselinum* (Apiaceae), *Medicago sativa* (Fabaceae) (Noyes, 2019).
Hosts: *Formica sp., Formica rufa, Messor sp., Messor barbarous* (Hymenoptera: Formicidae) (Noyes, 2019).

Distribution: Austria, Belarus, Bosnia Hercegovina, Croatia, Czech Republic, Czechoslovakia, Europe, Germany, Hungary, Italy, Kazakhstan, Lithuania, Moldova, Poland, Romania, Russia, Serbia, Slovakia, Transcaucasus, Tselinograd Obl., Turkey, Ukraine, United Kingdom, USSR (Noyes, 2019).

Eucharis anatolica (Boucek, 1952)
Publications with Turkey records: Noyes (2019), Öncüer (1991).
Associates: Unknown (Noyes, 2019).
Hosts: Unknown (Noyes, 2019).
Distribution: Turkey (Noyes, 2019).

Eucharis punctata (Foerster, 1859)
Publications with Turkey records: Noyes (2019), Öncüer (1991).
Hosts: *Messor sp., Messor barbarous, Messor concolor* (Hymenoptera: Formicidae) (Noyes, 2019).
Distribution: Algeria, Austria, Bulgaria, Croatia, Egypt, Europe, Macedonia, Saudi Arabia, Tunisia, Turkey, Yugoslavia (pre-1991) (Noyes, 2019).

Eucharis turca (Nikolskaya, 1952)
Publications with Turkey records: Noyes (2019), Öncüer (1991).
Associates: Unknown (Noyes, 2019).
Hosts: Unknown (Noyes, 2019).
Distribution: Turkey (Noyes, 2019).

Genus *Stilbula* (Spinola, 1811)
Stilbula cyniformis (Rossi, 1792)
Publications with Turkey records: Noyes (2019), Öncüer (1991).
Associates: *Cerceris arenaria* (Sphecidae), *Picris sp.* (Asteraceae), *Betula sp.* (Betulaceae), *Cladrastis sp.* (Fabaceae), *Castanea sp., Quercus sp.* (Fagaceae), *Morus sp.* (Moraceae) (Noyes, 2019).
Hosts: *Camponotus sp., Camponotus aethiops, Camponotus herculeanus, Camponotus maculatus, Camponotus marginatus, Rhytidoponera violacea* (Hymenoptera: Formicidae) (Noyes, 2019).
Distribution: Austria, Azerbaijan, Bosnia Hercegovina, Caucasus, Croatia, Czech Republic, Czechoslovakia, Europe, France, Germany, Hungary, Italy, Japan, Kazakhstan, Korea, Macedonia, Moldova, People's Republic of China, Romania, Russia, Slovakia, Tselinograd Obl., Turkey, Ukraine, United States of America (Noyes, 2019).

Stilbula vitripennis (Masi, 1934)
Publications with Turkey records: Noyes (2019), Öncüer (1991).
Associates: Unknown (Noyes, 2019).
Hosts: Unknown (Noyes, 2019).

References

Noyes, John S. 2019. *Universal Chalcidoidea Database*. Available at: ww.nhm.ac.uk/chalcidoids (accessed January 2022).

Öncüer, Cezmi. 1991. "A catologue of the parasites and predators of insect pests of Turkey". *Ege University Agricultural Faculty Publication 505*. İzmir. 354 pp.

Stringer, Danielle N., and Mantel, Sarah, and Jennings, John T., and Austin, Andrew D. 2012. "Family EUCHARITIDAE", Australian Centre for Evolutionary Biology and Biodiversity, and the School of Earth and Environmental Science. The University of Adelaide, Australia (Available from: https://www.researchgate.net /publication/268522095_Family_EUCHARITIDAE).

CHAPTER SEVEN

FAMILY: EULOPHIDAE WESTWOOD, 1829

The Eulophidae is a large family of ecto- and endo-parasitoid wasps, characterized by their small size (0.5 -2.0 mm). The group is found in all zoogeographical regions and in all land ecosystems. Individual species often occur in great numbers, significantly contributing to the biodiversity. Most Eulophidae are entomophagous, serving as parasitoids of a wide range of hosts from more than 100 families in the orders Lepidoptera, Diptera, Hymenoptera and Coleoptera. Some species are arephytophagous either gall-makers or inquilines within galls made by other insects.

The definition of Eulophidae is relatively straightforward: the tarsi in both sexes have four segments, the antennae generally have up to four funicular segments, and the gaster has a petiole, creating a distinct narrowing between propodeum and gaster (Yefremova, 2008).

The family currently includes 297 genera and 4472 species in four subfamilies. There are five subfamilies, 48 genera and 245 species in Turkey (Noyes, 2019).

Subfamily Entedoninae (Förster, 1856)
Genus *Ceranisus* (Walker, 1842)
Ceranisus adiyamanensis (Doğanlar and Doğanlar, 2013)
Publications with Turkey records: Doğanlar and Doğanlar (2013), Noyes (2019).
Associates: *Astragalus sp.* (Fabaceae) (Noyes, 2019).
Distribution: Turkey (Noyes, 2019).

Ceranisus amanosus (Doğanlar, Gumovsky and Doğanlar, 2009)
Publications with Turkey records: Doğanlar *et al*. (2009), Noyes (2019).
Associates: Unknown (Noyes, 2019).
Hosts: Unknown (Noyes, 2019).

Distribution: Turkey (Noyes, 2019).

Ceranisus antalyacus (Triapitsyn, 2004)
Publications with Turkey records: Noyes (2019).
Associates: *Aeolothrips glorious, Aeolothrips versicolor, Melanthrips ficalbii, Melanthrips pallidior (Aeolothripidae), Haplothrips andresi, Haplothrips reuteri, Neoheegeria dalmatica* (Phlaeothripidae), *Ceratothrips versicolor, Frankliniella occidentalis, Isoneurothrips australsi, Oxythrips ajugae, Taeniothrips inconsequens, Taeniothrips meridionalis, Thrips minutissimus* (Thripidae), *Arbutus andrachne* (Ericaceae), *Pyrus communis* (Rosaceae) (Noyes, 2019).
Hosts: *Thrips major* (Thysanoptera: Thripidae) (Noyes, 2019).
Distribution: Turkey (Noyes, 2019).

Ceranisus atcanacus (Doğanlar and Doğanlar, 2013)
Publications with Turkey records: Doğanlar and Doğanlar (2013), Noyes (2019).
Associates: *Sinapis sp.* (Brassicaceae) (Noyes, 2019).
Distribution: Turkey (Noyes, 2019).

Ceranisus bozovaensis (Doğanlar, 2003)
Publications with Turkey records: Doğanlar (2003), Doğanlar and Triapitsyn (2007), Noyes (2019).
Associates: *Haplothrips tritici* (Thysanoptera: Phlaeothripidae) (Noyes, 2019).
Distribution: Turkey (Noyes, 2019).

Ceranisus filizinancae (Doğanlar and Doğanlar, 2014)
Publications with Turkey records: Doğanlar and Doğanlar (2014), Noyes (2019).
Associates: Unknown (Noyes, 2019).
Hosts: Unknown (Noyes, 2019).
Distribution: Turkey (Noyes, 2019).

Ceranisus hirsutus (Doğanlar and Triapitsyn, 2007)
Publications with Turkey records: Doğanlar and Triapitsyn (2007), Noyes (2019).
Associates: Unknown (Noyes, 2019).
Hosts: Unknown (Noyes, 2019).
Distribution: Turkey (Noyes, 2019).

Ceranisus karacadagi (Doğanlar and Doğanlar, 2013)
Publications with Turkey records: Doğanlar and Doğanlar (2014), Noyes (2019).
Associates: Unknown (Noyes, 2019).
Hosts: Unknown (Noyes, 2019).
Distribution: Turkey (Noyes, 2019).

Ceranisus karkamisus (Doğanlar and Doğanlar, 2013)
Publications with Turkey records: Doğanlar and Doğanlar (2013), Noyes (2019).
Associates: *Narcissus pseudonarsissus* (Amaryllidaceae), *Triticum aestivum* (Poaceae) (Noyes, 2019).
Distribution: Turkey (Noyes, 2019).

Ceranisus kocaki (Doğanlar and Doğanlar, 2014)
Publications with Turkey records: Doğanlar and Doğanlar (2014), Noyes (2019).
Associates: Unknown (Noyes, 2019).
Hosts: Unknown (Noyes, 2019).
Distribution: Turkey (Noyes, 2019).

Ceranisus komurcukurus (Doğanlar and Doğanlar, 2013)
Publications with Turkey records: Doğanlar and Doğanlar (2013), Noyes (2019).
Associates: Unknown (Noyes, 2019).
Hosts: Unknown (Noyes, 2019).
Distribution: Turkey (Noyes, 2019).

Ceranisus menes (Walker, 1839)
Publications with Turkey records: Doğanlar and Doğanlar (2014), Doğanlar et al. (2009), Doğanlar and Triapitsyn (2007), Noyes (2019).
Associates: *Kakothrips robustus, Odontothrips meliloti, Taeniothrips primulae, Thrips tabaci* (Thysanoptera: Thripidae), *Rhus laurina* (Anacardiaceae), *Elaeis guineensis* (Arecaceae), *Rapistrum rugosum* (Brassicaceae), *Garcinia sp.* (Clusiaceae), *Cucumis sp.* (Cucurbitaceae), *Calluna sp.* (Ericaceae), *Acacia terminalis, Cajanus cajan, Medicago sativa, Melilotus alba, Tephrosia candida, Vigna angularis* (Fabaceae), *Rosmarinus officinalis* (Lamiaceae), *Persea americana, Umbellularia californica* (Lauraceae), *Allium cepa, Allium sphaerocephalon* (Liliaceae), *Digitalis sp., Linaria sp.* (Scrophulariaceae), *Lycopersicon esculentum,*

Solanum melongena, Solanum tuberosum (Solanaceae), *Vitis sp.* (Vitaceae) (Noyes, 2019).
Hosts: *Braueriella phillyreae* (Diptera: Cecidomyiidae), *Callirhytis glandium* (Hymenoptera: Cynipidae), *Ceratothripoides claratris, Frankliniella sp., Frankliniella intonsa, Frankliniella occidentalis, Frankliniella schultzei, Isoneurothrips fullawayi, Kakothrips sp., Kakothrips pisivorus, Kakothrips robustus, Megalurothrips sjostedti, Megalurothrips usitatus, Microcephalothrips abdominalis, Pseudodendrothrips mori, Scirtothrips sp., Scirtothrips citri, Scirtothrips perseae, Taeniothrips sp., Taeniothrips alliorum, Taeniothrips longistylus, Thrips sp., Thrips flavus, Thrips palmi, Thrips tabaci, Toxothrips ricinus* (Thysanoptera:Thripidae) (Noyes, 2019).
Distribution: Argentina, Australia, Benin, Brazil, Bulgaria, Canada, Canary Islands, Chile, Czech Republic, Czechoslovakia, Democratic Republic of Congo (Zaire), Dominican Republic, England, France, Germany, Greece, Hawaii, Hungary, India, Indonesia, Ireland, Israel, Italy, Japan, Kenya, Korea, Malaysia, Mexico, Micronesia, Moldova, Montenegro, Netherlands, New Zealand, Papua New Guinea, Philippines, Romania, Russia, Slovakia, Spain, Sri Lanka, Sweden, Switzerland, Taiwan, Thailand, Turkey, Ukraine, United States of America, USSR (Noyes, 2019).

Ceranisus nizipus (Doğanlar and Doğanlar, 2013)
Publications with Turkey records: Doğanlar and Doğanlar (2013), Noyes (2019).
Associates: *Narcissus pseudonarsissus* (Amaryllidaceae), *Sinapis arvensis* (Brassicaceae), *Triticum aestivum* (Poaceae) (Noyes, 2019).
Distribution: Turkey (Noyes, 2019).

Ceranisus oezdikmeni (Doğanlar and Doğanlar, 2014)
Publications with Turkey records: Doğanlar and Doğanlar (2014), Noyes (2019).
Associates: Unknown (Noyes, 2019).
Hosts: Unknown (Noyes, 2019).
Distribution: Turkey (Noyes, 2019).
Ceranisus oguzeliensis (Doğanlar and Doğanlar, 2013)
Publications with Turkey records: Doğanlar and Doğanlar (2013), Noyes (2019).
Associates: Unknown (Noyes, 2019).
Hosts: Unknown (Noyes, 2019).
Distribution: Turkey (Noyes, 2019).

Ceranisus onuri (Doğanlar, 2010)
Publications with Turkey records: Doğanlar et al. (2010), Noyes (2019).
Associates: *Asphodeline damascene* (Asphodelaceae) (Noyes, 2019).
Hosts: Thrips meridionalis (Thysanoptera: Thripidae) (Noyes, 2019).
Distribution: Turkey (Noyes, 2019).

Ceranisus pacuvius (Walker, 1838)
Publications with Turkey records: Doğanlar et al. (2011), Doğanlar and Triapitsyn (2007), Noyes (2019).
Associates: *Aelothrips intermedius, Aptinothrips rufus, Chirothrips hamatus, Frankliniella intonsa, Kakothrips robustus, Odontothrips meliloti* (Thysanoptera: Thripidae), *Lathyrus tuberosus, Lens culinaris, Medicago sativa, Melilotus alba, Sarothamnus scoparius* (Fabaceae), *Fagus sylvatica* (Fagaceae), *Picea excelsa* (Pinaceae) (Noyes, 2019).
Hosts: *Aleyrodes lonicerae, Asterobemisia carpini* (Hemiptera: Aleyrodidae), *Kakothrips pisivorus, Kakothrips robustus* (Thysanoptera: Thripidae) (Noyes, 2019).
Parasitoid hosts: *Angitia cerophaga, Pimpla alternans* (Hymenoptera: Ichneumonidae) (Noyes, 2019).
Distribution: Afghanistan, Austria, Czech Republic, Czechoslovakia, Denmark, Germany, Hungary, Ireland, Moldova, Romania, Sweden, Switzerland, Turkey, Ukraine, United Kingdom, Yugoslavia (pre-1991) (Noyes, 2019).

Ceranisus planitianus (Erdos, 1966)
Publications with Turkey records: Doğanlar and Doğanlar (2014), Noyes (2019).
Associates: *Lepidium subulatum* (Brassicaceae), *Juniperus phoeniciae* (Cupressaceae), *Rosmarinus officinalis* (Lamiaceae), *Tamarix canariensis* (Tamaricaceae) (Noyes, 2019).
Distribution: Canada, Czech Republic, Czechoslovakia, Hungary, Israel, Moldova, Slovakia, Spain, Turkey, Ukraine, United Kingdom, United States of America (Noyes, 2019).

Ceranisus sekilinensis (Doğanlar and Doğanlar, 2013)
Publications with Turkey records: Doğanlar and Doğanlar (2013), Noyes (2019).
Associates: Unknown (Noyes, 2019).
Hosts: Unknown (Noyes, 2019).
Distribution: Turkey (Noyes, 2019).

Ceranisus sincanus (Doğanlar and Doğanlar, 2013)
Publications with Turkey records: Doğanlar and Doğanlar (2013), Noyes (2019).
Associates: *Sinapis sp*. (Brassicaceae) (Noyes, 2019).
Distribution: Turkey (Noyes, 2019).

Genus *Chrysocharis* (Foerster, 1856)
Chrysocharis amanus (Walker, 1839)
Publications with Turkey records: Noyes (2019), Öncüer (1991).
Associates: *Acer campestre, Acer pseudoplatanus* (Aceraceae), *Cotinus coggygria* (Anacardiaceae), *Crataegus sp., Crataegus laevigata, Crataegus monogyna, Prunus spinosa, Rosa sp., Rosa sempervirens, Sorbus aria* (Rosaceae), *Ulmus sp.* (Ulmaceae) (Noyes, 2019).
Hosts: *Leucoptera lotella* (Lepidoptera: Lyonetiidae), *Nepticula sp., Nepticula aceris, Nepticula marginicolella, Nepticula plagicolella, Nepticula promissa, Nepticula sanguisorbae, Stigmella sp., Stigmella aceris, Stigmella Ariella, Stigmella catharticella, Stigmella centifoliella, Stigmella crataegella, Stigmella freyella, Stigmella fulvomacula, Stigmella marginicolella, Stigmella plagicolella, Stigmella pomella, Stigmella promissa, Stigmella prunetorum, Stigmella pygmaeella, Stigmella rhamnella, Stigmella sanguisorbae, Stigmella speciosa* (Nepticulidae) (Noyes, 2019).
Distribution: Algeria, Bulgaria, Czech Republic, Czechoslovakia, England, France, Greece, Hungary, Moldova, Montenegro, Netherlands, Poland, Romania, Serbia, Slovakia, Sweden, Turkey, USSR, Yugoslavia (pre-1991) (Noyes, 2019).

Chrysocharis debussyi (Hansson, 1985)
Publications with Turkey records: Noyes (2019).
Associates: Unknown (Noyes, 2019).
Hosts: Unknown (Noyes, 2019).
Distribution: Greece, Turkey (Noyes, 2019).

Chrysocharis gemma (Walker, 1839)
Publications with Turkey records: Bulut and Gocmen (2000), Noyes (2019).
Associates: *Ilex sp., Ilex aquifolium* (Aquifoliaceae), *Senecio squalidus* (Asteraceae), *Betula pendula* (Betulaceae), *Lonicera sp.* (Caprifoliaceae), *Sarothamnus scoparius* (Fabaceae), *Quercu coccifera, Quercus sp., Quercus ilex, Quercus robur, Quercus suber* (Fagaceae), *Olea europaea, Phillyrea*

sp. (Oleaceae), *Abies alba, Abies nordmanniana* (Pinaceae), *Brachypodium sp.* (Poaceae), *Tamarix canariensis* (Tamaricaceae) (Noyes, 2019).
Parasitoid hosts: *Colastes braconius, Pholetesor sp.* (Hymenoptera: Braconidae), *Microterys mayri* (Encyrtidae), *Sympiesis sericeicornis* (Eulophidae) (Noyes, 2019).
Hosts: *Agromyza sp., Cerodontha pygmaea, Chromatomyia horticola, Napomyza lateralis, Napomyza lonicerella, Phytomyza sp., Phytomyza ilicis, Phytomyza messaniella* (Diptera: Agromyzidae), *Asterolecanium ilicicola* (Hemiptera: Asterolecaniidae), *Silliana lhommei* (Hymenoptera: Tenthredinidae), *Lithocolletis messaniella, Metriochroa latifoliella, Phyllonorycter joviella, Phyllonorycter messaniella* (Lepidoptera: Gracillariidae), *Crinopteryx familiella* (Incurvariidae), *Leucoptera spartifoliella* (Lyonetiidae), *Cosmopterix pulchrimella* (Momphidae), *Nepticula heringella, Stigmella aurella, Stigmella heringella, Stigmella suberivora* (Nepticulidae), *Argyresthia fundella, Prays oleae* (Yponomeutidae) (Noyes, 2019).
Distribution: Austria, Canada, Canary Islands, Central African Republic, Channel Islands, Croatia, Cyprus, Europe, France, Germany, Greece, Hawaii, Israel, Italy, Madeira, Montenegro, Morocco, Nearctic, Netherlands, New Zealand, Portugal, Russia, Spain, Sweden, Switzerland, Turkey, United Kingdom, Yugoslavia (pre-1991) (Noyes, 2019).

Chrysocharis idyia (Walker, 1839)
Publications with Turkey records: Noyes (2019), Öncüer (1991).
Associates: *Sisymbrium irio* (Brassicaceae), *Pinus halepensis* (Pinaceae), *Triticum sp.* (Poaceae), *Reseda lutea* (Resedaceae), *Malus sp.* (Rosaceae), *Tamarix canariensis* (Tamaricaceae) (Noyes, 2019).
Hosts: *Agromyza nana, Phytobia artemisiae, Phytobia gyrans, Phytobia morionella, Phytobia pygmaea, Phytobia verbasci, Phytomyza myosotica, Phytomyza ranunculi* (Diptera: Agromyzidae), *Phyllonorycter blancardella, Phyllonorycter coryllfoliella, Phyllonorycter medicaginella* (Lepidoptera: Gracillariidae) (Noyes, 2019).
Distribution: Bulgaria, Czech Republic, Czechoslovakia, England, Europe, Germany, Hungary, Moldova, Netherlands, Poland, Russia, Serbia, Spain, Sweden, Switzerland, Turkey (Noyes, 2019).

Chrysocharis illustris (Graham, 1963)
Publications with Turkey records: Noyes (2019), Yefremova et al. (2010).
Associates: *Artemisia vulgaris, Tussilago farfara* (Asteraceae) (Noyes, 2019).

Hosts: *Philophylla heracleid, Prionimera cognata, Trypeta artemisiae* (Diptera: Tephritidae) (Noyes, 2019).
Distribution: Czech Republic, Czechoslovakia, Denmark, England, Finland, Germany, Poland, Sweden, Turkey (Noyes, 2019).

Chrysocharis laomedon (Walker, 1839)
Publications with Turkey records: Noyes (2019), Öncüer (1991).
Associates: *Acer sp., Acer campestre, Acer platanoides, Acer pseudoplatanus, Acer rubrum* (Aceraceae), *Sonchus sp.* (Asteraceae), *Alnus sp., Alnus glutinosa, Alnus hirsute, Betula sp., Carpinus betulus, Corylus sp., Corylus avellana, Ostrya virginiana* (Betulaceae), *Lonicera periclymenum, Viburnum sp., Viburnum furcatum, Viburnum lantana* (Caprifoliaceae), *Elaeagnus umbellate* (Elaeagnaceae), *Medicago sativa, Robinia sp.* (Fabaceae), *Castanea sp., Fagus sp., Fagus sylvatica, Quercus sp., Quercus cerris, Quercus dentata, Quercus mongolica, Quercus petraea, Quercus robur* (Fagaceae), *Carya illinoinensis* (Juglandaceae), *Syringa sp.* (Oleaceae), *Crataegus sp., Crataegus arnoldi, Crataegus laevigata, Crataegus monogyna, Crataegus oxyacanthoides, Malus sp., Malus pumila, Malus sieboldii, Malus sylvestris, Prunus sp., Prunus avium, Prunus padus, Prunus pennsylvanica, Prunus serotina, Prunus spinosa, Pyrus sp., Pyrus communis, Pyrus malus, Sorbus sp., Sorbus aucuparia, Sorbus intermedia, Sorbus torminalis* (Rosaceae), *Populus tremula, Salix sp., Salix alba, Salix caprea, Salix cinerea, Salix purpurea, Salix viminalis* (Salicaceae), *Tilia americana* (Tiliaceae), *Ulmus sp., Ulmus americana, Ulmus glabra* (Ulmaceae), *Stigmella sp.* (Lepidoptera: Nepticulidae) (Noyes, 2019).
Hosts: *Rhynchaenus sp.* (Coleoptera: Curculionidae), *Phytomyza sp.* (Diptera: Agromyzidae), *Cameraria sp., Gracillaria syringella, Lithocolletis sp., Lithocolletis acerifoliella, Lithocolletis alnicolella, Lithocolletis blancardella, Lithocolletis carpinicolella, Lithocolletis cerasicolella, Lithocolletis cydoniella, Lithocolletis harrisella, Lithocolletis hauderiella, Lithocolletis heegeriella, Lithocolletis lantanella, Lithocolletis maestingella, Lithocolletis messaniella, Lithocolletis pomifoliella, Lithocolletis pyrifoliella, Lithocolletis quercifoliella, Lithocolletis quinnata, Lithocolletis sorbi, Lithocolletis spinicolella, Parornix betulae, Parornix scoticella, Phyllocnistis sp., Phyllocnistis saligna, Phyllonorycter sp., Phyllonorycter acernella, Phyllonorycter alnifoliella, Phyllonorycter bicinctella, Phyllonorycter blancardella, Phyllonorycter carpini, Phyllonorycter cavella, Phyllonorycter cerasicolella, Phyllonorycter concomitella, Phyllonorycter coryli, Phyllonorycter crataegella, Phyllonorycter cydoniella,*

Phyllonorycter dubitella, Phyllonorycter elmaella, Phyllonorycter emberizaepennella, Phyllonorycter fagifolia, Phyllonorycter geniculella, Phyllonorycter hancola, Phyllonorycter harrisella, Phyllonorycter issikii, Phyllonorycter japonica, Phyllonorycter jezoniella, Phyllonorycter joannisi, Phyllonorycter jozanae, Phyllonorycter laciniatae, Phyllonorycter lantanella, Phyllonorycter lucetiella, Phyllonorycter maestingella, Phyllonorycter nicellii, Phyllonorycter orientalis, Phyllonorycter oxyacanthae, Phyllonorycter padella, Phyllonorycter pomonella, Phyllonorycter propinquinella, Phyllonorycter quercifoliella, Phyllonorycter rajella, Phyllonorycter ringoniella, Phyllonorycter robiniella, Phyllonorycter salicicolella, Phyllonorycter salictella, Phyllonorycter schreberella, Phyllonorycter sorbi, Phyllonorycter sorbicola, Phyllonorycter stettinensis, Phyllonorycter strigulatella, Phyllonorycter sylvella, Phyllonorycter tenerella, Phyllonorycter tristrigella, Phyllonorycter turugisana, Phyllonorycter ulmi, Phyllonorycter ulmifoliella, Phyllonorycter viburni, Phyllonorycter viminetorum, Phyllonorycter viminiella, Phyllonorycter watanabei (Lepidoptera: Gracillariidae), *Coptodisca lucifluella* (Heliozelidae), *Leucoptera scitella* (Lyonetiidae), *Nepticula argyropeza, Stigmella sp., Stigmella argyropeza, Stigmella aucupariae, Stigmella juglandifoliella, Stigmella nylandriella, Stigmella ruficapitella* (Nepticulidae), *Tischeria sp., Tischeria ekebladella* (Tischeriidae) (Noyes, 2019).
Parasitoid hosts: *Apanteles circumscriptus, Colastes braconius* (Hymenoptera: Braconidae), *Achrysocharoides niveipes, Enaysma sp.* (Eulophidae) (Noyes, 2019).
Distribution: Bulgaria, Canada, Croatia, Czech Republic, Czechoslovakia, Denmark, England, Europe, Finland, France, Germany, Hungary, Iran, Ireland, Israel, Italy, Japan, Moldova, Netherlands, Poland, Romania, Serbia, Slovakia, Sweden, Switzerland, Turkey, United States of America, USSR, Yugoslavia (pre-1991) (Noyes, 2019).

Chrysocharis latifossa (Hansson, 1985)
Publications with Turkey records: Noyes (2019).
Associates: Unknown (Noyes, 2019).
Hosts: Unknown (Noyes, 2019).
Distribution: Cyprus, Lebanon, Syria, Turkey (Noyes, 2019).

Chrysocharis liriomyzae (Delucchi, 1954)
Publications with Turkey records: Cikman (2012), Öncüer (1991), Noyes (2019).

Associates: *Pastinaca sativa, Peucedanum palustre, Sium latifolium* (Apiaceae), *Centaurea jacea, Eupatorium cannabinum, Solidago virgaurea, Sonchus sp., Sonchus oleraceus, Taraxacum sp., Xanthium sp.* (Asteraceae), *Lonicera sp.* (Caprifoliaceae), *Laburnum anagyroides, Medicago sativa* (Fabaceae), *Aquilegia vulgaris* (Ranunculaceae) (Noyes, 2019).
Hosts: *Agromyza sp., Agromyza frontella, Agromyza nana, Amauromyza gyrans, Chromatomyia horticola, Liriomyza sp., Liriomyza congesta, Liriomyza eupatorii, Liriomyza pusilla, Liriomyza sonchi, Liriomyza taraxaci, Liriomyza trifoliearum, Liriomyza trifolii, Paraphytomyza hendeliana, Phytoliriomyza melampyga, Phytomyza sp., Phytomyza atricornis, Phytomyza chaerophylli, Phytomyza citisi, Phytomyza horticola, Phytomyza lonicerae, Phytomyza minuscula, Phytomyza mylini, Phytomyza pastinacae, Phytomyza sii, Phytomyza solidaginis, Phytomyza symphyti, Phytomyza syngenesiae, Phytomyza thysselini, Phytomyza thysselinivora* (Diptera: Agromyzidae) (Noyes, 2019).
Distribution: Bulgaria, Canada, Croatia, Czech Republic, Czechoslovakia, Denmark, England, Europe, Finland, France, Germany, Hungary, Italy, Japan, Korea, Moldova, Netherlands, North Africa, Norway, People's Republic of China, Poland, Romania, Serbia, Slovakia, Slovenia, Sweden, Turkey, United States of America, USSR, Yugoslavia (pre-1991) (Noyes, 2019).

Chrysocharis mediana (Foerster, 1861)
Publications with Turkey records: Öncüer (1991), Noyes (2019).
Hosts: *Rhynchaenus fagi* (Coleoptera: Curculionidae) (Noyes, 2019).
Distribution: Bulgaria, Canada, Croatia, Czech Republic, Czechoslovakia, Denmark, Europe, Finland, France, Germany, Hungary, Italy, Netherlands, Romania, Serbia, Slovakia, Spain, Sweden, Switzerland, Turkey, United Kingdom, Yugoslavia (pre-1991) (Noyes, 2019).

Chrysocharis nephereus (Walker, 1839)
Publications with Turkey records: Öncüer (1991), Noyes (2019).
Associates: *Formica aquilonia* (Formicidae), *Acer sp., Acer platanoides, Acer pseudoplatanus* (Aceraceae), *Alnus sp., Alnus glutinosa, Alnus hirsute, Alnus incana, Alnus japonica, Betula sp., Betula grossa, Betula papyrifera, Betula pendula, Betula pubescens, Betula verrucose, Carpinus betulus, Corylus sp., Corylus avellana* (Betulaceae), *Alliaria officinalis* (Brassicaceae), *Lonicera sp.* (Caprifoliaceae), *Ledum sp.* (Ericaceae), *Laburnum anagyroides* (Fabaceae), *Castanea sp., Castanea sativa, Fagus*

sp., Fagus sylvatica, Quercus sp., Quercus cerris, Quercus dumosa, Quercus petraea, Quercus robur, Quercus velutina (Fagaceae), *Aesculus hippocastanum* (Hippocastanaceae), *Myrica asplenifolia* (Myricaceae), *Platanus occidentalis* (Platanaceae), *Alopecurus myosuroides, Lolium perenne* (Poaceae), *Clematis vitalba* (Ranunculaceae), *Ceanothus crassifolius* (Rhamnaceae), *Crataegus sp., Crataegus laevigata, Crataegus monogyna, Malus sp., Malus pumila, Malus sylvestris, Prunus sp., Prunus padus, Prunus pennsylvanica, Prunus serotina, Sorbus sp., Sorbus americana, Sorbus aucuparia* (Rosaceae), *Rubus caesius* (Rubiaceae), *Populus sp., Populus alba, Populus balsamifera, Populus italicus, Populus nigra, Populus tremula, Populus tremuloides, Populus virginiana, Salix sp., Salix alba, Salix aurita, Salix caprea, Salix cinerea, Salix fragilis x pentandra, Salix nigricans, Salix pentandra, Salix purpurea, Salix rubens, Salix triandra, Salix triandra x pentandra, Salix viminalis* (Salicaceae), *Tilia americana* (Tiliaceae), *Ulmus sp.* (Ulmaceae) (Noyes, 2019).

Hosts: *Zeugophora sp., Zeugophora flavicollis* (Coleoptera: Chrysomelidae), *Ceutorhynchus contractus, Orchestes fagi, Orchestes quercus, Orchestes rusci, Rhamphus oxyacanthae, Rhamphus pulicarius, Rhynchaenus alni, Rhynchaenus aterrhimus, Rhynchaenus avellanae, Rhynchaenus fagi, Rhynchaenus populi, Rhynchaenus quercus, Rhynchaenus rufipes, Rhynchaenus rusci, Rhynchaenus salicis, Rhynchaenus subfasciatus, Rhynchaenus takabayashii, Rhynchaenus testaceus* (Curculionidae), *Agromyza demeijerei, Chromatomyia horticola, Paraphytomyza fulvicornis, Paraphytomyza hendeliana, Paraphytomyza populi, Paraphytomyza populicola, Phytomyza fuscula, Phytomyza ilicis, Phytomyza nigra, Phytomyza rajella, Phytomyza vitalbae* (Diptera: Agromyzidae), *Cimbex quadrimaculata* (Hymenoptera: Cimbicidae), (Cynipidae) *Fenusa dohrnii, Fenusa pusilla, Fenusa ulmi, Heterarthrus aceris, Heterarthrus nemoratus, Heterarthrus vagans, Messa sp., Messa glaucopis, Messa hortulana, Messa nana, Messa populifoliella, Nematus sp., Phyllocolpa sp., Profenusa pygmaea, Profenusa thomsoni, Scolioneura betulae* (Tenthredinidae), *Bucculatrix canadensisella* (Lepidoptera: Bucculatricidae), *Coleophora flavipennella, Coleophora fuscedinella, Coleophora laricella, Coleophora salmani, Coleophora serratella* (Coleophoridae), *Eriocrania sp., Eriocrania haworthi, Eriocrania sangii, Eriocrania semipurpurella, Eriocrania sparmannella, Eriocrania subpurpurella, Eriocrania unimaculella* (Eriocraniidae), *Anacampsis populella, Chrysoesthia sexguttella* (Gelechiidae), *Acrocercops brongniardella, Callisto denticulella, Cameraria jacintoensis, Cameraria ohridella, Lithocolletis sp., Lithocolletis acerifoliella, Lithocolletis anderidae, Lithocolletis blancardella,*

Lithocolletis cavella, Lithocolletis comparella, Lithocolletis connexella, Lithocolletis dubitella, Lithocolletis emberizaepennella, Lithocolletis faginella, Lithocolletis froelichiella, Lithocolletis harrisella, Lithocolletis hauderiella, Lithocolletis heegeriella, Lithocolletis hortella, Lithocolletis joannisi, Lithocolletis maestingella, Lithocolletis messaniella, Lithocolletis nicellii, Lithocolletis platani, Lithocolletis platanoidella, Lithocolletis populifoliella, Lithocolletis quercifoliella, Lithocolletis quinnata, Lithocolletis rajella, Lithocolletis salicifoliella, Lithocolletis schreberella, Lithocolletis sorbi, Lithocolletis spinolella, Lithocolletis strigulatella, Lithocolletis tenerella, Lithocolletis tremuloidiella, Lithocolletis ulmifoliella, Lithocolletis viminella, Parornix alni, Parornix scoticella, Phyllocnistis sp., Phyllocnistis klemannella, Phyllocnistis labyrinthella, Phyllocnistis populiella, Phyllocnistis saligna, Phyllocnistis suffusella, Phyllocnistis unipunctella, Phyllocnistis vitegenella, Phyllocnistis xenia, Phyllonorycter sp., Phyllonorycter acerifoliella, Phyllonorycter acernella, Phyllonorycter alnifoliella, Phyllonorycter anderidae, Phyllonorycter blancardella, Phyllonorycter cavella, Phyllonorycter corylifoliella, Phyllonorycter dubitella, Phyllonorycter elmaella, Phyllonorycter froelichiella, Phyllonorycter geniculella, Phyllonorycter harrisella, Phyllonorycter joannisi, Phyllonorycter maestingella, Phyllonorycter messaniella, Phyllonorycter nemoris, Phyllonorycter nicellii, Phyllonorycter oxyacanthae, Phyllonorycter padella, Phyllonorycter platani, Phyllonorycter pyrifoliella, Phyllonorycter quercifoliella, Phyllonorycter rajella, Phyllonorycter roboris, Phyllonorycter schreberella, Phyllonorycter sorbi, Phyllonorycter stettinensis, Phyllonorycter strigulatella, Phyllonorycter tenerella, Phyllonorycter tristrigella, Phyllonorycter ulmifoliella, Phyllonorycter viminetorum (Gracillariidae), *Coptodisca sp., Coptodisca arbutiella, Coptodisca splendoriferella* (Heliozelidae), *Paraclemensia sp.* (Incurvariidae), *Leucoptera laburnella, Leucoptera malifoliella, Leucoptera scitella, Leucoptera sinuella, Lyonetia clerkella, Paraleucoptera albella, Paraleucoptera sinuella* (Lyonetiidae), *Ectoedemia pulverosella, Ectoedemia subbimaculella, Nepticula sp., Nepticula gratiosella, Nepticula sorbi, Nepticula tityrella, Nepticula vimineticola, Stigmella sp., Stigmella argentipedella, Stigmella assimilella, Stigmella atricapitella, Stigmella aucupariae, Stigmella aurella, Stigmella auritella, Stigmella basalella, Stigmella centifoliella, Stigmella confusella, Stigmella crataegella, Stigmella floslactella, Stigmella hemargyrella, Stigmella ignobiliella, Stigmella lapponica, Stigmella malella, Stigmella myrtillella, Stigmella obliquella, Stigmella oxyacanthella, Stigmella plagicolella, Stigmella pulverosella, Stigmella ruficapitella, Stigmella salicis, Stigmella sorbi, Stigmella subbimaculella, Stigmella tityrella, Stigmella trimaculella, Stigmella vimineticola, Stigmella woolhopiella* (Nepticulidae), *Tischeria sp., Tischeria angusticollella, Tischeria decidua, Tischeria dodonaea, Tischeria ekebladella, Tischeria heinemanni* (Tischeriidae), *Prays oleae* (Yponomeutidae) (Noyes, 2019).

Parasitoid hosts: *Apanteles circumscriptus, Colastes braconius* (Hymenoptera: Braconidae), *Achrysocharoides latreillei, Cirrospilus vittatus* (Eulophidae) (Noyes, 2019).

Parasitoids: *Closterocerus sp., Pediobius saulius* (Eulophidae) (Noyes, 2019).

Distribution: Austria, Bosnia Hercegovina, Bulgaria, Canada, Croatia, Czech Republic, Czechoslovakia, Denmark, Europe, Finland, France, Germany, Greece, Hungary, Italy, Japan, Macedonia, Montenegro, Netherlands, Norway, Poland, Portugal, Romania, Russia, Serbia, Slovakia, Spain, Sweden, Switzerland, Turkey, Ukraine, United Kingdom, United States of America, USSR, Yugoslavia (pre-1991) (Noyes, 2019).

Chrysocharis pentheus (Walker, 1839)

Publications with Turkey records: Gencer (2009), Yefremova et al. (2010), Noyes (2019).

Associates: *Acer pseudoplatanus* (Aceraceae), *Aegopodium podagraria, Angelica archangelica, Angelica sylvestris, Anthriscus sylvestris* (Apiaceae), *Kalopanax septemlobus* (Araliaceae), *Arctium sp., Arctium nemorosum, Artemisia montana, Artemisia vulgaris, Chrysanthemum sp., Cirsium sp., Erechtites hieracifolia, Eupatorium cannabinum, Gerbera sp., Gerbera jamesoni, Lapsana communis, Petasites hybridus, Reichardia sp., Sonchus arvensis, Sonchus asper, Tanacetum vulgare* (Asteraceae), *Impatiens noli-tangere, Impatiens parviflora* (Balsaminaceae), *Alnus sp., Betula sp., Carpinus sp., Carpinus betulus, Corylus avellana* (Betulaceae), *Catalpa sp.* (Bignoniaceae), *Alliaria officinalis, Brassica sp., Brassica oleracea, Cardaria draba* (Brassicaceae), *Humulus lupulus* (Cannabidaceae), *Lonicera sp., Lonicera gracilipes, Lonicera periclymenum* (Caprifoliaceae), *Gypsophila sp., Melandrium rubrum* (Caryophyllaceae), *Chenopodium album* (Chenopodiaceae), *Elaeagnus macrophylla* (Elaeagnaceae), *Kalmia sp., Kalmia angustifolia, Lyonia ovalifolia* (Ericaceae), *Astragalus glycyphyllos, Laburnum sp., Laburnum anagyroides, Lathyrus odoratus, Lathyrus vernus, Phaseolus vulgaris, Pisum sativum* (Fabaceae), *Fagus sp., Fagus sylvatica, Quercus sp., Quercus acutissima, Quercus alba, Quercus dentata, Quercus mongolica, Quercus pedunculiflora, Quercus robur, Quercus serrata, Quercus* *velutina* (Fagaceae), *Gentiana sp.* (Gentianaceae), *Aesculus hippocastanum* (Hippocastanaceae), *Papaver rhoeas, Papaver somniferum* (Papaveraceae), *Cedrus sp., Larix decidua* (Pinaceae), *Hordeum vulgare, Milium effusum, Phalaris arundinacea* (Poaceae), *Aquilegia vulgaris, Caltha palustris, Clematis vitalba, Ranunculus sp., Ranunculus flammula* (Ranunculaceae), *Crataegus sp., Crataegus laevigata, Crataegus monogyna, Malus sp., Malus pumila, Prunus pennsylvanica, Prunus persica, Prunus serotina, Prunus spinosa, Pyrus sp., Rosa sp., Rosa acicularis, Rosa rubiginosa, Sorbus sp., Sorbus aucuparia* (Rosaceae), *Rubus sp., Rubus fruticosus* (Rubiaceae), *Citrus sp.* (Rutaceae), *Populus nigra, Populus tremuloides, Salix sp., Salix cinerea, Salix nigri*cans (Salicaceae), *Lycopersicon esculentum, Solanum melongena, Solanum nigrum* (Solanaceae), *Tropaeolum majus* (Tropaeolaceae), *Celtis occidentalis, Ulmus sp., Ulmus americana, Ulmus glabra, Ulmus rubra* (Ulmaceae), *Urtica sp.* (Urticaceae) (Noyes, 2019).

Hosts: *Bruchus rufipes* (Coleoptera: Bruchidae), *Ceutorhynchus contractus, Orchestes quercus, Rhamphus oxyacanthae, Rhamphus pulicarius, Rhynchaenus japonicus, Rhynchaenus mutabilis, Rhynchaenus pallicornis, Rhynchaenus populi, Rhynchaenus takabayashii* (Curculionidae), *Agromyza sp., Agromyza albipennis, Agromyza alnibetulae, Agromyza aristate, Agromyza demeijerei, Agromyza flaviceps, Agromyza hiemalis, Agromyza nana, Agromyza orobi, Agromyza oryzae, Agromyza rufipes, Amauromyza flavifrons, Calycomyza humeralis, Chromatomyia fuscula, Chromatomyia horticola, Liriomyza sp., Liriomyza bryoniae, Liriomyza eupatorii, Liriomyza huidobrensis, Liriomyza impatientis, Liriomyza sativae, Liriomyza trifolii, Paraphytomyza cornigera, Paraphytomyza hendeliana, Paraphytomyza orbitalis, Paraphytomyza populi, Paraphytomyza populicola, Phytoliriomyza melampyga, Phytoliriomyza variegate, Phytomyza sp., Phytomyza angelicae, Phytomyza angelicastri, Phytomyza anthrisci, Phytomyza aquilegiae, Phytomyza artemisivora, Phytomyza calthophila, Phytomyza chaerophylli, Phytomyza citisi, Phytomyza deciduae, Phytomyza horticola, Phytomyza ilicis, Phytomyza lappae, Phytomyza lonicerae, Phytomyza marginella, Phytomyza milii, Phytomyza minuscula, Phytomyza obscurella, Phytomyza paniculatae, Phytomyza periclymeni, Phytomyza pubicornis, Phytomyza ranunculi, Phytomyza sonchi, Phytomyza sorbi, Phytomyza sphondylii, Phytomyza syngenesiae, Phytomyza tanaceti, Phytomyza tussilaginis, Phytomyza vitalbae* (Diptera: Agromyzidae), *Mayetiola destructor, Phytophaga destructor* (Cecidomyiidae), *Diplolepis lens, Diplolepis rosaefolii* (Hymenoptera: Cynipidae), *Fenusa pusilla* (Tenthredinidae), *Bucculatrix sp.* (Lepidoptera: Bucculatricidae), *Coleophora laricella, Coleophora pruniella*

(Coleophoridae), *Biselachista fulgens, Elachista cerusella* (Elachistidae), *Eriocrania subpurpurella* (Eriocraniidae), *Chrysoesthia sexguttella, Chrysoesthis drurella, Microsetia sexguttella* (Gelechiidae), *Callisto denticulella, Cameraria sp., Cameraria hikosanensis, Cameraria ohridella, Lithocolletis sp., Lithocolletis corylifoliella, Lithocolletis maestingella, Phyllocnistis citrella, Phyllocnistis labyrinthella, Phyllocnistis populiella, Phyllocnistis vitegenella, Phyllonorycter sp., Phyllonorycter blancardella, Phyllonorycter cavella, Phyllonorycter celtisella, Phyllonorycter corylifoliella, Phyllonorycter crataegella, Phyllonorycter lyoniae, Phyllonorycter malimalifoliella, Phyllonorycter oxyacanthae, Phyllonorycter propinquinella, Phyllonorycter spinolella, Phyllonorycter tristrigella, Phyllonorycter ulmifoliella* (Gracillariidae), *Coptodisca sp., Holocacista rivillei* (Heliozelidae), *Leucoptera laburnella, Leucoptera malifoliella, Leucoptera orobi, Leucoptera scitella, Leucoptera wailesella, Lyonetia clerkella, Lyonetia prunifoliella* (Lyonetiidae), *Cosmopterix pulchrimella* (Momphidae), *Ectoedemia agrimoniae, Ectoedemia pulverosella, Fomoria septembrella, Nepticula floslactella, Nepticula lapponica, Nepticula lindquisti, Nepticula malella, Nepticula microtheriella, Nepticula salicis, Nepticula sorbi, Nepticula tityrella, Stigmella sp., Stigmella aceris, Stigmella anomalella, Stigmella atricapitella, Stigmella atricollis, Stigmella aucupariae, Stigmella aurella, Stigmella centifoliella, Stigmella floslactella, Stigmella freyella, Stigmella hemargyrella, Stigmella hybnerella, Stigmella ignobiliella, Stigmella lapponica, Stigmella malella, Stigmella marginicolella, Stigmella microtheriella, Stigmella minusculella, Stigmella myrtillella, Stigmella nylandriella, Stigmella oxyacanthella, Stigmella plagicolella, Stigmella platanella, Stigmella pulverosella, Stigmella rubivora, Stigmella ruficapitella, Stigmella salicis, Stigmella sorbi, Stigmella splendidissimella, Stigmella tiliae, Stigmella tityrella, Stigmella trimaculella* (Nepticulidae), *Tischeria sp., Tischeria angusticollella, Tischeria dodonaea, Tischeria ekebladella, Tischeria heinemanni, Tischeria malifoliella* (Tischeriidae), *Argyresthia sp.* (Yponomeutidae) (Noyes, 2019).

Distribution: Austria, Bulgaria, Canada, Canary Islands, Channel Islands (British Is), Croatia, Cyprus, Czech Republic, Czechoslovakia, Denmark, England, Europe, Finland, France, Germany, Greece, Hungary, Israel, Italy, Japan, Korea, Macedonia, Malaysia, Moldova, Montenegro, Nearctic, Netherlands, Norway, People's Republic of China, Poland, Romania, Serbia, Slovakia, Spain, Sweden, Switzerland, Taiwan, Turkey, Ukraine, United States of America, USSR, Yugoslavia (pre-1991) (Noyes, 2019).

Chrysocharis phryne (Walker, 1839)
Publications with Turkey records: Cebeci et al. (2011), Gencer (2009), Noyes (2019), Öncüer (1991), Yefremova et al. (2010).
Associates: *Lonicera periclymenum* (Caprifoliaceae), *Medicago lupulina* (Fabaceae), *Fagus sylvatica, Quercus sp., Quercus petraea, Quercus robur* (Fagaceae), *Gentiana sp.* (Gentianaceae), *Aesculus hippocastanum* (Hippocastanaceae), *Crataegus oxyacanthoides, Sorbus aucuparia* (Rosaceae), *Ulmus sp.* (Ulmaceae) (Noyes, 2019).
Hosts: *Chromatomyia horticola, Phytomyza horticola* (Diptera: Agromyzidae), *Cameraria ohridella, Lithocolletis sp., Lithocolletis anderidae, Lithocolletis emberizaepennella, Lithocolletis helianthemella, Lithocolletis lautella, Lithocolletis platani, Lithocolletis quercifoliella, Lithocolletis roboris, Lithocolletis schreberella, Lithocolletis sorbi, Parornix avellanella, Parornix scoticella, Phyllonorycter sp., Phyllonorycter acernella, Phyllonorycter anderidae, Phyllonorycter emberizaepennella, Phyllonorycter insignitella, Phyllonorycter joannisi, Phyllonorycter lautella, Phyllonorycter maestingella, Phyllonorycter nigrescentella, Phyllonorycter oxyacanthae, Phyllonorycter quercifoliella, Phyllonorycter salictella, Phyllonorycter schreberella, Phyllonorycter sorbi, Phyllonorycter stettinensis, Phyllonorycter tenerella, Phyllonorycter tristrigella, Phyllonorycter ulmifoliella* (Lepidoptera: Gracillariidae), *Leucoptera lustratella* (Lyonetiidae), *Stigmella sp., Stigmella marginicolella* (Nepticulidae) (Noyes, 2019).
Parasitoids: *Closterocerus trifasciatus* (Hymenoptera: Eulophidae) (Noyes, 2019).
Distribution: Austria, Bulgaria, Czech Republic, Czechoslovakia, Denmark, England, Europe, Finland, Germany, Hungary, Japan, Moldova, Montenegro, Netherlands, Norway, People's Republic of China, Poland, Russia, Serbia, Sweden, Switzerland, Turkey, Ukraine, USSR, Yugoslavia (pre-1991) (Noyes, 2019).

Chrysocharis polyzo (Walker, 1839)
Publications with Turkey records: Noyes (2019), Öncüer (1991), Yefremova et al. (2010).
Associates: *Carex sp., Scirpus microcarpus* (Cyperaceae), *Melilotus sp.* (Fabaceae), *Iris sp., Iris pseudacorus, Iris setosa* (Iridaceae), *Alopecurus myosuroides, Arrhenatherum elatius, Bromus ciliates, Calamagrostis canadensis, Elytrigia repens, Hordeum vulgare, Phalaris arundinacea, Phleum nodosum, Phragmites sp., Phragmites communis, Setaria viridis, Triticum aestivum, Zea mays* (Poaceae), *Malus sp.* (Rosaceae) (Noyes, 2019).

Hosts: *Agromyza sp., Agromyza angulate, Agromyza nana, Agromyza nigripes, Agromyza yanonis, Ceradontha incisa, Cerodontha sp., Cerodontha caricicola, Cerodontha caricivora, Cerodontha incisa, Cerodontha ireos, Cerodontha okazakii, Cerodontha pygmaea, Cerodontha sasakawai, Cerodontha viridis, Chromatomyia fuscula, Dizygomyza atra, Dizygomyza caricivora, Dizygomyza incisa, Dizygomyza iraeos, Dizygomyza morosa, Dizygomyza pygmaea, Phytobia angulate, Phytobia atra, Phytobia caricivora, Phytobia incisa, Phytobia pygmaea, Phytomyza fuscula, Phytomyza obscurella, Pseudonapomyza atra* (Diptera: Agromyzidae), *Phyllonorycter sp.* (Lepidoptera: Gracillariidae) (Noyes, 2019).
Distribution: Austria, Bulgaria, Canada, Croatia, Czech Republic, Czechoslovakia, Denmark, Dominica, Europe, Finland, France, Germany, Hungary, Ireland, Italy, Japan, Korea, Macedonia, Moldova, Mongolia, Montenegro, Nearctic, Netherlands, Norway, Poland, Romania, Russia, Serbia, Slovakia, Sweden, Transcaucasus, Turkey, United Kingdom, United States of America, USSR, Yugoslavia (pre-1991) (Noyes, 2019).

Chrysocharis pubens Delucchi, 1954
Publications with Turkey records: Noyes (2019), Yefremova et al. (2010).
Associates: *Alocasia macrorhiza* (Araceae), *Acanthopanax sp.* (Arialiaceae), *Aster indicus* (Asteraceae), *Abelia spanthulata, Lonicera japonica, Viburnum furcatum* (Caprifoliaceae), *Carex sp.* (Cyperaceae), *Dioscorea quinqueloba* (Dioscoreaceae), *Hydrangea macrophylla, Hydrangea paniculata* (Hydrangeaceae), *Akebia sp.* (Lardizabalaceae), *Arundinaria pygmaea, Milium effusum* (Poaceae), *Malus sp., Malus pumila* (Rosaceae), *Camellia sp.* (Theaceae) (Noyes, 2019).
Hosts: *Cerodontha bisetiorbita, Cerodontha caricicola, Chromatomyia gentianae, Chromatomyia suikazurae, Dizygomyza bimaculate, Dizygomyza soenderupi, Ophiomyia maura, Phytobia bimaculate, Phytobia sonderupi, Phytomyza sp., Phytomyza dryoptericola, Phytomyza heringiana, Phytomyza jucunda, Phytomyza milii, Tropicomyia sp., Tropicomyia styricicola* (Diptera: Agromyzidae), *Lithocolletis blancardella* (Lepidoptera: Gracillariidae) (Noyes, 2019).
Distribution: Canary Islands, Czech Republic, Czechoslovakia, England, Europe, Finland, Germany, Italy, Japan, Korea, Moldova, Netherlands, People's Republic of China, Poland, Sweden, Switzerland, Taiwan, Turkey, Ukraine, USSR (Noyes, 2019).

Chrysocharis pubicornis (Zetterstedt, 1838)
Publications with Turkey records: Gencer (2009), Noyes (2019), Uygun et al. (1995), Yefremova et al. (2010).

Associates: *Ilex sp., Ilex aquifolium* (Aquifoliaceae), *Achillea millefolium, Arctium sp., Aster indicus, Carduus lanceolata, Chrysanthemum sp., Cineraria sp., Crepis biennis, Helianthus annuus, Lactuca sativa, Senecio jacobaea, Sonchus asper, Sonchus oleraceus, Sonchus palustris, Taraxacum officinale* (Asteraceae), *Betula pendula* (Betulaceae), *Brassica campestris, Cardaria draba* (Brassicaceae), *Lonicera periclymenum* (Caprifoliaceae), *Chenopodium sp.* (Chenopodiaceae), *Laburnum anagyroides, Lathyrus odoratus, Medicago hispida, Medicago sativa, Phaseolus vulgaris, Pisum sativum, Sarothamnus scoparius* (Fabaceae), *Gentiana sp., Gentiana asclepadea* (Gentianaceae), *Althaea rosea* (Malvaceae), *Fraxinus excelsior* (Oleaceae), *Papaver sp.* (Papaveraceae), *Alopecurus myosuroides, Holcus sp., Hordeum vulgare, Triticum aestivum* (Poaceae), *Rubus sp.* (Rubiaceae), *Populus sp.* (Salicaceae)(Noyes, 2019).
Hosts: *Agromyza atricornis, Agromyza myosotidis, Agromyza nana, Agromyza rufipes, Amauromyza flavifrons, Amauromyza morionella, Amauromyza verbasci, Calycomyza artemisiae, Chromatomyia sp., Chromatomyia fuscula, Chromatomyia horticola, Chromatomyia suikazurae, Dizygomyza sp., Dizygomyza verbasci, Liriomyza sp., Liriomyza bryoniae, Liriomyza congesta, Liriomyza flaveola, Liriomyza huidobrensis, Liriomyza katoi, Liriomyza strigata, Liriomyza trifolii, Napomyza sp., Napomyza lonicerella, Napomyza xylostei, Opiomyza maura, Paraphytomyza hendeliana, Paraphytomyza heringi, Paraphytomyza populi, Phytagromyza populi, Phytobia verbasci, Phytoliriomyza variegate, Phytomyza sp., Phytomyza affinis, Phytomyza angelicae, Phytomyza aprilina, Phytomyza atricornis, Phytomyza chaerophylli, Phytomyza cicutae, Phytomyza citisi, Phytomyza conyzae, Phytomyza crassiseta, Phytomyza fuscula, Phytomyza geniculate, Phytomyza gentianae, Phytomyza harlemensis, Phytomyza heringiana, Phytomyza horticola, Phytomyza ilicicola, Phytomyza ilicis, Phytomyza lonicerae, Phytomyza melana, Phytomyza milii, Phytomyza myosotica, Phytomyza nigra, Phytomyza obscura, Phytomyza origani, Phytomyza periclymeni, Phytomyza petoei, Phytomyza plantaginis, Phytomyza pulvinariae, Phytomyza ranunculi, Phytomyza sii, Phytomyza silai, Phytomyza spinaciae, Phytomyza symphyti, Phytomyza syngenesiae, Phytomyza vitalbae* (Diptera: Agromyzidae), *Cystiphora sonchi, Cystiphora taraxaci* (Cecidomyiidae), *Scaptomyza sp., Scaptomyza apicalis* (Drosophilidae), *Tephritis crepidis* (Tephritidae), *Lithocolletis blancardella, Lithocolletis corylifoliella* (Lepidoptera: Gracillariidae), *Lyonetia clerkella* (Lyonetiidae), *Nepticula aeneella, Nepticula catharticella, Nepticula malella, Nepticula splendidissimella*

(Nepticulidae), *Tischeria sp., Tischeria marginea* (Tischeriidae), *Zeiraphera diniana* (Tortricidae) (Noyes, 2019).
Parasitoids: *Pediobius metallicus* (Hymenoptera: Eulophidae) (Noyes, 2019).
Distribution: Australasian, Australia, Austria, Bulgaria, Canada, Croatia, Cyprus, Czech Republic, Czechoslovakia, Denmark, Europe, Finland, France, Germany, Greece, Greenland, Hungary, India, Italy, Japan, Jordan, Korea, Libya, Malta, Moldova, Mongolia, Montenegro, Nearctic, Netherlands, New Zealand, Norway, Oriental, Pakistan, Poland, Romania, Russia, Slovakia, Slovenia, Spain, Sweden, Switzerland, Turkey, Ukraine, United Kingdom, United States of America, USSR, Yemen, Yugoslavia (pre-1991) (Noyes, 2019).

Chrysocharis viridis (Nees, 1834)
Publications with Turkey records: Gencer (2009), Noyes (2019), Yefremova et al. (2010).
Associates: *Aegopodium sp., Aegopodium podagraria, Angelica sp., Anthriscus sylvestris, Astrantia major, Heracleum lanatum, Oenanthe sarmentosa, Osmorhiza depauperate, Pastinaca sativa, Sanicula chinensis, Sanicula europaea, Sanicula marilandica* (Apiaceae), *Aralia nudicaulis* (Araliaceae), *Acanthopanax sp.* (Arialiaceae), *Achillea sp., Achillea borealis, Achillea millefolium, Adenocaulon bicolor, Arctium tomentosum, Arnica alpina, Arnica cordifolia, Artemisia sp., Artemisia vulgaris, Aster sp., Aster indicus, Cacalia hastata, Cirsium sp., Cirsium arvense, Cirsium kamtschaticum, Eupatorium chinense, Kalimeris yomena, Petasites frigidus, Petasites hyperboreus, Petasites japonicus, Prenanthes alata, Saussurea angustifolia, Senecio cannabifolius, Senecio pauperculus, Solidago multiradiata, Sonchus arvensis, Synurus pungens, Taraxacum officinale* (Asteraceae), *Impatiens capensis* (Balsaminaceae), *Mertensia paniculata* (Boraginaceae), *Lonicera sp., Lonicera involucrate, Lonicera periclymenum, Sambucus nigra, Symphoricarpos alba, Symphoricarpos occidentalis* (Caprifoliaceae), *Melandrium rubrum* (Caryophyllaceae), *Cornus canadensis, Cornus stolonifera* (Cornaceae), *Juniperus thurifera* (Cupressaceae), *Carex sp.* (Cyperaceae), *Pteridium aquilinum* (Dennstaedtiaceae), *Laburnum sp., Laburnum anagyroides, Lupinus nootkatensis, Medicago sativa* (Fabaceae), *Plantago major* (Plantaginaceae), *Anemone canadensis, Aquilegia brevistyla, Caltha palustris, Clematis vitalba, Coptis aspleniifolia, Delphinium sp.* (Ranunculaceae), *Aruncus vulgaris* (Rosaceae), *Asperula dolorata* (Rubiaceae), *Populus sp., Populus nigra, Populus tremula, Salix alba, Salix rubens, Salix viminalis* (Salicaceae), *Mitella nuda, Saxifraga fusca,*

Tiarella trifoliata (Saxifragaceae), *Valeriana capitata* (Valerianaceae), *Viola nipponica* (Violaceae) (Noyes, 2019).

Hosts: *Agromyza sp., Agromyza demeijerei, Agromyza fragariae, Agromyza spiraeoidearum, Amauromyza flavifrons, Cerodontha incisa, Chromatomyia sp., Chromatomyia horticola, Chromatomyia kluanensis, Chromatomyia mitellae, Chromatomyia tiarellae, Dizygomyza flavifrons, Liriomyza sp., Liriomyza amoena, Liriomyza asterivora, Liriomyza congesta, Liriomyza impatientis, Liriomyza imurae, Liriomyza millefolii, Liriomyza pusilla, Liriomyza sonchi, Liriomyza takakoae, Napomyza sp., Napomyza glechomae, Napomyza rydeni, Napomyza xylostei, Nemorimyza posticata, Paraphytomyza cornigera, Paraphytomyza heringi, Paraphytomyza lonicerae, Paraphytomyza populi, Paraphytomyza populicola, Paraphytomyza spenceri, Paraphytomyza tremulae, Paraphytomyza tridentata, Phytagromyza sp., Phytagromyza hendeliana, Phytagromyza populi, Phytagromyza xylostei, Phytobia flavifrons, Phytoliriomyza hilariella, Phytoliriomyza melampyga, Phytomyza sp., Phytomyza aconiti, Phytomyza aegopodii, Phytomyza agromyzina, Phytomyza albimargo, Phytomyza alpina, Phytomyza angelicae, Phytomyza angelicastri, Phytomyza anthrisci, Phytomyza aprilina, Phytomyza aquilegivora, Phytomyza aralivora, Phytomyza archangelicae, Phytomyza artemisivora, Phytomyza atrantiae, Phytomyza atricornis, Phytomyza brunnipes, Phytomyza calthivora, Phytomyza calthophila, Phytomyza chaerophylli, Phytomyza ciliolate, Phytomyza cirsii, Phytomyza citisi, Phytomyza columbiana, Phytomyza delphinivora, Phytomyza eupatorii, Phytomyza glechomae, Phytomyza horticola, Phytomyza japonica, Phytomyza lappae, Phytomyza lappina, Phytomyza lonicerae, Phytomyza lycopi, Phytomyza matricariae, Phytomyza myosotica, Phytomyza obscurella, Phytomyza ovalis, Phytomyza pastinacae, Phytomyza plantaginis, Phytomyza prava, Phytomyza primulae, Phytomyza pubicornis, Phytomyza ramose, Phytomyza ranunculi, Phytomyza saniculae, Phytomyza saxifragae, Phytomyza scopulina, Phytomyza sphondylii, Phytomyza syngenesiae, Phytomyza tamui, Phytomyza tanaceti, Phytomyza tetrasticha, Phytomyza trollii, Phytomyza tundrensis, Phytomyza tussilaginis, Phytomyza vitalbae, Tropicomyia sp.* (Diptera: Agromyzidae), *Phyllonorycter medicaginella* (Lepidoptera: Gracillariidae), *Leucoptera laburnella* (Lyonetiidae), *Ectoedemia argentipedella* (Nepticulidae) (Noyes, 2019).

Distribution: Austria, Bulgaria, Canada, Channel Islands, Croatia, Czech Republic, Czechoslovakia, Denmark, Europe, Finland, Germany, Hawaii, Hungary, Israel, Italy, Japan, Korea, Moldova, Mongolia, Montenegro, Netherlands, Norway, People's Republic of China, Poland, Romania,

Russia, Serbia, Slovakia, Slovenia, Spain, Sweden, Switzerland, Taiwan, Turkey, United Kingdom, United States of America, USSR, Yugoslavia (pre-1991) (Noyes, 2019).

Genus *Closterocerus* (Westwood, 1833)
Closterocerus chamaeleon (Girault, 1922)
Publications with Turkey records: Aytar et al. (2011), Noyes (2019).
Hosts: *Ophelimus eucalypti, Ophelimus maskelli* (Hymenoptera: Eulophidae), *Eucalyptus sp., Eucalyptus amplifolia, Eucalyptus camaldulensis, Eucalyptus cloeziana, Eucalyptus globosus, Eucalyptus rudis, Eucalyptus tereticornis* (Myrtaceae) (Noyes, 2019).
Distribution: Algeria, Argentina, Australia, France, Israel, Italy, Portugal, Spain, Tunisia, Turkey, United States of America (Noyes, 2019).

Closterocerus lyonetiae (Ferriere, 1952)
Publications with Turkey records: Noyes (2019), Uygun et al. (1995).
Associates: *Quercus lanuginosa* (Fagaceae), *Rosmarinus officinalis* (Lamiaceae), *Clematis vitalba* (Ranunculaceae), *Malus pumila, Prunus persica* (Rosaceae) (Noyes, 2019).
Hosts: *Phytomyza horticola, Phytomyza vitalbae* (Diptera: Agromyzidae), *Targionia vitis* (Hemiptera: Diaspididae), *Cameraria ohridella, Phyllonorycter blancardella, Phyllonorycter corylifoliella* (Lepidoptera: Gracillariidae), *Leucoptera malifoliella, Lyonetia clerkella* (Lyonetiidae), *Stigmella malella* (Nepticulidae) (Noyes, 2019).
Distribution: Croatia, Finland, Germany, Hungary, Italy, Japan, Romania, Serbia, Spain, Sweden, Switzerland, Turkey, Ukraine, USSR, Yugoslavia (pre-1991) (Noyes, 2019).

Closterocerus trifasciatus (Westwood, 1833)
Publications with Turkey records: Cebeci et al. (2011), Noyes (2019).
Associates: *Neuroterus numismalis* (Cynipidae), *Rhus sp.* (Anacardiaceae), *Ilex aquifolium, Ilex opaca* (Aquifoliaceae), *Eupatorium maculatum* (Asteraceae), *Alnus glutinosa, Betula sp., Betula platyphylla, Betula verrucose, Carpinus betulus* (Betulaceae), *Catalpa sp., Tecoma radicans* (Bignoniaceae), *Lonicera sp.* (Caprifoliaceae), *Chloranthus japonicus* (Chloranthaceae), *Carex sp., Carex vesicaria* (Cyperaceae), *Arbutus menziesii, Rhododendron sp.* (Ericaceae), *Amphicarpaea sp., Bauhinia japonica, Laburnum anagyroides, Robinia pseudoacacia* (Fabaceae), *Castanea sativa, Fagus sylvatica, Quercus sp., Quercus alba, Quercus cerris, Quercus dentata, Quercus glaucoides, Quercus laevis, Quercus lyrate, Quercus mongolica, Quercus petraea, Quercus robur,*

Quercus rubra (Fagaceae), *Aesculus hippocastanum* (Hippocastanaceae), *Carya sp.*, *Carya illinoinensis* (Juglandaceae), *Phillyrea latifolia*, *Syringa sp.* (Oleaceae), *Platanus sp.* (Platanaceae), *Glyceria maxima*, *Sasa sp.* (Poaceae), *Aquilegia sp.*, *Thalictrum sp.* (Ranunculaceae), *Ceanothus sp.* (Rhamnaceae), *Amelanchier sp.*, *Crataegus monogyna*, *Malus pumila*, *Prunus sp.*, *Prunus persica*, *Pyrus sp.* (Rosaceae), *Citrus sp.* (Rutaceae), *Populus sp.*, *Populus canadensis*, *Populus italicus*, *Salix cinerea*, *Salix triandra* (Salicaceae), *Dirca palustris* (Thymelaeaceae), *Tilia sp.* (Tiliaceae), *Ulmus sp.*, *Ulmus laciniata* (Ulmaceae) (Noyes, 2019).

Hosts: *Brachys tessellatus* (Coleoptera: Buprestidae), *Baliosus ruber*, *Microrhopala xerene*, *Zeugophora sp.* (Chrysomelidae), *Orchestes avellanae*, *Orchestes fagi*, *Rhynchaenus alni*, *Rhynchaenus populi*, *Rhynchaenus quercus*, *Rhynchaenus rufipes* (Curculionidae), *Agromyza sp.*, *Agromyza clara*, *Agromyza melampyga*, *Agromyza oryzae*, *Agromyza parvicornis*, *Liriomyza melampyga*, *Liriomyza pictella*, *Liriomyza sativae*, *Meoneura obscurella*, *Napomyza lonicerella*, *Phytagromyza populi*, *Phytagromyza populicola*, *Phytomyza sp.*, *Phytomyza aquilegiae*, *Phytomyza heringiana*, *Phytomyza horticola*, *Phytomyza ilicicola*, *Phytomyza ilicis*, *Phytomyza minuscula*, *Phytomyza obscurella*, *Phytomyza phillyreae* (Diptera: Agromyzidae), *Arge pectoralis* (Hymenoptera: Argidae), *Fenusa pusilla*, *Heterarthrus nemoratus*, *Kaliofenusa ulmi*, *Messa hortulana*, *Messa nana*, *Metallus bethunei*, *Profenusa pygmaea* (Tenthredinidae), *Coleophora sp.*, *Coleophora fuscedinella*, *Coleophora kroneella*, *Coleophora lutipennella*, *Coleophora pruniella*, *Coleophora prunifoliae*, *Coleophora serratella* (Lepidoptera: Coleophoridae), *Biselachista fulgens*, *Elachista sp.*, *Elachista poae* (Elachistidae), *Eriocrania semipurpurella*, *Eriocrania sparmannella* (Eriocraniidae), *Coleotechnites thujaella*, *Exoteleia pinifoliella*, *Recurvaria thujaella* (Gelechiidae), *Callicercops sp.*, *Caloptilia syringella*, *Cameraria sp.*, *Cameraria guttifinitella*, *Cameraria hamadryadella*, *Cameraria ohridella*, *Gracillaria syringella*, *Leucanthiza dircella*, *Liocrobyla lobata*, *Lithocolletis sp.*, *Lithocolletis blancardella*, *Lithocolletis cerasicolella*, *Lithocolletis distentella*, *Lithocolletis froelichiella*, *Lithocolletis guttifinitella*, *Lithocolletis lucetiella*, *Lithocolletis malimalifoliella*, *Lithocolletis manni*, *Lithocolletis messaniella*, *Lithocolletis platani*, *Lithocolletis populiella*, *Lithocolletis populifoliella*, *Lithocolletis quercifoliella*, *Lithocolletis tremuloidiella*, *Parectopa robiniella*, *Parornix geminatella*, *Phyllocnistis sp.*, *Phyllocnistis citrella*, *Phyllocnistis insignis*, *Phyllocnistis labyrinthella*, *Phyllocnistis magnoliella*, *Phyllocnistis populiella*, *Phyllocnistis salicifoliella*, *Phyllocnistis suffusella*, *Phyllocnistis unipunctella*, *Phyllocnistis vitegenella*, *Phyllonorycter sp.*, *Phyllonorycter blancardella*,

Phyllonorycter caryaefoliella, Phyllonorycter celtidis, Phyllonorycter cerasicolella, Phyllonorycter corylifoliella, Phyllonorycter crataegella, Phyllonorycter hamameliella, Phyllonorycter maestingella, Phyllonorycter malimalifoliella, Phyllonorycter ostryarella, Phyllonorycter propinquinella, Phyllonorycter quercifoliella, Phyllonorycter salicifoliella, Phyllonorycter tenerella, Phyllonorycter tremuloidiella (Gracillariidae), *Antispila nydaefoliella, Coptodisca arbutiella, Coptodisca splendoriferella* (Heliozelidae), *Paraclemensia acerifoliella* (Incurvariidae), *Leucoptera coffeella, Leucoptera laburnella, Leucoptera malifoliella, Leucoptera scitella, Leucoptera sinuella, Lyonetia sp., Lyonetia clerkella, Lyonetia latistrigella, Paraleucoptera sinuella* (Lyonetiidae), *Chrysopeleia ostryaeella, Cosmopteryx sp.* (Momphidae), *Nepticula sp., Nepticula lapponica, Nepticula malella, Stigmella malella* (Nepticulidae), *Tischeria sp., Tischeria complanella, Tischeria decidua, Tischeria ekebladella, Tischeria malifoliella, Tischeria purinosella* (Tischeriidae), *Argyresthia thuiella* (Yponomeutidae) (Noyes, 2019).
Parasitoid hosts: *Apanteles sp., Apanteles circumscriptus, Apanteles lautellus, Pholetesor ornigis, Pholetesor phaetusa* (Hymenoptera: Braconidae), *Achrysocharoides latreillei, Chrysocharis sp., Chrysocharis Phryne, Pediobius Alcaeus, Pnigalio mediterraneus, Winnemana argei* (Eulophidae), *Exephanes sp.* (Ichneumonidae) (Noyes, 2019).
Parasitoids: *Pnigalio agraules* (Eulophidae) (Noyes, 2019).
Distribution: Australia, Austria, Belgium, Bosnia Hercegovina, Bulgaria, Canada, Croatia, Czech Republic, Czechoslovakia, England, Europe, France, Germany, Hungary, Israel, Italy, Japan, Macedonia, Moldova, Nearctic, Netherlands, Norway, People's Republic of China, Poland, Romania, Russia, Serbia, Slovakia, Sweden, Switzerland, Thailand, Turkey, Ukraine, United States of America, USSR, Yemen, Yugoslavia (pre-1991) (Noyes, 2019).

Genus *Entedon* (Dalman, 1820)
Entedon abdera (Walker, 1839)
Publications with Turkey records: Öncüer (1991), Noyes (2019).
Associates: *Linaria dalmatica* (Scrophulariaceae) (Noyes, 2019).
Hosts: *Gymnetron sp.* (Coleoptera: Curculionidae) (Noyes, 2019).
Distribution: Bulgaria, Czech Republic, Czechoslovakia, England, Germany, Hungary, Ireland, Macedonia, Netherlands, People's Republic of China, Romania, Russia, Slovakia, Sweden, Turkey, Ukraine (Noyes, 2019).

Entedon adiyamanicus (Doğanlar, Doğanlar and Boyadzhiev, 2013)
Publications with Turkey records: Doğanlar et al. (2013), Noyes (2019).
Associates: Unknown (Noyes, 2019).
Hosts: Unknown (Noyes, 2019).
Distribution: Turkey (Noyes, 2019).

Entedon akdagicus (Doğanlar and Doğanlar, 2013)
Publications with Turkey records: Doğanlar and Doğanlar (2013), Noyes (2019).
Associates: Unknown (Noyes, 2019).
Hosts: Unknown (Noyes, 2019).
Distribution: Turkey (Noyes, 2019).

Entedon angorensis (Gumovsky, 1999)
Publications with Turkey records: Noyes (2019).
Associates: Unknown (Noyes, 2019).
Hosts: Unknown (Noyes, 2019).
Distribution: Turkey (Noyes, 2019).

Entedon arabanensis (Doğanlar and Doğanlar, 2013)
Publications with Turkey records: Doğanlar and Doğanlar (2013), Noyes (2019).
Associates: *Narcissus pseudonarsissus* (Amaryllidaceae), *Vicia sp.* (Fabaceae) (Noyes, 2019).
Distribution: Turkey (Noyes, 2019).

Entedon bakacakicus (Doğanlar, 2013)
Publications with Turkey records: Doğanlar (2013), Doğanlar et al. (2013), Noyes (2019).
Associates: *Cirsium sp.* (Asteraceae) (Noyes, 2019).
Distribution: Turkey (Noyes, 2019).

Entedon biroi (Erdos, 1944)
Publications with Turkey records: Noyes (2019), Öncüer (1991).
Associates: *Robinia pseudoacacia* (Fabaceae) (Noyes, 2019).
Distribution: Bulgaria, Czech Republic, Czechoslovakia, Hungary, Iran, Kazakhstan, Moldova, Mongolia, Slovakia, Tselinograd Obl., Turkey, Ukraine, United States of America, USSR, Uzbekistan (Noyes, 2019).

Entedon calcicola (Graham, 1971)
Publications with Turkey records: Noyes (2019), Öncüer (1991).

Associates: Unknown (Noyes, 2019).
Hosts: Unknown (Noyes, 2019).
Distribution: England, Germany, Montenegro, Slovakia, Sweden, Turkey, Yugoslavia (pre-1991) (Noyes, 2019).

Entedon cionobius (Thomson, 1878)
Publications with Turkey records: Noyes (2019), Öncüer (1991).
Associates: *Onopordum nervosum* (Asteraceae), *Fraxinus sp.* (Oleaceae), *Scrophularia sp., Scrophularia aquatica, Scrophularia nodosa, Verbascum sp., Verbascum thapsus* (Scrophulariaceae) (Noyes, 2019).
Hosts: *Cionus sp., Cionus clairvillei, Cionus hortulanus, Cionus longicollis, Cionus nigritarsis, Cionus scrophulariae, Cionus tuberculosus, Cleopus pulchellus, Rhynchaenus fagi, Stereonychus sp., Stereonychus fraxini* (Coleoptera: Curculionidae) (Noyes, 2019).
Distribution: Bulgaria, Canary Islands, Croatia, England, Europe, France, Germany, Hungary, Montenegro, Netherlands, Spain, Sweden, Turkey, Ukraine, Yugoslavia (pre-1991) (Noyes, 2019).

Entedon fufius (Walker, 1846)
Publications with Turkey records: Noyes (2019), Öncüer (1991).
Associates: *Lotus corniculatus* (Fabaceae), *Rosmarinus officinalis* (Lamiaceae) (Noyes, 2019).
Hosts: *Apion sp.* (Coleoptera: Apionidae), *Meligethes bidentatus* (Nitidulidae) (Noyes, 2019).
Distribution: Bulgaria, England, Europe, Hungary, Mongolia, Romania, Russia, Spain, Sweden, Turkey (Noyes, 2019).
Entedon gumushanensis (Doğanlar and Doğanlar, 2013)
Publications with Turkey records: Doğanlar and Doğanlar (2013), Noyes (2019).
Associates: Unknown (Noyes, 2019).
Hosts: Unknown (Noyes, 2019).
Distribution: Turkey (Noyes, 2019).

Entedon hercyna (Walker, 1839)
Publications with Turkey records: Noyes (2019), Öncüer (1991).
Associates: *Verbascum phlomoides* (Scrophulariaceae) (Noyes, 2019).
Hosts: *Gymnetron asellus* (Coleoptera: Curculionidae) (Noyes, 2019).
Distribution: Bulgaria, Czech Republic, Czechoslovakia, England, Hungary, Netherlands, Norway, Russia, Slovakia, Sweden, Turkey (Noyes, 2019).

Entedon insignis (Erdos, 1944)
Publications with Turkey records: Noyes (2019), Öncüer (1991).
Associates: *Larinus turbinatus* (Curculionidae), *Carduus sp., Carduus bourgeanus, Carlina sp., Centaurea sp., Centaurea scabiosa, Cirsium sp., Cirsium arvense* (Asteraceae) (Noyes, 2019).
Hosts: *Meliboeoides amethystinus* (Coleoptera: Buprestidae), *Larinus sp., Larinus sturnus, Larinus turbinatus, Lixus sp., Lixus elongatus, Lixus filiformis, Rhinocyllus sp.* (Curculionidae) (Noyes, 2019).
Distribution: Bulgaria, Croatia, Czech Republic, Czechoslovakia, Europe, France, Germany, Hungary, Italy, Macedonia, Montenegro, Netherlands, Spain, Sweden, Turkey (Noyes, 2019).

Entedon lixi (Erdos, 1951)
Publications with Turkey records: Noyes (2019), Öncüer (1991).
Associates: *Carduus acanthoides, Carduus nutans* (Asteraceae) (Noyes, 2019).
Hosts: *Lixus elongatus* (Coleoptera: Curculionidae) (Noyes, 2019).
Distribution: Austria, Czech Republic, Czechoslovakia, England, France, Germany, Hungary, Netherlands, Russia, Slovakia, Sweden, Turkey (Noyes, 2019).

Entedon longus (Boucek, 1968)
Publications with Turkey records: Noyes (2019), Öncüer (1991).
Associates: *Rumex sp.* (Polygonaceae) (Noyes, 2019).
Hosts: *Apion sp., Apion arrogans* (Coleoptera: Apionidae) (Noyes, 2019).
Distribution: Bulgaria, Germany, Hungary, Israel, Sweden, Turkey, United Kingdom (Noyes, 2019).

Entedon nevsehiricus (Doğanlar and Doğanlar, 2013)
Publications with Turkey records: Doğanlar and Doğanlar (2013), Noyes (2019).
Associates: Unknown (Noyes, 2019).
Hosts: Unknown (Noyes, 2019).
Distribution: Turkey (Noyes, 2019).

Entedon nizipicus (Doğanlar, Doğanlar and Boyadzhiev, 2013)
Publications with Turkey records: Doğanlar et al. (2013), Noyes (2019).
Associates: *Narcissus pseudonarsissus* (Amaryllidaceae) (Noyes, 2019).
Distribution: Turkey (Noyes, 2019).

Entedon pallicrus (Erdos, 1944)
Publications with Turkey records: Noyes (2019), Öncüer (1991).
Associates: *Apion aestivum, Apion apricans, Apion pavidum* (Coleoptera: Apionidae), *Coronilla varia, Lotus corniculatus, Medicago sativa, Melilotus latissimus, Trifolium sp., Trifolium pratense* (Fabaceae) (Noyes, 2019).
Hosts: *Apion apricans, Apion pavidum, Apion trifolii* (Coleoptera: Apionidae) (Noyes, 2019).
Distribution: Bulgaria, Czech Republic, England, France, Greece, Hungary, Kazakhstan, Moldova, Netherlands, Russia, Slovakia, Sweden, Tselinograd Obl., Turkey, Ukraine, Uzbekistan (Noyes, 2019).

Entedon parvicalcar (Thomson, 1878)
Publications with Turkey records: Noyes (2019), Öncüer (1991).
Associates: *Carduus bourgeanus, Cirsium sp.* (Asteraceae), *Eruca vesicaria* (Brassicaceae), *Lotus corniculatus, Medicago sativa, Trifolium sp.* (Fabaceae), *Festuca pratensis* (Poaceae), *Pyrus sp.* (Rosaceae) (Noyes, 2019).
Hosts: *Leucoptera spartifoliella* (Lepidoptera: Lyonetiidae) (Noyes, 2019).
Distribution: Bulgaria, Croatia, Czech Republic, Czechoslovakia, England, Germany, Hungary, Kazakhstan, Moldova, Montenegro, Netherlands, Romania, Russia, Serbia, Spain, Sweden, Tselinograd Obl., Turkey, Ukraine, USSR, Yugoslavia (pre-1991) (Noyes, 2019).

Entedon pinetorum (Ratzeburg, 1852)
Publications with Turkey records: Noyes (2019).
Hosts: *Carphoborus minimus, Ips sexdentatus* (Coleoptera: Scolytidae) (Noyes, 2019).
Distribution: Germany, Hungary, Turkey (Noyes, 2019).

Entedon procioni (Erdos, 1944)
Publications with Turkey records: Noyes (2019), Öncüer (1991).
Associates: *Apion meliloti, Cytisus scoparius, Melilotus sp., Retama sphaerocarpa* (Fabaceae), *Urtica sp., Urtica dioica* (Urticaceae).
Hosts: *Apion seniculus, Apion urticarium* (Coleoptera: Apionidae) (Noyes, 2019).
Distribution: Bulgaria, Czech Republic, England, France, Germany, Hungary, Netherlands, Russia, Slovakia, Spain, Sweden, Turkey, Ukraine (Noyes, 2019).

Entedon sarkislaensis (Doğanlar and Doğanlar, 2013)
Publications with Turkey records: Doğanlar and Doğanlar (2013), Noyes (2019).
Associates: *Narcissus pseudonarsissus* (Amaryllidaceae), *Onobrychis sp., Vicia sativa* (Fabaceae) (Noyes, 2019).
Distribution: Turkey (Noyes, 2019).

Entedon susuzensis (Doğanlar and Doğanlar, 2013)
Publications with Turkey records: Doğanlar and Doğanlar (2013), Noyes (2019).
Associates: Unknown (Noyes, 2019).
Hosts: Unknown (Noyes, 2019).
Distribution: Turkey (Noyes, 2019).

Entedon thomsonianus (Erdos, 1944)
Publications with Turkey records: Noyes (2019), Öncüer (1991).
Associates: *Carduus sp., Carlina sp., Onopordum sp., Onopordum acanthium, Onopordum nervosum* (Asteraceae) (Noyes, 2019).
Hosts: *Meliboeus amethystinus* (Coleoptera: Buprestidae), *Larinus sp., Lixus sp., Lixus cardui* (Curculionidae) (Noyes, 2019).
Distribution: Azerbaijan, Bulgaria, Europe, France, Georgia, Germany, Greece, Hungary, Iran, Israel, Macedonia, Romania, Russia, Spain, Turkey, Turkmenistan, Ukraine (Noyes, 2019).

Genus *Neochrysocharis* (Kurdjumov, 1912)
Neochrysocharis aratus (Walker, 1838)
Publications with Turkey records: Noyes (2019), Yefremova (2010).
Associates: *Hordeum vulgare, Triticum aestivum* (Poaceae), *Coffea sp.* (Rubiaceae) (Noyes, 2019).
Hosts: *Chromatomyia fuscula* (Diptera: Agromyzidae), *Oscinella frit* (Chloropidae), *Leucoptera coffeella* (Lepidoptera: Lyonetiidae) (Noyes, 2019).
Distribution: Bulgaria, Canada, Canary Islands, Czech Republic, Czechoslovakia, Denmark, Finland, Germany, Greece, Hungary, Iran, Ireland, Israel, Moldova, Montenegro, Netherlands, Norway, Peru, Poland, Portugal, Russia, Slovakia, Spain, Sweden, Turkey, Ukraine, United Kingdom, United States of America, USSR, Yugoslavia (pre-1991) (Noyes, 2019).

Neochrysocharis arvensis (Graham, 1963)
Publications with Turkey records: Gencer (2009), Noyes (2019).

Associates: *Helianthus annuus* (Asteraceae), *Coffea arabica* (Rubiaceae) (Noyes, 2019).
Hosts: *Agromyza arenaria, Chromatomyia horticola* (Diptera: Agromyzidae), *Leucoptera coffeella* (Lepidoptera: Lyonetiidae) (Noyes, 2019).
Distribution: Czech Republic, Denmark, England, Finland, Greece, Mexico, Montenegro, Romania, Russia, Sweden, Turkey, Yugoslavia (pre-1991) (Noyes, 2019).

Neochrysocharis chlorogaster (Erdos, 1966)
Publications with Turkey records: Gencer (2009), Noyes (2019), Ulubilir and Yabas (2000), Yabas and Ulubilir (1995).
Associates: *Angelica sylvestris* (Apiaceae), *Brassica oleracea* (Brassicaceae), *Lonicera sp.* (Caprifoliaceae), *Crataegus laevigata, Crataegus monogyna, Malus pumila, Sorbus aucuparia* (Rosaceae), *Lycopersicon esculentum, Solanum melongena* (Solanaceae) (Noyes, 2019).
Hosts: *Malus pumila* (Rosaceae), *Rhamphus oxyacanthae* (Coleoptera: Curculionidae), *Chromatomyia horticola, Liriomyza trifolii, Paraphytomyza hendeliana, Phytomyza angelicastri* (Diptera: Agromyzidae), *Cameraria ohridella, Phyllonorycter blancardella, Phyllonorycter corylifoliella* (Lepidoptera: Gracillariidae), *Leucoptera malifoliella, Leucoptera scitella, Lyonetia clerkella* (Lyonetiidae), *Nepticula lapponica, Nepticula microtheriella, Nepticula salicis, Nepticula sorbi, Stigmella sp., Stigmella albifasciella, Stigmella atricapitella, Stigmella aucupariae, Stigmella caradjai, Stigmella carpinella, Stigmella centifoliella, Stigmella crataegella, Stigmella floslactella, Stigmella hahniella, Stigmella hemargyrella, Stigmella hybnerella, Stigmella ignobiliella, Stigmella loranthella, Stigmella malella, Stigmella microtheriella, Stigmella minusculella, Stigmella oxyacanthella, Stigmella roborella, Stigmella ruficapitella, Stigmella tiliae* (Nepticulidae), *Tischeria ekebladella* (Tischeriidae) (Noyes, 2019).
Distribution: Croatia, Czech Republic, Czechoslovakia, Denmark, England, Europe, Germany, Hungary, Moldova, Netherlands, Sweden, Turkey, Yugoslavia (pre-1991) (Noyes, 2019).

Neochrysocharis clarus (Szelenyi, 1977)
Publications with Turkey records: Gencer (2009), Noyes (2019).
Associates: *Helianthus annuus, Lactuca sativa, Lapsana communis* (Asteraceae), *Brassica oleracea, Cardaria draba* (Brassicaceae), *Cucumis sativus* (Cucurbitaceae) (Noyes, 2019).

Hosts: *Chromatomyia horticola* (Diptera: Agromyzidae) (Noyes, 2019).
Distribution: Hungary, Mongolia, Russia, Turkey (Noyes, 2019).

Neochrysocharis cuprifrons (Erdos, 1954)
Publications with Turkey records: Noyes (2019), Öncüer (1991).
Associates: *Agropyron intermedium, Brachypodium pinnatum, Festuca pratensis* (Poaceae), *Tamarix canariensis* (Tamaricaceae) (Noyes, 2019).
Distribution: Croatia, Czech Republic, Czechoslovakia, England, Finland, Germany, Greece, Hungary, Israel, Moldova, Netherlands, Palearctic, Romania, Russia, Serbia, Slovakia, Spain, Sweden, Turkey, Yugoslavia (pre-1991) (Noyes, 2019).

Neochrysocharis formosus (Westwood, 1833)
Publications with Turkey records Bulut and Gocmen (2000), Cikman (2012), Civelek (2002), Gencer (2009), Noyes (2019), Ulubilir and Yabas (2000), Uygun et al. (1995), Uygun et al. (1997), Yabas and Ulubilir (1995), Yefremova et al. (2010).
Associates: *Aegopodium podagraria, Apium graveolens, Eryngium campestre, Thapsia villosa* (Apiaceae), *Ilex sp.* (Aquifoliaceae), Asclepias syriaca (Asclepiadaceae), Artemisia herba-alba, Bidens sp., Erechtites hieracifolia, Gerbera sp., *Gerbera jamesoni, Lactuca sp., Lactuca sativa, Reichardia sp.* (Asteraceae), *Brassica oleracea, Eruca vesicaria, Lepidium subulatum* (Brassicaceae), *Gypsophila sp., Gypsophila struthium* (Caryophyllaceae), *Atriplex halimus, Suaeda vera* (Chenopodiaceae), *Citrullus lanatus, Cucumis melo, Cucumis sativus* (Cucurbitaceae), *Juniperus thurifera, Thuja sp., Thuja occidentalis* (Cupressaceae), *Euphorbia obtusifolia, Euphorbia regisjubae* (Euphorbiaceae), *Colutea sp., Colutea arborescens, Crotalaria laburnifolia, Laburnum anagyroides, Medicago sativa, Melilotus officinalis, Phaseolus vulgaris, Pisum sativum, Retama sphaerocarpa, Trifolium pratense, Trifolium repens, Vicia faba* (Fabaceae), *Quercus sp., Quercus borealis* (Fagaceae), *Asparagus acutifolius* (Liliaceae), *Olea europaea* (Oleaceae), *Pinus sp., Pinus banksiana, Pinus mugo, Pinus resinosa, Pinus strobus* (Pinaceae), *Agropyron sp., Hordeum vulgare, Lygeum sp.* (Poaceae), *Portulaca oleracea* (Portulaceae), *Clematis vitalba* (Ranunculaceae), *Reseda lutea* (Resedaceae), *Malus sp., Malus pumila, Prunus sp.,Prunus dulcis, Pyrus communis, Pyrus malus* (Rosaceae), *Coffea arabica, Rubus allegheniensis* (Rubiaceae), *Citrus sp., Citrus clementina, Citrus limon* (Rutaceae), *Populus sp.* (Salicaceae), *Capsicum sp., Lycopersicon esculentum, Solanum melongena, Solanum nigrum, Solanum tuberosum* (Solanaceae),

Tamarix canariensis (Tamaricaceae), *Urtica sp.* (Urticaceae) (Noyes, 2019).

Hosts: *Cassida nebulosa, Phyllotreta aeneicollis* (Coleoptera: Chrysomelidae), *Rhynchaenus quercus* (Curculionidae), *Agromyza sp., Agromyza demeijerei, Agromyza hiemalis, Agromyza parvicornis, Agromyza pusilla, Calycomyza humeralis, Chromatomyia fuscula, Chromatomyia horticola, Chromatomyia syngenesiae, Liriomyza sp., Liriomyza bryoniae, Liriomyza cicerina, Liriomyza huidobrensis, Liriomyza munda, Liriomyza pictella, Liriomyza propepusilla, Liriomyza pusilla, Liriomyza sativae, Liriomyza strigata, Liriomyza subpusilla, Liriomyza trifolii, Napomyza lonicerella, Ophiomyia phaseoli, Phytagromyza dianthicola, Phytomyza aquilegiae, Phytomyza atricornis, Phytomyza heringiana, Phytomyza horticola, Phytomyza minuscula, Phytomyza pubicornis, Phytomyza vitalbae, Pseudonapomyza dianthicola, Pseudonapomyza spicata* (Diptera: Agromyzidae), *Asphondylia punica, Cystiphora schmidti* (Cecidomyiidae), *Scaptomyza pallida* (Drosophilidae), *Dacus oleae* (Tephritidae), *Aleyrodes proletella, Bemisia tabaci, Tetralicia erianthi* (Hemiptera: Aleyrodidae), *Myzocallis trifolii* (Aphididae), *Cimbex quadrimaculata, Palaeocimbex quadrimaculatus* (Hymenoptera: Cimbicidae), *Diprion pini, Diprion simile, Neodiprion sp., Neodiprion fulviceps, Neodiprion sertifer* (Diprionidae), *Acantholyda erythrocephala, Acantholyda nemoralis, Acantholyda posticalis, Acantholyda stellata, Cephalcia signata* (Pamphiliidae), *Caliroa cerasi, Fenusa pumila, Fenusa pusilla, Profenusa sp.* (Tenthredinidae), *Bucculatrix crataegi* (Lepidoptera: Bucculatricidae), *Coleophora deauratella, Coleophora laricella (Coleophoridae), Donacivola fullawayi, Donacivola saccharella* (Elachistidae), *Chrysoesthia sexguttella, Coleotechnites sp., Coleotechnites apicitripunctella, Coleotechnites laricella, Coleotechnites milleri, Coleotechnites thujaella, Evagora apicitripunctella, Exoteleia pinifoliella, Recurvaria sp., Tuta absoluta* (Gelechiidae), *Dialectica scalariella, Lithocolletis sp., Lithocolletis blancardella, Lithocolletis cavella, Lithocolletis cerasicolella, Lithocolletis corylifoliella, Lithocolletis messaniella, Lithocolletis millierella, Lithocolletis platani, Lithocolletis roboris, Metriochroa latifoliella, Phyllocnistis citrella, Phyllonorycter sp., Phyllonorycter blancardella, Phyllonorycter celtisella, Phyllonorycter cerasicolella, Phyllonorycter corylifoliella, Phyllonorycter medicaginella* (Gracillariidae), *Antispila rivillei, Holocacista rivillei* (Heliozelidae), *Malacosoma disstria* (Lasiocampidae), *Bedellia orchilella*, *Leucoptera coffeella, Leucoptera malifoliella, Leucoptera scitella, Leucoptera sinuella, Paraleucoptera sinuella* (Lyonetiidae), *Cosmopterix pulchrimella*

(Momphidae), *Nepticula sp., Nepticula malella, Nepticula pomella, Stigmella sp., Stigmella aurella, Stigmella hahniella, Stigmella helianthemella, Stigmella jubae, Stigmella malella, Stigmella pomella, Stigmella pyri, Stigmella thuringiaca, Stigmella thymi* (Nepticulidae), *Tetralopha robustella* (Pyralidae), *Argyresthia sp., Argyresthia thuiella, Ocnerostoma sp.* (Yponomeutidae) (Noyes, 2019).
Parasitoid hosts: *Diglyphus isaea* (Hymenoptera: Eulophidae), *Trichogramma embryophagum, Trichogramma evanescens* (Trichogrammatidae) (Noyes, 2019).
Distribution: Afrotropical, Argentina, Austria, Bulgaria, Canada, Canary Islands, Croatia, Czech Republic, Czechoslovakia, Denmark, Egypt, England, Eritrea, Ethiopia, Europe, Finland, France, Germany, Greece, Guam, Hawaii, Hungary, Indonesia, Iran, Iraq, Israel, Italy, Jamaica, Japan, Jordan, Kazakhstan, Macedonia, Malaysia, Mexico, Moldova, Montenegro, Nearctic, Netherlands, North Africa, Norway, Palearctic, People's Republic of China, Poland, Romania, Russia, Serbia, Slovakia, South Africa, Spain, Sudan, Sweden, Switzerland, Taiwan, Tselinograd Obl., Turkey, Turkmenistan, Ukraine, United Arab Emirates, United States of America, USSR, Yemen, Yugoslavia (Federal Republic), Yugoslavia (pre-1991) (Noyes, 2019).

Genus *Pediobius* (Walker, 1846)
Pediobius brachycerus (Thomson, 1878)
Publications with Turkey records: Noyes (2019), Yefremova et al. (2010).
Associates: *Phragmites vulgaris* (Poaceae) (Noyes, 2019).
Hosts: *Araneus solitaria, Argiope sp., Argiope aurantia, Argiope bruennichi, Metargiope bruennichi* (Araneae: Araneidae), *Clubiona japonicola* (Clubionidae), *Tetragnatha extensa* (Tetragnathidae), *Latrodectus tredecimguttatus* (Theridiidae), *Tromatobia sp.* (Hymenoptera: Ichneumonidae) (Noyes, 2019).
Parasitoid hosts: *Arachnoleter stagnalis, Gelis sp., Pimpla sp., Pimpla aquilonia, Polysphincta sp., Tromatobia sp., Tromatobia oculatoria, Tromatobia rufopectus, Zaglyptus iwatai* (Hymenoptera: Ichneumonidae) (Noyes, 2019).
Distribution: Canada, Croatia, Czech Republic, Czechoslovakia, England, Europe, Germany, Hungary, Japan, Moldova, Montenegro, Nearctic, Netherlands, People's Republic of China, Romania, Russia, Slovakia, Sweden, Turkey, Ukraine, United States of America (Noyes, 2019).

Pediobius bruchicida (Rondani, 1872)
Publications with Turkey records: Doğanlar (1985), Noyes (2019), Öncüer (1991), Öncüer et al. (1978), Polat and Tozlu (2010).
Associates: *Gypsophila struthium* (Caryophyllaceae), *Retama sphaerocarpa, Spartium sp*. (Fabaceae), *Quercus infectoria* (Fagaceae), *Gossypium sp*. (Malvaceae), *Olea europaea* (Oleaceae), *Pinus sp., Pinus nigra* (Pinaceae), *Oryza sativa, Sorghum bicolor* (Poaceae), *Malus pumila, Prunus sp., Rosa sp*. (Rosaceae), *Salix sp*. (Salicaceae) (Noyes, 2019).
Hosts: *Bruchus sp.* (Coleoptera: Bruchidae), *Plagiodera versicolora* (Chrysomelidae), *Rhamphus oxyacanthae* (Curculionidae), *Dasineura gleditchiae* (Diptera: Cecidomyiidae), *Diplolepis mayri* (Hymenoptera: Cynipidae), *Hyphantria cunea* (Lepidoptera: Arctiidae), *Pectinophora gossypiella, Pexicopia malvella* (Gelechiidae), *Apocheima cinerarius* (Geometridae), *Euproctis sp., Euproctis chrysorrhoea, Euproctis phaeorrhoea, Leucoma salicis, Lymantria dispar, Ocneria dispar, Orgyia dubia, Stilpnotia salicis, Teia dubia* (Lymantriidae), *Phalaenoides glycinae, Sesamia cretica, Spodoptera exempta* (Noctuidae), *Thaumetopoea pityocampa, Thaumetopoea wilkinsoni* (Notodontidae), *Pieris rapae* (Pieridae), *Cnaphalocrocis medinalis* (Pyralidae), *Apotomis lutozana, Archips rosanus, Cacoecimorpha pronubana* (Tortricidae), *Plutella xylostella, Prays oleae, Yponomeuta malinellus, Yponomeuta padellus, Yponomeuta rorellus* (Yponomeutidae) (Noyes, 2019).
Parasitoid hosts: (Diptera: Tachinidae), *Apanteles sp., Apanteles kazak, Bracon brevicornis* (Hymenoptera: Braconidae), *Elasmus flabellatus* (Elasmidae), *Ooencyrtus pityocampae* (Encyrtidae), *Euplectrus laphygmae, Tetrastichus evonymellae* (Eulophidae), *Angitia armillata, Diadegma armillatum* (Ichneumonidae), *Prays oleae* (Lepidoptera: Yponomeutidae), *Cynips divisa* (Hymenoptera: Cynipidae) (Noyes, 2019).
Distribution: Afrotropical, Algeria, Australia, Bulgaria, Czech Republic, Czechoslovakia, Egypt, Europe, Greece, Hungary, India, Iran, Israel, Italy, Lebanon, Libya, Macedonia, Moldova, New Zealand, Palearctic, Papua New Guinea, Romania, Russia, Serbia, Slovakia, Spain, Sweden, Tadzhikistan, Tanzania, Turkey, Turkmenistan, Ukraine, Yemen (Noyes, 2019).

Pediobius cassidae (Erdos, 1958)
Publications with Turkey records: Noyes (2019), Öncüer (1991), Yefremova (2010).
Associates: *Carduus sp*. (Asteraceae), *Fagus sylvatica* (Fagaceae)(Noyes, 2019).
Parasitoid hosts: *Apanteles sp., Apanteles melanoscelus, Macrocentrus abdominalis, Microgaster meridiana, Microgaster tibialis* (Hymenoptera:

Braconidae), *Apechthis rufata, Diadegma trochanterata, Ephialtes rufatus, Phytodietus polyzonias, Scambus calobatus* (Ichneumonidae) (Noyes, 2019).
Hosts: *Leiopus sp.* (Coleoptera: Cerambycidae), *Cassida sp., Cassida deflorate, Cassida nebulosa, Cassida rubiginosa, Cassida viridis* (Chrysomelidae), *Apechthis rufata, Phytodietus polyzonias* (Hymenoptera: Ichneumonidae), *Phyllonorycter medicaginella* (Lepidoptera: Gracillariidae), *Lymantria dispar* (Lymantriidae), *Haritala ruralis, Sylepta ruralis* (Pyralidae), *Archips sp., Archips rosanus, Archips xylosteanus, Cacoecia rosanus, Tortrix sp., Tortrix viridana* (Tortricidae), *Yponomeuta cognatellus* (Yponomeutidae), *Zygaena sp.* (Zygaenidae) (Noyes, 2019).
Distribution: Austria, Croatia, Czech Republic, Czechoslovakia, England, France, Germany, Hungary, Iran, Israel, Italy, Macedonia, Moldova, Netherlands, Palearctic, Romania, Russia, Serbia, Slovakia, Slovenia, Sweden, Transcaucasus, Turkey, Ukraine, Yemen, Yugoslavia (Federal Republic) (Noyes, 2019).

Pediobius epigonus (Walker, 1839)
Publications with Turkey records: Noyes (2019), Öncüer (1991).
Associates: *Melica nutans, Triticum aestivum* (Poaceae) (Noyes, 2019).
Hosts: *Agromyza simplex, Melanagromyza simplex* (Diptera: Agromyzidae), *Mayetiola sp., Mayetiola destructor, Mayetiola poae, Phytophaga destructor* (Cecidomyiidae), *Oscinella sp., Oscinella frit, Oscinella pusilla* (Chloropidae), *Clanoneurum menozzii, Hydrellia griseola* (Ephydridae), *Elachista atricomella* (Lepidoptera: Elachistidae), *Tortrix viridana* (Tortricidae) (Noyes, 2019).
Distribution: Austria, Canada, Czech Republic, Czechoslovakia, Europe, Finland, France, Germany, Hungary, Israel, Italy, Moldova, Montenegro, Nearctic, Netherlands, New Zealand, Norway, Russia, Slovakia, Slovenia, Sweden, Switzerland, Turkey, Ukraine, United Kingdom, United States of America, USSR (Noyes, 2019).

Pediobius eubius (Walker, 1839)
Publications with Turkey records: Korkmaz et al. (2010), Noyes (2019), Yefremova et al. (2010).
Associates: *Harmolita longicornis* (Eurytomidae), *Betula pendula* (Betulaceae), *Agropyron sp., Alopecurus pratensis, Calamagrostis epigeios, Phalaris sp., Triticum aestivum* (Poaceae) (Noyes, 2019).
Hosts: *Hylesinus crenatus* (Coleoptera: Scolytidae), *Agromyza parvicornis* (Diptera: Agromyzidae), *Mayetiola sp.* (Cecidomyiidae), *Cephus cinctus, Cephus infuscatus* (Hymenoptera: Cephidae), *Eurytoma*

collaris, Eurytoma suecica, Eurytoma tapio, Tetramesa sp., Tetramesa angustipennis, Tetramesa calamagrostidis, Tetramesa eximia, Tetramesa fulvicollis, Tetramesa hyalipennis, Tetramesa linearis, Tetramesa petiolata, Eurytoma collaris, Eurytoma suecica, Eurytoma tapio, Tetramesa angustipennis (Eurytomidae)(Noyes, 2019).
Parasitoids: *Mesopolobus graminum* (Pteromalidae) (Noyes, 2019).
Distribution: Afghanistan, Austria, Bulgaria, Canada, Croatia, Czech Republic, Czechoslovakia, Europe, Finland, Germany, Hungary, Japan, Korea, Macedonia, Moldova, Mongolia, Montenegro, Nearctic, Netherlands, People's Republic of China, Russia, Slovakia, Slovenia, Spain, Sweden, Turkey, United Kingdom, United States of America, USSR, Yemen, Yugoslavia (Federal Republic), Yugoslavia (pre-1991) (Noyes, 2019).

Pediobius flaviscapus (Thomson, 1878)
Publications with Turkey records: Noyes (2019), Yefremova (2010).
Associates: *Rubus sp.* (Rubiaceae) (Noyes, 2019).
Hosts: *Saperda sp.* (Coleoptera: Cerambycidae), *Phytobia flavifrons, Phytobia incisa, Phytomyza ranunculi, Phytomyza rufipes, Phytomyza sphondylii* (Diptera: Agromyzidae), *Scaptomyza flava* (Drosophilidae), *Coleophora sp.* (Lepidoptera: Coleophoridae), *Eupithecia tripunctaria* (Geometridae), *Nepticula sp., Nepticula arifoliella* (Nepticulidae), *Spilonota sp.* (Tortricidae) (Noyes, 2019).
Distribution: Austria, Denmark, Europe, Germany, Israel, Japan, Korea, Moldova, Sweden, Turkey (Noyes, 2019).

Pediobius lysis (Walker, 1839)
Publications with Turkey records: Noyes (2019).
Associates: *Andricus coriarius, Andricus quercustozae, Cerroneuroterus lanuginosus, Neuroterus albipes, Neuroterus numismalis, Neuroterus quercusbaccarum, Neuroterus tricolor, Trigonaspis megaptera* (Cynipidae), *Quercus sp., Quercus cerris, Quercus faginea, Quercus infectoria, Quercus lusitanica, Quercus petraea, Quercus pyrenaica, Quercus robur* (Fagaceae) (Noyes, 2019).
Hosts: *Galerucella lineola* (Coleoptera: Chrysomelidae), *Adleria coriaria, Andricus sp., Andricus curvator, Andricus grossulariae, Andricus testaceipes, Chilaspis nitida, Cynipidae unspecified, Cynips longiventris, Neuroterus sp., Neuroterus albipes, Neuroterus lanuginosus, Neuroterus numismalis, Neuroterus quercusbaccarum, Neuroterus saliens, Neuroterus saltans, Trigonaspis sp., Trigonaspis megaptera* (Hymenoptera: Cynipidae),

Lithocolletis acerifoliella, Lithocolletis platani, Phyllonorycter sp. (Lepidoptera: Gracillariidae), *Orgyia antiqua* (Lymantriidae) (Noyes, 2019).
Parasitoids: *Mesopolobus tibialis* (Hymenoptera: Pteromalidae), *Torymus auratus* (Torymidae) (Noyes, 2019).
Distribution: Croatia, Czech Republic, Czechoslovakia, England, France, Germany, Hungary, Iran, Israel, Italy, Jordan, Moldova, Netherlands, Palearctic, Slovakia, Spain, Sweden, Turkey, Ukraine, Yugoslavia (pre-1991) (Noyes, 2019).

Pediobius metallicus (Nees, 1834)
Publications with Turkey records: Cebeci et al. (2011), Civelek (2002), Gencer (2009), Noyes (2019), Öncüer (1991), Uygun et al. (1995), Yefremova et al. (2010).
Associates: *Formica aquilonia* (Formicidae), *Heracleum sp.* (Apiaceae), *Ilex aquifolium* (Aquifoliaceae), *Cirsium sp., Cynara scolymus, Helianthus annuus, Leontodon sp., Reichardia sp., Sonchus sp., Sonchus arvensis, Sonchus oleraceus* (Asteraceae), *Betula pendula, Betula pubescens* (Betulaceae), *Brassica sp., Brassica oleracea, Cardaria draba, Raphanus sativus* (Brassicaceae), *Lonicera sp., Lonicera periclymenum* (Caprifoliaceae), *Spinacia oleracea* (Chenopodiaceae), *Cucumis sativus* (Cucurbitaceae), *Crotalaria laburnifolia, Dorycnium pentaphyllum, Medicago sativa, Pisum sativum, Retama sphaerocarpa, Tetragonolobus siliquosus* (Fabaceae), *Quercus ithaburensis* (Fagaceae), *Aesculus hippocastanum* (Hippocastanaceae), *Iris pseudacorus* (Iridaceae), *Phillyrea sp., Phillyrea latifolia* (Oleaceae), *Papaver somniferum* (Papaveraceae), *Glyceria maxima, Holcus sp., Milium effusum, Sorghum vulgare* (Poaceae), *Primula veris* (Primulaceae), *Ranunculus flammula* (Ranunculaceae), *Populus sp., Salix sp., Salix alba, Salix rubens, Salix viminalis* (Salicaceae), *Antirrhinum majus* (Scrophulariaceae), *Tamarix canariensis* (Tamaricaceae) (Noyes, 2019).
Hosts: *Agromyza sp., Agromyza angulate, Agromyza pusilla, Agromyza scutellate, Cerodontha ireos, Chromatomyia horticola, Chromatomyia syngenesiae, Dizygomyza sp., Dizygomyza incisa, Dizygomyza verbasci, Liriomyza sp., Liriomyza bryoniae, Liriomyza cicerina, Liriomyza huidobrensis, Liriomyza pusilla, Liriomyza strigata, Liriomyza trifolii, Napomyza lonicerella, Ophiomyia phaseoli, Paraphytomyza hendeliana, Paraphytomyza populi, Paraphytomyza tridentata, Phytagromyza populi, Phytagromyza populicola, Phytagromyza tridentata, Phytobia angulate, Phytomyza sp., Phytomyza atricornis, Phytomyza fuscula, Phytomyza horticola, Phytomyza ilicis, Phytomyza milii, Phytomyza obscurella, Phytomyza periclymeni, Phytomyza phillyreae, Phytomyza populi,*

Phytomyza primulae, Phytomyza ranunculi (Diptera: Agromyzidae), *Pegomya hyoscyami, Pegomya tenera* (Anthomyiidae), *Mayetiola destructor, Mayetiola Phalaris, Rabdophaga sp., Rabdophaga justini, Rabdophaga purpureaperda* (Cecidomyiidae), *Oscinella frit* (Chloropidae), *Syringophilus lhommei* (Hymenoptera: Tenthredinidae), *Elachista nigrella, Elachista poae* (Lepidoptera: Elachistidae), *Cameraria ohridella, Lithocolletis sp., Lithocolletis pastorella, Lithocolletis pomifoliella, Lithocolletis pyrifoliella, Lithocolletis salictella, Lithocolletis scitulella, Micrurapteryx kollariella, Parectopa kollariella, Parornix anguliferella, Phyllonorycter medicaginella, Phyllonorycter quercifoliella, Phyllonorycter salictella (Gracillariidae), Nepticula dorycniella, Nepticula gozmanyi, Nepticula helianthemella, Stigmella sp., Stigmella dorycniella, Stigmella gozmanyi, Stigmella helianthemella* (Nepticulidae), *Marasmia trapezalis* (Pyralidae), *Argyroploce oblongana, Endothenia gentianaeana* (Tortricidae) (Noyes, 2019).
Parasitoid hosts: *Pholetesor phaetusa* (Hymenoptera: Braconidae), *Chrysocharis pubicornis* (Eulophidae), *Itoplectis alternans* (Ichneumonidae), *Cyrtogaster vulgaris, Sphegigaster flavicornis* (Pteromalidae) (Noyes, 2019).
Parasitoids: *Eupelmus allynii* (Eupelmidae) (Noyes, 2019).
Distribution: Austria, Azerbaijan, Bulgaria, Canada, Canary Islands, Croatia, Czech Republic, Czechoslovakia, Ethiopia, Europe, Germany, Hungary, India, Iraq, Israel, Italy, Japan, Jordan, Korea, Macedonia, Madeira, Moldova, Montenegro, Nearctic, Netherlands, New Zealand, Pakistan, People's Republic of China, Romania, Russia, Serbia, Slovakia, Slovenia, Spain, Sweden, Turkey, Ukraine, United Kingdom, United States of America, USSR, Yemen, Yugoslavia (pre-1991) (Noyes, 2019).

Pediobius nigritarsis (Thomson, 1878)
Publications with Turkey records: Noyes (2019), Öncüer (1991), Yefremova et al. (2010).
Hosts: *Mayetiola destructor, Phytophaga destructor* (Diptera: Cecidomyiidae), *Cephus sp., Cephus cinctus, Cephus pygmaeus, Cephus tabidus* (Hymenoptera: Cephidae), *Trachelus tabidus* (Tenthredinidae), *Euplexia lucipara* (Lepidoptera: Noctuidae) (Noyes, 2019).
Distribution: Canada, Czech Republic, Czechoslovakia, England, Finland, France, Germany, Hungary, India, Israel, Italy, Nearctic, Netherlands, Russia, Sweden, Switzerland, Turkey, United States of America, USSR (Noyes, 2019).

Pediobius pyrgo (Walker, 1839)
Publications with Turkey records: Noyes (2019), Öncüer (1991).

Associates: *Andricus caputmedusae* (Cynipidae), *Quercus sp., Quercus pubescens* (Fagaceae), *Amygdalus communis, Crataegus sp., Cydonia vulgaris, Malus sp., Malus communis, Mespilus germanica, Prunus armeniaca, Prunus avium, Prunus cerasus, Prunus domestica, Prunus spinosa, Pyrus communis, Pyrus elaeagrifolia, Pyrus malus, Rosa canina* (Rosaceae), *Populus sp., Populus alba, Populus nigra* (Salicaceae), *Ulmus campestris* (Ulmaceae) (Noyes, 2019).

Hosts: *Malus pumila* (Rosaceae), *Ceutorhynchus maculalba, Rhynchaenus quercus* (Coleoptera: Curculionidae), *Forficula auricularia* (Dermaptera: Forficulidae), *Contarinia sorghicola* (Diptera: Cecidomyiidae), *Nemorilla floralis* (Tachinidae), *Biorhiza sp., Biorhiza pallida* (Hymenoptera: Cynipidae), *Hyphantria cunea* (Lepidoptera: Arctiidae), *Brachmia macroscopa, Coleotechnites piceaella, Recurvaria leucatella* (Gelechiidae), *Lithocolletis blancardella, Lithocolletis corylifoliella, Phyllocnistis citrella, Phyllonorycter sp., Phyllonorycter blancardella* (Gracillariidae), *Thymelicus lineola* (Hesperiidae), *Euproctis chrysorrhoea, Euproctis Melania, Euproctis phaeorrhoea, Leucoma salicis, Lymantria dispar, Nygmia phaeorrhoea, Ocneria terebynthina, Orgyia dubia, Porthetria dispar, Stilpnotia salicis* (Lymantriidae), *Leucoptera malifoliella, Leucoptera scitella, Leucoptera susinella* (Lyonetiidae), *Blepharita adusta, Orthosia carnipennis* (Noctuidae), *Clostera anachoreta* (Notodontidae), *Aporia crataegi* (Pieridae), *Fumaria casta, Fumea casta, Solenobia sp.* (Psychidae), *Etiella zinckenella, Glyphodes pyloalis* (Pyralidae), *Archippus abiephaga, Archips sp., Archips pronubana, Archips rosanus, Cacoecia murinana, Cacoecia rosanus, Cacoecia sorbiana, Cacoecimorpha pronubana, Choristoneura diversana, Choristoneura murinana, Choristoneura sorbiana, Grapholita molesta, Gypsonoma minutana, Lozotaenia coniferana, Pandemis cerasana, Pandemis ribeana, Sparganothis pilleriana, Tortrix sp., Tortrix viridana* (Tortricidae), *Cerostoma coriacellum, Yponomeuta sp., Yponomeuta evonymellus, Yponomeuta malinellus, Yponomeuta padellus, Ypsolophus coriacellus* (Yponomeutidae), *Illiberis pruni* (Zygaenidae) (Noyes, 2019).

Parasitoid hosts: *Compsilura concinnata, Nemorilla floralis, Triarthria setipennis* (Diptera: Tachinidae), *Goniozus japonicus, Perisierola sp.* (Hymenoptera: Bethylidae), *Apanteles sp., Apanteles albipennis, Apanteles fulvipes, Apanteles glomeratus, Apanteles liparidis, Apanteles melanoscelus, Apanteles rectinervis, Apanteles solitarius, Bracon sp., Bracon piger, Bracon variegator, Macrocentrus sp., Macrocentrus abdominalis, Macrocentrus resinellae, Meteorus adoxophyesi, Meteorus versicolor, Microgaster globate* (Braconidae), *Elasmus nudus, Necremnus*

metalarus (Eulophidae), *Angitia sp., Diadegma apostatae, Lissonota sp., Tranosema Arenicola, Tranosema nigricans, Zoophthorus palpator* (Ichneumonidae), *Cyclogastrella deplanata, Dibrachys sp., Dibrachys cavus, Eupteromalus hemipterus, Eupteromalus nidulans, Eupteromalus peregrinus, Psychophagus omnivorus* (Pteromalidae) (Noyes, 2019).
Distribution: Bulgaria, Canada, Canary Islands, Czech Republic, England, Europe, France, Germany, Honduras, Hungary, Iran, Iraq, Israel, Italy, Japan, Korea, Moldova, Nearctic, Netherlands, Oriental, Palearctic, People's Republic of China, Poland, Russia, Serbia, Slovakia, Spain, Sweden, Taiwan, Turkey, Ukraine, United Arab Emirates, United States of America, USSR, Yugoslavia (pre-1991) (Noyes, 2019).

Pediobius rotundatus (Fonscolombe, 1832)
Publications with Turkey records: Noyes (2019).
Associates: *Andricus burgundus, Andricus grossulariae, Andricus pseudoinflator, Chilaspis nitida, Neuroterus quercusbaccarum, Plagiotrochus amenti, Plagiotrochus australis, Plagiotrochus cardiguensis, Plagiotrochus coriaceus, Plagiotrochus panteli, Plagiotrochus quercusilicis, Plagiotrochus suberi, Plagiotrochus vilagelui, Pseudoneuroterus saliens* (Hymenoptera: Cynipidae), *Quercu coccifera, Quercus sp., Quercus faginea, Quercus ilex, Quercus ithaburensis, Quercus suber* (Fagaceae), *Tamarix canariensis* (Tamaricaceae) (Noyes, 2019).
Hosts: *Andricus burgundus, Plagiotrochus sp., Plagiotrochus fusifex, Plagiotrochus kiefferianus, Plagiotrochus quercusilicis* (Hymenoptera: Cynipidae), *Tortrix viridana* (Lepidoptera: Tortricidae) (Noyes, 2019).
Distribution: Algeria, Andorra, France, Italy, Jordan, Portugal, Spain, Tunisia, Turkey, Yemen (Noyes, 2019).

Pediobius saulius (Walker, 1839)
Publications with Turkey records: Cebeci et al. (2011), Noyes (2019), Önceler (1991).
Associates: *Andricus caputmedusae, Andricus hungaricus, Dryocosmus kuriphilus, Neuroterus albipes, Neuroterus numismalis* (Cynipidae), *Alnus sp., Alnus glutinosa, Betula verrucose, Carpinus betulus, Corylus avellana* (Betulaceae), *Laburnum anagyroides, Robinia sp.* (Fabaceae), *Quercus sp., Quercus ithaburensis, Quercus petraea, Quercus robur* (Fagaceae), *Aesculus hippocastanum* (Hippocastanaceae), *Platanus sp., Platanus orientalis* (Platanaceae), *Malus sp., Malus pumila, Prunus sp.* (Rosaceae), *Populus alba, Salix caprea, Salix cinerea* (Salicaceae), *Celtis australis, Ulmus sp.* (Ulmaceae) (Noyes, 2019).

Hosts: *Orchestes sp., Orchestes alni, Orchestes avellanae, Orchestes fagi, Orchestes quercus, Orchestes testaceus, Rhamphus sp., Rhamphus oxyacanthae, Rhynchaenus sp., Rhynchaenus alni, Rhynchaenus fagi, Rhynchaenus populi, Rhynchaenus quercus* (Coleoptera: Curculionidae), *Apanteles circumscriptus, Apanteles fulvipes, Calyptus minutus, Triaspis pallidipes* (Hymenoptera: Braconidae), *Bucculatrix sp., Bucculatrix crataegi, Bucculatrix ulmella* (Lepidoptera: Bucculatricidae), *Zagulajevia tadzhikiella* (Coleophoridae), *Eriocrania sparmannella* (Eriocraniidae), *Recurvaria leucatella* (Gelechiidae), *Callisto denticulella, Cameraria ohridella, Lithocolletis sp., Lithocolletis acerifoliella, Lithocolletis agilella, Lithocolletis blancardella, Lithocolletis carpinicolella, Lithocolletis cavella, Lithocolletis comparella, Lithocolletis coryli, Lithocolletis corylifoliella, Lithocolletis delitella, Lithocolletis distentella, Lithocolletis lautella, Lithocolletis malella, Lithocolletis manni, Lithocolletis messaniella, Lithocolletis millierella, Lithocolletis platani, Lithocolletis populifoliella, Lithocolletis pyrifoliella, Lithocolletis quercifoliella, Lithocolletis quinnata, Lithocolletis roboris, Lithocolletis sagitella, Lithocolletis salicicolella, Lithocolletis salictella, Lithocolletis schreberella, Lithocolletis stettinensis, Lithocolletis tremulae, Parornix persicella, Phyllocnistis sp., Phyllocnistis saligna, Phyllocnistis xenia, Phyllonorycter sp., Phyllonorycter alnifoliella, Phyllonorycter bicinctella, Phyllonorycter blancardella, Phyllonorycter cerasicolella, Phyllonorycter coryli, Phyllonorycter corylifoliella, Phyllonorycter dubitella, Phyllonorycter harrisella, Phyllonorycter issikii, Phyllonorycter lantanella, Phyllonorycter millierella, Phyllonorycter nipponicella, Phyllonorycter platani, Phyllonorycter quercifoliella, Phyllonorycter rajella, Phyllonorycter ringoniella, Phyllonorycter robiniella, Phyllonorycter sagitella, Phyllonorycter saliciphaga, Phyllonorycter sorbi, Phyllonorycter stettinensis, Phyllonorycter tenerella, Phyllonorycter tritorrhecta, Phyllonorycter ulmi, Phyllonorycter viminetorum* (Gracillariidae), *Holocasta rivillei* (Heliozelidae), *Leucoptera laburnella, Leucoptera malifoliella, Leucoptera scitella* (Lyonetiidae), *Stigmella malella* (Nepticulidae), *Tischeria sp., Tischeria dodonaea, Tischeria ekebladella, Tischeria heinemanni* (Tischeriidae), *Ancylis upupana, Apotomis lutozana, Tortrix viridana* (Tortricidae), *Yponomeuta malinellus* (Yponomeutidae) (Noyes, 2019).

Parasitoid hosts: *Apanteles sp., Apanteles circumscriptus, Apanteles fulvipes, Apanteles kuwayamai, Calyptus minutus, Triaspis pallidipes* (Hymenoptera: Braconidae), *Ageniaspis testaceipes, Holcothorax testaceipes* (Encyrtidae), *Chrysocharis orchestidis, Chrysocharis orchestis, Enaysma albiscapus, Enaysma splendens, Eulophus sp.,*

Pnigalio sp., Pnigalio pectinicornis, Sympiesis sp., Sympiesis sericeicornis, Tetrastichus ecus, Tetrastichus miser (Eulophidae), *Habrocytus diachymatis* (Pteromalidae) (Noyes, 2019).
Distribution: Austria, Bulgaria, Canary Islands, Croatia, Czech Republic, Czechoslovakia, England, Europe, France, Germany, Greece, Hungary, Iran, Israel, Italy, Japan, Kazakhstan, Korea, Moldova, Montenegro, Netherlands, Palearctic, Poland, Romania, Russia, Serbia, Slovakia, Sweden, Switzerland, Tselinograd Obl., Turkey, Turkmenistan, Ukraine, United Arab Emirates, USSR, Yemen, Yugoslavia (pre-1991) (Noyes, 2019).

Subfamily Entiinae (Hedqvist, 1974)
Genus *Euderus* (Haliday, 1844)
Euderus agrili (Boucek, 1963)
Publications with Turkey records: Noyes (2019), Öncüer (1991).
Hosts: *Agrilus sp., Agrilus aurichalceus, Agrilus suvorovi, Agrilus viridis* (Coleoptera: Buprestidae) (Noyes, 2019).
Distribution: Bulgaria, Czech Republic, Czechoslovakia, Germany, Greece, Hungary, Israel, Italy, Kirgizia, Moldova, Netherlands, Poland, Romania, Russia, Slovakia, Sweden, Tadzhikistan, Turkey, Ukraine, USSR (Noyes, 2019).
Euderus brevicornis (Boucek, 1963)
Publications with Turkey records: Noyes (2019), Öncüer (1991).
Associates: *Carduus pycnocephalus* (Asteraceae), *Ephedra nebrodensis* (Gnetaceae), *Marrubium vulgare, Salvia officinalis, Teucrium capitatum* (Lamiaceae), *Hordeum leporinum* (Poaceae), *Reseda lutea* (Resedaceae) (Noyes, 2019).
Distribution: Bulgaria, Canary Islands, Croatia, Czech Republic, Czechoslovakia, Europe, Georgia, Greece, Hungary, Iceland, Israel, Kirgizia, Moldova, Romania, Russia, Serbia, Slovakia, Spain, Transcaucasus, Turkey (Noyes, 2019).

Euderus palustris (Erdos, 1951)
Publications with Turkey records: Noyes (2019), Öncüer (1991).
Associates: Unknown (Noyes, 2019).
Hosts: Unknown (Noyes, 2019).
Distribution: Hungary, Kirgizia, Russia, Tadzhikistan, Turkey, Turkmenistan, Yemen (Noyes, 2019).

Genus *Wichmannia* (Ruschka, 1916)
Wichmannia decorata (Ruschka, 1916)

Publications with Turkey records: Noyes (2019).
Associates: *Liparthrum sp., Liparthrum albidum* (Coleoptera: Scolytidae), *Genista sp., Spartium junceum* (Fabaceae) (Noyes, 2019).
Hosts: *Liparthrum albidum* (Coleoptera: Scolytidae) (Noyes, 2019).
Distribution: Algeria, Croatia, France, Israel, North Africa, Slovenia, Turkey, Yugoslavia (pre-1991) (Noyes, 2019).

Subfamily Eulophinae
Genus *Aulogymnus* (Foerster, 1851)
Aulogymnus arsames (Walker, 1838)
Publications with Turkey records: Noyes (2019).
Associates: *Andricus conglomeratus, Andricus crispator, Andricus curvator, Andricus cydoniae, Andricus fecundatrix, Andricus inflator, Andricus kollari, Andricus lucidus, Andricus multiplicatus, Andricus pseudoinflator, Andricus quadrilineatus, Andricus quercusramuli, Andricus schroeckingeri, Andricus singularis, Andricus vindobonensis, Aphelonyx cerricola, Biorhiza pallida, Cynips disticha, Cynips quercus, Dryocosmus kuriphilus, Neuroterus albipes, Neuroterus anthracinus, Neuroterus numismalis, Neuroterus quercusbaccarum, Neuroterus tricolor, Plagiotrochus australis, Plagiotrochus quercusilicis* (Cynipidae), *Fagus sylvatica, Quercus sp., Quercus cerris, Quercus faginea, Quercus frainetto, Quercus ilex, Quercus lusitanica, Quercus pedunculiflora, Quercus petraea, Quercus pubescens, Quercus pyrenaica, Quercus robur, Quercus trojana* (Fagaceae) (Noyes, 2019).
Hosts: *Andricus crispator, Andricus curvator, Andricus infectorius, Andricus inflator, Andricus ostreus, Andricus quadrilineatus, Andricus quercusramuli, Andricus vindobonensis, Cynipidae unspecified sp., Neuroterus albipes, Neuroterus numismalis, Neuroterus quercusbaccarum, Plagiotrochus quercusilicis* (Hymenoptera: Cynipidae) (Noyes, 2019).
Parasitoids: *Torymus auratus* (Torymidae) (Noyes, 2019).
Distribution: Andorra, Austria, Channel Islands, Czech Republic, Czechoslovakia, Denmark, Finland, France, Germany, Hungary, Italy, Netherlands, Norway, Romania, Russia, Slovenia, Spain, Sweden, Turkey, Ukraine, United Kingdom (Noyes, 2019).

Aulogymnus gallarum (Linnaeus, 1761)
Publications with Turkey records: Askew et al. (2013), Noyes (2019).
Associates: *Quercus sp., Quercus calliprinos, Quercus cerris, Quercus dalechampi, Quercus faginea, Quercus infectoria, Quercus lusitanica, Quercus pedunculiflora, Quercus petraea, Quercus pubescens, Quercus pyrenaica, Quercus robur, Quercus suber, Quercus trojana* (Fagaceae);

Andricus amenti, Andricus burgundus, Andricus callidoma, Andricus caputmedusae, Andricus cecconi, Andricus conglomeratus, Andricus coriarius, Andricus crispator, Andricus curvator, Andricus cydoniae, Andricus galeatus, Andricus gallaeurnaeformis, Andricus grossulariae, Andricus kollari, Andricus lucidus, Andricus multiplicatus, Andricus quadrilineatus, Andricus quercuscalicis, Andricus quercusramuli, Andricus quercustozae, Andricus schroeckingeri, Andricus singularis, Andricus solitarius, Andricus vindobonensis, Aphelonyx cerricola, Biorhiza pallida, Callirhytis glandium, Chilaspis nitida, Cynips disticha, Cynips quercusfolii, Neuroterus albipes, Neuroterus anthracinus, Neuroterus numismalis, Neuroterus quercusbaccarum, Plagiotrochus amenti, Pseudoneuroterus saliens (Hymenoptera: Cynipidae) (Noyes, 2019).

Hosts: *Adleria sp., Adleria coriaria, Andricus sp., Andricus aestivalis, Andricus amenti, Andricus callidoma, Andricus coriarius, Andricus curvator, Andricus fecundatrix, Andricus grossulariae, Andricus kollari, Andricus multiplicatus, Andricus nudus, Andricus quadrilineatus, Andricus quercusramuli, Andricus ramuli, Biorhiza sp., Biorhiza pallida, Callirhytis glandium, Cynipidae unspecified, Cynips sp., Cynips coriaria, Cynips kollari, Cynips quercusfolii, Neuroterus sp., Neuroterus albipes, Neuroterus glandiformis, Neuroterus numismalis, Neuroterus quercusbaccarum, Trigonaspis mendesi* (Hymenoptera: Cynipidae) (Noyes, 2019).

Distribution: Andorra, Austria, Belgium, Croatia, Czech Republic, Czechoslovakia, Europe, Finland, France, Germany, Greece, Hungary, Iran, Israel, Italy, Moldova, Netherlands, Poland, Romania, Russia, Slovenia, Spain, Sweden, Turkey, Ukraine, United Kingdom, USSR, Yugoslavia (pre-1991) (Noyes, 2019).

Aulogymnus skianeuros (Ratzeburg, 1844)

Publications with Turkey records: Noyes (2019), Önuçar (1991).

Associates: *Andricus coriarius, Andricus coronatus, Andricus crispator, Andricus curvator, Andricus cydoniae, Andricus dentimitratus, Andricus grossulariae, Andricus infectorius, Andricus lignicolus, Andricus multiplicatus, Andricus quercuscalicis, Andricus quercusramuli, Andricus quercustozae, Andricus vindobonensis, Biorhiza pallida, Callirhytis glandium, Chilaspis nitida, Cynips divisa, Neuroterus numismalis, Neuroterus quercusbaccarum, Pseudoneuroterus saliens* (Cynipidae), *Baryscapus anasillus* (Eulophidae), *Quercus sp., Quercus cerris, Quercus faginea, Quercus lusitanica, Quercus pedunculiflora, Quercus petraea,*

Quercus pubescens, Quercus pyrenaica, Quercus roboris, Quercus robur (Fagaceae), *Rosa sp.* (Rosaceae) (Noyes, 2019).
Parasitoids: *Cecidostiba leucopeza* (Pteromalidae) (Noyes, 2019).
Hosts: *Andricus sp., Andricus curvator, Andricus lignicolus, Andricus quercusradicis, Andricus quercusramuli, Biorhiza pallida, Cynipidae unspecified, Diplolepis rosae, Neuroterus glandiformis, Synergus gallaepomiformis* (Hymenoptera: Cynipidae) (Noyes, 2019).
Distribution: Andorra, Austria, Belgium, Bulgaria, Croatia, Czech Republic, Czechoslovakia, Denmark, France, Germany, Greece, Hungary, Italy, Moldova, Netherlands, Poland, Romania, Russia, Slovenia, Spain, Sweden, Turkey, Ukraine, United Kingdom, Yugoslavia (pre-1991) (Noyes, 2019).

Aulogymnus trilineatus (Mayr, 1877)
Publications with Turkey records: Noyes (2019).
Associates: *Andricus caputmedusae, Andricus conglomeratus, Andricus coriarius, Andricus coronatus, Andricus dentimitratus, Andricus fecundatrix, Andricus glutinosus, Andricus grossulariae, Andricus hungaricus, Andricus kollari, Andricus lignicolus, Andricus lucidus, Andricus mitratus, Andricus pictus, Andricus polycerus, Andricus quadrilineatus, Andricus quercuscalicis, Andricus quercustozae, Andricus seckendorffi, Andricus solitarius, Andricus tomentosus, Cynips quercus, Cynips quercusfolii, Dryocosmus cerriphilus, Synophrus politus, Trigonaspis megaptera* (Hymenoptera: Cynipidae), *Eudecatoma biguttata* (Eurytomidae), *Quercus sp., Quercus cerris, Quercus faginea, Quercus frainetto, Quercus infectoria, Quercus lusitanica, Quercus petraea, Quercus pubescens, Quercus pyrenaica, Quercus robur, Quercus suber* (Fagaceae) (Noyes, 2019).
Hosts: *Adleria coriaria, Adleria coronate, Adleria glutinosa, Adleria kollari, Andricus sp., Andricus albopunctatus, Andricus caputmedusae, Andricus clementinae, Andricus coriarius, Andricus coronarius, Andricus fecundatrix, Andricus gemmeus, Andricus grossulariae, Andricus hungaricus, Andricus lignicolus, Andricus lucidus, Andricus quadrilineatus, Andricus quercuscalicis, Andricus quercusfolii, Andricus solitarius, Andricus testaceipes, Biorhiza pallida, Cynipidae unspecified sp., Cynips calycis, Cynips coronaria, Cynips quercusfolii, Diplolepis folii, Synergus pallicornis, Synophrus politus, Trigonaspis megaptera* (Hymenoptera: Cynipidae) (Noyes, 2019).
Distribution: Austria, Bulgaria, Croatia, Czech Republic, Czechoslovakia, England, France, Greece, Hungary, Italy, Netherlands,

Poland, Portugal, Romania, Russia, Slovakia, Slovenia, Spain, Sweden, Turkey, Ukraine, Yugoslavia (pre-1991) (Noyes, 2019).

Genus *Cirrospilus* (Westwood, 1832)
Cirrospilus elegantissimus (Westwood, 1832)
Publications with Turkey records: Gencer (2012), Noyes (2019).
Associates: *Alnus sp.* (Betulaceae), *Robinia sp.* (Fabaceae), *Quercus sp.* (Fagaceae), *Aesculus hippocastanum* (Hippocastanaceae), *Crataegus sp., Malus sp., Malus pumila, Prunus sp.* (Rosaceae), *Populus sp., Populus nigra, Salix alba* (Salicaceae) (Noyes, 2019).
Hosts: *Rhynchaenus sp.* (Coleoptera: Curculionidae), *Cameraria ohridella, Lithocolletis sp., Lithocolletis cerasicolella, Lithocolletis cerrutiella, Lithocolletis corylifoliella, Lithocolletis messaniella, Lithocolletis platani, Lithocolletis populiella, Lithocolletis populifoliella, Lithocolletis pyrifoliella, Lithocolletis tremulae, Lithocolletis turanica, Phyllocnistis sp., Phyllocnistis labyrinthella, Phyllocnistis suffusella, Phyllocnistis unipunctella, Phyllonorycter alnifoliella, Phyllonorycter blancardella, Phyllonorycter cerasicolella, Phyllonorycter platani, Phyllonorycter rajella, Phyllonorycter robiniella, Phyllonorycter salictella, Phyllonorycter stettinensis* (Lepidoptera: Gracillariidae), *Leucoptera scitella, Lyonetia sp.* (Lyonetiidae) (Noyes, 2019).
Distribution: Bulgaria, Czech Republic, Czechoslovakia, Europe, France, Germany, Hungary, Ireland, Israel, Italy, Macedonia, Moldova, Netherlands, People's Republic of China, Romania, Russia, Serbia, Sweden, Switzerland, Turkey, Turkmenistan, United Kingdom, Yugoslavia (pre-1991) (Noyes, 2019).

Cirrospilus ingenuus (Gahan, 1932)
Publications with Turkey records: Noyes (2019).
Associates: *Quercus ithaburensis* (Fagaceae), *Citrus sp., Citrus aurantifolia, Citrus jambhiri, Citrus limon, Citrus limonia, Citrus reticulata* (Rutaceae), *Salix sp.* (Salicaceae) (Noyes, 2019).
Hosts: (Diptera: Agromyzidae), *Phyllocnistis citrella, Phyllonorycter sp., Phyllonorycter quercifoliella* (Lepidoptera: Gracillariidae), *Leucoptera coffeella* (Lyonetiidae) (Noyes, 2019).
Parasitoid hosts: *Ageniaspis citricola* (Hymenoptera: Encyrtidae) (Noyes, 2019).
Distribution: Australia, Cyprus, India, Indonesia, Iran, Israel, Japan, Jordan, Malaysia, Mexico, Morocco, Oman, People's Republic of China, Philippines, Spain, Syria, Taiwan, Thailand, Tunisia, Turkey, United States of America, Yemen (Noyes, 2019)

Cirrospilus lyncus (Walker, 1838)
Publications with Turkey records: Cikman (2012), Noyes (2019), Öncüer (1991).
Associates: *Acer sp., Acer platanoides* (Aceraceae), *Alnus sp., Alnus glutinosa, Betula sp., Corylus avellana* (Betulaceae), *Viburnum lantana* (Caprifoliaceae), *Fagus sylvatica, Quercus sp., Quercus acutissima, Quercus dentata, Quercus ithaburensis, Quercus mongolica, Quercus robur, Quercus variabilis* (Fagaceae), *Triticum sp.* (Poaceae), *Rhamnus cathartica* (Rhamnaceae), *Cerasus sp., Crataegus sp., Crataegus laevigata, Crataegus monogyna, Malus sp., Malus domestica, Malus pumila, Malus sylvestris, Prunus cerasus, Prunus domestica, Prunus spinosa, Sorbus sp., Sorbus aucuparia, Sorbus torminalis* (Rosaceae), *Citrus sp.* (Rutaceae), *Populus alba, Populus nigra, Salix caprea, Salix cinerea* (Salicaceae), *Ulmus sp.* (Ulmaceae) (Noyes, 2019).
Hosts: *Agromyza albitarsis* (Diptera: Agromyzidae), *Enaysma splendens* (Hymenoptera: Eulophidae), *Eriocrania semipurpurella* (Lepidoptera: Eriocraniidae), *Callisto denticulella, Lithocolletis sp., Lithocolletis abrasella, Lithocolletis acerifoliella, Lithocolletis agilella, Lithocolletis blancardella, Lithocolletis cerasicolella, Lithocolletis comparella, Lithocolletis coryli, Lithocolletis corylifoliella, Lithocolletis delitella, Lithocolletis distentella, Lithocolletis dubitella, Lithocolletis lantanella, Lithocolletis messaniella, Lithocolletis oxyacanthae, Lithocolletis platani, Lithocolletis populifoliella, Lithocolletis pyrifoliella, Lithocolletis quercifoliella, Lithocolletis roboris, Lithocolletis schreberella, Lithocolletis spinicolella, Lithocolletis spinolella, Lithocolletis strigulatella, Lithocolletis ulmifoliella, Parornix anglicella, Parornix finitimella, Parornix petiolella, Phyllocnistis sp., Phyllocnistis citrella, Phyllocnistis suffusella, Phyllocnistis unipunctella, Phyllonorycter sp., Phyllonorycter blancardella, Phyllonorycter cerasicolella, Phyllonorycter coryli, Phyllonorycter corylifoliella, Phyllonorycter harrisella, Phyllonorycter hilarella, Phyllonorycter klemannella, Phyllonorycter lantanella, Phyllonorycter lautella, Phyllonorycter oxyacanthae, Phyllonorycter quercifoliella, Phyllonorycter sorbi, Phyllonorycter spinicolella* (Gracillariidae), *Leucoptera malifoliella* (Lyonetiidae), *Nepticula sp. Nepticula basalella, Nepticula catharticella, Nepticula hemargyrella, Nepticula salicis, Nepticula sorbi, Stigmella sp. Stigmella catharticella, Stigmella hemargyrella, Stigmella malella, Stigmella oxyacanthella* (Nepticulidae), *Tischeria sp., Tischeria complanella, Tischeria dodonaea, Tischeria ekebladella* (Tischeriidae) (Noyes, 2019).
Parasitoid hosts: *Apanteles sp., Apanteles circumscriptus* (Hymenoptera: Braconidae), *Chrysocharis nautius* (Eulophidae) (Noyes, 2019).

Distribution: Austria, Azerbaijan, Bosnia Hercegovina, Bulgaria, Croatia, Czech Republic, Czechoslovakia, England, Europe, Germany, Hungary, Iran, Ireland, Israel, Italy, Japan, Korea, Lithuania, Moldova, Montenegro, Netherlands, People's Republic of China, Poland, Romania, Russia, Slovakia, Spain, Sweden, Turkey, Ukraine, USSR, Yugoslavia (pre-1991) (Noyes, 2019).

Cirrospilus pictus (Nees, 1834)
Publications with Turkey records: Çikman and LaSalle (2011), Noyes (2019).
Associates: *Acer platanoides* (Aceraceae), *Schinus molle* (Anacardiaceae), *Alnus sp., Alnus glutinosa, Alnus japonica, Betula sp., Betula verrucosa* (Betulaceae), *Chamaecytisus proliferus, Laburnum anagyroides* (Fabaceae), *Fagus sylvatica, Quercus sp., Quercus petraea, Quercus robur, Quercus serrata* (Fagaceae), *Aesculus hippocastanum* (Hippocastanaceae), *Carya illinoinensis* (Juglandaceae), *Larix sp., Larix decidua* (Pinaceae), *Phalaris arundinacea* (Poaceae), *Cerasus sp., Crataegus monogyna, Malus sp., Malus pumila, Prunus sp., Pyrus communis, Pyrus pyraster, Rosa canina, Sorbus aucuparia* (Rosaceae), *Citrus sp., Citrus aurantium, Citrus clementina, Citrus limon, Citrus paradisi* (Rutaceae), *Populus sp., Populus canadensis, Populus nigra, Populus tremula, Salix sp., Salix alba, Salix caprea, Salix cuspidate, Salix purpurea, Salix rubens* (Salicaceae), *Ulmus sp., Ulmus glabra* (Ulmaceae), *Urtica sp.* (Urticaceae) (Noyes, 2019).
Hosts: *Rhamphus oxyacanthae, Rhynchaenus sp., Rhynchaenus alni, Rhynchaenus populi, Rhynchaenus quercus, Rhynchaenus rusci, Rhynchaenus testaceus, Rhynchaenus ulmi, Stereonychus fraxini* (Coleoptera: Curculionidae), *Agromyza demeijerei, Agromyza hiemalis, Japanagromyza salicifolii, Paraphytomyza populi* (Diptera: Agromyzidae), *Apanteles sp.* (Hymenoptera: Braconidae), *Aylax papaveris, Aylax rhoeadis* (Cynipidae), *Fenusa sp., Fenusa pusilla, Fenusa ulmi, Heterarthrus sp., Heterarthrus aceris, Heterarthrus nemoratus, Heterarthrus vagans, Hoplocampa sp., Messa hortulana, Nematus sp., Nematus salicis, Profenusa pygmaea, Scolioneura betulae, Scolioneura betuleti, Trichiocampus ulmi* (Tenthredinidae), *Bucculatrix crataegi, Bucculatrix ulmella* (Lepidoptera: Bucculatricidae), *Coleophora sp., Coleophora binderella, Coleophora fuscedinella, Coleophora ibipennella, Coleophora laricella, Coleophora serratella* (Coleophoridae), *Elachista cerusella* (Elachistidae), *Eriocrania semipurpurella, Eriocrania sparmannella* (Eriocraniidae), *Chrysoesthia sexguttella, Chrysoesthis drurella, Chrysopora bosae* (Gelechiidae), *Callisto denticulella, Caloptilia*

sp., Caloptilia elongella, Caloptilia stigmatella, Caloptilia syringella, Cameraria ohridella, Gracillaria sp., Gracillaria syringella, Lithocolletis sp., Lithocolletis blancardella, Lithocolletis cerasicolella, Lithocolletis connexella, Lithocolletis corylifoliella, Lithocolletis cytisella, Lithocolletis fuscedinella, Lithocolletis malella, Lithocolletis oxyacanthae, Lithocolletis pomifoliella, Lithocolletis populiella, Lithocolletis populifoliella, Lithocolletis pyrifoliella, Lithocolletis ringoniella, Lithocolletis salictella, Lithocolletis schreberella, Lithocolletis sorbi, Lithocolletis spinicolella, Lithocolletis strigulatella, Micrurapteryx sophorivora, Parornix anguliferella, Phyllocnistis sp., Phyllocnistis citrella, Phyllocnistis labyrinthella, Phyllocnistis saligna, Phyllocnistis suffusella, Phyllocnistis unipunctella, Phyllonorycter sp., Phyllonorycter alnifoliella, Phyllonorycter blancardella, Phyllonorycter cerasicolella, Phyllonorycter connexella, Phyllonorycter corylifoliella, Phyllonorycter cydoniella, Phyllonorycter cytisifoliae, Phyllonorycter dubitella, Phyllonorycter maestingella, Phyllonorycter platanoidella, Phyllonorycter pomonella, Phyllonorycter populifoliella, Phyllonorycter quercifoliella, Phyllonorycter rajella, Phyllonorycter salictella, Phyllonorycter stettinensis, Phyllonorycter strigulatella, Phyllonorycter tristrigella, Phyllonorycter ulmifoliella (Gracillariidae), *Coptodisca lucifluella, Holocacista rivillei* (Heliozelidae), *Dendrolimus spectabilis, Malacosoma sp.* (Lasiocampidae), *Ivela auripes, Leucoma salicis, Lymantria dispar, Porthetria dispar, Stilpnotia salicis* (Lymantriidae), *Cemiostoma scitella, Leucoptera sp., Leucoptera laburnella, Leucoptera malifoliella, Leucoptera scitella, Leucoptera sinuella, Lyonetia sp., Lyonetia clerkella, Paraleucoptera sinuella* (Lyonetiidae), *Cosmopterix pulchrimella* (Momphidae), *Nepticula confusella, Nepticula malella, Nepticula marginicolella, Nepticula salicis, Stigmella sp., Stigmella aurella, Stigmella incognitella, Stigmella juglandifoliella, Stigmella malella, Stigmella pyri, Stigmella trimaculella* (Nepticulidae), *Nycteola asiatica* (Noctuidae), *Cerura menciana* (Notodontidae), *Tischeria angusticollella, Tischeria ekebladella* (Tischeriidae), *Zeiraphera diniana* (Tortricidae) (Noyes, 2019).

Parasitoid hosts: *Apanteles sp., Apanteles chinensis, Apanteles circumscriptus, Apanteles euchaetis, Apanteles glomeratus, Apanteles liparidis, Rhogas pallidator* (Hymenoptera: Braconidae), *Chrysocharis nitetis, Kratoysma usticrus, Pnigalio sp.* (Eulophidae), *Cephaloglypta murinanae, Gregopimpla kuwanae* (Ichneumonidae) (Noyes, 2019).

Parasitoids: *Pnigalio agraules* (Eulophidae) (Noyes, 2019).

Distribution: Algeria, Austria, Bulgaria, Canada, Canary Islands, Croatia, Czech Republic, Czechoslovakia, Europe, Finland, France, Germany,

Hungary, Iran, Ireland, Italy, Japan, Jordan, Kazakhstan, Korea, Libya, Madeira, Malta, Moldova, Montenegro, Morocco, Nearctic, Netherlands, Norway, People's Republic of China, Romania, Russia, Serbia, Slovakia, Spain, Sweden, Switzerland, Tselinograd Obl., Turkey, United Kingdom, United States of America, USSR, Yugoslavia (pre-1991) (Noyes, 2019).

Cirrospilus staryi (Boucek, 1959)
Publications with Turkey records: Noyes (2019), Öncüer (1991).
Associates: *Malus sp., Malus pumila* (Rosaceae) (Noyes, 2019).
Hosts: *Aspilapteryx limosella, Gracillaria limosella, Lithocolletis helianthemella, Phyllocnistis citrella, Phyllonorycter sp., Phyllonorycter blancardella, Phyllonorycter corylifoliella, Phyllonorycter pyrifoliella* (Lepidoptera: Gracillariidae), *Nepticula microtheriella, Stigmella sp., Stigmella microtheriella* (Nepticulidae) (Noyes, 2019).
Distribution: Algeria, Austria, Bulgaria, Croatia, Czech Republic, Czechoslovakia, France, Georgia, Germany, Iran, Lebanon, Montenegro, North Africa, Serbia, Sweden, Transcaucasus, Turkey, Yugoslavia (pre-1991) (Noyes, 2019).

Cirrospilus viticola (Rondani, 1877)
Publications with Turkey records: Cebeci et al. (2011), Noyes (2019).
Associates: *Neuroterus numismalis* (Cynipidae), *Cornus mas* (Cornaceae), *Quercus sp., Quercus petraea, Quercus robur* (Fagaceae), *Aesculus hippocastanum* (Hippocastanaceae), *Olea europaea* (Oleaceae), *Malus pumila, Prunus sp.* (Rosaceae), *Populus italica, Populus italicus* (Salicaceae) (Noyes, 2019).
Hosts: *Orchestes quercus, Rhynchaenus sp., Rhynchaenus quercus* (Coleoptera: Curculionidae), *Biselachista fulgens* (Lepidoptera: Elachistidae), *Recurvaria leucatella* (Gelechiidae), *Aspilapteryx limosella, Cameraria ohridella, Lithocolletis sp., Lithocolletis blancardella, Lithocolletis corylifoliella, Lithocolletis delitella, Lithocolletis hortella, Lithocolletis manni, Lithocolletis messaniella, Lithocolletis populifoliella, Lithocolletis pyrifoliella, Lithocolletis quercifoliella, Lithocolletis sagitella, Lithocolletis schreberella, Lithocolletis tremulae, Lithocolletis ulmifoliella, Parornix petiolella, Phyllocnistis sp., Phyllocnistis citrella, Phyllocnistis labyrinthella, Phyllonorycter sp., Phyllonorycter blancardella, Phyllonorycter corylifoliella, Phyllonorycter robiniella* (Gracillariidae), *Antispila rivillei, Holocacista rivillei* (Heliozelidae), *Leucoptera coronillae, Leucoptera malifoliella* (Lyonetiidae), *Stigmella sp., Stigmella malella, Stigmella nitidella* (Nepticulidae), *Eublemma*

scitula (Noctuidae), *Tischeria sp., Tischeria complanella, Tischeria ekebladella* (Tischeriidae) (Noyes, 2019).
Parasitoid hosts: *Apanteles circumscriptus* (Hymenoptera: Braconidae) (Noyes, 2019).
Distribution: Austria, Azerbaijan, Bulgaria, Croatia, Czech Republic, Czechoslovakia, England, Europe, France, Germany, Hungary, Iran, Israel, Italy, Moldova, Montenegro, Netherlands, Norway, Poland, Romania, Russia, Serbia, Slovakia, Sweden, Switzerland, Turkey, Ukraine, USSR (Noyes, 2019).

Cirrospilus vittatus (Walker, 1838)
Publications with Turkey records: Noyes (2019), Öncüer (1991).
Associates: *Formica aquilonia* (Formicidae), *Acer sp.* (Aceraceae), *Pistacia lentiscus* (Anacardiaceae), *Angelica archangelica, Anthriscus sp., Anthriscus sylvestris, Foeniculum vulgare* (Apiaceae), *Artemisia vulgaris, Chrysanthemum sp., Launaea nudicaulis, Senecio sp., Sonchus arvensis* (Asteraceae), *Berberis thunbergii* (Berberidaceae), *Alnus sp., Alnus cordata, Alnus glutinosa, Alnus incana, Betula sp., Betula pendula, Betula pubescens, Betula verrucose, Carpinus betulus, Corylus avellana* (Betulaceae), *Jacaranda mimosifolia* (Bignoniaceae), *Alliaria officinalis* (Brassicaceae), *Lonicera sp., Symphoricarpos rivularis* (Caprifoliaceae), *Chenopodium album* (Chenopodiaceae), *Hypericum maculatum* (Clusiaceae), *Cornus mas* (Cornaceae), *Thuja sp., Thuja occidentalis* (Cupressaceae), *Euphorbia dendroides* (Euphorbiaceae), *Astragalus glycyphyllos, Coronilla emerus, Laburnum anagyroides, Medicago sativa* (Fabaceae), *Castanea sativa, Fagus sylvatica, Quercus sp., Quercus cerris, Quercus ilex, Quercus petraea, Quercus pubescens, Quercus robur, Quercus suber* (Fagaceae), *Ephedra nebrodensis* (Gnetaceae), *Aesculus hippocastanum* (Hippocastanaceae), *Carya illinoinensis* (Juglandaceae), *Fraxinus excelsior* (Oleaceae), A*bies alba, Abies nordmanniana, Larix sp., Pinus sp., Pinus resinosa, Pinus sylvestris* (Pinaceae), *Hordeum vulgare, Milium effusum, Phalaris arundinacea, Secale sp., Triticum aestivum* (Poaceae), *Ranunculus sp., Ranunculus flammula (Ranunculaceae), Rhamnus alaternus, Rhamnus alpina, Rhamnus cathartica, Rhamnus saxatilis* (Rhamnaceae), *Cerasus sp., Cotoneaster integerrimus, Crataegus sp., Crataegus heldreichi, Crataegus laevigata, Crataegus monogyna, Geum sp., Malus sp., Malus communis, Malus domestica, Malus pumila, Malus sylvestris, Prunus sp., Prunus amygdaliformis, Prunus cocomilia, Prunus dulcis, Prunus spinosa, Pyrus malus, Rosa sp., Rosa canina, Sorbus aucuparia* (Rosaceae), *Rubus sp.* (Rubiaceae), *Citrus sp., Citrus limon* (Rutaceae), *Populus sp., Populus*

canadensis, Populus euramericana, Populus nigra, Populus tremula, Salix sp., Salix alba, Salix caprea, Salix cinerea, Salix pentandra, Salix rubens, Salix triandra (Salicaceae), *Ulmus sp., Ulmus campestris, Ulmus glabra* (Ulmaceae), *Vitis sp.* (Vitaceae), *Phytomyza atricornis* (Diptera: Agromyzidae) (Noyes, 2019).

Hosts: *Malus pumila* (Rosaceae), *Ceutorhynchus contractus, Orchestes sp., Orchestes fagi, Rhamphus sp., Rhamphus fagi, Rhamphus oxyacanthae, Rhamphus pulicarius, Rhamphus rusci, Rhynchaenus sp., Rhynchaenus alni, Rhynchaenus fagi, Rhynchaenus populi, Rhynchaenus quercus* (Coleoptera: Curculionidae), *Agromyza sp., Agromyza albipennis, Agromyza alnibetulae, Agromyza alnivora, Agromyza demeijerei, Agromyza frontella, Agromyza nana, Chromatomyia fuscula, Chromatomyia horticola, Dizygomyza crucifericola, Dizygomyza humeralis, Liriomyza sp., Liriomyza huidobrensis, Liriomyza sativae, Liriomyza trifolii, Paraphytomyza hendeliana, Paraphytomyza heringi, Paraphytomyza populi, Paraphytomyza populicola, Paraphytomyza tridentata, Phytobia sp., Phytobia crucifericola, Phytoliriomyza jacarandae, Phytoliriomyza variegate, Phytomyza sp., Phytomyza angelicae, Phytomyza anthrisci, Phytomyza artemisivora, Phytomyza atricornis, Phytomyza chaerophylli, Phytomyza citisi, Phytomyza fuscula, Phytomyza harlemensis, Phytomyza heringiana, Phytomyza horticola, Phytomyza lonicerae, Phytomyza milii, Phytomyza obscurella, Phytomyza ranunculi, Phytomyza syngenesiae* (Diptera: Agromyzidae), *Massalongia betulifolia* (Cecidomyiidae), (Diptera), *Aleyrodes sp.* (Hemiptera: Aleyrodidae), *Asterolecanium ilicicola* (Asterolecaniidae), *Arge berberidis* (Hymenoptera: Argidae), *Phanomeris phyllotomae* (Braconidae), *Cimbex quadrimaculata, Palaeocimbex quadrimaculatus* (Cimbicidae), *Neodiprion abietis, Neodiprion nanulus* (Diprionidae), *Anoplonyx destructor, Caliroa cerasi, Fenusa ulmi, Heterarthrus sp., Heterarthrus nemoratus, Messa hortulana, Metallus gei, Pristiphora laricis, Profenusa pygmaea, Scolioneura sp., Scolioneura betulae, Scolioneura betuleti, Syringophilus sp., Syringophilus lhommei* (Tenthredinidae), *Bucculatrix crataegi* (Lepidoptera: Bucculatricidae), *Coleophora sp.* (Coleophoridae), *Elachista sp., Elachista adscitella, Elachista argentella, Elachista cerusella* (Elachistidae), *Dyseriocrania subpurpurella, Eriocrania unimaculella* (Eriocraniidae), *Evagora starki, Exoteleia dodecella, Microsetia sexguttella, Recurvaria milleri* (Gelechiidae), *Caloptilia syringella, Cameraria caryaefoliella, Cameraria ohridella, Gracillaria syringella, Lithocolletis sp., Lithocolletis anderidae, Lithocolletis blancardella, Lithocolletis corylifoliella, Lithocolletis gerasimowi, Lithocolletis nicellii, Lithocolletis pomifoliella, Lithocolletis populiella, Lithocolletis populifoliella, Lithocolletis pyrifoliella, Lithocolletis salictella, Lithocolletis strigulatella,*

Parornix anglicella, Parornix avellanella, Parornix finitimella, Parornix scoticella, Parornix torquillella, Phyllocnistis sp., Phyllocnistis citrella, Phyllocnistis labyrinthella, Phyllocnistis suffusella, Phyllocnistis unipunctella, Phyllocnistis vitegenella, Phyllonorycter sp.m Phyllonorycter alnifoliella, Phyllonorycter anderidae, Phyllonorycter blancardella, Phyllonorycter caryaealbella, Phyllonorycter cerasicolella, Phyllonorycter corylifoliella, Phyllonorycter crataegella, Phyllonorycter cydoniella, Phyllonorycter dubitella, Phyllonorycter harrisella, Phyllonorycter lautella, Phyllonorycter maestingella, Phyllonorycter messaniella, Phyllonorycter oxyacanthae, Phyllonorycter quercifoliella, Phyllonorycter quinnata, Phyllonorycter rajella, Phyllonorycter ringoniella, Phyllonorycter sorbi, Phyllonorycter spinicolella, Phyllonorycter stettinensis, Phyllonorycter strigulatella, Phyllonorycter tenerella, Phyllonorycter ulmifoliella (Gracillariidae), *Antispila sp., Antispila rivillei, Coptodisca lucifluella, Holocacista rivillei* (Heliozelidae), *Bedellia somnulentella, Leucoptera laburnella, Leucoptera malifoliella, Leucoptera scitella, Lyonetia sp., Lyonetia clerkella, Lyonetia prunifoliella* (Lyonetiidae), *Mompha raschkiella* (Momphidae), *Ectoedemia angulifasciella, Ectoedemia argentipedella, Ectoedemia euphorbiella, Ectoedemia mediofasciella, Ectoedemia spinosella, Ectoedemia subbimaculella, Ectoedemia suberis, Fedalmia zollikofferiella, Fomoria septembrella, Nepticula sp., Nepticula confusella, Nepticula heringella, Nepticula juglandifoliella, Nepticula lapponica, Nepticula malella, Nepticula mali, Nepticula pomella, Nepticula tityrella, Niepeltia minimella, Stigmella sp., Stigmella alternella, Stigmella amygdali, Stigmella anomalella, Stigmella argentipedella, Stigmella assimilella, Stigmella aucupariae, Stigmella aurella, Stigmella basiguttella, Stigmella betulicola, Stigmella betulicosa, Stigmella catharticella, Stigmella centifoliella, Stigmella confusella, Stigmella floslactella, Stigmella glutinosae, Stigmella hemargyrella, Stigmella heringella, Stigmella juglandifoliella, Stigmella lapponica, Stigmella luteella, Stigmella malella, Stigmella microtheriella, Stigmella minusculella, Stigmella myrtillella, Stigmella nanivora, Stigmella obliquella, Stigmella oxyacanthella, Stigmella perpygmaeella, Stigmella pomella, Stigmella pubescivora, Stigmella rhamnophila, Stigmella roborella, Stigmella ruficapitella, Stigmella salicis, Stigmella sorbi, Stigmella suberivora, Stigmella tityrella, Stigmella trimaculella, Stigmella ulmivora, Trifurcula cryptella* (Nepticulidae), *Tischeria angusticollella* (Tischeriidae), *Argyresthia sp., Argyresthia fundella, Argyresthia thuiella, Ocnerostoma friesei, Ocnerostoma piniariella* (Yponomeutidae) (Noyes, 2019).

Parasitoid hosts: *Colastes braconius, Mirax minuta, Phanomeris phyllotomae* (Hymenoptera: Braconidae), *Achrysocharoides latreillei,*

Chrysocharis nitetis, Pnigalio sp., Pnigalio soemius (Eulophidae) (Noyes, 2019).

Parasitoids: *Chrysocharis nephereus* (Eulophidae) (Noyes, 2019).

Distribution: Algeria, Austria, Azerbaijan, Bosnia Hercegovina, Bulgaria, Canada, Canary Islands, Croatia, Czech Republic, Czechoslovakia, Europe, France, Germany, Greece, Hungary, Iceland, Iran, Iraq, Ireland, Italy, Japan, Jordan, Latvia, Macedonia, Moldova, Montenegro, Nearctic, Netherlands, North Africa, Norway, People's Republic of China, Poland, Romania, Russia, Serbia, Slovakia, Spain, Sweden, Switzerland, Tunisia, Turkey, Ukraine, United Kingdom, United States of America, USSR, Yugoslavia (Federal Republic), Yugoslavia (pre-1991) (Noyes, 2019).

Genus *Colpoclypeus* (Lucchese, 1941)

Colpoclypeus florus (Walker, 1839)

Publications with Turkey records: Noyes (2019), Öncüer (1991).

Associates: *Diplolepis rosae* (Hymenoptera: Cynipidae), *Corylus avellana* (Betulaceae), *Glycine max* (Fabaceae), *Quercus sp.* (Fagaceae), *Fragaria sp., Fragaria ananassa, Malus sp., Malus pumila, Prunus persica, Pyrus communis, Rosa sp., Rosa damascene, Rosa woodsia* (Rosaceae), *Citrus sp.* (Rutaceae), *Populus balsamifera, Populus euramericana* (Salicaceae), *Vitis vinifera* (Vitaceae), *Sparganothis pilleriana* (Lepidoptera: Tortricidae) (Noyes, 2019).

Hosts: *Rabdophaga rosaria* (Diptera: Cecidomyiidae), *Filatima sp.* (Lepidoptera: Gelechiidae), *Lyonetia clerkella* (Lyonetiidae), *Acalla sp., Acalla logiana, Acleris sp., Acleris hastiana, Acleris holmiana, Acleris lipsiana, Acleris maccana, Acleris rhombana, Acleris schalleriana, Acleris variegana, Acroclita sp., Acroclita naevana, Adoxophyes sp., Adoxophyes orana, Adoxophyes reticulana, Ancylis comptana, Ancylis myrtillana, Aphelia paleana, Archips crataeganus, Archips oporanus, Archips podanus, Archips rosanus, Archips xylosteanus, Argyroploce lacunana, Argyrotaenia pulchellana, Argyrotaenia velutinana, Cacoecia sp., Cacoecia musculana, Cacoecia pronubana, Cacoecia xylosteana, Choristoneura lafauryana, Choristoneura rosaceana, Cnephasia chrysantheana, Cydia pomonella, Eulia ministrana, Eupoecilia ambiguella, Griselda myrtillana, Gypsonoma minutana, Hedya dimiodalba, Hedya pruniana, Hedya variegana, Lobesia botrana, Lozotaenia forsterana, Pandemis sp., Pandemis cerasana, Pandemis corylana, Pandemis dumetana, Pandemis heparana, Pandemis pyrusana, Pandemis ribeana, Pardia cynosbatella, Philedonides prodromanus, Rhopobota naevana, Sparganothis pilleriana, Spilonota ocellana,*

Syndemis sp., Syndemis musculana (Tortricidae), *Procris budensis* (Zygaenidae) (Noyes, 2019).
Distribution: Armenia, Azerbaijan, Bulgaria, Canada, Czech Republic, Czechoslovakia, Europe, France, Germany, Hungary, Italy, Moldova, Morocco, Netherlands, Poland, Romania, Russia, Slovakia, Spain, Sweden, Transcaucasus, Turkey, Ukraine, United Kingdom, United States of America, USSR (Noyes, 2019).

Genus *Dahlbominus* (Hincks, 1945)
Dahlbominus fuscipennis (Zetterstedt, 1838)
Publications with Turkey records: Noyes (2019), Öncüer (1991).
Associates: *Wolbachia sp.* (Bacteria: Wolbachiaceae), *Picea sp., Picea abies, Pinus sp., Pinus banksiana, Pinus contorta, Pinus nigra, Pinus resinosa, Pinus strobus, Pinus sylvestris* (Pinaceae) (Noyes, 2019).
Parasitoid hosts: *Banchus femoralis, Microcryptus basizonus, Pleolophus basizonus* (Hymenoptera: Ichneumonidae) (Noyes, 2019).
Parasitoids: *Eupelmella vesicularis* (Eupelmidae) (Noyes, 2019).
Hosts: *Glyptoscelis sp.* (Coleoptera: Chrysomelidae), *Pityogenes bidentatus, Scolytus mali* (Scolytidae), *Diprion sp., Diprion frutetorum, Diprion hercyniae, Diprion pallidum, Diprion pini, Diprion polytomum, Diprion simile, Gilpinia sp., Gilpinia frutetorum, Gilpinia hercyniae, Gilpinia pallida, Gilpinia polytoma, Gilpinia socia, Gilpinia virens, Neodiprion sp., Neodiprion lecontei, Neodiprion annulus, Neodiprion pinetum, Neodiprion pratti, Neodiprion sertifer, Neodiprion swainei, Neodiprion tsugae* (Hymenoptera: Diprionidae), *Anoplonyx destructor, Pristiphora abietina, Pristiphora erichsonii, Pristiphora wesmaeli* (Tenthredinidae), *Panolis flammea* (Lepidoptera: Noctuidae), *Archips xylosteanus, Tortrix viridana* (Tortricidae) (Noyes, 2019).
Distribution: Austria, Belarus, Belgium, Bosnia Hercegovina, Bulgaria, Canada, Croatia, Czech Republic, Czechoslovakia, England, Europe, Finland, Georgia, Germany, Hungary, Italy, Moldova, Montenegro, Nearctic, Netherlands, Poland, Russia, Serbia, Slovakia, Slovenia, Spain, Sweden, Turkey, United States of America, USSR, Yugoslavia (pre-1991) (Noyes, 2019).

Genus *Dicladocerus* (Westwood, 1832)
Dicladocerus westwoodii (Westwood, 1832)
Publications with Turkey records: Noyes (2019), Yefremova (2010).
Associates: *Formica aquilonia* (Hymenoptera: Formicidae), *Aster sp., Aster tripolium* (Asteraceae), *Betula pubescens* (Betulaceae), *Juniperus thurifera* (Cupressaceae), *Laburnum anagyroides* (Fabaceae), *Origanum*

sp. (Lamiaceae), *Olea europaea* (Oleaceae), *Larix sp., Larix decidua, Larix occidentalis, Pinus sp.* (Pinaceae), *Rosa canina* (Rosaceae) (Noyes, 2019).

Hosts: *Agromyza sp., Phytomyza sp., Phytomyza atricornis, Phytomyza geniculata* (Diptera: Agromyzidae), *Panaxia dominula* (Lepidoptera: Arctiidae), *Carposina scirrhosella* (Carposinidae), *Coleophora caespititiella, Coleophora dahurica, Coleophora laricella, Coleophora taeniipennella, Coleophora virgaureae* (Coleophoridae), *Evagora starki, Recurvaria nanella* (Gelechiidae), *Thaumetopoea pityocampa* (Notodontidae), *Archips oporanus, Croesia bergmanniana, Grapholita tenebrosana, Gypsonoma aceriana, Gypsonoma sociana, Laspeyresia roseticolana, Tortrix viridana, Zeiraphera diniana* (Tortricidae), *Argyresthia conjugella, Cedestis gysselinella, Prays oleae* (Yponomeutidae) (Noyes, 2019).

Distribution: Austria, Canada, Canary Islands, Croatia, Czech Republic, Czechoslovakia, Denmark, Europe, France, Germany, Greece, Hungary, Ireland, Italy, Japan, Madeira, Netherlands, People's Republic of China, Portugal, Russia, Serbia, Slovakia, Slovenia, Spain, Sweden, Switzerland, Syria, Turkey, United Kingdom, United States of America, Yugoslavia (pre-1991) (Noyes, 2019).

Genus *Diglyphus* (Walker, 1844)

Diglyphus anadolucus (Doğanlar, 1982)

Publications with Turkey records: Doğanlar (1982), Noyes (2019).

Associates: Unknown (Noyes, 2019).

Hosts: Unknown (Noyes, 2019).

Distribution: Turkey (Noyes, 2019).

Diglyphus begini (Ashmead, 1904)

Publications with Turkey records: Gencer (2009), Noyes (2019).

Associates: *Steinernema carpocapsae* (Nematoda), *Heracleum lanatum* (Apiaceae), *Bidens pilosa, Chrysanthemum sp., Cynara scolymus, Dendranthema grandiflora, Dendranthema morifolium, Jaegeria hirta, Lactuca sp., Lactuca sativa, Silybum marianum, Sonchus oleraceus, Tagetes sp., Tagetes erecta, Venegasia carpesioides* (Asteraceae), *Brassica oleracea, Hirschfeldia incana* (Brassicaceae), *Lonicera subspicata* (Caprifoliaceae), *Gypsophila paniculata* (Caryophyllaceae), *Commelina benghalensis* (Commelinaceae), *Citrullus lanatus, Cucumis melo, Cucumis sativus, Cucurbita foetidissima* (Cucurbitaceae), *Medicago sativa, Phaseolus sp., Phaseolus lunatus, Phaseolus vulgaris* (Fabaceae), *Eriodictyon crassifolium, Eriodictyon trichocalyx* (Hydrophyllaceae),

Salvia mellifera (Lamiaceae), *Allium cepa* (Liliaceae), *Platanus racemosa* (Platanaceae), *Hordeum vulgare, Triticum aestivum* (Poaceae), *Richardia mucronate* (Rubiaceae), *Citrus sp.* (Rutaceae), *Populus sp.* (Salicaceae), *Lycopersicon esculentum* (Solanaceae), *Tropaeolum majus, Tropaeolum nasturtium* (Tropaeolaceae), *Lantana camara* (Verbenaceae) (Noyes, 2019).

Hosts: *Agromyza sp., Agromyza frontella, Agromyza parvicornis, Agromyza pusilla, Agromyza scutellate, Calycomyza playptera, Chromatomyia fuscula, Chromatomyia syngenesiae, Liriomyza sp., Liriomyza ativae, Liriomyza helianthin, Liriomyza huidobrensis, Liriomyza munda, Liriomyza pictella, Liriomyza pusilla, Liriomyza sativae, Liriomyza trifolii, Phytomyza sp., Phytomyza albiceps, Phytomyza aquilegiae, Phytomyza atricornis, Phytomyza lanate, Phytomyza minuscula, Phytomyza syngenesiae* (Diptera: Agromyzidae), *Coelopoeta glutinosi* (Lepidoptera: Elachistidae), *Phyllocnistis citrella, Phyllonorycter felinelle, Nepticula juglandifoliella, Stigmella sp.* (Gracillariidae: Nepticulidae) (Noyes, 2019).

Distribution: Brazil, Canada, Canary Islands, Colombia, Czech Republic, Hawaii, Israel, Mexico, Nearctic, Norway People's Republic of China, Peru, Puerto Rico, Slovakia, Turkey, United States of America, Yemen (Noyes, 2019).

Diglyphus chabrias (Walker, 1838)

Publications with Turkey records: Noyes (2019).

Associates: *Aegopodium podagraria, Aegopodium vulgaris, Anthriscus sylvestris* (Apiaceae), *Artemisia vulgaris, Carduus bourgeanus, Lapsana communis, Senecio sp., Solidago virgaurea, Sonchus arvensis, Tanacetum vulgare, Tussilago farfara* (Asteraceae), *Impatiens parviflora* (Balsaminaceae), *Alnus incana* (Betulaceae), *Lonicera sp., Lonicera periclymenum* (Caprifoliaceae), *Astragalus glycyphyllos, Pisum sativum, Retama sphaerocarpa* (Fabaceae), *Hordeum vulgare* (Poaceae), *Aquilegia vulgaris, Ranunculus flammula* (Ranunculaceae), *Populus nigra* (Salicaceae), *Tamarix canariensis* (Tamaricaceae) (Noyes, 2019).

Hosts: *Nanophyes niger* (Coleoptera: Nanophyidae), *Agromyza sp., Agromyza alnivora, Agromyza rufipes, Amauromyza flavifrons, Chromatomyia fuscula, Chromatomyia horticola, Liriomyza sp., Liriomyza sativae, Liriomyza strigata, Paraphytomyza sp., Paraphytomyza hendeliana, Paraphytomyza populi, Phytoliriomyza sp., Phytoliriomyza melampyga, Phytoliriomyza populicola, Phytoliriomyza variegate, Phytomyza sp., Phytomyza chaerophylli, Phytomyza horticola, Phytomyza marginella, Phytomyza minuscula, Phytomyza periclymeni, Phytomyza*

pubicornis, Phytomyza ranunculi, Phytomyza solidaginis (Diptera: Agromyzidae), *Prionimera cognata, Trypeta artemisiae* (Tephritidae), *Syringopais temperatella* (Lepidoptera: Scythrididae) (Noyes, 2019).
Distribution: Bulgaria, Canary Islands, Czech Republic, Czechoslovakia, England, Germany, Hungary, Iran, Israel, Madeira, Netherlands, Norway, Pakistan, Russia, Slovakia, Spain, Sweden, Turkey (Noyes, 2019).

Diglyphus crassinervis (Erdos, 1958)
Publications with Turkey records: Bulut and Gocmen (2000), Civelek et al. (2002), Gencer (2009; 2012), Noyes (2019).
Associates: *Lactuca sativa, Tanacetum sp.* (Asteraceae), *Brassica sp.* (Brassicaceae), *Cucumis sativus* (Cucurbitaceae), *Dipsacus sylvestris* (Dipsacaceae), *Phaseolus sp., Pisum sativum, Vicia faba* (Fabaceae), *Papaver somniferum* (Papaveraceae), *Populus sp.* (Salicaceae), *Lycopersicon esculentum* (Solanaceae) (Noyes, 2019).
Hosts: *Agromyza sp., Chromatomyia horticola, Liriomyza sp., Liriomyza huidobrensis, Liriomyza sativae, Liriomyza trifolii, Paraphytomyza populi, Phytomyza sp., Phytomyza atricornis* (Diptera: Agromyzidae) (Noyes, 2019).
Distribution: Bulgaria, Canary Islands, Czech Republic, Czechoslovakia, Egypt, England, France, Germany, Hungary, Iran, Iraq, Israel, Italy, Jordan, Moldova, Morocco, People's Republic of China, Portugal, Slovakia, Spain, Sweden, Turkey, Yemen (Noyes, 2019).

Diglyphus isaea (Walker, 1838)
Publications with Turkey records: Bulut and Gocmen (2000), Civelek et al. (2002), Cikman (2012), Çikman and LaSalle, (2011), Civelek et al. (2002), Gencer (2009; 2012), Noyes (2019), Öncüer (1991), Ulubilir and Sekeroglu (1997), Ulubilir and Yabas (2000), Uygun et al. (1995), Yabas and Ulubilir (1995), Yasarakinci and Hincal (1997; 2000).
Associates: *Glomus calodonicum, Glomus fasciculatum, Glomus mosseae* (Fungus), *Diglyphus intermedius* (Hymenoptera: Eulophidae), *Steinerma feltiae* (Nematoda: Steinermidae), *Anthriscus sylvestris, Apium graveolens, Heracleum sphondylium, Pastinaca sativa* (Apiaceae), *Achillea millefolium, Anacyclus clavatus, Artemisia sp., Artemisia vulgaris, Carduus bourgeanus, Centaurea sp., Chamomilla suaveolens, Chrysanthemum sp., Cirsium lappaceum, Gerbera sp., Helianthus sp., Helianthus annuus, Lactuca sativa, Leucanthemum vulgare, Onopordum nervosum, Pulicaria sp., Pyrethrum sp., Reichardia sp., Solidago virgaurea, Sonchus sp., Sonchus arvensis, Sonchus oleraceus, Tagetes sp., Tanacetum vulgare, Xanthium sp., Zinnia sp.* (Asteraceae), *Impatiens*

parviflora (Balsaminaceae), *Brassica sp., Brassica napus, Brassica oleracea, Brassica parachinensis, Brassica rapa, Cardaria draba, Eruca sativa, Eruca vesicaria, Lepidium subulatum* (Brassicaceae), *Melandrium rubrum* (Caryophyllaceae), *Citrullus lanatus, Cucumis melo, Cucumis sativus, Cucurbita sp.* (Cucurbitaceae), *Juniperus thurifera* (Cupressaceae), *Caragana sp., Cicer arietinum, Laburnum anagyroides, Lathyrus odoratus, Medicago sativa, Melilotus officinalis, Phaseolus sp., Phaseolus vulgaris, Pisum sp., Pisum sativum, Retama sphaerocarpa, Trifolium medium, Trifolium pratense, Trifolium repens, Vicia faba* (Fabaceae), *Rosmarinus officinalis* (Lamiaceae), *Alopecurus myosuroides, Holcus mollis, Hordeum vulgare, Triticum sp.* (Poaceae), *Portulaca oleracea* (Portulaceae), *Aquilegia vulgaris, Ranunculus sp., Ranunculus asiaticus, Ranunculus flammula* (Ranunculaceae), *Malus pumila* (Rosaceae), *Citrus sp.* (Rutaceae), *Populus sp.* (Salicaceae), *Osyris alba* (Santalaceae), *Antirrhinum majus* (Scrophulariaceae), *Capsicum sp., Lycopersicon esculentum, Solanum melongena, Solanum tuberosum* (Solanaceae), *Tamarix canariensis* (Tamaricaceae), *Urtica sp.* (Urticaceae) (Noyes, 2019).

Hosts: *Lycopersicon esculentum* (Solanaceae), *Agromyza sp., Agromyza albipennis, Agromyza albitarsis, Agromyza frontella, Agromyza hiemalis, Agromyza nana, Agromyza oryzae, Amauromyza flavifrons, Cerodontha sp., Chromatomyia sp., Chromatomyia fuscula, Chromatomyia horticola, Chromatomyia syngenesiae, Dizygomyza crucifericola, Dizygomyza humeralis, Dizygomyza lateralis, Japanagromyza sp., Liriomyza sp., Liriomyza bryoniae, Liriomyza cicerina, Liriomyza congesta, Liriomyza Fasciola, Liriomyza huidobrensis, Liriomyza impatientis, Liriomyza pusilla, Liriomyza pusio, Liriomyza sativae, Liriomyza strigata, Liriomyza trifolii, Napomyza sp., Paraphytomyza populi, Phytagromyza dianthicola, Phytagromyza jannonei, Phytoagromyza sp., Phytobia crucifericola, Phytomyza sp., Phytomyza affinis, Phytomyza angelicae, Phytomyza anthrisci, Phytomyza atricornis, Phytomyza caulinaris, Phytomyza chaerophylli, Phytomyza citisi, Phytomyza echinopis, Phytomyza fuscula, Phytomyza horticola, Phytomyza lappina, Phytomyza melana, Phytomyza minuscula, Phytomyza mylini, Phytomyza orobanchia, Phytomyza petoei, Phytomyza plantaginis, Phytomyza ranunculi, Phytomyza solidaginis, Phytomyza sphondylii, Phytomyza syngenesiae, Phytomyza tetrasticha, Phytomyza vitalbae, Pseudonapomyza dianthicola* (Diptera: Agromyzidae), *Diptera indet., Trypeta artemisiae* (Tephritidae), *Phyllocnistis citrella, Phyllonorycter blancardella, Phyllonorycter corylifoliella* (Lepidoptera: Gracillariidae), *Leucoptera malifoliella,*

Lyonetia clerkella (Lyonetiidae), *Stigmella malella* (Nepticulidae) (Noyes, 2019).
Parasitoids: *Neochrysocharis formosa* (Hymenoptera: Eulophidae) (Noyes, 2019).
Distribution: Australia, Austria, Azerbaijan, Belgium, Brazil, Bulgaria, Canada, Canary Islands, Channel Islands, Croatia, Czech Republic, Denmark, Eritrea, Ethiopia, Europe, Finland, France, Germany, Greece, Greenland, Hungary, Iceland, India, Iran, Iraq, Israel, Italy, Japan, Jordan, Korea, Lithuania, Macedonia, Madeira, Malta, Moldova, Mongolia, Montenegro, Morocco, Netherlands, New Zealand, North Africa, Norway, Pakistan, People's Republic of China, Poland, Portugal, Romania, Russia, Serbia, Slovakia, Slovenia, South Africa, Spain, Sweden, Switzerland, Syria, Turkey, Ukraine, United Arab Emirates, United Kingdom, United States of America, USSR, Yemen, Yugoslavia (pre-1991) (Noyes, 2019).

Diglyphus minoeus (Walker, 1838)
Publications with Turkey records: Cikman (2012), Noyes (2019).
Associates: *Ilex radicans* (Aquifoliaceae), *Artemisia vulgaris, Sonchus sp., Sonchus arvensis, Tagetes sp.* (Asteraceae), *Lonicera sp., Lonicera alpigena, Lonicera gracilipes, Symphoricarpos rivularis* (Caprifoliaceae), *Cucumis sativus* (Cucurbitaceae), *Laburnum anagyroides, Medicago sativa, Pisum sativum* (Fabaceae), *Fagus sylvatica* (Fagaceae), *Fraxinus excelsior* (Oleaceae), *Alopecurus myosuroides, Hordeum vulgare, Milium effusum* (Poaceae), *Clematis vitalba, Thalictrum aquilegifolium* (Ranunculaceae), *Prunus sp.* (Rosaceae), *Populus sp., Populus nigra, Salix fragilis* (Salicaceae), *Lycopersicon esculentum* (Solanaceae) (Noyes, 2019).
Hosts: *Agromyza sp., Agromyza demeijerei, Calycomyza artemisiae, Chromatomyia fuscula, Chromatomyia horticola, Chromatomyia suikazurae, Liriomyza sp., Liriomyza bryoniae, Liriomyza congestaLiriomyza huidobrensis, Liriomyza sativae, Liriomyza strigata, Liriomyza trifolii, Napomyza lonicerella, Napomyza xylostei, Paraphytomyza hendeliana, Paraphytomyza heringi, Paraphytomyza populi, Paraphytomyza tridentata, Phytagromyza atricornis, Phytagromyza cytisi, Phytagromyza populi, Phytagromyza vitalbae, Phytomyza atricornis, Phytomyza citisi, Phytomyza fuscula, Phytomyza harlemensis, Phytomyza horticola, Phytomyza jucunda, Phytomyza lonicerae, Phytomyza milii, Phytomyza ranunculi, Phytomyza vitalbae* (Diptera: Agromyzidae), *Lithocolletis cerasicolella, Lithocolletis millierella, Phyllonorycter sp., Phyllonorycter cerasicolella* (Lepidoptera: Gracillariidae), *Stigmella aucupariae* (Nepticulidae) (Noyes, 2019).

Distribution: Austria, Barbados, Bulgaria, Croatia, Czech Republic, Czechoslovakia, England, Europe, Germany, Hungary, Iran, Israel, Italy, Japan, Korea, Malta, Moldova, Netherlands, Norway, Pakistan, People's Republic of China, Romania, Russia, Slovakia, Sweden, Switzerland, Turkey, Turkmenistan, Yemen (Noyes, 2019).

Diglyphus pachyneurus (Graham, 1963)
Publications with Turkey records: Cikman (2012), Çikman and LaSalle (2011), Gencer (2009; 2012), Noyes (2019), Öncüer (1991).
Associates: *Xanthium sp.* (Asteraceae), *Cardaria draba* (Brassicaceae), *Portulaca oleracea* (Portulaceae), *Populus sp.* (Salicaceae) (Noyes, 2019).
Hosts: *Agromyza salicifolii, Chromatomyia horticola, Japanagromyza salicifolii, Liriomyza trifolii* (Diptera: Agromyzidae) (Noyes, 2019).
Distribution: Croatia, Germany, Italy, Moldova, Montenegro, Netherlands, People's Republic of China, Russia, Sweden, Turkey, United Kingdom, Yemen (Noyes, 2019).

Diglyphus poppoea (Walker, 1848)
Publications with Turkey records: Noyes (2019).
Associates: *Anthriscus sylvestris* (Apiaceae), *Carduus lanceolata, Chrysanthemum sp., Lactuca sativa* (Asteraceae), *Betula pendula* (Betulaceae), *Dianthus barbatus* (Caryophyllaceae), *Cucumis sativus* (Cucurbitaceae), *Phaseolus sp.* (Fabaceae), *Alopecurus myosuroides, Elytrigia repens* (Poaceae), *Lycopersicon esculentum* (Solanaceae)(Noyes, 2019).
Hosts: *Agromyza sp., Chromatomyia horticola, Liriomyza sp., Liriomyza bryoniae, Liriomyza huidobrensis, Liriomyza trifolii, Phytomyza atricornis, Phytomyza fuscula, Phytomyza syngenesiae, Phytomyza vitalbae* (Diptera: Agromyzidae) (Noyes, 2019).
Distribution: Canary Islands, Czech Republic, Finland, Germany, Hungary, Israel, Italy, Moldova, Morocco, Netherlands, Portugal, Romania, Russia, Spain, Sweden, Switzerland, Turkey, United Kingdom, Yemen (Noyes, 2019).

Diglyphus pusztensis (Erdos and Novicky, 1951)
Publications with Turkey records: Gencer (2009), Noyes (2019).
Associates: *Cardaria draba* (Brassicaceae), *Solanum melongena* (Solanaceae) (Noyes, 2019).
Hosts: *Agromyza sp., Agromyza albipennis, Chromatomyia horticola, Liriomyza sp., Liriomyza trifolii, Phytomyza sp., Phytomyza horticola, Phytomyza ranunculi* (Diptera: Agromyzidae) (Noyes, 2019).

Distribution: Azerbaijan, Croatia, Czech Republic, Czechoslovakia, England, Europe, Germany, Hungary, Israel, Japan, Poland, Romania, Russia, Slovakia, Sweden, Transcaucasus, Turkey, Yemen, Yugoslavia (pre-1991) (Noyes, 2019).

Diglyphus sabulosus (Erdos, 1951)
Publications with Turkey records: Noyes (2019).
Associates: *Colutea sp., Colutea arborescens* (Fabaceae), *Liriomyza sp.* (Diptera: Agromyzidae) (Noyes, 2019).
Hosts: *Liriomyza sp.* (Diptera: Agromyzidae) (Noyes, 2019).
Distribution: Czech Republic, Czechoslovakia, Hungary, Israel, Romania, Slovakia, Sweden, Turkey, Ukraine (Noyes, 2019).

Diglyphus sensilis (Yefremova, 2011)
Publications with Turkey records: Noyes (2019).
Associates: *Trifolium repens* (Fabaceae) (Noyes, 2019).
Hosts: *Liriomyza cicerina, Liriomyza sativae* (Diptera: Agromyzidae) (Noyes, 2019).
Distribution: Israel, Turkey (Noyes, 2019).

Genus *Elachertus* (Spinola, 1811)
Elachertus fenestratus (Nees, 1834)
Publications with Turkey records: Noyes (2019), Öncüer (1991).
Associates: *Rhus sp.* (Anacardiaceae), *Betula sp.* (Betulaceae), *Medicago sativa* (Fabaceae), *Quercus acutissima, Quercus dentata, Quercus mongolica, Quercus pubescens, Quercus variabilis* (Fagaceae), *Persea americana* (Lauraceae), *Larix sp., Larix occidentalis, Pinus sylvestris* (Pinaceae), *Malus pumila, Rosa sp.* (Rosaceae), *Populus sp.* (Salicaceae), *Tamarix sp.* (Tamaricaceae), *Vitis vinifera* (Vitaceae) (Noyes, 2019).
Parasitoids: *Baryscapus pospelovi, Baryscapus szocsi* (Hymenoptera: Eulophidae) (Noyes, 2019).
Hosts: *Gymnetron villosulum* (Coleoptera: Curculionidae), *Cephus pygmaeus* (Hymenoptera: Cephidae), *Fenusa pusilla* (Tenthredinidae), *Valentinia glandulella* (Lepidoptera: Blastobasidae), *Coleophora sp., Coleophora fuscedinella, Coleophora laricella, Coleophora limosipennella, Coleophora pruniella* (Coleophoridae), *Exoteleia dodecella, Phthorimaea operculella* (Gelechiidae), *Callisto geminatella, Caloptilia packardella, Parornix geminatella, Phyllonorycter sp., Phyllonorycter blancardella* (Gracillariidae), *Palthis angulalis* (Noctuidae), *Stenoma algidella* (Oecophoridae), *Acrobasis caryivorella, Dioryctria sp., Dioryctria zimmermanni, Omiodes indicata* (Pyralidae),

Amorbia cuneana, Archips argyrospilus, Archips rosanus, Archips xylosteanus, Barbara colfaxiana, Endopizia viteana, Epinotia fraternana, Epinotia laracana, Epinotia proximana, Lobesia botrana, Notocelia roborana, Paralobesia viteana, Platynota idaeudalis, Platynota stultana, Polychrosis viteana, Proteoteras aesculana, Rhyacionia buoliana, Rhyacionia frustrana, Sparganothis pettitana, Sparganothis sulfureana, Spilonota ocellana, Tortrix viridana, Zeiraphera diniana (Tortricidae) (Noyes, 2019).

Distribution: Austria, Bosnia Hercegovina, Canada, Croatia, Czech Republic, Czechoslovakia, England, Europe, France, Germany, Greece, Greenland, Hungary, Iran, Israel, Italy, Japan, Korea, Mexico, Nearctic, Neotropical, Netherlands, Palearctic, People's Republic of China, Poland, Romania, Russia, Serbia, Slovakia, Sweden, Switzerland, Tadzhikistan, Turkey, Turkmenistan, United States of America, USSR, Yemen (Noyes, 2019).

Elachertus inunctus (Nees, 1834)

Publications with Turkey records: Gencer (2012), Noyes (2019).

Associates: *Acer platanoides, Acer pseudoplatanus* (Aceraceae), *Carlina brevibracteata, Centaurea sadleriana* (Asteraceae), *Alnus sp., Alnus glutinosa, Betula sp., Carpinus betulus, Corylus avellana* (Betulaceae), *Lonicera periclymenum, Symphoricarpos rivularis* (Caprifoliaceae), *Laburnum anagyroides, Lathyrus vernus, Robinia pseudoacacia* (Fabaceae), *Fagus sylvatica, Quercus sp., Quercus robur* (Fagaceae), *Frangula alnus* (Rhamnaceae), *Crataegus sp., Crataegus laevigata, Crataegus monogyna, Malus sp., Malus sylvestris, Prunus cerasus, Prunus padus, Prunus spinosa, Sorbus aucuparia, Sorbus rupicola* (Rosaceae), *Salix sp. Salix nigricans, Salix repens* (Salicaceae), *Lycopersicon esculentum* (Solanaceae), *Ulmus sp.* (Ulmaceae) (Noyes, 2019).

Hosts: *Diplolepis eglanteriae* (Hymenoptera: Cynipidae), *Perittia obscurepunctella* (Lepidoptera: Elachistidae), *Epermenia illigerella* (Epermeniidae), *Tuta absoluta* (Gelechiidae), *Callisto denticulella, Caloptilia cuculipennella, Cameraria ohridella, Euspilapterix auroguttella, Gracillaria cuculipennella, Gracillaria syringella, Lithocolletis blancardella, Lithocolletis coryli, Lithocolletis messaniella, Lithocolletis nicellii, Lithocolletis nigrescentella, Lithocolletis populiella, Parectopa robiniella, Parornix sp., Parornix anglicella, Parornix avellanella, Parornix finitimella, Parornix scoticella, Phyllonorycter sp., Phyllonorycter alnifoliella, Phyllonorycter blancardella, Phyllonorycter cerasicolella, Phyllonorycter coryli, Phyllonorycter emberizaepennella, Phyllonorycter nicellii, Phyllonorycter nigrescentella, Phyllonorycter*

oxyacanthae, Phyllonorycter platanoidella, Phyllonorycter pomonella, Phyllonorycter robiniella, Phyllonorycter schreberella, Phyllonorycter sorbi, Phyllonorycter stettinensis, Phyllonorycter strigulatella, Phyllonorycter tenerella, Phyllonorycter tristrigella, Phyllonorycter ulmifoliella (Gracillariidae), *Leucoptera laburnella, Leucoptera malifoliella, Leucoptera scitella* (Lyonetiidae), *Cosmopterix pulchrimella* (Momphidae), *Nepticula salicis, Stigmella sp.* (Nepticulidae), *Agonopterix heracliana* (Oecophoridae), *Adoxophyes reticulana, Ancylis mitterbachiana, Lathronympha hypericana* (Tortricidae) (Noyes, 2019).
Distribution: Austria, Bosnia Hercegovina, Bulgaria, Czech Republic, Czechoslovakia, England, Germany, Hungary, Iran, Israel, Italy, Japan, Korea, Lithuania, Moldova, Montenegro, Netherlands, Norway, People's Republic of China, Poland, Romania, Russia, Slovakia, Slovenia, Sweden, Switzerland, Taiwan, Turkey, Turkmenistan, Ukraine, USSR, Yugoslavia (pre-1991) (Noyes, 2019).

Elachertus pulcher (Erdos, 1961)
Publications with Turkey records: Noyes (2019), Öncüer (1991).
Associates: *Populus sp.* (Salicaceae), *Lycopersicon esculentum* (Solanaceae) (Noyes, 2019).
Hosts: *Tuta absoluta* (Lepidoptera: Gelechiidae) (Noyes, 2019).
Distribution: Austria, Azerbaijan, Croatia, Czech Republic, Czechoslovakia, France, Georgia, Germany, Hungary, Iran, Israel, Madeira, Moldova, Russia, Slovakia, Spain, Taiwan, Turkey, Turkmenistan, Yemen, Yugoslavia (pre-1991) (Noyes, 2019).

Genus *Elasmus* (Westwood, 1833)
Elasmus cyprianus (Ferriere, 1947)
Publications with Turkey records: Noyes (2019), Öncüer (1991).
Associates: Unknown (Noyes, 2019).
Hosts: Unknown (Noyes, 2019).
Distribution: Cyprus, Spain, Turkey (Noyes, 2019).

Elasmus flabellatus (Fonscolombe, 1832)
Publications with Turkey records: Noyes (2019), Öncüer (1991).
Associates: *Thapsia villosa* (Apiaceae), *Onopordum nervosum* (Asteraceae), *Olea europaea* (Oleaceae), *Citrus sp.* (Rutaceae) (Noyes, 2019).
Parasitoid hosts: *Goniozus claripennis* (Hymenoptera: Bethylidae), *Apanteles glomeratus, Hormius sp., Hormius moniliatus, Macrocentrus sp., Macrocentrus abdominalis, Microgaster sp., Microgaster globate*

(Braconidae), *Angitia sp., Angitia armillata, Angitia fenestralis, Eulimneria sp., Limneria sp., Mesochorus sp.* (Ichneumonidae) (Noyes, 2019).
Parasitoids: *Pediobius routensis* (Eulophidae) (Noyes, 2019).
Hosts: *Janus compressus* (Hymenoptera: Cephidae), *Anarsia lineatella, Platyedra malvella* (Lepidoptera: Gelechiidae), *Antispila rivillei* (Heliozelidae), *Eublemma scitula* (Noctuidae), *Acanthopsyche atra, Acanthopsyche opacella, Apterona sp., Apterona crenulella, Canephora unicolor, Deuterohyalina albida, Luffia sp., Luffia lapidella, Oreopsyche tabanivicinella, Pachythelia unicolor, Phalacropteryx apiformis, Psyche constancella* (Psychidae), *Etiella zinckenella, Pyrausta aurata* (Pyralidae), *Clysia ambiguella, Eudemis botrana, Lobesia botrana, Polychrosis botrana, Polychrosis semblidis* (Tortricidae), *Prays citri, Prays oleae, Prays oleellus, Yponomeuta malinellus, Yponomeuta padellus* (Yponomeutidae) (Noyes, 2019).
Distribution: Austria, Bosnia Hercegovina, Bulgaria, Croatia, Czech Republic, Europe, France, Germany, Greece, Hungary, Israel, Italy, Lebanon, Moldova, Morocco, Netherlands, Portugal, Romania, Russia, Serbia, Slovakia, Spain, Sweden, Turkey, United Kingdom, USSR, Yugoslavia (Federal Republic), Yugoslavia (pre-1991) (Noyes, 2019).

Elasmus nudus (Nees, 1834)
Publications with Turkey records: Noyes (2019), Öncüer (1991).
Associates: *Arbutus unedo* (Ericaceae), *Olea europaea* (Oleaceae), *Picea sp.* (Pinaceae), *Pyrus malus* (Rosaceae).
Parasitoid hosts: *Apanteles sp., Apanteles lacteicolor, Apanteles melanoscelus* (Hymenoptera: Braconidae), *Angitia armillata* (Ichneumonidae) (Noyes, 2019).
Parasitoids: *Pediobius pyrgo* (Eulophidae) (Noyes, 2019).
Hosts: *Apanteles sp.* (Hymenoptera: Braconidae), *Leucospis sp.* (Leucospidae), *Choreutis bjerkandrella* (Lepidoptera: Choreutidae), *Simaethis sp., Simaethis pariana* (Glyphipterygidae), *Euproctis chrysorrhoea, Euproctis phaeorrhoea, Lymantria dispar, Porthesia chrysorrhoea* (Lymantriidae), *Vanessa cardui* (Nymphalidae), *Ephestia kuehniella, Etiella sp., Galleria mellonella, Ostrinia nubilalis, Psammotis hyalinalis* (Pyralidae), *Ancylis achatina, Archips murinana, Archips rosanus, Cydia pomonella, Epinotia nigricana, Laspeyresia pomonella, Lobesia botrana, Semasia sp., Semasia rufimitrana, Sparganothis pilleriana, Zeiraphera rufimitrana* (Tortricidae), *Prays oleae, Scythropia crataegella, Yponomeuta sp., Yponomeuta evonymellus, Yponomeuta malinellus* (Yponomeutidae) (Noyes, 2019).

Distribution: Azerbaijan, Bulgaria, Canary Islands, Caucasus, Croatia, Czech Republic, Czechoslovakia, Europe, Finland, France, Georgia, Germany, Hungary, Iran, Israel, Italy, Japan, Kazakhstan, Moldova, Poland, Russia, Serbia, Slovakia, Spain, Sweden, Switzerland, Tadzhikistan, Tselinograd Obl., Turkey, Turkmenistan, Ukraine, United Arab Emirates, United Kingdom, Uzbekistan, Vietnam, Yemen, Yugoslavia (pre-1991) (Noyes, 2019).

Elasmus phthorimaeae (Ferriere, 1947)
Publications with Turkey records: Noyes (2019), Öncüer (1991).
Hosts: *Phthorimaea operculella* (Lepidoptera: Gelechiidae) (Noyes, 2019).
Distribution: Cyprus, Europe, Israel, Spain, Turkey, United Arab Emirates, Yemen (Noyes, 2019).

Elasmus platyedrae (Ferriere, 1935)
Publications with Turkey records: Doğanlar (1985), Noyes (2019), Öncüer (1991).
Associates: *Eryngium campestre* (Apiaceae), *Carduus pycnocephalus, Onopordum corymbosum, Onopordum nervosum* (Asteraceae), *Gypsophila struthium* (Caryophyllaceae), *Medicago sativa, Retama sphaerocarpa* (Fabaceae), *Gossypium sp.* (Malvaceae), *Hordeum leporinum* (Poaceae), *Reseda lutea* (Resedaceae) (Noyes, 2019).
Hosts: *Ceroplastes sp.* (Hemiptera: Coccidae), *Zeuzera coffea* (Lepidoptera: Cossidae), *Pectinophora gossypiella, Pectinophora malvella, Pexicopia malvella, Platyedra gossypiella* (Gelechiidae), *Euproctis chrysorrhoea* (Lymantriidae), *Parametriotes theae* (Momphidae), *Prophantis smaragdina* (Pyralidae) (Noyes, 2019).
Distribution: Armenia, Austria, Bulgaria, Canary Islands, Croatia, Cyprus, Czech Republic, Czechoslovakia, Egypt, Europe, France, Georgia, Greece, Hungary, India, Israel, Italy, Macedonia, Madeira, North Africa, Portugal, Slovakia, Spain, Sweden, Turkey, United Arab Emirates, United States of America, Yemen, Yugoslavia (pre-1991) (Noyes, 2019).

Elasmus steffani (Viggiani, 1967)
Publications with Turkey records: Noyes (2019), Öncüer (1991), Tozlu and Coruh (2011).
Associates: *Anchusa leptophylla* (Boraginaceae), *Olea europaea* (Oleaceae), *Citrus sp*. (Rutaceae) (Noyes, 2019).
Hosts: *Cynaeda gigantea* (Lepidoptera: Crambidae), *Tachyptilia disquei* (Gelechiidae), *Eudemis botrana, Lobesia botrana* (Tortricidae), *Prays citri,*

Prays oleae, Prays oleellus, Yponomeuta malinellus (Yponomeutidae) (Noyes, 2019).
Parasitoid hosts: *Apanteles sp., Apanteles glomeratus* (Hymenoptera: Braconidae) (Noyes, 2019).
Distribution: Bulgaria, Croatia, France, Greece, Israel, Italy, Spain, Syria, Turkey, United Arab Emirates (Noyes, 2019).

Elasmus unicolor (Rondani, 1877)
Publications with Turkey records: Noyes (2019), Öncüer (1991).
Hosts: *Coleophora vibicella* (Lepidoptera: Coleophoridae), *Galleria mellonella* (Pyralidae) (Noyes, 2019).
Distribution: Bulgaria, Caucasus, Czech Republic, Europe, France, Germany, Hungary, Israel, Italy, Moldova, Russia, Slovakia, Spain, Sweden, Switzerland, Turkey, Turkmenistan (Noyes, 2019).

Elasmus westwoodi (Giraud, 1856)
Publications with Turkey records: Noyes (2019).
Associates: *Thapsia villosa* (Apiaceae), *Gypsophila struthium* (Caryophyllaceae), *Atriplex halimus* (Chenopodiaceae), *Ephedra nebrodensis* (Gnetaceae), *Rosmarinus officinalis* (Lamiaceae), *Olea europaea* (Oleaceae), *Pinus halepensis* (Pinaceae), *Reseda lutea* (Resedaceae), *Malus pumila* (Rosaceae) (Noyes, 2019).
Hosts: *Pseudopterna pruinata* (Lepidoptera: Geometridae), *Lithocolletis malella* (Gracillariidae), *Apterona sp., Apterona crenulella, Apterona helicoidella, Canephora unicolor, Psyche turatii* (Psychidae), *Prays oleae* (Yponomeutidae) (Noyes, 2019).
Parasitoid hosts: *Apanteles sp.* (Hymenoptera: Braconidae) (Noyes, 2019).
Distribution: Austria, Bulgaria, Europe, Finland, France, Germany, Greece, Hungary, Israel, Italy, Kazakhstan, Moldova, Slovakia, Spain, Sweden, Switzerland, Tselinograd Obl., Turkey, Yugoslavia (Federal Republic) (Noyes, 2019).

Genus *Eulophus* (Geoffroy, 1762)
Eulophus larvarum (Linnaeus, 1758)
Publications with Turkey records: Noyes (2019), Öncüer (1991).
Associates: *Alnus incana, Betula pubescens* (Betulaceae), *Fagus sylvatica, Quercus sp., Quercus ilex, Quercus robur* (Fagaceae), *Malus domestica, Prunus sp.* (Rosaceae), *Populus euramericana, Salix sp.* (Salicaceae), *Coryssomerus capucinus* (Coleoptera: Curculionidae) (Noyes, 2019).

Hosts: *Cassida viridis* (Coleoptera: Chrysomelidae), *Rhynchaenus pilosus, Rhynchaenus quercus* (Curculionidae), *Dacus oleae* (Diptera: Tephritidae), *Achlya flavicornis, Polyploca flavicornis* (Lepidoptera: Drepanidae), *Agriopis leucophaearia, Alsophila aescularia, Alsophila quadripunctaria, Anisopteryx aceriana, Archiearis parthenias, Brephos parthenias, Ennomos fuscantaria, Epirrita autumnata, Geometra papilionaria, Operophtera brumata, Operophtera fagata, Perizoma flavofasciata* (Geometridae), *Lithocolletis blancardella* (Gracillariidae), *Eriogaster lanestris* (Lepidoptera: Lasiocampidae), *Dasychira selenitica, Lymantria dispar, Lymantria monachal, Orgyia antiqua, Orgyia gonostigma, Orgyia recens, Orgyia thyellina* (Lymantriidae), *Acronicta sp., Acronicta aceris, Acronicta Hercules, Acronicta leporine, Acronicta megacephala, Acronicta psi, Acronicta tridens, Acronycta sp., Amphipyra pyramidea, Apatele aceris, Apatele incretata, Apatele leporine, Apatele megacephala, Apatele psi, Apatele tridens, Barathra brassicae, Bena prasinana, Brachionycha sp., Brachionycha sphinx, Calymnia sp., Calymnia affinis, Calymnia trapezina, Colocasia coryli, Cosmia affinis, Cosmia trapezina, Demas coryli, Diataraxia sp., Diataraxia oleracea, Diloba coeruleocephala, Dryobotodes eremita, Dryobotodes monochroma, Dryobotodes protea, Episema coeruleocephala, Lithophane ornitopus, Mamestra brassicae, Mamestra oleracea, Orthosia carnipennis, Orthosia cruda, Orthosia gothica, Orthosia gracilis, Orthosia incerta, Orthosia munda, Orthosia stabilis, Polia oleracea, Pseudoips fagana, Scoliopteryx libatrix, Taeniocampa sp., Taeniocampa pulverulenta, Thecophora fovea, Trisateles emortualis* (Noctuidae), *Cerura vinula, Clostera anachoreta, Clostera anastomosis, Clostera curtula, Clostera pigra, Dicranura vinula, Drymonia sp., Lophopteryx camelina, Lophopteryx capucina, Lophopteryx cucullina, Ptilodon capucina, Ptilophora plumigera* (Notodontidae), *Depressaria sp.* (Oecophoridae), *Pieris sp., Pieris brassicae* (Pieridae), *Laothoe populi* (Sphingidae), *Archips histrionana, Archips rosanus, Cochylis atricapitana, Epinotia nigricana, Pandemis cerasana, Pandemis ribeana, Tortrix viridana, Zeiraphera isertana* (Tortricidae) (Noyes, 2019).

Parasitoid hosts: *Apanteles sp., Apanteles eulipis* (Hymenoptera: Braconidae) (Noyes, 2019).

Parasitoids: *Pediobius cothurnatus, Pediobius crassicornis* (Eulophidae) (Noyes, 2019).

Distribution: Armenia, Bulgaria, Croatia, Czech Republic, Czechoslovakia, England, Europe, Finland, France, Germany, Hungary, Italy, Japan, Kirgizia, Korea, Moldova, Montenegro, Netherlands, Norway, Palearctic, People's Republic of China, Portugal, Romania, Russia, Serbia, Slovakia,

Sweden, Switzerland, Tadzhikistan, Transcaucasus, Turkey, USSR (Noyes, 2019).

Genus *Euplectrus* (Westwood, 1832)
Euplectrus bicolor (Swederus, 1795)
Publications with Turkey records: Noyes (2019), Öncüer (1991).
Associates: *Brassica oleracea* (Brassicaceae), *Juniperus phoeniciae, Juniperus thurifera* (Cupressaceae), *Quercus sp.* (Fagaceae), *Ephedra nebrodensis* (Gnetaceae), *Aesculus hippocastanum* (Hippocastanaceae), *Gossypium sp.* (Malvaceae), *Pinus halepensis* (Pinaceae), *Panicum miliaceum, Zea mays* (Poaceae), *Populus sp.* (Salicaceae), *Tamarix canariensis* (Tamaricaceae) (Noyes, 2019).
Parasitoids: *Pediobius atamiensis* (Hymenoptera: Eulophidae) (Noyes, 2019).
Hosts: *Nanophyes telephii* (Coleoptera: Nanophyidae), *Asphondylia ignorata* (Diptera: Cecidomyiidae), *Ceroplastes rusci, Lecanium sp.* (Hemiptera: Coccidae), *Tetramesa stipae* (Hymenoptera: Eurytomidae), *Agriopis leucophaearia, Biston betularia, Crocalis elinguaria, Crocallis elinguaria, Itame fulvaria, Tephronia codetaria* (Lepidoptera: Geometridae), *Cameraria ohridella, Lithocolletis populifoliella* (Gracillariidae), *Abagrotis alternata, Amathes c-nigrum, Anaplectoides prasine, Anarta myrtillin, Autographa gamma, Autographa pulchrina, Blepharita satura, Brachionycha sphinx, Cosmia trapezina, Crino satura, Diarsia brunnea, Dichonia aeruginea, Epiglaea apiata, Euplexia lucipara, Eurois occulta, Heliothis armigera, Hypena rostralis, Laphygma exigua, Leucania sp., Lithomoia solidaginis, Mamestra brassicae, Mamestra contigua, Mamestra peregrina, Mamestra persicariae, Mamestra thalassina, Morrisonia confuse, Mythimna unipuncta, Orthosia gracilis, Plusia gamma, Polia bombycine, Polia hepatica, Polia nebulosa, Polia tincta, Scotia sp., Spodoptera exigua* (Noctuidae), *Etiella zinckenella* (Pyralidae), *Archips crataeganus* (Tortricidae) (Noyes, 2019).
Distribution: Australia, Bulgaria, Canada, Caribbean, Croatia, Czech Republic, Czechoslovakia, Denmark, England, Europe, France, Germany, Hungary, Israel, Italy, Japan, Korea, Moldova, Montenegro, Nearctic, Netherlands, North Africa, People's Republic of China, Russia, Serbia, Seychelles, Slovakia, Somalia, Spain, Sweden, Turkey, Turkmenistan, United States of America, USSR (Noyes, 2019).
Euplectrus ceresensis (Ferriere, 1941)
Publications with Turkey records: Noyes (2019), Öncüer (1991).
Hosts: *Plusia sp.* (Lepidoptera: Noctuidae) (Noyes, 2019).
Distribution: South Africa, Turkey (Noyes, 2019).

Euplectrus flavipes (Fonscolombe, 1832)
Publications with Turkey records: Gencer (2012), Noyes (2019), Öncüer (1991).
Associates: *Lactuca sativa, Sonchus oleraceus* (Asteraceae), *Cucumis sativus* (Cucurbitaceae), *Phaseolus sp.* (Fabaceae), *Capsicum annuum, Lycopersicon esculentum* (Solanaceae), *Isodromus niger* (Hymenoptera: Encyrtidae) (Noyes, 2019).
Hosts: *Autographa gamma, Catocala nymphagoga, Ephesia nymphaea, Laphygma sp., Plusia sp., Prodenia litura* (Lepidoptera: Noctuidae), *Archips rosanus, Cacoecia rosanus* (Tortricidae) (Noyes, 2019).
Distribution: Bulgaria, Canary Islands, Croatia, Cyprus, Dominican Republic, Europe, France, Hungary, Israel, Italy, Mexico, Moldova, Montenegro, People's Republic of China, Portugal, Serbia, Slovakia, Slovenia, Spain, Sweden, Syria, Taiwan, Turkey, Turkmenistan, Yugoslavia (Federal Republic) (Noyes, 2019).

Euplectrus phthorimaeae (Ferriere, 1941)
Publications with Turkey records: Noyes (2019), Öncüer (1991).
Hosts: *Pristiphora conjugata* (Hymenoptera: Tenthredinidae), *Phthorimaea terrella* (Lepidoptera: Gelechiidae), *Euproctis chrysorrhoea, Euproctis phaeorrhoea, Porthesia chrysorrhoea* (Lymantriidae) (Noyes, 2019).
Distribution: Cyprus, Hungary, Israel, Turkey (Noyes, 2019).

Genus *Hemiptarsenus* (Westwood, 1833)
Hemiptarsenus ornatus (Nees, 1834)
Publications with Turkey records: Noyes (2019), Yefremova et al. (2010).
Associates: *Sonchus sp.* (Asteraceae), *Betula sp., Betula nana, Carpinus betulus* (Betulaceae), *Gypsophila sp.* (Caryophyllaceae), *Helianthemum sp.* (Cistaceae), *Astragalus glycyphyllos, Genista scorpius, Lotus sp.* (Fabaceae), *Aesculus hippocastanum* (Hippocastanaceae), *Ranunculus sp.* (Ranunculaceae), *Crataegus laevigata, Dryas octopetala, Pyrus sp.* (Rosaceae), *Populus sp., Populus nigra* (Salicaceae), *Tamarix canariensis* (Tamaricaceae) (Noyes, 2019).
Hosts: *Hypurus bertrandi, Rhamphus oxyacanthae* (Coleoptera: Curculionidae), *Chromatomyia horticola, Liriomyza sp., Liriomyza huidobrensis, Liriomyza trifolii, Phytoliriomyza variegate, Phytomyza sp., Phytomyza atricornis, Phytomyza fuscula, Phytomyza horticola, Phytomyza melana, Phytomyza plantaginis, Phytomyza ranunculi* (Diptera: Agromyzidae), *Heterarthrus nemoratus, Heterarthrus*

ochropodus (Hymenoptera: Tenthredinidae), *Elachista klimeschi, Elachista occulta* (Lepidoptera: Elachistidae), *Chrysoesthia sexguttella* (Gelechiidae), *Cameraria ohridella, Lithocolletis sp., Lithocolletis anderidae, Lithocolletis helianthemella, Lithocolletis nigrescentella, Lithocolletis sorbi, Phyllonorycter anderidae, Phyllonorycter medicaginella, Phyllonorycter nigrescentella* (Gracillariidae), *Stigmella aeneofasciella, Stigmella aurella, Stigmella cryptella, Stigmella dryadella, Stigmella nanivora* (Nepticulidae) (Noyes, 2019).

Distribution: Bulgaria, Canary Islands, Croatia, Czech Republic, Czechoslovakia, England, Europe, France, Germany, Hungary, Israel, Italy, Japan, Jordan, Korea, Madeira, Moldova, Morocco, Nearctic, Netherlands, North Africa, People's Republic of China, Poland, Portugal, Romania, Russia, Slovakia, Slovenia, Spain, Sweden, Switzerland, Syria, Taiwan, Turkey, Turkmenistan, United Arab Emirates, United States of America, USSR, Yemen, Yugoslavia (pre-1991) (Noyes, 2019).

Hemiptarsenus unguicellus (Zetterstedt, 1838)

Publications with Turkey records: Gencer (2012), Noyes (2019), Yefremova et al. (2010).

Associates: *Leucopis silesiaca* (Diptera: Chamaemyiidae), *Cirsium dissectum, Senecio sp.* (Asteraceae), *Carex sp.* (Cyperaceae), *Fagus sylvatica* (Fagaceae), *Iris pseudacorus* (Iridaceae), *Alopecurus myosuroides, Deschampsia cespitosa, Glyceria maxima, Hordeum vulgare, Milium effusum, Oryza sativa, Phalaris arundinacea, Phragmites communis, Sasa sp.* (Poaceae), *Aquilegia vulgaris, Ranunculus flammula, Ranunculus repens* (Ranunculaceae), *Osyris alba* (Santalaceae), *Tamarix canariensis* (Tamaricaceae) (Noyes, 2019).

Hosts: *Ceutorhynchus assimilis* (Coleoptera: Curculionidae), *Blastophagus piniperda, Myelophilus piniperda* (Scolytidae), *Agromyza mobilis, Agromyza oryzae, Cerodontha ireos, Cerodontha pygmaea, Chromatomyia fuscula, Dizygomyza iridis, Dizygomyza pygmaea, Phytobia sp., Phytobia iridis, Phytomyza fuscula, Phytomyza milii, Phytomyza minuscula, Phytomyza nigra, Phytomyza ranunculi, Phytomyza spinaciae* (Diptera: Agromyzidae), *Hydrellia griseola* (Ephydridae), *Eriopeltis sp.* (Hemiptera: Coccidae), *Biselachista fulgens, Elachista sp., Elachista adscitella, Elachista bisulcella, Elachista cerusella, Elachista griseella, Elachista megerlella, Elachista poae, Elachista pollinariella* (Lepidoptera: Elachistidae), *Gortyna flavago, Xanthoecia flavago* (Noctuidae), *Ostrinia nubilalis, Pyrausta nubilalis* (Pyralidae), *Prays oleae* (Yponomeutidae) (Noyes, 2019).

Distribution: Afghanistan, Azerbaijan, Bulgaria, Canada, Canary Islands, Croatia, Czech Republic, Czechoslovakia, Europe, Faeroe Islands, France, Germany, Hungary, India, Iran, Italy, Japan, Korea, Macedonia, Madeira, Moldova, Mongolia, Montenegro, Morocco, Netherlands, North Africa, Norway, People's Republic of China, Poland, Romania, Russia, Serbia, Slovakia, Slovenia, Spain, Sweden, Taiwan, Turkey, United Kingdom, United States of America, USSR, Yugoslavia (pre-1991) (Noyes, 2019).

Hemiptarsenus varicornis (Girault, 1913)
Publications with Turkey records: Bulut and Gocmen (2000), Noyes (2019).
Associates: *Erechtites hieracifolia, Gerbera jamesoni* (Asteraceae), *Brassica oleracea* (Brassicaceae), *Citrullus lanatus* (Cucurbitaceae), *Phaseolus vulgaris, Pisum sativum, Vicia faba, Vigna unguiculata* (Fabaceae), *Allium fistulosum* (Liliaceae), *Althaea rosea* (Malvaceae), *Lycopersicon esculentum, Solanum nigrum, Solanum tuberosum* (Solanaceae) (Noyes, 2019).
Hosts: *Agromyza sp., Agromyza phaseoli, Chromatomyia horticola, Liriomyza sp., Liriomyza bryoniae, Liriomyza huidobrensis, Liriomyza pusilla, Liriomyza sativae, Liriomyza trifolii, Melanagromyza phaseoli, Ophiomyia sp., Ophiomyia phaseoli, Pseudonapomyza spicata* (Diptera: Agromyzidae) (Noyes, 2019).
Distribution: Australia, Canary Islands, Ethiopia, Fiji, Ghana, Guam, Hawaii, India, Indonesia, Iran, Japan, Kenya, Madeira, Malaysia, New Caledonia, New Zealand, Pakistan, People's Republic of China, Saudi Arabia, Senegal, Seychelles, Spain, Sri Lanka, Sudan, Taiwan, Tanzania, Turkey, Vanuatu (Noyes, 2019).

Hemiptarsenus zilahisebessi (Erdos, 1951)
Publications with Turkey records: Gencer (2009; 2012), Noyes (2019), Öncüer (1991), Yasarakinci and Ulucal (1997).
Associates: *Anacyclus clavatus* (Asteraceae), *Gypsophila struthium* (Caryophyllaceae), *Beta vulgaris* (Chenopodiaceae), *Cucumis sativus* (Cucurbitaceae), *Pisum sativum* (Fabaceae), *Zea mays* (Poaceae), *Lycopersicon esculentum* (Solanaceae) (Noyes, 2019).
Hosts: *Hypurus sp., Hypurus bertrandi* (Coleoptera: Curculionidae), *Chromatomyia horticola, Liriomyza sp., Liriomyza bryoniae, Liriomyza congesta, Liriomyza huidobrensis, Liriomyza sativae, Liriomyza trifolii* (Agromyzidae), *Stigmella sp., Stigmella agrimoniae* (Lepidoptera: Nepticulidae) (Noyes, 2019).

Distribution: Bulgaria, Czechoslovakia, Egypt, France, Hungary, Iran, Israel, Jordan, Korea, Moldova, People's Republic of China, Poland, Russia, Serbia, Slovakia, Spain, Turkey (Noyes, 2019).

Genus *Hyssopus* (Girault, 1916)
Hyssopus geniculatus (Hartig, 1838)
Publications with Turkey records: Noyes (2019), Öncüer (1991).
Associates: *Centaurea scabiosa* (Asteraceae), *Larix sp., Picea sp., Picea abies, Pinus sp., Pseudotsuga taxifolia* (Pinaceae) (Noyes, 2019).
Hosts: *Pityogenes bidentatus, Pityogenes chalcographus* (Coleoptera: Scolytidae), *Thecodiplosis brachyntera* (Diptera: Cecidomyiidae), *Exoteleia sp., Exoteleia dodecella, Metzneria sp.* (Lepidoptera: Gelechiidae), *Lithocolletis sp., Phyllocnistis citrella, Phyllonorycter sp.* (Gracillariidae), *Dioryctria pineae, Dioryctria sylvestrella, Doryctria splendidella, Galleria mellonella* (Pyralidae), *Cacoecimorpha pronubana, Cydia strobilella, Laspeyresia strobilella, Petrova sp., Petrova resinella, Pseudococcyx tessulatana, Pseudotomoides strobilellus, Rhyacionia sp., Rhyacionia buoliana* (Tortricidae) (Noyes, 2019).
Distribution: Bulgaria, Croatia, Czech Republic, Czechoslovakia, England, Europe, Finland, France, Germany, Hungary, Iran, Israel, Italy, Japan, Korea, Macedonia, Netherlands, Norway, Palearctic, People's Republic of China, Poland, Romania, Russia, Slovakia, Spain, Sweden, Tadzhikistan, Turkey, Turkmenistan, Yemen (Noyes, 2019).

Hyssopus nigritulus (Zetterstedt, 1838)
Publications with Turkey records: Gencer (2003), Noyes (2019), Öncüer (1991).
Associates: *Phalonia roseana* (Tortricidae), *Daucus sp.* (Apiaceae), *Carduus bourgeanus* (Asteraceae), *Alnus sp.* (Betulaceae), *Calystegia sepium* (Convolvulaceae), *Cupressus sp.* (Cupressaceae), *Dipsacus fullonum, Dipsacus sylvestris* (Dipsacaceae), *Alhagi pseudalhagi, Retama sphaerocarpa, Sophora japonica* (Fabaceae), *Rosmarinus officinalis* (Lamiaceae), *Ficus carica* (Moraceae), *Olea sp.* (Oleaceae), *Picea sp., Picea abies, Picea crassifolia* (Pinaceae), *Reseda lutea* (Resedaceae), *Rosa sp., Rosa canina* (Rosaceae), *Populus sp.* (Salicaceae), *Osyris alba* (Santalaceae), *Tamarix canariensis* (Tamaricaceae) (Noyes, 2019).
Hosts: *Trupanea amoena* (Diptera: Tephritidae), *Diplolepis mayri* (Hymenoptera: Cynipidae), *Ancistrocerus sp.* (Vespidae), *Zeuzera pyrina* (Lepidoptera: Cossidae), *Exoteleia dodecella* (Gelechiidae), *Bastodacna sp.* (Momphidae), *Paranthrene tabaniformis* (Sesiidae), *Barbara tessulatana, Cnephasia chrysantheana, Cochylis roseana, Cydia*

funebrana, Cydia strobilella, Grapholita funebrana, Laspeyresia strobilella, Petrova resinella, Phalonia roseana, Pseudococcyx tessulatana, Pseudotomoides strobilellus (Tortricidae) (Noyes, 2019).
Distribution: Armenia, Bulgaria, Canary Islands, Croatia, Czech Republic, Czechoslovakia, Europe, Finland, France, Germany, Hungary, Ireland, Israel, Italy, Japan, Korea, Moldova, Netherlands, Norway, Palearctic, People's Republic of China, Poland, Romania, Russia, Serbia, Slovakia, Slovenia, Spain, Sweden, Transcaucasus, Turkey, Turkmenistan, United Kingdom, USSR, Yemen (Noyes, 2019).

Genus *Microlycus* (Thomson, 1878)
Microlycus erdoesi (Boucek, 1959)
Publications with Turkey records: Noyes (2019), Öncüer (1991).
Associates: Unknown (Noyes, 2019).
Hosts: Unknown (Noyes, 2019).
Distribution: Austria, Bulgaria, Czech Republic, Czechoslovakia, England, Hungary, Israel, Moldova, Russia, Slovakia, Sweden, Turkey (Noyes, 2019).

Genus *Miotropis* (Thomson, 1878)
Miotropis unipuncta (Nees, 1834)
Publications with Turkey records: Noyes (2019).
Associates: *Betula sp., Betula verrucosa* (Betulaceae), *Quercus sp.* (Fagaceae) (Noyes, 2019).
Parasitoid hosts: *Agathis sp.* (Hymenoptera: Braconidae) (Noyes, 2019).
Hosts: *Coleophora sp., Coleophora caespititiella, Coleophora fuscedinella, Coleophora glaucicolella, Coleophora lutipennella, Coleophora serratella, Coleophora troglodytella* (Lepidoptera: Coleophoridae), *Elachista quadripunctella* (Elachistidae) (Noyes, 2019).
Distribution: Austria, Bulgaria, Croatia, Czech Republic, Czechoslovakia, England, Germany, Hungary, Israel, Italy, Madeira, Moldova, Mongolia, Morocco, Netherlands, North Africa, Romania, Serbia, Slovakia, Slovenia, Spain, Sweden, Turkey, USSR (Noyes, 2019).

Genus *Necremnus* (Thomson, 1878)
Necremnus artynes (Walker, 1839)
Publications with Turkey records: Noyes (2019), Öncüer (1991).
Associates: *Eryngium campestre* (Apiaceae), *Lepidium subulatum* (Brassicaceae), *Salvia aethiopis* (Lamiaceae) (Noyes, 2019).
Hosts: *Cosmopterix pulchrimella, Vulkaniella pomposella* (Lepidoptera: Momphidae) (Noyes, 2019).

Distribution: Austria, Canary Islands, Croatia, Czech Republic, Czechoslovakia, England, France, Germany, Italy, Macedonia, Madeira, Moldova, Mongolia, Montenegro, Netherlands, Slovakia, Spain, Sweden, Turkey, United States of America (Noyes, 2019).

Necremnus croton (Walker, 1839)
Publications with Turkey records: Noyes (2019), Öncüer (1991).
Hosts: *Pseudobankesia macedoniella* (Lepidoptera: Psychidae) (Noyes, 2019).
Distribution: England, Macedonia, Sweden, Turkey (Noyes, 2019).

Necremnus leucarthros (Nees, 1834)
Publications with Turkey records: Gencer (2012), Noyes (2019), Öncüer (1991).
Associates: *Centaurea nigra* (Asteraceae), *Juniperus sp.* (Cupressaceae), *Cicer arietinum, Medicago sativa* (Fabaceae), *Fagus sylvatica* (Fagaceae), *Picea sp., Pinus strobus* (Pinaceae), *Triticum sp., Triticum aestivum* (Poaceae) (Noyes, 2019).
Hosts: *Choragus piceus* (Coleoptera: Anthribidae), *Apion apricans, Apion flavipes, Apion pisi* (Apionidae), *Crioceris asparagi, Lema sp., Lema cyanella, Lema lichenis, Lema melanopa, Mantura obtusata, Oulema sp., Oulema gallaeciana, Oulema lichenis, Oulema melanopus* (Chrysomelidae), *Ceutorhynchus sp., Ceutorhynchus posthumus, Hypera sp., Hypera adspersa, Hypera arator, Hypera murina, Hypera polygoni, Hypera postica, Hypera rumicis, Hypera variabilis, Hypurus bertrandi, Phytonomus sp., Phytonomus arator, Phytonomus nigrirostris, Phytonomus variabilis, Sirocalus posthumus* (Curculionidae), *Quedius brevis* (Staphylinidae), *Phytomyza obscurella* (Diptera: Agromyzidae), *Dasineura brassicae* (Cecidomyiidae), *Oscinella frit* (Chloropidae), *Coleophora laricella* (Lepidoptera: Coleophoridae), *Eriocrania sp.* (Eriocraniidae), *Archips murinana, Cacoecia murinana, Choristoneura murinana* (Tortricidae), *Yponomeuta padellus* (Yponomeutidae) (Noyes, 2019).
Distribution: Algeria, Armenia, Bulgaria, Czech Republic, Czechoslovakia, Denmark, England, Europe, France, Germany, Greece, Hungary, India, Italy, Macedonia, Moldova, Montenegro, Netherlands, Palearctic, Poland, Romania, Russia, Slovakia, Sweden, Switzerland, Transcaucasus, Turkey, United States of America, USSR, Yugoslavia (pre-1991) (Noyes, 2019).

Necremnus tidius (Walker, 1839)
Publications with Turkey records: Gencer (2012), Noyes (2019), Öncüer (1991).
Associates: *Ceutorhynchus obstrictus* (Coleoptera: Curculionidae), *Daucus carota* (Apiaceae), *Carduus bourgeanus, Santolina chamaecyparissus, Senecio sp.* (Asteraceae), *Alliaria officinalis, Brassica sp., Brassica campestris, Brassica napus, Brassica rapa, Capsella bursa-pastoris, Descurainia Sophia, Eruca vesicaria, Hormathophylla spinosa, Sinapis arvensis, Sisymbrium altissimum, Sisymbrium irio, Sophia filipes* (Brassicaceae), *Salsola pestifer* (Chenopodiaceae), *Helianthemum squamatum* (Cistaceae), *Juniperus phoeniciae, Juniperus thurifera* (Cupressaceae), *Ephedra nebrodensis* (Gnetaceae), *Rosmarinus officinalis* (Lamiaceae), *Abies alba, Abies nordmanniana, Pinus sp.* (Pinaceae), *Lolium sp.* (Poaceae), *Tamarix canariensis* (Tamaricaceae) (Noyes, 2019).
Hosts: *Apion sp., Apion arrogans* (Coleoptera: Apionidae), *Psylliodes marcida, Psylliodes napi* (Chrysomelidae), *Ceutorhynchus sp., Ceutorhynchus assimilis, Ceutorhynchus floralis, Ceutorhynchus obstrictus, Ceutorhynchus typhae* (Curculionidae), *Phytomyza sp., Phytomyza ranunculi* (Diptera: Agromyzidae), *Chlorops pumilionis* (Chloropidae), *Cephus pygmaeus* (Hymenoptera: Cephidae), *Coleotechnites piceaella* (Lepidoptera: Gelechiidae), *Syringopais temperatella* (Scythrididae), *Argyresthia fundella, Ochsenheimeria vacculella* (Yponomeutidae) (Noyes, 2019).
Distribution: Bulgaria, Canada, Canary Islands, Croatia, Czech Republic, Czechoslovakia, England, Europe, France, Germany, Iran, Israel, Montenegro, Nearctic, Netherlands, Palearctic, Poland, Romania, Russia, Serbia, Slovakia, Spain, Sweden, Turkey, United States of America (Noyes, 2019).

Genus *Platyplectrus* (Ferriere, 1941)
Platyplectrus bouceki (Erdos, 1966)
Publications with Turkey records: Noyes (2019), Öncüer (1991).
Associates: *Artemisia herba-alba* (Asteraceae), *Juniperus thurifera* (Cupressaceae), *Retama sphaerocarpa* (Fabaceae), *Rhamnus lycioides* (Rhamnaceae), *Tamarix canariensis* (Tamaricaceae) (Noyes, 2019).
Distribution: Bulgaria, Croatia, Czech Republic, Czechoslovakia, Europe, Hungary, Italy, Moldova, Montenegro, Serbia, Slovakia, Turkey (Noyes, 2019).

Platyplectrus chlorocephalus (Nees, 1834)
Publications with Turkey records: Gencer (2012), Noyes (2019).

Associates: *Calluna vulgaris* (Ericaceae) (Noyes, 2019).
Hosts: *Rhagades pruni* (Lepidoptera: Zygaenidae) (Noyes, 2019).
Distribution: Bulgaria, Czechoslovakia, Germany, Hungary, Poland, Slovenia, Turkey (Noyes, 2019).

Platyplectrus pannonica (Erdos, 1966)
Publications with Turkey records: Gencer (2012), Noyes (2019).
Hosts: *Cifuna locuples* (Lepidoptera: Lymantriidae) (Noyes, 2019).
Distribution: Croatia, Europe, France, Germany, Hungary, Italy, Netherlands, People's Republic of China, Spain, Turkey, Yugoslavia (pre-1991) (Noyes, 2019).

Genus *Pnigalio* (Schrank, 1802)
Pnigalio agraules (Walker, 1839)
Publications with Turkey records: Noyes (2019), Öncüer (1991).
Associates: *Alnus glutinosa, Betula sp., Betula verrucosa* (Betulaceae), *Anagyris foetida, Lathyrus vernus (Fabaceae), Fagus sylvatica, Quercus sp., Quercus robur* (Fagaceae), *Aesculus hippocastanum* (Hippocastanaceae), *Olea europaea* (Oleaceae), *Phalaris arundinacea* (Poaceae), *Malus communis, Rosa sp., Rosa canina* (Rosaceae), *Citrus sp., Citrus limon* (Rutaceae), *Populus sp., Populus canadensis, Populus italicus* (Salicaceae), *Ulmus sp.* (Ulmaceae), *Phyllocnistis citrella* (Lepidoptera: Gracillariidae) (Noyes, 2019).
Hosts: *Apion croceifemoratum* (Coleoptera: Apionidae), *Orchestes alni, Orchestes avellanae, Orchestes fagi, Orchestes rusci, Rhynchaenus alni, Rhynchaenus fagi, Rhynchaenus populi, Rhynchaenus quercus, Rhynchaenus rusci, Rhynchaenus salicis, Rhynchaenus testaceus* (Curculionidae), *Phytagromyza populi* (Diptera: Agromyzidae), *Bactrocera oleae, Dacus oleae* (Tephritidae), *Neuroterus numismalis* (Hymenoptera: Cynipidae), *Fenusa dohrnii, Heterarthrus aceris, Messa hortulana, Profenusa pygmaea* (Tenthredinidae), *Elachista cerusella* (Lepidoptera: Elachistidae), *Eriocrania haworthi, Eriocrania sparmannella* (Eriocraniidae), *Chrysoesthia sexguttella* (Gelechiidae), *Caloptilia stigmatella, Cameraria ohridella, Lithocolletis cerasicolella, Lithocolletis froelichiella, Lithocolletis klemannella, Lithocolletis nigrescentella, Lithocolletis populifoliella, Lithocolletis schreberella, Metriochroa latifoliella, Phyllocnistis citrella, Phyllonorycter sp., Phyllonorycter blancardella, Phyllonorycter cerasicolella, Phyllonorycter corylifoliella, Phyllonorycter gerasimowi, Phyllonorycter millierella, Phyllonorycter pomonella, Phyllonorycter quercifoliella, Phyllonorycter rajella, Phyllonorycter tristrigella* (Gracillariidae), *Leucoptera scitella,*

Leucoptera sinuella, Paraleucoptera sinuella (Lyonetiidae), *Stigmella atricapitella, Stigmella aurella* (Nepticulidae), *Tischeria sp., Tischeria angusticollella, Tischeria dodonaea, Tischeria ekebladella* (Tischeriidae) (Noyes, 2019).
Distribution: Austria, Bosnia Hercegovina, Bulgaria, Czech Republic, Czechoslovakia, Denmark, Egypt, England, Europe, France, Germany, Greece, Hungary, Iran, Israel, Italy, Jordan, Macedonia, Malta, Netherlands, Palearctic, People's Republic of China, Romania, Russia, Serbia, Slovenia, Spain, Sweden, Switzerland, Turkey, Ukraine (Noyes, 2019).

Pnigalio cristatus (Ratzeburg, 1848)
Publications with Turkey records: Noyes (2019), Öncüer (1991).
Hosts: *Liriomyza congesta* (Diptera: Agromyzidae), *Chrysoesthia sexguttella* (Lepidoptera: Gelechiidae), *Lithocolletis nigrescentella, Phyllonorycter sp.* (Gracillariidae), *Antispila sp.* (Heliozelidae), *Cosmopterix pulchrimella* (Momphidae), *Stigmella aurella* (Nepticulidae), *Calymnia trapezina, Cosmia trapezina* (Noctuidae) (Noyes, 2019).
Distribution: Bulgaria, Croatia, Czech Republic, Czechoslovakia, England, Europe, France, Germany, Hungary, Italy, Macedonia, Moldova, Montenegro, Romania, Russia, Slovakia, Spain, Sweden, Turkey, Yugoslavia (pre-1991) (Noyes, 2019).

Pnigalio incompletus (Boucek, 1971)
Publications with Turkey records: Noyes (2019), Öncüer (1991), Uygun (1997).
Associates: *Citrus sp., Citrus limon* (Rutaceae), *Urtica sp.* (Urticaceae), *Vitis sp.* (Vitaceae) (Noyes, 2019).
Hosts: *Agromyza hiemalis, Chromatomyia horticola, Liriomyza sp., Liriomyza huidobrensis* (Diptera: Agromyzidae), *Chrysoesthia sexguttella* (Lepidoptera: Gelechiidae), *Phyllocnistis citrella, Phyllonorycter corylifoliella* (Gracillariidae), *Antispila rivillei, Holocacista rivillei* (Heliozelidae), *Cosmopterix pulchrimella* (Momphidae) (Noyes, 2019).
Distribution: Azerbaijan, Bulgaria, Czech Republic, Czechoslovakia, Israel, Italy, Jordan, Moldova, Slovakia, Syria, Turkey (Noyes, 2019).

Pnigalio longulus (Zetterstedt, 1838)
Publications with Turkey records: Noyes (2019), Öncüer (1991).
Associates: *Formica aquilonia* (Formicidae), *Alnus sp., Alnus glutinosa,Betula sp., Betula pendula, Betula pubescens, Betula verrucosa* (Betulaceae), *Fagus sylvatica, Quercus sp., Quercus petraea, Quercus robur* (Fagaceae), *Aesculus hippocastanum* (Hippocastanaceae), *Carya*

illinoinensis (Juglandaceae), *Olea europaea* (Oleaceae), *Phalaris arundinacea* (Poaceae), *Malus sp., Malus pumila, Malus sylvestris, Pyrus malus, Rosa canina, Sorbus aria, Sorbus aucuparia* (Rosaceae), *Populus tremula, Salix caprea, Salix cinerea, Salix viminalis* (Salicaceae), *Ulmus sp., Ulmus glabra* (Ulmaceae) (Noyes, 2019).

Hosts: *Zeugophora flavicollis* (Coleoptera: Chrysomelidae), *Orchestes quercus, Rhynchaenus fagi, Rhynchaenus rusci* (Curculionidae), *Dacus oleae* (Tephritidae), *Fenusa ulmi, Heterarthrus vagans, Messa nana, Scolioneura betulae* (Hymenoptera: Tenthredinidae), *Elachista cerusella* (Lepidoptera: Elachistidae), *Dyseriocrania subpurpurella, Eriocrania haworthi, Eriocrania sangii, Eriocrania semipurpurella, Eriocrania sparmannella, Eriocrania unimaculella (Eriocraniidae), Cameraria caryaefoliella, Cameraria ohridella, Gracillaria syringella, Lithocolletis sp., Lithocolletis alpina, Lithocolletis blancardella, Lithocolletis corylifoliella, Lithocolletis froelichiella, Lithocolletis klemannella, Lithocolletis populifoliella, Lithocolletis quercifoliella, Lithocolletis rajella, Lithocolletis ringoniella, Lithocolletis spinolella, Lithocolletis viminella, Metriochroa latifoliella, Parornix betulae, Pecophyllembius neglectus, Phyllocnistis labyrinthella, Phyllonorycter alnifoliella, Phyllonorycter blancardella, Phyllonorycter caryaealbella, Phyllonorycter cerasicolella, Phyllonorycter cydoniella, Phyllonorycter froelichiella, Phyllonorycter hilarella, Phyllonorycter klemannella, Phyllonorycter pyrifoliella, Phyllonorycter quercifoliella, Phyllonorycter rajella, Phyllonorycter spinolella, Phyllonorycter stettinensis, Phyllonorycter tristrigella, Phyllonorycter ulmifoliella* (Gracillariidae), *Coptodisca lucifluella* (Heliozelidae), *Leucoptera scitella, Lyonetia clerkella* (Lyonetiidae), *Ectoedemia argentipedella, Nepticula sorbi, Nepticula tityrella, Stigmella ruficapitella, Stigmella tityrella* (Nepticulidae), *Nycteola revayana* (Noctuidae), *Tischeria angusticollella, Tischeria ekebladella* (Tischeriidae), *Prays oleae* (Yponomeutidae) (Noyes, 2019).

Parasitoid hosts: *Colastes braconius* (Hymenoptera: Braconidae), *Achrysocharoides niveipes* (Eulophidae) (Noyes, 2019).

Distribution: Austria, Bulgaria, Canada, Czech Republic, Czechoslovakia, Denmark, France, Germany, Greece, Hungary, Italy, Lithuania, Netherlands, Norway, People's Republic of China, Poland, Russia, Serbia, Spain, Sweden, Switzerland, Turkey, Ukraine, United Kingdom, United States of America, USSR, Yugoslavia (pre-1991) (Noyes, 2019).

Pnigalio mediterraneus (Ferriere and Delucchi, 1957)
Publications with Turkey records: Cebeci et al. (2011), Noyes (2019), Öncüer (1991).
Hosts: *Inula cretica* (Asteraceae), *Chilocorus bipustulatus* (Coleoptera: Coccinellidae), *Orchestes sp.* (Curculionidae), *Phytomyza populi* (Diptera: Agromyzidae), *Bactrocera oleae, Dacus oleae, Myopites sp., Myopites stylata* (Tephritidae), *Messa hortulana* (Hymenoptera: Tenthredinidae), *Cameraria ohridella, Lithocolletis sp., Lithocolletis blancardella, Lithocolletis corylifoliella, Lithocolletis millierella, Metriochroa latifoliella, Phyllocnistis citrella, Phyllonorycter sp., Phyllonorycter medicaginella* (Lepidoptera: Gracillariidae), *Leucoptera scitella* (Lyonetiidae), *Tischeria dodonaea, Tischeria ekebladella* (Tischeriidae), *Prays oleae* (Yponomeutidae) (Noyes, 2019).
Parasitoid hosts: *Apanteles circumscriptus, Apanteles cricumscriptus* (Hymenoptera: Braconidae), *Chrysocharis nautius* (Eulophidae), *Eupelmus martelli, Eupelmus urozonus* (Eupelmidae) (Noyes, 2019).
Parasitoids: *Quercus sp., Quercus ithaburensis* (Fagaceae), *Aesculus hippocastanum* (Hippocastanaceae), *Olea europaea* (Oleaceae), *Malus pumila* (Rosaceae), *Citrus sp., Citrus limon* (Rutaceae), *Populus sp.* (Salicaceae) (Noyes, 2019).
Distribution: Algeria, Austria, Croatia, Europe, France, Greece, Israel, Italy, Libya, Moldova, North Africa, Portugal, Russia, Spain, Tunisia, Turkey, Yugoslavia (pre-1991) (Noyes, 2019).

Pnigalio nemati (Westwood, 1838)
Publications with Turkey records: Noyes (2019), Yefremova et al. (2010).
Associates: *Salix sp., Salix alba, Salix fragilis* (Salicaceae) (Noyes, 2019).
Hosts: *Curculio salicivorus* (Coleoptera: Curculionidae), *Euura sp., Euura mucronate, Pontania sp., Pontania bridgmanii, Pontania proxima, Pontania triandrae, Pontania viminalis* (Hymenoptera: Tenthredinidae), *Phyllonorycter medicaginella* (Lepidoptera: Gracillariidae) (Noyes, 2019).
Distribution: Bulgaria, Canada, Czech Republic, Czechoslovakia, England, Europe, Finland, France, Germany, Hungary, Israel, Netherlands, Russia, Slovakia, Sweden, Turkey (Noyes, 2019).

Pnigalio pectinicornis (Linnaeus, 1758)
Publications with Turkey records: Cebeci et al. (2011), Gencer (2012), Noyes (2019), Öncüer (1991).
Associates: *Formica aquilonia* (Formicidae), *Acer platanoides, Acer pseudoplatanus* (Aceraceae), *Artemisia glutinosa, Artemisia vulgaris*

(Asteraceae), *Alnus sp., Alnus glutinosa, Alnus incana, Betula sp., Betula pubescens, Betula verrucose, Carpinus betulus, Corylus avellana* (Betulaceae), *Laburnum anagyroides, Retama sphaerocarpa, Trifolium pratense* (Fabaceae), *Fagus sylvatica, Quercus sp., Quercus cerris, Quercus conferta, Quercus ilex, Quercus petraea, Quercus robur* (Fagaceae), *Aesculus hippocastanum* (Hippocastanaceae), *Platanus sp.* (Platanaceae), *Alopecurus myosuroides, Avena sativa, Elytrigia repens, Festuca pratensis, Hordeum vulgare, Melica nutans, Phalaris arundinacea, Sesleria uliginosa, Triticum aestivum, Triticum vulgare* (Poaceae), *Cerasus sp., Crataegus laevigata, Crataegus monogyna, Crataegus oxyacanthoides, Malus sp., Malus pumila, Malus sylvestris, Prunus sp., Prunus padus, Prunus spinosa, Sorbus aria, Sorbus aucuparia, Sorbus intermedia, Sorbus torminalis* (Rosaceae), *Rubus sp., Rubus caesius* (Rubiaceae), *Citrus sp., Citrus clementina, Citrus limon* (Rutaceae), *Populus sp., Populus tremula, Salix cinerea, Salix nigricans, Salix repens, Salix viminalis* (Salicaceae), *Verbascum nigrum* (Scrophulariaceae), *Solanum tuberosum* (Solanaceae), *Tamarix canariensis* (Tamaricaceae), *Celtis australis, Ulmus sp., Ulmus glabra* (Ulmaceae) (Noyes, 2019).

Hosts: *Stegobium paniceum* (Coleoptera: Anobiidae), *Zeugophora flavicollis* (Chrysomelidae), *Anoplus plantaris, Hypera arator, Hypurus bertrandi, Orchestes avellanae, Orchestes fagi, Orchestes quercus, Orchestes rusci, Orchestes salicis, Rhynchaenus sp., Rhynchaenus alni, Rhynchaenus ermischi, Rhynchaenus fagi, Rhynchaenus quercus, Rhynchaenus testaceus* (Curculionidae), *Agromyza sp., Agromyza alnivora, Agromyza demeijerei, Agromyza frontella, Agromyza mobilis, Agromyza nana, Agromyza spiraeae, Phytomyza sp., Phytomyza fuscula, Phytomyza obscurella, Phytomyza plantaginis, Phytomyza verbasci* (Diptera: Agromyzidae), *Trypeta artemisiae, Dacus oleae* (Tephritidae), *Fenusa dohrnii, Fenusa ulmi, Heterarthrus aceris, Heterarthrus vagans, Profenusa pygmaea* (Hymenoptera: Tenthredinidae), *Elachista sp., Elachista albifrontella, Elachista bisulcella, Elachista cerusella, Elachista dimicatella, Elachista perplexella, Elachista pollinariella, Elachista sparmannella* (Lepidoptera: Elachistidae), *Dyseriocrania fastuosella, Eriocrania fastuosella, Eriocrania haworthi, Eriocrania unimaculella* (Eriocraniidae), *Phthorimaea operculella, Recurvaria leucatella* (Gelechiidae), *Acrocercops omissella, Callisto denticulella, Caloptilia hemidactylella, Cameraria ohridella, Euspilapteryx auroguttella, Leucospilapteryx omissella, Lithocolletis sp., Lithocolletis acerifoliella, Lithocolletis anderidae, Lithocolletis blancardella, Lithocolletis cerasicolella, Lithocolletis comparella, Lithocolletis corylifoliella,*

Lithocolletis froelichiella, Lithocolletis heegeriella, Lithocolletis junoniella, Lithocolletis klemannella, Lithocolletis messaniella, Lithocolletis millierella, Lithocolletis nicellii, Lithocolletis nigrescentella, Lithocolletis oxyacanthae, Lithocolletis platani, Lithocolletis pomifoliella, Lithocolletis populifoliella, Lithocolletis pyrifoliella, Lithocolletis quercifoliella, Lithocolletis quinqueguttella, Lithocolletis rajella, Lithocolletis roboris, Lithocolletis salictella, Lithocolletis schreberella, Lithocolletis spinicolella, Lithocolletis tenerella, Lithocolletis ulmifoliella, Lithocolletis viminella, Parornix scoticella, Phyllocnistis sp., Phyllocnistis citrella, Phyllocnistis labyrinthella, Phyllonorycter sp., Phyllonorycter alnifoliella, Phyllonorycter blancardella, Phyllonorycter cerasicolella, Phyllonorycter corylifoliella, Phyllonorycter cydoniella, Phyllonorycter froelichiella, Phyllonorycter harrisella, Phyllonorycter joviella, Phyllonorycter lautella, Phyllonorycter maestingella, Phyllonorycter medicaginella, Phyllonorycter messaniella, Phyllonorycter oxyacanthae, Phyllonorycter platani, Phyllonorycter platanoidella, Phyllonorycter pomonella, Phyllonorycter pyrifoliella, Phyllonorycter quercifoliella, Phyllonorycter quinnata, Phyllonorycter rajella, Phyllonorycter roboris, Phyllonorycter salictella, Phyllonorycter schreberella, Phyllonorycter sorbi, Phyllonorycter spinolella, Phyllonorycter stettinensis, Phyllonorycter strigulatella, Phyllonorycter tenerella, Phyllonorycter trifasciella, Phyllonorycter tristrigella, Phyllonorycter ulmifoliella (Gracillariidae), *Antispila treitschkiella, Holocacista rivillei* (Heliozelidae), *Leucoptera sp., Leucoptera aceris, Leucoptera lustratella, Leucoptera malifoliella, Leucoptera scitella, Lyonetia sp., Lyonetia clerkella, Paraleucoptera sinuella* (Lyonetiidae), *Nepticula sp., Nepticula lapponica, Nepticula malella, Nepticula oxyacanthella, Nepticula plagicolella, Nepticula salicis, Nepticula septembrella, Stigmella sp., Stigmella aceris, Stigmella agrimoniae, Stigmella aucupariae, Stigmella catharticella, Stigmella hemargyrella, Stigmella malella, Stigmella myrtillella, Stigmella oxyacanthella, Stigmella plagicolella, Stigmella ruficapitella* (Nepticulidae), *Barathra brassicae, Polia oleracea* (Noctuidae), *Scythris temperatella, Syringopais temperatella* (Scythrididae), *Tischeria sp., Tischeria angusticollella, Tischeria complanella, Tischeria decidua, Tischeria dodonaea, Tischeria ekebladella, Tischeria heinemanni, Tischeria marginea* (Tischeriidae), *Prays oleae* (Yponomeutidae) (Noyes, 2019).

Parasitoid hosts: *Apanteles circumscriptus* (Hymenoptera: Braconidae), *Achrysocharoides latreillei, Enaysma splendens* (Eulophidae) (Noyes, 2019).

Distribution: Australia, Austria, Bosnia Hercegovina, Bulgaria, Canary Islands, Croatia, Czech Republic, Europe, France, Germany, Greece, Hungary, Iran, Israel, Italy, Macedonia, Moldova, Montenegro, Netherlands, New Zealand, Norway, Palearctic, People's Republic of China, Romania, Russia, Serbia, Slovakia, Slovenia, Spain, Sweden, Switzerland, Turkey, Ukraine, United Kingdom, USSR, Yugoslavia (pre-1991) (Noyes, 2019).

Pnigalio soemius (Walker, 1839)
Publications with Turkey records: Çam (2011), Doğanlar (1985), Gencer (2009), Noyes (2019), Yefremova et al. (2010).
Associates: *Acer platanoides, Acer pseudoplatanus* (Aceraceae), *Aegopodium podagraria, Angelica archangelica, Pastinaca sativa* (Apiaceae), *Artemisia podagraria, Artemisia vulgaris, Cynara cardunculus, Eupatorium cannabinum, Lapsana communis, Solidago virgaurea, Sonchus palustris, Taraxacum officinale, Tussilago farfara* (Asteraceae), *Impatiens parviflora* (Balsaminaceae), *Alnus sp., Alnus glutinosa, Alnus incana, Betula sp., Betula pendula, Betula verrucose, Carpinus betulus, Corylus avellana* (Betulaceae), *Cerinthe minor* (Boraginaceae), *Alliaria officinalis Humulus lupulus* (Brassicaceae), *Lonicera periclymenum* (Caprifoliaceae), *Chenopodium album, Suaeda vera* (Chenopodiaceae), *Hypericum maculatum* (Clusiaceae), *Pteridium aquilinum* (Dennstaedtiaceae), *Astragalus glycyphyllos, Laburnum anagyroides, Lathyrus vernus, Medicago falcata, Pisum sativum, Robinia pseudoacacia, Sarothamnus scoparius* (Fabaceae), *Fagus sylvatica, Quercus robur* (Fagaceae), *Uromyces trifolii* (Fungus), *Aesculus hippocastanum* (Hippocastanaceae), *Ballota nigra, Lamium galeobdolon* (Lamiaceae), *Gossypium sp.* (Malvaceae), *Olea europaea* (Oleaceae), *Agropyron sp., Deschampsia cespitosa, Hordeum vulgare, Lygeum sp., Milium effusum, Oryza sativa, Phalaris arundinacea* (Poaceae), *Aquilegia vulgaris, Ranunculus repens, Thalictrum flavum* (Ranunculaceae), *Cerasus sp., Crataegus laevigata, Crataegus monogyna, Filipendula ulmaria, Geum urbanum, Malus sp., Malus pumila, Potentilla repens, Prunus padus, Rosa canina, Sorbus aucuparia* (Rosaceae), *Citrus sp.* (Rutaceae), *Populus sp., Populus nigra, Populus tremula, Salix alba, Salix cinerea, Salix fragilis, Salix pentandra, Salix rubens, Salix triandra* (Salicaceae), *Ulmus sp., Ulmus glabra, Ulmus procera* (Ulmaceae) (Noyes, 2019).
Parasitoids: *Cirrospilus donatellae* (Hymenoptera: Eulophidae) (Noyes, 2019).

Hosts: *Dicladispa armigera, Lema cyanella, Sphaeroderma rubidum* (Coleoptera: Chrysomelidae), *Anoplus plantaris, Ceutorhynchus contractus, Orchestes avellanae, Orchestes fagi, Orchestes quercus, Rhamphus oxyacanthae, Rhamphus pulicarius, Rhynchaenus sp., Rhynchaenus fagi, Rhynchaenus populi, Rhynchaenus quercus, Rhynchaenus rusci, Rhynchaenus testaceus* (Curculionidae), *Agromyza sp., Agromyza albitarsis, Agromyza alnibetulae, Agromyza demeijerei, Agromyza flaviceps, Agromyza frontella, Agromyza potentillae, Amauromyza morionella, Cerodontha incisa, Chromatomyia fuscula, Japanagromyza salicifolii, Liriomyza sp., Liriomyza eupatorii, Liriomyza huidobrensis, Liriomyza impatientis, Liriomyza strigata, Paraphytomyza hendeliana, Paraphytomyza populi, Paraphytomyza tridentata, Phytoliriomyza melampyga, Phytoliriomyza variegate, Phytomyza sp., Phytomyza angelicae, Phytomyza artemisivora, Phytomyza atricornis, Phytomyza heringiana, Phytomyza horticola, Phytomyza milii, Phytomyza minuscula, Phytomyza obscurella, Phytomyza periclymeni, Phytomyza pubicornis, Phytomyza ranunculi, Phytomyza solidaginis, Phytomyza sphondylii, Pycnoglossa hystrix, Trilobomyza artemisiae* (Diptera: Agromyzidae), *Chirosia parvicornis* (Anthomyiidae), *Cystiphora sonchi, Cystiphora taraxaci, Mycodiplosis sp., Rhopalomyia ptarmicae* (Cecidomyiidae), *Dacus oleae, Prionimera cognata* (Tephritidae), *Fenusa dohrnii, Fenusa ulmi, Messa nana, Metallus gei, Profenusa pygmaea, Scolioneura betulae* (Hymenoptera: Tenthredinidae), *Elachista cerusella, Elachista klimeschi, Elachista occulta, Perittia herrichiella, Perittia weberella* (Lepidoptera: Elachistidae), *Eriocrania haworthi, Eriocrania semipurpurella, Eriocrania sparmannella, Eriocrania unimaculella* (Eriocraniidae), *Chrysoesthia sexguttella, Chrysoesthis drurella, Microsetia sexguttella, Pexicopia malvella* (Gelechiidae), *Callisto denticulella, Caloptilia syringella, Cameraria ohridella, Gracillaria syringella, Lithocolletis sp., Lithocolletis nigrescentella, Lithocolletis populifoliella, Lithocolletis schreberella, Parectopa robiniella, Phyllocnistis citrella, Phyllocnistis labyrinthella, Phyllonorycter sp., Phyllonorycter alnifoliella, Phyllonorycter anderidae, Phyllonorycter blancardella, Phyllonorycter cavella, Phyllonorycter cerasicolella, Phyllonorycter corylifoliella, Phyllonorycter cydoniella, Phyllonorycter dubitella, Phyllonorycter emberizaepennella, Phyllonorycter maestingella, Phyllonorycter medicaginella, Phyllonorycter nigrescentella, Phyllonorycter oxyacanthae, Phyllonorycter platanoidella, Phyllonorycter pomonella, Phyllonorycter rajella, Phyllonorycter salictella, Phyllonorycter sorbi, Phyllonorycter tenerella, Phyllonorycter tristrigella, Phyllonorycter ulmifoliella* (Gracillariidae), *Leucoptera sp., Leucoptera*

laburnella, Leucoptera lotella, Leucoptera lustratella, Leucoptera malifoliella, Leucoptera spartifoliella, Lyonetia clerkella (Lyonetiidae), *Cosmopterix pulchrimella* (Momphidae), *Ectoedemia argentipedella, Nepticula sp., Nepticula lapponica, Nepticula sorbi, Stigmella sp., Stigmella argentipedella, Stigmella aucupariae, Stigmella aurella, Stigmella crataegella, Stigmella floslactella, Stigmella hemargyrella, Stigmella hybnerella, Stigmella malella, Stigmella oxyacanthella, Stigmella ruficapitella, Stigmella salicis, Stigmella septembrella, Stigmella sorbi, Stigmella staphyleae* (Nepticulidae), *Tischeria angusticollella, Tischeria ekebladella* (Tischeriidae), *Kessleria saxifragae* (Yponomeutidae) (Noyes, 2019).
Distribution: Austria, Bangladesh, Bulgaria, Canary Islands, Croatia, Czech Republic, Czechoslovakia, Europe, France, Germany, Hungary, Iran, Iraq, Israel, Italy, Korea, Macedonia, Madeira, Moldova, Montenegro, Netherlands, Norway, Pakistan, Palearctic, People's Republic of China, Romania, Russia, Serbia, Slovakia, Spain, Sweden, Switzerland, Syria, Turkey, Ukraine, United Kingdom, USSR, Yugoslavia (pre-1991) (Noyes, 2019).

Genus *Rhicnopelte* (Foerster, 1878)
Rhicnopelte crassicornis (Nees, 1834)
Publications with Turkey records: Noyes (2019), Öncüer (1991).
Hosts: *Hyphantria cunea* (Lepidoptera: Arctiidae), *Phyllonorycter medicaginella* (Gracillariidae), *Agrotis sp., Eugraphe sigma, Scotia sp.* (Noctuidae) (Noyes, 2019).
Distribution: Austria, Bulgaria, Czech Republic, Czechoslovakia, Europe, France, Germany, Hungary, Iran, Israel, Italy, Kazakhstan, Macedonia, Moldova, Netherlands, Palearctic, Poland, Russia, Slovakia, Tselinograd Obl., Turkey, Yemen (Noyes, 2019).

Genus *Semielacher* (Boucek, 1988)
Semielacher petiolata (Girault, 1915)
Publications with Turkey records: Noyes (2019).
Associates: *Sonchus sp.* (Asteraceae), *Echium sp.* (Boraginaceae), *Mercurialis annua* (Euphorbiaceae), *Rubus ulmifolius* (Rubiaceae), *Citrus sp., Citrus aurantium, Citrus clementina, Citrus limon, Citrus paradisi, Citrus sinensis, Poncirus trifoliatus* (Rutaceae), *Parietaria diffusa, Urtica sp.* (Urticaceae) (Noyes, 2019).
Hosts: *Agromyza hiemalis, Chromatomyia horticola, Chromatomyia syngenesiae, Liriomyza sp., Liriomyza bryoniae* (Diptera: Agromyzidae), *Dialectica scalariella, Phyllocnistis citrella* (Lepidoptera: Gracillariidae),

Cosmopterix pulchrimella (Momphidae), *Stigmella aurella* (Nepticulidae) (Noyes, 2019).
Distribution: Algeria, Australia, Cyprus, Greece, Israel, Italy, Jordan, Morocco, North Africa, Oman, Solomon Islands, Spain, Syria, Tunisia, Turkey (Noyes, 2019).

Genus *Stenomesius* (Westwood, 1833)
Stenomesius rufescens (Retzius, 1783)
Publications with Turkey records: Noyes (2019), Öncüer (1991).
Associates: *Ficus sp., Ficus carica* (Moraceae) (Noyes, 2019).
Hosts: *Phloeotribus scarabaeoides* (Coleoptera: Scolytidae), *Anthophila nemorana, Simaethis nemorana* (Lepidoptera: Glyphipterygidae), *Phyllocnistis citrella* (Gracillariidae), *Ectoedemia sp., Nepticula argyropeza, Stigmella sp., Stigmella argyropeza, Stigmella aurella* (Nepticulidae), *Cnephasia longana* (Tortricidae) (Noyes, 2019).
Distribution: Armenia, Austria, Bulgaria, Czech Republic, Czechoslovakia, England, Europe, France, Georgia, Germany, Hungary, Iran, Italy, Montenegro, Nearctic, Netherlands, Norway, Palearctic, Slovakia, Sweden, Transcaucasus, Turkey, United States of America (Noyes, 2019).

Genus *Sympiesis* (Foerster, 1856)
Sympiesis acalle (Walker, 1848)
Publications with Turkey records: Gencer (2012), Noyes (2019), Öncüer (1991).
Associates: *Juniperus excelsa* (Cupressaceae), *Vaccinium macrocarpon* (Ericaceae), *Colutea sp., Robinia sp.* (Fabaceae), *Larix sp., Larix europaeus* (Pinaceae), *Crataegus sp., Malus sp., Malus domestica, Malus pumila, Prunus sp., Prunus cerasus, Prunus spinosa, Rosa sp.* (Rosaceae), *Salix sp., Salix viminalis* (Salicaceae) (Noyes, 2019).
Parasitoid hosts: *Apanteles sp., Apanteles circumscriptus, Apanteles lacteus, Pholetesor ornigis* (Hymenoptera: Braconidae), *Angitia fenestralis, Diadegma fenestrale* (Ichneumonidae) (Noyes, 2019).
Hosts: *Liriomyza variegata* (Diptera: Agromyzidae), *Spilochalcis sp.* (Hymenoptera: Chalcididae), *Blastobasis aurantiaca* (Lepidoptera: Blastobasidae), *Elachista sp.* (Elachistidae), *Anacampsis populella, Gelechia mulinella, Gelechia senticetella, Mirificarma mulinella, Recurvaria leucatella, Tachyptilia populella* (Gelechiidae), *Anthophila pariana* (Glyphipterygidae), *Callisto denticulella, Callisto geminatella, Caloptilia sp., Caloptilia alchimiella, Caloptilia elongella, Caloptilia palustriella, Caloptilia roscipennella, Caloptilia stigmatella, Lithocolletis*

sp., Lithocolletis blancardella, Lithocolletis corylifoliella, Lithocolletis hortella, Lithocolletis lantanella, Lithocolletis nigrescentella, Lithocolletis pomifoliella, Lithocolletis populiella, Lithocolletis populifoliella, Lithocolletis pyrifoliella, Lithocolletis spinicolella, Ornix sp., Parornix sp., Parornix anglicella, Parornix anguliferella, Parornix avellanella, Parornix finitimella, Parornix geminatella, Parornix periolella, Parornix petiolella, Phyllonorycter sp., Phyllonorycter blancardella, Phyllonorycter corylifoliella, Phyllonorycter crataegella, Phyllonorycter mespilella, Phyllonorycter ringoniella, Phyllonorycter robiniella, Phyllonorycter spinicolella, Phyllonorycter trifasciella (Gracillariidae), *Leucoptera scitella* (Lyonetiidae), *Stigmella plagicolella* (Nepticulidae), *Agonopterix posticella* (Oecophoridae), *Tischeria sp., Tischeria malifoliella* (Tischeriidae), *Acleris minuta, Ancylis comptana, Apotomis lutozana, Archips rosanus, Cacoecia rosanus, Choristoneura rosaceana, Epinotia laracana, Rhopobota naevana, Spilonota lariciana, Spilonota ocellana, Zeiraphera diniana* (Tortricidae) (Noyes, 2019).
Distribution: Azerbaijan, Bosnia Hercegovina, Bulgaria, Canada, Canary Islands, Croatia, Czech Republic, Czechoslovakia, England, Europe, Germany, Hungary, Israel, Italy, Japan, Kirgizia, Korea, Moldova, Montenegro, Nearctic, Netherlands, Romania, Russia, Serbia, Slovakia, Spain, Sweden, Switzerland, Tadzhikistan, Transcaucasus, Turkey, Turkmenistan, Ukraine, United States of America, Yemen (Noyes, 2019).

Sympiesis albiflava (Fahringer, 1944)
Publications with Turkey records: Noyes (2019), Öncüer (1991).
Associates: *Fraxinus sp.* (Oleaceae) (Noyes, 2019).
Hosts: *Caloptilia cuculipennella* (Lepidoptera: Gracillariidae) (Noyes, 2019).
Distribution: Turkey (Noyes, 2019).

Sympiesis euspilapterygis (Erdos, 1958)
Publications with Turkey records: Noyes (2019), Öncüer (1991).
Associates: *Erysimum sp.* (Brassicaceae), *Polygonum sp., Polygonum lapathifolium* (Polygonaceae), *Malus sp.* (Rosaceae), *Salix sp., Salix rosmarinifolia* (Salicaceae) (Noyes, 2019).
Hosts: *Aspilapteryx limosella, Euspilapteryx phasianipennella, Gracillaria limosella, Gracillaria phasianipennella, Lithocolletis blancardella, Lithocolletis quinqueguttella, Lithocolletis ulmifoliella, Phyllonorycter blancardella, Phyllonorycter pyrifoliella* (Lepidoptera: Gracillariidae) (Noyes, 2019).

Distribution: Austria, Bulgaria, Czech Republic, Czechoslovakia, Germany, Hungary, Montenegro, Poland, Sweden, Turkey (Noyes, 2019).

Sympiesis flavopicta (Boucek, 1959)
Publications with Turkey records: Noyes (2019).
Hosts: *Medicago sativa* (Fabaceae) (Noyes, 2019).
Distribution: Armenia, Austria, Bulgaria, Croatia, Czech Republic, Czechoslovakia, Germany, Hungary, Iraq, Israel, Kirgizia, Moldova, Montenegro, Pakistan, Romania, Russia, Serbia, Slovakia, Slovenia, Spain, Sweden, Tadzhikistan, Turkey, Turkmenistan, USSR, Yugoslavia (Federal Republic), Yugoslavia (pre-1991) (Noyes, 2019).

Sympiesis sericeicornis (Nees, 1834)
Publications with Turkey records: Gencer (2012), Noyes (2019), Öncüer (1991).
Associates: *Formica aquilonia* (Formicidae), *Acer sp., Acer platanoides, Acer pseudoplatanus, Acer tatarica* (Aceraceae), *Ambrosia artemisiifolia* (Asteraceae), *Impatiens balsamina* (Balsaminaceae), *Alnus sp., Alnus glutinosa, Alnus hirsute, Alnus incana, Betula sp., Betula pendula, Betula pubescens, Betula verrucose, Carpinus betulus, Corylus avellana* (Betulaceae), *Viburnum lantana* (Caprifoliaceae), *Castanea sativa, Fagus sp., Fagus sylvatica, Quercu coccifera, Quercus sp., Quercus acutissima, Quercus cerris, Quercus dentata, Quercus ilex, Quercus ithaburensis, Quercus macrocarpa, Quercus mongolica, Quercus petraea, Quercus robur, Quercus variabilis* (Fagaceae), *Ephedra nebrodensis* (Gnetaceae), *Aesculus hippocastanum* (Hippocastanaceae), *Carya illinoinensis* (Juglandaceae), *Fraxinus excelsior, Olea europaea* (Oleaceae), *Platanus sp., Platanus orientalis* (Platanaceae), *Amelanchier sp., Cerasus sp., Crataegus sp., Crataegus laevigata, Crataegus monogyna, Crataegus oxyacantha, Crataegus oxyacanthoides, Cydonia oblonga, Fragaria sp., Malus sp., Malus communis, Malus domestica, Malus pumila, Malus sylvestris, Prunus sp., Prunus avium, Prunus cerasus, Prunus padus, Prunus spinosa, Pyrus sp., Pyrus communis, Pyrus malus, Pyrus pyraster, Sorbus aria, Sorbus aucuparia, Sorbus intermedia, Sorbus torminalis* (Rosaceae), *Populus sp., Populus alba, Populus italicus, Populus nigra, Populus tremula, Salix sp., Salix alba, Salix caprea, Salix cinerea, Salix fragilis x pentandra, Salix nigricans, Salix purpurea, Salix rubens, Salix viminalis* (Salicaceae), *Simmondsia chinensis* (Simmondsaceae), *Celtis australis, Ulmus sp., Ulmus foliacea, Ulmus glabra* (Ulmaceae) (Noyes, 2019).

Hosts: *Hypera adspersa, Orchestes avellanae, Rhamphus oxyacanthae, Rhynchaenus salicis* (Coleoptera: Curculionidae), *Heterarthrus sp., Heterarthrus aceris, Heterarthrus nemoratus, Heterarthrus vagans* (Hymenoptera: Tenthredinidae), *Bucculatrix sp.* (Lepidoptera: Bucculatricidae), *Coleophora flaginella, Coleophora flavaginella, Coleophora pruniella* (Coleophoridae), *Anacampsis populella, Coleotechnites milleri* (Gelechiidae), *Callisto denticulella, Callisto geminatella, Caloptilia sp., Caloptilia alchimiella, Caloptilia elongella, Caloptilia hemidactylella, Caloptilia palustriella, Caloptilia rhombifoliella, Caloptilia roscipennella, Caloptilia stigmatella, Caloptilia syringella, Cameraria caryaefoliella, Cameraria hamadryadella, Cameraria ohridella, Cameraria tristrigella, Gracillaria alnivorella, Gracillaria azaleella, Gracillaria negundella, Gracillaria stigmatella, Gracillaria syringella, Lithocolletis sp., Lithocolletis acerifoliella, Lithocolletis acernella, Lithocolletis agilella, Lithocolletis alnicolella, Lithocolletis alpina, Lithocolletis bicinctella, Lithocolletis blancardella, Lithocolletis caryaefoliella, Lithocolletis cavella, Lithocolletis cerasicolella, Lithocolletis cincinnatiella, Lithocolletis comparella, Lithocolletis corylifoliella, Lithocolletis cramerella, Lithocolletis crataegella, Lithocolletis cytisella, Lithocolletis dubitella, Lithocolletis emberizaepennella, Lithocolletis endryella, Lithocolletis ermani, Lithocolletis faginella, Lithocolletis froelichiella, Lithocolletis hamadryadella, Lithocolletis hancola, Lithocolletis harrisella, Lithocolletis hauderiella, Lithocolletis heegeriella, Lithocolletis hortella, Lithocolletis issikii, Lithocolletis juglandis, Lithocolletis klemannella, Lithocolletis lucetiella, Lithocolletis maestingella, Lithocolletis malella, Lithocolletis manni, Lithocolletis mespilella, Lithocolletis messaniella, Lithocolletis millierella, Lithocolletis nicellii, Lithocolletis nipponicella, Lithocolletis oxyacanthae, Lithocolletis pastorella, Lithocolletis platani, Lithocolletis pomifoliella, Lithocolletis populiella, Lithocolletis populifoliella, Lithocolletis pygmaea, Lithocolletis pyrifoliella, Lithocolletis quercifoliella, Lithocolletis quinnata, Lithocolletis rajella, Lithocolletis ringoniella, Lithocolletis sagitella, Lithocolletis salicifoliella, Lithocolletis salictella, Lithocolletis schreberella, Lithocolletis scitulella, Lithocolletis sorbi, Lithocolletis spinicolella, Lithocolletis spinolella, Lithocolletis stettinensis, Lithocolletis strigulatella, Lithocolletis sylvella, Lithocolletis tenerella, Lithocolletis tremulae, Lithocolletis tremuloidiella, Lithocolletis tristrigella, Lithocolletis ulmifoliella, Metriochroa latifoliella, Oecophyllembius sp., Parornix sp., Parornix anglicella, Parornix anguliferella, Parornix avellanella, Parornix devoniella, Parornix geminatella, Parornix petiolella, Parornix scoticella, Phyllocnistis sp., Phyllocnistis labyrinthella, Phyllocnistis saligna,*

Phyllocnistis sorghageniella, Phyllocnistis suffusella, Phyllocnistis unipunctella, Phyllonorycter sp., Phyllonorycter acerifoliella, Phyllonorycter alnifoliella, Phyllonorycter anderidae, Phyllonorycter blancardella, Phyllonorycter caryaealbella, Phyllonorycter cerasicolella, Phyllonorycter connexella, Phyllonorycter coryli, Phyllonorycter corylifoliella, Phyllonorycter crataegella, Phyllonorycter cydoniella, Phyllonorycter dubitella, Phyllonorycter elmaella, Phyllonorycter harrisella, Phyllonorycter klemannella, Phyllonorycter lautella, Phyllonorycter maestingella, Phyllonorycter malella, Phyllonorycter mespilella, Phyllonorycter nipigon, Phyllonorycter oxyacanthae, Phyllonorycter platani, Phyllonorycter platanoidella, Phyllonorycter pomonella, Phyllonorycter populifoliella, Phyllonorycter propinquinella, Phyllonorycter pyrifoliella, Phyllonorycter quercifoliella, Phyllonorycter quinqueguttella, Phyllonorycter rajella, Phyllonorycter roboris, Phyllonorycter salicifoliella, Phyllonorycter salictella, Phyllonorycter schreberella, Phyllonorycter sorbi, Phyllonorycter spinicolella, Phyllonorycter spinolella, Phyllonorycter stettinensis, Phyllonorycter strigulatella, Phyllonorycter tenerella, Phyllonorycter tristrigella, Phyllonorycter ulmifoliella, Phyllonorycter viminetorum (Gracillariidae), *Porthetria dispar* (Lymantriidae), *Leucoptera malifoliella, Leucoptera scitella, Lyonetia sp., Lyonetia clerkella* (Lyonetiidae), *Periploca sp.* (Momphidae), *Nepticula sp., Nepticula malella, Stigmella malella, Stigmella nylandriella, Stigmella turbidella* (Nepticulidae), *Tischeria sp., Tischeria decidua, Tischeria dodonaea, Tischeria malifoliella* (Tischeriidae), *Ancylis comptana, Argyroploce lacunana, Choristoneura conflictana, Choristoneura lafauryana, Epinotia laracana, Epinotia sordidana, Pandemis dumetana, Sciaphila sp., Tortrix viridana* (Tortricidae), *Scythropia crataegella* (Yponomeutidae) (Noyes, 2019).

Parasitoid hosts: *Apanteles sp., Apanteles albipennis, Apanteles circumscriptus, Apanteles lacteus, Apanteles lautellus, Apanteles melanoscelus, Apanteles ornigis, Colastes braconius, Microgaster subcompleta, Pholetesor ornigis* (Hymenoptera: Braconidae), *Enaysma splendens, Pnigalio sp.* (Eulophidae) (Noyes, 2019).

Distribution: Armenia, Austria, Azerbaijan, Bosnia Hercegovina, Bulgaria, Canada, Canary Islands, Caucasus, Croatia, Czech Republic, Czechoslovakia, Europe, France, Germany, Greece, Hungary, Iran, Israel, Italy, Japan, Kazakhstan, Korea, Macedonia, Moldova, Montenegro, Nearctic, Netherlands, Norway, Palearctic, People's Republic of China, Poland, Romania, Russia, Serbia, Slovakia, Slovenia, Spain, Sweden, Switzerland, Tadzhikistan, Taiwan, Transcaucasus, Tselinograd Obl.,

Turkey, Ukraine, United Kingdom, United States of America, USSR, Yugoslavia (pre-1991) (Noyes, 2019).

Sympiesis turcica (Fahringer, 1944)
Publications with Turkey records: Noyes (2019), Öncüer (1991).
Hosts: *Lithocolletis platani* (Lepidoptera: Gracillariidae) (Noyes, 2019).
Distribution: Turkey (Noyes, 2019).

Sympiesis viridula (Thomson, 1878)
Publications with Turkey records: Noyes (2019), Yefremova et al. (2010).
Associates: *Ambrosia artemisiifolia, Arctium lappa* (Asteraceae), *Betula sp.* (Betulaceae), *Cannabis sativa* (Cannabidaceae), *Quercus laevis* (Fagaceae), *Zea mays* (Poaceae), *Malus sp., Malus pumila* (Rosaceae), *Solanum tuberosum* (Solanaceae) (Noyes, 2019).
Hosts: *Brachys tessellatus* (Coleoptera: Buprestidae), *Anacampsis populella, Phthorimaea operculella, Scrobipalpa suasella, Tachyptilia populella* (Lepidoptera: Gelechiidae), *Lithocolletis blancardella, Phyllonorycter blancardella* (Gracillariidae), *Leucoma salicis* (Lymantriidae), *Gortyna flavago, Papaipema nebris, Sesamia sp., Sesamia cretica* (Noctuidae), *Ostrinia nubilalis, Ostrinia obumbratalis, Ostrinia penitalis, Pyrausta ainsliei, Pyrausta nubilalis, Pyrausta penitalis* (Pyralidae), *Cochylis posterana, Phalonia sp., Phalonia posterana, Tortrix viridana* (Tortricidae) (Noyes, 2019).
Parasitoid hosts: *Cratotechus longicornis, Eulophus abdominalis* (Hymenoptera: Eulophidae) (Noyes, 2019).
Distribution: Bosnia Hercegovina, Bulgaria, Canada, Croatia, Czech Republic, Czechoslovakia, England, Europe, France, Germany, Hungary, Israel, Italy, Moldova, Nearctic, Netherlands, Poland, Romania, Russia, Slovakia, Sweden, Turkey, Ukraine, United States of America, USSR, Yugoslavia (pre-1991) (Noyes, 2019).

Genus *Zagrammosoma* (Ashmead, 1904)
Zagrammosoma talitzkii (Boucek, 1961)
Publications with Turkey records: Çikman and LaSalle (2011), Noyes (2019).
Associates: *Medicago sativa* (Fabaceae), *Aesculus hippocastanum* (Hippocastanaceae), *Malus pumila* (Rosaceae), *Lycopersicon esculentum, Solanum melongena* (Solanaceae) (Noyes, 2019).
Hosts: *Liriomyza congesta, Liriomyza pseudopygmina, Liriomyza trifolii* (Diptera: Agromyzidae), *Bucculatrix crataegi* (Lepidoptera: Bucculatricidae),

Cameraria ohridella, Lithocolletis cerasicolella, Lithocolletis connexella, Lithocolletis corylifoliella, Lithocolletis malella, Lithocolletis saliciphaga, Lithocolletis sorbi, Parornix persicella, Phyllonorycter sp. (Gracillariidae), *Holocacista sp., Holocacista rivillei* (Heliozelidae), *Leucoptera sp., Leucoptera scitella* (Lyonetiidae) (Noyes, 2019).
Distribution: Bulgaria, Iran, Israel, Italy, Kazakhstan, Moldova, Palearctic, Russia, Serbia, Tselinograd Obl., Turkey, Turkmenistan, Ukraine, USSR (Noyes, 2019).

Zagrammosoma variegatum (Masi, 1907)
Publications with Turkey records: Noyes (2019), Öncüer (1991).
Associates: *Jatropha curcas, Ricinus communis* (Euphorbiaceae), *Aesculus hippocastanum* (Hippocastanaceae), *Malus sp., Malus pumila, Pyrus malus* (Rosaceae), *Coffea sp., Coffea arabica* (Rubiaceae), *Citrus sp.* (Rutaceae), *Lycopersicon esculentum* (Solanaceae) (Noyes, 2019).
Hosts: *Chromatomyia horticola, Liriomyza congesta, Liriomyza sativae, Liriomyza trifolii* (Diptera: Agromyzidae), *Dacus oleae, Euleia heraclei* (Tephritidae), *Bucculatrix thurberiella* (Lepidoptera: Bucculatricidae), *Cameraria ohridella, Lithocolletis sp., Lithocolletis corylifoliella, Lithocolletis messaniella, Lithocolletis pyrifoliella, Metriochroa inferior, Metriochroa latifoliella, Metriochroa unferior, Oecophyllembius sp., Oecophyllembius inferior, Oecophyllembius neglectus, Phyllocnistis citrella, Phyllonorycter sp., Phyllonorycter blancardella, Phyllonorycter corylifoliella, Stomphastis thraustica* (Gracillariidae), *Antispila sp., Antispila rivillei, Holocacista sp., Holocacista rivillei* (Heliozelidae), *Leucoptera sp., Leucoptera caffeine, Leucoptera coffeella, Leucoptera malifoliella, Leucoptera meyricki, Leucoptera scitella, Lyonetia sp., Lyonetia clerkella* (Lyonetiidae), *Nepticula malella, Stigmella sp., Stigmella ariella, Stigmella malella* (Nepticulidae) (Noyes, 2019).
Distribution: Afrotropical, Armenia, Australia, Azerbaijan, Barbados, Bulgaria, Cameroon, Canary Islands, Croatia, Czechoslovakia, Eritrea, Ethiopia, Europe, Hungary, India, Iran, Italy, Kenya, Libya, Moldova, Morocco, New Zealand, North Africa, Pakistan, People's Republic of China, Peru, Romania, Russia, Serbia, Slovakia, Spain, Sweden, Switzerland, Tanzania, Turkey, Turkmenistan, United Arab Emirates, Uzbekistan, Yemen, Yugoslavia (pre-1991) (Noyes, 2019).

Subfamily Opheliminae (Westwood, 1833)
Genus *Ophelimus* (Haliday, 1844)
Ophelimus maskelli (Ashmead, 1900)
Publications with Turkey records: Aytar et al. (2011), Noyes (2019).

Associates: *Eucalyptus sp*. (Myrtaceae) (Noyes, 2019).
Hosts: *Eurhinocola eucalypti, Rhinocola eucalypti* (Hemiptera: Psyllidae), *Eucalyptus sp., Eucalyptus albens, Eucalyptus angulosa, Eucalyptus auringens, Eucalyptus bockwayi, Eucalyptus botryoides, Eucalyptus camaldulensis, Eucalyptus dundasii, Eucalyptus gillii, Eucalyptus globulus, Eucalyptus gomphocephala, Eucalyptus kruseana, Eucalyptus loxophleba, Eucalyptus maculate, Eucalyptus melanophloeia, Eucalyptus melliodora, Eucalyptus saligna, Eucalyptus tereticornis, Eucalyptus torelliana, Eucalyptus trabuti, Eucalyptus woodwardia* (Myrtaceae) (Noyes, 2019).
Parasitoids: *Closterocerus sp., Closterocerus chamaeleon* (Hymenoptera: Eulophidae), *Stethynium sp., Stethynium brevioviipositor, Stethynium ophelimi* (Mymaridae) (Noyes, 2019).
Distribution: Algeria, Argentina, Australia, Chile, England, France, Greece, Indonesia, Israel, Italy, Malta, New Zealand, Portugal, South Africa, Spain, Tunisia, Turkey, United States of America, Vietnam (Noyes, 2019).

Subfamily Tetrastichinae (Graham, 1987)
Genus *Anaprostocetus* (Graham, 1987)
Anaprostocetus acuminatus (Ratzeburg, 1848)
Publications with Turkey records: Doğanlar (2011), Doğanlar et al. (2011), Noyes (2019), Sakaltas and Gençer (2005).
Associates: *Dasineura oleae, Lasioptera oleicola* (Diptera: Cecidomyiidae), *Salix sp., Salix blanda* (Salicaceae) (Noyes, 2019).
Hosts: *Cecidomyia salicis, Dasineura oleae, Lasioptera oleicola* (Diptera: Cecidomyiidae), *Euura atra, Euura laeta* (Hymenoptera: Tenthredinidae), *Gypsonoma dealban* (Lepidoptera: Tortricidae), *Yponomeuta malinellus* (Yponomeutidae) (Noyes, 2019).
Distribution: Austria, Canada, Czech Republic, Czechoslovakia, England, Europe, France, Germany, Hungary, India, Italy, Japan, Nearctic, Netherlands, Sweden, Turkey, Ukraine (Noyes, 2019).

Anaprostocetus ankarensis (Gencer, 2010)
Publications with Turkey records: Gencer (2010), Noyes (2019).
Associates: Unknown (Noyes, 2019).
Hosts: Unknown (Noyes, 2019).
Distribution: Turkey (Noyes, 2019).

Genus *Aprostocetus* (Westwood, 1833)
Aprostocetus agevilleae (Graham, 1987)

Publications with Turkey records: Sakaltas and Gençer (2005), Noyes (2019).
Associates: *Abies alba* (Pinaceae) (Noyes, 2019).
Hosts: *Agevillea abietis* (Diptera: Cecidomyiidae) (Noyes, 2019).
Distribution: Czechoslovakia, Italy, Slovakia, Turkey (Noyes, 2019).

Aprostocetus agrus (Walker, 1839)
Publications with Turkey records: Sakaltas and Gençer (2005), Noyes (2019).
Associates: Unknown (Noyes, 2019).
Hosts: Unknown (Noyes, 2019).
Distribution: Czech Republic, Czechoslovakia, England, France, Greece, Hungary, Ireland, Netherlands, Russia, Slovakia, Sweden, Turkey (Noyes, 2019).

Aprostocetus bucculentus (Kostjukov, 1978)
Publications with Turkey records: Noyes (2019).
Associates: *Amygdalus sp., Amygdalus communis, Amygdalus fenzliana* (Rosaceae) (Noyes, 2019).
Hosts: *Eurytoma amygdali* (Hymenoptera: Eurytomidae) (Noyes, 2019).
Distribution: Armenia, Iran, Israel, Palearctic, Turkey (Noyes, 2019).

Aprostocetus catius (Walker, 1839)
Publications with Turkey records: Sakaltas and Gençer (2005), Noyes (2019).
Associates: Unknown (Noyes, 2019).
Hosts: Unknown (Noyes, 2019).
Distribution: Czech Republic, Czechoslovakia, England, Ireland, Russia, Sweden, Turkey (Noyes, 2019).

Aprostocetus caudatus (Westwood, 1833)
Publications with Turkey records: Sakaltas and Gençer (2005), Noyes (2019).
Associates: *Elymus farctus, Elymus repens* (Poaceae) (Noyes, 2019).
Hosts: *Dasineura alopecuri* (Diptera: Cecidomyiidae) (Noyes, 2019).
Distribution: Austria, Bulgaria, Czech Republic, Czechoslovakia, England, France, Germany, Greece, Hungary, Ireland, Italy, Netherlands, People's Republic of China, Russia, Slovakia, Sweden, Turkey, Yugoslavia (pre-1991) (Noyes, 2019).

Aprostocetus ceroplastae (Girault, 1916)
Publications with Turkey records: Graham (1987), Noyes (2019), Öncüer (1991), Ozsemerci and Aksit (2003).
Associates: *Maytenus senegalensis* (Celastraceae), *Ficus sp., Ficus carica* (Moraceae), *Myrtus communis, Syzygium malaccensis* (Myrtaceae), *Saccharum officinarum* (Poaceae), *Cydonia oblonga* (Rosaceae), *Citrus sp., Citrus nobilis, Citrus paradisi, Citrus reticulata, Citrus sinensis, Toddalia sp.* (Rutaceae) (Noyes, 2019).
Hosts: *Ceroplastes sp., Ceroplastes brevicauda, Ceroplastes cirripediformis, Ceroplastes destructor, Ceroplastes floridensis, Ceroplastes galeatus, Ceroplastes japonicus, Ceroplastes rubens, Ceroplastes rusci, Ceroplastes sinensis, Gascardia brevicauda, Gascardia destructor, Pulvinaria elongate, Pulvinaria iceryi* (Hemiptera: Coccidae) (Noyes, 2019).
Parasitoids: *Baryscapus ceroplastophilus, Tetrastichus ceroplastophilus (Hymenoptera: Eulophidae), Pachyneuron siculum* (Pteromalidae) (Noyes, 2019).
Distribution: Australia, Egypt, France, Greece, Israel, Italy, Lebanon, Libya, Madagascar, Mauritius, Morocco, Peoples' Republic of, China, South Africa, Turkey, Uganda, United States of America (Noyes, 2019).

Aprostocetus cracens (Graham, 1987)
Publications with Turkey records: Kaya and Kovanci (2001), Noyes (2019).
Hosts: *Coraebus rubi* (Coleoptera: Buprestidae) (Noyes, 2019).
Distribution: Greece, Italy, Turkey (Noyes, 2019).

Aprostocetus crino (Walker, 1838)
Publications with Turkey records: Sakaltas and Gençer (2005), Noyes (2019).
Associates: *Oecanthus sp.* (Orthoptera: Gryllidae), *Padus racemose* (Rosaceae) (Noyes, 2019).
Hosts: *Yponomeuta padellus* (Lepidoptera: Yponomeutidae), *Oecanthus sp., Oecanthus nigricornis, Oecanthus pellucens, Oecanthus quadripunctatus* (Orthoptera: Gryllidae) (Noyes, 2019).
Distribution: Andorra, Bulgaria, Croatia, Czech Republic, Czechoslovakia, Denmark, England, France, Germany, Hungary, Iceland, Ireland, Italy, Moldova, Montenegro, Nearctic, Netherlands, People's Republic of China, Romania, Russia, Slovakia, Sweden, Turkey, United States of America, Yugoslavia (Federal Republic), Yugoslavia (pre-1991) (Noyes, 2019).

Aprostocetus eriophyes (Taylor, 1909)
Publications with Turkey records: Noyes (2019), Öncüer (1991).
Associates: *Buxus sempervirens* (Buxaceae), *Ribes nigrum* (Grossulariaceae) (Noyes, 2019).
Hosts: *Aceria rudis, Aceria unguiculata, Cecidophyes ribis, Cecidophyopsis ribis, Eriophyes sp., Eriophyes buxi, Eriophyes canestrinii, Eriophyes ribis, Phytoptus avellanae, Phytoptus tiliae* (Acarina: Eriophyidae) (Noyes, 2019).
Distribution: England, Finland, Germany, Italy, Netherlands, Poland, Sweden, Turkey (Noyes, 2019).

Aprostocetus forsteri (Walker, 1847)
Publications with Turkey records: Sakaltas and Gençer (2005), Noyes (2019).
Associates: *Urophora sp.* (Diptera: Tephritidae), *Centaurea jacea, Centaurea nigra, Centaurea paniculate, Centaurea pannonica, Centaurea sadleriana, Centaurea scabiosa, Leucanthemum vulgare* (Asteraceae), *Salvia lavandulifolia, Salvia nemorosa, Salvia officinalis, Salvia pratensis* (Lamiaceae), *Papaver dubium, Papaver rhoeas* (Papaveraceae), *Rosa sp.* (Rosaceae) (Noyes, 2019).
Hosts: *Tephritis neesii* (Diptera: Tephritidae), *Aulax rogenhoferi, Aulax salvia, Aylax jaceae, Aylax rogenhoferi, Aylax salvia, Barbotinia oraniensis, Diplolepis mayri, Isocolus jaceae, Isocolus rogenhoferi, Isocolus scabiosae* (Hymenoptera: Cynipidae) (Noyes, 2019).
Distribution: Austria, Bulgaria, Cyprus, Czech Republic, Czechoslovakia, England, France, Hungary, Iran, Italy, Palearctic, Russia, Slovakia, Spain, Sweden, Switzerland, Turkey, Yugoslavia (pre-1991) (Noyes, 2019).

Aprostocetus holoxanthus (Graham, 1991)
Publications with Turkey records: Noyes (2019).
Associates: Unknown (Noyes, 2019).
Hosts: Unknown (Noyes, 2019).
Distribution: People's Republic of China, Turkey (Noyes, 2019).

Aprostocetus lasiopterus (Doğanlar, 2011)
Publications with Turkey records: Doğanlar (2011), Doğanlar et al. (2011), Noyes (2019).
Associates: *Dasineura oleae, Lasioptera oleicola* (Diptera: Cecidomyiidae), *Olea europaea* (Oleaceae) (Noyes, 2019).
Hosts: *Dasineura oleae, Lasioptera oleicola* (Diptera: Cecidomyiidae) (Noyes, 2019).

Distribution: Turkey (Noyes, 2019).

Aprostocetus levadiensis (Graham, 1987)
Publications with Turkey records: Sakaltas and Gençer (2005), Noyes (2019).
Associates: Unknown (Noyes, 2019).
Hosts: Unknown (Noyes, 2019).
Distribution: Greece, Sweden, Turkey (Noyes, 2019).

Aprostocetus longicauda (Thomson, 1878)
Publications with Turkey records: Noyes (2019).
Associates: Unknown (Noyes, 2019).
Hosts: Unknown (Noyes, 2019).
Distribution: Bulgaria, Czech Republic, Czechoslovakia, France, Germany, Greece, Hungary, Italy, Netherlands, Russia, Slovakia, Sweden, Turkey, United Kingdom, United States of America (Noyes, 2019).

Aprostocetus lycidas (Walker, 1839)
Publications with Turkey records: Noyes (2019).
Associates: *Fagus sp., Fagus sylvatica* (Fagaceae), *Salix sp*. (Salicaceae) (Noyes, 2019).
Hosts: *Cecidomyia sp., Dasineura epilobii, Dasineura leguminicola, Hartigiola annulipes, Kaltenbachiola strobe, Mayetiola destructor, Mikiola fagi, Rabdophaga sp., Rabdophaga heterobia, Thomasiniana theobaldi* (Diptera: Cecidomyiidae).
Distribution: Czech Republic, Czechoslovakia, Denmark, England, Finland, France, Germany, Italy, Montenegro, Morocco, Netherlands, Poland, Slovakia, Sweden, Turkey, Yugoslavia (pre-1991).

Aprostocetus neglectus (Domenichini, 1957)
Publications with Turkey records: Noyes (2019), Öncüer (1991).
Associates: *Quadraspidiotus perniciosus* (Diaspididae), *Acer pseudoplatanus* (Aceraceae), *Fraxinus excelsior* (Oleaceae), *Malus pumila* (Rosaceae), *Tilia sp.* (Tiliaceae) (Noyes, 2019).
Hosts: *Adalia bipunctata, Chilocorus sp., Chilocorus bijugus, Chilocorus bipustulatus, Chilocorus kuwanae, Coccinella sp., Coccinella septempunctata, Exochomus sp., Exochomus quadripustulatus, Scymnus sp., Scymnus subvillosus* (Coleoptera: Coccinellidae), *Myzus cerasi* (Hemiptera: Aphididae) (Noyes, 2019).
Distribution: Austria, Czech Republic, Czechoslovakia, England, Europe, France, Georgia, Germany, Greece, Hungary, India, Iran, Israel, Italy,

Moldova, Morocco, Nearctic, Pakistan, Russia, Sweden, Turkey, USSR (Noyes, 2019).

Aprostocetus pausiris (Walker, 1839)
Publications with Turkey records: Sakaltas and Gençer (2005), Noyes (2019).
Associates: *Phragmites australis* (Poaceae) (Noyes, 2019).
Hosts: *Dasineura leguminicola* (Diptera: Cecidomyiidae), *Lipara sp.* (Chloropidae) (Noyes, 2019).
Distribution: Canada, Czech Republic, Czechoslovakia, Denmark, England, France, Germany, Greece, Hungary, Ireland, Madeira, Nearctic, Netherlands, Norway, Portugal, Romania, Russia, Spain, Sweden, Turkey, USSR, Yugoslavia (pre-1991) (Noyes, 2019).

Aprostocetus productus (Graham, 1987)
Publications with Turkey records: Sakaltas and Gençer (2005), Noyes (2019).
Associates: Unknown (Noyes, 2019).
Hosts: Unknown (Noyes, 2019).
Distribution: Turkey (Noyes, 2019).

Aprostocetus pygmaeus (Zetterstedt, 1838)
Publications with Turkey records: Sakaltas and Gençer (2005), Noyes (2019).
Associates: *Alopecurus pratensis* (Poaceae), *Rosa sp.* (Rosaceae) (Noyes, 2019).
Hosts: *Cecidomyiidae unspecified, Dasineura alopecuri* (Diptera: Cecidomyiidae), *Megachile centuncularis, Osmia coerulescens* (Hymenoptera: Apidae), *Diplolepis rosae* (Cynipidae) (Noyes, 2019).
Distribution: Canada, Czech Republic, Czechoslovakia, Denmark, England, Europe, Finland, France, Germany, Greece, Hungary, Italy, Netherlands, Norway, Russia, Sweden, Turkey, Ukraine (Noyes, 2019).

Aprostocetus samandagus (Doğanlar, 2011)
Publications with Turkey records: Doğanlar (2011), Doğanlar et al. (2011), Noyes (2019).
Associates: *Olea europaea* (Oleaceae) (Noyes, 2019).
Hosts: *Dasineura oleae, Lasioptera oleicola* (Diptera: Cecidomyiidae) (Noyes, 2019).
Distribution: Turkey (Noyes, 2019).

Aprostocetus terebrans (Erdos, 1954)
Publications with Turkey records: Sakaltas and Gençer (2005), Noyes (2019).
Associates: Unknown (Noyes, 2019).
Hosts: Unknown (Noyes, 2019).
Distribution: Bulgaria, Czech Republic, Czechoslovakia, England, France, Germany, Hungary, Ireland, Netherlands, Norway, Russia, Slovakia, Sweden, Turkey, United States of America (Noyes, 2019).

Aprostocetus toddaliae (Risbec, 1958)
Publications with Turkey records: Noyes (2019).
Hosts: *Ceroplastes sp*. (Hemiptera: Coccidae) (Noyes, 2019).
Distribution: Israel, Madagascar, Turkey (Noyes, 2019).

Aprostocetus trjapitzini (Kostjukov, 1976)
Publications with Turkey records: Karaca et al. (2003), Noyes (2019), Ulgenturk (2001), Ulgenturk and Toros (2001).
Associates: *Eulecanium sp.* (Hemiptera: Coccidae), *Maackia sp., Robinia pseudoacacia* (Fabaceae), *Picea abies* (Pinaceae), *Dasiphora fruticose, Prunus sp., Prunus domestica, Prunus spinosa* (Rosaceae), *Ulmus sp.* (Ulmaceae).
Hosts: *Eulecanium sp., Eulecanium ciliatum, Eulecanium rugulosum, Eulecanium secretum, Eulecanium tiliae, Parthenolecanium corni, Physokermes hemicryphus, Physokermes piceae, Physokermes sugonjaevi, Sphaerolecanium prunastri* (Hemiptera: Coccidae) (Noyes, 2019).
Parasitoid hosts: *Microterys hortulanus* (Hymenoptera: Encyrtidae) (Noyes, 2019).
Distribution: Germany, Kazakhstan, Moldova, Netherlands, Romania, Russia, Sweden, Tadzhikistan, Tselinograd Obl., Turkey, United Kingdom, Uzbekistan (Noyes, 2019).

Aprostocetus tymber (Walker, 1839)
Publications with Turkey records: Sakaltas and Gençer (2005), Noyes (2019).
Associates: *Rubus sp., Rubus idaeus* (Rubiaceae), *Salix sp.* (Salicaceae) (Noyes, 2019).
Hosts: *Helicomyia saliciperda, Lasioptera rubi* (Diptera: Cecidomyiidae) (Noyes, 2019).
Distribution: Czech Republic, England, France, Ireland, Italy, Netherlands, Poland, Russia, Serbia, Slovakia, Sweden, Turkey, Yugoslavia (pre-1991) (Noyes, 2019).

Genus *Baryscapus* (Foerster, 1856)
Baryscapus adalia (Walker, 1839)
Publications with Turkey records: Sakaltas and Gençer (2005), Noyes (2019).
Associates: *Angelica sp*. (Apiaceae) (Noyes, 2019).
Hosts: *Cionus scrophulariae* (Coleoptera: Curculionidae), *Phytomyza orobanchia* (Diptera: Agromyzidae), *Leucoptera spartifoliella* (Lepidoptera: Lyonetiidae), *Yponomeuta malinellus, Yponomeuta padellus* (Yponomeutidae).
Parasitoid hosts: *Discochaeta hyponomeutae* (Diptera: Tachinidae), *Entedon sp*. (Hymenoptera: Eulophidae), *Angitia armillata, Herpestomus bruneicornis, Mesochorus confuses, Pimpla examinator* (Ichneumonidae) (Noyes, 2019).
Distribution: Bulgaria, Czech Republic, Czechoslovakia, England, France, Germany, Hungary, Italy, Russia, Serbia, Sweden, Turkey, Yugoslavia (pre-1991) (Noyes, 2019).

Baryscapus babiyi (Doğanlar, 1993)
Publications with Turkey records: Doğanlar (1993), Noyes (2019), Yegorenkova and Yefremova (2011).
Associates: *Cirsium vulgare* (Asteraceae), *Lepidium latifolium* (Brassicaceae) (Noyes, 2019).
Hosts: *Lasiosina devitata* (Diptera: Chloropidae) (Noyes, 2019).
Distribution: Turkey (Noyes, 2019).

Baryscapus berhidanus (Erdos, 1954)
Publications with Turkey records: Noyes (2019).
Associates: *Andricus caputmedusae, Andricus glutinosus, Andricus grossulariae, Andricus kollari, Andricus lucidus, Andricus pictus, Andricus quercuscalicis, Andricus quercustozae, Biorhiza pallida, Chilaspis nitida* (Hymenoptera: Cynipidae), *Quercus cerris, Quercus faginea, Quercus infectoria, Quercus lanuginose, Quercus petraea, Quercus pubescens, Quercus pyrenaica, Quercus robur* (Fagaceae) (Noyes, 2019).
Hosts: *Andricus glutinosus, Andricus hispanicus, Andricus kollari, Andricus quercuscalicis, Andricus quercustozae, Cynips quercustozae* (Hymenoptera: Cynipidae) (Noyes, 2019).
Distribution: Bulgaria, Czech Republic, Czechoslovakia, Europe, France, Hungary, Italy, Jordan, Spain, Turkey (Noyes, 2019).

Baryscapus bruchidii (Erdos, 1951)
Publications with Turkey records: Sakaltas and Gençer (2005), Noyes (2019).
Associates: *Glycyrrhiza echinate* (Fabaceae) (Noyes, 2019).
Parasitoid hosts: *Oncophanes lanceolator* (Hymenoptera: Braconidae), *Scambus calobatus* (Ichneumonidae) (Noyes, 2019).
Hosts: *Bruchidius cinerascens, Bruchidius peregii, Bruchidius unicolor* (Coleoptera: Bruchidae), *Archips rosanus* (Lepidoptera: Tortricidae) (Noyes, 2019).
Distribution: Hungary, Moldova, Russia, Turkey (Noyes, 2019).

Baryscapus bruchivorus (Gahan, 1942)
Publications with Turkey records: Noyes (2019).
Associates: *Coronilla emerus* (Fabaceae) (Noyes, 2019).
Hosts: *Bruchidius sp., Bruchidius ater, Bruchidius fasciatus, Bruchus brachialis, Bruchus lentis, Bruchus pallicornis, Bruchus rufipes, Bruchus ulicis* (Coleoptera: Bruchidae) (Noyes, 2019).
Distribution: France, Italy, Nearctic, Spain, Turkey, United States of America (Noyes, 2019).

Baryscapus bruchophagi (Gahan, 1913)
Publications with Turkey records: Sakaltas and Gençer (2005), Noyes (2019).
Associates: *Lotus corniculatus, Lotus tenuis, Medicago sp., Medicago hispida, Medicago sativa, Onobrychis viciifolia, Trifolium sp., Trifolium pratense* (Fabaceae) (Noyes, 2019).
Hosts: *Apion seniculus* (Coleoptera: Apionidae), *Hypera postica* (Curculionidae), *Tropicomyia atomella* (Diptera: Agromyzidae), *Asphondylia websteri* (Cecidomyiidae), *Bruchophagus sp., Bruchophagus gibbus, Bruchophagus kolobovae, Bruchophagus platypterus, Bruchophagus roddi, Eurytoma onobrychidis, Systole sp., Systole geniculate* (Hymenoptera: Eurytomidae) (Noyes, 2019).
Parasitoid hosts: *Bruchophagus sp.* (Hymenoptera: Eurytomidae), *Bathyplectes curculionis* (Ichneumonidae) (Noyes, 2019).
Distribution: Afrotropical, Argentina, Armenia, Austria, Belgium, Bulgaria, Canada, Germany, Hungary, India, Kirgizia, Mexico, Nearctic, Neotropical, New Zealand, Russia, Sweden, Turkey, United States of America, Uruguay, USSR, Uzbekistan (Noyes, 2019).

Baryscapus ceroplastophilus (Domenichini, 1965)
Publications with Turkey records: Noyes (2019), Öncüer (1974; 1991), Ozsemerci and Aksit (2003).
Associates: *Ficus sp., Ficus carica* (Moraceae), *Citrus sp., Citrus limon, Citrus nobilis, Citrus paradisi* (Rutaceae) (Noyes, 2019).
Hosts: *Ceroplastes sp., Ceroplastes floridensis, Ceroplastes rusci, Coccus hesperidum, Coccus pseudomagnoliarum, Saissetia oleae* (Hemiptera: Coccidae), *Trioza ceroplastae* (Triozidae), *Scutellista cyanea* (Hymenoptera: Pteromalidae) (Noyes, 2019).
Parasitoid hosts: *Microterys flavus* (Encyrtidae), *Aprostocetus ceroplastae, Tetrastichus ceroplastae, Tetrastichus gibbus* (Eulophidae), *Scutellista coerulea, Scutellista cyanea* (Pteromalidae) (Noyes, 2019).
Parasitoids: *Ceroplastes rusci* (Hemiptera: Coccidae) (Noyes, 2019).
Distribution: France, Greece, Israel, Italy, Lebanon, Turkey (Noyes, 2019).

Baryscapus cirsiicola (Graham, 1991)
Publications with Turkey records: Sakaltas and Gençer (2005), Noyes (2019).
Associates: *Cirsium vulgare* (Asteraceae) (Noyes, 2019).
Distribution: England, France, Russia, Turkey (Noyes, 2019).

Baryscapus crassicornis (Erdos, 1954)
Publications with Turkey records: Koçak and Ozdemir (2012), Noyes (2019), Sakaltas and Gençer (2005).
Associates: *Heracleum platytaenium* (Apiaceae), *Carduus acanthoides, Cirsium canum, Cirsium vulgare, Crepis conyzifolia* (Asteraceae) (Noyes, 2019).
Hosts: *Larinus sp., Larinus jaceae, Larinus turbinatus, Lixus nordmanni* (Coleoptera: Curculionidae), *Tephritis conycifoliae* (Diptera: Tephritidae) (Noyes, 2019).
Distribution: Czech Republic, Czechoslovakia, France, Germany, Hungary, Israel, Russia, Spain, Turkey (Noyes, 2019).

Baryscapus daira (Walker, 1839)
Publications with Turkey records: Noyes (2019), Sakaltas and Gençer (2005), Öncüer (1991).
Associates: *Centaurea sp., Centaurea scabiosa, Cirsium sp., Cirsium arvense, Cirsium eriophorum, Cirsium vulgare, Cynara scolymus* (Asteraceae), *Olea europaea* (Oleaceae) (Noyes, 2019).

Hosts: *Cecidomyia sp., Dasineura oleae, Perrisia oleae* (Diptera: Cecidomyiidae), *Chaetorellia jaceae, Chaetorellia loricate, Orellia colon, Orellia ruficauda, Terellia florescentiae, Terellia fuscicornis, Terellia longicauda, Terellia serratulae, Trypeta colon, Trypeta florescentiae, Urophora cuspidate, Urophora jaceana, Urophora stylata* (Tephritidae), *Megachile rotundata* (Hymenoptera: Apidae) (Noyes, 2019).
Parasitoid hosts*:* *Crataepus marbis* (Eulophidae), *Eurytoma sp.* (Eurytomidae) (Noyes, 2019).
Distribution: Argentina, Austria, Bulgaria, Canada, Chile, Czech Republic, Czechoslovakia, England, France, Germany, Greece, Hungary, Italy, Moldova, Nearctic, Netherlands, Russia, Slovakia, Spain, Sweden, Switzerland, Turkey, United States of America, USSR (Noyes, 2019).

Baryscapus endemus (Walker, 1839)
Publications with Turkey records: Noyes (2019), Sakaltas and Gençer (2005).
Associates: *Lotus corniculatus, Medicago sativa* (Fabaceae), *Juncus squarrosus* (Juncaceae), *Pinus sp.* (Pinaceae), *Plantago lanceolata* (Plantaginaceae), *Scrophularia sp., Scrophularia nodosa, Verbascum lychnitis* (Scrophulariaceae) (Noyes, 2019).
Hosts: *Cionus sp., Cionus hortulanus, Cionus scrophulariae, Hypera postica, Orchestes fagi, Orchestes quercus, Rhynchaenus fagi, Rhynchaenus quercus, Stereonychus fraxini* (Coleoptera: Curculionidae), *Sphaerolecanium prunastri* (Hemiptera: Coccidae), *Kermococcus quercus (Kermesidae), Bruchophagus gibbus, Bruchophagus platypterus, Eurytoma onobrychidis* (Hymenoptera: Eurytomidae), *Coleophora alticolella, Coleophora deauratella* (Lepidoptera: Coleophoridae), *Peribatodes rhomboidaria* (Geometridae), *Lithocolletis corylifoliella, Lithocolletis pyrifoliella, Parornix periolella, Parornix petiolella, Phyllocnistis citrella* (Gracillariidae), *Heterogynis penella* (Heterogynidae), *Enargia ypsillon* (Noctuidae), *Thaumetopoea pityocampa* (Notodontidae), *Agonopterix alpigena, Depressaria alpigenella* (Oecophoridae), *Clavigestia purdeyi, Rhyacionia buoliana* (Tortricidae), *Argyresthia conjugella, Scythropia crataegella* (Yponomeutidae) (Noyes, 2019).
Parasitoid hosts: *Apanteles sp., Apanteles circumscriptus, Apanteles lacteus, Eubadizon minutus, Protomicroplitis alvearius, Sigalphus fagi* (Hymenoptera: Braconidae), *Holcothorax testaceipes, Litomastix kriechbaumeri* (Encyrtidae), *Elachertus olivaceus* (Eulophidae), *Angitia sp.* (Ichneumonidae), *Mesopolobus subfumatus, Pteromalus sp., Trichomalus sp.* (Pteromalidae) (Noyes, 2019).

Distribution: Austria, Bulgaria, Czechoslovakia, Denmark, England, France, Germany, Hungary, Iran, Ireland, Moldova, Netherlands, Norway, Palearctic, Poland, Russia, Spain, Sweden, Switzerland, Turkey, Ukraine, USSR, Yugoslavia (pre-1991) (Noyes, 2019).

Baryscapus euphorbiae (Graham, 1991)
Publications with Turkey records: Noyes (2019), Sakaltas and Gençer (2005).
Associates: *Euphorbia sp.* (Euphorbiaceae) (Noyes, 2019).
Hosts: *Bayeria capitigena* (Diptera: Cecidomyiidae) (Noyes, 2019).
Distribution: Netherlands, Russia, Sweden, Turkey (Noyes, 2019).

Baryscapus evonymellae (Bouche, 1834)
Publications with Turkey records: Noyes (2019), Öncüer (1991), Özbek and Coruh (2010).
Associates: *Euonymus verrucose* (Celastraceae), *Astragalus meridionalis, Cytisus scoparius, Sarothamnus scoparius* (Fabaceae), *Crataegus azarolus, Malus communis, Malus domestica, Malus pumila, Padus racemose, Prunus armeniaca, Prunus padus, Pyrus malus* (Rosaceae), *Salix sp., Salix alba* (Salicaceae) (Noyes, 2019).
Parasitoid hosts: *Angitia armillata, Diadegma armillatum, Mesochorus confuses* (Hymenoptera: Ichneumonidae) (Noyes, 2019).
Hosts: *Diplolepis spinosa, Diplolepis spinosissimae* (Hymenoptera: Cynipidae), *Malacosoma neustria* (Lepidoptera: Lasiocampidae), *Leucoptera spartifoliella* (Lyonetiidae), *Myelois cribrella, Myelois cribrumella* (Pyralidae), *Yponomeuta sp., Yponomeuta cognatellus, Yponomeuta evonymellus, Yponomeuta malinellus, Yponomeuta padellus, Yponomeuta rorellus* (Yponomeutidae) (Noyes, 2019).
Distribution: Armenia, Azerbaijan, Belgium, Bosnia Hercegovina, Bulgaria, Czech Republic, Czechoslovakia, England, France, Germany, Hungary, Iran, Italy, Kirgizia, Lithuania, Moldova, Nearctic, Netherlands, Palearctic, Poland, Romania, Russia, Serbia, Spain, Sudan, Sweden, Switzerland, Transcaucasus, Turkey, Ukraine, United States of America, USSR, Yugoslavia (pre-1991) (Noyes, 2019).

Baryscapus fossarum (Graham, 1991)
Publications with Turkey records: Noyes (2019), Sakaltas and Gençer (2005).
Associates: *Epilobium hirsutum* (Onagraceae) (Noyes, 2019).

Hosts: *Mompha epilobiella, Mompha fulvescens* (Lepidoptera: Momphidae) (Noyes, 2019).
Distribution: Czechoslovakia, England, France, Italy, Netherlands, Russia, Sweden, Turkey (Noyes, 2019).

Baryscapus galactopus (Ratzeburg, 1844)
Publications with Turkey records: Avci and Özbek (1990), Noyes (2019), Öncüer (1991).
Associates: *Brassica sp., Brassica oleracea* (Brassicaceae), *Malus pumila* (Rosaceae) (Noyes, 2019).
Parasitoid hosts: *Oncodes fumatus* (Diptera: Acroceridae), *Apanteles sp., Apanteles affinis, Apanteles glomeratus, Apanteles limbatus, Apanteles pieridis, Apanteles rubecula, Apanteles spurius, Cotesia sp., Cotesia glomerat, Cotesia rubecula, Microgaster sp., Microgaster vinulae, Rhogas pallidator* (Hymenoptera: Braconidae), *Anilastus ebeninus, Hyposoter sp., Hyposoter ebeninus* (Ichneumonidae) (Noyes, 2019).
Hosts: *Clubiona sp.* (Araneae: Clubionidae), *Phytomyza orobanchia* (Diptera: Agromyzidae), *Leucoma salicis* (Lepidoptera: Lymantriidae), *Cerura sp., Cerura vinula, Dicranura vinula, Thaumetopoea pityocampa* (Notodontidae), *Aporia crataegi, Artogeia sp., Artogeia rapae, Colias sp., Pieris sp., Pieris brassicae, Pieris rapae, Pontia daplidice* (Pieridae), *Ancylis minimana* (Tortricidae), *Plutella maculipennis, Yponomeuta evonymellus, Yponomeuta malinellus* (Yponomeutidae) (Noyes, 2019).
Distribution: Australia, Azores, Bulgaria, Canada, Canary Islands, Czech Republic, Czechoslovakia, Denmark, England, Finland, France, Germany, Hungary, India, Italy, Lithuania, Moldova, Nearctic, Netherlands, New Zealand, North Africa, Romania, Russia, Serbia, Spain, Sweden, Switzerland, Turkey, Ukraine, United States of America, USSR, Yugoslavia (pre-1991) (Noyes, 2019).

Baryscapus garganus (Domenichini, 1958)
Publications with Turkey records: Noyes (2019), Sakaltas and Gençer (2005).
Associates: *Sonchus oleraceus* (Asteraceae) (Noyes, 2019).
Parasitoid hosts: *Habrocytus fenomenalis* (Hymenoptera: Pteromalidae) (Noyes, 2019).
Hosts: *Lixus iridis* (Coleoptera: Curculionidae), *Tephritis formosa* (Diptera: Tephritidae) (Noyes, 2019).
Distribution: Germany, Italy, Turkey (Noyes, 2019).

Baryscapus gaziantepensis Doğanlar, 1993
Publications with Turkey records: Doğanlar (1993), Noyes (2019).
Associates: *Pistacia vera* (Anacardiaceae) (Noyes, 2019).
Hosts: *Agonoscena succincta* (Hemiptera: Psyllidae) (Noyes, 2019).
Distribution: Turkey (Noyes, 2019).

Baryscapus impeditus (Nees, 1834)
Publications with Turkey records: Noyes (2019), Öncüer (1991), Sakaltas and Gençer (2005).
Associates: *Azadirachta indica* (Meliaceae) (Noyes, 2019).
Hosts: *Diuraphis noxia* (Hemiptera: Aphididae), *Pulvinaria sp.* (Coccidae), *Anisochrysa carnea, Chrysopa sp., Chrysopa carnea, Chrysopa flavifrons, Chrysoperla carnea* (Neuroptera: Chrysopidae) (Noyes, 2019).
Distribution: Andorra, Bulgaria, Croatia, Czech Republic, Czechoslovakia, France, Germany, India, Italy, Moldova, Netherlands, Pakistan, Russia, Serbia, Slovakia, Sweden, Turkey (Noyes, 2019).
Baryscapus nigroviolaceus (Nees, 1834)
Publications with Turkey records: Cebeci (2011), Özbek and Coruh (2010), Noyes (2019), Sakaltas and Gençer (2005).
Associates: *Alnus glutinosa* (Betulaceae), *Viburnum lantana* (Caprifoliaceae), *Laburnum anagyroides* (Fabaceae), *Aesculus hippocastanum* (Hippocastanaceae), *Malus sp., Malus pumila, Pyrus malus* (Rosaceae), *Populus alba* (Salicaceae), *Vitis vinifera* (Vitaceae) (Noyes, 2019).
Parasitoid hosts: *Elasmus flabellatus* (Hymenoptera: Elasmidae), *Pediobius saulius, Pnigalio mediterraneus, Sympiesis sericeicornis, Tetrastichus platanellus* (Eulophidae) (Noyes, 2019).
Hosts: *Hypera postica, Orchestes quercus, Rhynchaenus quercus* (Coleoptera: Curculionidae), *Dasineura gleditchiae* (Diptera: Cecidomyiidae), *Cameraria ohridella, Lithocolletis sp., Lithocolletis blancardella, Lithocolletis cavella, Lithocolletis comparella, Lithocolletis corylifoliella, Lithocolletis froelichiella, Lithocolletis lantanella, Micrurapteryx sophorivora, Parornix petiolella, Phyllocnistis suffusella, Phyllonorycter sp., Phyllonorycter blancardella, Phyllonorycter corylifoliella, Phyllonorycter robiniella* (Lepidoptera: Gracillariidae), *Antispila rivillei, Holocacista rivillei, Holocasta rivillei* (Heliozelidae), *Malacosoma neustria* (Lasiocampidae), *Leucoptera laburnella, Leucoptera lotella, Leucoptera malifoliella, Leucoptera scitella, Leucoptera spartifoliella, Lyonetia clerkella* (Lyonetiidae), *Tinea cavella* (Tineidae), *Lobesia botrana, Spilonota ocellana* (Tortricidae), *Argyresthia sp., Argyresthia fundella, Prays oleae* (Yponomeutidae) (Noyes, 2019).

Distribution: Austria, Bulgaria, Czech Republic, Czechoslovakia, England, France, Germany, Hungary, Iran, Italy, Moldova, Montenegro, Palearctic, Russia, Serbia, Spain, Sweden, Switzerland, Turkey, Ukraine, USSR, Yugoslavia (pre-1991) (Noyes, 2019).

Baryscapus oophagus (Otten, 1942)
Publications with Turkey records: Noyes (2019), Özbek and Coruh (2010).
Associates: *Pinus sp.* (Pinaceae) (Noyes, 2019).
Hosts: *Diprion pini, Neodiprion sertifer* (Hymenoptera: Diprionidae), *Phyllocnistis citrella* (Lepidoptera: Gracillariidae), *Malacosoma neustria* (Lasiocampidae) (Noyes, 2019).
Distribution: Bulgaria, Czech Republic, Czechoslovakia, Europe, France, Germany, Iran, Palearctic, Poland, Russia, Sweden, Turkey (Noyes, 2019).

Baryscapus pallidae (Graham, 1991)
Publications with Turkey records: Noyes (2019), Sakaltas and Gençer (2005).
Associates: *Andricus burgundus, Andricus caputmedusae, Andricus conificus, Andricus cydoniae, Andricus grossulariae, Andricus kollari, Andricus lucidus, Andricus quercuscalicis, Andricus schroeckingeri, Biorhiza pallida, Chilaspis nitida, Cynips cornifex, Cynips longiventris, Cynips quercus, Cynips quercusfolii, Dryocosmus kuriphilus, Neuroterus anthracinus, Neuroterus quercusbaccarum, Pseudoneuroterus macropterus, Pseudoneuroterus saliens, Quercus sp., Quercus cerris, Quercus faginea, Quercus petraea, Quercus pubescens, Quercus robur* (Fagaceae), *Rosa sp.* (Rosaceae) (Noyes, 2019).
Hosts: *Biorhiza pallida, Diplolepis mayri, Dryocosmus kuriphilus* (Hymenoptera: Cynipidae) (Noyes, 2019).
Distribution: England, Finland, France, Germany, Hungary, Iran, Italy, Spain, Sweden, Turkey (Noyes, 2019).

Baryscapus protasis (Graham, 1991)
Publications with Turkey records: Noyes (2019).
Associates: *Medicago sativa* (Fabaceae) (Noyes, 2019).
Hosts: *Bruchus sp.* (Coleoptera: Bruchidae) (Noyes, 2019).
Distribution: France, Turkey (Noyes, 2019).

Baryscapus servadeii (Domenichini, 1965)
Publications with Turkey records: Avci (2000), Can and Duzbastilar (2002), Noyes (2019), Öncüer (1991).

Associates: *Ooencyrtus pityocampae* (Encyrtidae), *Pinus sp., Pinus brutia, Pinus halepensis, Pinus insignis, Pinus nigra, Pinus pinaster* (Pinaceae) (Noyes, 2019).
Hosts: *Thaumetopoea bonjeani, Thaumetopoea pityocampa, Thaumetopoea wilkinsoni, Traumatocampa pityocampa* (Lepidoptera: Notodontidae) (Noyes, 2019).
Parasitoid hosts: *Ooencyrtus pityocampae* (Hymenoptera: Encyrtidae) (Noyes, 2019).
Parasitoids: *Baryscapus transversalis* (Eulophidae) (Noyes, 2019).
Distribution: Albania, Algeria, Bosnia Hercegovina, Bulgaria, Croatia, Europe, France, Germany, Greece, Israel, Italy, Morocco, North Africa, Portugal, Russia, Spain, Syria, Tunisia, Turkey, Yugoslavia (pre-1991) (Noyes, 2019).

Baryscapus spartifoliellae (Graham, 1991)
Publications with Turkey records: Noyes (2019), Sakaltas and Gençer (2005).
Associates: *Sarothamnus sp., Sarothamnus scoparius* (Fabaceae) (Noyes, 2019).
Hosts: *Leucoptera spartifoliella* (Lepidoptera: Lyonetiidae) (Noyes, 2019).
Distribution: Czech Republic, England, Germany, Netherlands, Turkey (Noyes, 2019).

Baryscapus transversalis (Graham, 1991)
Publications with Turkey records: Noyes (2019).
Associates: *Pinus sp., Pinus nigra* (Pinaceae) (Noyes, 2019).
Hosts: *Thaumetopoea pityocampa, Thaumetopoea wilkinsoni* (Lepidoptera: Notodontidae) (Noyes, 2019).
Parasitoid hosts: *Ooencyrtus pityocampae* (Hymenoptera: Encyrtidae), *Baryscapus servadeii* (Eulophidae) (Noyes, 2019).
Distribution: Albania, Bosnia Hercegovina, Bulgaria, Greece, Turkey (Noyes, 2019).

Baryscapus turionum (Hartig, 1838)
Publications with Turkey records: Noyes (2019), Sakaltas and Gençer (2005).
Associates: *Pinus sp., Pinus nigra, Pinus radiata, Pinus sylvestris* (Pinaceae) (Noyes, 2019).
Hosts: *Exoteleia dodecella* (Lepidoptera: Gelechiidae), *Aporia crataegi* (Pieridae), *Blastesthia turionella, Rhyacionia buoliana, Rhyacionia duplana* (Tortricidae) (Noyes, 2019).

Parasitoid hosts: *Actia nudibasis* (Diptera: Tachinidae), *Apanteles glomeratus* (Hymenoptera: Braconidae), *Copidosoma geniculatum* (Encyrtidae) (Noyes, 2019).
Distribution: Austria, Belgium, Bulgaria, Canada, Czech Republic, Czechoslovakia, England, Europe, Germany, Hungary, Italy, Nearctic, Netherlands, Poland, Romania, Russia, Serbia, Spain, Sweden, Switzerland, Turkey, United States of America, Yugoslavia (pre-1991) (Noyes, 2019).

Genus *Crataepus* (Foerster, 1878)
Crataepus marbis (Walker, 1839)
Publications with Turkey records: Noyes (2019), Öncüer (1991), Sakaltas and Gençer (2005).
Associates: *Carduus sp., Cirsium sp., Cirsium arvense, Cirsium vulgare* (Asteraceae) (Noyes, 2019).
Hosts: *Cecidomyia sp.* (Diptera: Cecidomyiidae), *Orellia sp., Orellia ruficauda, Tephritis conura, Terellia sp., Terellia flavicauda, Terellia florescentiae, Terellia fuscicornis, Terellia longicauda, Terellia serratulae, Trypeta sp., Urophora sp., Urophora stylata* (Tephritidae) (Noyes, 2019).
Distribution: Austria, Bulgaria, Canada, Czech Republic, Czechoslovakia, England, Europe, France, Germany, Hungary, Italy, Moldova, Nearctic, Netherlands, Poland, Russia, Slovakia, Spain, Sweden, Turkey, United States of America, USSR, Uzbekistan (Noyes, 2019).

Genus *Leptocybe* (Fisher and La Salle, 2004)
Leptocybe invasa (Fisher and La Salle, 2004)
Publications with Turkey records: Aytar et al. (2011), Noyes (2019).
Associates: *Eucalyptus camaldulensis* (Myrtaceae) (Noyes, 2019).
Hosts: *Eucalyptus sp., Eucalyptus botryoides, Eucalyptus bridgesiana, Eucalyptus camaldulensis, Eucalyptus cinerea, Eucalyptus deanei, Eucalyptus eurograndis, Eucalyptus globulus, Eucalyptus grandis, Eucalyptus grandis x camaldulensis, Eucalyptus gunii, Eucalyptus nicholii, Eucalyptus nitens, Eucalyptus pellita, Eucalyptus pulverulenta, Eucalyptus robusta, Eucalyptus rudis, Eucalyptus saligna, Eucalyptus tereticornis, Eucalyptus urophylla, Eucalyptus urophylla x grandis, Eucalyptus viminalis* (Myrtaceae) (Noyes, 2019).
Parasitoids: *Coccobius furviflagellatus* (Hymenoptera: Aphelinidae), *Aprostocetus causalis* (Eulophidae), *Megastigmus sp., Megastigmus brasiliensis, Megastigmus viggianii* (Torymidae) (Noyes, 2019).

Distribution: Algeria, Argentina, Australia, Brazil, Canary Islands, Chile, Ethiopia, France, Greece, India, Iran, Iraq, Israel, Italy, Jordan, Kenya, Lebanon, Malta, Mexico, Morocco, Mozambique, New Zealand, Paraguay, People's Republic of China, Portugal, South Africa, Spain, Syria, Taiwan, Tanzania, Thailand, Tunisia, Turkey, Uganda, United States of America, Uruguay, Vietnam, Zimbabwe (Noyes, 2019).

Genus *Melittobia* (Westwood, 1848)
Melittobia acasta (Walker, 1839)
Publications with Turkey records: Güler and Kodan (2010), Noyes (2019).
Associates: *Ensliniella parasitica* (Acarina: Winterschmidtiidae), *Osmia sp.* (Hymenoptera: Apidae), *Allodynerus delphinalis* (Vespidae), *Artemisia taurica* (Asteraceae), *Phragmites vulgaris* (Poaceae), *Rubus sp.* (Rubiaceae) (Noyes, 2019).
Hosts: *Coroebus undatus* (Coleoptera: Buprestidae), *Donacia sp.* (Chrysomelidae), *Calliphora sp., Calliphora vomitoria* (Diptera: Calliphoridae), Giraudiella inclusa, Lasioptera arundinis, Thomasiella arundinis (Cecidomyiidae), *Musca sp., Musca domestica* (Muscidae), *Diplostichus janithrix, Lydella grisescens* (Tachinidae), *Urophora solstitialis* (Tephritidae), *Anthidium florentinum, Anthophora abrupta, Anthophorabia sp., Apis mellifera, Bombus sp., Bombus hortorum, Bombus hypnorum, Bombus ruderatus, Bombus terrestris, Chalicodoma muraria, Chalicodoma spissula, Heriades crenulatus, Heriades truncorum, Hylaeus sp., Megachile sp., Megachile bombycine, Megachile centuncularis, Megachile leachella, Megachile muraria, Megachile pacifica, Megachile relativa, Megachile rotundata, Megachile willughbiella, Osmia sp., Osmia adunca, Osmia bicornis, Osmia coerulescens, Osmia leucomelena, Osmia rufa, Prosopis sp., Stelis nasuta, Xylocopa turanica* (Hymenoptera: Apidae), *Chrysis cyanea, Chrysis ignita* (Chrysididae), *Diprion pini, Diprion similo, Neodiprion lecontei* (Diprionidae), *Formica fusca, Formica pratensis, Formica sanguinea* (Formicidae), *Corynopus coarctatus, Isodontia Mexicana, Pison morosum, Pison spinolae, Psen sp., Psenulus fuscipennis, Psenulus pallipes, Psenulus schenski, Sceliphron destillatorium, Sceliphron figulus, Trypoxylon sp., Trypoxylon figulus* (Sphecidae), *Ancistrocerus nigricornis, Anterhynchium sp., Discoelius sp., Gymnomerus laevipes, Odynerus antilope, Odynerus callosus, Odynerus renformis, Odynerus spinipes, Rhynchium delphinale, Symmorphus sp., Symmorphus debilitates, Vespa sp.* (Vespidae), *Malacosoma neustria* (Lepidoptera: Lasiocampidae), *Cnidocampa flavescens* (Limacodidae), *Lymantria dispar, Porthetria*

dispar (Lymantriidae), *Amphipyra pyramidea* (Noctuidae), *Pieris brassicae* (Pieridae), *Carpocapsa pomonella, Laspeyresia pomonella* (Tortricidae) (Noyes, 2019).

Parasitoid hosts: *Pachyophthalmus signatus* (Diptera: Sarcophagidae), *Compsilura concinnata, Diplostichus janithrix, Lydella grisescens* (Tachinidae), *Chrysis cyanea* (Hymenoptera: Chrysididae), *Aritranis sp., Campoplex alkae, Eulimneria alkae, Pimpla roborator, Sphecophaga vesparum* (Ichneumonidae), *Leucospis intermedia* (Leucospidae), *Monodontomerus obsoletus* (Torymidae) (Noyes, 2019).

Distribution: Argentina, Austria, Azores, Bulgaria, Canada, Caucasus, Czech Republic, Czechoslovakia, Denmark, England, Europe, Finland, France, Germany, Guam, Hungary, India, Iran, Italy, Japan, Kazakhstan, Madeira, Moldova, Nearctic, Netherlands, New Zealand, North Africa, Palearctic, People's Republic of China, Poland, Russia, Serbia, Spain, Sri Lanka, Sweden, Switzerland, Tselinograd Obl., Turkey, Ukraine, United States of America, USSR, Venezuela, Yugoslavia (pre-1991) (Noyes, 2019).

Genus *Minotetrastichus* (Kostjukov, 1977)

Minotetrastichus frontalis (Nees, 1834)

Publications with Turkey records: Cebeci et al. (2011), Noyes (2019).

Associates: *Andricus schroeckingeri, Neuroterus numismalis, Neuroterus quercusbaccarum* (Cynipidae), *Alnus sp., Alnus glutinosa, Betula sp.* (Betulaceae), *Robinia pseudoacacia* (Fabaceae), *Fagus sylvatica, Quercus cerris, Quercus roboris, Quercus robur* (Fagaceae), *Aesculus hippocastanum* (Hippocastanaceae), *Larix sp.* (Pinaceae), *Platanus orientalis* (Platanaceae), *Malus sp., Malus pumila, Prunus sp., Pyrus malus, Sorbus aucuparia* (Rosaceae), *Rubus ulmifolius* (Rubiaceae), *Populus sp., Salix purpurea* (Salicaceae), *Ulmus sp.* (Ulmaceae) (Noyes, 2019).

Hosts: *Orchestes alni, Orchestes fagi, Orchestes pallicornis, Orchestes populi, Orchestes quercus, Rhynchaenus sp., Rhynchaenus alni, Rhynchaenus fagi, Rhynchaenus pallicornis, Rhynchaenus populi, Rhynchaenus testaceus* (Coleoptera: Curculionidae), *Cimbex femorata* (Hymenoptera: Cimbicidae), *Diplolepis rosae* (Cynipidae), *Fenusa sp., Fenusa pusilla, Fenusa ulmi, Heterarthrus sp., Heterarthrus nemoratus, Scolioneura sp., Scolioneura betulae* (Tenthredinidae), *Coleophora fuscedinella, Coleophora laricella, Coleophora pruniella* (Lepidoptera: Coleophoridae), *Eriocrania haworthi, Eriocrania semipurpurella, Eriocrania unimaculella* (Eriocraniidae), *Caloptilia sp., Caloptilia alchimiella, Caloptilia syringella, Cameraria ohridella, Gracillaria* *alchimiella, Gracillaria syringella, Lithocolletis sp., Lithocolletis blancardella, Lithocolletis cerasicolella, Lithocolletis corylifoliella, Lithocolletis delitella, Lithocolletis distentella, Lithocolletis lautella, Lithocolletis maestingella, Lithocolletis messaniella, Lithocolletis platani, Lithocolletis populifoliella, Lithocolletis pyrifoliella, Lithocolletis quercifoliella, Lithocolletis roboris, Lithocolletis schreberella, Lithocolletis tremulae, Parectopa robiniella, Parornix finitimella, Parornix petiolella, Phyllocnistis labyrinthella, Phyllocnistis saligna, Phyllocnistis vitegenella, Phyllonorycter sp., Phyllonorycter alnifoliella, Phyllonorycter blancardella, Phyllonorycter cerasicolella, Phyllonorycter corylifoliella, Phyllonorycter crataegella, Phyllonorycter harrisella, Phyllonorycter heegeriella, Phyllonorycter issikii, Phyllonorycter lautella, Phyllonorycter messaniella, Phyllonorycter platani, Phyllonorycter propinquinella, Phyllonorycter rajella, Phyllonorycter robiniella, Phyllonorycter salictella, Phyllonorycter stettinensis, Phyllonorycter strigulatella* (Gracillariidae), *Antispila rivillei, Coptodisca splendoriferella, Holocacista rivillei* (Heliozelidae), *Leucoptera laburnella, Leucoptera malifoliella, Leucoptera scitella, Leucoptera sinuella, Lyonetia sp., Lyonetia clerkella* (Lyonetiidae), *Nepticula malella, Stigmella sp., Stigmella hemargyrella, Stigmella malella, Stigmella sorbi* (Nepticulidae), *Cerura vinula* (Notodontidae), *Tischeria sp., Tischeria complanella, Tischeria ekebladella, Tischeria longiciliata* (Tischeriidae) (Noyes, 2019).

Parasitoid hosts: *Apanteles circumscriptus, Apanteles spurius, Phanomeris phyllotomae, Pholetesor ornigis* (Hymenoptera: Braconidae), *Atoposomoidea unifasciata, Chrysocharis sp., Enaysma albiscapus, Enaysma splendens, Pediobius saulius, Pnigalio agraules, Pnigalio pectinicornis* (Eulophidae) (Noyes, 2019).

Parasitoids: *Pediobius saulius* (Eulophidae) (Noyes, 2019).

Distribution: Austria, Bosnia Hercegovina, Bulgaria, Canada, Canary Islands, Croatia, Czech Republic, Czechoslovakia, Denmark, England, Europe, Finland, France, Germany, Hungary, Ireland, Italy, Macedonia, Moldova, Nearctic, Netherlands, Norway, Pakistan, Romania, Russia, Serbia, Slovakia, Slovenia, Spain, Sweden, Switzerland, Turkey, Ukraine, United Arab Emirates, United States of America, USSR, Yugoslavia (pre-1991) (Noyes, 2019).

Minotetrastichus platanellus (Mercet, 1922)

Publications with Turkey records: Cebeci et al. (2011), Noyes (2019).

Associates: *Quercus sp., Quercus ithaburensis* (Fagaceae), *Aesculus hippocastanum* (Hippocastanaceae), *Platanus sp., Platanus occidentalis*

(Platanaceae), *Prunus sp.* (Rosaceae), *Populus italicus* (Salicaceae), *Lithocolletis platani* (Lepidoptera: Gracillariidae) (Noyes, 2019).
Hosts: *Cameraria ohridella, Lithocolletis sp., Lithocolletis blancardella, Lithocolletis cerasicolella, Lithocolletis millierella, Lithocolletis platani, Lithocolletis populiella, Lithocolletis populifoliella, Lithocolletis quercifoliella, Lithocolletis ulmifoliella, Phyllonorycter blancardella, Phyllonorycter cerasicolella, Phyllonorycter millierella, Phyllonorycter platani, Phyllonorycter populifoliella* (Lepidoptera: Gracillariidae) (Noyes, 2019).
Parasitoid hosts: *Apanteles sp., Apanteles circumscriptus* (Hymenoptera: Braconidae), *Sympiesis sericeicornis* (Eulophidae) (Noyes, 2019).
Distribution: Bulgaria, Czech Republic, Czechoslovakia, England, France, Georgia, Germany, Greece, Hungary, Iran, Israel, Italy, Netherlands, Serbia, Slovakia, Turkey, Ukraine, Yugoslavia (pre-1991) (Noyes, 2019).

Genus *Neotrichoporoides* (Girault, 1913)
Neotrichoporoides beyarslani (Doğanlar, 1993)
Publications with Turkey records: Doğanlar (1993), Noyes (2019).
Associates: Unknown (Noyes, 2019).
Hosts: Unknown (Noyes, 2019).
Distribution: Turkey (Noyes, 2019).

Neotrichoporoides biogradensis (Graham, 1987)
Publications with Turkey records: Noyes (2019), Sakaltas and Gençer (2005).
Associates: Unknown (Noyes, 2019).
Hosts: Unknown (Noyes, 2019).
Distribution: Czechoslovakia, Moldova, Slovakia, Sweden, Turkey, Yugoslavia (pre-1991) (Noyes, 2019).

Neotrichoporoides cavigena Graham, 1987
Publications with Turkey records: Noyes (2019), Sakaltas and Gençer (2005).
Associates: Unknown (Noyes, 2019).
Hosts: Unknown (Noyes, 2019).
Distribution: Bulgaria, Czechoslovakia, France, Russia, Slovakia, Turkey (Noyes, 2019).

Neotrichoporoides cynodontis (Domenichini, 1967)
Publications with Turkey records: Noyes (2019), Sakaltas and Gençer (2005).
Associates: *Cynodon dactylon* (Poaceae) (Noyes, 2019).
Hosts: *Dasiops latifrons* (Diptera: Lonchaeidae) (Noyes, 2019).
Distribution: Algeria, Cyprus, France, Italy, Spain, Turkey (Noyes, 2019).

Neotrichoporoides mediterraneus (Graham, 1986)
Publications with Turkey records: Noyes (2019), Sakaltas and Gençer (2005).
Associates: *Oryza sativa* (Poaceae) (Noyes, 2019).
Distribution: Australia, Bulgaria, Canary Islands, Czech Republic, Czechoslovakia, France, India, Italy, Madeira, People's Republic of China, Romania, Russia, Slovakia, Spain, Turkey, USSR (Noyes, 2019).

Neotrichoporoides szelenyii (Erdos, 1951)
Publications with Turkey records: Noyes (2019), Sakaltas and Gençer (2005).
Associates: *Andropogon ischaemum, Triticum aestivum* (Poaceae) (Noyes, 2019).
Distribution: Azerbaijan, Bulgaria, Czechoslovakia, Greece, Hungary, Iran, Italy, Moldova, Portugal, Romania, Russia, Saudi Arabia, Turkey, United Arab Emirates (Noyes, 2019).

Neotrichoporoides tokatensis (Doğanlar, 1993)
Publications with Turkey records: Doğanlar (1993), Noyes (2019).
Associates: Unknown (Noyes, 2019).
Hosts: Unknown (Noyes, 2019).
Distribution: Turkey (Noyes, 2019).

Neotrichoporoides viridimaculatus (Fullaway, 1955)
Publications with Turkey records: Noyes (2019), Sakaltas and Gençer (2005).
Associates: *Cynodon dactylon, Hyparrhenia hirta* (Poaceae) (Noyes, 2019).
Hosts: *Pseudaletia unipuncta* (Lepidoptera: Noctuidae) (Noyes, 2019).
Distribution: Argentina, Bermuda, Bulgaria, Caribbean, Colombia, Cuba, Czechoslovakia, Europe, France, Hawaii, Hungary, India, Italy, Madeira, Nearctic, People's Republic of China, Portugal, Russia, Slovakia, Sweden,

Turkey, United States of America, USSR, Yugoslavia (pre-1991) (Noyes, 2019).

Genus *Oomyzus* (Rondani, 1870)
Oomyzus scaposus (Thomson, 1878)
Publications with Turkey records: Noyes (2019), Öncüer (1991).
Associates: *Brevicoryne brassicae* (Hemiptera: Aphididae), *Parlatoria pergandii* (Diaspididae), *Medicago sativa, Vigna unguiculata* (Fabaceae), *Sorghum sp.* (Poaceae), *Citrus aurantium* (Rutaceae) (Noyes, 2019).
Hosts: *Adalia bipunctata, Calvia quattuordecimguttata, Cheilomenes sexmaculata, Chilocorus sp., Chilocorus bipustulatus, Chilocorus renipustulatus, Coccinella sp., Coccinella quinquenotata, Coccinella quinquepunctata, Coccinella septempunctata, Coccinella transveroguttata, Coccinella transversalis, Coccinella undecimpunctata, Exochomus quadripustulatus, Harmonia quadripunctata, Menochilus sexmaculatus, Propylea quattuordecimpunctata, Scymnus sp., Synharmonia conglobata* (Coleoptera: Coccinellidae), *Chrysopa carnea, Chrysopa septempunctata, Chrysopa ventralis* (Neuroptera: Chrysopidae) (Noyes, 2019).
Distribution: Afghanistan, Australia, Austria, Bahamas, Bulgaria, Canada, Canary Islands, Colombia, Czech Republic, Czechoslovakia, Egypt, England, Europe, France, Germany, Greece, Hawaii, Hungary, India, Indonesia, Iraq, Italy, Korea, Moldova, Nearctic, Netherlands, New Zealand, Pakistan, People's Republic of China, Russia, Serbia, Slovakia, Sri Lanka, Sweden, Switzerland, Syria, Turkey, Turkmenistan, Ukraine, United States of America, USSR, Uzbekistan (Noyes, 2019).

Oomyzus sempronius (Erdos, 1954)
Publications with Turkey records: Noyes (2019), Öncüer (1991).
Associates: *Carduus nutans* (Asteraceae), *Abies alba* (Pinaceae) (Noyes, 2019).
Hosts: *Chilocorus bipustulatus* (Coleoptera: Coccinellidae), *Anisochrysa sp., Anisochrysa carnea, Chrysopa sp., Chrysopa carnea, Chrysopa flavifrons, Chrysopa ventralis* (Neuroptera: Chrysopidae) (Noyes, 2019).
Distribution: Croatia, Czech Republic, Czechoslovakia, Egypt, France, Germany, Greece, Hungary, Italy, Moldova, Poland, Russia, Slovakia, Sweden, Switzerland, Turkey, USSR, Yemen (Noyes, 2019).

Genus *Pronotalia* (Gradwell, 1957)
Pronotalia carlinarum (Szelenyi and Erdos, 1951)

Publications with Turkey records: Koçak and Ozdemir (2012), Noyes (2019), Öncüer (1991), Özbek and Coruh (2010), Sakaltas and Gençer (2005).
Associates: *Heracleum platytaenium* (Apiaceae), *Arnica montana, Carduus crispus, Carlina brevibracteata, Centaurea sp., Centaurea cetia, Centaurea cyanus, Centaurea sadleriana, Centaurea scabiosa, Cirsium oleraceum, Cirsium vulgare, Hieracium piloselloides, Solidago californica* (Asteraceae), *Orobanche sp*. (Orobanchaceae) (Noyes, 2019).
Hosts: *Melanagromyza heracleana, Phytomyza sp., Phytomyza orobanchia* (Diptera: Agromyzidae), *Setulia convergens* (Calliphoridae), *Calamoncosis minima, Lipara lucens* (Chloropidae), *Chaetorellia jaceae, Chaetorellia loricate, Chaetostomella cylindrica, Noeeta bisetosa, Orellia colon, Procecidochares anthracina, Tephritis arnica, Tephritis conura, Terellia winthemi, Urophora quadrifasciata, Urophora stylata* (Tephritidae), *Malacosoma neustria* (Lepidoptera: Lasiocampidae) (Noyes, 2019).
Distribution: Armenia, Bulgaria, Czech Republic, England, Finland, France, Germany, Hungary, Israel, Italy, Moldova, Netherlands, Russia, Slovakia, Sweden, Switzerland, Turkey, Ukraine, United States of America (Noyes, 2019).

Pronotalia erzurumica (Doğanlar, 1993)
Publications with Turkey records: Doğanlar (1993), Noyes (2019).
Associates: *Centaurea solstitialis* (Asteraceae) (Noyes, 2019).
Distribution: Turkey (Noyes, 2019).

Pronotalia fiorii (Domenichini, 1958)
Publications with Turkey records: Noyes (2019), Sakaltas and Gençer (2005).
Hosts: *Lixus iridis* (Coleoptera: Curculionidae) (Noyes, 2019).
Parasitoid hosts: *Habrocytus fenomenalis* (Hymenoptera: Pteromalidae) (Noyes, 2019).
Distribution: Czechoslovakia, Hungary, Italy, Turkey (Noyes, 2019).

Pronotalia hungarica (Erdos, 1955)
Publications with Turkey records: Noyes (2019), Öncüer (1991).
Associates: *Centaurea aspera, Cirsium sp*. (Asteraceae) (Noyes, 2019).
Distribution: Bulgaria, Czech Republic, Czechoslovakia, France, Hungary, Italy, Russia, Slovakia, Sweden, Turkey (Noyes, 2019).

Pronotalia tortumensis (Doğanlar, 1993)
Publications with Turkey records: Doğanlar (1993), Noyes (2019).

Hosts: *Apterona helix* (Lepidoptera: Psychidae) (Noyes, 2019).
Distribution: Turkey (Noyes, 2019).

Genus *Puklina* (Graham, 1991)
Puklina gelincika (Doğanlar, 1993)
Publications with Turkey records: Doğanlar (1993), Noyes (2019).
Associates: *Asphodelus sp*. (Asphodelaceae) (Noyes, 2019).
Hosts: *Bruchophagus sp.* (Hymenoptera: Eurytomidae) (Noyes, 2019).
Distribution: Turkey (Noyes, 2019).

Genus *Quadrastichodella* (Girault, 1913)
Quadrastichodella nova (Girault, 1922)
Publications with Turkey records: Aytar et al. (2011), Noyes (2019).
Associates: *Eucalyptus sp., Eucalyptus resinifera, Eucalyptus umbellatus* (Myrtaceae) (Noyes, 2019).
Hosts: *Eucalyptus sp., Eucalyptus resinifera, Eucalyptus umbellatus* (Myrtaceae) (Noyes, 2019).
Distribution: Afrotropical, Argentina, Australia, Canary Islands, Europe, Israel, Italy, Nearctic, Neotropical, New Zealand, South Africa, Spain, Turkey, United States of America (Noyes, 2019).

Genus *Quadrastichus* (Girault, 1913)
Quadrastichus dasineurae Doğanlar, LaSalle, Sertkaya and Doğanlar, 2009
Publications with Turkey records: Doğanlar (2011), Doğanlar et al. (2009), Doğanlar et al. (2011), Noyes (2019).
Associates: *Dasineura oleae, Lasioptera oleicola* (Diptera: Cecidomyiidae), *Olea europaea* (Oleaceae) (Noyes, 2019).
Hosts: *Dasineura oleae, Lasioptera oleicola* (Diptera: Cecidomyiidae) (Noyes, 2019).
Distribution: Turkey (Noyes, 2019).
Genus *Sigmophora* (Rondani, 1867)
Sigmophora brevicornis (Panzer, 1804)
Publications with Turkey records: Noyes (2019), Sakaltas and Gençer (2005).
Associates: *Eupelmus urozonus* (Eupelmidae), *Eurytoma rosae* (Eurytomidae), *Pteromalus bedeguaris* (Pteromalidae), *Daucus carota* (Apiaceae), *Capparis spinosa* (Capparidaceae), *Coronilla varia, Cytisus scoparius, Lens esculenta, Medicago sativa, Sarothamnus scoparius, Ulex europaeus* (Fabaceae), *Fraxinus sp.* (Oleaceae), *Papaver sp.* (Papaveraceae), *Rosa canina* (Rosaceae), *Salix sp.* (Salicaceae),

Verbascum austriacum, Verbascum pulverulenti (Scrophulariaceae) (Noyes, 2019).

Hosts: *Mesocoelopus niger* (Coleoptera: Anobiidae), *Apion trifolii* (Apionidae), *Asphondylia sp., Asphondylia calycotomae, Asphondylia coronillae, Asphondylia cytisi, Asphondylia dufouri, Asphondylia ervi, Asphondylia melanopus, Asphondylia mikii, Asphondylia ononidis, Asphondylia pruniperda, Asphondylia sarothamni, Asphondylia ulicis, Asphondylia verbasci, Contarinia sp., Contarinia lentis, Contarinia medicaginis, Contarinia tritici, Cystiphora sonchi, Dasineura affinis, Dasineura brassicae, Dasineura papaveris, Dasineura pyri, Eumarchalia gennadii, Ischnonyx verbasci, Kiefferia pericarpiicola, Kiefferia pimpinellae, Lasioptera rubi, Rabdophaga sp., Schizomyia galiorum* (Cecidomyiidae), *Urophora jaceana* (Tephritidae), *Diplolepis rosae* (Hymenoptera: Cynipidae), *Bruchophagus gibbus, Eurytoma onobrychidis* (Eurytomidae), *Euura atra* (Tenthredinidae), *Cochylidia implicitana* (Lepidoptera: Tortricidae), *Yponomeuta evonymellus* (Yponomeutidae) (Noyes, 2019).

Parasitoid hosts: *Pseudocatolaccus asphondyliae* (Hymenoptera: Pteromalidae), *Torymus dauci* (Torymidae) (Noyes, 2019).

Distribution: Andorra, Austria, Bulgaria, Canada, Croatia, Cyprus, Czech Republic, Czechoslovakia, Denmark, England, Finland, France, Georgia, Germany, Greece, Hungary, India, Iran, Ireland, Israel, Italy, Korea, Lebanon, Macedonia, Moldova, Montenegro, Morocco, Netherlands, Norway, Pakistan, Portugal, Romania, Russia, Serbia, Slovakia, Spain, Sweden, Switzerland, Thailand, Turkey, Ukraine, USSR, Vietnam, Yugoslavia (Federal Republic), Yugoslavia (pre-1991) (Noyes, 2019).

Genus *Stepanovia* (Kostjukov, 2004)

Stepanovia eurytomae (Nees, 1834)

Publications with Turkey records: Bayram et al. (1998), Gencer (2003), Mete and Mergen (2017), Noyes (2019), Ozbek et al. (1999).

Associates: *Rosa sp., Rosa agrestis, Rosa canina, Rosa pimpinellifolia* (Rosaceae), *Sphaeropyx bicolor* (Hymenoptera: Braconidae) (Noyes, 2019).

Hosts: *Diplolepis sp., Diplolepis centifoliae, Diplolepis eglanteriae, Diplolepis fructuum, Diplolepis mayri, Diplolepis nervosa, Diplolepis rosae, Diplolepis spinosissimae* (Hymenoptera: Cynipidae) (Noyes, 2019).

Parasitoid hosts: *Eurytoma rosae* (Eurytomidae) (Noyes, 2019).

Distribution: Armenia, Czech Republic, Czechoslovakia, England, Finland, France, Germany, Iran, Morocco, Netherlands, Norway, Poland, Russia, Spain, Sweden, Turkey (Noyes, 2019).

Stepanovia fructirosae (Boyadzhiev, Yefremova and Tozlu, 2017)
Publications with Turkey records: Noyes (2019).
Associates: *Diplolepis fructuum* (Hymenoptera: Cynipidae), *Rosa sp.* (Rosaceae) (Noyes, 2019).
Distribution: Turkey (Noyes, 2019).

Genus *Tetrastichus* (Haliday, 1844)
Tetrastichus coelarchus (Graham, 1991)
Publications with Turkey records: Noyes (2019), Sakaltas and Gençer (2005).
Associates: Unknown (Noyes, 2019).
Hosts: Unknown (Noyes, 2019).
Distribution: Czechoslovakia, England, People's Republic of China, Sweden, Turkey, Yugoslavia (pre-1991) (Noyes, 2019).

Tetrastichus decrescens (Graham, 1991)
Publications with Turkey records: Noyes (2019), Sakaltas and Gençer (2005).
Hosts: *Henosepilachna vigintioctomaculata* (Coleoptera: Coccinellidae) (Noyes, 2019).
Distribution: England, France, People's Republic of China, Sweden, Turkey (Noyes, 2019).

Tetrastichus melasomae (Graham, 1991)
Publications with Turkey records: Noyes (2019), Sakaltas and Gençer (2005).
Hosts: *Chrysomela vigintipunctata* (Coleoptera: Chrysomelidae) (Noyes, 2019).
Distribution: Czech Republic, Sweden, Turkey (Noyes, 2019).

Tetrastichus sinope (Walker, 1839)
Publications with Turkey records: Kilinçer (1982), Noyes (2019).
Parasitoid hosts: *Erynnia nitida* (Diptera: Tachinidae), *Apanteles affinis, Apanteles glomeratus, Apanteles rubecula, Microgaster sp., Rogas circumscriptus* (Hymenoptera: Braconidae), *Anilastus ebeninus* (Ichneumonidae) (Noyes, 2019).
Hosts: *Brevicoryne brassicae* (Hemiptera: Aphididae), *Trichoplusia ni* (Lepidoptera: Noctuidae), *Pieris sp., Pieris brassicae, Pieris rapae* (Pieridae) (Noyes, 2019).

Distribution: Algeria, Canada, England, France, Germany, Hungary, Ireland, Italy, Lithuania, Moldova, Nearctic, Netherlands, New Zealand, Sweden, Turkey, Ukraine, United States of America, Yugoslavia (pre-1991) (Noyes, 2019).

Genus *Zeytinus* (Doğanlar, 2011)
Zeytinus hatayensis (Doğanlar, 2011)
Catalagues (Journals) with Turkiye records: Doğanlar (2011), Doğanlar et al. (2011), Noyes (2019).
Associates: *Dasineura oleae, Lasioptera oleicola* (Diptera: Cecidomyiidae), *Olea europaea* (Oleaceae) (Noyes, 2019).
Hosts: *Dasineura oleae, Lasioptera oleicola* (Diptera: Cecidomyiidae) (Noyes, 2019).
Distribution: Turkey (Noyes, 2019).

References

Avcı, Mustafa. 2000. "Investigations on structure of egg-batches, parasitism and egg layinh habits of *Thaumetopoea pityocampa* (Den. & Schiff.) (Lepidoptera: Thaumetopoeidae) in various regions of Turkey". *Turkish Journal of Entomology*, 24(3):167-178.

Avcı, Utku, and Özbek, Hikmet. 1990. "Cabbage lepidopterous pests and their parasitoids in Erzurum". *Proceedings of the Second Turkish National Congress of Biological Control*, 26-29 September 1990, pp. 319-330.

Aytar, Fatih, and Dağdaş, Said, and Duran, Celalettin. 2011. "Australian insects affecting *Eucalyptus* species in Turkey". *Silva Lusitana*, 19, 41-47.

Bayram, Şerife, and Ülgentürk, Selma, and Toros, Seval. 1998. "Research on the insects causing galls on dog rose (*Rosa spp.*) and their parasitoids in Ankara province". *Turkish Journal of Entomology*, 22(4):259-268.

Bulut, Ebrucan, and Göçmen, Hüseyin. 2000. "Pests and their natural enemies on greenhouse vegetables in Antalya". *Bulletin. Section Regionale Ouest Palaearctique, Organisation Internationale de Lutte Biologique*, 23(1):33-37.

Çam, Halit. 2011. "A checklist of the Eurytomidae (Hymenoptera, Chalcidoidea) species of Turkey". *Zootaxa*, 3113:53-64.

Can, P. & Duzbastilar, I.M. 2002. "Determination of egg parasitoids of pine processionary caterpillar (*Thaumetopoea pityocampa* Den. & Schiff.) in pine populations in the Aegean region". *(Abstract 120)*

Abstracts, VII th European Congress of Entomology, October 7-13 2002, Thessaloniki, Greece pp.118 Hellenic Entomological Society.

Cebeci, Hacı Hüseyin, and Grabenweger, Giselher, and Ayberk, Hamit. 2011. "Eulophid parasitoids (Hymenoptera: Eulophidae) of the horse chestnut leafminer, *Cameraria ohridella* (Lepidoptera: Gracillariidae), from Istanbul, Turkey". *Turkish Journal of Zoology*, 35(5):777-780.

Çıkman, Emine, and La Salle, John. 2011. "Parasitoids of the leafminers (Diptera: Agromyzidae) in Malatya, Turkey". *Turkiye Entomoloji Dergisi*, 35(3):475-484.

Çıkman, Emine. 2012. "Parasitoids of the leafminers (Diptera: Agromyzidae) from Elazig Province, Turkey". *African Journal of Agricultural Research*, 7(12):1937-1943.

Civelek, Hasan Sungur, and Yoldaş, Zeynep, and Weintraub, Phyllis. 2002. "The parasitoid complex of *Liriomyza huidobrensis* in cucumber greenhouses in Izmir Province, western Turkey". *Phytoparasitica*, 30(3):285-287.

Civelek, Hasan Sungur. 2002. "A new record for the parasitoid fauna of *Chromatomyia horticola* (Goureau, 1851) (Diptera: Agromyzidae): *Pediobius metallicus* (Nees, 11834) (Hymenoptera: Eulophidae)". *Turkish Journal of Entomology*, 26(2):155-159

Doğanlar, Mikdat, and Doğanlar, Oğuzhan. & Boyadzhiev, S. Boyadziev. 2013. "Hypopygia of the sparetus species group of Entedon Dalman (Hymenoptera: Eulophidae), with descriptions of two new species". *Munis Entomology and Zoology*, 8(2):831-851.

Doğanlar, Mikdat, and Doğanlar, Oğuzhan. 2013. "Systematics of Urfacus Doğanlar, 2003 (Hymenoptera, Eulophidae, Ceranisinae), with descriptions of new species in Turkey". *Journal of Biological Sciences*, 7(1):76-90.

—. 2013a. "Hypopygia of the hercyna species group of Entedon Dalman (Hymenoptera: Eulophidae), with descriptions of new species". *Munis Entomology and Zoology*, 8 (2):812-824.

—. 2013b. "Revision of the cyanellus species group of Entedon Dalman (Hymenoptera: Eulophidae), with descriptions of new species". *Munis Entomology and Zoology*, 8 (2):532-548.

—. 2014a. "A review of some species of *Epomphale* Girajult (Hymenoptera: Eulophidae: Ceranisinae) from several parts of the world, with description of new species". *Munis Entomology and Zoology*, 9(2):727-755.

Doğanlar, Mikdat, and Gumovsky, Alex, and Doğanlar, Oğuzhan. 2009. "A new species of the menes species group of the genus *Ceranisus*

(Hymenoptera: Eulophidae) from Turkey". Turkish Journal of Entomology, 33(4):265-278.

Doğanlar, Mikdat, and LaSalle, John, and Sertkaya, Erdal, and Doğanlar, Oğuzhan. 2009. "A new species of Quadrastichus Girault, 1913 (Hymenoptera: Eulophidae: Tetrastichinae) from Hatay, Turkey". Turkish Journal of Zoology 33(3):309-314.

Doğanlar, Mikdat, and Sertkaya, Erdal, and & Skuhravá, Marcela. 2011a. "Pest status of olive leaf gall midge *Dasineura oleae* (Angelini, 1831), description of Lasioptera oleicola Skuhravá sp. new (Diptera: Cecidomyiidae) and effectiveness of parasitoids on their populations in Hatay, Turkey". *Turkish Journal of Entomology*, 35(2):265-284.

Doğanlar, Mikdat, and Triapitsyn, Serguei. 2007. "Review of *Ceranisus* (Hymenoptera: Eulophidae) of Turkey, with description of a new species". *European Journal of Entomology*, 104(1):105-110.

Doğanlar, Mikdat, Gumovsky, Alex, and Doğanlar, Oğuzhan. 2011b. "A review of *Ceranisus* (Hymenoptera: Eulophidae) of Ukraine, with description of two new species". *Turkish Journal of Entomology*, 35(2):215-229.

Doğanlar, Mikdat. 1985a. "Notes on Chalcidoidea of Turkey. III. Encyrtidae, Tetracampidae, Aphelinidae, Eulophidae and Elasmidae". *Türkiye Bitki Koruma Dergisi*, 9(2):91-103.

—. 1985b. "A new cotton pest of Turkey from Igdir Plain, *Pexicopia malvella* (Hb.) (Lepidoptera; Gelechiidae) and its parasites". *Türkiye Bitki Koruma Dergisi*, 9(3):165-172.

—. 1993a. "Notes on *Baryscapus* Förster, 1856, with description of four new species (Hymenoptera, Eulophidae, Tetrastichinae)". *Entomofauna, Zeitschrift für Entomologie*, 14(23):381-389.

—. 1993b. "Notes on the species of *Neotrichoporoides* Girault, 1913, with descriptions of some new species from Ghana and Turkey (Hymenoptera, Eulophidae, Tetrastichinae)". *Entomofauna, Zeitschrift für Entomologie*, 14(23):369-380.

—. 2003. "A new genus and a new species of Entedontinae (Hymenoptera, Eulophidae) from southeastern Anatolia, Turkey". *Turkish Journal of Zoology*, 27(3):181-185.

—. 2011b. "Parastioid complex of the olive leaf gall midges, *Dasineura oleae* (Angelini 1831) and *Lasioptera oleicola* Skuhravá, 2011 (Diptera: Cecidomyiidae) in Hatay Turkey, with descriptions of new genus and species from Tetrastichinae (Hymenoptera: Eulophidae)". *Turkish Journal of Entomology*, 35(2):245-264

—. 2013. "A peculiar new species of *Entedon* Dalman, 1820 (Hymenoptera: Eulophidae) from Sanliurfa Province, Turkey". *Munis Entomology and Zoology*, 8(2):654-658.

Doğanlar, Oğuzhan, and Doğanlar, Mikdat, and Frary, Anne. 2010. "Preliminary phylogeny of thrips parasitoids of Turkey based on some morphological scales and 28S D2 rDNA, with description of a new species". Turkish Journal of Zoology 34(3):279-289.

Gençer, Lütfiye. 2003. "The chalcidoid parasitoids of *Diplolepis mayri* Schld. (Hymenoptera: Cynipidae), a pest of *Rosa canina* L. in Sivas province". *Turkish Journal of Entomology*, 27(2):107-111.

—. 2009. "A new species of *Anaprostocteus* Graham 1987 (Hymenoptera: Eulophidae) from Turkey". *Entomologica Fennica*, 20: 225—227.

—. 2009. "Contribution to the knowledge of the chalcid parasitoid complex (Hymenoptera: Chalcidoidea) of agromyzid leafminers (Diptera: Agromizidae) from Turkey, with new hosts and records". *Journal of Plant Protection Research*, 49(2):158-161.

—. 2012. "Contributions to the knowledge of the Eulophinae (Hymenoptera: Eulophidae) from Sivas, Turkey, with some new records". *Journal of the Entomological Research Society*, 14(3):83-89.

Güler, Yasemin, and Kodan, Münevver. 2010, "A new record of ectoparasitoid *Melittobia acasta* (Walker, 1839) (Hymenoptera: Eulophidae) from Turkey". *Mellifera*, 10(19):34-36.

Karaca, İsmail, and Japoshvili, George, and Demirozer, Ozan. 2003. "The chalcid parasitoid complex (Hymenoptera: Chalcidoidea) associated with the globose scale (*Sphaerolecanium prunasttri* Fonscolombe) (Hemiptera: Coccoidea) in Isparta Province, Turkey and some European countries". *Zeitschrift für Pflanzenkrankheiten und Pflanzenschutz*, 110(5):505-511.

Kaya, Mehmet, and Kovancı, Bahattin. 2001. "Investigations on the biology of *Coroebus rubi* (L.) (Coleoptera: Buprestidae) in Bursa province of Turkey". *Turkish Journal of Entomology*, 25(4):275-285.

Kılınçer, Neşet. 1982. "Investigations on the parasite complex of the small cabbage white butterfly (*Pieris rapae* L.) (Lepidoptera: Pieridae) in the Ankara region". *Bitki Koruma Bülteni 22*: 107-119.

Koçak, Erhan, and Özdemir, Mustafa. 2012. "Parasitic Hymenoptera reared from the insects on *Heracleum platytaenium* Boiss. (Apiaceae) with new faunistic and biological records". *Turkish Journal of Zoology*, 36(2):201-208.

Korkmaz, Ertan Mahir, and Hastaoğlu Örgen, Sevda, and Gençer, Lütfiye, and Ülgentürk, Selma, and Başıbüyük, Hasan Hüseyin. 2010. "Determination of some wheat pests and their parasitoids in wheat fields of Central Anatolia Region". *Türkiye Entomoloji Dergisi*, 34(3):361-377.

Mete, Özlem, and Mergen, Orhan. 2017. "The community components associated with two common rose gall wasps (Hymenoptera: Cynipidae: Diplolepidini) in Turkey". *Turkish Journal of Zoology*, 41:696-701.

Noyes, John S. 2019. *Universal Chalcidoidea Database*. Available at: www.nhm.ac.uk/chalcidoids (accessed January 2022).

Öncüer Cezmi. 1991. "A catalogue of the parasites and predators of insect pests of Turkey". – *E.Ü. Ziraat Fakültesi Yayınları No. 505*, Bornova-İzmir, pp. 354.

Öncüer, Cezmi, and Yalçın, E. & Erkin, E. 1978. "The natural enemies of the pupae of *Euproctis chrysorrhoea* L. (Lepidoptera: Lymantriidae), which damages fruit trees in the Aegean region, and their effectiveness". *Türkiye Bitki Koruma Dergisi*, 2(1):31-36.

Öncüer, Cezmi. 1974. "Ege bölgesinde turunçgil bahçelerinde zararli *Coccus* (Homoptera: Coccidae) türlerinin taninmasi yayilisi ve dogal düsmanlari üzerinde arastirmalar". *Bitki Koruma Bülteni*, Supplement, 1:1-61.

Özbek, Hikmet, and Çoruh, Saliha. 2010. "Egg parasitoids of *Malacosoma neustria* (Linnaeus, 1758) (Lepidoptera: Lasiocampidae) in Erzurum province of Turkey". *Turkish Journal of Entomology*, 34(4):551-560.

Özbek, Hikmet, and Güçlü, Şaban, and Tozlu, Göksel. 1999. "Biology and natural enemies of *Diplolepis mayri* Schld. (Hymenoptera: Cynipidae), a pest of Rosa canina L. in Erzurum Province". *Turkish Journal of Entomology*, 23(1):39-50.

Özsemerci, Fatma, and Aksit, Tülin. 2003. "Investigations on some biological characteristics and populations fluctuation of *Ceroplastes rusci* L. (Homoptera: Coccidae) harmful to fig trees in Aydin province". *Turkish Journal of Entomology*, 27(1):13-25.

Polat, Alper, and Tozlu, Göksel. 2010. "Investigations on the brief biology, host plants and parasitoids of *Archips rosana* (Linnaeus, 1758) (Lepidoptera: Tortricidae) in Erzurum". *Turkish Journal of Entomology*, 34(4):529-542.

Sakaltaş, Elif, and Gençer, Lütfiye. 2005. "Contribution to the knowledge of the Tetrastichinae (Hymenoptera: Eulophidae) from Ankara, Turkey, with Some New Records". *Acta Phytopathologica et Entomologica Hungarica*, 40 (3-4): 383-390.

Tozlu, Göksel, and Çoruh, Saliha. 2011. "Parasitoids of *Cynaeda gigantea* (Wocke, 1871) (Lepidoptera: Crambidae), a pest of *Anchusa leptophylla* Roemer and Schultes (Boraginacaea) from the East

Anatolia region of Turkey". *Journal of the Entomological Research Society*, 13(3): 117-124.

Ülgenturk, Selma, and Toros, Seval. 2001. "Natural enemies of the oak scale insect, *Eulecanium ciliatum* (Douglas) (Hemiptera: Coccidae) in Turkey". *Entomologica*, Bari 33:219-224.

Ülgentürk, Selma. 2001. "Parasitoids and predators of Coccidae (Homoptera: Coccoidea) species on ornamental plants in Ankara, Turkey". *Acta Phytopathologica et Entomologica Hungarica*, 36(3/4):369-375.

Ulubilir, Asuman, and Şekeroğlu, Erdal. 1997. "Biological control of *Liriomyza trifolii* by *Diglyphus isaea* on unheated greenhouse tomatoes in Adana, Turkey". *Bulletin. Section Regionale Ouest Palaearctique, Organisation Internationale de Lutte Biologique*, 20(4): 232-235.

Ulubilir, Asuman, and Yabaş, Cahide. 2000. "Studies on population development of leafminers (*Liriomyza spp.*) and parasitation situation". *Bulletin. Section Regionale Ouest Palaearctique, Organisation Internationale de Lutte Biologique*, 23(1):151-156.

Uygun, Nedim, and Elekcioğlu, Naime, and Erkilic, Lerzan, and Karaca, İsmail, and Kersting, Ulrich, and Vacante, V. 1997. "Studies on biological control of *Phyllocnistis citrella* Stainton (Lepidoptera: Gracillariidae) in Turkey". *Bulletin. Section Regionale Ouest Palaearctique, Organisation Internationale de Lutte Biologique*. 20(7):96-101.

Uygun, Nedim, and Polatöz, Zühal, and Başpınar, Hüseyin. 1995. "Faunistic studies on Agromyzidae (Diptera) in the south east Mediterranean rigion of Turkey". *Turkish Journal of Entomology*, 19(2):123-136.

Yabaş, Cahide, and Ulubilir, Asuman. 1995. "Investigations on the population fluctuations and parasitization of the leaf miner (*Liriomyza trifolii* Burgess)". *Bitki Koruma Bülteni*, 35(1-2):35-44.

Yaşarakıncı, Nilgün, and Hıncal, Pervin. 1997. "Determining the pests and beneficial species and their population densities on tomato, cucumber, pepper and lettuce greenhouses in İzmir". *Bitki Koruma Bülteni*, 37(1/2):79-89.

Yaşarakıncı, Nilgün, and Hıncal, Pervin. 2000. "The studies on the pests, their natural enemies and population developments on protected pepper in Izmir province". *Bitki Koruma Bülteni*, 40(3/4):135-152.

Yefremova, Zoya A., and Civelek, Hasan Sungur, and Boyadzhiev, Peter, and Dursun, Oktay, and Eskin, Ata. 2010. "Contributions to the Turkish Eulophidae (Hymenoptera, Chalcidoidea) with new records". *Turkish Journal of Entomology*, 34 (4): 447-463.

Yefremova, Zoya. 2008. Order Hymenoptera, family Eulophidae. Arthropod fauna of the UAE, 1: 345–360. Available from: https://www.researchgate.net/publication/261988579_Order_Hymenoptera_family_Eulophidae [accessed: Dec 12, 2022].

Yegorenkova, Ekaterina N. & Yefremova, Zoya A. 2011. "*Baryscapus babiyi* Doğanlar, 1993 (Hymenoptera: Eulophidae) is parasitoid of *Lasiosina devitata* (Diptera: Chloropidae) in Turkey: description of male and discovery of host". Zoosystematica Rossica, 20(2):325-329.

CHAPTER EIGHT

FAMILY: EUPELMIDAE WALKER, 1833

The Eupelmidae is a medium-sized family, characterized by several features: the mesopleuron is large, convex and longer than its height; the spur of the mid tibia is usually somewhat enlarged; the midcoxa is inserted anterior to the midline of the mesopleuron and at the posterior margin of mesopleuron; the mesoscutum often has a large concave median groove or furrow; and cerci are usually positioned at or near the apex of the metasoma.

For the most part, Eupelmids are primary parasitoids of a variety of insects, often associated with hosts concealed in plant tissue. Calosotinae and Neanastatinae primarily develop on wood-boring beetles, although some species also attack various gall-midges (Diptera: Cecidomyiidae). In contrast, Eupelminae display a much broader host range, including species of Lepidoptera (moths & butterflies), Coleoptera (beetles), Hemiptera (Coccoidea scale insects), Orthoptera (crickets and grasshoppers), Blattodea (cockroaches), Mantodea (mantids), and Diptera (flies). They can act as either primary parasitoids or hyperparasitoids, often targeting the cocoons of Braconids. One genus, *Anastatus*, specifically attacks large eggs of Lepidoptera, Orthoptera, Mantodea and Blattodea (Stringer et al., 2012).

The family currently includes 45 genera and 907 species placed in three subfamilies, and there are two subfamilies, four genera and 17 species in Turkey (Noyes, 2019).

Subfamily Calosotinae (Boucek, 1958)
Genus *Calosota* (Curtis, 1836)
Calosota aestivalis (Curtis, 1836)
Publications with Turkey records: Askew and Nieves-Aldrey (2006), Noyes (2019).

Associates: *Ptinomorphus imperialis, Xestobium plumbeum* (Coleoptera: Anobiidae), *Anthaxia sp.* (Buprestidae), *Chlorophorus sartor* (Cerambycidae), *Magdalis memnonia, Magdalis rufa* (Curculionidae), *Carpinus sp.* (Betulaceae), *Chilopsis linearis* (Bignoniaceae), *Retama sphaerocarpa* (Fabaceae), *Fagus sp., Quercus sp.* (Fagaceae), *Fraxinus ornus* (Oleaceae), *Larix laricina, Pinus sp., Pinus contorta, Pinus coulteri, Pinus flexilis, Pinus halepensis, Pinus nigra, Pinus sabiniana, Pseudotsuga menziesii* (Pinaceae), *Tamarix canariensis* (Tamaricaceae), *Taxodium distichum* (Taxodiaceae) (Noyes, 2019).
Hosts: *Hedobia sp., Ptilinus sp.* (Coleoptera: Anobiidae), *Anthaxia sp.* (Buprestidae), *Pogonocherus fasciculatus, Tetrops praeusta* (Cerambycidae), *Trichodes leucopsideus* (Cleridae), *Pissodes sp.* (Curculionidae), *Chaetoptelius vestitus, Myelophilus minor, Orthotomicus erosus, Phloeosinus sp., Phloeosinus armatus, Phloeosinus aubei, Pityogenes calcaratus, Scolytus sp.* (Scolytidae), *Megachile sp.* (Hymenoptera: Apidae) (Noyes, 2019).
Distribution: Canada, Croatia, Cyprus, Czech Republic, Czechoslovakia, England, Europe, France, Hungary, Israel, Jordan, Morocco, Netherlands, Poland, Romania, Russia, Slovakia, Spain, Sweden, Turkey, United States of America, USSR (Noyes, 2019).

Calosota metallica (Gahan, 1922)
Publications with Turkey records: Gibson (2010), Noyes (2019).
Associates: *Foeniculum vulgare, Heracleum sp.* (Apiaceae), *Encoelia sp.* (Fungus), *Umbellularia californica* (Lauraceae), *Agropyron sp., Elymus repens, Festuca sp., Lygeum sp., Triticum sp., Triticum aestivum* (Poaceae), *Tamarix canariensis* (Tamaricaceae) (Noyes, 2019).
Hosts: *Scolytus rugulosus* (Coleoptera: Scolytidae), *Mayetiola destructor, Phytophaga destructor* (Diptera: Cecidomyiidae), *Dacus sp., Dacus pusillus* (Tephritidae), *Harmolita grandis, Harmolita tritici, Tetramesa sp., Tetramesa grandis, Tetramesa holci, Tetramesa tritici* (Hymenoptera: Eurytomidae), *Ancylis comptana, Choristoneura rosaceana* (Lepidoptera: Tortricidae) (Noyes, 2019).
Parasitoid hosts: *Eurytoma parva, Harmolita sp., Isosoma sp., Isosoma grandis, Tetramesa sp., Tetramesa brevicornis, Tetramesa cereipes, Tetramesa grandis, Tetramesa tritici* (Hymenoptera: Eurytomidae), *Ditropinotus aureoviridis* (Torymidae) (Noyes, 2019).
Distribution: Bosnia Hercegovina, Bulgaria, Canada, Croatia, Cyprus, Czech Republic, Czechoslovakia, Europe, France, Germany, Hungary, Iran, Italy, Nearctic, North Africa, Romania, Russia, Slovakia, Spain, Tadzhikistan, Turkey, United States of America, Uzbekistan (Noyes, 2019).

Genus *Pentacladia* (Westwood, 1835)
Pentacladia hatayensis (Doğanlar, 2003)
Publications with Turkey records: Doğanlar (2003), Noyes (2019).
Associates: Unknown (Noyes, 2019).
Hosts: Unknown (Noyes, 2019).
Distribution: Turkey (Noyes, 2019).

Subfamily Eupelminae (Walker, 1833)
Genus *Anastatus* (Motschulsky, 1859)
Anastatus bifasciatus (Geoffroy, 1785)
Publications with Turkey records: Avci (2000), Noyes (2019), Öncüer (1991), Özbek and Coruh (2010).
Associates: *Corylus avellana* (Betulaceae), *Quercus sp.* (Fagaceae), *Pinus sp., Pinus brutia, Pinus halepensis, Pinus insignis, Pinus nigra, Pinus pinaster* (Pinaceae), *Triticum aestivum* (Poaceae), *Crataegus monogyna, Prunus domestica, Prunus persica, Prunus spinosa* (Rosaceae), *Populus sp.* (Salicaceae) (Noyes, 2019).
Hosts: *Cinara schimitscheki* (Hemiptera: Aphididae), *Gonocerus acuteangulatus, Leptoglossus occidentalis* (Coreidae), *Antestia lineaticollis, Eurydema festivum, Eurydema ornatum, Nezara viridula, Rhaphigaster nebulosa* (Pentatomidae), *Psylla pruni, Psylla pyri* (Psyllidae), *Eurygaster sp., Eurygaster integriceps* (Scutelleridae), *Dendrolimus pini, Dendrolimus punctatus, Dendrolimus spectabilis, Gastropacha quercifolia, Macrothylacia rubi, Malacosoma neustria* (Lepidoptera: Lasiocampidae), *Lymantria dispar, Nygmia phaeorrhoea, Porthetria dispar, Stilpnotia salicis* (Lymantriidae), *Thaumetopoea pityocampa, Thaumetopoea processionea, Thaumetopoea wilkinsoni, Traumatocampa pityocampa* (Notodontidae), *Charaxes jasius* (Nymphalidae), *Iphiclides podalirius* (Papilionidae), *Philosamia Cynthia, Saturnia pyri* (Saturniidae), *Polyptychus dentatus* (Sphingidae), *Oxya velox* (Orthoptera: Acrididae), *Phaneroptera sp.* (Tettigoniidae) (Noyes, 2019).
Parasitoid hosts: *Meteorus sp., Pauesia pini* (Hymenoptera: Braconidae) (Noyes, 2019).
Distribution: Bosnia Hercegovina, Bulgaria, Croatia, Cyprus, Czech Republic, Czechoslovakia, Europe, France, Germany, Greece, Hungary, India, Israel, Italy, Japan, Kazakhstan, Kenya, Kirgizia, Korea, Lebanon, Moldova, Montenegro, Morocco, North Africa, People's Republic of China, Portugal, Romania, Russia, Serbia, Slovakia, Spain, Sweden, Taiwan, Transcaucasus, Tselinograd Obl., Turkey, Ukraine, United States of America, USSR, Yugoslavia (pre-1991) (Noyes, 2019).

Anastatus japonicus (Ashmead, 1904)
Publications with Turkey records: Avci (2009), Noyes (2019).
Associates: *Nosema serbicae* (Nosematidae), *Trichogramma dendrolimi* (Hymenoptera: Trichogrammatidae), *Glycine max, Robinia pseudoacacia* (Fabaceae), *Quercus sp., Quercus suber* (Fagaceae), *Eucalyptus camaldulensis* (Myrtaceae), *Pinus sp.* (Pinaceae), *Platanus sp.* (Platanaceae), *Populus sp., Salix sp., Salix alba* (Salicaceae), *Dimocarpus longan, Litchi chinensis* (Sapindaceae) (Noyes, 2019).
Hosts: *Riptortus clavatus* (Hemiptera: Alydidae), *Nezara viridula, Palomena prasine, Tessaratoma papillosa* (Pentatomidae), *Dendrolimus sibiricus, Dendrolimus spectabilis, Malacosoma neustria* (Lepidoptera: Lasiocampidae), *Lymantria dispar, Lymantria obfuscate, Nygmia phaeorrhoea, Porthetria dispar* (Lymantriidae), *Cerura vinula, Dicranura vinula* (Notodontidae), *Iphiclides podalirius* (Papilionidae), *Antheraea pernyi, Philosamia cynthia* (Saturniidae) (Noyes, 2019).
Parasitoid hosts: *Apanteles melanoscelus, Rogas dendrolimi* (Hymenoptera: Braconidae), *Ooencyrtus pityocampae, Tyndarichus navae* (Encyrtidae) (Noyes, 2019).
Distribution: Austria, Belarus, Bosnia Hercegovina, Bulgaria, Canada, Croatia, Czechoslovakia, Europe, France, Germany, Hungary, India, Italy, Japan, Kazakhstan, Kirgizia, Korea, Macedonia, Nearctic, North Africa, Oriental, People's Republic of China, Romania, Russia, Serbia, Slovakia, Spain, Tselinograd Obl., Turkey, Ukraine, United States of America, USSR, Uzbekistan, Yugoslavia (pre-1991) (Noyes, 2019).

Genus *Eupelmus* (Dalman, 1820)
Eupelmus annulatus (Nees, 1834)
Publications with Turkey records: Gibson (2011), Noyes (2019).
Associates: *Tomicus piniperda* (Coleoptera: Scolytidae), *Dryocosmus kuriphilus* (Hymenoptera: Cynipidae), *Quercus sp., Quercus cerris, Quercus faginea, Quercus pyrenaica* (Fagaceae), *Pinus contorta, Pinus halepensis, Pinus sylvestris* (Pinaceae), *Malus pumila* (Rosaceae) (Noyes, 2019).
Parasitoid hosts: *Apanteles sp., Apanteles melanoscelus, Apanteles solitarius, Cotesia melanoscela, Macrocentrus sp., Macrocentrus ancylivorus, Meteorus versicolor* (Hymenoptera: Braconidae), *Synergus pallicornis, Synergus umbraculus* (Cynipidae), *Cremastus minor, Ephialtes caudatus, Glypta sp., Glypta rufiscutellaris, Liotryphon caudatus, Omorgus borealis* (Ichneumonidae) (Noyes, 2019).
Hosts: *Agrilus planipennis* (Coleoptera: Buprestidae), *Pissodes sitchenisis, Pissodes strobi* (Curculionidae), *Andricus aestivalis, Andricus*

coriarius, Andricus grossulariae, Andricus hispanicus, Andricus kollari, Andricus legitimus, Andricus lignicolus, Andricus multiplicatus, Andricus pictus, Andricus polycerus, Andricus quercustozae, Aphelonyx cerricola, Cynips quercusfolii, Dryocosmus kuriphilus, Synophrus politus (Hymenoptera: Cynipidae), *Diprion simile (Diprionidae), Lymantria dispar, Nygmia phaeorrhoea, Porthetria dispar, Stilpnotia salicis* (Lepidoptera: Lymantriidae), *Chrysoclista lariella* (Momphidae), *Apterona crenulella* (Psychidae), *Carpocapsa pomonella, Grapholita molesta, Rhyacionia buoliana* (Tortricidae) (Noyes, 2019).
Distribution: Austria, Bulgaria, Canada, Croatia, Czech Republic, Czechoslovakia, England, Europe, France, Germany, Greece, Hungary, Iran, Italy, Moldova, Netherlands, Romania, Russia, Serbia, Slovakia, Spain, Sweden, Switzerland, Turkey, Ukraine, United States of America (Noyes, 2019).

Eupelmus aseculatus (Kalina, 1981)
Publications with Turkey records: Fusu (2017), Noyes (2019).
Associates: *Andricus quercusradicis, Callirhytis glandium, Callirhytis rufescens* (Hymenoptera: Cynipidae), *Quercus pyrenaica, Quercus suber* (Fagaceae), *Hypecoum imberbe, Papaver dubium* (Papaveraceae) (Noyes, 2019).
Hosts: *Andricus quercusradicis, Aylax hypecoi, Aylax papaveris, Callirhytis glandium, Callirhytis rufescens* (Hymenoptera: Cynipidae) (Noyes, 2019).
Distribution: Azerbaijan, Bulgaria, Croatia, Cyprus, France, Greece, Israel, Italy, Romania, Spain, Turkey, Yugoslavia (pre-1991) (Noyes, 2019).

Eupelmus azureus (Ratzeburg, 1844)
Publications with Turkey records: Askew et al. (2013), Noyes (2019), Öncüer (1991).
Associates: *Andricus aries, Andricus caliciformis, Andricus callidoma, Andricus caputmedusae, Andricus conglomeratus, Andricus conificus, Andricus coriarius, Andricus coronatus, Andricus corruptrix, Andricus curvator, Andricus cydoniae, Andricus fecundatrix, Andricus galeatus, Andricus gallaeurnaeformis, Andricus glutinosus, Andricus grossulariae, Andricus hartigi, Andricus hispanicus, Andricus hungaricus, Andricus hystrix, Andricus infectorius, Andricus inflator, Andricus kollari, Andricus legitimus, Andricus lignicolus, Andricus lucidus, Andricus megalucidus, Andricus multiplicatus, Andricus pictus, Andricus polycerus, Andricus quercuscalicis, Andricus quercusradicis, Andricus quercusramuli,*

Andricus quercustozae, Andricus seckendorffi, Andricus seminationis, Andricus sieboldin, Andricus singularis, Andricus solitarius, Andricus tomentosus, Andricus truncicolus, Aphelonyx cerricola, Biorhiza pallida, Chilaspis nitida, Cynips agama, Cynips cornifex, Cynips divisa, Cynips quercus, Cynips quercusfolii, Dryocosmus cerriphilus, Dryocosmus kuriphilus, Neuroterus quercusbaccarum, Plagiotrochus quercusilicis, Pseudoneuroterus macropterus, Pseudoneuroterus saliens, Synophrus politus (Cynipidae), *Quercu coccifera, Quercus sp., Quercus cerris, Quercus frainetto, Quercus infectoria, Quercus lusitanica, Quercus petraea, Quercus pubescens, Quercus robur, Quercus suber* (Fagaceae), *Pinus strobus* (Pinaceae) (Noyes, 2019).

Hosts: *Scolytus ratzeburgi* (Coleoptera: Scolytidae), *Urophora cardui* (Diptera: Tephritidae), *Andricus caputmedusae, Andricus coriarius, Andricus fecundatrix, Andricus glutinosus, Andricus grossulariae, Andricus hungaricus, Andricus kollari, Andricus lucidus, Andricus quercustozae, Biorhiza pallida, Cynips flosculi, Cynips kollari, Cynips lignicola, Cynips polycera, Neuroterus glandiformis, Neuroterus saltans, Synergus pallicornis, Synergus umbraculus* (Hymenoptera: Cynipidae), *Diprion simile* (Diprionidae), *Exoteleia dodecella* (Lepidoptera: Gelechiidae), *Leucoma salicis, Nygmia phaeorrhoea, Porthetria dispar, Stilpnotia salicis* (Lymantriidae), *Luffia lapidella* (Psychidae), *Etiella zinckenella* (Pyralidae), *Grapholita molesta* (Tortricidae) (Noyes, 2019).

Parasitoid hosts: *Apanteles sp., Apanteles melanoscelus, Apanteles solitarius, Bracon piger, Macrocentrus ancylivorus, Macrocentrus delicatus, Meteorus versicolor* (Hymenoptera: Braconidae), *Cremastus minor, Glypta rufiscutellaris* (Ichneumonidae), *Eupelmella vesicularis* (Eupelmidae) (Noyes, 2019).

Distribution: Austria, Bulgaria, Canada, Caucasus, Croatia, Czechoslovakia, Europe, France, Germany, Greece, Hungary, Italy, Moldova, Nearctic, Poland, Romania, Spain, Turkey, Ukraine, United Kingdom, United States of America (Noyes, 2019)

Eupelmus impennis (Nikolskaya, 1952)

Publications with Turkey records: Fusu (2017), Noyes (2019).

Associates: *Eryngium campestre* (Apiaceae) (Noyes, 2019).

Hosts: *Boucheella artemisiae, Lasioptera eryngii* (Diptera: Cecidomyiidae), *Diplolepis fructuum* (Hymenoptera: Cynipidae) (Noyes, 2019).

Distribution: Bulgaria, Greece, Hungary, Iran, Romania, Slovakia, Tadzhikistan, Turkey, USSR, Uzbekistan (Noyes, 2019).

Eupelmus lanceolatus (Gibson and Fusu, 2016)
Publications with Turkey records: Gibson and Fusu (2016), Noyes (2019).
Associates: Unknown (Noyes, 2019).
Hosts: Unknown (Noyes, 2019).
Distribution: Bulgaria, Cyprus, France, Greece, Israel, Italy, Kenya, Malta, Somalia, Turkey (Noyes, 2019).

Eupelmus messene (Walker, 1839)
Publications with Turkey records: Fusu (2017), Noyes (2019).
Hosts: *Anthonomus signatus* (Coleoptera: Curculionidae) (Noyes, 2019).
Distribution: Argentina, Armenia, Australia, Austria, Azores, Bulgaria, Canada, Colombia, Czech Republic, England, France, Greece, Hungary, Iran, Italy, Japan, Kirgizia, Korea, Madeira, Moldova, New Zealand, People's Republic of China, Portugal, Romania, Russia, Serbia, Spain, Switzerland, Turkey, Ukraine, United States of America (Noyes, 2019).

Eupelmus microzonus (Foerster, 1860)
Publications with Turkey records: Askew et al. (2006), Korkmaz et al. (2010), Noyes (2019), Öncüer (1991).
Associates: *Acroptilon repens, Centaurea sp., Centaurea aspera, Centaurea melitensis, Centaurea nicaeensis, Centaurea scabiosa, Centaurea solstitialis, Centaurea sphaerocephala, Cichorium intybus, Inula crithmoides, Notobasis syriaca, Podospermum laciniatum, Scorzonera laciniata, Silybum marianum, Tragopogon sp.* (Asteraceae), *Medicago sativa, Onobrychis viciifolia* (Fabaceae), *Quercus sp.* (Fagaceae), *Nepeta latifolia, Phlomis tuberosa, Salvia lavandulifolia, Salvia nemorosa, Salvia officinalis, Salvia pratensis* (Lamiaceae), *Hypecoum imberbe, Papaver sp., Papaver dubium, Papaver rhoeas, Papaver somniferum* (Papaveraceae), *Triticum aestivum* (Poaceae), *Potentilla argentea, Potentilla reptans, Rosa sp.* (Rosaceae), *Tribulus terrestris* (Zygophyllaceae) (Noyes, 2019).
Hosts: *Apion seniculus* (Coleoptera: Apionidae), *Bruchus affinis, Bruchus emarginatus, Bruchus rufipes, Bruchus tristiculus* (Bruchidae), *Microlarinus lareynii* (Curculionidae), *Mayetiola destructor, Mayetiola mimeuri* (Diptera: Cecidomyiidae), *Chlorops taeniopus, Lasiosina cinctipes* (Chloropidae), *Myopites sp.* (Tephritidae), *Aulacidea sp., Aulacidea acroptilonica, Aulacidea freesei, Aulacidea laurae, Aulacidea tragopogonis, Aylax sp., Aylax hypecoi, Aylax minor, Aylax papaveris, Barbotinia sp., Barbotinia oraniensis, Diplolepis sp., Diplolepis mayri, Iraella luteipes, Isocolus sp., Isocolus centaureae, Isocolus lichtensteini,*

Isocolus tauricus, Liposthenes sp., Liposthenes kerneri, Neaylax sp., Neaylax salvia, Phanacis sp., Phanacis caulicola, Phanacis centaureae, Phanacis zwoelferi, Timaspis cichorii, Timaspis lampsanae, Xestophanes sp., Xestophanes potentillae (Hymenoptera: Cynipidae), *Bruchophagus caraganae, Bruchophagus gibbus, Bruchophagus roddi, Eurytoma onobrychidis, Eurytoma roddi* (Eurytomidae), *Dendrolimus sibiricus* (Lepidoptera: Lasiocampidae), *Fumea danielis* (Psychidae), *Etiella zinckenella* (Pyralidae) (Noyes, 2019).
Parasitoid hosts: *Rhogas dendrolimi* (Hymenoptera: Braconidae) (Noyes, 2019).
Distribution: Bulgaria, Canada, Canary Islands, Croatia, Cyprus, Czech Republic, Czechoslovakia, Europe, France, Germany, Greece, Hungary, Iran, Italy, Kazakhstan, Kirgizia, Macedonia, Moldova, Montenegro, Morocco, Netherlands, North Africa, Portugal, Romania, Russia, Serbia, Slovakia, Spain, Transcaucasus, Tselinograd Obl., Turkey, Turkmenistan, Ukraine, USSR, Uzbekistan (Noyes, 2019).

Eupelmus pistaciae (Al khatib, 2015)
Publications with Turkey records: Al khatib et al. (2014), Noyes (2019).
Associates: *Pistacia terebinthus, Pistacia vera* (Anacardiaceae) (Noyes, 2019).
Hosts: *Megastigmus sp., Megastigmus pistaciae* (Hymenoptera: Torymidae) (Noyes, 2019).
Distribution: Cyprus, France, Greece, Iran, Tunisia, Turkey (Noyes, 2019).

Eupelmus seculatus (Ferriere, 1954)
Publications with Turkey records: Askew et al. (2006), Noyes (2019).
Associates: *Plagiotrochus australis, Plagiotrochus quercusilicis, Plagiotrochus yeusei* (Hymenoptera: Cynipidae), *Acroptilon repens, Artemisia sp., Artemisia herba-alba, Lapsana communis, Podospermum laciniatum, Scorzonera laciniata* (Asteraceae), *Gypsophila struthium* (Caryophyllaceae), *Atriplex halimus* (Chenopodiaceae), *Medicago sativa, Ononis sp., Retama sphaerocarpa* (Fabaceae), *Quercus ilex* (Fagaceae), *Ephedra nebrodensis* (Gnetaceae), *Potentilla reptans* (Rosaceae) (Noyes, 2019).
Hosts: *Aulacidea laurae, Plagiotrochus sp., Plagiotrochus australis, Plagiotrochus quercusilicis, Timaspis sp., Timaspis lampsanae, Xestophanes sp., Xestophanes potentillae* (Hymenoptera: Cynipidae), *Bruchophagus roddi (Eurytomidae), Thaumetopoea pityocampa* (Lepidoptera: Notodontidae) (Noyes, 2019).

Distribution: Azerbaijan, Croatia, France, Italy, Macedonia, Monaco, Palearctic, Spain, Turkey (Noyes, 2019).

Eupelmus urozonus (Dalman, 1820)
Publications with Turkey records: Askew et al. (2013), Bayram et al. (1998), Doğanlar (2011; 2014), Doğanlar et al. (2011), Gencer (2003), Mete and Mergen (2017), Noyes (2019), Öncüer (1991), Özbek et al. (1999).
Associates: *Bruchidius chloroticus* (Coleoptera: Bruchidae), *Phloeosinus aubei* (Scolytidae), *Dasineura oleae, Lasioptera oleicola* (Diptera: Cecidomyiidae), *Paracolopha morrisoni* (Hemiptera: Aphididae), *Andricus amblycerus, Andricus amenti, Andricus aries, Andricus burgundus, Andricus callidoma, Andricus caputmedusae, Andricus conglomeratus, Andricus conificus, Andricus coriarius, Andricus coronatus, Andricus corruptrix, Andricus curtisii, Andricus curvator, Andricus cydoniae, Andricus fecundatrix, Andricus galeatus, Andricus gallaeurnaeformis, Andricus gemmeus, Andricus glutinosus, Andricus grossulariae, Andricus hartigi, Andricus hispanicus, Andricus hungaricus, Andricus infectorius, Andricus inflator, Andricus kollari, Andricus lignicolus, Andricus lucidus, Andricus malphigii, Andricus multiplicatus, Andricus paradoxus, Andricus pictus, Andricus polycerus, Andricus pseudoinflator, Andricus quadrilineatus, Andricus quercuscalicis, Andricus quercusradicis, Andricus quercusramuli, Andricus quercustozae, Andricus schroeckingeri, Andricus seckendorffi, Andricus sieboldin, Andricus singularis, Andricus solitarius, Andricus testaceipes, Andricus truncicolus, Andricus vindobonensis, Aphelonyx cerricola, Biorhiza pallida, Callirhytis glandium, Cerroneuroterus lanuginosus, Chilaspis nitida, Cynips agama, Cynips cornifex, Cynips disticha, Cynips divisa, Cynips longiventris, Cynips quercus, Cynips quercusfolii, Dryocosmus cerriphilus, Dryocosmus kuriphilus, Neuroterus albipes, Neuroterus anthracinus, Neuroterus numismalis, Neuroterus quercusbaccarum, Neuroterus tricolor, Plagiotrochus amenti, Plagiotrochus australis, Plagiotrochus panteli, Plagiotrochus quercusilicis, Plagiotrochus razeti, Pseudoneuroterus macropterus, Pseudoneuroterus saliens, Synophrus politus, Trigonaspis brunneicornis, Trigonaspis synaspis* (Hymenoptera: Cynipidae), *Sigmophora brevicornis* (Eulophidae), *Eupelmus kamijoi* (Eupelmidae), *Eurytoma rosae* (Eurytomidae), *Pteromalus bedeguaris* (Pteromalidae), *Acer sp., Acer monspessulanum, Acer pseudoplatanus* (Aceraceae), *Rhus lucida, Rhus undulata* (Anacardiaceae), *Centaurea sadleriana, Centaurea scabiosa, Centaurea trinervia, Hieracium sp., Hieracium pilosella, Inula sp., Inula viscosa, Lapsana communis*

(Asteraceae), *Betula sp.* (Betulaceae), *Capparis spinosa* (Capparidaceae), *Cupressus funebris, Cupressus torulosa, Juniperus excelsa, Juniperus phoeniciae, Juniperus thurifera, Platycladus orientalis* (Cupressaceae), *Diospyros kaki* (Ebenaceae), *Astragalus glycyphyllos, Cyamopsis tetragonoloba, Cytisus scoparius, Medicago sativa, Phaseolus sp., Sesbania aculeata, Vigna unguiculata* (Fabaceae), *Castanea sp., Castanea crenata, Castanea sativa, Fagus sp., Fagus sylvatica, Quercu coccifera, Quercus sp., Quercus cerris, Quercus dentata, Quercus faginea, Quercus frainetto, Quercus ilex, Quercus infectoria, Quercus lusitanica, Quercus mongolica, Quercus pedunculiflora, Quercus petraea, Quercus pubescens, Quercus pyrenaica, Quercus robur, Quercus suber, Quercus trojana* (Fagaceae), *Ephedra foliate, Ephedra fragilis, Ephedra nebrodensis* (Gnetaceae), *Aesculus hippocastanum* (Hippocastanaceae), *Phlomis tuberosa, Salvia fruticose, Salvia triloba* (Lamiaceae), *Fraxinus sp., Fraxinus excelsior, Olea sp., Olea europaea* (Oleaceae), *Larix sp., Larix decidua, Larix polonica, Pinus sp., Pinus bungeana, Pinus sylvestris, Pseudotsuga menziesii* (Pinaceae), *Calamagrostis epigeios, Phyllostachys sp.* (Poaceae), *Clematis vitalba* (Ranunculaceae), *Prunus sp., Prunus amygdalus, Prunus armeniaca, Rosa sp., Rosa canina, Rosa damascene, Rosa rubiginosa* (Rosaceae), *Rubus sp., Rubus caesius, Rubus fruticosus, Rubus idaeus* (Rubiaceae), *Citrus nobilis, Citrus sinensis* (Rutaceae), *Salix sp., Salix pentandra* (Salicaceae), *Tamarix sp.* (Tamaricaceae), *Corchorus capsularis* (Tiliaceae), *Celtis sp., Ulmus pumila, Zelkova serrata* (Ulmaceae), *Vitis vinifera* (Vitaceae) (Noyes, 2019).

Hosts: *Apion corchori, Apion fuscirostre, Apion holosericeum* (Coleoptera: Apionidae), *Atellabus nitens* (Attelabidae), *Bruchidius sp., Bruchidius chloroticus, Bruchidius marginalis, Bruchidius unicolor, Bruchus lentis, Bruchus lividimanus, Bruchus rufipes* (Bruchidae), *Plagiodera versicolora* (Chrysomelidae), *Chilocorus bipustulatus* (Coccinellidae), *Brachonyx pineti, Curculio vicetinus, Hypolixus truncatulus, Lixus juncii, Lixus truncatulus, Pempheres affinis, Pempherulus affinis, Pissodes castaneus, Pissodes notatus, Rhynchaenus hustachei, Stereonychus fraxini, Xylocleptis bispinus* (Curculionidae), *Blastophagus minor, Chaetoptelius vestitus, Hylesinus fraxini, Hylesinus vestitus, Ips acuminatus, Ips typographus, Ips vorontzovi, Phloeotribus scarabaeoides, Pityophthorus polonicus, Polygraphus poligraphus, Scolytus amygdali, Scolytus ensifer, Scolytus intricatus, Scolytus rugulosus* (Scolytidae), *Melanagromyza phaseoli* (Diptera: Agromyzidae), *Asphondylia sp., Asphondylia Capparis, Asphondylia pruniperda, Asphondylia sarothamni, Celticecis japonica, Dasineura gleditchiae, Dasineura oleae, Dryomyia circinnans, Etsuhoa thuriferae, Eumarchalia* *gennadii, Janetia cerris, Kaltenbachiola strobi, Lasioptera berlesiana, Lasioptera oleicola, Lasioptera rubi, Mikiola fagi, Prolasioptera berlesiana, Psectrosema reticulatum,, Pseudasphondylia diospyri, Rabdophaga sp.* (Cecidomyiidae), *Bactrocera oleae, Carpomya incomplete, Ceratitis capitata, Dacus oleae, Myopites sp., Myopites limbardae, Myopites olivieri, Myopites stylata* (Tephritidae), *Toxoptera aurantia, Toxoptera theaicola* (Hemiptera: Aphididae), *Ceroplastes floridensis, Ceroplastes rubens, Eulecanium corni, Parthenolecanium corni, Parthenolecanium rufulum, Sphaerolecanium prunastri* (Coccidae), *Aonidiella aurantii* (Diaspididae), *Andricus aestivalis, Andricus albopunctatus, Andricus amenti, Andricus coriarius, Andricus corruptrix, Andricus curvator, Andricus fecundatrix, Andricus galeata, Andricus grossulariae, Andricus hispanicus, Andricus infectorius, Andricus kollari, Andricus lignicolus, Andricus lucidus, Andricus multiplicatus, Andricus ostreus, Andricus paradoxus, Andricus quercuscalicis, Andricus quercusradicis, Andricus quercusramuli, Andricus singulus, Andricus solitarius, Andricus testaceipes, Andricus vindobonensis, Aphelonyx cerricola, Aulacidea sp., Aulacidea hieracia, Biorhiza pallida, Cynips disticha, Cynips divisa, Cynips longiventris, Cynips quercus, Cynips quercusfolii, Diastrophus rubi, Diplolepis sp., Diplolepis centifoliae, Diplolepis eglanteriae, Diplolepis fructuum, Diplolepis mayri, Diplolepis nervosa, Diplolepis rosae, Diplolepis spinosissimae, Dryocosmus kuriphilus, Isocolus sp., Isocolus rogenhoferi, Isocolus scabiosae, Neuroterus glandiformis, Neuroterus quercusbaccarum, Neuroterus saltans, Pediaspis aceris, Plagiotrochus australis, Plagiotrochus quercusilicis, Rhodites rosae, Rhodus cyprius, Rhoophilus loewi, Salviela kezivi, Synergus pallicornis, Synophrus politus, Timaspis lampsanae, Trigonaspis mendesi* (Hymenoptera: Cynipidae), *Diprion pini, Neodiprion sertifer* (Diprionidae), *Bruchophagus roddi, Eurytoma sp., Eurytoma appendigaster, Eurytoma gallephedrae, Tetramesa eximia* (Eurytomidae), *Blascoa ephedrae* (Pteromalidae), *Pseudoxiphydria betulae* (Siricidae), *Euura amerinae, Pontania viminalis* (Tenthredinidae), *Megastigmus amicorum, Megastigmus pictus, Megastigmus pistaciae, Megastigmus seitneri, Megastigmus spermotrophus* (Torymidae), *Scyrotis sp.* (Lepidoptera: Cecidosidae), *Coleophora hemerobiella, Coleophora serratella* (Coleophoridae), *Coleotechnites piceaella, Dichomeris marginella, Exoteleia dodecella, Gelechia senticetella* (Gelechiidae), *Callisto denticulella, Cameraria ohridella, Lithocolletis corylifoliella, Lithocolletis pyrifoliella, Phyllonorycter sp.* (Gracillariidae), *Dendrolimus pini* (Lasiocampidae), *Lymantria dispar* (Lymantriidae), *Parametriotes theae* (Momphidae), *Thaumetopoea pityocampa* (Notodontidae), *Aporia*

crataegi, Pieris brassicae (Pieridae), *Etiella zinckenella* (Pyralidae), *Archips rosanus, Carpocapsa pomonella, Epinotia festivana, Grapholita funebrana, Lobesia botrana, Pandemis ribeana, Polychrosis botrana, Rhyacionia buoliana, Sparganothis pilleriana, Tortrix viridana* (Tortricidae), *Prays oleae* (Yponomeutidae) (Noyes, 2019).
Parasitoid hosts: *Anthomyiopsis plagioderae, Eurysthaea scutellaris, Nemorilla floralis* (Diptera: Tachinidae), *Goniozus claripennis* (Hymenoptera: Bethylidae), *Apanteles sp., Apanteles circumscriptus, Apanteles glomeratus, Apanteles melanoscelus, Bracon intercessor, Bracon piger, Meteorus sp., Meteorus versicolor* (Braconidae), *Synergus gallaepomiformis, Synergus reinhardi, Synergus umbraculus* (Cynipidae), *Copidosoma geniculatum* (Encyrtidae), *Dahlbominus fuscipennis, Dichatomus acerinus, Olynx arsames, Pnigalio mediterraneus* (Eulophidae), *Eupelmus urozonus* (Eupelmidae), *Eurytoma brunniventris, Eurytoma gallephedrae* (Eurytomidae), *Angitia punctoria* (Ichneumonidae), *Blascoa ephedrae, Mesopolobus sp., Mesopolobus tibialis, Pteromalus sp.* (Pteromalidae), *Megastigmus pictus, Torymus sp., Torymus beneficus, Torymus cingulatus, Torymus nigricornis, Torymus sinensis* (Torymidae) (Noyes, 2019).
Distribution: Afghanistan, Algeria, Andorra, Armenia, Australia, Austria, Belgium, Bosnia Hercegovina, Bulgaria, Canary Islands, Croatia, Cyprus, Czech Republic, Czechoslovakia, Egypt, England, Europe, Finland, France, Germany, Greece, Hungary, India, Iran, Israel, Italy, Japan, Jordan, Kazakhstan, Korea, Lebanon, Luxembourg, Moldova, Montenegro, Morocco, Netherlands, North Africa, Norway, Pakistan, People's Republic of China, Poland, Portugal, Romania, Russia, Serbia, Slovakia, Slovenia, South Africa, Spain, Sweden, Switzerland, Syria, Transcaucasus, Tselinograd Obl., Turkey, Ukraine, United States of America, USSR, Yugoslavia (Federal Republic), Yugoslavia (pre-1991) (Noyes, 2019).

Eupelmus vesicularis (Retzius, 1783)
Publications with Turkey records: Askew et al. (2006), Noyes (2019).
Associates: *Andricus burgundus, Andricus caputmedusae, Andricus conglomeratus, Andricus coriarius, Andricus crispator, Andricus curvator, Andricus grossulariae, Andricus hispanicus, Andricus hungaricus, Andricus kollari, Andricus quercuscalicis, Andricus quercusradicis, Andricus schroeckingeri, Andricus sieboldin, Biorhiza pallida, Chilaspis nitida, Cynips agama, Cynips disticha, Cynips divisa, Cynips quercus, Cynips quercusfolii, Neuroterus albipes, Neuroterus anthracinus, Neuroterus quercusbaccarum, Neuroterus tricolor, Plagiotrochus australis, Plagiotrochus quercusilicis, Trigonaspis*

brunneicornis (Cynipidae), *Aprostocetus monacoi* (Eulophidae), *Eurytoma centaureae, Eurytoma punctatella, Eurytoma solenosopheriae, Sycophila vacciniicola* (Eurytomidae), *Ormyrus nitidulus, Ormyrus papaveris, Ormyrus pomaceus, Ormyrus vacciniicola* (Ormyridae), *Hemadas nubilipennis, Homoporus subniger, Pteromalus sp., Pteromalus hieracia, Pteromalus vibulenus, Stinoplus jenningsi* (Pteromalidae).
Associates: *Eryngium campestre* (Apiaceae), *Acroptilon repens, Ambrosia artemisiifolia, Artemisia sp., Carduus nutans, Centaurea sp., Centaurea nigra, Centaurea paniculate, Centaurea sadleriana, Centaurea scabiosa, Cirsium arvense, Crepis biennis, Dittrichia sp., Hieracium sp., Hypochaeris radicata, Inula viscosa, Lactuca viminea, Lapsana communis, Picris sp., Picris echioides, Senecio jacobaea, Sonchus asper, Tanacetum vulgare, Tragopogon sp.* (Asteraceae), *Alnus sp.* (Betulaceae), *Brassica napus* (Brassicaceae), *Convolvulus arvensis* (Convolvulaceae), *Thuja sp., Thuja occidentalis* (Cupressaceae), *Carex lasiocarpa* (Cyperaceae), *Dipsacus fullonum* (Dipsacaceae), *Vaccinium sp., Vaccinium angustifolium* (Ericaceae), *Astragalus glycyphyllos, Lotus corniculatus, Medicago sativa, Melilotus indicus, Trifolium hybridum* (Fabaceae), *Fagus sp., Quercu coccifera, Quercus sp., Quercus cerris, Quercus faginea, Quercus ilex, Quercus lusitanica, Quercus petraea, Quercus pubescens, Quercus pyrenaica, Quercus robur, Quercus suber* (Fagaceae), *Aesculus hippocastanum* (Hippocastanaceae), *Glechoma hederacea, Nepeta latifolia, Salvia fruticose, Salvia lavandulifolia, Salvia nemorosa, Salvia officinalis, Salvia pratensis, Salvia triloba* (Lamiaceae), *Hypecoum imberbe, Papaver sp., Papaver dubium, Papaver rhoeas* (Papaveraceae), *Pinus sp., Pinus nigra, Pinus strobus* (Pinaceae), *Agropyron tenerum, Alopecurus myosuroides, Alopecurus pratensis, Ammophila arenaria, Calamagrostis epigeios, Elymus sp., Elymus repens, Festuca rubra, Phleum sp.* (Poaceae), *Rumex sp., Rumex crispus* (Polygonaceae), *Malus pumila, Potentilla argentea, Potentilla erecta, Potentilla reptans, Rosa sp., Rosa blanda* (Rosaceae), *Rubus sp., Rubus caesius, Rubus fruticosus, Rubus idaeus* (Rubiaceae), *Salix repens* (Salicaceae), *Linaria dalmatica* (Scrophulariaceae), *Ulmus carpinifolia, Ulmus pumila* (Ulmaceae), *Sphaeropyx bicolor* (Hymenoptera: Braconidae) (Noyes, 2019).
Hosts: *Apion sp., Apion gibbirostre, Apion meieri, Apion tubiferum, Apion violaceum* (Coleoptera: Apionidae), *Bruchus brachialis, Bruchus lentis, Bruchus rufipes* (Bruchidae), *Agrilus rubicola* (Buprestidae), *Cassida rubiginosa, Chlamisus cribripennis* (Chrysomelidae), *Anthonomus signatus, Ceutorhynchus assimilis, Ceutorhynchus obstrictus, Cylindrocopturus furnissi, Gymnetron sp., Hypera postica, Hypera*

rumicis, Lixus juncii, Mecinus collaris, Rhinocyllus conicus, Tachypterellus consors (Curculionidae), *Gyrinus sp.* (Gyrinidae), *Nanophyes flavidus* (Nanophyidae), *Rhynchites bacchus* (Rhynchitidae), *Hylesinus fraxini, Hylesinus oleiperda, Ips typographus, Phloeotribus scarabaeoides, Pityogenes pilidens, Scolytus butovitschi, Scolytus pygmaeus* (Scolytidae), *Melanagromyza albocilia, Phytomyza fuscula* (Diptera: Agromyzidae), *Botanophila seneciella, Hylemya sp.* (Anthomyiidae), *Isosandalum dentipes, Lasioptera eryngii, Lasioptera rubi, Mayetiola avenae, Mayetiola destructor, Mikiola fagi, Phytophaga destructor, Planetella sp., Rhopalomyia sp., Rhopalomyia baccarum* (Cecidomyiidae), *Chlorops taeniopus* (Chlor opidae), *Musca domestica* (Muscidae), *Chaetorellia jaceae, Dacus oleae, Myopites olivieri, Tephritis stictica, Urophora jaceana, Urophora quadrifasciata* (Tephritidae), *Ceratina dupla* (Hymenoptera: Apidae), *Cephus cinctus, Cephus pygmaeus, Janus integer* (Cephidae), *Andricus caputmedusae, Andricus kollari, Andricus paradoxus, Andricus quercusradicis, Aulacidea sp., Aulacidea acroptilonica, Aulacidea follioti, Aulacidea hieracia, Aulacidea podagrae, Aulacidea scorzonerae, Aulacidea tragopogonis, Aulacidea tumida, Aylax hypecoi, Aylax minor, Aylax papaveris, Aylax salvia, Cynips disticha, Cynips quercusfolii, Diastrophus sp., Diastrophus mayri, Diastrophus rubi, Diplolepis sp., Diplolepis fructuum, Diplolepis mayri, Diplolepis nervosa, Diplolepis nodulosa, Diplolepis rosae, Diplolepis spinosissimae, Isocolus sp., Isocolus rogenhoferi, Isocolus scabiosae, Liposthenes glechomae, Liposthenes kerneri, Phanacis sp., Phanacis caulicola, Phanacis centaureae, Phanacis hypochoeridis, Plagiotrochus quercusilicis, Rhodus cyprius, Salviela kezivi, Synergus gallaepomiformis, Synergus hayneanus, Timaspis lampsanae, Timaspis lusitanica, Timaspis phoenixopodos, Trigonaspis brunneicornis, Trigonaspis Megaptera, Trigonaspis synaspis, Xestophanes brevitarsis, Xestophanes potentillae* (Cynipidae), *Diprion pini, Diprion simile, Neodiprion abietis, Neodiprion lecontei, Neodiprion annulus, Neodiprion pratti, Neodiprion sertifer* (Diprionidae), *Bruchophagus sp., Bruchophagus caraganae, Bruchophagus gibbus, Bruchophagus kolobovae, Bruchophagus platypterus, Bruchophagus roddi, Eurytoma onobrychidis, Eurytoma parva, Eurytoma roddi, Harmolita sp., Harmolita albomaculata, Harmolita elymi, Harmolita festucae, Harmolita grandis, Harmolita hordei, Harmolita linearis, Harmolita longula, Harmolita maculate, Harmolita tritici, Harmolita vaginicola, Philachyra aptera, Tetramesa sp., Tetramesa albomaculatum, Tetramesa atlanticum, Tetramesa cylindrica, Tetramesa dactylicola, Tetramesa elymi, Tetramesa eximia, Tetramesa festucae, Tetramesa grandis, Tetramesa holci, Tetramesa hordei,*

Tetramesa maculatum, Tetramesa oregon, Tetramesa phalaridis, Tetramesa tritici, Tetramesa vaginicolum (Eurytomidae), *Hemadas nubilipennis* (Pteromalidae), *Spilomena enslini* (Sphecidae), *Pontania viminalis* (Tenthredinidae), *Coleophora sp., Coleophora franki, Coleophora fuscedinella, Coleophora laricella, Coleophora murinipennella, Coleophora pruniella, Coleophora salmani* (Lepidoptera: Coleophoridae), *Elachista adscitella, Elachista argentella, Elachista bisulcella, Elachista littoricola* (Elachistidae), *Coleotechnites piceaella, Exoteleia dodecella, Isophrictis striatella, Metzneria metzneriella* (Gelechiidae), *Cameraria ohridella, Lithocolletis blancardella, Phyllonorycter sp., Phyllonorycter quinqueguttella* (Gracillariidae), *Dendrolimus sibiricus, Macrothylacia rubi, Malacosoma disstria, Malacosoma pluviale* (Lasiocampidae*), Leucoma salicis, Porthetria dispar, Stilpnotia salicis* (Lymantriidae), *Thaumetopoea pityocampa, Thaumetopoea wilkinsoni* (Notodontidae), *Aporia crataegi* (Pieridae), *Eurukuttarus confederata* (Psychidae), *Acrobasis rubrifasciella, Etiella zinckenella* (Pyralidae), *Archippus packardianus, Archips packardiana, Grapholita molesta, Laspeyresia nigricana, Rhyacionia buoliana* (Tortricidae), *Argyresthia sp., Argyresthia laricella, Argyresthia thuiella* (Yponomeutidae), *Zygaena sp., Zygaena trifolii* (Zygaenidae) (Noyes, 2019).

Parasitoid hosts: *Agathis calcarata, Apanteles laricellae, Apanteles melanoscelus, Apanteles solitarius, Bracon sp., Bracon cephi, Bracon intercessor, Macrocentrus peroneae, Rhogas dendrolimi, Rogas sp., Triaspis thoracicus* (Hymenoptera: Braconidae), *Aulacidea tumida* (Cynipidae), *Copidosoma geniculatum* (Encyrtidae), *Dahlbominus fuscipennis, Microplectron fuscipenne* (Eulophidae), *Eupelmus allynii, Eupelmus spongipartus* (Eupelmidae), *Ahtola atra, Eurytoma sp., Eurytoma parva, Tetramesa angustipennis, Tetramesa brevicollis, Tetramesa brevicornis, Tetramesa calamagrostidis, Tetramesa eximia, Tetramesa linearis* (Eurytomidae), *Bathyplectes curculionis, Exenterus amictorius, Glypta rufiscutellaris* (Ichneumonidae), *Platygaster zosine* (Platygastridae), *Dinarmus acutus, Dinarmus laticeps, Homoporus sp., Homoporus luniger, Merisoporus chalcidiphagus* (Pteromalidae), *Ditropinotus aureoviridis, Syntomaspis baudysi* (Torymidae) (Noyes, 2019).

Parasitoids: *Pediobius tarsalis* (Eulophidae) (Noyes, 2019).

Distribution: Algeria, Andorra, Austria, Belgium, Bulgaria, Canada, Caucasus, Croatia, Czech Republic, Czechoslovakia, Denmark, Estonia, Europe, Finland, France, Germany, Greece, Hungary, Iran, Israel, Italy, Kazakhstan, Kirgizia, Macedonia, Middle East, Moldova, Morocco,

Nearctic, Netherlands, North Africa, People's Republic of China, Poland, Romania, Russia, Serbia, Slovakia, Slovenia, Spain, Sweden, Switzerland, Transcaucasus, Tselinograd Obl., Turkey, Turkmenistan, Ukraine, United Kingdom, United States of America, USSR, Yugoslavia (pre-1991) (Noyes, 2019).

Eupelmus vladimiri (Fusu, 2017)
Publications with Turkey records: Fusu (2017), Noyes (2019).
Associates: *Apomyelois ceratoniae* (Lepidoptera: Pyralidae), *Acroptilon repens, Artemisia sp.* (Asteraceae), *Ceratonia siliqua* (Fabaceae), *Rosa canina* (Rosaceae) (Noyes, 2019).
Hosts: *Asphondylia gennadii, Rhopalomyia artemisiae* (Diptera: Cecidomyiidae), *Aulacidea acroptilonica, Diplolepis sp.* (Hymenoptera: Cynipidae) (Noyes, 2019).
Distribution: Albania, Bulgaria, Croatia, Cyprus, Greece, Iran, Montenegro, Turkey (Noyes, 2019).

References

Al Khatib, Fadel, and Fusu, Lucian, and Cruaud, Astrid, and Gibson, Gary A.P., and Borowiec, Nicholas, and Rasplus, Jean-Yves, and Ris, Nicholas, and Delvare, Gerard. 2014. "An integrative approach to the species discrimination in the Euplemus urozonus complex (Hymenoptera, Eupelmidae), with the description of 11 new species from the Western Palaearctic". *Systematic Entomology*, 39: 806–862.

Askew, Richard R., and Melika, Gerotge, and Pujade-Villar, Juli, and Schönrogge, Karsten, and Stone, Graham N., and Nieves-Aldrey, José Luis. 2013. "Catalogue of parasitoids and inquilines in cynipid oak galls in the west Palaearctic". *Zootaxa*, 3643(1):1-133.

Askew, Richard Robinson, and Nieves-Aldrey, Jose-Luis. 2006. "Calosotinae and Neanastatinae in the Iberian peninsula and Canary Islands, with descriptions of new species and a supplementary note on *Brasema* Cameron, 1884 (Hymenoptera, Chalcidoidea, Eupelmidae)". *Graellsia*, 62(1):87-100.

Askew, Richard Robinson, and Plantard, Oliver, and Gómez, José F., and Hernandez Nieves, Maria, and Nieves-Aldrey, José Luis. 2006. "Catalogue of parasitoids and inquilines in galls of Aylacini, Diplolepini and Pediaspidini (Hym., Cynipidae) in West Palaearctic". *Zootaxa*, 1301:7.

Avcı, Mustafa. 2000. "Investigations on structure of egg-batches, parasitism and egg layinh habits of *Thaumetopoea pityocampa* (Den. &

Schiff.) (Lep.: Thaumetopoeidae) in various regions of Turkey". *Turkish Journal of Entomology*, 24(3):167-178.

—. 2009. "Parasitoid complex and new host plants of the gypsy moth, *Lymantria dispar* L. in the Lakes District, Turkey". *Journal of Animal and Veterinary Advances*, 8(7):1402-1405.

Bayram, Şerife, and Ülgentürk, Selma, and Toros, Seval. 1998. "Research on the insects causing galls on dog rose (*Rosa spp.*) and their parasitoids in Ankara province". *Turkish Journal of Entomology*, 22(4):259-268.

Doğanlar, Mikdat, and Sertkaya, Erdal, and & Skuhravá, Marcela. 2011a. "Pest status of olive leaf gall midge *Dasineura oleae* (Angelini, 1831), description of *Lasioptera oleicola* Skuhravá sp. new (Diptera: Cecidomyiidae) and effectiveness of parasitoids on their populations in Hatay, Turkey". *Turkish Journal of Entomology*, 35(2):265-284.

Doğanlar, Mikdat. 2003. "A new species of *Pentacladia* Westwood, 1835 (Hymenoptera, Chalcidoidea, Euplemidae) from Yayladagi, Hatay, Turkey". *Turkish Journal of Entomology*, 27(2):83-90.

—. 2011b. "Parastioid complex of the olive leaf gall midges, Dasineura oleae (Angelini 1831) and *Lasioptera oleicola* Skuhravá, 2011 (Diptera: Cecidomyiidae) in Hatay Turkey, with descriptions of new genus and species from Tetrastichinae (Hymenoptera: Eulophidae)". *Turkish Journal of Entomology*, 35(2):245-264.

—. 2014. "First reports on the natural enemy fauna of the chestnut gall wasp, *Dryocosmus kuriphilus* Yasumatsu (Hymenoptera: Cynipidae) in Yalova, Turkey". *Türkiye Biyolojik Mucadele Dergisi*, 5(1):67-74.

Fusu, Lucian. 2017. "An integrative taxonomic study of European Eupelmus (Macroneura) (Hymenoptera: Chalcidoidea: Eupelmidae), with molecular and cytogenetic analysis of *Eupelmus* (Macroneura) vesicularis: several species hiding under one name for 240 years". *Zoological Journal of the Linnean Society*, 181(3):519-603.

Gençer, Lütfiye. 2003. "The chalcidoid parasitoids of *Diplolepis mayri* Schld. (Hymenoptera: Cynipidae), a pest of *Rosa canina* L. in Sivas province". *Turkish Journal of Entomology*, 27(2):107-111.

Gibson, Gary A.P. & Fusu, Lucian. 2016. "Revision of the Palaearctic species of *Eupelmus* (Eupelmus) Dalman (Hymenoptera: Chalcidoidea: Eupelmidae)". *Zootaxa*, 4081(1):1-331.

Gibson, Gary A.P. 2010. "*Calosota curtis* (Hymenoptera, Chalcidoidea, Eupelmidae) - review of the New World and European fauna including revision of species from the West Indies and Central and North America". *ZooKeys*, 55:1-75.

—. 2011. "The species of *Eupelmus* (Eupelmus) Dalman and Eupelmus (Episolindelia) Girault (Hymenoptera: Eupelmidae) in North America north of Mexico". *Zootaxa*, 2951:1-97.

Korkmaz, Ertan Mahir, and Hastaoğlu Örgen, Sevda, and Gençer, Lütfiye, and Ülgentürk, Selma, and Başıbüyük, Hasan Hüseyin. 2010. "Determination of some wheat pests and their parasitoids in wheat fields of Central Anatolia Region". *Türkiye Entomoloji Dergisi*, 34(3):361-377.

Mete, Özlem, and Mergen, Orhan. 2017. "The community components associated with two common rose gall wasps (Hymenoptera: Cynipidae: Diplolepidini) in Turkey". *Turkish Journal of Zoology*, 41:696-701.

Noyes, John S. 2019. *Universal Chalcidoidea Database*. Available at: www.nhm.ac.uk/chalcidoids (accessed January 2022).

Öncüer, Cezmi. 1991. "A Catalogue of the Parasites and Predators of Insect Pests of Turkey". - *E.Ü.Ziraat Fakültesi Yayınları No.: 505*, Bornova-İzmir: pp. 354.

Özbek, Hikmet, and Çoruh, Saliha. 2010. "Egg parasitoids of *Malacosoma neustria* (Linnaeus, 1758) (Lepidoptera: Lasiocampidae) in Erzurum province of Turkey". *Turkish Journal of Entomology*, 34(4):551-560.

Özbek, Hikmet, and Güçlü, Şaban, and Tozlu, Göksel. 1999. "Biology and natural enemies of *Diplolepis mayri* Schld. (Hymenoptera: Cynipidae), a pest of *Rosa canina* L. in Erzurum Province". *Turkish Journal of Entomology*, 23(1):39-50.

Stringer, Danielle N., and Mantel, Sarah, and Jennings, John T., and Austin, Andrew D. 2012. "Family Aphelinidae", *Australian Centre for Evolutionary Biology and Biodiversity, and the School of Earth and Environmental Science*. The University of Adelaide, Australia (Available from:
https://www.researchgate.net/publication/
262008500_Family_APHELINIDAE).

CHAPTER NINE

FAMILY: EURYTOMIDAE WALKER, 1832

The Eurytomidae is a medium-sized family characterized by the following features: the pronotum is large, rectangular, generally at least half the medial length of the mesoscutum and has prominent anterior corners; the dorsal thorax is generally marked with coarse setiferous punctures; and the metasoma is convex dorsally and well-sclerotised (non-collapsing). Most species in this family are black or dark in color and non-metallic.

The majority of eurytomids are parasitic, although a significant number are phytophagous, feeding on seeds or boring in stems. Parasitic species are often associated with insect larval hosts hidden in plant tissue, such as stem-borers, seed-eaters or gall-formers. Several species have been recorded from spider (Araneae) and cicada (Hemiptera) eggs. Some species display a remarkable mixture of biologies, with their larvae starting development as parasitoids, but completing their developlement by feeding on plant tissue once their host is consumed. Other species are facultative hyperparasitoids, and often have a broad host range.

Eurytomidae is an interesting family within Chalcidoidea because it includes many pest species. For example, the almond wasp, *Eurytoma amygdali*, is a significant pest of almonds in the Palaearctic region, certain Tetramesa species are cereal pests, and some *Bruchophagus* species may be pests of legume crops. However, some of the phytophagous species can be beneficial, such as *Eurytoma attiva*, which has been used successfully for controlling of the weed black sage (*Cordia macrostachia*) in Mauritius. A few Eurytomid parasitoids have been employed in biological control programs, but without notable success (Stringer et al., 2012).

The family currently includes 88 genera and 1424 species placed in three subfamilies. In Turkey, there are two subfamilies, nine genera and 98 species (Noyes, 2019).

Subfamily Eurytominae (Walker, 1832)
Genus *Aximopsis* (Ashmead, 1904)
Aximopsis augasmae (Zerova, 1977)
Publications with Turkey records: Stojanova et al. (2012), Noyes (2019).
Associates: *Atraphaxis spinosa* (Polygonaceae), *Zygophyllum sp.* (Zygophyllaceae) (Noyes, 2019).
Hosts: *Augasma atraphaxidellum* (Lepidoptera: Coleophoridae) (Noyes, 2019).
Distribution: Armenia, Azerbaijan, Bulgaria, Georgia, Turkey, Turkmenistan, USSR (Noyes, 2019).

Aximopsis collina (Zerova, 1984)
Publications with Turkey records: Mete and Mergen (2017), Noyes (2019).
Associates: *Diplolepis fructuum, Diplolepis rosae* (Hymenoptera: Cynipidae), *Rosa sp., Rosa canina* (Rosaceae) (Noyes, 2019).
Hosts: *Diplolepis sp., Diplolepis fructuum, Diplolepis mayri* (Hymenoptera: Cynipidae) (Noyes, 2019).
Distribution: Bulgaria, Iran, Tadzhikistan, Turkey, Turkmenistan, Ukraine, Uzbekistan (Noyes, 2019).

Aximopsis nodularis (Boheman, 1836)
Publications with Turkey records: Cam (2011; 2012), Noyes (2019).
Associates: *Prosopis sp.* (Fabaceae), *Pinus sp.* (Pinaceae), *Rosa sp.* (Rosaceae), *Rubus sp.* (Rubiaceae) (Noyes, 2019).
Hosts: *Pissodes notatus* (Coleoptera: Curculionidae), *Ceratina sp., Ceratina callosa, Heriades sp., Megachile sp., Osmia adunca, Osmia coerulescens, Osmia parvula, Prosopis sp.* (Hymenoptera: Apidae), *Cynips quercusfolii, Diplolepis mayri, Diplolepis rosae* (Cynipidae), *Nitela spinolai, Pemphredon lethifer, Psenulus schencki, Trypoxylon attenuatum, Trypoxylon figulus* (Sphecidae), *Symmorphus debilitatus* (Vespidae) (Noyes, 2019).
Parasitoid hosts: *Xylophurus sp.* (Ichneumonidae) (Noyes, 2019).
Distribution: Belgium, Bosnia Hercegovina, Bulgaria, Croatia, Czech Republic, Europe, Finland, France, Germany, Hungary, Italy, Moldova, Netherlands, Romania, Slovakia, Sweden, Turkey, United Kingdom, USSR, Yugoslavia (pre-1991) (Noyes, 2019).

Genus *Bruchophagus* (Ashmead, 1888)
Bruchophagus astragali (Fedoseeva, 1954)
Publications with Turkey records: Cam (2011; 2012), Noyes (2019), Stojanova *et al.* (2012).
Hosts: *Astragalus sp., Astragalus albicaulis, Astragalus asper, Astragalus caucasica, Astragalus caucasicus, Astragalus cicer, Astragalus davuricus, Astragalus glycyphyllos, Halimodendron halodendron* (Fabaceae) (Noyes, 2019).
Parasitoids: *Spintherus dubius* (Hymenoptera: Pteromalidae) (Noyes, 2019).
Distribution: Bulgaria, Caucasus, Georgia, Germany, Hungary, Iran, Kazakhstan, Mongolia, Romania, Russia, Sweden, Tselinograd Obl., Turkey, Turkmenistan, USSR (Noyes, 2019).

Bruchophagus bajariae (Erdos, 1957)
Catalagues (Journals) with Turkiye records: Cam (2011; 2012), Noyes (2019), Stojanova et al. (2012).
Hosts: *Euphorbia sp*. (Euphorbiaceae) (Noyes, 2019).
Distribution: Europe, France, Hungary, Iran, Turkey, USSR (Noyes, 2019).

Bruchophagus coluteae (Boucek, 1954)
Publications with Turkey records: Cam (2011; 2012), Noyes (2019), Stojanova et al. (2012).
Hosts: *Colutea sp., Colutea arborescens* (Fabaceae) (Noyes, 2019).
Distribution: Bulgaria, Czech Republic, Czechoslovakia, Moldova, Turkey, USSR, Uzbekistan (Noyes, 2019).

Bruchophagus coronillae (Erdelyi and Szelenyi, 1975)
Publications with Turkey records: Cam (2011; 2012), Noyes (2019), Stojanova et al. (2012).
Associates: *Coronilla varia* (Fabaceae) (Noyes, 2019).
Distribution: Hungary, Turkey (Noyes, 2019).

Bruchophagus gibbus (Boheman, 1836)
Publications with Turkey records: Cam (2011; 2012), Lotfalizadeh et al. (2007), Noyes (2019), Stojanova et al. (2012).
Associates: *Astragalus sp., Lotus sp., Medicago sp., Medicago sativa, Trifolium sp., Trifolium incarnatum, Trifolium pratense* (Fabaceae) (Noyes, 2019).

Hosts: *Apion apricans* (Coleoptera: Apionidae), *Asphondylia websteri, Dasineura leguminicola* (Diptera: Cecidomyiidae), *Trachyspermum ammi* (Apiaceae), *Anisolotus puberula, Astragalus sp., Astragalus gouglasii, Lotus sp., Medicago sativa, Melilotus sp., Oxytropis lambertii, Trifolium sp., Trifolium alpestrea, Trifolium pratense* (Fabaceae) (Noyes, 2019).
Parasitoid hosts: *Apanteles glomeratus* (Hymenoptera: Braconidae) (Noyes, 2019).
Parasitoids: *Aprostocetus venustus* (Eulophidae), *Idiomacromerus terebrator, Liodontomerus insuetus, Liodontomerus longfellowi* (Torymidae) (Noyes, 2019).
Distribution: Argentina, Australia, Austria, Azerbaijan, Bulgaria, Canada, Caucasus, Chile, Finland, Germany, Hungary, India, Iran, Iraq, Italy, Nearctic, Neotropical, Netherlands, New Zealand, People's Republic of China, Peru, Romania, Russia, Sweden, Turkey, Ukraine, United Kingdom, United States of America, USSR (Noyes, 2019).

Bruchophagus mutabilis (Nikolskaya, 1952)
Publications with Turkey records: Cam (2011; 2012), Noyes (2019), Stojanova et al. (2012).
Associates: *Astragalus sp., Colutea arborescens* (Fabaceae) (Noyes, 2019).
Distribution: Bulgaria, Caucasus, France, Iran, Kazakhstan, Turkey, USSR (Noyes, 2019).

Bruchophagus platypterus (Walker, 1834)
Publications with Turkey records: Noyes (2019), Stojanova et al. (2012).
Associates: *Apion loti* (Coleoptera: Apionidae), *Aprostocetus bruchophagi* (Hymenoptera: Eulophidae), *Lotus sp., Lotus corniculatus* (Fabaceae) (Noyes, 2019).
Hosts: *Lotus sp., Lotus corniculatus, Lotus hispidus, Lotus rectum, Lotus tenuis, Lotus uliginosus, Melilotus sp., Trifolium sp., Trifolium pratense* (Fabaceae) (Noyes, 2019).
Parasitoids: *Idiomacromerus perplexus, Liodontomerus insuetus, Liodontomerus longfellowi, Liodontomerus perplexus* (Hymenoptera: Torymidae) (Noyes, 2019).
Distribution: Argentina, Brazil, Bulgaria, Canada, Chile, England, France, Georgia, Germany, Ireland, Kazakhstan, Nearctic, Netherlands, Romania, Russia, Serbia, Sweden, Tselinograd Obl., Turkey, Ukraine, United States of America, Uruguay (Noyes, 2019).

Bruchophagus robiniae (Zerova, 1970)
Publications with Turkey records: Cam (2011; 2012), Noyes (2019), Stojanova et al. (2012).
Hosts: Robinia pseudoacacia (Fabaceae) (Noyes, 2019).
Distribution: Armenia, Bulgaria, Caucasus, Iran, Turkey, Ukraine, USSR, Uzbekistan (Noyes, 2019).

Bruchophagus roddi (Gussakovskiy, 1933)
Publications with Turkey records: Cam (2011; 2012), Noyes (2019), Öncüer (1991), Stojanova et al. (2012).
Associates: *Habrocytus medicaginis, Pteromalus sequester* (Pteromalidae), *Liodontomerus perplexus* (Torymidae), *Astragalus sp., Lotus sp., Medicago sp., Medicago arabica, Medicago coerulea, Medicago falcata, Medicago glutinosa, Medicago hispida, Medicago minima, Medicago sativa, Melilotus sp.* (Fabaceae) (Noyes, 2019).
Hosts: *Medicago sativa, Astragalus sp., Lotus sp., Medicago sp., Medicago arabica, Medicago coerulea, Medicago falcata, Medicago glutinosa, Medicago hispida, Medicago sativa, Melilotus sp.* (Fabaceae) (Noyes, 2019).
Parasitoids: *Aprostocetus venustus, Tetrastichus bruchophagi* (Hymenoptera: Eulophidae), *Habrocytus medicaginis* (Pteromalidae), *Idiomacromerus pannonicus, Idiomacromerus perplexus, Liodontomerus insuetus, Liodontomerus perplexus* (Torymidae) (Noyes, 2019).
Distribution: Argentina, Australia, Austria, Bulgaria, Canada, Caucasus, Chile, Czech Republic, Europe, France, Germany, Hungary, India, Iran, Iraq, Israel, Kazakhstan, Mexico, Moldova, Nearctic, Netherlands, New Zealand, Peru, Poland, Romania, Russia, Serbia, Slovakia, Tselinograd Obl., Turkey, Ukraine, United States of America, USSR, Uzbekistan (Noyes, 2019).

Genus *Eurytoma* (Illiger, 1807)
Eurytoma adleriae (Zerova, 1995)
Publications with Turkey records: Askew (2013), Doğanlar (2014), Noyes (2019).
Associates: *Andricus aries, Andricus caliciformis, Andricus caputmedusae, Andricus conglomeratus, Andricus coriarius, Andricus coronatus, Andricus grossulariae, Andricus hystrix, Andricus infectorius, Andricus kollari, Andricus lignicolus, Andricus lucidus, Andricus polycerus, Andricus quercuscalicis, Andricus quercusradicis, Andricus quercusramuli, Andricus seckendorffi, Andricus stefanii, Andricus sternlichti, Andricus tomentosus, Andricus truncicolus, Aphelonyx*

cerricola, Biorhiza pallida, Cynips disticha, Cynips longiventris, Cynips quercus, Cynips quercusfolii, Dryocosmus kuriphilus, Plagiotrochus quercusilicis (Hymenoptera: Cynipidae), *Quercus sp., Quercus frainetto, Quercus infectoria, Quercus pubescens, Quercus robur* (Fagaceae) (Noyes, 2019).
Hosts: *Adleria conglomerate, Cynips conglomerate, Diplolepis fructuum, Dryocosmus kuriphilus* (Hymenoptera: Cynipidae) (Noyes, 2019).
Distribution: Austria, Bulgaria, Croatia, France, Germany, Greece, Hungary, Iran, Italy, Lebanon, Morocco, Slovakia, Spain, Turkey, Ukraine (Noyes, 2019).

Eurytoma aemula (Szelenyi, 1974)
Publications with Turkey records: Cam (2011; 2012), Noyes (2019), Stojanova et al. (2012).
Associates: *Centaurea sp.* (Asteraceae), *Hypecoum imberbe* (Papaveraceae) (Noyes, 2019).
Hosts: *Aylax hypecoi* (Hymenoptera: Cynipidae) (Noyes, 2019).
Distribution: Armenia, Bulgaria, Mongolia, Turkey (Noyes, 2019).

Eurytoma amygdali (Enderlein, 1907)
Publications with Turkey records: Cam (2011; 2012), Noyes (2019), Zerova and Fursov (1991).
Associates: *Pyemotes muraiae* (Acarina: Pyemotidae), *Amygdalus communis, Prunus dulcis* (Rosaceae) (Noyes, 2019).
Hosts: *Prunus dulcis, Amygdalus sp., Amygdalus bucharica, Amygdalus communis, Amygdalus fenzliana, Prunus sp., Prunus armeniaca, Prunus dulcis* (Rosaceae) (Noyes, 2019).
Parasitoids: *Aprostocetus bucculentus* (Hymenoptera: Eulophidae), *Eurytoma armenica, Gugolzia bademia* (Eurytomidae) (Noyes, 2019).
Distribution: Armenia, Azerbaijan, Bosnia Hercegovina, Bulgaria, Caucasus, Cyprus, France, Georgia, Greece, Hungary, Iran, Israel, Jordan, Lebanon, Macedonia, Russia, Syria, Turkey, Ukraine, USSR, Yugoslavia (pre-1991) (Noyes, 2019).

Eurytoma appendigaster (Swederus, 1795)
Publications with Turkey records: Cam (2011; 2012), Noyes (2019), Stojanova et al. (2012).
Associates: *Deschampsia cespitosa* (Poaceae) (Noyes, 2019).
Hosts: *Tetramesa sp.* (Hymenoptera: Eurytomidae), *Hyphantria cunea* (Lepidoptera: Arctiidae), *Coleophora pruniella* (Coleophoridae), *Lymantria dispar, Porthetria dispar* (Lymantriidae), *Blastodacna*

curvilineella (Momphidae), *Antheraea pernyi* (Saturniidae), *Carpocapsa pomonella, Grapholita molesta, Rhyacionia buoliana* (Tortricidae), *Chrysopa sp.* (Neuroptera: Chrysopidae) (Noyes, 2019).
Parasitoid hosts: *Arrhinomyia tragica, Compsilura concinnata* (Diptera: Tachinidae), *Apanteles sp., Apanteles fulvipes, Apanteles glomeratus, Apanteles melanoscelus, Apanteles solitarius, Cotesia melanoscela, Hormius moniliatus, Meteorus versicolor, Microbracon terebella, Microgaster globate, Petalodes unicolor* (Hymenoptera: Braconidae), *Tetramesa petiolata* (Eurytomidae), *Cremastus minor, Trichomma enecator* (Ichneumonidae) (Noyes, 2019).
Parasitoids: *Pediobius deschampsiae* (Eulophidae), *Endromopoda sp.* (Ichneumonidae), *Chlorocytus deschampsiae* (Pteromalidae) (Noyes, 2019).
Distribution: Bulgaria, Canada, Czech Republic, Czechoslovakia, Europe, France, Germany, Hungary, Italy, Moldova, Netherlands, North Africa, People's Republic of China, Poland, Romania, Sweden, Turkey, United Kingdom, United States of America, Yugoslavia (pre-1991) (Noyes, 2019).

Eurytoma aspila (Walker, 1836)
Publications with Turkey records: Askew et al. (2006), Cam (2011; 2012), Noyes (2019), Stojanova et al. (2012).
Associates: *Acroptilon repens, Centaurea sp., Centaurea aspera, Centaurea diffusa, Centaurea nicaeensis, Centaurea scabiosa, Centaurea sphaerocephala, Cichorium intybus, Lactuca viminea, Lapsana communis, Notobasis syriaca, Picris sp., Silybum marianum, Sonchus asper* (Asteraceae) (Noyes, 2019).
Hosts: *Agromyza affinis, Phytomyza affinis* (Diptera: Agromyzidae), *Aulacidea acroptilonica, Aulacidea follioti, Aulacidea freesei, Aulacidea tragopogonis, Aylax sp., Isocolus lichtensteini, Phanacis sp., Phanacis caulicola, Phanacis centaureae, Phanacis zwoelferi, Timaspis cichorii, Timaspis lampsanae, Timaspis phoenixopodos* (Hymenoptera: Cynipidae), *Oecanthus pellucens* (Orthoptera: Gryllidae) (Noyes, 2019).
Distribution: Bulgaria, Caucasus, Croatia, England, Europe, France, Germany, Greece, Hungary, Italy, Kazakhstan, Moldova, Mongolia, Serbia, Spain, Tselinograd Obl., Turkey, Ukraine, USSR (Noyes, 2019).

Eurytoma brevicoxa (Zerova and Cam, 2003)
Publications with Turkey records: Askew et al. (2006), Cam (2011; 2012), Noyes (2019), Stojanova et al. (2012), Zerova et al. (2003).
Associates: Unknown (Noyes, 2019).

Hosts: Unknown (Noyes, 2019).
Distribution: Turkey (Noyes, 2019).
Eurytoma brunniventris (Ratzeburg, 1852)
Publications with Turkey records: Askew et al. (2013), Cam (2011; 2012), Noyes (2019), Stojanova et al. (2012).
Associates: *Andricus amenti, Andricus aries, Andricus askewi, Andricus burgundus, Andricus caliciformis, Andricus callidoma, Andricus caputmedusae, Andricus cecconi, Andricus conglomeratus, Andricus conificus, Andricus coriarius, Andricus coronatus, Andricus corruptrix, Andricus crispator, Andricus curtisii, Andricus curvator, Andricus cydoniae, Andricus fecundatrix, Andricus galeatus, Andricus gallaeurnaeformis, Andricus gemmeus, Andricus glandulae, Andricus glutinosus, Andricus grossulariae, Andricus hartigi, Andricus hispanicus, Andricus hungaricus, Andricus hystrix, Andricus infectorius, Andricus inflator, Andricus kollari, Andricus legitimus, Andricus lignicolus, Andricus lucidus, Andricus malphigii, Andricus megalucidus, Andricus mitratus, Andricus multiplicatus, Andricus paradoxus, Andricus pictus, Andricus polycerus, Andricus pseudoinflator, Andricus quadrilineatus, Andricus quercuscalicis, Andricus quercuscorticis, Andricus quercusradicis, Andricus quercusramuli, Andricus quercustozae, Andricus seckendorffi, Andricus seminationis, Andricus sieboldin, Andricus singularis, Andricus solitarius, Andricus stefanii, Andricus sternlichti, Andricus superfetationis, Andricus testaceipes, Andricus tomentosus, Andricus truncicolus, Andricus vindobonensis, Aphelonyx cerricola, Biorhiza pallida, Callirhytis erythrocephala, Callirhytis glandium, Callirhytis rufescens, Cerroneuroterus lanuginosus, Chilaspis nitida, Cynips agama, Cynips cornifex, Cynips disticha, Cynips divisa, Cynips longiventris, Cynips quercus, Cynips quercusfolii, Dryocosmus cerriphilus, Dryocosmus kuriphilus, Dryocosmus mayri, Neuroterus albipes, Neuroterus anthracinus, Neuroterus numismalis, Neuroterus quercusbaccarum, Neuroterus saliens, Neuroterus tricolor, Plagiotrochus australis, Plagiotrochus burnayi, Plagiotrochus gallaeramulorum, Plagiotrochus panteli, Plagiotrochus quercusilicis, Pseudoneuroterus macropterus, Pseudoneuroterus saliens, Trigonaspis synaspis (Cynipidae), Castanea sp., Castanea crenata, Castanea sativa, Quercu coccifera, Quercus sp., Quercus cerris, Quercus faginea, Quercus frainetto, Quercus ilex, Quercus infectoria, Quercus lusitanica, Quercus pedunculiflora, Quercus petraea, Quercus pubescens, Quercus pyrenaica, Quercus robur* (Fagaceae) (Noyes, 2019).
Hosts: *Andricus sp., Andricus albopunctatus, Andricus argenteus, Andricus caliciformis, Andricus callidoma, Andricus caputmedusae,*

Andricus conglomeratus, Andricus coriarius, Andricus coronatus, Andricus corruptrix, Andricus curvator, Andricus fecundatrix, Andricus galeata, Andricus glutinosus, Andricus grossulariae, Andricus hispanicus, Andricus hungaricus, Andricus infectorius, Andricus inflator, Andricus kollari, Andricus legitimus, Andricus lignicolus, Andricus lucidus, Andricus mitratus, Andricus multiplicatus, Andricus nudus, Andricus ostreus, Andricus paradoxus, Andricus quercuscalicis, Andricus quercusradicis, Andricus quercusramuli, Andricus quercustozae, Andricus seminationis, Andricus seminatorius, Andricus solitarius, Andricus stefanii, Andricus testaceipes, Andricus vindobonensis, Aphelonyx cerricola, Biorhiza sp., Biorhiza pallida, Callirhytis erythrocephala, Chilaspis nitida, Cynips sp., Cynips agama, Cynips disticha, Cynips divisa, Cynips longiventris, Cynips quercus, Cynips quercusfolii, Dryocosmus kuriphilus, Neuroterus sp., Neuroterus albipes, Neuroterus anthracinus, Neuroterus glandiformis, Neuroterus lanuginosus, Neuroterus numismalis, Neuroterus quercusbaccarum, Neuroterus saltans, Neuroterus tricolor, Plagiotrochus australis, Plagiotrochus burnayi, Plagiotrochus quercusilicis, Trigonaspis synaspis (Hymenoptera: Cynipidae), *Caenacis divisa* (Pteromalidae) (Noyes, 2019).
Parasitoid hosts: *Synergus sp., Synergus albipes, Synergus gallaepomiformis, Synergus nervosus, Synergus pallicornis, Synergus reinhardi, Synergus umbraculus* (Cynipidae), *Olynx arsames* (Eulophidae), *Eudecatoma biguttata, Eurytoma brunniventris* (Eurytomidae), *Caenacis divisa, Cecidostiba semifascia, Mesopolobus sp., Mesopolobus tibialis* (Pteromalidae), *Syntomaspis cyanea, Torymus sp., Torymus cingulatus, Torymus cyaneus* (Torymidae) (Noyes, 2019).
Parasitoids: *Mesopolobus fasciiventris* (Pteromalidae), *Torymus auratus, Torymus cingulatus, Torymus nigricornis* (Torymidae) (Noyes, 2019).
Distribution: Andorra, Austria, Belgium, Bosnia Hercegovina, Bulgaria, Caucasus, Croatia, Czech Republic, Czechoslovakia, Denmark, Europe, Finland, France, Germany, Greece, Hungary, Iran, Israel, Italy, Japan, Jordan, Korea, Lebanon, Morocco, Netherlands, People's Republic of China, Portugal, Romania, Serbia, Slovenia, Spain, Sweden, Switzerland, Turkey, Ukraine, United Kingdom, USSR (Noyes, 2019).

Eurytoma centaureae (Claridge, 1960)
Publications with Turkey records: Cam (2011; 2012), Noyes (2019), Stojanova et al. (2012).
Associates: *Eupelmus vesicularis* (Eupelmidae), *Eurytoma punctatella* (Eurytomidae), *Homoporus subniger, Pteromalus hieracia, Pteromalus vibulenus, Stinoplus jenningsi* (Pteromalidae), *Achillea sp., Centaurea sp.,*

Centaurea scabiosa, Cichorium intybus, Cirsium acanthoides, Cirsium vulgare, Crepis sp., Crepis biennis, Picris sp., Sonchus asper (Asteraceae) (Noyes, 2019).
Hosts: *Lixus elongatus* (Coleoptera: Curculionidae), *Melanagromyza sp., Melanagromyza aeneoventris* (Diptera: Agromyzidae), *Aulacidea follioti, Aylax sp., Isocolus sp., Phanacis sp., Phanacis caulicola, Phanacis centaureae, Timaspis cichorii, Timaspis lusitanica* (Hymenoptera: Cynipidae) (Noyes, 2019).
Distribution: Austria, Bulgaria, England, Europe, France, Germany, Spain, Turkey, Ukraine, USSR (Noyes, 2019).

Eurytoma collaris (Walker, 1832)
Publications with Turkey records: Cam (2011; 2012), Noyes (2019), Stojanova et al. (2012).
Associates: *Brachypodium sylvaticum* (Poaceae) (Noyes, 2019).
Hosts: *Tetramesa fulvicollis* (Hymenoptera: Eurytomidae) (Noyes, 2019).
Parasitoid hosts: *Pediobius eubius* (Eulophidae) (Noyes, 2019).
Distribution: Bulgaria, Czech Republic, Europe, France, Germany, Hungary, Kazakhstan, Slovakia, Sweden, Tselinograd Obl., Turkey, United Kingdom (Noyes, 2019).

Eurytoma compressa (Fabricius, 1794)
Publications with Turkey records: Cam (2011; 2012), Noyes (2019), Stojanova et al. (2012).
Associates: *Carduus sp., Centaurea sp., Centaurea calcitrapa, Centaurea iberica, Centaurea scabiosa, Cirsium sp., Cirsium vulgare, Jurinea sp., Onopordum corymbosum* (Asteraceae) (Noyes, 2019).
Hosts: *Chaetorellia sp., Chaetorellia loricate, Chaetostomella onotrophes, Ensina sonchi, Orellia sp., Orellia colon, Tephritis conura, Terellia sp., Urophora sp., Urophora cardui, Urophora cuspidate, Urophora jaceana, Urophora quadrifasciata, Urophora solstitialis, Urophora stylata* (Diptera: Tephritidae), *Isocolus rogenhoferi, Isocolus scabiosae* (Hymenoptera: Cynipidae) (Noyes, 2019).
Parasitoids: *Baryscapus gradwelli* (Eulophidae) (Noyes, 2019).
Distribution: Belgium, Bulgaria, Czech Republic, Europe, France, Germany, Hungary, Iran, Italy, Netherlands, Romania, Russia, Serbia, Slovakia, Spain, Sweden, Turkey, United Kingdom, USSR (Noyes, 2019).

Eurytoma crambeae (Zerova, 1978)
Publications with Turkey records: Cam (2011; 2012), Noyes (2019), Stojanova et al. (2012).

Associates: *Crambe orientalis, Crambe tataria* (Brassicaceae) (Noyes, 2019).
Hosts: *Crambe abyssinica* (Brassicaceae) (Noyes, 2019).
Distribution: Romania, Turkey, Ukraine (Noyes, 2019).

Eurytoma curculionum (Mayr, 1878)
Publications with Turkey records: Cam (2011; 2012), Noyes (2019), Stojanova et al. (2012).
Associates: *Brassica napus* (Brassicaceae), *Rumex sp., Rumex crispus* (Polygonaceae), *Rubus idaeus* (Rubiaceae), *Linaria dalmatica, Linaria genistifolia* (Scrophulariaceae) (Noyes, 2019).
Hosts: *Apion sp., Apion violaceum* (Coleoptera: Apionidae), *Agrilus rubicola* (Buprestidae), *Baris chlorizans, Baris cuprirostris, Baris laticollis, Brachonyx pineti, Ceutorhynchus assimilis, Ceutorhynchus obstrictus, Ceutorhynchus verrucatus, Cleopus campanulae, Gymnetron sp., Gymnetron asellus, Gymnetron campanulae, Lixus juncii, Lixus scabricollis, Lixus truncatulus, Mecinus janthinus, Pissodes validirostris* (Curculionidae), *Lasioptera rubi* (Diptera: Cecidomyiidae) (Noyes, 2019).
Parasitoid hosts: *Bracon intercessor* (Hymenoptera: Braconidae), *Picroscytoides cerasiops* (Pteromalidae) (Noyes, 2019).
Distribution: Austria, Bosnia Hercegovina, Europe, Finland, France, Germany, Hungary, India, Iran, Italy, Macedonia, Morocco, Norway, Poland, Sweden, Turkey, United Kingdom, USSR, Yugoslavia (pre-1991) (Noyes, 2019).

Eurytoma curta (Walker, 1832)
Publications with Turkey records: Cam (2011; 2012), Noyes (2019), Öncüer (1991), Stojanova et al. (2012).
Hosts: *Euribia jaceana, Urophora sp., Urophora jaceana* (Diptera: Tephritidae), *Isocolus gratiosus* (Hymenoptera: Cynipidae), *Carduus sp., Centaurea nigra, Cirsium sp.* (Asteraceae) (Noyes, 2019).
Distribution: Austria, England, Europe, France, Germany, Hungary, Moldova, Netherlands, Sweden, Turkey (Noyes, 2019).

Eurytoma cypriaca Masi, 1934
Publications with Turkey records: Cam (2011; 2012), Noyes (2019).
Associates: *Foeniculum vulgare* (Apiaceae) (Noyes, 2019).
Distribution: Cyprus, Turkey (Noyes, 2019).

Eurytoma dentata (Mayr, 1878)
Publications with Turkey records: Cam (2011; 2012), Noyes (2019), Stojanova et al. (2012).
Associates: *Capparis spinosa* (Capparidaceae), *Atriplex halimus* (Chenopodiaceae), *Astragalus lagurus, Sarothamnus scoparius* (Fabaceae), *Sesamum indicum* (Pedaliaceae), *Zea mays* (Poaceae), *Verbascum sp.* (Scrophulariaceae), *Capsicum sp.* (Solanaceae) (Noyes, 2019).
Hosts: *Asphondylia sp., Asphondylia anatolica, Asphondylia capparis, Asphondylia capsica, Asphondylia coronillae, Asphondylia gennadii, Asphondylia mikii, Asphondylia punica, Asphondylia sarothamni, Asphondylia scrophulariae, Asphondylia sesame, Asphondylia trabuti, Asphondylia verbasci, Contarinia medicaginis, Eumarchalia gennadii* (Diptera: Cecidomyiidae) (Noyes, 2019).
Distribution: Austria, Belgium, Bosnia Hercegovina, Bulgaria, Caucasus, Croatia, Cyprus, Egypt, England, Europe, France, Germany, Greece, Hungary, India, Italy, Philippines, Romania, Sweden, Turkey, USSR (Noyes, 2019).

Eurytoma erdoesi (Szelenyi, 1974)
Publications with Turkey records: Cam (2011; 2012), Noyes (2019).
Associates: *Dactylis glomerata* (Poaceae) (Noyes, 2019).
Hosts: *Tetramesa longula* (Hymenoptera: Eurytomidae) (Noyes, 2019).
Parasitoids: *Mesopolobus graminum* (Pteromalidae) (Noyes, 2019).
Distribution: Bulgaria, Europe, Germany, Hungary, Turkey, Ukraine, United Kingdom, USSR (Noyes, 2019).

Eurytoma flavimana (Boheman, 1836)
Publications with Turkey records: Cam (2011; 2012), Korkmaz et al. (2010), Noyes (2019), Stojanova et al. (2012).
Associates: *Agropyron sp., Agropyron repens, Elymus repens, Triticum sp., Triticum aestivum* (Poaceae) (Noyes, 2019).
Hosts: *Magdalis violacea* (Coleoptera: Curculionidae), *Tetramesa hordei, Tetramesa linearis, Tetramesa tritici* (Hymenoptera: Eurytomidae) (Noyes, 2019).
Parasitoid hosts: *Tetramesa linearis* (Hymenoptera: Eurytomidae) (Noyes, 2019).
Distribution: Bulgaria, Czech Republic, Europe, Finland, France, Germany, Hungary, Netherlands, Romania, Sweden, Turkey, Ukraine, United Kingdom, USSR (Noyes, 2019).
Eurytoma goidanichi (Boucek, 1970)

Publications with Turkey records: Cam (2011; 2012), Noyes (2019), Stojanova et al. (2012).
Hosts: *Callisto denticulella, Phyllonorycter robiniella* (Lepidoptera: Gracillariidae), *Lymantria dispar, Ocneria dispar* (Lymantriidae), *Nycteola asiatica* (Noctuidae), *Cerura vinula* (Notodontidae), *Pieris brassicae* (Pieridae), *Marumba quercus* (Sphingidae), *Apotomis lutozana* (Tortricidae) (Noyes, 2019).
Parasitoid hosts: *Apanteles sp., Apanteles glomeratus, Apanteles orobena, Apanteles porthetriae, Apanteles solitarius, Cotesia glomerata, Meteorus sp., Meteorus pulchricornis, Rhogas sp.* (Hymenoptera: Braconidae), *Angitia sp.* (Ichneumonidae) (Noyes, 2019).
Distribution: Croatia, Czechoslovakia, Europe, Germany, Iran, Italy, Macedonia, Moldova, Poland, Russia, Serbia, Slovakia, Turkey, Turkmenistan, Ukraine, USSR, Yugoslavia (Federal Republic), Yugoslavia (pre-1991) (Noyes, 2019).

Eurytoma harmolitarum (Erdos, 1957)
Publications with Turkey records: Cam (2011; 2012), Noyes (2019), Stojanova et al. (2012).
Associates: *Stipa sp., Stipa capillata* (Poaceae) (Noyes, 2019).
Hosts: *Tetramesa aciculata* (Hymenoptera: Eurytomidae) (Noyes, 2019).
Distribution: Bulgaria, Caucasus, Hungary, Turkey, USSR (Noyes, 2019).

Eurytoma herbaria (Zerova, 1994)
Publications with Turkey records: Noyes (2019), Stojanova et al. (2012).
Associates: *Picris sp., Picris koreana, Tragopogon sp., Tragopogon dubius* (Asteraceae) (Noyes, 2019).
Hosts: Diptera: Tephritidae (Noyes, 2019).
Distribution: Russia, Turkey, Ukraine (Noyes, 2019).

Eurytoma infracta (Mayr, 1904)
Publications with Turkey records: Cam (2011; 2012), Noyes (2019), Stojanova et al. (2012).
Associates: *Cichorium intybus* (Asteraceae), *Phlomis tuberosa, Salvia sp., Salvia lavandulifolia, Salvia nemorosa, Salvia officinalis, Salvia pratensis, Salvia sclarea, Salvia triloba, Salvia verbenaca* (Lamiaceae), *Papaver dubium, Papaver rhoeas* (Papaveraceae) (Noyes, 2019).
Hosts: *Oxyna parietina* (Diptera: Tephritidae), *Aulacidea sp., Aylax minor, Aylax salvia, Barbotinia oraniensis, Hedickiana levantina, Neaylax*

sp., Neaylax salvia, Neaylax verbenacus, Timaspis cichorii (Hymenoptera: Cynipidae) (Noyes, 2019).
Distribution: Caucasus, Croatia, Europe, France, Jordan, Romania, Spain, Turkey, Ukraine, USSR, Yugoslavia (pre-1991) (Noyes, 2019).

Eurytoma koeleriae (Erdos, 1969)
Publications with Turkey records: Noyes (2019), Stojanova et al. (2012).
Associates: *Koeleria dactyloides, Koeleria glauca* (Poaceae) (Noyes, 2019).
Hosts: *Tetramesa schlechtendali* (Hymenoptera: Eurytomidae) (Noyes, 2019).
Distribution: Bulgaria, Hungary, Turkey (Noyes, 2019).

Eurytoma laserpitii (Mayr, 1878)
Publications with Turkey records: Cam (2011; 2012), Noyes (2019), Stojanova et al. (2012).
Associates: *Daucus carota, Eryngium campestre, Petroselinum sativum* (Apiaceae) (Noyes, 2019).
Hosts: *Lasioptera carophila, Lasioptera eryngii* (Diptera: Cecidomyiidae) (Noyes, 2019).
Distribution: Austria, Bulgaria, England, Europe, Hungary, Romania, Transcaucasus, Turkey, United Kingdom, USSR (Noyes, 2019).

Eurytoma lobopterae (Erdos, 1964)
Publications with Turkey records: Cam (2011; 2012), Noyes (2019), Öncüer (1991), Stojanova et al. (2012).
Hosts: *Loboptera decipiens, Phyllodromia decipiens* (Dictyoptera: Blattellidae) (Noyes, 2019).
Distribution: Croatia, France, Spain, Turkey (Noyes, 2019).

Eurytoma martellii (Domenichini, 1960)
Publications with Turkey records: Cam (2011; 2012), Noyes (2019), Stojanova et al. (2012).
Associates: *Olea europaea* (Oleaceae) (Noyes, 2019).
Hosts: *Bactrocera oleae, Dacus oleae* (Diptera: Tephritidae) (Noyes, 2019).
Distribution: Egypt, France, Greece, Italy, North Africa, Spain, Syria, Turkey (Noyes, 2019).

Eurytoma mayri (Ashmead, 1887)
Publications with Turkey records: Cam (2011; 2012), Noyes (2019), Stojanova et al. (2012).
Associates: *Potentilla argentea* (Rosaceae), *Rubus sp., Rubus caesius, Rubus fruticosus, Rubus idaeus* (Rubiaceae) (Noyes, 2019).
Hosts: *Diastrophus mayri, Diastrophus rubi* (Hymenoptera: Cynipidae) (Noyes, 2019).
Distribution: Austria, Belgium, Croatia, Czech Republic, Europe, Finland, France, Germany, Hungary, Italy, Netherlands, Poland, Romania, Spain, Sweden, Turkey, Ukraine, United Kingdom, USSR (Noyes, 2019).

Eurytoma morio (Boheman, 1836)
Publications with Turkey records: Cam (2011; 2012), Noyes (2019), Stojanova et al. (2012).
Associates: *Thuja occidentalis* (Cupressaceae), *Olea europaea* (Oleaceae), *Picea sp., Picea abies, Pinus sp.* (Pinaceae), *Malus sp., Malus pumila, Malus sylvestris, Prunus sp., Prunus armeniaca, Prunus avium, Prunus glandulosa, Prunus persica, Prunus pisardii, Pyrus sp., Pyrus bretschneideri, Pyrus malus* (Rosaceae), *Ulmus carpinifolia, Ulmus pumila* (Ulmaceae) (Noyes, 2019).
Hosts: *Pogonochaerus hispidus, Pogonocherus hispidus, Tetrops praeusta* (Coleoptera: Cerambycidae), *Magdalis armigera, Magdalis frontalis, Magdalis violacea, Pissodes notatus* (Curculionidae), *Blastophagus minor, Carphoborus minimus, Chaetoptelius vestitus, Dryocoetes autographus, Hylesinus crenatus, Hylesinus fraxini, Hylesinus oleiperda, Hylesinus orni, Hylesinus varius, Ips acuminatus, Ips amitinus, Ips chalcographus, Ips typographus, Leperesinus varius, Orthotomicus erosus, Phloeosinus sp., Phloeosinus armatus,Phloeosinus aubei, Phloeosinus thujae, Phloeotribus scarabaeoides, Pityogenes bidentatus, Pityogenes calcaratus, Pityogenes chalcographus, Pityogenes quadridens, Polygraphus subopacus, Scolytus sp., Scolytus carpini, Scolytus ensifer, Scolytus intricatus, Scolytus japonicus, Scolytus laevis, Scolytus mali, Scolytus multistriatus, Scolytus nitidus, Scolytus pygmaeus, Scolytus rugulosus, Scolytus schevyrewi, Scolytus seulensis, Xyleborus dispar* (Scolytidae), *Exoteleia dodecella* (Lepidoptera: Gelechiidae), *Aporia crataegi* (Pieridae) (Noyes, 2019).
Parasitoid hosts: *Coeloides bostrychorum, Coeloides filiformis, Coeloides melanotus* (Hymenoptera: Braconidae), *Cheiropachus quadrum* (Pteromalidae) (Noyes, 2019).
Distribution: Austria, Belgium, Croatia, Czech Republic, Czechoslovakia, Denmark, England, Europe, Finland, France, Germany, Hungary, India,

Iran, Israel, Italy, Kazakhstan, Lebanon, Lithuania, Moldova, Mongolia, Morocco, Netherlands, Norway, People's Republic of China, Poland, Romania, Russia, Serbia, Slovakia, Spain, Sweden, Switzerland, Tselinograd Obl., Tunisia, Turkey, Turkmenistan, Ukraine, USSR, Yemen (Noyes, 2019).

Eurytoma ochraceipes (Kalina, 1970)
Publications with Turkey records: Cam (2011; 2012), Noyes (2019), Stojanova et al. (2012).
Associates: *Ceratoides papposa* (Chenopodiaceae) (Noyes, 2019).
Distribution: Bulgaria, Croatia, Europe, France, Israel, Turkey, USSR, Vietnam, Yemen, Yugoslavia (pre-1991) (Noyes, 2019).

Eurytoma onobrychidis (Nikolskaya, 1933)
Publications with Turkey records: Cam (2011; 2012), Noyes (2019), Stojanova et al. (2012).
Associates: *Habrocytus medicaginis* (Pteromalidae), *Onobrychis sativa, Onobrychis viciifolia* (Fabaceae) (Noyes, 2019).
Hosts: *Onobrychis sp., Onobrychis viciaefolia, Onobrychis viciifolia* (Fabaceae) (Noyes, 2019).
Parasitoids: *Idiomacromerus perplexus* (Hymenoptera: Torymidae) (Noyes, 2019).
Distribution: Bulgaria, Canada, Caucasus, Czechoslovakia, Europe, Germany, Hungary, Kirgizia, Nearctic, Sweden, Turkey, Ukraine, USSR (Noyes, 2019).

Eurytoma oophaga (Silvestri, 1920)
Publications with Turkey records: Cam (2011; 2012), Noyes (2019).
Associates: *Eryngium sp.* (Apiaceae), *Centaurea sp., Cirsium sp., Inula sp., Solidago sp.* (Asteraceae), *Corylus avellana* (Betulaceae), *Hypericum sp.* (Clusiaceae), *Cephalaria uralensis* (Dipsacaceae), *Melissa officinalis, Spartium junceum* (Fabaceae), *Mentha sp.* (Lamiaceae), *Althaea sp.* (Malvaceae), *Rumex sp.* (Polygonaceae), *Spiraea sp.* (Rosaceae), *Verbena sp.* (Verbenaceae) (Noyes, 2019).
Hosts: *Oecanthus pellucens* (Orthoptera: Gryllidae) (Noyes, 2019).
Distribution: Caucasus, Europe, France, Germany, Italy, Romania, Serbia, Transcaucasus, Turkey (Noyes, 2019).

Eurytoma phalaridis (Graham, 1974)
Publications with Turkey records: Cam (2011; 2012), Noyes (2019).
Associates: *Phalaris arundinacea* (Poaceae) (Noyes, 2019).

Parasitoid hosts: *Tetramesa longicornis* (Hymenoptera: Eurytomidae) (Noyes, 2019).
Parasitoids: *Pediobius phalaridis* (Eulophidae), *Mesopolobus graminum* (Pteromalidae) (Noyes, 2019).
Distribution: England, Germany, Netherlands, Sweden, Turkey (Noyes, 2019).

Eurytoma pistaciae (Rondani, 1877)
Publications with Turkey records: Askew et al. (2006), Cam (2011; 2012), Gencer (2003), Mete and Mergen (2017), Noyes (2019), Ozbek et al. (1999), Stojanova et al. (2012).
Associates: *Diplolepis mayri, Neuroterus macropterus* (Cynipidae), *Gugolzia oezdenoerneki* (Pteromalidae), *Megastigmus pistaciae* (Torymidae), *Pistacia sp., Pistacia mutica, Pistacia palaestina, Pistacia terebinthus, Pistacia vera* (Anacardiaceae), *Hieracium virosum* (Asteraceae), *Quercus sp., Quercus cerris* (Fagaceae), *Rosa sp., Rosa canina* (Rosaceae) (Noyes, 2019).
Hosts: *Andricus aestivalis, Andricus conglomeratus, Andricus grossulariae, Andricus kollari, Andricus vindobonensis, Aphelonyx cerricola, Aulacidea hieracia, Biorhiza pallida, Diplolepis mayri, Diplolepis rosae, Dryocosmus kuriphilus* (Hymenoptera: Cynipidae), *Megastigmus pistaciae* (Torymidae) (Noyes, 2019).
Parasitoid hosts: *Synergus pallicornis, Synergus umbraculus* (Cynipidae), *Torymus sinensis* (Torymidae) (Noyes, 2019).
Distribution: Austria, Bulgaria, Croatia, England, France, Germany, Hungary, Iran, Italy, Japan, Korea, Romania, Russia, Spain, Switzerland, Turkey, Ukraine, USSR (Noyes, 2019).

Eurytoma plotnikovi (Nikolskaya, 1934)
Publications with Turkey records: Cam (2012), Doğanlar and Doğanlar (2010), Noyes (2019), Stojanova et al. (2012).
Associates: *Pistacia vera* (Anacardiaceae) (Noyes, 2019).
Hosts: *Pistacia chinensis, Pistacia palaestina, Pistacia vera* (Anacardiaceae) (Noyes, 2019).
Parasitoids: *Gugolzia karadagae* (Hymenoptera: Pteromalidae) (Noyes, 2019).
Distribution: Greece, Iran, Israel, Italy, Kirgizia, People's Republic of China, Tadzhikistan, Tunisia, Turkey, Turkmenistan (Noyes, 2019).

Eurytoma pollux (Claridge, 1959)
Publications with Turkey records: Cam (2011; 2012), Noyes (2019).

Associates: *Ammophila arenaria, Calamagrostis sp., Calamagrostis epigeios* (Poaceae) (Noyes, 2019).
Hosts: *Tetramesa calamagrostidis, Tetramesa eximia* (Hymenoptera: Eurytomidae) (Noyes, 2019).
Parasitoid hosts: *Pediobius calamagrostidis* (Eulophidae), *Tetramesa calamagrostidis* (Eurytomidae) (Noyes, 2019).
Parasitoids: *Mesopolobus graminum* (Pteromalidae) (Noyes, 2019).
Distribution: Bulgaria, England, Europe, Germany, Hungary, Sweden, Turkey, United Kingdom, USSR (Noyes, 2019).

Eurytoma robusta (Mayr, 1878)
Publications with Turkey records: Cam (2012), Noyes (2019), Öncüer (1991), Stojanova et al. (2012).
Associates: *Carduus sp., Centaurea sp., Centaurea iberica, Centaurea scabiosa, Cirsium sp., Cirsium arvense, Cirsium lanceolatum, Cirsium vulgare, Crepis conyzifolia, Echinops sp., Jurinea sp., Saussurea sp., Senecio sp.* (Asteraceae), *Medicago sativa* (Fabaceae), *Papaver dubium, Papaver rhoeas* (Papaveraceae) (Noyes, 2019).
Hosts: *Rhinocyllus conicus* (Coleoptera: Curculionidae), *Chastorellia jaceae, Euribia sp., Euribia cardui, Euribia jaceana, Ptiloedaspis tavaresaiana, Tephritis conycifoliae, Terellia virens, Trypeta sp., Urophora sp., Urophora affinis, Urophora cardui, Urophora cuspidate, Urophora jaceana, Urophora quadrifasciata, Urophora solstitialis, Urophora stylata* (Diptera: Tephritidae), *Aphelonyx cerricola, Aylax sp., Aylax papaveris, Aylax rhoeadis, Isocolus scabiosae* (Hymenoptera: Cynipidae), *Evetria buoliana* (Lepidoptera: Tortricidae) (Noyes, 2019).
Parasitoids: *Baryscapus gradwelli* (Hymenoptera: Eulophidae) (Noyes, 2019).
Distribution: Austria, Bulgaria, Czech Republic, Denmark, Europe, France, Germany, Greece, Hungary, Iran, Italy, Moldova, Netherlands, Palearctic, Romania, Serbia, Slovakia, Spain, Sweden, Switzerland, Turkey, United Kingdom, USSR (Noyes, 2019).

Eurytoma rosae (Nees, 1834)
Publications with Turkey records: Bayram et al. (1998), Cam (2011; 2012), Gencer (2003), Mete and Mergen (2017), Noyes (2019), Öncüer (1991), Ozbek et al. (1999), Stojanova et al. (2012).
Associates: *Dasineura rosarum* (Diptera: Cecidomyiidae), *Andricus kollari, Diplolepis eglanteriae, Diplolepis mayri, Diplolepis rosae* (Hymenoptera: Cynipidae), *Sigmophora brevicornis* (Eulophidae), *Eupelmus urozonus* (Eupelmidae), *Pteromalus bedeguaris* (Pteromalidae),

Acer sp. (Aceraceae), *Centaurea sp.* (Asteraceae), *Castanea sp., Quercus sp.* (Fagaceae), *Olea europaea* (Oleaceae), *Rosa sp., Rosa canina, Rosa damascene, Rosa pimpinellifolia* (Rosaceae), *Rubus idaeus* (Rubiaceae) (Noyes, 2019).

Hosts: *Agrilus rubicola* (Coleoptera: Buprestidae), *Scolytus mali* (Scolytidae), *Lasioptera rubi, Oligotrophus lyciicola* (Diptera: Cecidomyiidae), *Dacus oleae, Myopites limbardae* (Tephritidae), *Diplolepis sp., Diplolepis centifoliae, Diplolepis eglanteriae, Diplolepis fructuum, Diplolepis mayri, Diplolepis nervosa, Diplolepis nervulosa, Diplolepis rosae, Diplolepis rosarum, Diplolepis spinosissimae, Dryocosmus kuriphilus, Rhodites rosae, Synergus pallicornis, Synergus reinhardi, Timaspis lampsanae* (Hymenoptera: Cynipidae), *Hyphantria cunea* (Lepidoptera: Arctiidae), *Anthophila nemorana* (Glyphipterygidae), *Clysia ambiguella, Polychrosis botrana* (Tortricidae) (Noyes, 2019).

Parasitoid hosts: *Periclistus sp., Periclistus brandtii, Periclistus caninae, Periclistus spinosissimae* (Hymenoptera: Cynipidae) (Noyes, 2019).

Distribution: Andorra, Argentina, Armenia, Austria, Belgium, Bulgaria, Caucasus, Czech Republic, Czechoslovakia, England, Europe, Finland, France, Germany, Greece, Hungary, Iran, Italy, Kazakhstan, Malaysia, Netherlands, North Africa, Palearctic, People's Republic of China, Poland, Romania, Serbia, Slovakia, Spain, Sweden, Switzerland, Tselinograd Obl., Turkey, USSR, Yugoslavia (Federal Republic), Yugoslavia (pre-1991) (Noyes, 2019).

Eurytoma rufipes (Walker, 1832)

Publications with Turkey records: Cam (2011; 2012), Noyes (2019), Stojanova et al. (2012).

Associates: *Potentilla sp., Potentilla erecta, Potentilla reptans* (Rosaceae) (Noyes, 2019).

Hosts: *Blastophagus piniperda, Hylurgops palliatus* (Coleoptera: Scolytidae), *Dacus oleae* (Diptera: Tephritidae), *Xestophanes brevitarsis, Xestophanes potentillae* (Hymenoptera: Cynipidae) (Noyes, 2019).

Parasitoids: *Habrocytus isarchus* (Pteromalidae) (Noyes, 2019).

Distribution: Austria, Czech Republic, Czechoslovakia, Europe, Finland, France, Germany, Hungary, Ireland, Italy, Poland, Spain, Sweden, Switzerland, Turkey, Ukraine, United Kingdom, USSR (Noyes, 2019).

Eurytoma schreineri (Schreiner, 1908)

Publications with Turkey records: Cam (2011; 2012), Noyes (2019), Ozbek et al. (1996), Stojanova et al. (2012).

Associates: *Prunus armeniaca* (Rosaceae) (Noyes, 2019).

Hosts: *Armeniaca vulgaris, Cerasus avium, Cerasus vulgaris, Prunus sp., Prunus armeniaca, Prunus divaricate, Prunus domestica, Prunus institia, Prunus spinosa* (Rosaceae) (Noyes, 2019).
Parasitoids: *Syntomaspis eurytomae* (Hymenoptera: Torymidae) (Noyes, 2019).
Distribution: Armenia, Georgia, Greece, Moldova, Romania, Russia, Turkey, Ukraine, USSR (Noyes, 2019).

Eurytoma serratulae (Fabricius, 1798)
Publications with Turkey records: Cam (2011; 2012), Noyes (2019), Öncüer (1991), Stojanova et al. (2012).
Associates: *Cirsium arvense* (Asteraceae), *Medicago sativa* (Fabaceae) (Noyes, 2019).
Hosts: *Lixus juncii* (Coleoptera: Curculionidae), *Myopites frauenfeldi, Tephritis stictica, Urophora sp., Urophora cardui* (Diptera: Tephritidae) (Noyes, 2019).
Parasitoid hosts: *Bracon intercessor* (Hymenoptera: Braconidae), *Picroscytoides cerasiops* (Pteromalidae) (Noyes, 2019).
Distribution: Austria, Bulgaria, Czech Republic, England, Europe, France, Germany, Hungary, Italy, Moldova, Morocco, Netherlands, Sweden, Turkey, USSR (Noyes, 2019).

Eurytoma setigera (Mayr, 1878)
Publications with Turkey records: Askew et al. (2013), Noyes (2019), Öncüer (1991).
Associates: *Andricus amblycerus, Andricus caliciformis, Andricus caputmedusae, Andricus conglomeratus, Andricus conificus, Andricus coriarius, Andricus corruptrix, Andricus curvator, Andricus cydoniae, Andricus fecundatrix, Andricus gallaeurnaeformis, Andricus grossulariae, Andricus hartigi, Andricus hispanicus, Andricus kollari, Andricus lignicolus, Andricus lucidus, Andricus multiplicatus, Andricus polycerus, Andricus pseudoinflator, Andricus schroeckingeri, Andricus seckendorffi, Andricus singularis, Andricus solitarius, Andricus vindobonensis, Aphelonyx cerricola, Biorhiza pallida, Chilaspis nitida, Cynips agama, Cynips cornifex, Cynips disticha, Cynips divisa, Cynips quercus, Cynips quercusfolii, Dryocosmus cerriphilus, Dryocosmus kuriphilus, Neuroterus numismalis, Neuroterus quercusbaccarum, Neuroterus tricolor, Plagiotrochus australis, Plagiotrochus panteli, Pseudoneuroterus macropterus, Pseudoneuroterus saliens* (Cynipidae), *Castanea sp., Castanea sativa, Quercu coccifera, Quercus cerris, Quercus ilex, Quercus*

infectoria, Quercus lusitanica, Quercus petraea, Quercus pubescens, Quercus robur (Fagaceae) (Noyes, 2019).
Hosts: *Andricus grossulariae, Biorhiza pallida, Diplolepis mayri, Dryocosmus kuriphilus, Synergus gallaepomiformis, Synergus pallicornis, Synergus reinhardi, Synergus umbraculus* (Hymenoptera: Cynipidae), *Megastigmus pistaciae* (Torymidae) (Noyes, 2019).
Parasitoid hosts: *Torymus beneficus* (Torymidae) (Noyes, 2019).
Distribution: Austria, Croatia, France, Germany, Hungary, Italy, Japan, Korea, People's Republic of China, Romania, Russia, Spain, Switzerland, Turkey, Ukraine, USSR (Noyes, 2019).

Eurytoma strigifrons (Thomson, 1876)
Publications with Turkey records: Cam (2011; 2012), Noyes (2019), Stojanova et al. (2012).
Associates: *Centaurea sp., Centaurea aspera, Centaurea iberica, Centaurea jacea, Centaurea melitensis, Centaurea nicaeensis, Centaurea nigra, Centaurea sadleriana, Centaurea scabiosa, Centaurea sphaerocephala, Leuzea conifera* (Asteraceae), *Salvia triloba* (Lamiaceae) (Noyes, 2019).
Hosts: *Aylax sp., Isocolus sp., Isocolus jaceae, Isocolus leuzeae, Isocolus lichtensteini, Isocolus rogenhoferi, Isocolus scabiosae, Rhodus cyprius* (Hymenoptera: Cynipidae), *Mamestra brassicae* (Lepidoptera: Noctuidae), *Aporia crataegi* (Pieridae), *Archips rosanus, Tortrix viridana* (Tortricidae), *Pieris brassicae, Pieris rapae* (Pieridae) (Noyes, 2019).
Parasitoid hosts: *Apanteles sp., Apanteles falcatus, Apanteles glomeratus, Apanteles nigerrimus, Apanteles rubecula, Apanteles spurius, Macrocentrus thoracicus* (Hymenoptera: Braconidae), *Anilastus ebeninus, Hyposoter ebeninus, Itoplectis maculator* (Ichneumonidae) (Noyes, 2019).
Distribution: Andorra, Bulgaria, Czech Republic, Europe, Finland, France, Germany, Greece, Hungary, Israel, Italy, Moldova, Romania, Russia, Spain, Sweden, Turkey, Ukraine, United Kingdom, USSR (Noyes, 2019).

Eurytoma striolata (Ratzeburg, 1848)
Publications with Turkey records: Cam (2011; 2012), Narendran et al. (1995), Noyes (2019), Stojanova et al. (2012).
Associates: *Olea europaea* (Oleaceae), *Prunus sp.* (Rosaceae) (Noyes, 2019).
Hosts: *Chaetoptelius vestitus, Chetoptelius vestitus, Hylesinus fraxini, Hylesinus oleiperda, Hylesinus vestitus, Phloeotribus scarabaeoides, Scolytus amygdali, Scolytus multistriatus, Scolytus rugulosus, Scolytus scolytus* (Coleoptera: Scolytidae) (Noyes, 2019).
Distribution: Germany, Italy, Turkey (Noyes, 2019).

Eurytoma terebinthi (Rondani, 1877)
Publications with Turkey records: Cam (2011; 2012), Noyes (2019).
Associates: *Pistacia terebinthus* (Anacardiaceae) (Noyes, 2019).
Distribution: Bulgaria, Italy, Turkey (Noyes, 2019).

Eurytoma tokatensis (Doğanlar and Çam, 1991)
Publications with Turkey records: Doğanlar and Cam (1991), Noyes (2019).
Associates: Unknown (Noyes, 2019).
Hosts: Unknown (Noyes, 2019).
Distribution: Turkey (Noyes, 2019).

Eurytoma turkezia (Zerova and Cam, 2003)
Publications with Turkey records: Cam (2011), Noyes (2019), Stojanova et al. (2012), Zerova et al. (2003).
Associates: *Verbascum sp.* (Scrophulariaceae) (Noyes, 2019).
Distribution: Turkey (Noyes, 2019).

Eurytoma verticillata (Fabricius, 1798)
Publications with Turkey records: Cam (2011; 2012), Noyes (2019), Stojanova et al. (2012).
Associates: *Gelis tenellus* (Ichneumonidae), *Brassica sp., Brassica oleracea* (Brassicaceae), *Medicago sativa* (Fabaceae), *Picea pungens* (Pinaceae), *Oryza sativa* (Poaceae), *Populus sp.* (Salicaceae) (Noyes, 2019).
Hosts: *Parthenolecanium corni* (Hemiptera: Coccidae), *Diprion pini* (Hymenoptera: Diprionidae), *Hyphantria cunea* (Lepidoptera: Arctiidae), *Coleophora pruniella, Coleophora serratella* (Coleophoridae), *Coleotechnites piceaella*, *Exoteleia dodecella* (Gelechiidae), *Parnara guttata* (Hesperiidae), *Dendrolimus spectabilis* (Lasiocampidae), *Leucoma salicis, Lymantria dispar, Lymantria monachal, Nygmia phaeorrhoea, Ocneria dispar, Porthetria dispar, Stilpnotia salicis* (Lymantriidae), *Blastodacna curvilineella* (Momphidae), *Sesamia inferens* (Noctuidae), *Thaumetopoea processionea* (Notodontidae), *Aporia crataegi, Pieris brassicae* (Pieridae), *Luffia lapidella* (Psychidae), *Etiella zinckenella* (Pyralidae), *Synanthedon tipuliformis* (Sesiidae), *Archips rosanus* (Tortricidae), *Grapholita funebrana, Grapholita molesta, Rhyacionia buoliana, Plutella xylostella, Yponomeuta evonymellus* (Yponomeutidae), *Chrysopa sp.* (Neuroptera: Chrysopidae) (Noyes, 2019).
Parasitoid hosts: *Compsilura concinnata* (Diptera: Tachinidae), *Apanteles sp., Apanteles baoris, Apanteles glomeratus, Apanteles*

lacteicolor, Apanteles liparidis, Apanteles melanoscelus, Apanteles pieridis, Apanteles porthetriae, Apanteles solitarius, Bracon onukii, Bracon piger, Cotesia plutellae, Hormius moniliatus, Macrocentrus sp., Macrocentrus ancylivorus, Macrocentrus instabilis, Meteorus sp., Meteorus pulchricornis, Meteorus versicolor, Microgaster globate, Microgaster meridiana, Phanerotoma grapholithae, Rhogas sp. (Hymenoptera: Braconidae), *Ephialtes sp., Eriborus molestee, Glypta sp., Glypta rufiscutellaris, Mastrus molestee, Phobocampe disparis, Pristomerus vulnerator* (Ichneumonidae) (Noyes, 2019).
Distribution: Austria, Bulgaria, Canada, Croatia, Czech Republic, Czechoslovakia, Denmark, Europe, France, Germany, Hungary, Italy, Japan, Korea, Macedonia, Moldova, Nearctic, Netherlands, People's Republic of China, Poland, Romania, Russia, Serbia, Slovakia, Sweden, Switzerland, Turkey, Ukraine, United Kingdom, United States of America, USSR, Yugoslavia (pre-1991) (Noyes, 2019).

Genus *Exeurytoma* (Burks, 1971)
Exeurytoma anatolica (Cam, 1998)
Publications with Turkey records: Cam (1998; 2011; 2012), Doğanlar and Çam (2005), Noyes (2019), Stojanova et al. (2012).
Hosts: *Astragalus lagurus* (Fabaceae) (Noyes, 2019).
Distribution: Turkey (Noyes, 2019).

Exeurytoma kebanensis (Doğanlar, 2005)
Publications with Turkey records: Cam (2011; 2012), Doğanlar and Çam (2005), Noyes (2019), Stojanova et al. (2012).
Associates: *Amygdalus communis* (Rosaceae) (Noyes, 2019).
Distribution: Turkey (Noyes, 2019).

Genus *Macrorileya* (Ashmead, 1900)
Macrorileya femorata (Boucek, 1952)
Publications with Turkey records: Cam (2011; 2012), Doğanlar and Çam (2005), Noyes (2019), Öncüer (1991), Stojanova et al. (2012).
Associates: Unknown (Noyes, 2019).
Hosts: Unknown (Noyes, 2019).
Distrubution: Europe, Hungary, Turkey, USSR (Noyes, 2019).

Macrorileya inopinata (Silvestri, 1920)
Publications with Turkey records: Cam (2011; 2012), Doğanlar and Çam (2005), Noyes (2019), Öncüer (1991), Stojanova et al. (2012).

Associates: *Eryngium sp.* (Apiaceae), *Centaurea sp., Inula sp.* (Asteraceae), *Melissa sp.* (Fabaceae), *Clinopodium sp., Mentha sp., Origanum vulgare, Satureja sp.* (Lamiaceae), *Andropogon halepensis* (Poaceae), *Verbena sp*. (Verbenaceae) (Noyes, 2019).

Hosts: *Cicada orni, Cicada plebeja, Lyristes gemelli, Lyristes plebeja, Tettigia orni* (Hemiptera: Cicadidae), *Oecanthus pellucens* (Orthoptera: Gryllidae) (Noyes, 2019).

Distribution: Bulgaria, Caucasus, Croatia, Czech Republic, Czechoslovakia, Europe, Georgia, Hungary, Italy, Kazakhstan, Mexico, Moldova, North Africa, Palearctic, Romania, Serbia, Slovakia, Spain, Transcaucasus, Tselinograd Obl., Tunisia, Turkey, United States of America, USSR (Noyes, 2019).

Genus *Sycophila* (Walker, 1871)

Sycophila biguttata (Swederus, 1795)

Publications with Turkey records: Askew et al. (2013), Cam (2011; 2012), Doğanlar and Çam (2005), Gencer (2003), Mete and Mergen (2017), Noyes (2019), Öncüer (1991), Stojanova et al. (2012).

Associates: *Andricus amblycerus, Andricus aries, Andricus askewi, Andricus burgundus, Andricus caliciformis, Andricus caputmedusae, Andricus clementinae, Andricus conglomeratus, Andricus coriarius, Andricus coronatus, Andricus corruptrix, Andricus crispator, Andricus curvator, Andricus cydoniae, Andricus dentimitratus, Andricus fecundatrix, Andricus floridus, Andricus galeatus, Andricus gallaeurnaeformis, Andricus gemmeus, Andricus glandulae, Andricus glutinosus, Andricus grossulariae, Andricus hispanicus, Andricus hungaricus, Andricus hystrix, Andricus infectorius, Andricus inflator, Andricus kollari, Andricus lignicolus, Andricus lucidus, Andricus mitratus, Andricus multiplicatus, Andricus polycerus, Andricus pseudoinflator, Andricus quadrilineatus, Andricus quercuscalicis, Andricus quercusramuli, Andricus quercustozae, Andricus seckendorffi, Andricus singularis, Andricus solitarius, Andricus sternlichti, Andricus testaceipes, Andricus vindobonensis, Aphelonyx cerricola, Biorhiza pallida, Callirhytis glandium, Cerroneuroterus lanuginosus, Chilaspis nitida, Cynips agama, Cynips cornifex, Cynips disticha, Cynips divisa, Cynips longiventris, Cynips quercus, Cynips quercusfolii, Dryocosmus cerriphilus, Dryocosmus kuriphilus, Neuroterus albipes, Neuroterus anthracinus, Neuroterus numismalis, Neuroterus quercusbaccarum, Neuroterus saliens, Neuroterus tricolor, Plagiotrochus australis, Plagiotrochus quercusilicis, Pseudoneuroterus macropterus, Pseudoneuroterus saliens, Synophrus olivieri, Synophrus politus, Trigonaspis Megaptera, Trigonaspis mendesi,*

Trigonaspis synaspis (Cynipidae), *Aulogymnus trilineatus* (Eulophidae), *Pistacia sp.* (Anacardiaceae), *Quercu coccifera, Quercus sp., Quercus brantii, Quercus cerrioides, Quercus cerris, Quercus faginea, Quercus frainetto, Quercus ilex, Quercus infectoria, Quercus ithaburensis, Quercus lusitanica, Quercus pedunculiflora, Quercus petraea, Quercus pubescens, Quercus pyrenaica, Quercus robur, Quercus suber* (Fagaceae), *Rosa sp., Rosa canina* (Rosaceae), *Rubus caesius, Rubus idaeus* (Rubiaceae) (Noyes, 2019).
Hosts: *Andricus sp., Andricus aestivalis, Andricus aries, Andricus callidoma, Andricus caputmedusae, Andricus conglomeratus, Andricus coriarius, Andricus corruptrix, Andricus gallaetincticorne, Andricus grossulariae, Andricus hungaricus, Andricus inflator, Andricus kollari, Andricus lignicolus, Andricus lucidus, Andricus multiplicatus, Andricus panteli, Andricus quercuscalicis, Andricus quercusfolii, Andricus quercusramuli, Andricus quercustozae, Andricus seckendorffi, Andricus testaceipes, Andricus trotter, Andricus vindobonensis, Aphelonyx cerricola, Aylax sp., Biorhiza sp., Biorhiza pallida, Callirhytis glandium, Chilaspis Israeli, Chilaspis nitida, Cynipidae unspecified, Cynips sp., Cynips divisa, Cynips longiventris, Cynips quercus, Cynips quercusfolii, Diastrophus rubi, Diplolepis sp., Diplolepis fructuum, Diplolepis mayri, Diplolepis rosae, Dryocosmus kuriphilus, Neuroterus albipes, Neuroterus glandiformis, Neuroterus lanuginosus, Neuroterus numismalis, Neuroterus quercusbaccarum, Neuroterus tricolor, Phanacis sp., Plagiotrochus sp., Plagiotrochus australis, Rhodites sp., Synophrus politus, Trigonaspis mendesi* (Hymenoptera: Cynipidae), *Megastigmus pistaciae* (Torymidae), *Quercus sp.* (Fagaceae) (Noyes, 2019).
Parasitoid hosts: *Aulogymnus aceris* (Hymenoptera: Eulophidae) (Noyes, 2019).
Parasitoids: *Torymus nigricornis* (Torymidae) (Noyes, 2019).
Distribution: Andorra, Austria, Belgium, Bosnia Hercegovina, Bulgaria, Caucasus, Croatia, Czech Republic, England, Europe, France, Germany, Greece, Hungary, Iran, Italy, Jordan, Kazakhstan, Macedonia, Moldova, Montenegro, Netherlands, North Africa, Poland, Romania, Russia, Serbia, Slovenia, Spain, Sweden, Tselinograd Obl., Tunisia, Turkey, Ukraine, USSR, Yugoslavia (pre-1991) (Noyes, 2019).

Sycophila flavicollis (Walker, 1834)
Publications with Turkey records: Cam (2011; 2012), Noyes (2019), Stojanova et al. (2012).
Associates: *Hieracium sp., Hieracium pilosella* (Asteraceae), *Quercu coccifera, Quercus sp., Quercus cerris, Quercus fagine, Quercus ilex,*

Quercus petraea, Quercus pubescens, Quercus pyrenaica, Quercus robur, Quercus suber (Fagaceae) (Noyes, 2019).
Hosts: *Andricus sp., Andricus aries, Andricus conglomeratus, Andricus quercusradicis, Andricus testaceipes, Aulacidea hieracia, Aulacidea subterminalis, Cynips longiventris, Cynips quercus, Diplolepis rosae, Plagiotrochus quercusilicis, Synophrus politus, Trigonaspis brunneicornis* (Hymenoptera: Cynipidae) (Noyes, 2019).
Associates: *Andricus amblycerus, Andricus aries, Andricus conglomeratus, Andricus coriarius, Andricus grossulariae, Andricus hispanicus, Andricus kollari, Andricus polycerus, Andricus quercusradicis, Andricus testaceipes, Cynips longiventris, Cynips quercus, Plagiotrochus amenti, Plagiotrochus gallaeramulorum, Plagiotrochus panteli, Plagiotrochus quercusilicis, Pseudoneuroterus macropterus, Pseudoneuroterus saliens, Synophrus politus* (Hymenoptera: Cynipidae) (Noyes, 2019).
Distribution: Andorra, Austria, Bulgaria, Caucasus, Croatia, Czech Republic, Czechoslovakia, England, Europe, Finland, France, Germany, Greece, Hungary, Iran, Montenegro, Netherlands, Poland, Romania, Slovakia, Spain, Sweden, Turkey (Noyes, 2019).

Sycophila iracemae (Nieves Aldrey, 1984)
Publications with Turkey records: Askew et al. (2013), Noyes (2019).
Associates: *Andricus coriarius, Andricus fecundatrix, Andricus hispanicus, Andricus inflator, Andricus pseudoinflator, Andricus quercustozae, Dryocosmus kuriphilus, Synophrus politus, Trigonaspis synaspis* (Hymenoptera: Cynipidae), *Acer monspessulanum* (Aceraceae), *Quercus faginea, Quercus frainetto, Quercus infectoria, Quercus petraea, Quercus pubescens, Quercus pyrenaica* (Fagaceae), *Rosa sp.* (Rosaceae) (Noyes, 2019).
Hosts: *Andricus coriarius, Andricus kollari, Cynips quercus, Diplolepis mayri, Dryocosmus kuriphilus, Pediaspis aceris* (Hymenoptera: Cynipidae) (Noyes, 2019).
Distribution: France, Greece, Hungary, Iran, Italy, Spain, Turkey (Noyes, 2019).

Sycophila mayri (Erdos, 1959)
Publications with Turkey records: Cam (2011; 2012), Noyes (2019), Stojanova et al. (2012).
Associates: *Centaurea sp., Centaurea ornate, Centaurea scabiosa, Hieracium pilosella, Picris sp., Tragopogon dubius* (Asteraceae) (Noyes, 2019).

Hosts: *Aulacidea subterminalis, Aulacidea tragopogonis, Isocolus rogenhoferi, Phanacis caulicola, Phanacis centaureae* (Hymenoptera: Cynipidae) (Noyes, 2019).
Distribution: Austria, Czech Republic, Czechoslovakia, England, Europe, Hungary, Moldova, Slovakia, Spain, Turkey, Ukraine, USSR (Noyes, 2019).

Sycophila mellea (Curtis, 1831)
Publications with Turkey records: Cam (2012), Noyes (2019).
Associates: *Papaver dubium* (Papaveraceae), *Ammophila arenaria, Calamagrostis epigeios, Elymus repens, Festuca rubra* (Poaceae), *Rosa sp.* (Rosaceae) (Noyes, 2019).
Hosts: *Mayetiola destructor, Phytophaga destructor* (Diptera: Cecidomyiidae), *Kermes sp.* (Hemiptera: Kermesidae), *Aylax papaveris* (Hymenoptera: Cynipidae), *Eurytoma pollux, Harmolita hordei, Harmolita linearis, Harmolita rossica, Harmolita tritici, Tetramesa sp., Tetramesa brevicornis, Tetramesa cylindrica, Tetramesa eximia, Tetramesa hordei, Tetramesa linearis, Tetramesa tritici* (Eurytomidae) (Noyes, 2019).
Parasitoid hosts: *Eurytoma parva, Tetramesa sp.* (Eurytomidae), *Ditropinotus aureoviridis* (Torymidae) (Noyes, 2019).
Distribution: Bulgaria, Canada, Caucasus, Croatia, Europe, France, Germany, Hungary, Kazakhstan, Moldova, Nearctic, Netherlands, Romania, Russia, Sweden, Tselinograd Obl., Turkey, United Kingdom, United States of America, USSR (Noyes, 2019).

Sycophila submutica (Thomson, 1876)
Publications with Turkey records: Askew et al. (2006), Cam (2011; 2012), Noyes (2019), Stojanova et al. (2012).
Associates: *Pistacia mutica* (Anacardiaceae), *Carthamus oxyacantha, Carthamus tinctorius, Centaurea sp., Centaurea adpressa, Centaurea aspera, Centaurea diffusa, Centaurea jacea, Centaurea maculosa, Centaurea melitensis, Centaurea nicaeensis, Centaurea nigra, Centaurea paniculate, Centaurea pseudomaculosa, Centaurea sadleriana, Centaurea scabiosa, Centaurea solstitialis, Centaurea sphaerocephala, Chartolepis intermedia, Cichorium sp., Cirsium sp., Hieracium sp., Hieracium vulgatum, Hypochaeris radicata, Leuzea conifera, Serratula heterophylla, Serratula xeranthemoides* (Asteraceae), *Glechoma hederacea, Salvia triloba* (Lamiaceae), *Potentilla erecta* (Rosaceae), *Linaria vulgaris, Verbascum phlomoides* (Scrophulariaceae) (Noyes, 2019).

Hosts: *Lasioptera rubi* (Diptera: Cecidomyiidae), *Andricus conglomeratus, Andricus kollari, Andricus lucidus, Aulacidea sp., Aulacidea ascanica, Aulacidea hieracia, Aulacidea hypochoeridis, Aulacidea macula, Aulacidea scorzonerae, Aulax sp., Aylax jaceae, Aylax lichtensteini, Aylax serratulae, Isocolus sp., Isocolus areolate, Isocolus gratiosus, Isocolus jaceae, Isocolus leuzeae, Isocolus lichtensteini, Isocolus rogenhoferi, Isocolus scabiosae, Isocolus serratulae, Isocolus tauricus, Liposthenes sp., Liposthenes glechomae, Liposthenes latreillei, Phanacis sp., Phanacis centaureae, Phanacis hypochoeridis, Rhodus cyprius, Xestophanes brevitarsis* (Hymenoptera: Cynipidae) (Noyes, 2019).
Parasitoid hosts: *Megastigmus pistaciae* (Torymidae) (Noyes, 2019).
Distribution: Andorra, Austria, Belgium, Bosnia Hercegovina, Bulgaria, Caucasus, Croatia, Czechoslovakia, Denmark, England, Europe, Finland, France, Germany, Greece, Hungary, Iran, Iraq, Kazakhstan, Macedonia, Moldova, Montenegro, Netherlands, Palearctic, Romania, Serbia, Spain, Sweden, Switzerland, Tselinograd Obl., Turkey, Ukraine, USSR (Noyes, 2019).

Sycophila variegata (Curtis, 1831)
Publications with Turkey records: Askew et al. (2013), Cam (2011; 2012), Noyes (2019), Öncüer (1991), Stojanova et al. (2012).
Associates: *Andricus amblycerus, Andricus burgundus, Andricus caputmedusae, Andricus conglomeratus, Andricus coriarius, Andricus corruptrix, Andricus curvator, Andricus fecundatrix, Andricus grossulariae, Andricus hispanicus, Andricus hungaricus, Andricus infectorius, Andricus kollari, Andricus lignicolus, Andricus lucidus, Andricus polycerus, Andricus pseudoinflator, Andricus quercuscalicis, Andricus quercusradicis, Andricus quercustozae, Andricus solitarius, Andricus testaceipes, Andricus vindobonensis, Biorhiza pallida, Callirhytis glandium, Callirhytis rufescens, Cerroneuroterus lanuginosus, Cynips cornifex, Cynips quercus, Cynips quercusfolii, Dryocosmus kuriphilus, Dryocosmus mayri, Neuroterus quercusbaccarum, Plagiotrochus amenti, Plagiotrochus australis, Plagiotrochus coriaceus, Plagiotrochus gallaeramulorum, Plagiotrochus panteli, Plagiotrochus quercusilicis, Pseudoneuroterus macropterus, Pseudoneuroterus saliens* (Hymenoptera: Cynipidae), *Pistacia lentiscus* (Anacardiaceae), *Castanea sp., Castanea sativa, Quercu coccifera, Quercus sp., Quercus brantii, Quercus cerris, Quercus faginea, Quercus frainetto, Quercus ilex, Quercus infectoria, Quercus ithaburensis, Quercus lanuginose, Quercus*

lusitanica, Quercus pedunculiflora, Quercus petraea, Quercus pubescens, Quercus pyrenaica, Quercus robur, Quercus suber (Fagaceae).

Hosts: *Andricus sp., Andricus coriarius, Andricus corruptrix, Andricus fecundatrix, Andricus grossulariae, Andricus hispanicus, Andricus kollari, Andricus lignicolus, Andricus pseudoinflator, Andricus quercustozae, Andricus vindobonensis, Biorhiza pallida, Callirhytis rufescens, Chilaspis Israeli, Cynips sp., Diplolepis sp., Dryocosmus kuriphilus, Neuroterus sp., Neuroterus glandiformis, Neuroterus lanuginosus, Neuroterus saliens, Plagiotrochus sp., Plagiotrochus australis, Plagiotrochus kiefferianus, Plagiotrochus panteli, Plagiotrochus quercusilicis, Synergus umbraculus, Trichagalma serrata, Trigonaspis brunneicornis* (Hymenoptera: Cynipidae) (Noyes, 2019).

Parasitoid hosts: *Synergus umbraculus* (Hymenoptera: Cynipidae) (Noyes, 2019).

Distribution: Andorra, Austria, Bulgaria, Caucasus, Croatia, Czechoslovakia, England, Europe, France, Greece, Hungary, Iran, Iraq, Israel, Italy, Japan, Jordan, Korea, Netherlands, People's Republic of China, Romania, Slovakia, Spain, Sweden, Switzerland, Turkey (Noyes, 2019).

Genus *Systole* (Walker, 1832)

Systole albipennis (Walker, 1832)

Publications with Turkey records: Cam (2011; 2012), Noyes (2019), Stojanova et al. (2012).

Associates: *Conium maculatum, Daucus carota, Foeniculum vulgare, Petroselinum crispum, Torilis japonica* (Apiaceae) (Noyes, 2019).

Hosts: *Anethum graveolens, Bupleurum fruticosum, Bupleurum rotundifolium, Carum sp., Carum carvi, Conium maculatum, Coriandrum sativum, Cuminum cyminum, Daucus carota, Ferula orientalis, Foeniculum sp., Foeniculum vulgare, Pastinaca sp., Petroselinum sativum, Pimpinella anisum, Torilis sp., Torilis japonica* (Apiaceae) (Noyes, 2019).

Distribution: Bulgaria, Canada, Caucasus, Chile, Croatia, Czech Republic, Czechoslovakia, England, Europe, France, Germany, Hungary, India, Iran, Italy, Moldova, Montenegro, Netherlands, Romania, Sweden, Switzerland, Turkey, Ukraine, USSR (Noyes, 2019).

Systole bipunctata (Erdos, 1952)

Publications with Turkey records: Cam (2011; 2012), Noyes (2019), Stojanova et al. (2012).

Hosts: *Eryngium sp.* (Apiaceae) (Noyes, 2019).

Distribution: Azerbaijan, Bulgaria, Croatia, Czechoslovakia, Europe, France, Hungary, Slovakia, Turkey, Ukraine, USSR (Noyes, 2019).

Systole brunnicornis (Zerova and Cam, 2003)
Publications with Turkey records: Cam (2011; 2012), Noyes (2019), Stojanova et al. (2012), Zerova et al. (2003; 2004).
Associates: Unknown (Noyes, 2019).
Hosts: Unknown (Noyes, 2019).
Distribution: Turkey (Noyes, 2019).

Systole conspicua (Erdos, 1951)
Publications with Turkey records: Noyes (2019), Stojanova et al. (2012).
Hosts: *Eupithecia pimpinellata* (Lepidoptera: Geometridae), *Anthriscus longirostris, Anthriscus scandicina* (Apiaceae) (Noyes, 2019).
Distribution: Bulgaria, Europe, Hungary, Tunisia, Turkey, Ukraine (Noyes, 2019).

Systole coriandri (Gussakovsky, 1933)
Publications with Turkey records: Cam (2011; 2012), Noyes (2019), Stojanova et al. (2012).
Hosts: *Coriandrum sativum* (Apiaceae) (Noyes, 2019).
Distribution: Afghanistan, Argentina, Brazil, Bulgaria, Caribbean, Caucasus, Central America, Chile, Europe, India, Iran, Japan, Malaysia, Moldova, North Africa, People's Republic of China, Romania, Russia, Turkey, United States of America, USSR, Uzbekistan (Noyes, 2019).

Systole minima (Zerova and Cam, 2003)
Publications with Turkey records: Cam (2011; 2012), Noyes (2019), Stojanova et al. (2012), Zerova et al. (2003).
Associates: Unknown (Noyes, 2019).
Hosts: Unknown (Noyes, 2019).
Distribution: Turkey (Noyes, 2019).

Systole plana (Zerova and Cam, 2004)
Publications with Turkey records: Cam (2011; 2012), Noyes (2019), Stojanova et al. (2012), Zerova et al. (2004).
Hosts: Unknown (Noyes, 2019).
Associates: Unknown (Noyes, 2019).
Distribution: Turkey (Noyes, 2019).

Systole salviae (Zerova, 1968)
Publications with Turkey records: Cam (2011; 2012), Noyes (2019), Stojanova et al. (2012).
Hosts: *Salvia glutinosa, Salvia grandiflora, Salvia scabiosifolia* (Lamiaceae) (Noyes, 2019).
Distribution: Turkey, Ukraine (Noyes, 2019).

Systole tokata (Zerova and Cam, 2004)
Publications with Turkey records: Cam (2011; 2012), Noyes (2019), Stojanova et al. (2012), Zerova et al. (2004).
Associates: Unknown (Noyes, 2019).
Hosts: Unknown (Noyes, 2019).
Distribution: Turkey (Noyes, 2019).

Genus *Tetramesa* (Walker, 1848)
Tetramesa agrostidis (Howard, 1896)
Publications with Turkey records: Cam (2011; 2012), Noyes (2019), Stojanova et al. (2012).
Associates: *Agrostis sp.* (Poaceae) (Noyes, 2019).
Hosts: *Agrostis sp., Agrostis vulgaris* (Poaceae) (Noyes, 2019).
Distribution: Bulgaria, Europe, Germany, Kazakhstan, Nearctic, Tselinograd Obl., Turkey, United States of America, USSR (Noyes, 2019).

Tetramesa anatolica (Zerova and Cam, 2003)
Publications with Turkey records: Cam (2012), Noyes (2019), Stojanova et al. (2012), Zerova et al. (2003).
Associates: Unknown (Noyes, 2019).
Hosts: Unknown (Noyes, 2019).
Distribution: Turkey (Noyes, 2019).

Tetramesa angustipenne (Walker, 1832)
Publications with Turkey records: Cam (2011; 2012), Noyes (2019), Stojanova et al. (2012).
Hosts: *Alopecurus pratensis* (Poaceae) (Noyes, 2019).
Parasitoid hosts: *Chlorocytus pulchripes* (Hymenoptera: Pteromalidae) (Noyes, 2019).
Parasitoids: *Pediobius eubius* (Eulophidae), *Eupelmus atropurpureus, Macroneura vesicularis* (Eupelmidae), *Eurytoma tapio, Sycophila sp.* (Eurytomidae), *Endromopoda sp.* (Ichneumonidae), *Chlorocytus planus, Homoporus febriculosus, Mesopolobus graminum* (Pteromalidae) (Noyes, 2019).

Distribution: Bulgaria, Czech Republic, Europe, Hungary, Romania, Sweden, Turkey, United Kingdom, USSR (Noyes, 2019).

Tetramesa brevicornis (Walker, 1832)
Publications with Turkey records: Cam (2011; 2012), Noyes (2019), Stojanova et al. (2012).
Associates: *Festuca sp.* (Poaceae) (Noyes, 2019).
Hosts: *Festuca sp., Festuca ovina, Festuca rubra, Festuca rupicola, Festuca sulcate, Festuca valesiaca, Festuca wagneri* (Poaceae) (Noyes, 2019).
Parasitoid hosts: *Homoporus sp.* (Hymenoptera: Pteromalidae) (Noyes, 2019).
Parasitoids: *Bracon sp.* (Braconidae), *Calosota viridis, Eupelmus atropurpureus, Macroneura vesicularis* (Eupelmidae), *Eurytoma sp., Sycophila sp., Sycophila mellea* (Eurytomidae), *Merisus splendidus* (Pteromalidae), *Eridontomerus laticornis, Glyphomerus isosomatis* (Torymidae) (Noyes, 2019).
Distribution: Czech Republic, Europe, Germany, Hungary, Netherlands, Romania, Serbia, Slovakia, Sweden, Turkey, United Kingdom, USSR (Noyes, 2019).

Tetramesa brischkei (Schlechtendal, 1891)
Publications with Turkey records: Cam (2011; 2012), Noyes (2019).
Associates: *Agropyron repens* (Poaceae) (Noyes, 2019).
Hosts: *Elymus sp., Leymus sabulosus* (Poaceae) (Noyes, 2019).
Parasitoids: *Eurytoma elymi* (Hymenoptera: Eurytomidae) (Noyes, 2019).
Distribution: Bulgaria, Europe, Germany, Kazakhstan, Russia, Sweden, Tselinograd Obl., Turkey, USSR (Noyes, 2019).

Tetramesa cereipes (Erdos, 1955)
Publications with Turkey records: Cam (2011; 2012), Noyes (2019).
Associates: *Elymus repens* (Poaceae) (Noyes, 2019).
Hosts: *Agropyron sp., Agropyron intermedium, Agropyron repens, Agropyron ruthenicum* (Poaceae) (Noyes, 2019).
Parasitoids: *Calosota viridis* (Hymenoptera: Eupelmidae) (Noyes, 2019).
Distribution: Azerbaijan, Bulgaria, Caucasus, Czech Republic, Czechoslovakia, Europe, Hungary, Moldova, Palearctic, Romania, Slovakia, Turkey, Ukraine, USSR (Noyes, 2019).

Tetramesa cornuta (Walker, 1832)
Publications with Turkey records: Cam (2011; 2012), Noyes (2019), Stojanova et al. (2012).
Associates: *Agropyron sp., Agropyron repens* (Poaceae) (Noyes, 2019).
Hosts: *Agropyron sp., Agropyron intermedium, Agropyron repens, Elymus hispidus, Elymus repens* (Poaceae) (Noyes, 2019).
Parasitoids: *Pediobius alaspharus* (Hymenoptera: Eulophidae), *Chlorocytus agropyri, Homoporus chalcidiphagus* (Pteromalidae) (Noyes, 2019).
Distribution: Bulgaria, Czech Republic, Europe, Hungary, Kazakhstan, Moldova, Nearctic, Romania, Slovakia, Sweden, Tselinograd Obl., Turkey, Ukraine, United Kingdom, United States of America, USSR, Yemen (Noyes, 2019).

Tetramesa crassicornis (Walker, 1832)
Publications with Turkey records: Cam (2011; 2012), Noyes (2019).
Associates: Unknown (Noyes, 2019).
Hosts: Unknown (Noyes, 2019).
Distribution: Europe, Hungary, Sweden, Turkey, United Kingdom (Noyes, 2019).

Tetramesa elongia (Zerova, 2007)
Publications with Turkey records: Cam (2011; 2012), Noyes (2019), Stojanova *et al.* (2012).
Hosts: *Agropyron pectiniforme* (Poaceae) (Noyes, 2019).
Distribution: Bulgaria, Caucasus, Kazakhstan, Mongolia, Tselinograd Obl., Turkey, Ukraine, USSR (Noyes, 2019).

Tetramesa eximia (Giraud, 1863)
Publications with Turkey records: Cam (2011; 2012), Noyes (2019).
Associates: *Homoporus febriculosus* (Pteromalidae), *Calamagrostis epigeios* (Poaceae) (Noyes, 2019).
Hosts: *Ammophila arenaria, Calamagrostis sp., Calamagrostis epigeios* (Poaceae) (Noyes, 2019).
Parasitoid hosts: *Eurytoma sp.* (Hymenoptera: Eurytomidae), *Homoporus fulviventris* (Pteromalidae) (Noyes, 2019).
Parasitoids: *Aprostocetus sp., Pediobius eubius* (Eulophidae), *Eupelmus atropurpureus, Eupelmus urozonus, Macroneura vesicularis* (Eupelmidae), *Eurytoma danuvica, Sycophila sp., Sycophila mellea* (Eurytomidae), *Chlorocytus harmolitae, Homoporus sp., Homoporus*

destructor, Homoporus luniger (Pteromalidae), *Pseudotorymus sp., Syntomaspis baudysi, Torymus baudysi* (Torymidae) (Noyes, 2019).
Distribution: Austria, Czech Republic, Czechoslovakia, Europe, Germany, Hungary, Netherlands, Romania, Slovakia, Sweden, Turkey, Ukraine, United Kingdom, USSR (Noyes, 2019).

Tetramesa fulvicollis (Walker, 1832)
Publications with Turkey records: Cam (2011; 2012), Noyes (2019), Stojanova et al. (2012).
Hosts: *Brachypodium sp., Brachypodium sylvaticum* (Poaceae) (Noyes, 2019).
Parasitoids: *Bracon sp.* (Hymenoptera: Braconidae), *Pediobius eubius* (Eulophidae), *Eurytoma collaris* (Eurytomidae) (Noyes, 2019).
Distribution: Bulgaria, Croatia, Czech Republic, Czechoslovakia, Europe, France, Germany, Hungary, Montenegro, Netherlands, Palearctic, Romania, Slovakia, Sweden, Turkey, United Kingdom, USSR (Noyes, 2019).

Tetramesa fumipennis (Walker, 1832)
Publications with Turkey records: Cam (2011; 2012), Noyes (2019), Stojanova et al. (2012).
Hosts: *Alopecurus geniculatus* (Poaceae) (Noyes, 2019).
Distribution: Caucasus, Europe, Germany, Hungary, Palearctic, Romania, Sweden, Turkey, United Kingdom, USSR (Noyes, 2019).

Tetramesa giraudi (Schlechtendal, 1891)
Publications with Turkey records: Cam (2011; 2012), Noyes (2019), Stojanova et al. (2012).
Hosts: *Festuca sp., Festuca gigantea, Festuca orientalis, Festuca pratensis, Oryzopsis virescens* (Poaceae) (Noyes, 2019).
Distribution: Bulgaria, Czech Republic, Europe, France, Germany, Hungary, Romania, Sweden, Turkey, USSR (Noyes, 2019).

Tetramesa linearis (Walker, 1832)
Publications with Turkey records: Cam (2011; 2012), Noyes (2019).
Associates: *Agropyron sp., Agropyron repens, Elymus repens* (Poaceae) (Noyes, 2019).
Hosts: *Agropyron sp., Agropyron intermedium, Agropyron repens, Agropyron trichophorum, Calamagrostis sp., Elymus junceus, Elymus repens, Elytrigia sp.* (Poaceae) (Noyes, 2019).

Parasitoids: *Pediobius sp., Pediobius eubius* (Hymenoptera: Eulophidae), *Eupelmella vesicularis, Eupelmus allynii, Macroneura vesicularis* (Eupelmidae), *Eudecatoma mellea, Eurytoma sp., Eurytoma differta, Eurytoma flavimana, Sycophila sp., Sycophila mellea* (Eurytomidae), *Homoporus sp., Homoporus chalcidiphagus* (Pteromalidae), *Ditropinotus aureoviridis, Eridontomerus biroi, Eridontomerus isosomatis, Eridontomerus laticornis* (Torymidae) (Noyes, 2019).
Distribution: Bulgaria, Caucasus, Czech Republic, Czechoslovakia, Europe, France, Germany, Hungary, Kazakhstan, Moldova, Mongolia, Nearctic, Neotropical, Netherlands, New Zealand, Romania, Russia, Serbia, Slovakia, Sweden, Tselinograd Obl., Turkey, Ukraine, United Kingdom, United States of America, USSR (Noyes, 2019).

Tetramesa longula (Dalman, 1820)
Publications with Turkey records: Cam (2011; 2012), Noyes (2019), Stojanova et al. (2012).
Associates: *Dactylis glomerata* (Poaceae) (Noyes, 2019).
Hosts: *Dactylis glomerata* (Poaceae) (Noyes, 2019).
Parasitoids: *Pediobius dactylicola* (Hymenoptera: Eulophidae), *Eupelmella vesicularis* (Eupelmidae), *Eurytoma erdoesi* (Eurytomidae), *Mesopolobus graminum* (Pteromalidae), *Eridontomerus isosomatis* (Torymidae) (Noyes, 2019).
Distribution: Caucasus, Europe, Germany, Hungary, Ireland, Nearctic, Romania, Sweden, Turkey, United Kingdom, United States of America, USSR (Noyes, 2019).

Tetramesa maderae (Walker, 1849)
Publications with Turkey records: Cam (2011; 2012), Korkmaz et al. (2010), Öncüer (1991), Noyes (2019), Stojanova et al. (2012).
Associates: *Bromus ciliates, Elymus glaucus, Hordeum vulgare, Triticum sp.* (Poaceae) (Noyes, 2019).
Hosts: *Bromus ciliates, Elymus glaucus, Hordeum sp., Hordeum vulgare, Triticum sp., Triticum aestivum* (Poaceae).
Parasitoids: *Calosota metallica, Eupelmella vesicularis, Eupelmus allynii* (Hymenoptera: Eupelmidae), *Homoporus chalcidiphagus, Homoporus febriculosus* (Pteromalidae), *Cryptopristus harrisii, Ditropinotus aureoviridis, Eridontomerus isosomatis* (Torymidae) (Noyes, 2019).
Distribution: Belarus, Canada, Caucasus, Europe, Hungary, Iran, Israel, Italy, Kazakhstan, Madeira, Mexico, Nearctic, Romania, Tadzhikistan, Tselinograd Obl., Turkey, Ukraine, United Kingdom, United States of America, USSR, Uzbekistan (Noyes, 2019).

Tetramesa romana (Walker, 1873)
Publications with Turkey records: Cam (2011; 2012), Doğanlar and Doğanlar (2010), Noyes (2019), Stojanova et al. (2012).
Associates: *Endobia donacisa* (Eurytomidae), *Arundo donax* (Poaceae) (Noyes, 2019).
Hosts: *Arundo donax* (Poaceae) (Noyes, 2019).
Parasitoids: *Gugolzia harmolitae* (Hymenoptera: Pteromalidae) (Noyes, 2019).
Distribution: France, Italy, Spain, Turkey (Noyes, 2019).

Tetramesa tritici (Fitch, 1859)
Publications with Turkey records: Cam (2011; 2012), Korkmaz et al. (2010), Noyes (2019), Stojanova et al. (2012).
Associates: *Bromus ciliates, Elymus glaucus, Elymus virginicus, Hordeum vulgare, Triticum sp., Triticum vulgare* (Poaceae) (Noyes, 2019).
Hosts: *Sitophilus granarius* (Coleoptera: Dryophthoridae), *Phytophaga destructor* (Diptera: Cecidomyiidae), *Hordeum vulgare, Triticum sp., Triticum aestivum* (Poaceae) (Noyes, 2019).
Parasitoids: *Centrodora speciosissima* (Hymenoptera: Aphelinidae), *Calosota metallica, Eupelmella vesicularis, Eupelmus allynii* (Eupelmidae), *Eudecatoma mellea, Eurytoma flavimana, Eurytoma parva (Eurytomidae), Homoporus chalcidiphagus, Homoporus febriculosus* (Pteromalidae), *Cryptopristus harrisii, Ditropinotus aureoviridis, Eridontomerus isosomatis* (Torymidae) (Noyes, 2019).
Distribution: Brazil, Canada, Colombia, Iran, Nearctic, Spain, Turkey, United States of America, USSR (Noyes, 2019).

Subfamily Rileyinae (Ashmead, 1904)
Genus *Rileya* (Ashmead, 1888)
Rileya asiatica (Zerova, 1976)
Publications with Turkey records: Noyes (2019), Stojanova et al. (2012).
Associates: *Astragalus sp.* (Fabaceae) (Noyes, 2019).
Hosts: *Astragalus caucasica, Astragalus caucasicus* (Fabaceae), *Asphondylia sp.* (Diptera: Cecidomyiidae) (Noyes, 2019).
Distribution: Georgia, Turkey, Turkmenistan (Noyes, 2019).

Rileya Doğanlari (Cam, 2003)
Publications with Turkey records: Cam (2003; 2011; 2012), Noyes (2019), Stojanova et al. (2012).
Associates: *Astragalus lagurus* (Fabaceae) (Noyes, 2019).

Hosts: *Asphondylia anatolica* (Diptera: Cecidomyiidae) (Noyes, 2019).
Distribution: Turkey (Noyes, 2019).

References

Askew, Richard Robinson, and Melika, Gerotge, and Pujade-Villar, Juli, and Schönrogge, Karsten, and Stone, Graham N., and Nieves-Aldrey, José Luis. 2013. "Catalogue of parasitoids and inquilines in cynipid oak galls in the west Palaearctic". *Zootaxa*, 3643(1):1-133.

Askew, Richard Robinson, and Plantard, Oliver, and Gómez, José F., and Hernandez Nieves, Maria, and Nieves-Aldrey, José Luis. 2006. "Catalogue of parasitoids and inquilines in galls of Aylacini, Diplolepini and Pediaspidini (Hym., Cynipidae) in West Palaearctic". *Zootaxa*, 1301:7.

Bayram, Şerife, and Ülgentürk, Selma, and Toros, Seval. 1998. "Research on the insects causing galls on dog rose (*Rosa spp.*) and their parasitoids in Ankara province". *Turkish Journal of Entomology*, 22(4):259-268.

Çam, Halit. 1998. "A new species of Exeurytoma Burks, 1971 from Turkey (Hymenoptera: Chalcidoidea; Eurytomidae)". *Turkish Journal of Zoology*, 22(3):191-193.

—. 2003. "A new species of Rileya Ashmead (Hymenoptera: Chalcidoidea: Eurytomidae) from Tokat, Turkey". *Annals of the Entomological Society of America*, 96(1):39-43.

—. 2011. "A checklist of the Eurytomidae (Hymenoptera, Chalcidoidea) species of Turkey". *Zootaxa*, 3113:53-64.

—. 2012. "Updated checklist of the Eurytomidae (Hymenoptera: Chalcidoidea) species". *Archives of Biological Sciences*, 64(2):667-674.

Doğanlar, Mikdat, and Çam, Halit. 1991. "The species of Eurytoma Ill. with two-three strong setae on hind tibiae from Türkiye, and description of a new species from Tokat, Türkiye (Hymenoptera, Eurytomidae)". *Turkish Journal of Entomology*, 15(3): 143–151.

—. 2005. "Morphology and systematics of Turkish species of Exeurytoma Burks, 1971 (Hymenoptera, Chalcidoidea, Eurytomidae), with description of a new species". *Turkish Journal of Zoology*, 29(3):243-248.

Doğanlar, Mikdat, and Doğanlar, Oğuzhan. 2010. "Review of the species of *Gugolzia* Delucchi and Steffan (Hymenoptera: Pteromalidae) in Europe and Turkey, with descriptions of new species". *Turkish Journal of Zoology*, 34(1):23-34.

Doğanlar, Mikdat. 2014. "First reports on the natural enemy fauna of the chestnut gall wasp, *Dryocosmus kuriphilus* Yasumatsu (Hymenoptera: Cynipidae) in Yalova, Turkey". *Türkiye Biyolojik Mucadele Dergisi*, 5(1):67-74.

Gençer, Lütfiye. 2003. "The chalcidoid parasitoids of *Diplolepis mayri* Schld. (Hymenoptera: Cynipidae), a pest of *Rosa canina* L. in Sivas province". *Turkish Journal of Entomology*, 27(2):107-111.

Korkmaz, Ertan Mahir, and Hastaoğlu Örgen, Sevda, and Gençer, Lütfiye, and Ülgentürk, Selma, and Başıbüyük, Hasan Hüseyin. 2010. "Determination of some wheat pests and their parasitoids in wheat fields of Central Anatolia Region". *Türkiye Entomoloji Dergisi*, 34(3):361-377.

Lotfalizadeh, Hosseinali, and Delvare, Gérard, and Rasplus, Jean-Yves. 2007. "Phylogenetic analysis of Eurytominae (Chalcidoidea: Eurytomidae) based on morphological characters". *Zoological Journal of the Linnean Society*, 151:441-510.

Mete, Özlem, and Mergen, Orhan. 2017. "The community components associated with two common rose gall wasps (Hymenoptera: Cynipidae: Diplolepidini) in Turkey". *Turkish Journal of Zoology*, 41:696-701.

Narendran, Thekke Curuppathe, and Tezcan, Serdar, and Civelek, Hasan Sungur. 1995. "A new species of *Eurytoma* Illiger (Hymenoptera, Eurytomidae) parasitic on *Scolytus rugulosus* Ratzeburg (Coleoptera, Scolytidae) in Turkey and some notes about it". *Turkish Journal of Entomology*, 19(2):81-86.

Noyes, John S. 2019. *Universal Chalcidoidea Database*. Available at: www.nhm.ac.uk/chalcidoids (accessed January 2022).

Öncüer, Cezmi. 1991. "A Catalogue of the Parasites and Predators of Insect Pests of Turkey". *Agricultural Faculty Paper No. 505*. Ege University, İzmir.

Özbek, Hikmet, and Güçlü, Şabam, and Tozlu, Göksel. 1999. "Biology and natural enemies of *Diplolepis mayri* Schld. (Hymenoptera: Cynipidae), a pest of *Rosa canina* L. in Erzurum Province". *Turkish Journal of Entomology*, 23(1):39-50.

Özbek, Hikmet, and Güçlü, Şaban, and Hayat, Rüstem. 1996. "Investigations on the phytophagous and predator insect species on stone fruits in the north-east agricultural region of Turkey". *Turkish Journal of Agriculture and Forestry*, 20(3):267-282.

Stojanova, Anelia, and Civelek, Hasan Sungur, and Yörük, Berna, and Sari, Sabahat, and Atahan, Tuğçe. 2012. "Checklists of Turkish Eurytomidae Walker, 1832 and Torymidae Walker, 1833

(Hymenoptera: Chalcidoidea)". *Turkish Journal of Entomology*, 36(1):69-82.

Stringer, Danielle N., and Mantel, Sarah, and Jennings, John T., and Austin, Andrew D. 2012. "Family Eurytomidae", *Australian Centre for Evolutionary Biology and Biodiversity, and the School of Earth and Environmental Science*. The University of Adelaide, Australia (Available from: https://biodiversity.org.au/afd/taxa/Eurytomidae, accessed October 2022).

Zerova, Marina Dmitrievna, and Fursov, Viktor N. 1991. "The Palaearctic species of Eurytoma (Hymenoptera: Eurytomidae) developing in stone fruits (Rosaceae: Prunoideae)". *Bulletin of Entomological Research*, 81(2):209-219.

Zerova, Marina Dmitrievna, and Seryogina, Ludmila Yakovlevna, and Çam, Halit. 2003. "New species of Eurytomidae (Hymenoptera: Chalcidoidea) from Turkey". *Zoologische Mededelingen*, Leiden, 77(1-4):259-266.

Zerova, Marina Dmitrievna, Seryogina, Ludmila Yakovlevna, and Çam, Halit. 2004. "New species of the genus Systole (Hymenoptera, Eurytomidae) from Turkey". *Vestnik Zoologii*, Kiev, 38(3):81-85.

CHAPTER TEN

FAMILY: LEUCOSPIDAE WALKER, 1834

Leucospids are among the largest of the Chalcidoids, with some species reaching up to 16 mm in length. They are characterized by a swollen hind femur with a toothed ventral surface, an elongated tegula that is more than twice as long as it is wide, and often longitudinally folded fore wings. The ovipositor is generally exserted and curves upward over the metasoma. The body is typically black, non-metallic and may have yellow or red markings. Leucospids are parasitoids of solitary aculeate Hymenoptera, predominantly bees (Stringer et al., 2012).

The family currently includes 4 genera and 134 species, and there are one genus and six species in Turkey (Noyes, 2019).

Genus *Leucospis* (Fabricius, 1775)
Leucospis bifasciata (Klug, 1812)
Publications with Turkey records: Baur and Amiet (2000), Noyes (2019), Öncüer (1991).
Hosts: *Anthidiellum strigatum, Anthidium strigatum* (Hymenoptera: Apidae) (Noyes, 2019).
Distributon: Albania, Azerbaijan, Bosnia Hercegovina, Bulgaria, Caucasus, Croatia, Cyprus, Europe, France, Greece, Iran, Israel, Italy, Jordan, Lebanon, Montenegro, Spain, Syria, Tadzhikistan, Turkey, Turkmenistan, Ukraine, USSR, Uzbekistan (Noyes, 2019).

Leucospis biguetina (Jurine, 1807)
Publications with Turkey records: Baur and Amiet (2000), Noyes (2019), Öncüer (1991).
Associates: *Euphorbia sp*. (Euphorbiaceae) (Noyes, 2019).
Hosts: *Hoplitis acuticornis, Hoplitis tridentata, Osmia acuticornis, Osmia tridentata* (Hymenoptera: Apidae), *Isodontia mexicana* (Sphecidae) (Noyes, 2019).

Distribution: Algeria, Austria, Bosnia Hercegovina, Croatia, Czech Republic, Europe, France, Hungary, Iran, Israel, Italy, Montenegro, Morocco, North Africa, Russia, Serbia, Slovakia, Slovenia, Spain, Switzerland, Turkey, USSR (Noyes, 2019).

Leucospis brevicauda (Fabricius, 1804)
Publications with Turkey records: Noyes (2019), Öncüer (1991).
Associates: Unknown (Noyes, 2019).
Hosts: Unknown (Noyes, 2019).
Distribution: Algeria, Italy, Morocco, Tunisia, Turkey (Noyes, 2019).

Leucospis dorsigera (Fabricius, 1775)
Publications with Turkey records: Noyes (2019), Öncüer (1991).
Associates: *Daucus carota, Eryngium sp.* (Apiaceae), *Carduus pycnocephalus* (Asteraceae), *Retama sphaerocarpa* (Fabaceae), *Allium sp.* (Liliaceae), *Hordeum leporinum* (Poaceae), *Prunus armeniaca* (Rosaceae), *Rubus sp.* (Rubiaceae) (Noyes, 2019).
Hosts: *Apate monacha* (Coleoptera: Bostrychidae), *Osphranteria coerulescens* (Cerambycidae), *Anthidiellum sp., Anthidiellum strigatum, Anthidium sp., Anthidium diadema, Anthidium strigatum, Megachile sp., Megachile ericetorum, Osmia sp., Osmia adunca, Osmia bicornis, Osmia cornuta, Osmia fedtschenkoi, Osmia fulviventris, Osmia ligurica, Osmia niveata, Osmia parietina, Osmia rufa, Osmia tricornis* (Hymenoptera: Apidae) (Noyes, 2019).
Parasitoid hosts: *Xorides corcyrensis* (Ichneumonidae) (Noyes, 2019).
Distribution: Afghanistan, Algeria, Austria, Bulgaria, Caucasus, Croatia, Cyprus, Czech Republic, Czechoslovakia, Egypt, Europe, France, Germany, Greece, Hungary, Iran, Israel, Italy, Kazakhstan, Macedonia, Moldova, Montenegro, North Africa, Poland, Romania, Serbia, Slovakia, Slovenia, Spain, Switzerland, Tselinograd Obl., Turkey, Turkmenistan, Ukraine, USSR (Noyes, 2019).

Leucospis gigas (Fabricius, 1793)
Publications with Turkey records: Baur and Amiet (2000), Noyes (2019), Öncüer (1991), Ye et al. (2017).
Associates: *Coelioxys sp.* (Apidae), *Sedum album* (Crassulaceae) (Noyes, 2019).
Hosts: *Anthophora sp., Anthophora garrulus, Chalicodoma sp., Chalicodoma muraria, Chalicodoma pyrenaica, Megachile sp., Megachile desertorum, Megachile hungarica, Megachile parietina, Megachile pyrenaica, Megachile sicula, Osmia sp., Osmia bicornis, Osmia coerulea,*

Osmia coerulescens, Osmia rufa (Hymenoptera: Apidae), *Katamenes flavigularis* (Vespidae) (Noyes, 2019).
Distribution: Afghanistan, Albania, Algeria, Armenia, Austria, Bosnia Hercegovina, Caucasus, Croatia, Czechoslovakia, Egypt, Europe, France, Germany, Gibraltar, Greece, Hungary, India, Iran, Israel, Italy, Jordan, Kazakhstan, Kirgizia, Lebanon, Libya, Macedonia, Malta, Morocco, Nearctic, North Africa, Pakistan, People's Republic of China, Romania, Russia, Slovakia, Slovenia, Spain, Switzerland, Syria, Tadzhikistan, Tselinograd Obl., Tunisia, Turkey, Turkmenistan, Ukraine, United States of America, USSR, Uzbekistan (Noyes, 2019).

Leucospis intermedia (Illiger, 1807)
Publications with Turkey records: Noyes (2019), Öncüer (1991), Ye et al. (2017).
Hosts: *Megachile sp., Osmia emarginata, Osmia mustelina* (Hymenoptera: Apidae) (Noyes, 2019).
Distribution: Afghanistan, Albania, Algeria, Armenia, Australia, Austria, Azerbaijan, Caucasus, Croatia, Cyprus, Czech Republic, Czechoslovakia, Egypt, Europe, France, Germany, Greece, Hungary, Iran, Israel, Italy, Jordan, Kazakhstan, Lebanon, Libya, Moldova, Morocco, North Africa, People's Republic of China, Russia, Slovakia, Spain, Switzerland, Syria, Tadzhikistan, Transcaucasus, Tselinograd Obl., Turkey, Ukraine, USSR, Uzbekistan (Noyes, 2019).

References

Baur, Hannes, and Amiet, Félix. 2000. "The Leucospidae (Hymenoptera: Chalcidoidea) of Switzerland, with a key and data on the European species". *Revue Suisse de Zoologie*, 107(2):359-388.

Noyes, John S. 2019. *Universal Chalcidoidea Database*. Available at: www.nhm.ac.uk/chalcidoids (accessed January 2022).

Öncüer, Cezmi. 1991. "A Catalogue of the Parasites and Predators of Insect Pests of Turkey". *Ege Üniversitesi Ziraat Fakültesi Yayınları* [Ege University Agriculture Faculty Publ.] 505.

Stringer, Danielle N., and Mantel, Sarah, and Jennings, John T., and Austin, Andrew D. 2012. "Family Leucospidae", *Australian Centre for Evolutionary Biology and Biodiversity, and the School of Earth and Environmental Science*. The University of Adelaide, Australia (Available from: https://www.researchgate.net/publication/265383166_Family_LEUCOSPIDAE, accessed October 2022).

Ye, Xin-hai, Achterberg, Cornelis van, and Yue, Qi, and Xu, Zai-fu. 2017. "Review of Chinese Leucospidae (Hymenoptera, Chalcidoidea)". *ZooKeys*, 651:107-157.

CHAPTER ELEVEN

FAMILY: MEGASTIGMIDAE THOMSON, 1876

Megastigmidae was once considered a subfamily of Torymidae. However, researchers have recognized Megastigmidae as its own family because it has been found that Torymidae and Megastigmidae are not as closely related as previously believed. Most species within Megastigmidae are phytophagous, feeding on seeds of various plants worldwide, including *Rhus* (Anacardiaceae), *Schinus* (Anacardiaceae), *Pandanus* (Pandanaceae), and conifers. While some phytophagous species are considered severe pests or parasitoids, two genera, Megastigmus and Mangostigmus, are known to attack *Eupelmus pistaciae* and Cecidomyiidae respectively.

Megastigmidae can be easily identified by the enlarged stigma in their fore wing. Many species have an orange-yellow body color, although some may be patterned black, entirely black, or exhibit metallic coloration. Their ovipositors are long and usually curve upwards, and their hind coxae are relatively small. Setae are symmetrically distributed throughout the thorax.

There is a total of 12 genera in the Megastigmidae, and 10 are biologically known (Jansta et al., 2018). There are three genera and 20 species in Turkey (Noyes, 2019)

Genus *Bootanomyia* (Girault, 1915)
Bootanomyia almusiensis (Doğanlar, 1989)
Publications with Turkey records: Doğanlar (1989), Stojanova et al. (2012).
Associates: *Pseudoneuroterus macropterus* (Hymenoptera: Cynipidae), *Quercus cerris* (Fagaceae) (Noyes, 2019).
Hosts: *Neuroterus macropterus* (Hymenoptera: Cynipidae) (Noyes, 2019).
Distribution: Iran, Turkey (Noyes, 2019).

Bootanomyia balikesirensis (Doğanlar, 2011)
Publications with Turkey records: Doğanlar (2011), Noyes (2019), Stojanova et al. (2012).
Associates: *Olea europaea* (Oleaceae) (Noyes, 2019).
Distribution: Turkey (Noyes, 2019).

Bootanomyia dorsalis (Fabricius, 1798)
Publications with Turkey records: Askew et al. (2013), Öncüer (1991), Noyes (2019), Stojanova et al. (2012).
Associates: *Andricus aries, Andricus askewi, Andricus burgundus, Andricus caliciformis, Andricus caputmedusae, Andricus conglomeratus, Andricus coriarius, Andricus coronatus, Andricus corruptrix, Andricus curvator, Andricus cydoniae, Andricus dentimitratus, Andricus fecundatrix, Andricus glutinosus, Andricus grossulariae, Andricus hispanicus, Andricus hungaricus, Andricus infectorius, Andricus inflator, Andricus kollari, Andricus legitimus, Andricus lignicolus, Andricus lucidus, Andricus multiplicatus, Andricus polycerus, Andricus pseudoinflator, Andricus quercuscalicis, Andricus quercusradicis, Andricus quercusramuli, Andricus quercustozae, Andricus seckendorffi, Andricus singularis, Andricus solitarius, Andricus stefanii, Andricus sternlichti, Andricus testaceipes, Andricus tomentosus, Andricus vindobonensis, Aphelonyx cerricola, Biorhiza pallida, Callirhytis erythrocephala, Callirhytis glandium, Callirhytis rufescens, Chilaspis nitida, Cynips disticha, Cynips longiventris, Cynips quercus, Cynips quercusfolii, Dryocosmus cerriphilus, Dryocosmus kuriphilus, Dryocosmus mayri, Neuroterus quercusbaccarum, Neuroterus saliens, Plagiotrochus amenti, Plagiotrochus gallaeramulorum, Plagiotrochus panteli, Plagiotrochus quercusilicis, Plagiotrochus yeusei, Pseudoneuroterus macropterus, Pseudoneuroterus saliens, Synophrus olivieri, Synophrus politus, Trigonaspis synaspis* (Hymenoptera: Cynipidae), *Fagus sylvatica, Quercu coccifera, Quercus sp., Quercus cerris, Quercus faginea, Quercus frainetto, Quercus infectoria, Quercus ithaburensis, Quercus lusitanica, Quercus pedunculiflora, Quercus persica, Quercus petraea, Quercus pubescens, Quercus pyrenaica, Quercus robur, Quercus semecarpifolia, Quercus suber* (Fagaceae) (Noyes, 2019).
Hosts: *Andricus sp., Andricus aestivalis, Andricus albopunctatus, Andricus aries, Andricus caputmedusae, Andricus conglomeratus, Andricus coriarius, Andricus corruptor, Andricus corruptrix, Andricus curvator, Andricus fecundatrix, Andricus grossulariae, Andricus hispanicus, Andricus hungaricus, Andricus infectorius, Andricus inflator, Andricus kollari, Andricus lignicolus, Andricus lucidus, Andricus multiplicatus, Andricus peyerimhoffi, Andricus quercuscorticis, Andricus*

quercusradicis, Andricus quercusramuli, Andricus stefanii, Andricus testaceipes, Andricus vindobonensis, Aphelonyx cerricola, Biorhiza sp., Biorhiza pallida, Callirhytis erythrocephala, Callirhytis glandium, Cynips sp., Cynips conglomerate, Cynips coriaria, Cynips divisa, Cynips kollari, Cynips lignicola, Cynips longiventris, Cynips quercusfolii, Diplolepis rosae, Dryocosmus kuriphilus, Neuroterus macropterus, Neuroterus quercusbaccarum, Synergus pallicornis, Synophrus politus (Hymenoptera: Cynipidae), *Syntomaspis sp.* (Torymidae), *Quercus semecarpifolia* (Fagaceae) (Noyes, 2019).
Parasitoid: *Synergus reinhardi, Synergus umbraculus* (Hymenoptera: Cynipidae) (Noyes, 2019).
Distribution: Andorra, Austria, Belgium, Bosnia Hercegovina, Bulgaria, Caucasus, Croatia, Czech Republic, Czechoslovakia, Denmark, England, Europe, Finland, France, Germany, Greece, Hungary, India, Iran, Israel, Italy, Jordan, Macedonia, Moldova, Netherlands, People's Republic of China, Portugal, Romania, Russia, Serbia, Slovakia, Spain, Sweden, Tunisia, Turkey, Ukraine, USSR (Noyes, 2019).

Bootanomyia emrezaferi (Doğanlar, 2011)
Publications with Turkey records: Doğanlar (2011), Noyes (2019), Stojanova et al. (2012).
Associates: *Prunus sp.* (Rosaceae) (Noyes, 2019).
Hosts: Hymenoptera: Cynipidae (Noyes, 2019).
Distribution: Turkey (Noyes, 2019).

Bootanomyia hepdurgunae (Doğanlar, 2011)
Publications with Turkey records: Doğanlar (2011), Noyes (2019), Stojanova et al. (2012).
Associates: *Olea europaea* (Oleaceae) (Noyes, 2019).
Distribution: Turkey (Noyes, 2019).

Bootanomyia mehmeti (Doğanlar, 2011)
Publications with Turkey records: Doğanlar (2011; 2014), Noyes (2019), Stojanova et al. (2012).
Associates: *Quercus sp*. (Fagaceae) (Noyes, 2019).
Hosts: *Dryocosmus kuriphilus* (Hymenoptera: Cynipidae) (Noyes, 2019).
Distribution: Turkey (Noyes, 2019).

Bootanomyia onuri (Doğanlar, 2011)
Publications with Turkey records: Doğanlar (2011), Noyes (2019), Stojanova et al. (2012).

Associates: *Quercus sp.* (Fagaceae) (Noyes, 2019).
Hosts: *Neuroterus sp.* (Hymenoptera: Cynipidae) (Noyes, 2019).
Distribution: Turkey (Noyes, 2019).

Bootanomyia shebnemae (Doğanlar, 2011)
Publications with Turkey records: Doğanlar (2011), Noyes (2019), Stojanova et al. (2012).
Associates: *Rosa sp.* (Rosaceae) (Noyes, 2019).
Hosts: Hymenoptera: Cynipidae (Noyes, 2019).
Distribution: Turkey (Noyes, 2019).

Bootanomyia stigmatizans (Fabricius, 1798)
Publications with Turkey records: Askew et al. (2013), Noyes (2019), Stojanova et al. (2012).
Associates: *Andricus aries, Andricus caputmedusae, Andricus coriarius, Andricus coronatus, Andricus dentimitratus, Andricus glutinosus, Andricus grossulariae, Andricus hispanicus, Andricus hungaricus, Andricus infectorius, Andricus kollari, Andricus lignicolus, Andricus lucidus, Andricus megalucidus, Andricus pictus, Andricus polycerus, Andricus quercuscalicis, Andricus quercustozae, Cynips longiventris, Cynips quercusfolii, Synophrus politus* (Cynipidae*), Quercus sp., Quercus cerris, Quercus faginea, Quercus frainetto, Quercus infectoria, Quercus lusitanica, Quercus petraea, Quercus pubescens, Quercus pyrenaica, Quercus robur, Quercus suber* (Fagaceae) (Noyes, 2019).
Hosts: *Andricus sp., Andricus caputmedusae, Andricus coriarius, Andricus curvator, Andricus hispanicus, Andricus kollari, Andricus lignicolus, Andricus lucidus, Andricus quercuscalicis, Andricus quercustozae, Cynips divisa, Neuroterus numismalis* (Hymenoptera: Cynipidae) (Noyes, 2019).
Parasitoid hosts: *Torymus nigricornis* (Torymidae) (Noyes, 2019).
Distribution: Austria, Bosnia Hercegovina, Bulgaria, Croatia, Czech Republic, Czechoslovakia, England, Europe, France, Germany, Greece, Hungary, Iran, Italy, Morocco, Portugal, Romania, Russia, Serbia, Slovakia, Slovenia, Spain, Sweden, Turkey, Ukraine, USSR (Noyes, 2019).

Bootanomyia synophri (Mayr, 1874)
Publications with Turkey records: Askew et al. (2013), Noyes (2019), Stojanova et al. (2012).
Associates: *Andricus glutinosus, Aphelonyx cerricola, Synophrus politus, Quercus sp., Quercus cerris, Quercus suber* (Fagaceae) (Noyes, 2019).
Hosts: *Synophrus politus* (Hymenoptera: Cynipidae) (Noyes, 2019).

Distribution: Algeria, Austria, Bulgaria, Europe, Greece, Hungary, Turkey (Noyes, 2019).

Genus *Megastigmus* (Dalman, 1820)
Megastigmus juniperi (Nikolskaya, 1952)
Publications with Turkey records: Noyes (2019), Stojanova et al. (2012).
Hosts: *Juniperus sp., Juniperus excelsa, Juniperus semiglobosa, Juniperus seravschanica, Juniperus turkestanica* (Cupressaceae) (Noyes, 2019).
Distribution: Armenia, Kazakhstan, Turkey, USSR, Uzbekistan (Noyes, 2019).

Megastigmus leptocybus (Doğanlar and Hassan, 2013)
Publications with Turkey records: Noyes (2019), Doğanlar and Hassan (2010).
Associates: *Eucalyptus camaldulensis* (Myrtaceae) (Noyes, 2019).
Hosts: *Leptocybe invasa* (Hymenoptera: Eulophidae) (Noyes, 2019).
Distribution: Israel, Italy, Turkey (Noyes, 2019).

Megastigmus pistaciae (Walker, 1871)
Publications with Turkey records: Noyes (2019).
Associates: *Eurytoma sp., Eurytoma pistaciae, Sycophila sp.* (Eurytomidae), *Gugolzia melengicia, Gugolzia oezdenoerneki* (Pteromalidae), *Pistacia atlantica, Pistacia chinensis, Pistacia integerrima, Pistacia lentiscus, Pistacia mutica, Pistacia palaestina, Pistacia terebinthus, Pistacia vera, Schinus terebinthifolius* (Anacardiaceae) (Noyes, 2019).
Hosts: *Pistacia atlantica x vera Pistacia sp., Pistacia atlantica, Pistacia chinensis, Pistacia integerrima, Pistacia lentiscus, Pistacia mutica, Pistacia terebinthus, Pistacia vera, Schinus mole* (Anacardiaceae), *Eucalyptus sp.* (Myrtaceae) (Noyes, 2019).
Parasitoids: *Eupelmus gemellus, Eupelmus pistaciae* (Hymenoptera: Eupelmidae), *Decatoma trogocarpi, Eurytoma rosae, Sycophila biguttata* (Eurytomidae) (Noyes, 2019).
Distribution: Afghanistan, Armenia, Australia, Bulgaria, Caucasus, Croatia, Cyprus, Europe, France, Georgia, Greece, Iran, Israel, Italy, Kenya, Kirgizia, Mexico, Montenegro, Morocco, People's Republic of China, Portugal, Russia, Spain, Syria, Tadzhikistan, Transcaucasus, Tunisia, Turkey, Turkmenistan, Ukraine, United States of America, USSR, Uzbekistan (Noyes, 2019).

Megastigmus rosae (Boucek, 1971)
Publications with Turkey records: Gencer (2003), Mete and Mergen (2017), Noyes (2019), Öncüer (1991), Stojanova et al. (2012).
Associates: *Diplolepis fructuum* (Hymenoptera: Cynipidae), *Rosa beggeriana, Rosa canina, Rosa kokanica* (Rosaceae) (Noyes, 2019).
Hosts: *Diplolepis fructuum, Diplolepis mayri* (Hymenoptera: Cynipidae), *Rosa sp., Rosa arvensis, Rosa canina, Rosa ferruginea, Rosa penduline, Rosa rubiginosa, Rosa tomentosa, Rosa tshatyrdagi, Rosa turkestanica* (Rosaceae) (Noyes, 2019).
Distribution: Armenia, Austria, Azerbaijan, Caucasus, Czech Republic, Czechoslovakia, France, Georgia, Germany, Iran, Kazakhstan, Russia, Switzerland, Tadzhikistan, Tselinograd Obl., Turkey, Turkmenistan, Ukraine, USSR (Noyes, 2019).

Megastigmus schimitscheki (Novicky, 1954)
Publications with Turkey records: Noyes (2019), Stojanova et al. (2012).
Hosts: *Abies pinsapo, Cedrus atlantica, Cedrus brevifolia, Cedrus libani* (Pinaceae) (Noyes, 2019).
Distribution: Cyprus, France, Greece, Lebanon, Syria, Turkey (Noyes, 2019).

Megastigmus suspectus (Borries, 1895)
Publications with Turkey records: Ayberk and Cebeci (2010), Noyes (2019), Stojanova et al. (2012).
Associates: *Mastrus clausus* (Ichneumonidae), *Mesopolobus pinus* (Pteromalidae), *Abies alba* (Pinaceae) (Noyes, 2019).
Hosts: *Abies sp., Abies alba, Abies amabilis, Abies borisii-regis, Abies bornmulleriana, Abies cephalonica, Abies cilicica, Abies concolor, Abies equi-trojani, Abies grandis, Abies homolepis, Abies nebrodensis, Abies nordmanniana, Abies numidica, Abies pinsapo, Abies sibirica, Cedrus atlantica, Cedrus brevifolia* (Pinaceae) (Noyes, 2019).
Parasitoids: *Mesopolobus pinus* (Hymenoptera: Pteromalidae) (Noyes, 2019).
Distribution: Algeria, Austria, Caucasus, Czech Republic, Czechoslovakia, Denmark, Europe, Finland, France, Georgia, Germany, Greece, Hungary, Italy, Kazakhstan, Montenegro, Netherlands, Poland, Romania, Slovakia, Slovenia, Spain, Sweden, Tselinograd Obl., Turkey, Ukraine, United Kingdom, USSR (Noyes, 2019).

Megastigmus usakensis (Doğanlar and Zengin, 2018)
Publications with Turkey records: Doğanlar and Zengin (2018), Noyes (2019).
Associates: Unknown (Noyes, 2019).
Hosts: Unknown (Noyes, 2019).
Distribution: Turkey (Noyes, 2019).

Megastigmus wachtli (Seitner, 1916)
Publications with Turkey records: Noyes (2019), Stojanova et al. (2012).
Hosts: *Cupressus abramsiana, Cupressus arizonica, Cupressus atlantica, Cupressus bakeri, Cupressus dupreziana, Cupressus goveniana, Cupressus lusitanica, Cupressus macrocarpa, Cupressus sempervirens* (Cupressaceae) (Noyes, 2019).
Distribution: Albania, Algeria, Croatia, Europe, France, Greece, Italy, Malta, Montenegro, Morocco, Portugal, Slovenia, Tunisia, Turkey, Yugoslavia (pre-1991) (Noyes, 2019).

Megastigmus zvimendeli (Doğanlar and Hassan, 2010)
Publications with Turkey records: Doğanlar (2015), Noyes (2019).
Hosts: *Leptocybe invasa* (Hymenoptera: Eulophidae) (Noyes, 2019).
Associates: *Eucalyptus camaldulensis, Eucalyptus grandis* (Myrtaceae) (Noyes, 2019).
Distribution: Australia, Israel, Turkey (Noyes, 2019).

Genus *Westralianus* (Boucek, 1988)
Westralianus altinoezus (Doğanlar, 2011)
Publications with Turkey records: Doğanlar (2011), Noyes (2019), Stojanova et al. (2012).
Associates: *Gymnosporangium confusum* (Uredinales: Pucciniaceae), *Crataegus monogyna* (Rosaceae) (Noyes, 2019).
Distribution: Turkey (Noyes, 2019).

References

Askew, Richard Robinson, and Melika, Gerotge, and Pujade-Villar, Juli, and Schönrogge, Karsten, and Stone, Graham N., and Nieves-Aldrey, José Luis. 2013. "Catalogue of parasitoids and inquilines in cynipid oak galls in the west Palaearctic". *Zootaxa*, 3643(1):1-133.

Ayberk, Hamit, and Cebeci, Hüseyin. 2010. "Survey of Trojan fir seed orchard in Edremit-Gurgendag Region of Turkey". *African Journal of Biotechnology*, 9(41): 6963-6965.

Doğanlar, Mikdat, and Hassan, Errol. 2010. "Review of Australian species of Megastigmus (Hymenoptera: Torymidae) associated with Eucalyptus, with descriptions of new species". *Australian Journal of Basic and Applied Sciences*, 4(10): 5059-5120.

Doğanlar, Mikdat, and Zengin, Erdal. 2018. "A new species of Megastigmus (Hymenoptera: Torymidae: Megastigminae) from Usak, Turkey". *Munis Entomology and Zoology*, 13(1):288-291.

Doğanlar, Mikdat. 1989. "A new parasitic species of *Megastigmus* Dalman (Hymenoptera: Torymidae) from Tokat, Turkey". *Turkish Journal of Entomology*, 13:197-203.

—. 2011a. "Review of Paleatcic and Australian species of *Bootanomyia* Girault 1915 (Hymenoptera: Torymidae: Megastigminae), with description of new species". *Turkish Journal of Zoology*, 35(2):123-157.

—. 2011b. "A new species of Westralianus Boucek 1988 (Hymenoptera: Torymidae: Megastigminae) from Turkey". *Turkish Journal of Zoology*, 35(1):23-27.

—. 2014. "First reports on the natural enemy fauna of the chestnut gall wasp, *Dryocosmus kuriphilus* Yasumatsu (Hymenoptera: Cynipidae) in Yalova, Turkey". *Türkiye Biyolojik Mücadele Dergisi*, 5(1):67-74.

—. 2015. "Diagnosis of *Megastigmus spp.* (Hymenoptera: Torymidae) reared from galls of *Leptocybe invasa* Fisher & Lasalle, 2004 (Hymenoptera: Eulophidae) on *Eucalyptus spp.* (Myrtaceae), with description of a new species from South Africa". *Entomofauna*, 36(43):561-580.

Gençer, Lütfiye. 2003. "The chalcidoid parasitoids of *Diplolepis mayri* Schld. (Hymenoptera: Cynipidae), a pest of Rosa canina L. in Sivas province". *Turkish Journal of Entomology*, 27(2):107-111.

Janšta Petr, and Cruaud Astrid, and Delvare Gérard, and Genson Guénaëlle, and Heraty John, and Křížková Barbora, and Rasplus, Jean-Yves. 2018. "Torymidae (Hymenoptera; Chalcidoidea) revised: molecular phylogeny, circumscription and reclassification of the familywith discussion of its biogeography and evolution of life–history traits". *Cladistics*. 34:627–651.doi:10.1111/cla.12228.

Mete, Özlem, and Mergen, Orhan. 2017. "The community components associated with two common rose gall wasps (Hymenoptera: Cynipidae: Diplolepidini) in Turkey". *Turkish Journal of Zoology*, 41:696-701.

Noyes, John S. 2019. *Universal Chalcidoidea Database*. Available at: www.nhm.ac.uk/chalcidoids (accessed January 2022).

Öncüer, Cezmi. 1991. "A Catalogue of the Parasites and Predators of Insect Pests of Turkey". *Ege Üniversitesi Ziraat Fakültesi Yayınları* [Ege University Agriculture Faculty Publ.] 505.

Stojanova, Anelia, and Civelek, Hasan Sungur, and Yörük, Berna, and Sari, Sabahat, and Atahan, Tuğçe. 2012. "Checklists of Turkish Eurytomidae Walker, 1832 and Torymidae Walker, 1833 (Hymenoptera: Chalcidoidea)". *Turkish Journal of Entomology*, 36(1):69-82.

CHAPTER TWELVE

FAMILY: MYMARIDAE HALIDAY, 1833

The family Mymaridae includes the smallest known insects, all of which are parasitoids in the eggs of other insects (Huber et al., 2009). The head has a transverse, membranous suture (sometimes referred to as a trabecula or carina) below the anterior ocellus and along the inner eye margins, with additonal sutures behind the ocelli being rare. The scutellum is typically divided transversely into an anterior and posterior part. The hind wing is petiolate, and the membrane of the disc does not extend to the base of the wing. These incests are minute to moderate in size, with length varying from 0.35-1.8mm. The antennal toruli is situated far apart, much closer to eye margin than to each other, and are positioned relatively high on the head, about midway between mouth margin and vertex.

The family currently comprises 103 genera and 1424 species, and there are nine genera and 19 species in Turkey (Noyes, 2019).

Genus *Anagrus* (Haliday, 1833)
Anagrus atomus (Linnaeus, 1767)
Publications with Turkey records: Noyes (2019), Öncüer (1991).
Associates: *Metarhizum anisopliae, Paecilomyces fumosoroseus* (Fungus), *Actinidia deliciosa* (Actinidiaceae), *Corylus avellana, Ostrya virginiana* (Betulaceae), *Sambucus nigra* (Caprifoliaceae), *Beta vulgaris, Chenopodium sp.* (Chenopodiaceae), *Cucumis sativus* (Cucurbitaceae), *Ricinus communis* (Euphorbiaceae), *Medicago sativa, Phaseolus vulgaris, Vicia faba* (Fabaceae), *Fagus sylvatica, Quercus ilex, Quercus suber* (Fagaceae), *Gossypium sp.* (Malvaceae), *Zea mays* (Poaceae), *Portulaca sp.* (Portulaceae), *Malus pumila, Prunus domestica, Rosa sp., Rosa canina, Rosa multiflora* (Rosaceae), *Rubus sp., Rubus fruticosus* (Rubiaceae), *Capsicum annuum, Lycopersicon esculentum* (Solanaceae), *Ulmus sp.* (Ulmaceae), *Vitis sp., Vitis vinifera* (Vitaceae) (Noyes, 2019).

Hosts: *Arboridia adanae, Arboridia Kermanshah, Asymmetrasca decedens, Cicadula sexnotata, Circulifer tenellus, Dikrella sp., Edwardsiana crataegi, Edwardsiana prunicola, Edwardsiana rosae, Empoasca sp., Empoasca decendens, Empoasca decipiens, Empoasca flavescens, Empoasca maligna, Empoasca vitis, Erythroneura eburnean, Erythroneura pallidifrons, Erythroneura rorida, Hauptidia maroccana, Jacobiasca lybica, Lindbergina aurovittata, Lindbergina spoliate, Scaphoideus titanus, Solenopyx sulphurellus, Tettigella viridis, Typhlocyba sp., Typhlocyba pomeria, Typhlocyba quercus, Typhlocyba rosae, Zygina flammigera, Zygina pallidifrons, Zygina rhamni, Zyginidia sp., Zyginidia pullula* (Hemiptera: Cicadellidae), *Conomelus anceps, Delphacodes fairmairei, Megamelus notula* (Delphacidae), *Quadraspidiotus perniciosus* (Diaspididae), *Panstenon oxylus* (Hymenoptera: Pteromalidae), *Malacosoma neustria* (Lepidoptera: Lasiocampidae), *Yponomeuta malinellus* (Yponomeutidae), *Triticum aestivum* (Poaceae) (Noyes, 2019).
Distribution: Argentina, Austria, Belgium, Bulgaria, Canada, Cape Verde Islands, Chile, Croatia, Denmark, Egypt, Europe, Finland, France, Germany, Greece, Iran, Israel, Italy, Kirgizia, Korea, Macedonia, Moldova, New Zealand, Pakistan, People's Republic of China, Poland, Romania, Russia, Serbia, Spain, Sweden, Switzerland, Turkey, Turkmenistan, United Kingdom, United States of America, Yemen, Yugoslavia (pre-1991) (Noyes, 2019).

Genus *Anaphes* (Haliday, 1833)
Anaphes diana (Girault, 1911)
Publications with Turkey records: Noyes (2019).
Associates: *Lens esculenta, Medicago sp., Medicago sativa* (Fabaceae), *Fagus sylvatica* (Fagaceae) (Noyes, 2019).
Hosts: *Hypera sp., Sitona sp., Sitona bicolor, Sitona callosus, Sitona crinitus, Sitona cylindricollis, Sitona discoideus, Sitona flavescens, Sitona hispidulus, Sitona humeralis, Sitona inops, Sitona lineatus, Sitona sulcifrons, Sitona tenuis* (Coleoptera: Curculionidae) (Noyes, 2019).
Distribution: Algeria, Australia, Austria, Belgium, Bulgaria, Canada, England, Europe, France, Germany, Greece, Italy, Moldova, Morocco, Netherlands, New Zealand, Poland, Romania, Spain, Syria, Turkey, United States of America (Noyes, 2019).

Genus *Cosmocomoidea* (Howard, 1908)
Cosmocomoidea tremulae (Bakkendorf, 1934)
Publications with Turkey records: Noyes (2019).

Associates: *Populus sp*. (Salicaceae) (Noyes, 2019).
Hosts: *Idiocerus populi, Rhytidodus decimusquartus, Rhytidodusa decimusquartus* (Hemiptera: Cicadellidae) (Noyes, 2019).
Distribution: Denmark, Italy, Netherlands, Romania, Turkey, United Kingdom (Noyes, 2019).

Genus *Gonatocerus* (Nees, 1834)
Gonatocerus fuscicornis (Walker, 1846)
Publications with Turkey records: Noyes (2019).
Associates: *Medicago sativa* (Fabaceae), *Baldingera arundincea* (Poaceae) (Noyes, 2019).
Hosts: *Oscinella frit* (Diptera: Chloropidae) (Noyes, 2019).
Distribution: Belgium, Bulgaria, Denmark, England, Europe, Germany, Japan, Norway, People's Republic of China, Poland, Romania, Spain, Sweden, Turkey (Noyes, 2019).

Gonatocerus longicornis (Nees, 1834)
Publications with Turkey records: Noyes (2019).
Associates: *Corylus avellana* (Betulaceae), *Medicago sativa* (Fabaceae) (Noyes, 2019).
Hosts: *Cicadella viridis, Rhytidodus decimusquartus, Rhytidodusa decimusquartus, Tettigella viridis, Tettigoniella spectra* (Hemiptera: Cicadellidae) (Noyes, 2019).
Distribution: Austria, Bulgaria, Czech Republic, Czechoslovakia, Europe, Finland, France, Germany, Greece, India, Italy, Japan, Kirgizia, Korea, Netherlands, Palearctic, People's Republic of China, Romania, Russia, Slovakia, Sweden, Thailand, Turkey, United Kingdom (Noyes, 2019).

Genus *Litus* (Haliday, 1833)
Litus cynipseus (Haliday, 1833)
Publications with Turkey records: Noyes (2019).
Associates: *Fagus sylvatica* (Fagaceae) (Noyes, 2019).
Hosts: *Ocypus sp., Ocypus olens, Staphylinus sp., Staphylinus olens* (Coleoptera: Staphylinidae), *Limnobates sp.* (Hemiptera: Hydrometridae) (Noyes, 2019).
Distribution: Austria, Belgium, Bulgaria, Canada, Czech Republic, Denmark, Europe, Finland, France, Germany, Greece, Hungary, Italy, Japan, Kirgizia, Korea, Mexico, Moldova, Montenegro, Morocco, Nepal, Netherlands, Romania, Russia, Slovenia, Spain, Sweden, Switzerland, Turkey, United Kingdom, United States of America, Yugoslavia (Federal Republic) (Noyes, 2019).

Genus *Lymaenon* (Walker, 1846)
Lymaenon aureus (Girault, 1911)
Publications with Turkey records: Noyes (2019).
Associates: *Medicago sativa* (Fabaceae) (Noyes, 2019).
Hosts: *Quadraspidiotus perniciosus* (Hemiptera: Diaspididae) (Noyes, 2019).
Distribution: Argentina, Belgium, Bermuda, Brazil, Bulgaria, Canada, Chile, Egypt, Finland, France, Greece, India, Japan, Romania, Russia, Spain, Sweden, Turkey, United Kingdom, United States of America (Noyes, 2019).

Lymaenon litoralis (Haliday, 1833)
Publications with Turkey records: Noyes (2019).
Associates: *Medicago sativa* (Fabaceae), *Juncus effusus* (Juncaceae), *Triticum aestivum, Zea mays* (Poaceae), *Crataegus sp.* (Rosaceae) (Noyes, 2019).
Hosts: *Anthonomus grandis, Anthonomus quadrigibbus, Microlarinus lypriformis, Tachypterellus quadrigibbus* (Coleoptera: Curculionidae), *Aphis pomi, Rhopalosiphum padi* (Hemiptera: Aphididae), *Acocephalus sp., Anoplotettix putoni, Aphrodes sp., Cicadula sexnotata, Macrosteles sexnotatus, Zyginidia sohrab* (Cicadellidae), *Neoaliturus tenellus* (Tingidae) (Noyes, 2019).
Distribution: Argentina, Australia, Austria, Bahamas, Belgium, Brazil, Bulgaria, Chile, Czech Republic, Denmark, Egypt, Europe, Finland, Germany, Greece, India, Iran, Ireland, Italy, Japan, Macedonia, Mexico, Nearctic, Netherlands, Norway, Poland, Romania, Serbia, Slovakia, Spain, Sweden, Switzerland, Turkey, United Kingdom, United States of America (Noyes, 2019).

Lymaenon longior (Soyka, 1946)
Publications with Turkey records: Noyes (2019).
Associates: Unknown (Noyes, 2019)
Hosts: Unknown (Noyes, 2019).
Distribution: Austria, Belgium, Bulgaria, England, Germany, Netherlands, Spain, Turkey (Noyes, 2019).

Lymaenon thyrides (Debauche, 1948)
Publications with Turkey records: Noyes (2019).
Hosts: Unknown (Noyes, 2019).
Associates: Unknown (Noyes, 2019).
Distribution: Belgium, Bulgaria, England, Romania, Spain, Turkey (Noyes, 2019).

Genus *Ooctonus* (Haliday, 1833)
Ooctonus hemipterus (Haliday, 1833)
Publications with Turkey records: Noyes (2019).
Associates: Unknown (Noyes, 2019).
Hosts: Unknown (Noyes, 2019).
Distribution: Austria, Belgium, Bulgaria, Canada, Croatia, Czech Republic, Europe, Finland, France, Germany, Greece, Hungary, Ireland, Italy, Kirgizia, Morocco, Netherlands, Palearctic, Poland, Romania, Russia, Slovakia, Slovenia, Sweden, Switzerland, Turkey, United Kingdom, United States of America (Noyes, 2019).

Genus *Polynema* (Haliday, 1833)
Polynema fennicum (Soyka, 1946)
Publications with Turkey records: Noyes (2019).
Associates: Unknown (Noyes, 2019).
Hosts: Unknown (Noyes, 2019).
Distribution: Bulgaria, Finland, Netherlands, Turkey (Noyes, 2019).

Polynema fuscipes (Haliday, 1833)
Publications with Turkey records: Noyes (2019).
Associates: Unknown (Noyes, 2019).
Hosts: Unknown (Noyes, 2019).
Distribution: Belgium, Bulgaria, Chile, Denmark, England, Germany, Greece, Norway, Slovakia, Sweden, Turkey (Noyes, 2019).

Polynema laetum (Foerster, 1847)
Publications with Turkey records: Noyes (2019).
Associates: Unknown (Noyes, 2019).
Hosts: Unknown (Noyes, 2019).
Distribution: Germany, Romania, Turkey (Noyes, 2019).

Polynema ovulorum (Linnaeus, 1758)
Publications with Turkey records: Noyes (2019).
Hosts: *Aphrodes bicinctus* (Hemiptera: Cicadellidae), *Pieris brassicae* (Lepidoptera: Pieridae) (Noyes, 2019).
Distribution: Austria, Belgium, Bulgaria, Europe, France, Germany, Greece, Romania, Sweden, Turkey, United Kingdom (Noyes, 2019).

Polynema pusillum (Haliday, 1833)
Publications with Turkey records: Noyes (2019).
Associates: Unknown (Noyes, 2019).

Hosts: Unknown (Noyes, 2019).
Distribution: Bulgaria, Czech Republic, England, Germany, Netherlands, Romania, Turkey (Noyes, 2019).

Polynema tenue (Soyka, 1956)
Publications with Turkey records: Noyes (2019).
Associates: Unknown (Noyes, 2019).
Hosts: Unknown (Noyes, 2019).
Distribution: Poland, Romania, Turkey (Noyes, 2019).

Polynema valkenburgense (Soyka, 1931)
Publications with Turkey records: Noyes (2019).
Associates: Unknown (Noyes, 2019).
Hosts: Unknown (Noyes, 2019).
Distribution: Belgium, Bulgaria, Netherlands, Romania, Sweden, Switzerland, Turkey, United Kingdom (Noyes, 2019).

Genus *Stephanodes* (Enock, 1909)
Stephanodes similis (Foerster, 1847)
Publications with Turkey records: Noyes (2019).
Associates: *Medicago sativa* (Fabaceae) (Noyes, 2019).
Distribution: Argentina, Austria, Belgium, Bulgaria, Canada, Cape Verde Islands, Croatia, Cyprus, Czech Republic, Democratic Republic of Congo, Denmark, Europe, Finland, France, Germany, Greece, Hungary, Italy, Kirgizia, Macedonia, Moldova, Norway, Paraguay, Poland, Romania, Russia, Slovenia, Spain, Sweden, Switzerland, Turkey, Turkmenistan, Ukraine, United Kingdom, United States of America (Noyes, 2019).

References

Huber, John Theodore, and Viggiani, Gennaro, and Jesu, Riccardo. 2009. "Order Hymenoptera, family Mymaridae", *Arthropod fauna of the UAE*, 2: 270–297.

Noyes, John S. 2019. *Universal Chalcidoidea Database*. Available at: www.nhm.ac.uk/chalcidoids (accessed January 2022).

Öncüer, Cezmi. 1991. "A Catalogue of the Parasites and Predators of Insect Pests of Turkey". *Ege Üniversitesi Ziraat Fakültesi Yayınları* [Ege University Agriculture Faculty Publ.] 505.

CHAPTER THIRTEEN

FAMILY: ORMYRIDAE FOERSTER, 1856

Many species of Ormyridae are parasitoids of various gall-forming insects. They are known to attack cecidogenic Cynipids, Chalcids and Diptera. A few species may also parasitize phytophagous Eurytomids in seeds. In the tropical regions, certain species are known to be parasitoids of the inhabitants of galls on figs.

Main characteristics of Ormyridae are as follows: the gaster is never petiolate, highly convex, and exhibits characteristic coarse, pitted sculpture; the stigmal vein is very short, although sometimes stigma may be enlarged, resulting in apex of the uncus almost touching the anterior margin of forewing; petiole is strongly transverse, and not distinct; in females, the ovipositor is either not or hardly exserted, with the exserted part being much shorter than abdomen; the body is generally not very elongate, measuring about 1.1-6.7 mm in length.

The family currently includes three genera and 125 species. In Turkey, there are 12 species of one genus (Noyes, 2019).

Genus *Ormyrus* (Westwood, 1832)
Ormyrus ardahanensis (Doğanlar, 1991)
Publications with Turkey records: Doğanlar (1991), Noyes (2019).
Associates: Unknown (Noyes, 2019).
Hosts: Unknown (Noyes, 2019).
Distribution: Turkey (Noyes, 2019).

Ormyrus bingoeliensis (Doğanlar, 1991)
Publications with Turkey records: Doğanlar (1991), Noyes (2019).
Associates: *Quercus sp., Quercus brantii* (Fagaceae) (Noyes, 2019).
Hosts: *Chilaspis israeli* (Hymenoptera: Cynipidae) (Noyes, 2019).
Distribution: Iran, Turkey (Noyes, 2019).

Ormyrus diffinis (Fonscolombe, 1832)
Publications with Turkey records: Noyes (2019), Öncüer (1991).
Associates: *Centaurea sp.* (Asteraceae), *Melissa officinalis* (Fabaceae), *Quercus sp.* (Fagaceae), *Nepeta latifolia, Phlomis sp., Salvia sp., Salvia aethiopis, Salvia fruticose, Salvia lavandulifolia, Salvia nemorosa, Salvia officinalis, Salvia pratensis, Salvia triloba* (Lamiaceae), *Potentilla sp.* (Rosaceae) (Noyes, 2019).
Hosts: *Dasineura potentillae* (Diptera: Cecidomyiidae), *Aulacidea sp., Aylax sp., Aylax nemorosae, Aylax salvia, Isocolus sp., Liposthenes kerneri, Neaylax sp., Neaylax nemorosae, Panteliella sp., Phanacis sp., Rhodus cyprius, Salviela kezivi, Xestophanes sp., Xestophanes potentillae* (Hymenoptera: Cynipidae) (Noyes, 2019).
Distribution: Bulgaria, Caucasus, Croatia, Czech Republic, Czechoslovakia, Europe, France, Germany, Hungary, Iran, Israel, Kazakhstan, Macedonia, Moldova, Montenegro, Russia, Serbia, Slovakia, Spain, Tadzhikistan, Transcaucasus, Tselinograd Obl., Turkey, Turkmenistan, Ukraine, USSR, Yugoslavia (pre-1991) (Noyes, 2019).

Ormyrus gratiosus (Foerster, 1860)
Publications with Turkey records: Noyes (2019), Öncüer (1991).
Associates: *Carthamus oxyacantha, Centaurea sp., Centaurea aspera, Centaurea jacea, Centaurea melitensis, Centaurea nicaeensis, Centaurea nigra, Centaurea paniculate, Centaurea sadleriana, Centaurea scabiosa, Centaurea solstitialis, Centaurea sphaerocephala, Hieracium sp., Serratula sp., Serratula heterophylla* (Asteraceae), *Scabiosa sp.* (Dipsacaceae), *Medicago sativa, Onobrychis sativa* (Fabaceae), *Phlomis sp.* (Lamiaceae), *Potentilla sp., Potentilla argentea, Rosa sp.* (Rosaceae) (Noyes, 2019).
Hosts: *Terellia sp., Urophora sp.* (Diptera: Tephritidae), *Aulacidea sp., Aylax sp., Aylax rogenhoferi, Aylax scabiosae, Diastrophus mayri, Diplolepis sp., Isocolus sp., Isocolus jaceae, Isocolus lichtensteini, Isocolus rogenhoferi, Isocolus scabiosae, Isocolus serratulae, Isocolus tauricus, Isocolus tinctorius, Xestophanes sp.* (Hymenoptera: Cynipidae) (Noyes, 2019).
Distribution: Armenia, Austria, Bulgaria, Caucasus, Czech Republic, England, Europe, Finland, France, Georgia, Germany, Hungary, Iran, Kazakhstan, Macedonia, Netherlands, Poland, Russia, Serbia, Slovakia, Spain, Sweden, Switzerland, Tselinograd Obl., Turkey, Turkmenistan, Ukraine, USSR, Uzbekistan (Noyes, 2019).

Ormyrus kazovaensis (Doğanlar, 1991)
Publications with Turkey records: Doğanlar (1991), Noyes (2019).
Associates: Unknown (Noyes, 2019).

Hosts: Unknown (Noyes, 2019).
Distribution: Turkey (Noyes, 2019).

Ormyrus longicornis (Boucek, 1969)
Publications with Turkey records: Noyes (2019).
Associates: Unknown (Noyes, 2019).
Hosts: Unknown (Noyes, 2019).
Distribution: Azerbaijan, Bulgaria, Europe, Georgia, Russia, Turkey (Noyes, 2019).

Ormyrus nitidulus (Fabricius, 1804)
Publications with Turkey records: Noyes (2019), Öncüer (1991).
Associates: *Andricus amblycerus, Andricus aries, Andricus askewi, Andricus caliciformis, Andricus caputmedusae, Andricus conglomeratus, Andricus conificus, Andricus coriarius, Andricus coronatus, Andricus corruptrix, Andricus curtisii, Andricus curvator, Andricus dentimitratus, Andricus fecundatrix, Andricus galeatus, Andricus gemmeus, Andricus glutinosus, Andricus grossulariae, Andricus hartigi, Andricus hispanicus, Andricus hungaricus, Andricus hystrix, Andricus infectorius, Andricus kollari, Andricus lignicolus, Andricus lucidus, Andricus megalucidus, Andricus mitratus, Andricus moreae, Andricus multiplicatus, Andricus polycerus, Andricus quercuscalicis, Andricus quercusradicis, Andricus quercustozae, Andricus seckendorffi, Andricus sieboldin, Andricus singularis, Andricus sternlichti, Andricus testaceipes, Andricus theophrastea, Andricus tomentosus, Andricus truncicolus, Aphelonyx cerricola, Biorhiza pallida, Cerroneuroterus lanuginosus, Cynips cornifex, Cynips disticha, Cynips divisa, Cynips longiventris, Cynips quercus, Cynips quercusfolii, Dryocosmus kuriphilus, Neuroterus anthracinus, Synophrus politus, Trigonaspis synaspis, Macroneura vesicularis* (Eupelmidae), *Pteromalus papaveris* (Pteromalidae), *Quercus sp., Quercus brantii, Quercus cerris, Quercus faginea, Quercus frainetto, Quercus infectoria, Quercus ithaburensis, Quercus lusitanica, Quercus macranthera, Quercus petraea, Quercus pubescens, Quercus pyrenaica, Quercus robur* (Fagaceae), *Rosa sp.* (Rosaceae) (Noyes, 2019).
Hosts: *Oligotrophus bergenstammi* (Diptera: Cecidomyiidae), *Adleria sp., Andricus sp., Andricus aestivalis, Andricus askewi, Andricus caputmedusae, Andricus conglomeratus, Andricus conificus, Andricus coriarius, Andricus coronatus, Andricus curtisii, Andricus curvator, Andricus fecundatrix, Andricus galeata, Andricus galeatus, Andricus gallaetincticorne, Andricus glutinosus, Andricus grossulariae, Andricus hartigi, Andricus hispanicus, Andricus kollari, Andricus lignicolus,*

Andricus lucidus, Andricus mayri, Andricus megalucidus, Andricus mitratus, Andricus multiplicatus, Andricus polycerus, Andricus pseudoaries, Andricus quercuscalicis, Andricus quercustozae, Andricus sternlichti, Andricus testaceipes, Andricus tomentosus, Andricus truncicolus, Aphelonyx cerricola, Aphelonyx persica, Biorhiza sp., Biorhiza pallida, Cynips sp., Cynips conglomerate, Cynips conifica, Cynips divisa, Cynips lignicola, Cynips longiventris, Cynips quercus, Cynips quercusfolii, Diplolepis sp., Diplolepis rosae, Rhodites rosae, Trigonaspis megaptera (Hymenoptera: Cynipidae) (Noyes, 2019).
Distribution: Algeria, Andorra, Austria, Azerbaijan, Belgium, Bosnia Hercegovina, Bulgaria, Caucasus, Croatia, Czech Republic, Czechoslovakia, Denmark, England, Europe, France, Georgia, Germany, Greece, Hungary, Iran, Italy, Jordan, Macedonia, Netherlands, North Africa, Romania, Serbia, Slovakia, Slovenia, Spain, Sweden, Switzerland, Turkey, Ukraine, United States of America, USSR, Yugoslavia (pre-1991) (Noyes, 2019).

Ormyrus orientalis (Walker, 1871)
Publications with Turkey records: Noyes (2019).
Associates: *Eryngium campestre* (Apiaceae), *Phoenix sylvestris* (Arecaceae), *Artemisia herba-alba, Carduus acanthoides, Carduus pycnocephalus, Carthamus oxyacantha, Carthamus tinctorius, Centaurea sp., Centaurea pannonica, Centaurea sadleriana, Inula viscosa, Onopordum nervosum, Serratula bracteifolia* (Asteraceae), *Gypsophila struthium* (Caryophyllaceae), *Cajanus cajan, Retama sphaerocarpa* (Fabaceae), *Quercu coccifera* (Fagaceae), *Hordeum leporinum* (Poaceae), *Reseda lutea* (Resedaceae), *Rosa sp.* (Rosaceae) (Noyes, 2019).
Hosts: *Melanagromyza obtuse, Phytomyza orobanchia* (Diptera: Agromyzidae), *Lasioptera eryngii* (Cecidomyiidae), *Lonchaea sp.* (Lonchaeidae), *Acanthiophilus helianthin, Myopites sp., Myopites frauenfeldi, Myopites inulaedyssentericae, Tephritis sp., Tephritis stictica* (Tephritidae*), Aylax sp., Diastrophus rubi, Diplolepis sp., Isocolus sp., Isocolus rogenhoferi, Neuroterus quercusbaccarum, Phanacis caulicola, Timaspis phoenixopodos* (Hymenoptera: Cynipidae) (Noyes, 2019).
Parasitoid hosts: *Eurytoma sp., Eurytoma tristis* (Eurytomidae) (Noyes, 2019).
Distribution: Afghanistan, Afrotropical, Azerbaijan, Belarus, Bulgaria, Canary Islands, Caucasus, Croatia, Czech, Republic, Czechoslovakia, Dominican Republic, Europe, France, Georgia, Germany, Greece, Hungary, India, Iran, Iraq, Italy, Macedonia, Moldova, Montenegro, Oriental,

Pakistan, Russia, Serbia, Slovakia, Spain, Sri Lanka, Tadzhikistan, Turkey, Turkmenistan, Ukraine, USSR, Uzbekistan (Noyes, 2019).

Ormyrus pomaceus (Geoffroy, 1785)
Publications with Turkey records: Doğanlar (2014), Noyes (2019).
Associates: *Andricus alniensis, Andricus amblycerus, Andricus aries, Andricus burgundus, Andricus caliciformis, Andricus callidoma, Andricus caputmedusae, Andricus cecconi, Andricus conglomeratus, Andricus conificus, Andricus coriarius, Andricus coronatus, Andricus corruptrix, Andricus curtisii, Andricus curvator, Andricus cydoniae, Andricus dentimitratus, Andricus fecundatrix, Andricus floridus, Andricus galeatus, Andricus gallaeurnaeformis, Andricus gemmeus, Andricus glandulae, Andricus glutinosus, Andricus grossulariae, Andricus hartigi, Andricus hispanicus, Andricus hungaricus, Andricus hystrix, Andricus infectorius, Andricus inflator, Andricus kollari, Andricus lignicolus, Andricus lucidus, Andricus megalucidus, Andricus melikai, Andricus mitratus, Andricus multiplicatus, Andricus pictus, Andricus polycerus, Andricus pseudoinflator, Andricus quadrilineatus, Andricus quercuscalicis, Andricus quercusradicis, Andricus quercusramuli, Andricus quercustozae, Andricus schroeckingeri, Andricus seckendorffi, Andricus seminationis, Andricus singularis, Andricus solitarius, Andricus stefanii, Andricus sternlichti, Andricus testaceipes, Andricus truncicolus, Andricus vindobonensis, Aphelonyx cerricola, Biorhiza pallida, Cerroneuroterus lanuginosus, Chilaspis nitida, Cynips agama, Cynips cornifex, Cynips disticha, Cynips divisa, Cynips longiventris, Cynips quercus, Cynips quercusfolii, Dryocosmus cerriphilus, Dryocosmus kuriphilus, Neuroterus albipes, Neuroterus anthracinus, Neuroterus numismalis, Neuroterus quercusbaccarum, Neuroterus tricolor, Plagiotrochus amenti, Plagiotrochus australis, Plagiotrochus britaniae, Plagiotrochus burnayi, Plagiotrochus coriaceus, Plagiotrochus gallaeramulorum, Plagiotrochus panteli, Plagiotrochus quercusilicis, Plagiotrochus razeti, Plagiotrochus suberi, Pseudoneuroterus macropterus, Pseudoneuroterus saliens, Trigonaspis brunneicornis, Trigonaspis mendesi, Trigonaspis synaspis* (Cynipidae), *Macroneura vesicularis* (Eupelmidae), *Pteromalus papaveris* (Pteromalidae), *Castanea sp., Castanea sativa, Quercu coccifera, Quercus sp., Quercus brantii, Quercus castaneifolii, Quercus cerris, Quercus dentata, Quercus faginea, Quercus frainetto, Quercus ilex, Quercus infectoria, Quercus ithaburensis, Quercus lusitanica, Quercus macranthera, Quercus mongolica, Quercus pedunculiflora, Quercus persica, Quercus petraea, Quercus pubescens, Quercus pyrenaica,*

Quercus robur, Quercus suber (Fagaceae), *Phyllostachys sp.* (Poaceae) (Noyes, 2019).
Hosts: *Andricus sp., Andricus aestivalis, Andricus callidoma, Andricus conglomeratus, Andricus coriarius, Andricus corruptor, Andricus corruptrix, Andricus curvator, Andricus cydoniae, Andricus fecundatrix, Andricus galeata, Andricus grossulariae, Andricus hartigi, Andricus hispanicus, Andricus hungaricus, Andricus infectorius, Andricus inflator, Andricus lignicolus, Andricus lucidus, Andricus mayri, Andricus megalucidus, Andricus multiplicatus, Andricus paradoxus, Andricus pictus, Andricus polycerus, Andricus pseudoaries, Andricus pseudoinflator, Andricus quercuscalicis, Andricus quercusradicis, Andricus quercusramuli, Andricus quercustozae, Andricus seckendorffi, Andricus singularis, Andricus solitarius, Andricus sternlichti, Andricus superfetationis, Andricus vindobonensis, Aphelonyx sp., Aphelonyx cerricola, Biorhiza sp., Biorhiza pallida, Callirhytis rufescens, Chilaspis Israeli, Cynips sp., Cynips amblycera, Cynips aries, Cynips calycis, Cynips conglomerate, Cynips coriaria, Cynips cornifex, Cynips coronaria, Cynips disticha, Cynips divisa, Cynips glutinosa, Cynips longiventris, Cynips quercus, Cynips quercusfolii, Diastrophus mayri, Diplolepis sp., Diplolepis divisa, Diplolepis folii, Dryocosmus kuriphilus, Neuroterus sp., Neuroterus lanuginosus, Neuroterus politus, Neuroterus quercusbaccarum, Neuroterus saliens, Neuroterus saltans, Plagiotrochus sp., Plagiotrochus australis, Plagiotrochus quercusilicis, Pseudoneuroterus macropterus, Synophrus sp., Trigonaspis brunneicornis, Trigonaspis mendesi, Trigonaspis synaspis* (Hymenoptera: Cynipidae) (Noyes, 2019).
Parasitoid hosts: *Torymus beneficus* (Torymidae) (Noyes, 2019).
Distribution: Andorra, Austria, Bulgaria, Caucasus, Croatia, Czech Republic, Denmark, England, Europe, Finland, France, Germany, Greece, Hungary, Iran, Ireland, Italy, Japan, Jordan, Korea, Moldova, Montenegro, Netherlands, Oriental, Palearctic, People's Republic of China, Poland, Romania, Russia, Serbia, Slovakia, Spain, Sweden, Turkey, Ukraine, USSR (Noyes, 2019).
Ormyrus tschami (Doğanlar, 1991)
Publications with Turkey records: Doğanlar (2014), Noyes (2019).
Associates: Unknown (Noyes, 2019).
Hosts: Unknown (Noyes, 2019).
Distribution: Turkey (Noyes, 2019).

Ormyrus wachtli (Mayr, 1904)
Publications with Turkey records: Noyes (2019).
Associates: *Centaurea sp., Centaurea jacea, Hieracium sp., Serratula sp.* (Asteraceae), *Quercus sp.* (Fagaceae), *Phlomis sp., Salvia sp., Salvia lavandulifolia, Salvia nemorosa, Salvia officinalis, Salvia pratensis, Salvia verbenaca* (Lamiaceae), *Papaver sp.* (Papaveraceae), *Salix sp.* (Salicaceae) (Noyes, 2019).
Hosts: *Centaurea jacea* (Asteraceae), *Rabdophaga dubia* (Diptera: Cecidomyiidae), *Aulacidea sp., Aulax salvia, Aylax salvia, Isocolus sp., Neaylax sp., Neaylax salvia, Neaylax verbenacus, Phanacis sp., Phanacis centaureae* (Hymenoptera: Cynipidae) (Noyes, 2019).
Distribution: Austria, Bulgaria, Croatia, Czech Republic, Europe, France, Hungary, Iran, Italy, Kazakhstan, Moldova, Romania, Russia, Serbia, Slovakia, Spain, Sweden, Tselinograd Obl., Turkey, Ukraine, USSR, Yugoslavia (pre-1991) (Noyes, 2019).

Ormyrus yeschilirmaka (Doğanlar, 1991)
Publications with Turkey records: Doğanlar (1991), Noyes (2019).
Associates: Unknown (Noyes, 2019).
Hosts: Unknown (Noyes, 2019).
Distribution: Turkey (Noyes, 2019).

References

Doğanlar, Mikdat. 1991. "Systematic studies on the species of *Cyrtosoma* Perris from Turkey and descriptions of some new species (Hymenoptera, Ormyridae)". *Turkish Journal of Entomology*, 15(2):71-87.

—. 2014. "First reports on the natural enemy fauna of the chestnut gall wasp, *Dryocosmus kuriphilus* Yasumatsu (Hymenoptera: Cynipidae) in Yalova, Turkey". *Türkiye Biyolojik Mücadele Dergisi*, 5(1):67-74.

Noyes, John S. 2019. *Universal Chalcidoidea Database*. Available at: www.nhm.ac.uk/chalcidoids (accessed January 2022).

Öncüer, Cezmi. 1991. "A Catalogue of the Parasites and Predators of Insect Pests of Turkey". *Ege Üniversitesi Ziraat Fakültesi Yayınları* [Ege University Agriculture Faculty Publ.] 505.

CHAPTER FOURTEEN

FAMILY: PERILAMPIDAE FORSTER, 1856

Perilampids are robust Chalcids with a large thorax that bears pockmarks. Their abdomen is small, shiny and triangular in shape. Some species exhibit a metallic green coloration resembling cuckoo wasps, while others are black. These wasps are commonly found on flowers, and most species are hyperparasitoids that attack Diptera and Hymenoptera which are parasitoids of caterpillars. Some species also attack free-living insects in various orders. For example, *Perilampus platygaster* Say is a hyperparasitoid that attacks *Meteorus dimididtus* Cresson, a braconid parasite of the grape leaf folder.

Similar to Eucharitids, the Perilampids lay their eggs on foliage, and the eggs hatch into planidium larvae. These planidia remain on the foliage, attaching to a passing host and penetrating its body cavity. If a hyperparasitic species enters a caterpillar that is not already parasitized, it usually does not develop. However, if the caterpillar is already parasitized, the Perilampid larva usually remains inactive within the caterpillar until the caterpillar's parasitoid has pupated, and then it attacks the parasitoid. Perilampids commonly target the pupae of Tachinids, Braconids, or Ichneumonids (see faculty.ucr.edu).

The main diagnostic characteristics of the Perilampids are as follows: the prepectus is in the same plane as the pronotum and fused with it. In the dorsal view, the pronotum has sides that are more or less parallel, and the medial portion is clearly visible behind the head. The fore wing has a moderately long marginal vein and short postmarginal and stigmal veins. Perilampids are moderately large, measuring about 1.3-5.5 mm in length.

The family currently includes 15 genera and 277 species placed in three subfamilies, and there are two subfamilies, five genera and 19 species in Turkey (Noyes, 2019).

Subfamily Chrysolampinae (Dalla Torre, 1898)
Genus *Brachyelatus* (Hoffer and Novicky, 1954)
Brachyelatus viridis (Hoffer and Novicky, 1954)
Publications with Turkey records: Doğanlar and Doğanlar (2014), Noyes (2019).
Associates: *Carduus bourgeanus* (Asteraceae) (Noyes, 2019).
Distribution: Austria, Azerbaijan, Czech Republic, Czechoslovakia, Hungary, Moldova, Spain, Turkey, Ukraine, USSR (Noyes, 2019).

Genus *Chrysolampus* (Spinola, 1811)
Chrysolampus adiyamanus (Doğanlar and Doğanlar, 2014)
Publications with Turkey records: Doğanlar and Doğanlar (2014), Noyes (2019).
Associates: *Cicer arietinum* (Fabaceae) (Noyes, 2019).
Distribution: Turkey (Noyes, 2019).

Chrysolampus akcakalenus (Doğanlar and Doğanlar, 2014)
Publications with Turkey records: Doğanlar and Doğanlar (2014), Noyes (2019).
Associates: *Triticum aestivum* (Poaceae) (Noyes, 2019).
Distribution: Turkey (Noyes, 2019).

Chrysolampus besnilus (Doğanlar and Doğanlar, 2014)
Publications with Turkey records: Doğanlar and Doğanlar (2014), Noyes (2019).
Associates: *Narcissus pseudonarsissus* (Amaryllidaceae) (Noyes, 2019).
Distribution: Turkey (Noyes, 2019).

Chrysolampus dortyolus (Doğanlar and Doğanlar, 2014)
Publications with Turkey records: Doğanlar and Doğanlar (2014), Noyes (2019).
Associates: Unknown (Noyes, 2019).
Hosts: Unknown (Noyes, 2019).
Distribution: Turkey (Noyes, 2019).

Chrysolampus gencerae (Doğanlar and Doğanlar, 2014)
Publications with Turkey records: Doğanlar and Doğanlar (2014), Noyes (2019).
Associates: Unknown (Noyes, 2019).
Hosts: Unknown (Noyes, 2019).
Distribution: Turkey (Noyes, 2019).

Chrysolampus pazarcikus (Doğanlar and Doğanlar, 2014)
Publications with Turkey records: Doğanlar and Doğanlar (2014), Noyes (2019).
Associates: *Narcissus pseudonarsissus* (Amaryllidaceae) (Noyes, 2019).
Distribution: Turkey (Noyes, 2019).

Chrysolampus splendidulus (Spinola, 1808)
Publications with Turkey records: Doğanlar and Doğanlar (2014), Noyes (2019).
Hosts: *Lyctus canaliculatus, Lyctus linearis* (Coleoptera: Lyctidae) (Noyes, 2019).
Distribution: Algeria, Caucasus, Czech Republic, Czechoslovakia, Europe, Germany, Hungary, Italy, Kazakhstan, Moldova, North Africa, Russia, Slovakia, Sweden, Transcaucasus, Tselinograd Obl., Tunisia, Turkey (Noyes, 2019).

Chrysolampus thenae (Walker, 1848)
Publications with Turkey records: Doğanlar and Doğanlar (2014), Noyes (2019), Öncüer (1991).
Associates: *Medicago sativa* (Fabaceae), *Betonica offinalis, Stachys officinalis* (Lamiaceae) (Noyes, 2019).
Hosts: *Meligethes sp., Meligethes pedicularis* (Coleoptera: Nitidulidae), *Macrosiphum rosae* (Hemiptera: Aphididae).
Distribution: Austria, Croatia, Czech Republic, Czechoslovakia, Europe, Finland, France, Germany, Hungary, Kazakhstan, Macedonia, Moldova, Montenegro, Netherlands, Poland, Russia, Serbia, Slovakia, Spain, Sweden, Tselinograd Obl., Turkey, Ukraine, United Kingdom, USSR (Noyes, 2019).

Chrysolampus yayladagus (Doğanlar and Doğanlar, 2014)
Publications with Turkey records: Doğanlar and Doğanlar (2014), Noyes (2019).
Associates: Unknown (Noyes, 2019).
Hosts: Unknown (Noyes, 2019).
Distribution: Turkey (Noyes, 2019).

Genus *Chrysomalla* (Foerster, 1859)
Chrysomalla gulsehirica (Doğanlar and Doğanlar, 2012)
Publications with Turkey records: Doğanlar and Doğanlar (2012), Noyes (2019).
Associates: Unknown (Noyes, 2019).

Hosts: Unknown (Noyes, 2019).
Distribution: Turkey (Noyes, 2019).

Chrysomalla nevsehirica (Doğanlar and Doğanlar, 2012)
Publications with Turkey records: Doğanlar and Doğanlar (2012), Noyes (2019).
Associates: Unknown (Noyes, 2019).
Hosts: Unknown (Noyes, 2019).
Distribution: Turkey (Noyes, 2019).

Chrysomalla roseri (Foerster, 1859)
Publications with Turkey records: Doğanlar and Doğanlar (2012), Noyes (2019), Öncüer (1991).
Associates: *Berteroa incana, Eruca vesicaria* (Brassicaceae), *Gypsophila arenaria* (Caryophyllaceae), *Medicago sativa* (Fabaceae), *Rosmarinus officinalis* (Lamiaceae) (Noyes, 2019).
Hosts: *Tychius flavus* (Coleoptera: Curculionidae) (Noyes, 2019).
Distribution: Austria, Azerbaijan, Czech Republic, Czechoslovakia, Europe, Germany, Hungary, Kazakhstan, Serbia, Spain, Tadzhikistan, Tselinograd Obl., Turkey, Ukraine, USSR, Yugoslavia (pre-1991) (Noyes, 2019).

Chrysomalla sekilica (Doğanlar and Doğanlar, 2012)
Publications with Turkey records: Doğanlar and Doğanlar (2012), Noyes (2019).
Associates: *Sinapis sp.* (Brassicaceae) (Noyes, 2019).
Distribution: Turkey (Noyes, 2019).

Chrysomalla stigmatica (Boucek, 1972)
Publications with Turkey records: Doğanlar and Doğanlar (2012), Noyes (2019).
Associates: Unknown (Noyes, 2019).
Hosts: Unknown (Noyes, 2019).
Distribution: Bulgaria, Jordan, Turkey (Noyes, 2019).
Chrysomalla turcica (Boucek, 1972)
Publications with Turkey records: Noyes (2019).
Associates: Unknown (Noyes, 2019).
Hosts: Unknown (Noyes, 2019).
Distribution: Turkey (Noyes, 2019).

Genus *Elatomorpha* (Zerova, 1970)
Elatomorpha nurdagica (Doğanlar and Doğanlar, 2014)

Publications with Turkey records: Doğanlar and Doğanlar (2014), Noyes (2019).
Associates: *Triticum aestivum* (Poaceae) (Noyes, 2019).
Distribution: Turkey (Noyes, 2019).

Subfamily Perilampinae (Foerster, 1856)
Genus *Perilampus* (Latreille, 1809)
Perilampus aeneus (Rossius, 1790)
Publications with Turkey records: Noyes (2019), Öncüer (1991).
Associates: *Brassica napus* (Brassicaceae), *Tamarix canariensis* (Tamaricaceae) (Noyes, 2019).
Hosts: *Lixus juncii* (Coleoptera: Curculionidae), *Athalia colibri, Athalia rosae, Tomostethus nigritus* (Hymenoptera: Tenthredinidae), *Carpocapsa pomonella* (Lepidoptera: Tortricidae) (Noyes, 2019).
Distribution: Bosnia Hercegovina, Croatia, Czech Republic, Europe, Germany, Hungary, Italy, Moldova, Netherlands, Russia, Serbia, Slovakia, Spain, Sweden, Turkey, United Kingdom, USSR (Noyes, 2019).

Perilampus catilus (Argaman, 1990)
Publications with Turkey records: Noyes (2019).
Associates: Unknown (Noyes, 2019).
Hosts: Unknown (Noyes, 2019).
Distribution: Turkey (Noyes, 2019).

References

Doğanlar, Mikdat, and Doğanlar, Oğuzhan. 2012. "Palearctic species of *Chrysomalla* Förster, 1859 (Hymenoptera: Perilampidae: Chrysolampinae), with descriptions of new species from Turkey". *Turkish Journal of Entomology*, 36(4):519-532.

—. 2014. "Review of Chrysolampinae Dalla Torre 1898 (Hymenoptera: Perilampidae) with descriptions of new species from Turkey". *Entomofauna*, 35(4):49-84.

Noyes, John S. 2019. *Universal Chalcidoidea Database*. Available at: www.nhm.ac.uk/chalcidoids (accessed January 2022).

Öncüer, Cezmi. 1991. "A Catalogue of the Parasites and Predators of Insect Pests of Turkey". *Ege Üniversitesi Ziraat Fakültesi Yayınları* [Ege University Agriculture Faculty Publ.] 505.

Perilampidae, https://faculty.ucr.edu/~legneref/identify/perilamp.htm (accessed October 2022).

CHAPTER FIFTEEN

FAMILY: PTEROMALIDAE DALMAN, 1820

The Pteromalidae is one of the three largest families of Chalcidoidea. The family does not have clearly defined characteristics, and its members are generally identified by the absence of distinctive features that are used to define other families. In fact, some subfamilies are included in Pteromalidae primarily because they do not fit into any other chalcid family. There is still much work to be done in establishing a more stable classification for the Chalcidoids, and reassessing the Pteromalid groups, will be an integral part of this process.

As currently defined, the family is characterized by having 5-segmented tarsi, usually 5–7 funicular segments in the antenna, rarely exserted ovipositors, well-developed postmarginal and stigmal veins in the fore wing, and a mesopleuron that is not swollen or convex- in general lacking distinctive features that are characteristics of other families.

The biology and host associations of pteromalids are extremely varied, with most species being idiobionts. Many species develop as ectoparasitoids of larvae and pupae of Lepidoptera (moths and butterflies), Diptera (flies), Coleoptera (beetles) and Hymenoptera. They commonly attack concealed hosts such as leaf-miners and gall-inducers. Other species are idiobiont endoparasitoids, frequently targeting Lepidopterous pupae.

Most members of the tribe Miscogasterini are koinobiont larval-pupal parasitoids of Diptera, specifically Agromyzidae. *Tomicobia* species, on the other hand, develop on adult Coleoptera. Additionally, some Pteromalids exhibits predatory behavior rather than being strictly parasitic, and others are phytophagous. As a result, the Pteromalidae encompass almost the entire range of biologies found within the Chalcidoidea.

Several species have been successfully used in biological control; however, they do not match the success of other parasitoid groups such as Aphelinidae, Encyrtidae and Braconidae (Stringer et al., 2012).

The family currently includes 588 genera and 3506 species placed in 31 subfamilies and there are 15 subfamilies, 67 genera and 179 species in Turkey (Noyes, 2019).

Subfamily Asaphinae (Ashmead, 1904)
Genus *Asaphes* (Walker, 1834)
Asaphes suspensus (Nees, 1834)
Publications with Turkey records: Öncüer (1991), Noyes (2019).
Associates: *Aphidencyrtus aphidivorus* (Encyrtidae), *Alloxysta megourae* (Figitidae), *Acer sp.* (Aceraceae), *Daucus carota* (Apiaceae), *Nerium indicum* (Apocynaceae), *Tetrapanax papyriferus* (Araliaceae), *Artemisia sp., Chrysanthemum sp., Cynara scolymus, Solidago sp., Sonchus sp., Sonchus oleraceus* (Asteraceae), *Betula sp.* (Betulaceae), *Brassica campestris, Brassica oleracea, Raphanus sativus* (Brassicaceae), *Sambucus sieboldiana* (Caprifoliaceae), *Euonymus europaea* (Celastraceae), *Chenopodium album* (Chenopodiaceae), *Elaeagnus umbellata* (Elaeagnaceae), *Medicago sativa, Pisum sativum, Robinia pseudoacacia, Trifolium pratense, Vicia faba* (Fabaceae), *Fagus sylvatica* (Fagaceae), *Ribes aureum* (Grossulariaceae), *Iris germanica* (Iridaceae), *Stauntonia hexaphylla* (Lardizabalaceae), *Parabenzoin praecox* (Lauraceae), *Magnolia kobus* (Magnoliaceae), *Ligustrum obtusifolium, Syringa emodi, Syringa reticulata* (Oleaceae), *Epilobium sp.* (Onagraceae), *Abies firma, Pinus densiflora, Pinus thunbergii* (Pinaceae), *Pittosporum tobira* (Pittosporaceae), *Limonium sp.* (Plumbaginaceae), *Sorghum sp., Sorghum bicolor, Triticum aestivum, Zea mays* (Poaceae), *Caltha palustris* (Ranunculaceae), *Chaenomeles japonica, Malus sp., Malus domestica, Malus pumila, Prunus sp., Prunus padus, Prunus persica, Rosa sp., Sorbaria sorbifolia, Spiraea thunbergia, Stephanandra incisa* (Rosaceae), *Salix sp., Salix vulpina* (Salicaceae), *Smilax china* (Smilacaceae), *Lycopersicon esculentum, Solanum tuberosum* (Solanaceae), *Staphylea bumalda* (Staphyleaceae), *Celtis sp.* (Ulmaceae) (Noyes, 2019).
Hosts: *Lysaphidus platensis, Liriomyza congesta* (Diptera: Agromyzidae), *Rabdophaga salicisstrobiloides* (Cecidomyiidae), *Acyrthosiphon ibotum, Acyrthosiphon kondoi, Acyrthosiphon magnoliae, Acyrthosiphon muradachi, Acyrthosiphon pisum, Acyrthosiphon porosum, Acyrthosiphon solani, Acyrthosiphon syringae, Anuraphis crataegifoliae, Aphis sp., Aphis*

armata, Aphis craccivora, Aphis fabae, Aphis gossypii, Aphis nasturtii, Aphis nerii, Aphis pomi, Aphis rosae, Aphis rumicis, Aphis sorbi, Aphis spiraecola, Aulacorthum solani, Ayrthosiphon pisum, Brachycaudus cardui, Brevicoryne brassicae, Calaphis betulaecolens, Callipterinella calliptera, Capitophorus sp., Capitophorus horni, Cavariella aegopodii, Cavariella araliae, Cavariella salicicola, Cinara sp., Cinara pineti, Coloradoa artemisicola, Cryptomyzus ribis, Dactynotus sp., Dactynotus rudbeckiae, Dysaphis plantaginea, Eriosoma lanigerum, Euceraphis punctipennis, Eulachnus thunbergia, Hayhurstia atriplicis, Hyalopterus pruni, Hyperomyzus lactucae, Indomegoura indica, Kakimia houghtonensis, Lipaphis erysimi, Lipaphis pseudobrassicae, Macrosiphoniella artemisiella, Macrosiphoniella sanborni, Macrosiphum sp., Macrosiphum akebiae, Macrosiphum avenae, Macrosiphum cornelli, Macrosiphum euphorbiae, Macrosiphum ibarae, Macrosiphum pisi, Macrosiphum rosae, Macrosiphum smilacifoliae, Melanaphis bambusae, Metopolophium dirhodum, Myzus sp., Myzus houghtonensis, Myzus malisuctus, Myzus persicae, Nearctaphis bakeri, Nectarosiphon persicae, Periphyllus sp., Periphyllus californiensis, Periphyllus negundinis, Pleotrichophorus glandulosus, Pterocomma sp., Rhodobium porosum, Rhopalosiphoninus calthae, Rhopalosiphum sp., Rhopalosiphum maidis, Rhopalosiphum padi, Rhopalosiphum pseudobrassicae, Schizaphis graminum, Schizolachnus sp., Shivaphis celti, Sipha sp., Sitobion avenae, Sitobion fragariae, Staticobium sp., Therioaphis sp., Therioaphis trifolii, Toxoptera aurantia, Toxoptera odinae, Tuberocephalus sakurae, Tuberolachnus salignus, Unisitobion sorbi, Uroleucon cirsii, Uroleucon sonchi (Hemiptera: Aphididae) (Noyes, 2019).

Parasitoid hosts: *Aphelinus sp., Aphelinus asychis, Aphelinus jucundus, Aphelinus mali, Aphelinus semiflavus* (Hymenoptera: Aphelinidae), *Aphidius sp., Aphidius areolatus, Aphidius avenae, Aphidius cingulatus, Aphidius ervi, Aphidius gifuensis, Aphidius matricariae, Aphidius nigripes, Aphidius phorodontis, Aphidius polygonaphis, Aphidius rhopalosiphi, Aphidius salicis, Aphidius sicarius, Aphidius smithi, Aphidius testaceipes, Binodoxys centaureae, Diaeretiella sp., Diaeretiella rapae, Diaeretus leucopterus, Diaeretus rapae, Diaretiella rapae, Dyscritulus sp., Ephedrus sp., Ephedrus californicus, Ephedrus nacheri, Ephedrus persicae, Ephedrus plagiator, Ephedrus salicicola, Lysaphidus pleotrichophori, Lysiphlebus japonicus, Lysiphlebus testaceipes, Pauesia abietis, Pauesia akamatsucola, Pauesia momicola, Pauesia salignae, Pauesia unilachni, Praon sp., Praon aguti, Praon barbatum, Praon capitophori, Praon dorsale, Praon exsoletum, Praon flavicorne, Praon flavinode, Praon simulans, Praon volucre, Toxares sp., Trioxys sp.,*

Trioxys shivaphis, Trioxys utilis (Braconidae), *Aphidencyrtus aphidivorus* (Encyrtidae), *Alloxysta sp.* (Figitidae), *Dendrocerus carpenteri* (Megaspilidae) (Noyes, 2019).
Distribution: Argentina, Belgium, Brazil, Bulgaria, Canada, Canary Islands, Chile, Croatia, Czech Republic, England, Europe, France, Germany, India, Iran, Israel, Italy, Japan, Kazakhstan, Kirgizia, Korea, Mexico, Morocco, Nearctic, Nepal, Netherlands, Norway, Pakistan, People's Republic of China, Poland, Portugal, Romania, Russia, Serbia, Slovakia, Spain, Sweden, Switzerland, Tselinograd Obl., Turkey, Ukraine, United States of America (Noyes, 2019).

Asaphes vulgaris (Walker, 1834)
Publications with Turkey records: Bulut and Gocmen (2000), Noyes (2019), Öncüer (1991).
Associates: *Sipha maydis* (Hemiptera: Aphididae), *Lasius niger, Myrmica laevinodis* (Hymenoptera: Formicidae), *Acer opulifolium, Acer platanoides, Acer pseudoplatanus* (Aceraceae), *Daucus carota, Pastinaca sp.* (Apiaceae), *Artemisia abrotanum, Artemisia herba-alba, Aster tripolium, Calendula sp., Centaurea nigra, Cirsium sp., Cirsium arvense, Senecio jacobaea, Sonchus sp., Tanacetum vulgare* (Asteraceae), *Alnus sp., Corylus avellana* (Betulaceae), *Sambucus nigra* (Caprifoliaceae), *Cerastium sp.* (Caryophyllaceae), *Euonymus atropurpurea, Euonymus europaea* (Celastraceae), *Atriplex halimus, Beta vulgaris, Chenopodium sp.* (Chenopodiaceae), *Cucurbita sativus* (Cucurbitaceae), *Juniperus thurifera* (Cupressaceae), *Succissa pratensis* (Dipsacaceae), *Vaccinium vitisidaea* (Ericaceae), *Lathyrus odoratus, Lens esculenta, Medicago sativa, Ononis sp., Ononis tridentata, Pisum sativum, Vicia sp.* (Fabaceae), *Fagus sylvatica* (Fagaceae), *Ephedra nebrodensis* (Gnetaceae), *Epilobium sp.* (Onagraceae), *Pinus sp., Pinus halepensis* (Pinaceae), *Limonium sp.* (Plumbaginaceae), *Molinia sp., Poa annua, Triticum aestivum, Zea mays* (Poaceae), *Caltha palustris* (Ranunculaceae), *Malus sp., Malus pumila, Prunus sp., Prunus cerasus, Pyrus communis, Pyrus malus* (Rosaceae), *Galium sp., Rubus sp., Rubus idaeus* (Rubiaceae), *Salix sp.* (Salicaceae), *Datura sp., Solanum tuberosum* (Solanaceae), *Tamarix canariensis* (Tamaricaceae), *Tilia sp.* (Tiliaceae), *Parietaria sp., Urtica dioica* (Urticaceae) (Noyes, 2019).
Hosts: *Lysaphidus platensis* (Coleoptera: Curculionidae), *Anthonomus pomorum, Tychius medicaginis, Liriomyza lutea* (Diptera: Agromyzidae), *Aphidoletes cucumeris, Dasineura ignorata* (Cecidomyiidae), *Syrphus ribesii* (Syrphidae), *Acyrthosiphon caraganae, Acyrthosiphon onobrychis, Acyrthosiphon pisum, Amphorophora ampullata, Amphorophora rubi,*

Anuraphis persicaeniger, Anuraphis schwartzi, Aphididae unspecified, Aphis sp., Aphis armata, Aphis craccivora, Aphis epilobii, Aphis euonymi, Aphis fabae, Aphis glycines, Aphis hederae, Aphis holci, Aphis maidis, Aphis parietariae, Aphis pomi, Aphis rumicis, Aphis sambuci, Aphis tripolii, Brachycaudus helichrysi, Brachycolus korotnewi, Brachycolus noxius, Brevicoryne brassicae, Capitophorus ribis, Cavariella aegopodii, Cinara cembrae, Cinara piceae, Cinara pinea, Corylobium avellanae, Dactynotus solidaginis, Drepanosiphum platanoidis, Dysaphis plantaginea, Elatobium abietinum, Eriosoma lanigerum, Eucallipterus tiliae, Eulachna sp., Hayhurstia atriplicis, Hyalopterus phragmiticola, Hyalopterus pruni, Hyperomyzus lactucae, Macrosiphum avenae, Macrosiphum euphorbiae, Macrosiphum funestum, Macrosiphum granarium, Macrosiphum hellebore, Macrosiphum rosae, Metopeurum fuscoviride, Metopolophium albidum, Myzocallis schreiberi, Myzus sp., Myzus ascalonicus, Myzus cerasi, Myzus persicae, Nasonovia ribisnigri, Nectarosiphon persicae, Periphyllus sp., Phorodon pruni, Rhopalosiphum insertum, Rhopalosiphum padi, Schizaphis graminum, Schizolachnus pineti, Sitobion sp., Sitobion avenae, Staticobium sp., Therioaphis trifolii, Toxoptera aurantia, Tuberculoides annulatus, Uroleucon cirsii, Uroleucon jaceae, Uroleucon sonchi, Wahlgreniella ossiannilssoni (Hemiptera: Aphididae), *Eulecanium fletcheri* (Coccidae), *Aulacaspis rosae* (Diaspididae), *Phenacoccus aceris, Pseudococcus adonidum* (Pseudococcidae), *Psylla mali, Psylla pyricola* (Psyllidae), *Phaenoglyphis villosa* (Hymenoptera: Cynipidae) (Noyes, 2019).

Parasitoid hosts: *Aphelinus flavus, Aphelinus mali, Aphelinus semiflavus, Aphelinus varipes* (Aphelinidae), *Aphidius sp., Aphidius avenae, Aphidius ervi, Aphidius funebris, Aphidius matricariae, Aphidius medicaginis, Aphidius nigripes, Aphidius picipes, Aphidius polygoni, Aphidius rhopalosiphi, Aphidius rosae, Aphidius smithi, Aphidius sonchi, Aphidius uzbekistanicus, Binodoxys angelicae, Diaeretiella sp., Diaeretiella rapae, Diaeretus rapae, Diaretiella rapae, Ephedrus sp., Ephedrus lacertosus, Ephedrus persicae, Ephedrus plagiator, Lysiphlebus sp., Lysiphlebus cardui, Lysiphlebus dissolutus, Lysiphlebus fabarum, Lysiphlebus testaceipes, Monoctonus cerasi, Pauesia unilachni, Praon sp., Praon dorsale, Praon exsoletum, Praon volucre, Trioxys sp., Trioxys aceris, Trioxys pallidus* (Braconidae), *Charips longicornis* (Charipidae), *Trechnites insidiosus* (Encyrtidae), *Alloxysta fractricornis* (Figitidae), *Diplazon laetatorius* (Ichneumonidae) (Noyes, 2019).

Parasitoids: *Dendrocerus carpenteri* (Megaspilidae), *Micradelus rotundus* (Pteromalidae) (Noyes, 2019).

Distribution: Algeria, Argentina, Armenia, Australia, Austria, Belgium, Brazil, Bulgaria, Canada, Caribbean, Channel, Islands, Chile, Croatia, Czech Republic, Czechoslovakia, Europe, Finland, France, Germany, Greece, Greenland, Hungary, Iceland, India, Iran, Italy, Kazakhstan, Kirgizia, Korea, Lithuania, Macedonia, Moldova, Montenegro, Morocco, Nearctic, Nepal, Netherlands, Norway, Pakistan, People's Republic of China, Poland, Portugal, Romania, Russia, Serbia, Slovakia, Slovenia, Spain, St Vincent & Grenadines, Sweden, Switzerland, Tselinograd Obl., Turkey, Turkmenistan, Ukraine, United Kingdom, United States of America, USSR, Uzbekistan, Yugoslavia (pre-1991) (Noyes, 2019).

Subfamily Ceinae (Boucek, 1961)
Genus *Cea* (Walker, 1837)
Cea pulicaris (Walker, 1837)
Publications with Turkey records: Noyes (2019), Öncüer (1991).
Associates: *Peucedanum oreoselinum* (Apiaceae) (Noyes, 2019).
Hosts: *Phytomyza pauliloewi* (Diptera: Agromyzidae) (Noyes, 2019).
Distribution: Algeria, Czech Republic, Czechoslovakia, England, Europe, Germany, Italy, Moldova, Netherlands, Romania, Russia, Spain, Sweden, Turkey (Noyes, 2019).

Subfamily Cerocephalinae (Gahan, 1946)
Genus *Theocolax* (Westwood, 1832)
Theocolax elegans (Westwood, 1874)
Publications with Turkey records: Noyes (2019), Öncüer (1991).
Associates: *Sitophilus oryzae* (Coleoptera: Dryophthoridae), *Anisopteromalus calandrae* (Hymenoptera: Pteromalidae), *Vigna unguiculata* (Fabaceae), *Oryza sativa, Triticum sp., Triticum aestivum, Zea mays* (Poaceae) (Noyes, 2019).
Hosts: *Lasioderma serricorne, Stegobium paniceum* (Coleoptera: Anobiidae), *Prostephanus truncates, Rhizopertha dominica, Rhyzopertha dominica* (Bostrychidae), *Acanthoscelides obtectus, Bruchus quadrimaculatus, Callosobruchus sp., Callosobruchus analis, Callosobruchus chinensis, Callosobruchus maculatus, Zaobrotes subfasciatus* (Bruchidae), *Cryptolestes ferrugineus* (Cucujidae), *Caulophilus latinasus, Caulophilus oryzae* (Curculionidae), *Calandra granaria, Calandra oryzae, Sitophilus sp., Sitophilus granarius, Sitophilus linearis, Sitophilus oryzae, Sitophilus zeamais* (Dryophthoridae), *Sitotroga cerealella* (Lepidoptera: Gelechiidae) (Noyes, 2019).
Distribution: Argentina, Australia, Belgium, Belize, Brazil, Canada, Congo, French Guiana, Germany, Greece, Guam, Guyana, Hawaii,

Honduras, India, Indonesia, Iran, Israel, Kirgizia, Madagascar, Malaysia, Mexico, Morocco, Nearctic, Netherlands, Nigeria, North Africa, Pakistan, Panama, People's Republic of China, Peru, Puerto Rico, Senegal, South Africa, Surinam, Sweden, Thailand, Togo, Turkey, United Kingdom, United States of America, Uruguay (Noyes, 2019).

Subfamily Cleonyminae (Walker, 1837)
Genus *Notanisus* (Walker, 1837)
Notanisus oulmesiensis (Delucchi, 1962)
Publications with Turkey records: Noyes (2019).
Associates: *Anthaxia sp.* (Buprestidae), *Scolytus mediterraneus* (Scolytidae), *Prunus sp.* (Rosaceae) (Noyes, 2019).
Hosts: *Anthaxia sp., Sphenoptera sp.* (Coleoptera: Buprestidae), *Scolytus rugulosus* (Scolytidae) (Noyes, 2019).
Distribution: Croatia, Cyprus, Greece, Iran, Montenegro, Morocco, Romania, Turkey, Uzbekistan, Yugoslavia (pre-1991) (Noyes, 2019).

Subfamily Colotrechninae (Thomson, 1876)
Genus *Colotrechnus* (Thomson, 1878)
Colotrechnus akifkansui (Doğanlar, 2018)
Publications with Turkey records: Doğanlar (2018a), Noyes (2019).
Associates: Unknown (Noyes, 2019).
Hosts: Unknown (Noyes, 2019).
Distribution: Turkey (Noyes, 2019).

Colotrechnus bekiralkani (Doğanlar, 2018)
Publications with Turkey records: Doğanlar (2018a), Noyes (2019).
Associates: Unknown (Noyes, 2019).
Hosts: Unknown (Noyes, 2019).
Distribution: Turkey (Noyes, 2019).

Colotrechnus birecikensis (Doğanlar, 2018)
Publications with Turkey records: Doğanlar (2018a), Noyes (2019).
Associates: Unknown (Noyes, 2019).
Hosts: Unknown (Noyes, 2019).
Distribution: Turkey (Noyes, 2019).

Colotrechnus hasangirayi (Doğanlar, 2018)
Publications with Turkey records: Doğanlar (2018a), Noyes (2019).
Associates: Unknown (Noyes, 2019).
Hosts: Unknown (Noyes, 2019).

Distribution: Turkey (Noyes, 2019).

Colotrechnus karatasensis (Doğanlar, 2018)
Publications with Turkey records: Doğanlar (2018a), Noyes (2019).
Associates: Unknown (Noyes, 2019).
Hosts: Unknown (Noyes, 2019).
Distribution: Turkey (Noyes, 2019).

Colotrechnus kemaliyenensis (Doğanlar, 2018)
Publications with Turkey records: Doğanlar (2018a), Noyes (2019).
Associates: Unknown (Noyes, 2019).
Hosts: Unknown (Noyes, 2019).
Distribution: Turkey (Noyes, 2019).

Colotrechnus mustafaoezeri (Doğanlar, 2018)
Publications with Turkey records: Doğanlar (2018a), Noyes (2019).
Associates: Unknown (Noyes, 2019).
Hosts: Unknown (Noyes, 2019).
Distribution: Turkey (Noyes, 2019).

Colotrechnus viridis (Masi, 1921)
Publications with Turkey records: Noyes (2019), Öncüer (1991).
Associates: *Onopordum nervosum* (Asteraceae), *Tribulus terrestris* (Zygophyllaceae) (Noyes, 2019).
Hosts: *Microlarinus lypriformis* (Coleoptera: Curculionidae) (Noyes, 2019).
Distribution: Azerbaijan, Canary Islands, Croatia, Czech Republic, Czechoslovakia, Europe, Hungary, Kazakhstan, Libya, Lithuania, Moldova, North Africa, Romania, Russia, Slovakia, Spain, Transcaucasus, Tselinograd Obl., Turkey, Turkmenistan, USSR, Uzbekistan (Noyes, 2019).

Colotrechnus zduzguenesae (Doğanlar, 2018)
Publications with Turkey records: Doğanlar (2018a), Noyes (2019).
Associates: Unknown (Noyes, 2019).
Hosts: Unknown (Noyes, 2019).
Distribution: Turkey (Noyes, 2019).

Subfamily Diparinae (Thomson, 1876)
Genus *Dipara* (Walker, 1833)
Dipara claviger (Kieffer, 1906)
Publications with Turkey records: Noyes (2019), Öncüer (1991).

Associates: Unknown (Noyes, 2019).
Hosts: Unknown (Noyes, 2019).
Distribution: Bosnia Hercegovina, Croatia, Europe, Greece, Italy, Romania, Turkey (Noyes, 2019).

Subfamily Epichrysomallinae (Hill and Riek, 1967)
Genus *Odontofroggatia* (Ishii, 1934)
Odontofroggatia ishii (Wiebes, 1980)
Publications with Turkey records: Doğanlar (2012), Noyes (2019).
Associates: *Ficus macrocarpa, Ficus macrocarpa, Ficus prasinicarpa* (Moraceae) (Noyes, 2019).
Hosts: *Ficus microcarpa* (Moraceae) (Noyes, 2019).
Distribution: Brazil, Hawaii, Malaysia, Papua New Guinea, People's Republic of China, Taiwan, Turkey, United States of America (Noyes, 2019).

Subfamily Eunotinae (Ashmead, 1904)
Genus *Eunotus* (Walker, 1834)
Eunotus areolatus (Ratzeburg, 1852)
Publications with Turkey records: Noyes (2019), Ulgenturk and Toros (2001).
Associates: *Pinus nigra* (Pinaceae), *Salix aurita, Salix repens* (Salicaceae) (Noyes, 2019).
Hosts: *Ceroplastes japonicus, Eulecanium ciliatum, Parthenolecanium persicae, Rhodococcus spiraeae* (Hemiptera: Coccidae) (Noyes, 2019).
Distribution: Czech Republic, Czechoslovakia, Denmark, Europe, France, Germany, Hungary, Italy, Kazakhstan, Netherlands, People's Republic of China, Romania, Spain, Sweden, Tselinograd Obl., Turkey (Noyes, 2019).

Genus *Scutellista* (Motschulsky, 1859)
Scutellista caerulea (Fonscolombe, 1832)
Publications with Turkey records: Noyes (2019), Öncüer (1991), Ozsemerci and Aksit (2003).
Associates: *Metaphycus lounsburyi* (Encyrtidae), *Baryscapus ceroplastophilus* (Eulophidae), *Aphelandra squarrosa* (Acanthaceae), *Mangifera indica* (Anacardiaceae), *Annona squamosa* (Annonaceae), *Nerium oleander* (Apocynaceae), *Baccharis pilularis* (Asteraceae), *Maytenus senegalensis* (Celastraceae), *Acacia tortilis* (Fabaceae), *Punica granatum* (Lythraceae), *Ficus sp., Ficus carica* (Moraceae), *Melaleuca hypericifolia, Psidium guajava* (Myrtaceae), *Olea europaea* (Oleaceae),

Cydonia oblonga (Rosaceae), *Citrus sp., Citrus deliciosa, Citrus limon, Citrus limonia, Citrus nobilis, Citrus unshiu* (Rutaceae), *Santalum album* (Santalaceae), *Leucophyllum taxanum* (Scrophulariaceae), *Vitis vinifera* (Vitaceae), *Saissetia miranda* (Hemiptera: Coccidae) (Noyes, 2019).

Hosts: *Cerococcus hibisci* (Hemiptera: Cerococcidae), *Ceroplastes sp., Ceroplastes actiniformis, Ceroplastes ceriferus, Ceroplastes cirripediformis, Ceroplastes destructor, Ceroplastes floridensis, Ceroplastes galeatus, Ceroplastes japonicus, Ceroplastes mimosa, Ceroplastes rubens, Ceroplastes rusci, Ceroplastes sinensis, Coccus hesperidum, Coccus oleae, Coccus viridis, Eulecanium corni, Gascardia brevicauda, Gascardia destructor, Kilifia acuminata, Lecanium sp., Lecanium corni, Lichtensia viburni, Parasaissetia nigra, Saissetia sp., Saissetia coffeae, Saissetia hemisphaerica, Saissetia neglecta, Saissetia nigra, Saissetia oleae, Waxiella mimosae* (Coccidae), *Quadraspidiotus perniciosus* (Diaspididae), *Phenacoccus artemisiae* (Pseudococcidae) (Noyes, 2019).

Parasitoids: *Quaylea whittieri* (Hymenoptera: Encyrtidae), *Baryscapus ceroplastophilus* (Eulophidae), *Lecaniobius cockerellii* (Eupelmidae) (Noyes, 2019).

Distribution: Afrotropical, Albania, Algeria, Argentina, Australia, Azerbaijan, Bolivia, Brazil, Canary Islands, Cayman Islands, Chile, Colombia, Croatia, Cuba, Cyprus, Czech Republic, Dominican Republic, Egypt, El Salvador, Eritrea, Ethiopia, France, Georgia, Greece, Hawaii, India, Israel, Italy, Ivory Coast, Japan, Kenya, Lebanon, Malta, Mexico, Morocco, Nearctic, Netherlands, New Zealand, North Africa, Oman, People's Republic of China, Peru, Puerto Rico, Senegal, South Africa, Spain, Sri Lanka, Trinidad & Tobago, Tunisia, Turkey, Uganda, United States of America, Venezuela (Noyes, 2019).

Subfamily Miscogastrinae (Walker, 1833)
Genus *Halticoptera* (Spinola, 1811)
Halticoptera adanacus (Doğanlar, 2006)
Publications with Turkey records: Doğanlar (2006), Noyes (2019).
Associates: Unknown (Noyes, 2019).
Hosts: Unknown (Noyes, 2019).
Distribution: Turkey (Noyes, 2019).

Halticoptera aenea (Walker, 1833)
Publications with Turkey records: Doğanlar (2006), Noyes (2019).
Associates: *Lactuca sativa* (Asteraceae), *Spergularia sp., Spergularia rupicola* (Caryophyllaceae), *Laburnum sp., Laburnum anagyroides,*

Medicago sativa (Fabaceae), *Fagus sylvatica* (Fagaceae), *Plantago sp.*, *Plantago coronopus* (Plantaginaceae), Triticum aestivum (Poaceae), *Crataegus sp.*, *Malus sylvestris*, *Prunus mahaleb* (Rosaceae), *Rubus idaeus* (Rubiaceae), *Salix sp.* (Salicaceae), *Tamarix canariensis* (Tamaricaceae) (Noyes, 2019).

Hosts: *Agromyza angulate, Agromyza demeijerei, Agromyza frontella, Cerodontha femoralis, Liriomyza congesta, Liriomyza sativae, Liriomyza trifolii, Phytomyza citisi, Phytomyza delphiniae, Phytomyza plantaginis, Phytomyza sedicola* (Diptera: Agromyzidae), *Mayetiola destructor* (Cecidomyiidae), *Oscinella frit, Oscinella pusilla, Oscinosoma frit* (Chloropidae), *Drosophila sp., Scaptomyza graminum* (Drosophilidae), *Eriogaster lanestris, Malacosoma neustria* (Lepidoptera: Lasiocampidae) (Noyes, 2019).

Distribution: Azerbaijan, Belgium, Bulgaria, Canary Islands, Czech Republic, Czechoslovakia, Europe, France, Germany, Hungary, Iran, Italy, Kazakhstan, Mexico, Moldova, Nearctic, Netherlands, Puerto Rico, Romania, Russia, Spain, Sweden, Switzerland, Tselinograd Obl., Turkey, Ukraine, United Kingdom, United States of America, USSR (Noyes, 2019).

Halticoptera circulus (Walker, 1833)

Publications with Turkey records: Cikman (2012), Doğanlar (2006), Noyes (2019), Öncüer (1991).

Associates: *Chasmodon apterus* (Hymenoptera: Braconidae), *Apium graveolens* (Apiaceae), *Arctium lappa, Lactuca sp., Solidago virgaurea, Sonchus sp., Sonchus oleraceus* (Asteraceae), *Brassica juncea* (Brassicaceae), *Citrullus lanatus, Cucumis melo* (Cucurbitaceae), *Ricinus communis* (Euphorbiaceae), *Laburnum anagyroides, Medicago sativa* (Fabaceae), *Mentha sp., Mentha spicata* (Lamiaceae), *Allium sp., Allium cepa, Allium schoenoprasum* (Liliaceae), *Alopecurus pratensis, Avena sativa, Hordeum sp., Hordeum vulgare, Lolium multiflorum, Lolium perenne, Triticum aestivum* (Poaceae), *Populus sp., Salix sp.* (Salicaceae), *Capsicum annuum, Solanum nigrum* (Solanaceae) (Noyes, 2019).

Hosts: *Agromyza frontella, Agromyza melampyga, Agromyza parvicornis, Cerodontha dorsalis, Chromatomyia fuscula, Japanagromyza inaequalis, Japanagromyza salicifolii, Liriomyza sp., Liriomyza bryoniae, Liriomyza congesta, Liriomyza huidobrensis, Liriomyza langei, Liriomyza melampyga, Liriomyza pictella, Liriomyza sativae, Liriomyza trifolii, Melanagromyza virens, Napomyza gymnostoma, Ophiomyia phaseoli, Phytobia angulate, Phytobia clara, Phytobia lateralis, Phytomyza sp., Phytomyza atricornis, Phytomyza citisi, Phytomyza delphiniae, Phytomyza*

horticola, Phytomyza petoei, Phytomyza solidaginis (Diptera: Agromyzidae), *Mayetiola destructor* (Cecidomyiidae), *Meromyza americana, Oscinella sp., Oscinella frit, Oscinella vastator, Oscinosoma frit* (Chloropidae), *Geomyza sp., Geomyza tripunctata* (Opomyzidae) (Noyes, 2019).
Distribution: Armenia, Austria, Azerbaijan, Belgium, Bulgaria, Canada, Canary Islands, Chile, Czech Republic, Dominican Republic, Ethiopia, Europe, France, Germany, Hawaii, Honduras, Hungary, Iran, Iraq, Italy, Japan, Jordan, Kazakhstan, Korea, Mexico, Moldova, Nearctic, Neotropical, Netherlands, Norway, People's Republic of China, Puerto Rico, Romania, Serbia, Slovakia, Spain, Sweden, Taiwan, Tselinograd Obl., Turkey, United Kingdom, United States of America (Noyes, 2019).

Halticoptera crius (Walker, 1839)
Publications with Turkey records: Doğanlar (2006), Noyes (2019), Öncüer (1991).
Associates: *Medicago sativa* (Fabaceae) (Noyes, 2019).
Distribution: Czech Republic, England, Europe, Germany, Hungary, Iran, Ireland, Italy, Netherlands, People's Republic of China, Romania, Russia, Sweden, Turkey (Noyes, 2019).

Halticoptera dimidiata (Foerster, 1841)
Publications with Turkey records: Noyes (2019), Öncüer (1991).
Associates: *Polygonum sp.* (Polygonaceae) (Noyes, 2019).
Distribution: Austria, Belgium, Bulgaria, Czech Republic, Czechoslovakia, England, Europe, Germany, Hungary, Ireland, Italy, Kazakhstan, Montenegro, Netherlands, Romania, Serbia, Slovakia, Slovenia, Spain, Sweden, Tselinograd Obl., Turkey, United Kingdom (Noyes, 2019).

Halticoptera kemaliyensis (Doğanlar, 2006)
Publications with Turkey records: Doğanlar (2006), Noyes (2019).
Associates: Unknown (Noyes, 2019).
Hosts: Unknown (Noyes, 2019).
Distribution: Turkey (Noyes, 2019).

Halticoptera laevigata (Thomson, 1876)
Publications with Turkey records: Noyes (2019), Öncüer (1991).
Associates: *Lonicera caprifolium, Lonicera steum, Lonicera xylosteum* (Caprifoliaceae), *Cotoneaster multiflora, Crataegus monogyna, Pyrus communis* (Rosaceae) (Noyes, 2019).

Hosts: *Anomoia permunda, Myoleja lucida, Phagocarpus permundus, Rhagoletis cerasi* (Diptera: Tephritidae) (Noyes, 2019).
Distribution: Belgium, Bulgaria, Czech Republic, Czechoslovakia, England, Europe, Germany, Kazakhstan, Kirgizia, Netherlands, People's Republic of China, Russia, Sweden, Switzerland, Tselinograd Obl., Turkey (Noyes, 2019).

Halticoptera longipetiolus (Hedqvist, 1975)
Publications with Turkey records: Doğanlar (2006), Noyes (2019).
Associates: *Dianthus caryophyllus* (Caryophyllaceae) (Noyes, 2019).
Hosts: *Liriomyza trifolii* (Diptera: Agromyzidae) (Noyes, 2019).
Distribution: Sweden, Turkey (Noyes, 2019).

Halticoptera moczari (Erdos, 1954)
Publications with Turkey records: Noyes (2019), Öncüer (1991).
Associates: *Medicago sativa* (Fabaceae) (Noyes, 2019).
Distribution: Hungary, Kazakhstan, Romania, Slovakia, Tselinograd Obl., Turkey (Noyes, 2019).

Halticoptera patellana (Dalman, 1818)
Publications with Turkey records: Cikman (2012), Doğanlar (2006), Noyes (2019), Öncüer (1991).
Associates: *Amaranthus sp.* (Amaranthaceae), *Lactuca sativa, Senecio sp.* (Asteraceae), *Corylus avellana* (Betulaceae), *Cucumis melo* (Cucurbitaceae), *Pteridium aquilinum* (Dennstaedtiaceae), *Laburnum sp., Laburnum anagyroides, Medicago sativa* (Fabaceae), *Allium sp.* (Liliaceae), *Triticum aestivum* (Poaceae), *Prunus mahaleb* (Rosaceae), *Solanum tuberosum* (Solanaceae) (Noyes, 2019).
Hosts: *Phloeosinus dentatus* (Coleoptera: Scolytidae), *Agromyza sp., Agromyza demeijerei, Agromyza parvicornis, Agromyza reptans, Cerodontha dorsalis, Cerodontha femoralis, Haplomyza sp., Japanagromyza inaequalis, Liriomyza sp., Liriomyza congesta, Liriomyza huidobrensis, Liriomyza langei, Liriomyza melampyga, Liriomyza munda, Liriomyza pictella, Liriomyza pusilla, Liriomyza trifolii, Melanagromyza virens, Ophiomyia phaseoli, Phytobia angulate, Phytobia cepae, Phytobia clara, Phytobia coquilletti, Phytobia crucifericola, Phytomyza sp., Phytomyza citisi, Phytomyza delphiniae, Phytomyza lappae, Phytomyza lappina, Phytomyza sedicola, Pseudonapomyza dianthicola, Pseudonapomyza spicata, Pycnoglossa hystrix* (Diptera: Agromyzidae), *Phytophaga destructor* (Cecidomyiidae), *Chlorops cucurbitae, Meromyza americana, Oscinella sp., Oscinella frit* (Chloropidae), *Scaptomyza flava*

(Drosophilidae), *Spilographa zoe, Trypeta zoe* (Tephritidae), *Leucoptera laburnella* (Lepidoptera: Lyonetiidae) (Noyes, 2019).
Distribution: Belgium, Brazil, Canada, Canary Islands, Chile, Czech Republic, Europe, France, Germany, Hawaii, Italy, Japan, Mexico, Montenegro, Nearctic, Neotropical, Netherlands, People's Republic of China, Peru, Romania, Russia, Sweden, Switzerland, Taiwan, Trinidad & Tobago, Turkey, United Kingdom, United States of America (Noyes, 2019).

Halticoptera polita (Walker, 1834)
Publications with Turkey records: Doğanlar (2006), Noyes (2019).
Associates: *Salix sp.* (Salicaceae) (Noyes, 2019).
Distribution: Belgium, Bulgaria, Czech Republic, Czechoslovakia, England, Europe, Germany, Hungary, Ireland, Netherlands, People's Republic of China, Romania, Sweden, Turkey (Noyes, 2019).

Halticoptera poreia (Walker, 1848)
Publications with Turkey records Doğanlar (2006), Noyes (2019).
Associates: *Medicago sativa* (Fabaceae), *Populus euramericana* (Salicaceae) (Noyes, 2019).
Distribution: Bulgaria, England, Italy, Montenegro, Netherlands, People's Republic of China, Sweden, Turkey (Noyes, 2019).

Halticoptera sakipagai (Doğanlar, 2006)
Publications with Turkey records: Doğanlar (2006), Noyes (2019).
Associates: Unknown (Noyes, 2019).
Hosts: Unknown (Noyes, 2019).
Distribution: Turkey (Noyes, 2019).

Halticoptera triannulata (Erdos, 1946)
Publications with Turkey records: Noyes (2019), Öncüer (1991).
Associates: *Medicago sativa* (Fabaceae) (Noyes, 2019).
Distribution: Austria, Czech Republic, Czechoslovakia, Germany, Hungary, Kazakhstan, Moldova, Palearctic, People's Republic of China, Romania, Slovakia, Sweden, Tselinograd Obl., Turkey, United Kingdom (Noyes, 2019).

Halticoptera yoncacus (Doğanlar, 2006)
Publications with Turkey records: Doğanlar (2006), Noyes (2019).
Associates: Unknown (Noyes, 2019).
Hosts: Unknown (Noyes, 2019).
Distribution: Turkey (Noyes, 2019).

Genus *Ksenoplata* (Boucek, 1965)
Ksenoplata quadrata (Boucek, 1965)
Publications with Turkey records: Noyes (2019), Öncüer (1991).
Associates: *Medicago lappacea* (Fabaceae) (Noyes, 2019).
Hosts: *Bruchidius sp.* (Coleoptera: Bruchidae) (Noyes, 2019).
Distribution: Algeria, Azerbaijan, Bulgaria, Czech Republic, Czechoslovakia, Macedonia, Serbia, Slovakia, Spain, Turkey (Noyes, 2019).

Genus *Lamprotatus* Westwood, 1833
Lamprotatus truncatus (Fonscolombe, 1832)
Publications with Turkey records: Noyes (2019), Öncüer (1991).
Associates: *Cirsium creticum* (Asteraceae), *Rumex sp., Rumex acetosa* (Polygonaceae) (Noyes, 2019).
Hosts: *Delia sp., Hylemya sp., Pegomya cirsicola, Pegomya nigricornis* (Diptera: Anthomyiidae) (Noyes, 2019).
Distribution: Croatia, England, Europe, France, Germany, Ireland, Italy, Kazakhstan, Netherlands, Sweden, Tselinograd Obl., Turkey (Noyes, 2019).

Genus *Miscogaster* Walker, 1833
Miscogaster elegans Walker, 1833
Publications with Turkey records: Noyes (2019), Öncüer (1991).
Associates: *Betula sp.* (Betulaceae), *Symphytum officinale* (Boraginaceae), *Luzula pilosa* (Juncaceae), *Lamium album, Teucrium scorodonia* (Lamiaceae), *Urtica dioica* (Urticaceae) (Noyes, 2019).
Hosts: *Agromyza anthracina, Agromyza reptans, Amauromyza labiatarum, Cerodontha bimaculate, Liriomyza impatientis, Phytobia crucifericola, Phytobia labiatarum, Phytomyza sp., Phytomyza lycopi, Phytomyza ranunculi, Phytomyza scotina, Phytomyza sonchi, Phytomyza symphyti* (Diptera: Agromyzidae) (Noyes, 2019).
Distribution: Belgium, Bulgaria, Czech Republic, Czechoslovakia, Europe, Germany, Hungary, Ireland, Italy, Netherlands, Poland, Sweden, Turkey, United Kingdom (Noyes, 2019).

Miscogaster hortensis (Walker, 1833)
Publications with Turkey records: Noyes (2019), Öncüer (1991).
Associates: *Medicago sativa* (Fabaceae) (Noyes, 2019).
Hosts: *Agromyza flavipennis, Agromyza frontella, Agromyza genistae, Agromyza nana, Agromyza nigrescens, Agromyza reptans, Liriomyza congesta, Phytagromyza hendeliana, Phytomyza angelicae, Phytomyza*

lappina, Phytomyza rufipes, Phytomyza sedicola (Diptera: Agromyzidae) (Noyes, 2019).
Distribution: Belgium, England, Europe, Hungary, India, Ireland, Kazakhstan, Netherlands, Romania, Spain, Sweden, Tselinograd Obl., Turkey, United States of America (Noyes, 2019).

Miscogaster rufipes (Walker, 1833)
Publications with Turkey records: Noyes (2019), Öncüer (1991).
Associates: *Anchusa arvensis, Anchusa italica, Borago officinalis, Cynoglossum officinale, Echium altissimum, Echium vulgare, Lycopsis arvensis, Myosotis sp., Symphytum officinale* (Boraginaceae), *Ranunculus sp.* (Ranunculaceae), *Urtica dioica* (Urticaceae) (Noyes, 2019).
Hosts: *Agromyza abiens, Agromyza pseudoreptans, Agromyza reptans, Agromyza rufipes, Phytomyza sp., Phytomyza atricornis, Phytomyza symphyti* (Diptera: Agromyzidae) (Noyes, 2019).
Distribution: Belgium, Bulgaria, Croatia, Czech Republic, England, Europe, France, Germany, Ireland, Kazakhstan, Moldova, Netherlands, Spain, Sweden, Tselinograd Obl., Turkey (Noyes, 2019).

Genus *Thinodytes* (Graham, 1956)
Thinodytes cyzicus (Walker, 1839)
Publications with Turkey records: Noyes (2019), Öncüer (1991).
Associates: *Senecio jacobaea* (Asteraceae), *Lathyrus maritimus, Medicago sativa, Pisum sativum, Retama sphaerocarpa* (Fabaceae), *Rhinanthus sp.* (Scrophulariaceae) (Noyes, 2019).
Hosts: *Liriomyza huidobrensis, Phytomyza atricornis, Phytomyza horticola* (Diptera: Agromyzidae) (Noyes, 2019).
Distribution: Bulgaria, Caucasus, Croatia, Czech Republic, Czechoslovakia, Europe, Germany, Hungary, India, Iraq, Ireland, Japan, Kazakhstan, Middle East, Moldova, Montenegro, Netherlands, People's Republic of China, Romania, Serbia, Slovakia, Slovenia, Spain, Sweden, Transcaucasus, Tselinograd Obl., Turkey, United Kingdom (Noyes, 2019).

Subfamily Ormocerinae (Walker, 1833)
Genus *Oxyglypta* (Foerster, 1856)
Oxyglypta rugosa (Ruschka, 1912)
Publications with Turkey records: Noyes (2019), Öncüer (1991).
Associates: *Pyrus communis* (Rosaceae) (Noyes, 2019).
Hosts: *Apiomyia bergenstammi, Oligotrophus bergenstammi* (Diptera: Cecidomyiidae), *Euproctis chrysorrhoea* (Lepidoptera: Lymantriidae) (Noyes, 2019).

Distribution: Austria, Europe, Israel, Turkey (Noyes, 2019).

Genus *Systasis* (Walker, 1834)
Systasis encyrtoides (Walker, 1834)
Publications with Turkey records: Noyes (2019), Öncüer (1991).
Associates: Diptera: Cecidomyiidae, *Carduus bourgeanus* (Asteraceae), *Berberis oblonga* (Berberidaceae), *Eruca vesicaria* (Brassicaceae), *Gypsophila struthium, Silene wolgensis* (Caryophyllaceae), *Atriplex halimus* (Chenopodiaceae), *Juniperus thurifera* (Cupressaceae), *Elaeagnus oxycarpa* (Elaeagnaceae), *Lathyrus pratensis, Lens esculenta, Medicago sp., Medicago sativa, Retama sphaerocarpa, Sarothamnus sp., Sarothamnus scoparius* (Fabaceae), *Ephedra nebrodensis* (Gnetaceae), *Mentha longifolia* (Lamiaceae), *Chamaenerion angustifolium, Epilobium angustifolium* (Onagraceae), *Citrus sp.* (Rutaceae), *Osyris alba* (Santalaceae), *Odontites sp., Odontites verna* (Scrophulariaceae), *Lycium dasystemum* (Solanaceae), *Tamarix canariensis* (Tamaricaceae), *Contarinia lentis* (Diptera: Cecidomyiidae) (Noyes, 2019).
Hosts: *Apion fuscirostre* (Coleoptera: Apionidae), *Bruchidius ater, Bruchidius fasciatus* (Bruchidae), *Tychius medicaginis* (Curculionidae), *Phytomyza isais* (Diptera: Agromyzidae), *Contarinia sp., Contarinia citri, Contarinia elaeagniflorae, Contarinia ilicis, Contarinia lentis, Contarinia lyciicola, Contarinia medicaginis, Contarinia pisi, Dasineura sp., Dasineura affinis, Dasineura epilobii, Dasineura gleditchiae, Dasineura ignorata, Diplosiola bursaria, Jaapiella floriperda, Jaapiella kovalevi, Jaapiella medicaginis, Stenodiplosis panici* (Cecidomyiidae), *Myopites stylata* (Tephritidae), *Cochylidia implicitana, Tortrix viridana* (Lepidoptera: Tortricidae), *Hormathophylla spinosa* (Brassicaceae) (Noyes, 2019).
Distribution: Austria, Azerbaijan, Belgium, Bulgaria, Canary Islands, Croatia, Czech Republic, Czechoslovakia, England, Europe, France, Germany, Greece, Hungary, Iran, Israel, Italy, Kazakhstan, Moldova, Montenegro, Morocco, Netherlands, Palearctic, People's Republic of China, Poland, Portugal, Romania, Russia, Serbia, Slovakia, Slovenia, Spain, Sweden, Transcaucasus, Tselinograd Obl., Turkey, Ukraine, United States of America, Yugoslavia (Federal Republic), Yugoslavia (pre-1991) (Noyes, 2019).

Subfamily Otitesellinae (Joseph, 1964)
Genus *Walkerella* (Westwood, 1883)
Walkerella microcarpae (Boucek, 1993)
Publications with Turkey records: Doğanlar (2012), Noyes (2019).

Associates: *Ficus macrocarpa, Ficus macrocarpa* (Moraceae) (Noyes, 2019).
Hosts: *Ficus microcarpa* (Moraceae) (Noyes, 2019).
Distribution: Bermuda, Brazil, Cayman Islands, Hawaii, Japan, People's Republic of China, Turkey, United States of America (Noyes, 2019).

Subfamily Pireninae (Haliday, 1844)
Genus *Gastrancistrus* (Westwood, 1833)
Gastrancistrus glabellus (Nees, 1834)
Publications with Turkey records: Noyes (2019), Öncüer (1991).
Associates: Unknown (Noyes, 2019).
Hosts: Unknown (Noyes, 2019).
Distribution: Bulgaria, Czech Republic, England, Europe, Germany, Hungary, Ireland, Kazakhstan, Montenegro, Netherlands, Romania, Spain, Sweden, Tselinograd Obl., Turkey (Noyes, 2019).

Gastrancistrus hirtulus (Graham, 1969)
Publications with Turkey records: Noyes (2019), Öncüer (1991).
Associates: *Betula sp.* (Betulaceae), *Larix decidua* (Pinaceae) (Noyes, 2019).
Distribution: England, Netherlands, Sweden, Turkey (Noyes, 2019).

Gastrancistrus latifrons (Thomson, 1876)
Publications with Turkey records: Noyes (2019), Öncüer (1991).
Associates: Unknown (Noyes, 2019).
Hosts: Unknown (Noyes, 2019).
Distribution: Croatia, England, Hungary, Netherlands, Serbia, Spain, Sweden, Turkey (Noyes, 2019).

Gastrancistrus pruniflorumus (Doğanlar and Yigit, 2019)
Publications with Turkey records: Doğanlar (2019), Noyes (2019).
Associates: *Prunus armeniaca* (Rosaceae) (Noyes, 2019).
Hosts: *Contarinia pruniflorum* (Diptera: Cecidomyiidae) (Noyes, 2019).
Distribution: Turkey (Noyes, 2019).

Genus *Macroglenes* (Westwood, 1832)
Macroglenes fidanlinensis (Doğanlar, 2018)
Publications with Turkey records: Doğanlar (2018a), Noyes (2019).
Associates: Unknown (Noyes, 2019).
Hosts: Unknown (Noyes, 2019).
Distribution: Turkey (Noyes, 2019).

Macroglenes varicornis (Haliday, 1833)
Publications with Turkey records: Noyes (2019), Öncüer (1991).
Associates: *Anthoxanthum sp.* (Poaceae) (Noyes, 2019).
Hosts: *Contarinia tritici* (Diptera: Cecidomyiidae) (Noyes, 2019).
Distribution: Belgium, Croatia, Czech Republic, Europe, France, Germany, Hungary, Ireland, Montenegro, Netherlands, People's Republic of China, Romania, Serbia, Slovakia, Spain, Sweden, Switzerland, Turkey, United Kingdom (Noyes, 2019).

Subfamily Pteromalinae (Dalman, 1820)
Genus *Anisopteromalus* (Ruschka, 1912)
Anisopteromalus calandrae (Howard, 1881)
Publications with Turkey records: Noyes (2019), Öncüer (1991).
Associates: *Heterospilus prosopidis* (Hymenoptera: Braconidae), *Choetospila elegans, Dinarmus basalis, Lariophagus distinguendus* (Pteromalidae), *Imbrasia belina* (Lepidoptera: Saturniidae), *Centella asiatica, Coriandrum sativum* (Apiaceae), *Cyperus rotundus* (Cyperaceae), *Ceratonia siliqua, Cicer arietinum, Phaseolus angularis, Phaseolus vulgaris, Pisum sp., Pisum sativum, Vigna angularis, Vigna radiata, Vigna unguiculata* (Fabaceae), *Ficus carica* (Moraceae), *Olea europaea* (Oleaceae), *Oryza sativa, Sorghum sp., Sorghum bicolor, Triticum sp., Triticum aestivum, Zea mays* (Poaceae) (Noyes, 2019).
Hosts: *Anobium sp., Catorama herbarium, Lasioderma serricorne, Sitodrepa sp., Stegobium paniceum* (Coleoptera: Anobiidae), *Araecerus fasciculatus* (Anthribidae), *Piezotrachelus varium* (Apionidae), *Prostephanus truncates, Rhizopertha dominica, Rhyzopertha dominica* (Bostrychidae), *Acanthoscelides obtectus, Bruchus sp., Bruchus chinensis, Bruchus obscurus, Bruchus pisorum, Bruchus quadrimaculatus, Bruchus rufimanus, Callosobruchus sp., Callosobruchus analis, Callosobruchus chinensis, Callosobruchus maculatus, Callosobruchus phaseoli, Callosobruchus quadrimaculatus, Pachymerus sp., Spermophagus subfasciatus, Zabrotes subfasciatus* (Bruchidae), *Rezotrachelus vorium* (Coleoptera), *Athesapeuta cyperi, Caulophilus latinasus, Ceutorhynchus assimilis, Pempheres affinis, Pempherulus affinis* (Curculionidae), *Trogoderma granarium* (Dermestidae), *Calandra granaria, Calandra oryzae, Sitophilus sp., Sitophilus granaries, Sitophilus oryzae, Sitophilus sasakii, Sitophilus zeamais* (Dryophthoridae), *Carpophilus obsoletus* (Nitidulidae), *Oryzaephilus surinamensis* (Silvanidae), *Tribolium castaneum* (Tenebrionidae), *Anisopteromalus calandrae* (Hymenoptera: Pteromalidae), *Sitotroga cerealella* (Lepidoptera: Gelechiidae), *Cadra*

cautella, Ectomyelois ceratoniae, Ephestia elutella, Ephestia kuehniella (Pyralidae) (Noyes, 2019).
Distribution: Argentina, Australia, Austria, Bangladesh, Belgium, Bermuda, Bolivia, Botswana, Brazil, Bulgaria, Burkina Faso, Colombia, Cuba, Czech Republic, Czechoslovakia, Egypt, El Salvador, France, Germany, Greece, Guyana, Hawaii, Hungary, India, Indonesia, Iran, Iraq, Israel, Italy, Jamaica, Japan, Kazakhstan, Kenya, Korea, Madeira, Malawi, Malaysia, Mauritius, Mexico, Moldova, Morocco, Myanmar, Nearctic, New Zealand, Nigeria, Pakistan, Papua New Guinea, Peru, Portugal, Puerto Rico, Romania, Russia, Saudi Arabia, Serbia, Seychelles, Slovakia, Solomon Islands, Sweden, Switzerland, Syria, Taiwan, Tanzania, Thailand, Togo, Tselinograd Obl., Turkey, United Kingdom, United States of America, Venezuela, West Africa (Noyes, 2019).

Genus *Arthrolytus* (Thomson, 1878)
Arthrolytus discoideus (Nees, 1834)
Publications with Turkey records: Noyes (2019), Öncüer (1991).
Associates: Unknown (Noyes, 2019).
Hosts: Unknown (Noyes, 2019).
Distribution: Belgium, Caucasus, Czech Republic, Europe, Germany, Netherlands, Romania, Slovenia, Spain, Sweden, Turkey, United Kingdom (Noyes, 2019).

Arthrolytus geyrazensis (Doğanlar, 2018)
Publications with Turkey records: Doğanlar (2018a), Noyes (2019).
Associates: Unknown (Noyes, 2019).
Hosts: Unknown (Noyes, 2019).
Distribution: Turkey (Noyes, 2019).

Arthrolytus gurderei (Doğanlar, 2018)
Publications with Turkey records: Doğanlar (2018a), Noyes (2019).
Associates: Unknown (Noyes, 2019).
Hosts: Unknown (Noyes, 2019).
Distribution: Turkey (Noyes, 2019).

Arthrolytus kahvecii (Doğanlar, 2018)
Publications with Turkey records: Doğanlar (2018a), Noyes (2019).
Associates: Unknown (Noyes, 2019).
Hosts: Unknown (Noyes, 2019).
Distribution: Turkey (Noyes, 2019).

Arthrolytus magarsosensis (Doğanlar, 2018)
Publications with Turkey records: Doğanlar (2018a), Noyes (2019).
Associates: Unknown (Noyes, 2019).
Hosts: Unknown (Noyes, 2019).
Distribution: Turkey (Noyes, 2019).

Arthrolytus nenehatunae (Doğanlar, 2018)
Publications with Turkey records: Doğanlar (2018a), Noyes (2019).
Associates: Unknown (Noyes, 2019).
Hosts: Unknown (Noyes, 2019).
Distribution: Turkey (Noyes, 2019).

Arthrolytus nuridemiragi (Doğanlar, 2018)
Publications with Turkey records: Doğanlar (2018a), Noyes (2019).
Associates: Unknown (Noyes, 2019).
Hosts: Unknown (Noyes, 2019).
Distribution: Turkey (Noyes, 2019).

Arthrolytus yazicioglui (Doğanlar, 2018)
Publications with Turkey records: Doğanlar (2018a), Noyes (2019).
Associates: Unknown (Noyes, 2019).
Hosts: Unknown (Noyes, 2019).
Distribution: Turkey (Noyes, 2019).

Genus *Caenocrepis* (Thomson, 1878)
Caenocrepis arenicola (Thomson, 1878)
Publications with Turkey records: Noyes (2019), Öncüer (1991).
Hosts: *Pachycerus madidus* (Coleoptera: Curculionidae) (Noyes, 2019).
Distribution: Austria, Azerbaijan, Caucasus, Croatia, Czech Republic, Czechoslovakia, Europe, France, Georgia, Greece, Hungary, Kazakhstan, Moldova, Montenegro, Morocco, North Africa, Romania, Serbia, Slovakia, Spain, Sweden, Transcaucasus, Tselinograd Obl., Turkey, USSR (Noyes, 2019).

Caenocrepis bothynoderi (Gromakov, 1940)
Publications with Turkey records: Noyes (2019), Öncüer (1991).
Associates: *Heliotropium europaeum* (Boraginaceae) (Noyes, 2019).
Hosts: *Bothynoderes punctiventris, Cleonus punctiventris, Pachycerus cordiger* (Coleoptera: Curculionidae).
Distribution: Azerbaijan, Europe, Iran, Kazakhstan, Kirgizia, Tselinograd Obl., Turkey, Turkmenistan, Ukraine, USSR (Noyes, 2019).

Genus *Catolaccus* (Thomson, 1878)
Catolaccus ater (Ratzeburg, 1852)
Publications with Turkey records: Noyes (2019), Öncüer (1991).
Associates: *Brassica oleracea* (Brassicaceae), *Onobrychis viciifolia* (Fabaceae), *Picea abies, Pinus sp.* (Pinaceae), *Populus sp.* (Salicaceae) (Noyes, 2019).
Hosts: *Hypera murina, Hypera variabilis, Rhynchaenus salicis* (Coleoptera: Curculionidae), *Contarinia pyrina, Contarinia pyrivora* (Diptera: Cecidomyiidae), *Apanteles sp.* (Hymenoptera: Braconidae), *Ocnogyna loewi* (Lepidoptera: Arctiidae), *Platyedra malvella* (Gelechiidae), *Apocheima cinerarius* (Geometridae), *Leucoma salicis, Lymantria dispar* (Lymantriidae), *Acronicta rumicis, Chamaepora rumicis, Symira dentinosa* (Noctuidae), *Vanessa urticae* (Nymphalidae), *Aporia crataegi, Pieris brassicae* (Pieridae), *Apotomis lutozana, Archips rosanus* (Tortricidae), *Yponomeuta malinellus* (Yponomeutidae) (Noyes, 2019).
Parasitoid hosts: *Goniozus claripennis* (Hymenoptera: Bethylidae), *Apanteles sp., Apanteles congestus, Apanteles enephes, Apanteles globatus, Apanteles glomeratus, Apanteles kazak, Apanteles melanoscelus, Apanteles pieridis, Apanteles spurius, Bracon brevicornis, Bracon hebetor, Cotesia sp., Rhogas pallidator* (Braconidae), *Gelis sp., Pimpla maculator* (Ichneumonidae) (Noyes, 2019).
Distribution: Armenia, Austria, Canary Islands, Croatia, Czech Republic, Czechoslovakia, Europe, France, Germany, Hungary, Iran, Italy, Kazakhstan, Kirgizia, Moldova, Poland, Romania, Russia, Serbia, Slovakia, Spain, Sweden, Transcaucasus, Tselinograd Obl., Turkey, Turkmenistan, Ukraine, United Kingdom, USSR (Noyes, 2019).

Genus *Cecidostiba* (Thomson, 1878)
Cecidostiba fungosa (Geoffroy, 1785)
Publications with Turkey records: Doğanlar (2014), Noyes (2019).
Associates: *Andricus askewi, Andricus burgundus, Andricus caliciformis, Andricus caputmedusae, Andricus conglomeratus, Andricus coriarius, Andricus coronatus, Andricus corruptrix, Andricus crispator, Andricus curtisii, Andricus curvator, Andricus cydoniae, Andricus dentimitratus, Andricus fecundatrix, Andricus galeatus, Andricus gallaeurnaeformis, Andricus glutinosus, Andricus grossulariae, Andricus hartigi, Andricus hispanicus, Andricus hungaricus, Andricus infectorius, Andricus kollari, Andricus lignicolus, Andricus lucidus, Andricus megalucidus, Andricus multiplicatus, Andricus pictus, Andricus polycerus, Andricus quercuscalicis, Andricus quercusradicis, Andricus quercusramuli,*

Andricus quercustozae, Andricus seckendorffi, Andricus solitarius, Andricus superfetationis, Andricus tomentosus, Andricus vindobonensis, Aphelonyx cerricola, Biorhiza pallida, Callirhytis erythrocephala, Callirhytis glandium, Cerroneuroterus lanuginosus, Chilaspis nitida, Cynips agama, Cynips cornifex, Cynips disticha, Cynips divisa, Cynips longiventris, Cynips quercus, Cynips quercusfolii, Dryocosmus cerriphilus, Dryocosmus mayri, Neuroterus anthracinus, Neuroterus numismalis, Neuroterus saliens, Pseudoneuroterus macropterus, Pseudoneuroterus saliens, Synophrus olivieri, Trigonaspis Megaptera, Trigonaspis synaspis (Cynipidae), *Quercus sp., Quercus cerris, Quercus faginea, Quercus frainetto, Quercus infectoria, Quercus ithaburensis, Quercus lusitanica, Quercus pedunculiflora, Quercus petraea, Quercus pubescens, Quercus pyrenaica, Quercus robur, Quercus suber* (Fagaceae) (Noyes, 2019).

Hosts: *Andricus sp., Andricus aestivalis, Andricus calicis, Andricus caputmedusae, Andricus corruptrix, Andricus curvator, Andricus glutinosus, Andricus grossulariae, Andricus hungaricus, Andricus kollari, Andricus multiplicatus, Andricus quercuscalicis, Andricus quercusradicis, Andricus quercusramuli, Andricus quercustozae, Andricus vindobonensis, Aphelonyx cerricola, Biorhiza pallida, Ceroptres arator, Cynips sp., Cynips longiventris, Cynips quercusfolii, Dryocosmus kuriphilus, Synergus sp., Synergus gallaepomiformis, Synergus pallicornis, Synergus umbraculus* (Hymenoptera: Cynipidae) (Noyes, 2019).

Parasitoid hosts: *Synergus gallaepomiformis, Synergus umbraculus, Olynx skianeuros* (Eulophidae) (Noyes, 2019).

Distribution: Andorra, Austria, Belgium, Bulgaria, Croatia, Czech Republic, Czechoslovakia, Denmark, Europe, France, Germany, Greece, Hungary, Israel, Italy, Jordan, Kazakhstan, Netherlands, Poland, Romania, Serbia, Slovakia, Slovenia, Spain, Sweden, Tselinograd Obl., Turkey, Ukraine, United Kingdom (Noyes, 2019).

Genus *Cheiropachus* (Westwood, 1829)

Cheiropachus quadrum (Fabricius, 1787)

Publications with Turkey records: Narendran et al. (1995), Noyes (2019).

Associates: *Apate sp., Sinoxylon sexdentatum* (Coleoptera: Bostrychidae), *Magdalis sp.* (Curculionidae), *Dryocoetes sp., Hylesinus sp., Ips sp., Leperesinus californicus, Leperesinus fraxini, Leperesinus oregonus, Phloeosinus sp., Scolytus sp., Scolytus laevis* (Scolytidae), *Quercus sp.* (Fagaceae), *Olea europaea* (Oleaceae), *Pinus halepensis* (Pinaceae), *Triticum aestivum* (Poaceae), *Cydonia oblonga, Malus domestica, Malus*

pumila, Malus sylvestris, Persica vulgaris, Prunus sp., Prunus amygdalus, Prunus armeniaca, Prunus glandulosa, Prunus persica, Pyrus communis (Rosaceae), *Ulmus propinqua, Ulmus pumila* (Ulmaceae) (Noyes, 2019).
Hosts: *Apate sp., Sinoxylon sexdentatum* (Coleoptera: Bostrychidae), *Tetrops sp., Tetrops praeusta* (Cerambycidae), *Magdalis sp., Magdalis aenescens, Magdalis armicollis, Magdalis armigera, Magdalis barbicornis, Magdalis barbita* (Curculionidae), *Chaetoptelius vestitus, Dryocoetes autographus, Hylesinus fraxini, Hylesinus oleiperda, Hylesinus orni, Hylesinus varius, Ips amitinus, Ips sexdentatus, Ips typographus, Leperesinus varius, Myelophilus minor, Myelophilus piniperda, Phloeosinus aubei, Phloeosinus canadensis, Phloeosinus punctatus, Phloeotribus oleae, Phloeotribus scarabaeoides, Pityogenes chalcographus, Scolytus sp., Scolytus amygdali, Scolytus carpini, Scolytus ensifer, Scolytus intricatus, Scolytus japonicus, Scolytus koenigi, Scolytus laevis, Scolytus mali, Scolytus multistriatus, Scolytus nitidus, Scolytus pygmaeus, Scolytus ratzeburgi, Scolytus rugulosus, Scolytus scolytus, Xyleborus xylographus* (Scolytidae), *Zeuzera pyrina* (Lepidoptera: Cossidae), *Lymantria dispar* (Lymantriidae) (Noyes, 2019).
Distribution: Argentina, Armenia, Austria, Belgium, Bosnia Hercegovina, Canada, Canary Islands, Caucasus, Chile, Croatia, Czech Republic, Czechoslovakia, Egypt, England, Europe, France, Germany, Hungary, India, Iran, Israel, Italy, Kazakhstan, Kirgizia, Lebanon, Macedonia, Moldova, Morocco, Nearctic, Neotropical, Netherlands, North Africa, Norway, Pakistan, People's Republic of China, Poland, Romania, Russia, Serbia, Slovakia, Slovenia, Spain, Sweden, Switzerland, Tselinograd Obl., Tunisia, Turkey, Turkmenistan, Ukraine, United States of America, USSR, Yugoslavia (pre-1991) (Noyes, 2019).

Genus *Chlorocytus* (Graham, 1956)
Chlorocytus breviscapus (Graham, 1965)
Publications with Turkey records: Korkmaz et al. (2010), Noyes (2019).
Associates: *Calamagrostis epigeios, Phalaris arundinacea, Triticum aestivum* (Poaceae) (Noyes, 2019).
Hosts: *Calameuta sp., Calameuta filiformis* (Hymenoptera: Cephidae) (Noyes, 2019).
Distribution: Czech Republic, England, Germany, Hungary, Netherlands, Sweden, Turkey (Noyes, 2019).

Chlorocytus diversus (Walker, 1836)
Publications with Turkey records: Noyes (2019), Öncüer (1991).

Associates: *Centaurea sp., Centaurea scabiosa* (Asteraceae), *Brassica napus* (Brassicaceae) (Noyes, 2019).
Hosts: *Ceutorhynchus obstrictus* (Coleoptera: Curculionidae), *Hartigia xanthostoma* (Hymenoptera: Cephidae), *Phanacis centaureae* (Cynipidae) (Noyes, 2019).
Distribution: Croatia, Czech Republic, England, Europe, France, Germany, Ireland, Kazakhstan, Kirgizia, Montenegro, Netherlands, Serbia, Slovakia, Spain, Sweden, Tselinograd Obl., Turkey, United Kingdom, Uzbekistan (Noyes, 2019).

Chlorocytus phalaridis (Graham, 1965)
Publications with Turkey records: Noyes (2019), Öncüer (1991).
Associates: *Phalaris arundinacea* (Poaceae) (Noyes, 2019).
Hosts: *Tetramesa longicornis* (Hymenoptera: Eurytomidae) (Noyes, 2019).
Parasitoid hosts: *Bracon sp.* (Braconidae), *Tetramesa longicornis* (Eurytomidae) (Noyes, 2019).
Parasitoids: *Mesopolobus graminum* (Pteromalidae) (Noyes, 2019).
Distribution: Belgium, Czech Republic, England, Europe, Germany, Ireland, Kazakhstan, Netherlands, People's Republic of China, Sweden, Tselinograd Obl., Turkey (Noyes, 2019).

Chlorocytus ultonicus (Graham, 1965)
Publications with Turkey records: Noyes (2019), Öncüer (1991).
Associates: *Fagus sylvatica* (Fagaceae), *Phleum pratense, Phleum ultonicus* (Poaceae) (Noyes, 2019).
Parasitoid hosts: *Pediobius claridgei* (Hymenoptera: Eulophidae), *Tetramesa phleicola* (Eurytomidae) (Noyes, 2019).
Distribution: Bulgaria, England, Europe, Germany, Ireland, Netherlands, Sweden, Turkey, United Kingdom (Noyes, 2019).

Genus *Coelopisthia* (Foerster, 1856)
Coelopisthia caledonica (Askew, 1980)
Publications with Turkey records: Noyes (2019), Öncüer (1991).
Associates: *Salix sp.* (Salicaceae) (Noyes, 2019).
Hosts: *Xestia alpicola* (Lepidoptera: Noctuidae), *Melitaea cinxia* (Nymphalidae) (Noyes, 2019).
Distribution: Bulgaria, Finland, People's Republic of China, Sweden, Switzerland, Turkey, United Kingdom (Noyes, 2019).

Genus *Conomorium* (Masi, 1924)
Conomorium patulum (Walker, 1835)
Publications with Turkey records: Noyes (2019), Öncüer (1991).
Associates: *Pistacia vera* (Anacardiaceae), *Gossypium sp.* (Malvaceae), *Abies sp., Larix sp., Pinus sp.* (Pinaceae), *Platanus sp.* (Platanaceae), *Populus sp., Populus canadensis* (Salicaceae) (Noyes, 2019).
Hosts: *Delia brassicae* (Diptera: Anthomyiidae), *Hyphantria cunea, Tyria jacobaeae* (Lepidoptera: Arctiidae), *Biston hirtaria, Erannis bajaria, Erannis defoliaria* (Geometridae), *Lithocolletis platani, Phyllonorycter platani* (Gracillariidae), *Malacosoma neustria* (Lasiocampidae), *Euproctis sp.* (Lymantriidae), *Leucoptera sinuella, Paraleucoptera sinuella* (Lyonetiidae), *Heliothis armigera, Mamestra brassicae, Oria musculosa, Spodoptera litura* (Noctuidae), *Thaumetopoea pityocampa, Thaumetopoea solitaria* (Notodontidae), *Tinea cloacella* (Tineidae) (Noyes, 2019).
Distribution: Armenia, Canary Islands, Caucasus, Croatia, Czech Republic, Czechoslovakia, Egypt, England, Europe, Finland, France, Germany, Greece, Hungary, Iran, Italy, Japan, Kazakhstan, Korea, Macedonia, Moldova, Morocco, North Africa, Romania, Russia, Serbia, Slovakia, Spain, Sweden, Switzerland, Transcaucasus, Tselinograd Obl., Turkey, Turkmenistan, Ukraine, USSR, Yugoslavia (pre-1991) (Noyes, 2019).

Conomorium pityocampae (Graham, 1992)
Publications with Turkey records: Noyes (2019).
Associates: *Upupa epops* (Birds), *Pinus halepensis, Pinus nigra* (Pinaceae), *Tamarix canariensis* (Tamaricaceae) (Noyes, 2019).
Hosts: *Thaumetopoea pityocampa, Thaumetopoea wilkinsoni* (Lepidoptera: Notodontidae) (Noyes, 2019).
Distribution: Croatia, Cyprus, France, Greece, Italy, Slovenia, Spain, Sweden, Turkey, Yugoslavia (Federal Republic), Yugoslavia (pre-1991) (Noyes, 2019).

Genus *Cyrtogaster* (Walker, 1833)
Cyrtogaster vulgaris (Walker, 1833)
Publications with Turkey records: Noyes (2019), Öncüer (1991).
Associates: *Daucus carota, Pastinaca sativa* (Apiaceae), *Ilex sp., Ilex aquifolium* (Aquifoliaceae), *Aster sp., Chrysanthemum sp., Gazania sp., Leucanthemum vulgare, Onopordum nervosum, Senecio jacobaea, Vernonia officinalis* (Asteraceae), *Brassica sp.* (Brassicaceae), *Carex sp.* (Cyperaceae), *Lupinus sp., Medicago sativa, Pisum sativum, Vicia sp., Vicia angustifolia* (Fabaceae), *Fagus sylvatica, Quercus sp.* (Fagaceae),

Luzula pilosa (Juncaceae), *Papaver sp.* (Papaveraceae), *Cedrus sp., Picea sp., Pinus sp.* (Pinaceae), *Alopecurus myosuroides, Avena sativa, Hordeum vulgare, Milium effusum, Triticum aestivum* (Poaceae), *Primula vulgaris* (Primulaceae), *Malus pumila, Prunus avium, Pyrus communis* (Rosaceae), *Salix sp., Salix alba* (Salicaceae), *Veronica officinalis* (Scrophulariaceae), *Lycopersicon esculentum* (Solanaceae), *Tamarix canariensis* (Tamaricaceae), *Ulmus campestris* (Ulmaceae) (Noyes, 2019).
Hosts: *Bruchidius marginalis* (Coleoptera: Bruchidae), *Agromyza sp., Agromyza frontella, Cerodontha bimaculate, Cerodontha caricicola, Chromatomyia fuscula, Chromatomyia horticola, Chromatomyia nigra, Chromatomyia paraciliata, Liriomyza bryoniae, Liriomyza huidobrensis, Liriomyza sativae, Paraphytomyza tridentata, Phytobia sonderupi, Phytomyza sp., Phytomyza atricornis, Phytomyza crassiseta, Phytomyza fuscula, Phytomyza ilicis, Phytomyza lateralis, Phytomyza milii, Phytomyza nigra, Phytomyza sonderupi, Phytomyza syngenesiae* (Diptera: Agromyzidae), *Chlorops pumilionis, Oscinella frit, Oscinosoma frit* (Chloropidae), *Scaptomyza flava* (Drosophilidae), *Lonchaea hirticeps* (Lonchaeidae), *Lonchoptera sp., Lonchoptera furcate, Lonchoptera lutea* (Lonchopteridae), *Opomyza florum* (Opomyzidae), *Melanostoma scalare* (Syrphidae), *Rhagoletis cerasi* (Tephritidae), *Macrosiphum euphorbiae* (Hemiptera: Aphididae), *Diplolepis rosae* (Hymenoptera: Cynipidae), *Eriogaster lanestris* (Lepidoptera: Lasiocampidae) (Noyes, 2019).
Parasitoid hosts: *Coelinius niger* (Hymenoptera: Braconidae) (Noyes, 2019).
Parasitoids: *Pediobius metallicus* (Eulophidae) (Noyes, 2019).
Distribution: Austria, Azerbaijan, Azores, Belgium, Bulgaria, Canada, Canary Islands, Croatia, Czech Republic, Czechoslovakia, Europe, France, Germany, Hungary, Iceland, Iran, Italy, Kazakhstan, Latvia, Lithuania, Madeira, Moldova, Montenegro, Morocco, Nearctic, Netherlands, North Africa, Norway, Romania, Russia, Serbia, Slovakia, Spain, St Vincent & Grenadines, Sweden, Switzerland, Transcaucasus, Tselinograd Obl., Turkey, United Kingdom, United States of America (Noyes, 2019).

Genus *Cyrtoptyx* (Delucchi, 1956)
Cyrtoptyx latipes (Rondani, 1874)
Publications with Turkey records: Noyes (2019).
Associates: *Amaranthus sp., Amaranthus spinosus* (Amaranthaceae), *Eryngium campestre* (Apiaceae), *Cannabis sativa* (Cannabidaceae), *Capparis decidua* (Capparidaceae), *Gypsophila struthium* (Caryophyllaceae), *Atriplex halimus* (Chenopodiaceae), *Retama*

sphaerocarpa (Fabaceae), *Olea europaea* (Oleaceae), *Verbascum sp.* (Scrophulariaceae), *Tribulus terrestris* (Zygophyllaceae) (Noyes, 2019).
Hosts: *Hypolixus sp., Hypolixus truncatulus, Pachytychius sp., Pachytychius hordei* (Coleoptera: Curculionidae), *Bactrocera oleae, Dacus sp., Dacus oleae* (Diptera: Tephritidae), *Coleophora stefanii* (Lepidoptera: Coleophoridae), *Cydia sinana* (Tortricidae) (Noyes, 2019).
Parasitoids: *Asecodes erxias, Teleopterus erxias* (Hymenoptera: Eulophidae) (Noyes, 2019).
Distribution: Azerbaijan, Bulgaria, Croatia, Cyprus, Egypt, Eritrea, Europe, Greece, India, Iran, Italy, Kazakhstan, Lebanon, Libya, Pakistan, People's Republic of China, Spain, Syria, Transcaucasus, Tselinograd Obl., Turkey, Yemen, Yugoslavia (pre-1991) (Noyes, 2019).

Cyrtoptyx robustus (Masi, 1907)
Publications with Turkey records: Noyes (2019).
Associates: *Andricus conglomeratus, Andricus coriarius, Andricus grossulariae, Andricus hispanicus, Andricus kollari, Andricus polycerus, Andricus quercustozae, Andricus tomentosus, Aphelonyx cerricola, Cynips disticha, Cynips divisa, Cynips quercus, Plagiotrochus australis, Trigonaspis synaspis, Quercus sp., Quercus faginea, Quercus frainetto, Quercus ilex, Quercus infectoria, Quercus ithaburensis, Quercus pubescens, Quercus pyrenaica* (Fagaceae) (Noyes, 2019).
Hosts: *Adleria sp., Andricus coriarius, Andricus hispanicus, Andricus kollari, Andricus polycerus, Andricus quercustozae, Andricus tomentosus, Cynips coriaria, Cynips disticha, Cynips kollari, Cynips polycera, Cynips quercus, Plagiotrochus australis* (Hymenoptera: Cynipidae) (Noyes, 2019).
Distribution: Andorra, Croatia, Greece, Hungary, Iran, Italy, Jordan, Romania, Spain, Turkey (Noyes, 2019).
Genus *Dibrachoides* (Kurdjumov, 1913)
Dibrachoides dynastes (Foerster, 1841)
Publications with Turkey records: Noyes (2019), Öncüer (1991).
Associates: *Pastinaca sativa* (Apiaceae), *Betula sp.* (Betulaceae), *Medicago sativa* (Fabaceae), *Ficus carica* (Moraceae), *Tamarix canariensis* (Tamaricaceae) (Noyes, 2019).
Hosts: *Hypera sp., Hypera brunneipennis, Hypera murina, Hypera nigrirostris, Hypera postica, Hypera punctata, Hypera rumicis, Hypera variabilis, Phytonomus sp., Phytonomus nigrirostris, Phytonomus posticus, Phytonomus rumicis* (Coleoptera: Curculionidae), *Homotoma ficus* (Hemiptera: Carsidaridae) (Noyes, 2019).

Distribution: Belgium, Canada, Caucasus, Croatia, Czech Republic, Czechoslovakia, Europe, France, Germany, Hungary, Iran, Italy, Kazakhstan, Kirgizia, Macedonia, Moldova, Morocco, Nearctic, Netherlands, North Africa, Romania, Russia, Serbia, Slovakia, Spain, Sweden, Transcaucasus, Tselinograd Obl., Turkey, United Kingdom, United States of America (Noyes, 2019).

Genus *Dibrachys* (Foerster, 1856)
Dibrachys affinis (Masi, 1907)
Publications with Turkey records: Noyes (2019), Öncüer (1991).
Associates: *Malus orientalis, Prunus persica, Pyrus malus* (Rosaceae), *Vitis vinifera* (Vitaceae) (Noyes, 2019).
Hosts: *Calliphora sp., Calliphora erythrocephala, Calliphora vomitoria* (Diptera: Calliphoridae), *Anarsia lineatella, Sitotroga cerealella* (Lepidoptera: Gelechiidae), *Leucoma salicis, Stilpnotia salicis* (Lymantriidae), *Synanthedon typhiaeformis* (Sesiidae), *Carpocapsa pomonella, Clysia ambiguella, Cydia pomonella, Enarmonia formosana, Eudemis botrana, Grapholita molesta, Laspeyresia pomonella, Lobesia botrana, Polychrosis botrana, Sparganothis pilleriana* (Tortricidae), *Yponomeuta sp.* (Yponomeutidae) (Noyes, 2019).
Parasitoid hosts: *Apanteles glomeratus, Apanteles solitarius* (Hymenoptera: Braconidae), *Anilastus ebeninus, Pristomerus vulnerator* (Ichneumonidae) (Noyes, 2019).
Distribution: Algeria, Austria, Bulgaria, Czech Republic, Europe, France, Germany, Greece, Hungary, Iran, Italy, Montenegro, Morocco, Nearctic, Netherlands, North Africa, Portugal, Romania, Serbia, Spain, Switzerland, Turkey, Ukraine, United Kingdom (Noyes, 2019).

Dibrachys microgastri (Bouche, 1834)
Publications with Turkey records: Gülel (1988), Kilinçer (1982), Noyes (2019), Polat and Tozlu (2010), Öncüer (1991).
Associates: *Apis mellifera, Bombus sp.* (Apidae), *Brachymeria intermedia* (Chalcididae), *Pachysomoides fulvus* (Ichneumonidae), *Polistes metricus* (Vespidae), *Ectropis deodarae* (Lepidoptera: Geometridae), *Chalcoela iphitalis* (Pyralidae), *Pistacia vera* (Anacardiaceae), *Betula sp.* (Betulaceae), *Brassica sp., Brassica oleracea* (Brassicaceae), *Juniperus thurifera* (Cupressaceae), *Medicago sativa, Prosopis tamarugo, Robinia pseudoacacia* (Fabaceae), *Fagus sylvatica, Quercus sp., Quercus persica, Quercus pubescens* (Fagaceae), *Ephedra nebrodensis* (Gnetaceae), *Juglans regia* (Juglandaceae), *Persea americana* (Lauraceae), *Gossypium sp.* (Malvaceae), *Cedrus atlantica, Pinus sp., Pinus strobus, Pinus*

sylvestris, Pinus taeda (Pinaceae), *Rhamnus lycioides* (Rhamnaceae), *Malus pumila, Prunus sp., Prunus persica* (Rosaceae), *Salix sp.* (Salicaceae), *Osyris alba* (Santalaceae), *Ulmus pumila* (Ulmaceae), *Vitis vinifera* (Vitaceae) (Noyes, 2019).

Hosts: *Epeira sp., Epeira foliata* (Araneae: Araneidae), *Philodromus aureolus* (Thomisidae), *Stegobium paniceum* (Coleoptera: Anobiidae), *Bruchus brachialis* (Bruchidae), *Pyrrhidium sanguineum* (Cerambycidae), *Coleomegilla maculata* (Coccinellidae), *Cryptolestes pusillus* (Cucujidae), *Hypera postica, Phytonomus variabilis* (Curculionidae), *Calandra granaria, Sitophilus granaries, Sitophilus oryzae* (Dryophthoridae), *Hypoborus ficus, Scolytus schevyrewi* (Scolytidae), *Tenebroides mauritanicus* (Trogossitidae), *Forficula auricularia* (Dermaptera: Forficulidae), *Phytomyza affinis* (Diptera: Agromyzidae), *Calliphora sp., Calliphora vomitoria, Phormia regina, Phormia terraenovae, Protocalliphora sp.* (Calliphoridae), *Crataerina pallida, Stenepteryx hirundinis* (Hippoboscidae), *Musca domestica, Muscina stabulans, Stomoxys calcitrans* (Muscidae), *Rhagoletis cingulata* (Tephritidae), *Pseudococcus citri* (Hemiptera: Pseudococcidae), *Apis mellifera, Megachile sp., Megachile argentata* (Hymenoptera: Apidae), *Cephus sp.* (Cephidae), *Diprion frutetorum, Diprion hercyniae, Diprion pini, Diprion polytomum, Diprion simile, Diprion virens, Gilpinia hercyniae, Gilpinia polytoma, Neodiprion sp., Neodiprion abbottii, Neodiprion abietis, Neodiprion excitans, Neodiprion fabricii, Neodiprion lecontei, Neodiprion pratti, Neodiprion sertifer* (Diprionidae), *Trypoxylon figulus* (Sphecidae), *Nematus sp., Nematus oligospilus, Nematus ribesii, Nematus turgaiensis, Neopareophora litura, Pachynematus montanus, Priophorus morio, Priophorus padi, Pristiphora bivittate, Pristiphora erichsonii* (Tenthredinidae), *Dolichovespula media, Polistes fuscatus, Vespula germanica* (Vespidae), *Halisidota argentata, Halysidota argentata, Hyphantria sp., Hyphantria cunea, Tyria jacobaeae* (Lepidoptera: Arctiidae), *Bucculatrix crataegi, Bucculatrix ulmella* (Bucculatricidae), *Eutromula pariana* (Choreutidae), *Coleophora sp., Coleophora fuscedinella, Coleophora malivorella, Zagulajevia tadzhikiella* (Coleophoridae), *Anarsia lineatella, Coleotechnites piceaella, Dichomeris marginella, Gnorimoschema gallaesolidaginis, Gnorimoschema operculella, Pectinophora gossypiella, Phthorimaea operculella, Scrobipalpula absoluta, Sitotroga sp., Sitotroga cerealella* (Gelechiidae), *Abraxas grossulariata, Apocheima cinerarius, Biston hirtaria, Cystidia couggaria, Ennomos subsignarius, Lambdina fiscellaria, Lambdina somniaria, Operophtera brumata, Sabulodes aegrotata* (Geometridae), *Anthophila pariana* (Glyphipterygidae), *Callisto denticulella, Lithocolletis* *corylifoliella, Lithocolletis pyrifoliella, Parornix periolella, Parornix petiolella, Phyllonorycter crataegella* (Gracillariidae), *Dendrolimus sibiricus, Dendrolimus spectabilis, Eriogaster lanestris, Gastropacha quercifolia, Malacosoma sp., Malacosoma Americanum, Malacosoma californicum, Malacosoma constrictum, Malacosoma disstria, Malacosoma neustria* (Lasiocampidae), *Euproctis chrysorrhoea, Euproctis phaeorrhoea, Hemerocampa leucostigma, Hemerocampa pseudotsugata, Hemerocampa vetusta, Leucoma salicis, Leucoma wiltshirei, Lymantria sp., Lymantria dispar, Lymantria monachal, Lymantria obfuscate, Notolophus antiquus, Nygmia phaeorrhoea, Ocneria dispar, Orgyia antiqua, Orgyia leucostigma, Orgyia pseudotsugata, Orgyia vetusta, Porthetria dispar, Stilpnotia salicis* (Lymantriidae), *Leucoptera malifoliella* (Lyonetiidae), *Acronicta aceris, Acronicta clarescens, Acronicta lepusculina, Acronicta populi, Acronicta psi, Actebia fennica, Amathes c-nigrum, Amathes collaris, Camptylochila aemula, Epizeuxis aemula, Mamestra purpurissata, Ochropleura fennica, Panolis flammea, Papaipema nebris, Peridroma margaritosa, Peridroma saucia, Polia legitima, Polia puprurissata* (Noctuidae), *Phryganidia californica, Thaumetopoea pityocampa, Thaumetopoea processionea, Thaumetopoea wilkinsoni* (Notodontidae), *Euphydryas aurinia, Oeneis melissa, Oeneis semidea, Vanessa polychloros* (Nymphalidae), *Depressaria heracliana, Hofmannophila pseudopratella* (Oecophoridae), *Aporia crataegi, Colias hyale, Pieris sp., Pieris brassicae, Pieris rapae* (Pieridae), *Clania variegate, Eurukuttarus confederate, Fumaria casta, Fumea casta, Luffia lapidella, Thyridopteryx ephemeraeformis* (Psychidae), *Platyptilia carduidactyla* (Pterophoridae), *Achroia grisella, Aphomia sociella, Dicymolomia pegasalis, Dioryctria disclusa, Etiella zinckenella, Euzophera bigella, Galleria sp., Galleria mellonella, Hypsopygia costalis, Ostrinia furnacalis, Ostrinia nubilalis, Ostrinia obumbratalis, Pempelia genistella, Phlyctaenia rubigalis, Plodia interpunctella, Pyrausta ainsliei, Pyrausta nubilalis, Salebria atrifasciella, Udea rubigalis* (Pyralidae), *Actias luna, Hyalophora cecropia, Platysamia cecropia* (Saturniidae), *Aegeria myopaeformis, Aegeria typhiaeformis, Synanthedon pictipes, Synanthedon typhiaeformis* (Sesiidae), *Kermania pistaciella, Nemapogon granella, Tinea columbariella, Tinea granella* (Tineidae), *Acleris variegana, Adoxophyes reticulana, Amorbia cuneana, Apotomis lutozana, Archippus packardianus, Archips cerasivoranus, Archips fervidanus, Archips murinana, Archips packardiana, Archips rosanus, Archips semiferanus, Archips xylosteanus, Argyroploce nubiferana, Argyrotaenia citrana, Argyrotaenia velutinana, Barbara colfaxiana, Cacoecia sp., Cacoecia rosanus, Carpocapsa pomonella,*

Choristoneura pinus, Choristoneura rosaceana, Clysia ambiguella, Cryptophlebia carpophagoides, Cydia sp., Cydia molesta, Cydia pomonella, Cymolomia hartigiana, Epiblema scudderiana, Epinotia pygmaeana, Epinotia tedella, Eupoecilia ambiguella, Grapholita funebrana, Grapholita molesta, Gypsonoma aceriana, Laspeyresia pomonella, Lobesia botrana, Pandemis ribeana, Polychrosis botrana, Ptycholomoides aeriferanus, Rhyacionia sp., Rhyacionia buoliana, Sparganothis pilleriana, Spilonota ocellana, Taniva albolineana, Taniva albolineata, Tortrix viridana, Zeiraphera diniana (Tortricidae), *Argyresthia nitidella, Plutella maculipennis, Plutella xylostella, Yponomeuta sp., Yponomeuta evonymellus, Yponomeuta malinellus, Yponomeuta rorellus* (Yponomeutidae), *Procris pruni* (Zygaenidae), *Chrysopa sp., Chrysopa californica, Chrysopa lateralis, Chrysopa majuscule, Chrysopa oculate, Chrysopa plorabunda, Chrysopa rufilabris, Chrysoperla carnea* (Neuroptera: Chrysopidae), *Wesmaelius subnebulosus* (Hemerobiidae), *Sympherobius californicus* (Sympherobiidae) (Noyes, 2019).

Parasitoid hosts: *Agria affinis, Sarcophaga uliginosa* (Diptera: Sarcophagidae), *Actia interrupta, Actia pilipennis, Actia rubigalis, Arrhinomyia tragica, Bigonicheta setipennis, Bigonicheta spinipennis, Carcelia guava, Carcelia laxifrons, Ceratochaeta caudata, Compsilura concinnata, Digonochaeta setipennis, Elodia sp., Ernestia rudis, Erynnia nitida, Erynnia tortricis, Eurysthaea scutellaris, Leskia aurea, Lydella nigripes, Lydella stabulans, Masicera zimini, Mericia aldrichi, Nemorilla floralis, Nemorilla maculosa, Nemorilla pyste, Ocytata pallipes, Panzeria penitalis, Panzeria rudis, Paraphorocera senilis, Phorocera tortricis, Phryxe caudata, Phryxe vulgaris, Pseudoperichaeta erecta, Pyraustomyia penitalis, Rhacodineura antiqua, Sturmia scutellate, Tachina sp., Tachina larvarum, Tortriciophaga tortricis, Triarthria setipennis, Tricholyga segregate, Zenillia separata* (Tachinidae), *Apanteles sp., Apanteles circumscriptus, Apanteles clisiocampae, Apanteles coleophorae, Apanteles congestus, Apanteles enephes, Apanteles fulvipes, Apanteles galleriae, Apanteles glomeratus, Apanteles hyphantriae, Apanteles kazak, Apanteles lacteicolor, Apanteles lacteus, Apanteles liparidis, Apanteles longicauda, Apanteles melanoscelus, Apanteles pieridis, Apanteles porthetriae, Apanteles rubripes, Apanteles solitarius, Apanteles spurius, Apanteles vitripennis, Apanteles zygaenarum, Ascogaster quadridentate, Bracon sp., Bracon gelechiae, Chelonus shoshoneanorum, Cotesia glomerata, Dendrosoter ferrugineus, Dinocampus coccinellae, Glyptapanteles liparidis, Hormius moniliatus, Macrocentrus sp., Macrocentrus ancylivorus, Macrocentrus delicatus, Macrocentrus peroneae,*

Macrocentrus philippinensis, Macrocentrus resinellae, Macrocentrus thoracicus, Meteorus sp., Meteorus acronyctae, Meteorus communis, Meteorus hyphantriae, Meteorus rubens, Meteorus versicolor, Microbracon gelechiae, Microchelonus luzhetzkji, Microdus rufipes, Microgaster sp., Orgilus longiceps, Perilitus coccinellae, Phanerotoma grapholithae, Rhogas dendrolimi, Rhogas hyphantriae, Rhogas pallidator, Rogas indiscretus, Rogas stigmator, Sigalphis bicolor, Sigalphus bicolor, Sphaeropyx bicolor, Sycosoter lavagnei (Hymenoptera: Braconidae), *Brachymeria rugulosa, Ceratosmicra debilis, Spilochalcis sp.* (Chalcididae), *Gilpinia hercyniae* (Diprionidae), *Pleurotropis tarsalis, Tetrastichus sp., Tetrastichus rapo* (Eulophidae), *Eurytoma sp.* (Eurytomidae), *Aenoplex sp., Angitia acuta, Angitia armillata, Angitia fenestralis, Aptesis basizona, Arenetra canadensis, Bathyplectes sp., Bathyplectes anurus, Bathyplectes curculionis, Bathythrix pimplae, Caenocryptus sexannulatus, Campoplex pallipes, Campoplex phthorimaeae, Campoplex Validus, Casinaria limenitidis, Casinaria tenuiventris, Diadegma acutum, Diadegma chrysosticta, Diadegma fenestrale, Diloa antipodialis, Dusona adjuncta, Enicospilus ramidulus, Ephialtes sp., Ephialtes caudatus, Ephialtes crassiseta, Ephialtes cydiae, Ephialtes extensor, Eriborus molestee, Erigorgus sp., Exenterus sp., Exenterus amictorius, Gambrus extrematis, Gelis sp., Gelis aerator, Glypta murinanae, Glypta rufiscutellaris, Gravenhorstia sp., Gregopimpla inquisitor, Hemiteles aerator, Hemiteles socialis, Horogenes acutus, Hyposoter ebeninus, Hyposoter fugitivus, Hyposoter lymantriae, Hyposoter pilosulus, Hyposoter takagii, Hyposoter validus, Iseropus coelebs, Itoplectis alternans, Itoplectis behrensii, Itoplectis conquisitor, Itoplectis maculator, Lamachus sp., Lamachus eques, Liotryphon caudatus, Lissonota artemisiae, Mastrus sp., Mastrus carpocapsae, Mastrus molestee, Mastrus pilifrons, Mastrus smithii, Mesostenus transfugia, Olesicampe ratzeburgi, Omorgus difformis, Pachysomoides fulvus, Phaeogenes walshiae, Phobocampe sp., Phobocampe clisiocampae, Phobocampe crassiuscula, Phobocampe pallipes, Phygadeuon sp., Phygadeuon scaposus, Phytodietus polyzonias, Phytodietus vulgaris, Pimpla sp., Pimpla aequalis, Pimpla alternans, Pimpla annulipes, Pimpla conquisitor, Pimpla inquisitor, Pimpla sanguinipes, Pimpla spuria, Polistiphaga arvalis, Polistiphaga fulva, Polyblastus pedalis, Pristomerus euryptychiae, Pristomerus vulnerator, Scambus nucum, Sphecophaga vesparum, Spilocryptus extrematis, Theronia atalantae, Tranosema nigiricans, Trichomma enecator* (Ichneumonidae), *Capellia lividicorpus, Dibrachys sp., Dibrachys cavus, Habrocytus sp., Habrocytus phycidis, Psychophagus omnivorus,*

Pteromalus sp., Pteromalus puparum, Sceptrothelys lividicorpus (Pteromalidae), Mengenilla nigritula, Mengenilla subnigrescens (Strepsiptera: Mengenillidae) (Noyes, 2019).

Parasitoids: *Baryscapus coerulescens, Pediobius pyrgo, Pediobius tarsalis, Tetrastichus coerulescens* (Hymenoptera: Eulophidae), *Pteromalus puparum* (Pteromalidae) (Noyes, 2019).

Distribution: Afghanistan, Algeria, Argentina, Australia, Austria, Belarus, Belgium, Bosnia Hercegovina, Brazil, Bulgaria, Canada, Chile, Croatia, Cyprus, Czech Republic, Czechoslovakia, Denmark, Egypt, England, Eritrea, Europe, Finland, France, Germany, Greece, Hungary, India, Iran, Italy, Japan, Kazakhstan, Kirgizia, Korea, Mexico, Moldova, Morocco, Nearctic, Neotropical, Netherlands, New Zealand, North Africa, Pakistan, People's Republic of China, Peru, Poland, Portugal, Romania, Russia, Serbia, Slovakia, South Africa, Spain, Sudan, Sweden, Switzerland, Syria, Tadzhikistan, Tselinograd Obl., Tunisia, Turkey, Turkmenistan, Ukraine, United States of America, Uruguay, USSR, Uzbekistan, Yugoslavia (pre-1991) (Noyes, 2019).

Dibrachys palandoekenus (Doğanlar, 1987)

Publications with Turkey records: Doğanlar (1987), Noyes (2019).

Hosts: *Pieris brassicae* (Lepidoptera: Pieridae) (Noyes, 2019).

Parasitoid hosts: *Apanteles glomeratus* (Hymenoptera: Braconidae) (Noyes, 2019).

Distribution: Turkey (Noyes, 2019).

Genus *Dinarmus* (Thomson, 1878)

Dinarmus acutus (Thomson, 1878)

Publications with Turkey records: Hizal and Parlak (2013), Noyes (2019), Öncüer (1991).

Associates: *Alhagi graecorum, Astragalus canadensis, Astragalus glycyphyllos, Carcis siliquastrum, Cicer arictinum, Coronilla emerus, Cytisus laburnum, Lathyrus sp., Lathyrus latifolius, Lathyrus sylvestris, Lens esculenta, Pisum sativum, Vicia sp., Vigna radiata* (Fabaceae), *Ficus sp., Ficus carica* (Moraceae), *Amygdalus communis* (Rosaceae) (Noyes, 2019).

Hosts: *Acanthoscelides perforates, Bruchidius sp., Bruchidius ater, Bruchidius fulvus, Bruchidius gilvus, Bruchidius lividimanus, Bruchidius marginalis, Bruchidius unicolor, Bruchus sp., Bruchus affinis, Bruchus brachialis, Bruchus emarginatus, Bruchus lentis, Bruchus lividimanus, Bruchus rufipes, Bruchus tristiculus, Bruchus viciae, Callosobruchus maculatus* (Coleoptera: Bruchidae), *Ceroplastes rusci* (Hemiptera: Coccidae), *Dinarmus rufimanus* (Hymenoptera: Pteromalidae) (Noyes, 2019).

Parasitoid hosts: *Dinarmus sp.* (Hymenoptera: Pteromalidae) (Noyes, 2019).
Distribution: Afrotropical, Algeria, Austria, Belgium, Bulgaria, Canada, Caucasus, Croatia, Czech Republic, Czechoslovakia, England, Europe, France, Gambia, Germany, Greece, Hungary, India, Iran, Iraq, Israel, Italy, Ivory Coast, Kazakhstan, Macedonia, Malta, Moldova, Montenegro, Morocco, Nearctic, Netherlands, North Africa, Oriental, Romania, Russia, Senegal, Slovakia, Spain, Sri Lanka, Sweden, Switzerland, Transcaucasus, Tselinograd Obl., Turkey, United States of America, USSR, Yugoslavia (Federal Republic) (Noyes, 2019).

Dinarmus italicus (Masi, 1922)
Publications with Turkey records: Noyes (2019), Öncüer (1991).
Associates: *Eryngium campestre* (Apiaceae), *Coronilla emerus* (Fabaceae) (Noyes, 2019).
Hosts: *Bruchidius sp., Bruchidius albosparsus, Bruchidius fasciatus, Bruchus brachialis, Bruchus lividimanus, Bruchus rufipes, Bruchus tristiculus, Bruchus tristis* (Coleoptera: Bruchidae), *Locusta sp.* (Orthoptera: Acrididae) (Noyes, 2019).
Parasitoids: *Pteromalus sequester* (Hymenoptera: Pteromalidae) (Noyes, 2019).
Distribution: Algeria, Croatia, France, Italy, Kazakhstan, Spain, Tselinograd Obl., Turkey, Turkmenistan, Uzbekistan (Noyes, 2019).

Genus *Dinotiscus* (Ghesquiere, 1946)
Dinotiscus eupterus (Walker, 1836)
Publications with Turkey records: Noyes (2019).
Associates: *Abies nephrolepis, Picea sp., Picea abies, Picea crassifolia, Picea excelsa, Picea wilsoni, Pinus sp., Pinus armandi, Pinus contorta, Pinus koraiensis, Pinus strobus, Pinus sylvestris* (Pinaceae) (Noyes, 2019).
Hosts: *Magdalis armicollis, Magdalis barbital, Pissodes harcyniae* (Coleoptera: Curculionidae), *Cryphalus sp., Cryphalus abietis, Cryphalus piceae, Cryphalus saltuarius, Dendroctonus monticolae, Dendroctonus ponderosae, Dryocoetes sp., Dryocoetes autographus, Dryocoetes hectographus, Hylastes ater, Hylesinus fraxini, Hylurgops palliates, Ips sp., Ips amitinus, Ips bidentatus, Ips bistridentatus, Ips cembrae, Ips chalcographus, Ips duplicatus, Ips monacensis, Ips pini, Ips sexdentatus, Ips spinidens, Ips typographus, Ips vorontzovi, Orthotomicus suturalis, Phloeophthorus spinulosus, Phthorophloeus sp., Phthorophloeus spinulosus, Pityogenes sp., Pityogenes bidentatus, Pityogenes*

bistridentatus, Pityogenes chalcographus, Pityogenes conjunctus, Pityogenes pilidens, Pityogenes quadridens, Pityophthorus sp., Pityophthorus bidentatus, Pityophthorus henscheli, Pityophthorus lichtensteini, Pityophthorus micrographus, Pityophthorus polonicus, Ploygraphus rufipennis, Polygraphus sp., Polygraphus jezoensis, Polygraphus poligraphus, Polygraphus Proximus, Polygraphus punctifrons, Polygraphus rufipennis, Polygraphus subopacus, Scolytus ensifer, Scolytus scolytus, Xylechinus pilosus (Scolytidae) (Noyes, 2019).
Parasitoid hosts: *Dendrosoter middendorfi* (Hymenoptera: Braconidae) (Noyes, 2019).
Distribution: Armenia, Austria, Belarus, Belgium, Bosnia Hercegovina, Canada, Czech Republic, Czechoslovakia, Europe, Finland, France, Germany, Hungary, Japan, Kazakhstan, Lithuania, Montenegro, Nearctic, Netherlands, New Zealand, Norway, People's Republic of China, Poland, Romania, Russia, Slovakia, Spain, Sweden, Transcaucasus, Tselinograd Obl., Turkey, Ukraine, United Kingdom, United States of America, USSR (Noyes, 2019).

Genus *Erdoesina* (Graham, 1957)
Erdoesina alboannulata (Ratzeburg, 1852)
Publications with Turkey records: Noyes (2019), Öncüer (1991).
Associates: *Pinus sp.* (Pinaceae) (Noyes, 2019).
Hosts: *Boarmia bistortata, Bupalus piniarius, Ellopia prosapiaria, Ematurga atomaria* (Lepidoptera: Geometridae), *Ceramica pisi, Chloridea obsolete, Mamestra pisi, Panolis flammea, Panolis griseovariegata* (Noctuidae), *Galleria mellonella* (Pyralidae), *Hyloicus pinastri, Sphinx pinastri* (Sphingidae), *Zeiraphera griseana* (Tortricidae) (Noyes, 2019).
Parasitoid hosts: *Phryxe erythrostoma* (Diptera: Tachinidae), *Banchus sp.* (Hymenoptera: Ichneumonidae) (Noyes, 2019).
Distribution: Azerbaijan, Czech Republic, Czechoslovakia, Europe, Germany, Kazakhstan, Poland, Romania, Russia Slovakia, Tselinograd Obl., Turkey, USSR, Uzbekistan (Noyes, 2019).

Genus *Goidanichium* (Boucek, 1970)
Goidanichium atrum (Boucek, 1970)
Publications with Turkey records: Noyes (2019), Öncüer (1991).
Associates: *Hibiscus rosa-sinensis* (Malvaceae), *Citrus sp.* (Rutaceae) (Noyes, 2019).

Hosts: *Maconellicoccus hirsutus, Planococcus citri, Pseudococcus sp.* (Hemiptera: Pseudococcidae), *Sympherobius sp.* (Neuroptera: Sympherobiidae) (Noyes, 2019).
Distribution: Italy, Turkey (Noyes, 2019).

Genus *Gugolzia* (Delucchi and Steffan, 1956)
Gugolzia bademia (Doğanlar, 2004)
Publications with Turkey records: Doğanlar and Bolu (2004), Doğanlar and Doğanlar (2010), Noyes (2019), Öncüer (1991).
Hosts: *Eurytoma amygdali* (Hymenoptera: Eurytomidae), *Amygdalus communis* (Rosaceae) (Noyes, 2019).
Distribution: Turkey (Noyes, 2019).

Gugolzia bandirmae (Doğanlar, 2019)
Publications with Turkey records: Doğanlar (2019), Noyes (2019).
Associates: *Pistacia terebinthus* (Anacardiaceae) (Noyes, 2019).
Distribution: Turkey (Noyes, 2019).

Gugolzia harmolitae (Delucchi and Steffan, 1956)
Publications with Turkey records: Doğanlar (2010), Noyes (2019).
Associates: *Arundo donax, Phragmites communis* (Poaceae) (Noyes, 2019).
Hosts: *Harmolita romana, Tetramesa romana* (Hymenoptera: Eurytomidae) (Noyes, 2019).
Distribution: France, Italy, Turkey (Noyes, 2019).

Gugolzia karadagae (Doğanlar and Doğanlar, 2010)
Publications with Turkey records: Doğanlar (2010), Noyes (2019).
Associates: *Pistacia vera* (Anacardiaceae) (Noyes, 2019).
Parasitoid hosts: *Eurytoma plotnikovi* (Hymenoptera: Eurytomidae) (Noyes, 2019).
Distribution: Turkey (Noyes, 2019).

Gugolzia melengicia (Doğanlar, 2010)
Publications with Turkey records: Doğanlar (2010), Noyes (2019).
Associates: *Megastigmus pistaciae* (Torymidae), *Pistacia terebinthus* (Anacardiaceae) (Noyes, 2019).
Parasitoid hosts: *Eurytoma sp., Sycophila sp.* (Hymenoptera: Eurytomidae) (Noyes, 2019).
Distribution: Turkey (Noyes, 2019).

Gugolzia oezdenoerneki (Doğanlar, 2019)
Publications with Turkey records: Doğanlar (2019), Noyes (2019).
Associates: *Eurytoma pistaciae* (Hymenoptera: Eurytomidae), *Megastigmus pistaciae* (Torymidae) *Pistacia palaestina, Pistacia terebinthus* (Anacardiaceae) (Noyes, 2019).
Distribution: Turkey (Noyes, 2019).

Gugolzia tuzlucanensis (Doğanlar, Gozuacik and Subasi, 2018)
Publications with Turkey records: Doğanlar *et al.* (2018), Noyes (2019).
Associates: *Sophora alopecuroides* (Fabaceae) (Noyes, 2019).
Distribution: Turkey (Noyes, 2019).

Genus *Hobbya* (Delucchi, 1957)
Hobbya stenonota (Ratzeburg, 1848)
Publications with Turkey records: Noyes (2019), Öncüer (1991).
Associates: *Andricus caputmedusae, Andricus conglomeratus, Andricus coriarius, Andricus curvator, Andricus cydoniae, Andricus glutinosus, Andricus grossulariae, Andricus hartigi, Andricus hispanicus, Andricus infectorius, Andricus kollari, Andricus legitimus, Andricus lucidus, Andricus multiplicatus, Andricus quercuscalicis, Andricus quercusradicis, Andricus quercusramuli, Andricus quercustozae, Andricus solitarius, Aphelonyx cerricola, Biorhiza pallida, Chilaspis nitida, Cynips divisa, Cynips quercus, Dryocosmus cerriphilus, Pseudoneuroterus macropterus* (Cynipidae), *Quercus sp., Quercus cerris, Quercus faginea, Quercus frainetto, Quercus ithaburensis, Quercus lusitanica, Quercus petraea, Quercus pubescens, Quercus pyrenaica, Quercus robur, Quercus trojana* (Fagaceae), *Rosa sp.* (Rosaceae) (Noyes, 2019).
Hosts: *Andricus aestivalis, Andricus coriarius, Andricus grossulariae, Andricus hispanicus, Andricus kollari, Andricus multiplicatus, Andricus pictus, Andricus quercusradicis, Andricus quercusramuli, Andricus quercustozae, Aphelonyx cerricola, Biorhiza sp., Biorhiza pallida, Diplolepis fructuum, Diplolepis mayri, Synergus sp., Synergus reinhardi* (Hymenoptera: Cynipidae) (Noyes, 2019).
Parasitoid hosts: *Torymus sp., Torymus cingulatus* (Torymidae) (Noyes, 2019).
Distribution: Austria, Belgium, Bulgaria, Croatia, Czech Republic, England, Europe, France, Germany, Greece, Hungary, Iran, Jordan, Netherlands, Romania, Russia, Slovenia, Spain, Sweden, Turkey, Ukraine (Noyes, 2019).

Genus *Homoporus* (Thomson, 1878)
Homoporus arestor (Walker, 1848)
Publications with Turkey records: Noyes (2019), Öncüer (1991).
Associates: Unknown (Noyes, 2019).
Hosts: Unknown (Noyes, 2019).
Distribution: Bulgaria, Czech Republic, Czechoslovakia, England, Europe, France, Germany, Hungary, Kazakhstan, Kirgizia, Moldova, People's Republic of China, Slovakia, Spain, Sweden, Tselinograd Obl., Turkey (Noyes, 2019).

Homoporus febriculosus (Girault, 1917)
Publications with Turkey records: Koçak and Ozdemir (2012), Noyes (2019).
Associates: *Heracleum platytaenium* (Apiaceae), *Alopecurus pratensis, Ammophila arenaria, Elymus canadensis, Elymus virginicus, Muhlenbergia sp., Phleum pratense* (Poaceae) (Noyes, 2019).
Hosts: *Melanagromyza heracleana* (Diptera: Agromyzidae), *Mayetiola destructor, Phytophaga destructor* (Cecidomyiidae), *Lasiosina cinctipes* (Chloropidae), (Hymenoptera: Braconidae), *Cephus cinctus* (Cephidae), *Harmolita sp., Harmolita albomaculata, Harmolita grandis, Harmolita hordei, Harmolita tritici, Harmolita vaginicola, Tetramesa sp., Tetramesa albomaculatum, Tetramesa eximia, Tetramesa grandis, Tetramesa hordei, Tetramesa longicornis, Tetramesa tritici, Tetramesa vaginicolum* (Eurytomidae) (Noyes, 2019).
Parasitoid hosts: *Bracon cephi* (Braconidae), *Eupelmus allynii (Eupelmidae), Platygaster hiemalis, Platygaster zosine* (Platygastridae) (Noyes, 2019).
Parasitoids: *Eupelmus allynii* (Eupelmidae), *Tetramesa angustipennis* (Eurytomidae) (Noyes, 2019).
Distribution: Belgium, Bulgaria, Canada, Czech Republic, Czechoslovakia, Europe, France, Germany, Hungary, Italy, Kazakhstan, Nearctic, Netherlands, Palearctic, Russia, Spain, Sweden, Tselinograd Obl., Turkey, United Kingdom, United States of America, USSR (Noyes, 2019).

Homoporus fulviventris (Walker, 1835)
Publications with Turkey records: Noyes (2019), Öncüer (1991).
Associates: *Phanacis centaureae, Timaspis cichorii* (Cynipidae), *Artemisia herba-alba, Centaurea sp., Centaurea aspera, Centaurea scabiosa, Cichorium sp., Cichorium intybus, Lactuca viminea*

(Asteraceae), *Agropyron sp., Ammophila arenaria, Lygeum sp.* (Poaceae), *Tamarix canariensis* (Tamaricaceae) (Noyes, 2019).
Hosts: *Phlomis tuberosa* (Lamiaceae), *Panteliella sp., Phanacis centaureae, Timaspis cichorii, Timaspis phoenixopodos* (Hymenoptera: Cynipidae), *Harmolita rossica, Phylachyra apterum, Tetramesa sp., Tetramesa aciculate, Tetramesa hyalipennis* (Eurytomidae) (Noyes, 2019).
Parasitoid hosts: *Eupelmus cicadae* (Eupelmidae), *Tetramesa eximia* (Eurytomidae) (Noyes, 2019).
Parasitoids: *Eupelmus atropurpureus* (Eupelmidae) (Noyes, 2019).
Distribution: Belgium, Bosnia Hercegovina, Bulgaria, Canary Islands, Croatia, Czech Republic, England, Europe, Germany, Hungary, Iran, Italy, Kazakhstan, Madeira, Moldova, Netherlands, People's Republic of China, Romania, Russia, Serbia, Slovakia, Spain, Sweden, Transcaucasus, Tselinograd Obl., Turkey, Ukraine (Noyes, 2019).

Homoporus subniger (Walker, 1835)
Publications with Turkey records: Noyes (2019), Öncüer (1991).
Associates: *Timaspis cichorii* (Hymenoptera: Cynipidae), *Calosota sp., Eupelmus vesicularis* (Eupelmidae), *Eurytoma sp., Eurytoma centaureae, Eurytoma punctatella* (Eurytomidae), *Pteromalus hieracia, Pteromalus vibulenus, Stinoplus cichorii, Stinoplus jenningsi* (Pteromalidae), *Centaurea sp., Centaurea aspera, Centaurea sadleriana, Centaurea scabiosa, Cichorium intybus, Crepis biennis, Hypochaeris radicata, Lactuca viminea, Lapsana communis, Launaea arborescens, Picris sp., Picris echioides, Silybum marianum, Sonchus asper, Tragopogon sp.* (Asteraceae), *Papaver sp., Papaver somniferum* (Papaveraceae), *Dactylis glomerata* (Poaceae), *Rumex sp.* (Polygonaceae), *Verbascum sp.* (Scrophulariaceae) (Noyes, 2019).
Hosts: *Apion sp.* (Coleoptera: Apionidae), *Aulacidea follioti, Aulacidea freesei, Aulacidea martae, Aulacidea tragopogonis, Aylax papaveris, Iraella luteipes, Phanacis caulicola, Phanacis centaureae, Phanacis hypochoeridis, Phanacis zwoelferi, Timaspis cichorii, Timaspis lampsanae, Timaspis lusitanica, Timaspis papaveris, Timaspis phoenixopodos* (Hymenoptera: Cynipidae), *Anacampsoides sp., Glyphipteryx simpliciella* (Lepidoptera: Glyphipterygidae) (Noyes, 2019).
Distribution: Austria, Bulgaria, Czech Republic, Czechoslovakia, Europe, France, Germany, Hungary, Iran, Kazakhstan, Moldova, Montenegro, Netherlands, People's Republic of China, Romania, Serbia, Slovakia, Spain, Sweden, Tselinograd Obl., Turkey, United Kingdom (Noyes, 2019).

Genus *Lampoterma* (Graham, 1956)
Lampoterma ceutorhynchae (Doğanlar, 2018)
Publications with Turkey records: Doğanlar (2018b), Noyes (2019).
Associates: *Lepidium draba* (Brassicaceae) (Noyes, 2019).
Hosts: *Ceutorhynchus assimilis* (Coleoptera: Curculionidae) (Noyes, 2019).
Distribution: Turkey (Noyes, 2019).

Lampoterma viride (Thomson, 1876)
Publications with Turkey records: Noyes (2019), Öncüer (1991).
Associates: *Lepidium draba* (Brassicaceae) (Noyes, 2019).
Distribution: Czech Republic, Czechoslovakia, England, Europe, Germany, Hungary, Moldova, Netherlands, Spain, Sweden, Turkey (Noyes, 2019).

Lampoterma yoncae (Doğanlar, 2018)
Publications with Turkey records: Doğanlar (2018c), Noyes (2019).
Associates: *Hypera postica* (Coleoptera: Curculionidae), *Medicago sativa* (Fabaceae) (Noyes, 2019).
Distribution: Turkey (Noyes, 2019).

Genus *Lariophagus* (Crawford, 1909)
Lariophagus distinguendus (Foerster, 1841)
Publications with Turkey records: Noyes (2019), Öncüer (1991).
Associates: *Anisopteromalus calandrae* (Pteromalidae), *Coriandrum sativum* (Apiaceae), *Phaseolus vulgaris, Vigna unguiculata* (Fabaceae), *Hordeum sp., Oryza sativa, Sorghum sp., Triticum sp., Triticum aestivum* (Poaceae) (Noyes, 2019).
Hosts: *Lasioderma serricorne, Nicobium schneideri, Oligomerus ptilinoides, Sitodrepa panicea, Stegobium paniceum* (Coleoptera: Anobiidae), *Rhizopertha dominica, Rhyzopertha dominica* (Bostrychidae), *Bruchus brachialis, Callosobruchus chinensis, Callosobruchus maculatus* (Bruchidae), *Pachytychius hordei* (Curculionidae), *Calandra granaria, Calandra oryzae, Sitophilus sp., Sitophilus granarius, Sitophilus oryzae, Sitophilus zeamais* (Dryophthoridae), *Gibbium psylloides, Ptinus fur, Ptinus tectus* (Ptinidae), *Tribolium sp.* (Tenebrionidae), *Mayetiola phalaris* (Diptera: Cecidomyiidae), *Lariophagus distinguendus* (Hymenoptera: Pteromalidae) (Noyes, 2019).
Parasitoid hosts: *Microbracon hebetor* (Braconidae), *Nemeritus canescens* (Ichneumonidae) (Noyes, 2019).

Distribution: Argentina, Australia, Austria, Belgium, Brazil, Bulgaria, Canada, Chile, Croatia, Czech Republic, Czechoslovakia, Egypt, Germany, Hungary, India, Israel, Italy, Japan, Kazakhstan, Korea, Lebanon, Madeira, Mexico, Morocco, Nearctic, Nepal, Netherlands, New Zealand, People's Republic of China, Portugal, Romania, Russia, Serbia, Slovakia, Sri Lanka, Sweden, Switzerland, Thailand, Tselinograd Obl., Turkey, Ukraine, United Kingdom, United States of America, Uruguay, USSR, Yugoslavia (pre-1991) (Noyes, 2019).

Genus *Merisus* (Walker, 1834)
Merisus splendidus (Walker, 1834)
Publications with Turkey records: Korkmaz et al. (2010), Noyes (2019), Öncüer (1991).
Associates: *Tetramesa brevicornis* (Hymenoptera: Eurytomidae), *Atriplex halimus* (Chenopodiaceae), *Retama sphaerocarpa* (Fabaceae), *Dactylis glomerata, Festuca rubra, Triticum aestivum* (Poaceae), *Reseda lutea* (Resedaceae) (Noyes, 2019).
Hosts: *Tetramesa sp., Tetramesa brevicornis, Tetramesa calamagrostidis* (Hymenoptera: Eurytomidae), *Glyphipteryx simpliciella* (Lepidoptera: Glyphipterygidae) (Noyes, 2019).
Distribution: Belgium, Bulgaria, Czech Republic, Czechoslovakia, Europe, France, Germany, Hungary, Iran, Italy, Kazakhstan, Macedonia, Moldova, Montenegro, Nearctic, Netherlands, North Africa, Romania, Serbia, Slovakia, Spain, Sweden, Tselinograd Obl., Turkey, United Kingdom (Noyes, 2019).
Genus *Mesopolobus* (Westwood, 1833)
Mesopolobus aequus (Walker, 1834)
Publications with Turkey records: Noyes (2019), Öncüer (1991).
Associates: *Achillea millefolium* (Asteraceae), *Juniperus thurifera* (Cupressaceae), *Medicago sativa* (Fabaceae), *Ephedra nebrodensis* (Gnetaceae), *Avena sp., Avena sativa, Triticum aestivum* (Poaceae), *Tamarix canariensis* (Tamaricaceae) (Noyes, 2019).
Hosts: *Calligypona pellucida, Dalphacodes pellucida, Delphacodes pellucida, Delphax pellucida, Javesella pellucida* (Hemiptera: Delphacidae) (Noyes, 2019).
Parasitoids: *Mesopolobus graminum* (Hymenoptera: Pteromalidae) (Noyes, 2019).
Distribution: Belgium, Bulgaria, Canary Islands, Czech Republic, Europe, Finland, Germany, Hungary, Kazakhstan, Madeira, Moldova, Nearctic, Netherlands, North Africa, Palearctic, People's Republic of

China, Serbia, Slovakia, Spain, Sweden, Tselinograd Obl., Turkey, United Kingdom, United States of America (Noyes, 2019).

Mesopolobus amaenus (Walker, 1834)

Publications with Turkey records: Noyes (2019).

Associates: *Andricus amenti, Andricus burgundus, Andricus coriarius, Andricus corruptrix, Andricus crispator, Andricus curvator, Andricus cydoniae, Andricus gallaeurnaeformis, Andricus grossulariae, Andricus hungaricus, Andricus kollari, Andricus lignicolus, Andricus multiplicatus, Andricus paradoxus, Andricus pictus, Andricus polycerus, Andricus pseudoinflator, Andricus quadrilineatus, Andricus quercuscalicis, Andricus quercusradicis, Andricus quercusramuli, Andricus schroeckingeri, Andricus singularis, Andricus solitarius, Andricus vindobonensis, Biorhiza pallida, Cerroneuroterus lanuginosus, Chilaspis nitida, Cynips divisa, Cynips longiventris, Cynips quercusfolii, Dryocosmus kuriphilus, Dryocosmus mayri, Neuroterus quercusbaccarum, Synergus umbraculus* (Hymenoptera: Cynipidae), *Quercus sp., Quercus cerris, Quercus faginea, Quercus ilex, Quercus infectoria, Quercus ithaburensis, Quercus persica, Quercus petraea, Quercus pubescens, Quercus pyrenaica, Quercus robur, Quercus suber* (Fagaceae), Rosa sp. (Rosaceae) (Noyes, 2019).

Hosts: *Dryomyia circinnans, Dryomyia lichtensteini* (Diptera: Cecidomyiidae), *Andricus sp., Andricus amenti, Andricus coriarius, Andricus corruptrix, Andricus curvator, Andricus grossulariae, Andricus inflator, Andricus kollari, Andricus lignicolus, Andricus multiplicatus, Andricus pictus, Andricus pseudoinflator, Andricus quercuscalicis, Andricus singularis, Andricus solitarius, Andricus vindobonensis, Biorhiza pallida, Cynips quercusfolii, Diplolepis mayri, Diplolepis quercusfolii, Dryocosmus kuriphilus, Neuroterus quercusbaccarum, Neuroterus saltans, Plagiotrochus ilicis, Synergus gallaepomiformis, Synergus umbraculus, Trigonaspis synaspis* (Hymenoptera: Cynipidae), *Leucoma wiltshirei* (Lepidoptera: Lymantriidae) (Noyes, 2019).

Parasitoid hosts: *Synergus umbraculus* (Hymenoptera: Cynipidae) (Noyes, 2019).

Distribution: Austria, Belgium, Croatia, Czech Republic, Europe, Finland, France, Germany, Hungary, Iran, Ireland, Israel, Italy, Japan, Jordan, Netherlands, People's Republic of China, Romania, Spain, Sweden, Tonga, Turkey, Ukraine, United Kingdom (Noyes, 2019).

Mesopolobus aspilus (Walker, 1835)

Publications with Turkey records: Doğanlar (2011), Doğanlar et al. (2011), Noyes (2019), Polat and Tozlu (2010).

Associates: *Dasineura oleae, Lasioptera oleicola* (Cecidomyiidae), *Juniperus sp., Juniperus communis* (Cupressaceae), *Quercus acutissima* (Fagaceae), *Ephedra nebrodensis* (Gnetaceae), *Salix sp.* (Salicaceae), *Taxus baccata* (Taxaceae) (Noyes, 2019).
Hosts: *Dasineura oleae, Lasioptera oleicola, Oligotrophus juniperinus, Taxomyia taxi* (Diptera: Cecidomyiidae), *Euura amerinae* (Hymenoptera: Tenthredinidae), *Dendrolimus pini* (Lepidoptera: Lasiocampidae), *Archips rosanus* (Tortricidae) (Noyes, 2019).
Parasitoid hosts: *Sarcophaga sp.* (Diptera: Sarcophagidae) (Noyes, 2019).
Distribution: Bulgaria, Czech Republic, Czechoslovakia, Denmark, England, Europe, Hungary, Kazakhstan, Moldova, Norway, People's Republic of China, Spain, Sweden, Tselinograd Obl., Turkey (Noyes, 2019).

Mesopolobus diffinis (Walker, 1834)
Publications with Turkey records: Doğanlar (2011), Doğanlar et al. (2011), Noyes (2019), Öncüer (1991).
Associates: *Dasineura oleae, Lasioptera oleicola* (Diptera: Cecidomyiidae), *Pastinaca sativa* (Apiaceae), *Achillea millefolium, Artemisia sp., Artemisia absinthium, Artemisia campestris, Artemisia dracunculus, Artemisia herba-alba, Artemisia juncea, Artemisia terrae-albae, Carduus bourgeanus, Chrysanthemum sp., Tanacetum vulgare* (Asteraceae), *Lepidium subulatum* (Brassicaceae), *Gypsophila struthium* (Caryophyllaceae), *Atriplex halimus* (Chenopodiaceae), *Juniperus communis* (Cupressaceae), *Retama sphaerocarpa, Vicia cracca* (Fabaceae), *Quercu coccifera* (Fagaceae), *Ephedra nebrodensis* (Gnetaceae), *Nepeta hederacea, Rosmarinus officinalis* (Lamiaceae), *Lythrum salicaria* (Lythraceae), *Abies alba, Abies nordmanniana* (Pinaceae), *Atraphaxis sp.* (Polygonaceae), *Rhamnus lycioides* (Rhamnaceae), *Galium verum* (Rubiaceae), *Tamarix sp., Tamarix canariensis* (Tamaricaceae), *Taxus baccata* (Taxaceae), *Bayeria thymicola* (Diptera: Cecidomyiidae) (Noyes, 2019).
Hosts: *Bayeriola salicariae, Dasineura glechomae, Dasineura oleae, Dasineura viciae, Diarthronomyia chrysanthemi, Dichelonyx terraealbae, Dracunculomyia ehstragoni, Geocrypta galii, Lasioptera oleicola, Misospatha tubifex, Rhopalomyia foliorum, Rhopalomyia millefolii, Rhopalomyia tanaceticola, Seriphidiomyia botryose, Seriphidiomyia juncea, Taxomyia taxi* (Diptera: Cecidomyiidae), *Argyresthia fundella* (Lepidoptera: Yponomeutidae) (Noyes, 2019).

Distribution: Belgium, Bulgaria, Croatia, Czech Republic, Denmark, Europe, Germany, Hungary, Isle of Man, Kazakhstan, Moldova, Montenegro, Netherlands, Slovakia, Spain, Sweden, Transcaucasus, Tselinograd Obl., Turkey, Turkmenistan, United Kingdom (Noyes, 2019).

Mesopolobus dubius (Walker, 1834)
Publications with Turkey records: Doğanlar (2014), Noyes (2019).
Associates: *Andricus burgundus, Andricus corruptrix, Andricus curvator, Andricus lignicolus, Andricus quercuscalicis, Andricus quercusradicis, Andricus quercusramuli, Andricus quercustozae, Biorhiza pallida, Callirhytis erythrocephala, Chilaspis nitida, Cynips divisa, Cynips quercusfolii, Neuroterus albipes, Neuroterus numismalis, Neuroterus quercusbaccarum, Neuroterus tricolor, Plagiotrochus amenti, Plagiotrochus panteli, Plagiotrochus quercusilicis, Plagiotrochus razeti, Pseudoneuroterus macropterus, Pseudoneuroterus saliens, Trigonaspis Megaptera, Trigonaspis mendesi* (Hymenoptera: Cynipidae), *Fagus sylvatica, Quercu coccifera, Quercus sp., Quercus cerris, Quercus faginea, Quercus ilex, Quercus petraea, Quercus pyrenaica, Quercus robur, Quercus suber* (Fagaceae) (Noyes, 2019).
Hosts: *Andricus sp., Andricus corruptrix, Andricus curvator, Andricus lignicolus, Andricus quercuscalicis, Andricus quercusramuli, Biorhiza sp., Biorhiza pallida, Cynips divisa, Dryocosmus kuriphilus, Neuroterus sp., Neuroterus numismalis, Neuroterus quercusbaccarum, Trigonaspis Megaptera, Trigonaspis mendesi* (Hymenoptera: Cynipidae), *Sparganothis pilleriana* (Lepidoptera: Tortricidae) (Noyes, 2019).
Distribution: Andorra, Belgium, Croatia, Czech Republic, Czechoslovakia, Europe, France, Germany, Hungary, Italy, Netherlands, Romania, Spain, Sweden, Turkey, Ukraine, United Kingdom (Noyes, 2019).

Mesopolobus graminum (Hardh, 1950)
Publications with Turkey records: Noyes (2019), Öncüer (1991).
Associates: *Trigonaspis synaspis* (Cynipidae), *Hieracium sp.* (Asteraceae), *Quercus sp.* (Fagaceae), *Agropyron repens, Avena sativa, Phalaris arundinacea, Triticum aestivum* (Poaceae) (Noyes, 2019).
Hosts: *Calligypona obscurella, Calligypona pellucida, Calligypona sordidula, Criomorphus bicarinate, Delphacodes pellucida, Javesella pellucida* (Hemiptera: Delphacidae), *Aulacidea hieracia, Trigonaspis synaspis* (Hymenoptera: Cynipidae), *Eurytoma suecica, Tetramesa angustipennis, Tetramesa calamagrostidis, Tetramesa hyalipennis, Tetramesa linearis, Tetramesa pilicornis* (Eurytomidae) (Noyes, 2019).

Parasitoid hosts: *Pediobius dactylicola, Pediobius eubius, Pediobius phalaridis* (Eulophidae*), Eurytoma sp., Eurytoma phalaridis, Eurytoma pollux, Tetramesa longicornis, Tetramesa longula* (Eurytomidae), *Chlorocytus phalaridis, Chlorocytus pulchripes, Chlrocytus pulchripes, Mesopolobus aequus, Panstenon oxylus* (Pteromalidae) (Noyes, 2019).
Parasitoids: *Tetramesa hyalipennis* (Eurytomidae) (Noyes, 2019).
Distribution: Bulgaria, Denmark, England, Europe, Finland, Germany, Kazakhstan, Netherlands, Spain, Sweden, Tselinograd Obl., Turkey, USSR (Noyes, 2019).

Mesopolobus incultus (Walker, 1834)
Publications with Turkey records: Noyes (2019), Öncüer (1991).
Associates: *Carduus bourgeanus* (Asteraceae), *Trifolium sp., Trifolium pratense, Trifolium repens* (Fabaceae), *Plantago sp., Plantago lanceolata* (Plantaginaceae) (Noyes, 2019).
Hosts: *Apion sp., Protapion fulvipes* (Coleoptera: Apionidae), *Gymnetron sp., Gymnetron pascuorum, Mecinus sp.* (Curculionidae), *Polygraphus poligraphus* (Scolytidae), *Phytobia humeralis* (Diptera: Agromyzidae), *Kaltenbachiola strobi* (Cecidomyiidae), *Bruchophagus gibbus* (Hymenoptera: Eurytomidae) (Noyes, 2019).
Parasitoid hosts: *Spintherus dubius* (Pteromalidae) (Noyes, 2019).
Distribution: Belgium, Bulgaria, Canada, Czech Republic, Europe, Germany, Hungary, Italy, Moldova, Morocco, Netherlands, New Zealand, North Africa, Romania, Russia, Slovakia, Spain, Sweden, Switzerland, Syria, Turkey, Ukraine, United Kingdom (Noyes, 2019).

Mesopolobus laticornis (Walker, 1834)
Publications with Turkey records: Noyes (2019), Öncüer (1991).
Associates: *Eryngium campestre* (Apiaceae), *Gypsophila struthium* (Caryophyllaceae), *Arrhenatherum sp., Arrhenatherum elatius, Avena elatior* (Poaceae) (Noyes, 2019).
Hosts: *Oscinella frit* (Diptera: Chloropidae) (Noyes, 2019).
Distribution: Bulgaria, Croatia, Czech Republic, England, Europe, Germany, Hungary, Ireland, Moldova, Montenegro, Netherlands, Russia, Spain, Sweden, Turkey, United Kingdom (Noyes, 2019).
Mesopolobus mediterraneus (Mayr, 1903)
Publications with Turkey records: Doğanlar (2011), Doğanlar et al. (2011), Noyes (2019), Öncüer (1991).
Associates: *Dasineura oleae, Lasioptera oleicola* (Diptera: Cecidomyiidae), *Andricus burgundus, Andricus floridus, Andricus pseudoinflator, Andricus quercuscalicis, Andricus quercusradicis,*

Andricus quercusramuli, Andricus solitarius, Biorhiza pallida, Chilaspis nitida, Cynips disticha, Cynips divisa, Dryocosmus kuriphilus, Neuroterus albipes, Neuroterus anthracinus, Neuroterus numismalis, Neuroterus quercusbaccarum, Plagiotrochus amenti, Plagiotrochus britaniae, Plagiotrochus cardiguensis, Plagiotrochus panteli, Plagiotrochus quercusilicis, Pseudoneuroterus saliens (Hymenoptera: Cynipidae), *Artemisia herba-alba, Hypochaeris radicata* (Asteraceae), *Juniperus communis, Juniperus thurifera* (Cupressaceae), *Cytisus scoparius, Genista Scorpius, Sarothamnus scoparius, Ulex europaeus* (Fabaceae), *Quercu coccifera, Quercus sp., Quercus cerris, Quercus faginea, Quercus ilex, Quercus pubescens, Quercus pyrenaica, Quercus robur, Quercus suber* (Fagaceae), *Olea europaea* (Oleaceae), *Abies alba, Abies nordmanniana, Pinus halepensis* (Pinaceae), *Malus communis* (Rosaceae), *Tamarix canariensis* (Tamaricaceae), *Vitis vinifera* (Vitaceae) (Noyes, 2019).

Hosts: *Apion fuscirostre, Apion ulicis* (Coleoptera: Apionidae), *Bruchidius ater* (Bruchidae), *Stereonychus fraxini* (Curculionidae), *Dasineura affinis, Dasineura gleditchiae, Dasineura oleae, Dryomyia lichtensteini, Lasioptera oleicola, Rhopalomyia producticeps, Schmidtiella gemmarum* (Diptera: Cecidomyiidae), *Ferrisia oleae* (Hemiptera: Pseudococcidae), *Andricus sp., Andricus pseudoinflator, Andricus quercusramuli, Biorhiza pallida, Cynips sp., Dryocosmus kuriphilus, Neuroterus sp., Phanacis hypochoeridis, Plagiotrochus sp., Plagiotrochus amenti, Plagiotrochus britaniae, Plagiotrochus cardiguensis, Plagiotrochus ilicis, Plagiotrochus quercusilicis* (Hymenoptera: Cynipidae), *Coleophora serratella* (Lepidoptera: Coleophoridae), *Coleotechnites piceaella* (Lepidoptera: Gelechiidae), *Archips murinana, Lobesia botrana, Rhyacionia buoliana, Tortrix viridana* (Tortricidae), *Argyresthia fundella, Prays oleellus, Yponomeuta sp., Yponomeuta malinellus, Yponomeuta padellus* (Yponomeutidae) (Noyes, 2019).

Parasitoid hosts: *Apanteles glomeratus, Macrocentrus thoracicus* (Hymenoptera: Braconidae), *Habrocytus sequester* (Pteromalidae) (Noyes, 2019).

Parasitoids: *Azotus chionaspidis* (Aphelinidae) (Noyes, 2019).

Distribution: Andorra, Bosnia Hercegovina, Bulgaria, Croatia, Czechoslovakia, England, Europe, France, Germany, Hungary, Isle of Man, Italy, Russia, Slovakia, Spain, Sweden, Tunisia, Turkey, Yugoslavia (pre-1991) (Noyes, 2019).

Mesopolobus morys (Walker, 1848)

Publications with Turkey records: Noyes (2019), Öncüer (1991).

Associates: *Brassica sp., Brassica napus, Capsella bursa-pastoris, Lepidium draba* (Brassicaceae), *Trifolium sp.* (Fabaceae), *Tamarix canariensis* (Tamaricaceae) (Noyes, 2019).
Hosts: *Apion apricans, Apion trifolii* (Coleoptera: Apionidae), *Ceutorhynchus assimilis, Ceutorhynchus floralis, Ceutorhynchus obstrictus, Ceutorhynchus turbatus* (Curculionidae), *Dasineura brassicae* (Diptera: Cecidomyiidae) (Noyes, 2019).
Distribution: Belgium, Bulgaria, Canada, Czech Republic, England, Europe, France, Germany, Hungary, Italy, Kazakhstan, Kirgizia, Lithuania, Moldova, Nearctic, Netherlands, Poland, Romania, Serbia, Slovakia, Spain, Sweden, Switzerland, Tselinograd Obl., Turkey, United States of America (Noyes, 2019).

Mesopolobus nobilis (Walker, 1834)
Publications with Turkey records: Noyes (2019), Öncüer (1991).
Associates: *Arrhenatherum elatius, Avena sp., Avena elatior, Bromus sp., Bromus inermis, Helictotrichon pratense, Helictotrichon pubescens, Zerna inermis* (Poaceae) (Noyes, 2019).
Hosts: *Apion sp.* (Coleoptera: Apionidae), *Avena sp., Bromus sp.* (Poaceae) (Noyes, 2019).
Distribution: Austria, Belgium, Bulgaria, Czech Republic, Czechoslovakia, England, Europe, France, Germany, Hungary, Ireland, Kazakhstan, Moldova, Netherlands, New Zealand, People's Republic of China, Slovakia, Spain, Sweden, Tselinograd Obl., Turkey, United States of America (Noyes, 2019).

Mesopolobus sericeus (Forster, 1770)
Publications with Turkey records: Gencer (2003), Noyes (2019), Ozbek et al. (1999).
Associates: *Andricus callidoma, Andricus conglomeratus, Andricus coronatus, Andricus corruptrix, Andricus curvator, Andricus dentimitratus, Andricus fecundatrix, Andricus grossulariae, Andricus hispanicus, Andricus inflator, Andricus kollari, Andricus lignicolus, Andricus lucidus, Andricus malphigii, Andricus mitratus, Andricus multiplicatus, Andricus paradoxus, Andricus polycerus, Andricus quadrilineatus, Andricus quercuscalicis, Andricus quercusradicis, Andricus quercusramuli, Andricus quercustozae, Andricus seminationis, Andricus solitarius, Aphelonyx cerricola, Biorhiza pallida, Callirhytis erythrocephala, Cerroneuroterus lanuginosus, Chilaspis nitida, Cynips disticha, Cynips divisa, Cynips longiventris, Cynips quercus, Cynips quercusfolii, Dryocosmus kuriphilus, Neuroterus albipes, Neuroterus*

anthracinus, Neuroterus numismalis, Neuroterus quercusbaccarum, Neuroterus tricolor, Plagiotrochus britaniae (Cynipidae), *Acer sp., Acer monspessulanum, Acer pseudoplatanus* (Aceraceae), *Quercus sp., Quercus cerris, Quercus faginea, Quercus ilex, Quercus petraea, Quercus pubescens, Quercus pyrenaica, Quercus robur* (Fagaceae), *Rosa sp., Rosa canina* (Rosaceae) (Noyes, 2019).
Hosts: *Andricus sp., Andricus callidoma, Andricus corruptrix, Andricus curvator, Andricus grossulariae, Andricus inflator, Andricus kollari, Andricus lignicolus, Andricus mayri, Andricus ostreus, Andricus pictus, Andricus quadrilineatus, Andricus quercuscalicis, Andricus solitarius, Biorhiza pallida, Cynips disticha, Cynips divisa, Cynips longiventris, Cynips quercusfolii, Diplolepis sp., Diplolepis mayri, Diplolepis rosae, Dryocosmus kuriphilus, Neuroterus sp., Neuroterus quercusbaccarum, Neuroterus saltans, Pediaspis aceris, Synergus gallaepomiformis, Synergus pallicornis, Synergus reinhardi, Synergus umbraculus* (Hymenoptera: Cynipidae) (Noyes, 2019).
Parasitoid hosts: *Synergus sp., Synergus gallaepomiformis, Synergus nervosus, Synergus pallicornis, Synergus reinhardi, Synergus umbraculus* (Hymenoptera: Cynipidae*), Eurytoma brunniventris* (Eurytomidae) (Noyes, 2019).
Distribution: Andorra, Austria, Belgium, Bulgaria, Croatia, Denmark, Europe, France, Germany, Greece, Hungary, Iran, Israel, Italy, Kazakhstan, Moldova, Netherlands, Romania, Spain, Sweden, Tselinograd Obl., Turkey, Ukraine, United Kingdom (Noyes, 2019).

Mesopolobus teliformis (Walker, 1834)
Publications with Turkey records: Noyes (2019), Öncüer (1991).
Associates: *Agropyron sp., Agropyron cristatum, Agropyron junceiforme, Agropyron repens* (Poaceae), *Tamarix ramosissima* (Tamaricaceae) (Noyes, 2019).
Hosts: Hemiptera: Diaspididae (Noyes, 2019).
Distribution: Afghanistan, Belgium, Bulgaria, Czech Republic, Czechoslovakia, England, Europe, France, Germany, Hungary, Kazakhstan, Netherlands, People's Republic of China, Slovakia, Sweden, Tselinograd Obl., Turkey, USSR (Noyes, 2019).

Mesopolobus tibialis (Westwood, 1833)
Publications with Turkey records Doğanlar (2014), Noyes (2019).
Associates: *Andricus alniensis, Andricus amenti, Andricus burgundus, Andricus callidoma, Andricus corruptrix, Andricus crispator, Andricus curvator, Andricus cydoniae, Andricus fecundatrix, Andricus floridus,*

Andricus glandulae, Andricus grossulariae, Andricus inflator, Andricus kollari, Andricus legitimus, Andricus lignicolus, Andricus malphigii, Andricus multiplicatus, Andricus paradoxus, Andricus pseudoinflator, Andricus quadrilineatus, Andricus quercuscalicis, Andricus quercusramuli, Andricus schroeckingeri, Andricus singularis, Andricus solitarius, Andricus vindobonensis, Biorhiza pallida, Cerroneuroterus obtectus, Chilaspis nitida, Cynips disticha, Cynips divisa, Cynips longiventris, Cynips quercus, Cynips quercusfolii, Dryocosmus kuriphilus, Neuroterus albipes, Neuroterus anthracinus, Neuroterus numismalis, Neuroterus politus, Neuroterus quercusbaccarum, Neuroterus saliens, Neuroterus tricolor, Plagiotrochus amenti, Plagiotrochus australis, Plagiotrochus cardiguensis, Plagiotrochus quercusilicis, Plagiotrochus razeti, Pseudoneuroterus saliens, Trigonaspis Megaptera, Trigonaspis synaspis (Cynipidae), *Eryngium maritimum* (Apiaceae), *Medicago sp.* (Fabaceae), *Fagus sylvatica, Quercu coccifera, Quercus sp., Quercus cerris, Quercus faginea, Quercus ilex, Quercus lusitanica, Quercus pedunculiflora, Quercus petraea, Quercus pubescens, Quercus pyrenaica, Quercus robur, Quercus suber* (Fagaceae), *Pinus halepensis* (Pinaceae), *Prunus sp.* (Rosaceae) (Noyes, 2019).

Hosts: *Kaltenbachiola strobi* (Diptera: Cecidomyiidae), *Andricus sp., Andricus albopunctatus, Andricus burgundus, Andricus curvator, Andricus grossulariae, Andricus kollari, Andricus legitimus, Andricus lignicolus, Andricus malpighi, Andricus ostreus, Andricus pseudoinflator, Andricus quercuscalicis, Andricus quercusramuli, Biorhiza pallida, Cynips sp., Cynips flosculi, Dryocosmus kuriphilus, Neuroterus sp., Neuroterus albipes, Neuroterus anthracinus, Neuroterus baccarum, Neuroterus numismalis, Neuroterus politus, Neuroterus quercusbaccarum, Neuroterus tricolor, Plagiotrochus sp., Plagiotrochus amenti, Plagiotrochus australis, Plagiotrochus cardiguensis, Plagiotrochus ilicis, Plagiotrochus quercusilicis, Plagiotrochus razeti, Synergus albipes, Synergus gallaepomiformis, Trigonaspis megaptera* (Hymenoptera: Cynipidae), *Tortrix viridana* (Lepidoptera: Tortricidae) (Noyes, 2019).

Parasitoid hosts: *Synergus albipes, Synergus gallaepomiformis* (Hymenoptera: Cynipidae), *Pediobius lysis* (Eulophidae), *Caenacis divisa* (Pteromalidae) (Noyes, 2019).

Parasitoids: *Torymus auratus* (Torymidae) (Noyes, 2019).

Distribution: Andorra, Austria, Belgium, Bulgaria, Canary Islands, Channel Islands, Croatia, Czech Republic, Europe, Finland, France, Germany, Greece, Hungary, Italy, Moldova, Montenegro, Netherlands, North Africa, People's Republic of China, Poland, Romania, Slovakia,

Spain, Sweden, Syria, Turkey, Ukraine, United Kingdom, USSR (Noyes, 2019).

Genus *Muscidifurax* (Girault and Sanders, 1910)
Muscidifurax adanacus (Doğanlar, 2007)
Publications with Turkey records: Doğanlar (2007), Noyes (2019).
Associates: *Fannia sp.* (Diptera: Muscidae), *Arundo donax* (Poaceae) (Noyes, 2019).
Distribution: Turkey (Noyes, 2019).

Genus *Nasonia* (Ashmead, 1904)
Nasonia vitripennis (Walker, 1836)
Publications with Turkey records: Noyes (2019), Öncüer (1991).
Associates: *Hirundo rustica, Parus major, Sialia sialia* (Birds), *Bacillus thuringiensis* (Bacteria: Bacillaceae), *Arsenophonus nasoniae* (Enterobacteriaceae), *Wolbachia sp., Wolbachia pipientis* (Wolbachiaceae), *Bombus hypnorum* (Hymenoptera: Apidae), *Nasonia giraulti, Nasonia longicornis, Nasonia vitripennis* (Pteromalidae), *Gypsophila struthium* (Caryophyllaceae), *Gossypium hirsutum* (Malvaceae) (Noyes, 2019).
Hosts: *Ips sp., Ips acuminatus, Ips typographus* (Coleoptera: Scolytidae), *Tenebrio molitor* (Tenebrionidae), *Euborrellia annulipes* (Dermaptera: Carcinophoridae), *Blattella germanica* (Dictyoptera: Blattellidae), *Periplaneta americana* (Blattidae), *Hydrotaea capensis* (Diptera: Anthomyiidae), *Aldrichina sp., Aldrichina graham, Apaulina sp., Apaulina avium, Apaulina metallica, Apaulina sialia, Calliphora sp., Calliphora augur, Calliphora erythrocephala, Calliphora graham, Calliphora oceaniae, Calliphora quadrimaculata, Calliphora stygia, Calliphora vicina, Calliphora vomitoria, Callitroga hominivorax, Callitroga macellaria, Chrysomya sp., Chrysomya albiceps, Chrysomya chloropyga, Chrysomya megacephala, Chrysomya putoria, Chrysomya rufifacies, Chrysomya varipes, Cochliomyia hominivorax, Cochliomyia macellaria, Lucilia sp., Lucilia bufonivora, Lucilia Caesar, Lucilia cuprina, Lucilia illustris, Lucilia sericata, Phaenicia sp., Phaenicia eximia, Phaenicia Mexicana, Phaenicia sericata, Phaenicia sericera, Phormia sp., Phormia azurea, Phormia regina, Phormia terraenovae, Pollenia stygia, Protocalliphora sp., Protocalliphora avium, Protocalliphora azurea, Protocalliphora metallica, Protocalliphora sialia, Protocalliphora sordida, Protophormia terraenovae, Pycnosoma rufifacies* (Calliphoridae), *Glossina morsitans* (Glossinidae), *Pseudolynchia canariensis* (Hippoboscidae), *Cyrtoneurina pararescita,*

Fannia sp., Fannia canicularis, Fannia femoralis, Fannia scalaris, Haematobia sp., Lispe orientalis, Musca sp., Musca domestica, Muscina prolapsa, Muscina stabulans, Ophyra aenescens, Ophyra leucostoma, Ophyra nigra, Polietes lardaria, Stomoxys sp., Stomoxys calcitrans, Synthesiomyia nudiseta (Muscidae), *Gasterophilus intestinalis, Hypoderma sp.* (Oestridae), *Piophila casei* (Piophilidae), *Agria affinis, Bellieria mellanura, Bercaea sp., Bercaea cruentata, Peckia abnormis, Sarcodexia sternodontis, Sarcophaga sp., Sarcophaga barbata, Sarcophaga bullata, Sarcophaga carnaria, Sarcophaga crassipalpis, Wohlfahrtia sp.* (Sarcophagidae), *Neottiophilum praeustum* (Sciomyzidae), *Eristalis tenax* (Syrphidae), *Hubneria estigmenensis, Omotoma fumiferanea, Sturmia schineri* (Tachinidae), *Aphis nerii* (Hemiptera: Aphididae), *Oncopeltus fasciatus* (Lygaeidae), *Apis mellifera* (Hymenoptera: Apidae), *Diprion pini, Neodiprion sertifer* (Diprionidae), *Dolichovespula saxonica* (Vespidae), *Reticulitermes flavipes* (Isoptera: Rhinotermitidae), *Malacosoma castrensis* (Lepidoptera: Lasiocampidae), *Lymantria dispar* (Lymantriidae), *Trichoplusia ni* (Noctuidae), *Achroia grisella, Plodia interpunctella, Pyrausta nubilalis* (Pyralidae), *Archips fumiferana, Choristoneura fumiferana* (Tortricidae), *Diapheromera femorata* (Orthoptera: Phasmatidae) (Noyes, 2019).

Parasitoid hosts: *Agria housei, Blaesoxipha sp., Parasarcophaga argyrostoma, Sarcophaga sp., Sarcophaga aurifrons, Sarcophaga barbata, Sarcophaga bullata, Sarcophaga carnaria, Sarcophaga crassipalpis, Sarcophaga haemorrhoidalis, Sarcophaga nodosa* (Diptera: Sarcophagidae), *Diplostichus janithrix, Lydella stabulans, Sturmia sericariae, Timavia fumiferanae, Tricholyga bombycis, Winthemia amoena* (Tachinidae), *Alysia manducatory, Apanteles sp., Phaenocarpa antipoda* (Hymenoptera: Braconidae), *Eulimneria alkae* (Ichneumonidae) (Noyes, 2019).

Distribution: Argentina, Australia, Austria, Belgium, Brazil, Canada, Chile, Croatia, Czech Republic, Czechoslovakia, Denmark, Egypt, England, Europe, France, Germany, Hawaii, India, Israel, Italy, Japan, Kazakhstan, Kirgizia, Korea, Madeira, Moldova, Nearctic, Netherlands, New Caledonia, New Zealand, North Africa, Norway, People's Republic of China, Poland, Romania, Senegal, Serbia, Slovakia, South Africa, Spain, Sri Lanka, Sweden, Tadzhikistan, Tselinograd Obl., Turkey, United States of America, Uzbekistan, Zimbabwe (Noyes, 2019).

Genus *Norbanus* (Walker, 1843)

Norbanus calabrus (Masi, 1942)

Publications with Turkey records: Noyes (2019), Öncüer (1991).

Associates: Unknown (Noyes, 2019).
Hosts: Unknown (Noyes, 2019).
Distribution: Azerbaijan, Bulgaria, Croatia, Cyprus, Czech Republic, Czechoslovakia, Europe, France, Iran, Italy, Kazakhstan, Serbia, Slovakia, Tadzhikistan, Tselinograd Obl., Turkey, Turkmenistan, Ukraine (Noyes, 2019).

Norbanus cerasiops (Masi, 1922)
Publications with Turkey records: Noyes (2019), Öncüer (1991).
Associates: *Amaranthus sp.* (Amaranthaceae), *Carduus sp., Carduus bourgeanus, Carduus pycnocephalus, Cirsium sp., Cirsium arvense, Onopordum sp., Onopordum nervosum* (Asteraceae), *Crambe sp., Crambe tataria* (Brassicaceae), *Hordeum leporinum* (Poaceae) (Noyes, 2019).
Hosts: *Larinus sp., Larinus planus, Larinus turbinatus, Lixus sp., Lixus ascanii, Lixus brevirostris, Lixus cardui, Lixus juncii, Lixus scabricollis* (Coleoptera: Curculionidae) (Noyes, 2019).
Parasitoid hosts: *Bracon intercessor* (Hymenoptera: Braconidae), *Eurytoma sp.* (Eurytomidae) (Noyes, 2019).
Distribution: Caucasus, Croatia, Cyprus, Czech Republic, Czechoslovakia, Europe, France, Greece, Hungary, Iran, Italy, Kazakhstan, Madeira, Moldova, Morocco, North Africa, People's Republic of China, Romania, Serbia, Slovakia, Spain, Tselinograd Obl., Turkey (Noyes, 2019).

Norbanus obscurus (Masi, 1922)
Publications with Turkey records: Korkmaz et al. (2010), Noyes (2019).
Associates: *Halogeton sp.* (Chenopodiaceae), *Triticum aestivum* (Poaceae) (Noyes, 2019).
Hosts: *Cephus pygmaeus* (Hymenoptera: Cephidae) (Noyes, 2019).
Distribution: Algeria, Azerbaijan, Bulgaria, Caucasus, Croatia, Czechoslovakia, Europe, France, Germany, Hungary, Iran, Italy, Kazakhstan, Macedonia, Romania, Russia, Serbia, Spain, Syria, Transcaucasus, Tselinograd Obl., Turkey, Ukraine (Noyes, 2019).

Genus *Pachycrepoideus* (Ashmead, 1904)
Pachycrepoideus vindemmiae (Rondani, 1875)
Publications with Turkey records: Noyes (2019), Öncüer (1991).
Associates: *Wolbachia sp.* (Bacteria: Wolbachiaceae), *Anastrepha sp., Ceratitis sp.* (Diptera: Tephritidae), *Asobara tabida* (Hymenoptera: Eucoilidae), *Spondias mombin* (Anacardiaceae), *Cynara scolymus* (Asteraceae), *Allium cepa* (Liliaceae), *Ficus sp., Ficus carica* (Moraceae),

Eugenia dysenterica, Eugenia uniflora (Myrtaceae), *Cydonia oblonga, Prunus sp.* (Rosaceae), *Tamarix canariensis* (Tamaricaceae) (Noyes, 2019).

Hosts: *Delia antiqua, Hylemya sp., Hylemya antiqua, Hylemya brassicae, Phorbia brassicae* (Diptera: Anthomyiidae), *Calliphora sp., Chrysomya albiceps, Chrysomya megacephala, Chrysomya putoria, Lucilia sp., Lucilia illustris, Phaenicia sericata, Phormia sp., Phormia regina* (Calliphoridae), *Orseolia oryzae* (Cecidomyiidae), *Drosophila sp., Drosophila melanogaster, Drosophila uvarum* (Drosophilidae), *Lonchaea aristella* (Lonchaeidae), *Fannia sp., Fannia canicularis, Fannia pusio, Fannia scalaris, Haematobia irritans, Lyperosia sp., Lyperosia exigua, Lyperosia irritans, Musca sp., Musca domestica, Muscina stabulans, Ophyra aenescens, Ophyra leucostoma, Stomoxys sp., Stomoxys calcitrans, Stomoxys niger* (Muscidae), *Megaselia scalaris* (Phoridae), *Piophila sp., Piophila casei* (Piophilidae), *Oxysarcodexia thornax, Peckia chrysostoma, Sarcodexia sp., Sarcodexia lambens, Sarcophagula sp.* (Sarcophagidae), *Poecilosomella sp., Poecilosomella angulate* (Sphaeroceridae), *Hermetia illucens* (Stratiomyiidae), *Ornidia obesa* (Syrphidae), *Anastrepha sp., Anastrepha fraterculus, Anastrepha ludens, Anastrepha mombinpraeoptans, Anastrepha obliqua, Anastrepha sororcula, Anastrepha suspensa, Ceratitis sp., Ceratitis capitata, Ceratitis rosa, Dacus ciliates, Dacus cucurbitae, Dacus dorsalis, Dacus oleae, Dacus passiflorae, Myiopardalis pardalina, Rhagoletis sp., Rhagoletis cingulate, Rhagoletis fausta, Rhagoletis indifferens, Terellia fuscicornis* (Tephritidae), *Anasa tristis* (Hemiptera: Coreidae), *Bombus sp.* (Hymenoptera: Apidae), *Bombyx mori* (Lepidoptera: Bombycidae), *Anagasta kuehniella, Cadra cautella, Diatraea saccharalis, Ephestia kuehniella, Plodia interpunctella* (Pyralidae) (Noyes, 2019).

Parasitoid hosts: *Sarcophaga sp., Sarcophagula occidua* (Diptera: Sarcophagidae), *Exorista sorbillans, Metagonistylum minense, Paratheresia brasiliensis, Paratheresia claripalpis, Pratheresia claripalpis, Trichopoda pennipes (Tachinidae), Diachasma fullawayi, Diachasma tryoni, Opion humilis* (Hymenoptera: Braconidae), *Galesus silvestrii* (Diapriidae), *Encyrtus sp.* (Encyrtidae), *Asobara tabida, Leptopilina heterodoma* (Eucoilidae), *Tetrastichus giffardianus* (Eulophidae), *Muscidifurax raptor* (Pteromalidae) (Noyes, 2019).

Distribution: Argentina, Australia, Barbados, Belgium, Belize, Bolivia, Brazil, Bulgaria, Canada, Canary Islands, Colombia, Costa Rica, Croatia, Czech Republic, Czechoslovakia, Denmark, Dominica, Europe, Fiji, Germany, Ghana, Greece, Hawaii, India, Indonesia, Iran, Israel, Italy, Jamaica, Japan, Kazakhstan, Kirgizia, Korea, Malaysia, Mauritius,

Mexico, Moldova, Morocco, Nearctic, Netherlands, New Zealand, North Africa, Panama, Philippines, Puerto Rico, Réunion, Romania, Serbia, Slovakia, Spain, St Christopher (Kitts) & Nevis, Sweden, Switzerland, Taiwan, Thailand, Trinidad & Tobago, Tselinograd Obl., Tunisia, Turkey, United Kingdom, United States of America, Uzbekistan, Venezuela (Noyes, 2019).

Genus *Pachyneuron* (Walker, 1833)
Pachyneuron ahlaense (Mani and Saraswat, 1974)
Publications with Turkey records: Noyes (2019), Öncüer (1991).
Associates: Unknown (Noyes, 2019).
Hosts: Unknown (Noyes, 2019).
Distribution: India, Turkey (Noyes, 2019).

Pachyneuron aphidis (Bouche, 1834)
Publications with Turkey records: Noyes (2019), Öncüer (1991).
Associates: *Phenacoccus manihoti* (Hemiptera: Pseudococcidae), *Aphidencyrtus aphidivorus* (Hymenoptera: Encyrtidae), *Lasius niger, Myrmica laevinodis* (Formicidae), *Acer sp.* (Aceraceae), *Daucus carota* (Apiaceae), *Nerium indicum* (Apocynaceae), *Asclepias curassavica* (Asclepiadaceae), *Artemisia sp., Artemisia herba-alba, Carduus pycnocephalus, Cirsium arvense, Picris sp., Sonchus sp., Sonchus oleraceus, Tanacetum vulgare* (Asteraceae), *Betula sp.* (Betulaceae), *Brassica oleracea, Brassica rapa, Cakile edentula, Raphanus sativus, Sisymbrium officinale* (Brassicaceae), *Gypsophila struthium* (Caryophyllaceae), *Euonymus europaea* (Celastraceae), *Chenopodium album* (Chenopodiaceae), *Juniperus thurifera* (Cupressaceae), *Elaeagnus umbellate, Hippophae rhamnoides* (Elaeagnaceae), *Manihot esculenta* (Euphorbiaceae), *Leucaena sp., Leucaena leucocephala, Medicago sativa, Retama sphaerocarpa, Robinia pseudoacacia, Trifolium pratense* (Fabaceae), *Quercus robur* (Fagaceae), *Ephedra nebrodensis* (Gnetaceae), *Ribes aureum* (Grossulariaceae), *Philadelphus coronarius* (Hydrangeaceae), *Juglans sp.* (Juglandaceae), *Marrubium vulgare* (Lamiaceae), *Punica granatum* (Lythraceae), *Liriodendron tulipifera* (Magnoliaceae), *Gossypium sp.* (Malvaceae), *Morus sp.* (Moraceae), *Ligustrum obtusifolium* (Oleaceae), *Pinus densiflora, Pinus halepensis, Pinus thunbergii* (Pinaceae), *Pittosporum tobira* (Pittosporaceae), *Limonium sp.* (Plumbaginaceae), *Hordeum leporinum, Sorghum sp., Sorghum bicolor, Triticum sp., Triticum aestivum, Zea mays* (Poaceae), *Reseda lutea* (Resedaceae), *Cerasus macrocarpa, Malus pumila, Prunus sp., Prunus domestica, Prunus persica, Pyrus communis* (Rosaceae),

Citrus sp., Citrus deliciosa (Rutaceae), *Salix sp.* (Salicaceae), *Solanum tuberosum* (Solanaceae), *Tamarix canariensis* (Tamaricaceae), *Rhopalomyia californica* (Diptera: Cecidomyiidae), *Aphis bambusae* (Hemiptera: Aphididae) (Noyes, 2019).

Hosts: *Hippodamia tibialis, Hippodamia tredecimpunctata* (Coleoptera: Coccinellidae), *Phytomyza atricornis* (Diptera: Agromyzidae), *Diadiplosis hirticornis* (Cecidomyiidae), *Baccha clavata* (Syrphidae), *Acyrthosiphon ibotum, Acyrthosiphon kondoi, Acyrthosiphon onobrychis, Acyrthosiphon pisum, Anuraphis crataegifoliae, Anuraphis persicaeniger, Anuraphis schwartzi, Aphis sp., Aphis craccivora, Aphis euphorbiae, Aphis fabae, Aphis gossypii, Aphis maidis, Aphis nerii, Aphis nigragregalis, Aphis pomi, Aphis punicae, Aphis rumicis, Aphis spiraecola, Ayrthosiphon pisum, Brachycaudus sp., Brachycaudus cardui, Brachycaudus helichrysi, Brachycolus korotnewi, Brachycolus noxius, Brevicoryne brassicae, Callipterinella calliptera, Capitophorus sp., Capitophorus bitrichus, Capitophorus elaeagni, Capitophorus hippophaes, Cavariella aegopodii, Cavariella archangelicae, Cavariella salicicola, Chaitophorus populeti, Chromaphis juglandicola, Cinara pineti, Cinara pini, Cryptomyzus ribis, Dactynotus erigeronensis, Dactynotus picridis, Diuraphis noxia, Dysaphis apiifolia, Dysaphis plantaginea, Eriosoma lanigerum, Eulachnus thunbergia, Hayhurstia atriplicis, Hyalopterus amygdali, Hyalopterus pruni, Hyperomyzus lactucae, Hysteroneura setariae, Illinoia liriodendri, Kakimia houghtonensis, Lipaphis erysimi, Lipaphis pseudobrassicae, Macrosiphoniella abrotani, Macrosiphoniella frigidicola, Macrosiphum sp., Macrosiphum avenae, Macrosiphum euphorbiae, Macrosiphum formosanum, Macrosiphum granarium, Macrosiphum rosae, Macrosiphum solanifolii, Melanaphis sacchari, Metopolophium albidum. Myzus sp., Myzus houghtonensis, Myzus nicotiaenae, Myzus persicae, Pemphigus bursarius, Periphyllus sp., Pleotrichophorus bitrichus, Pleotrichophorus glandulosus, Rhopalosiphum insertum, Rhopalosiphum maidis, Rhopalosiphum padi, Rhopalosiphum pseudobrassicae, Rhopalosiphum sacchari, Schizaphis graminum, Schizolachnus sp., Sipha flava, Staticobium sp., Therioaphis sp., Therioaphis maculate, Therioaphis trifolii, Toxoptera aurantia, Toxoptera citricidus, Toxoptera graminum, Toxoptera odinae, Tuberculoides annulatus, Tuberocephalus momonis, Uroleucon sonchi* (Hemiptera: Aphididae), *Didesmococcus koreanus, Lecanium sp., Saissetia oleae* (Coccidae), *Kermes pubescens* (Kermesidae), *Coccura ussuriensis, Phenacoccus manihoti* (Pseudococcidae), *Cacopsylla pyricola, Heteropsylla cubana, Psylla sp., Psylla pyri* (Psyllidae), *Phaenoglyphis villosa* (Hymenoptera: Cynipidae),

Sitotroga cerealella (Lepidoptera: Gelechiidae), *Archips murinana, Merophyas divulsana* (Tortricidae) (Noyes, 2019).

Parasitoid hosts: *Aphelinus sp., Aphelinus asychis, Aphelinus mali, Aphelinus semiflavus, Aphelinus toxopteraphidis* (Hymenoptera: Aphelinidae), *Aclitus obscuripennis, Aphidius sp., Aphidius alius, Aphidius colemani, Aphidius ervi, Aphidius funebris, Aphidius gifuensis, Aphidius lonicerae, Aphidius matricariae, Aphidius medicaginis, Aphidius picipes, Aphidius pisivorus, Aphidius rapae, Aphidius rosae, Aphidius salicis, Aphidius sicarius, Aphidius smithi, Aphidius sonchi, Aphidius testaceipes, Aphidius uzbekistanicus, Binodoxys angelicae, Diaeretiella rapae, Diaeretus chenopodiaphidis, Diaeretus leucopterus, Diaeretus rapae, Diaretiella rapae, Dyscritulus sp., Ephedrus sp., Ephedrus incompletes, Ephedrus lacertosus, Ephedrus nacheri, Ephedrus persicae, Ephedrus plagiator, Lipolexis gracilis, Lysaphidus pleotrichophori, Lysiphlebus cardui, Lysiphlebus dissolutus, Lysiphlebus fabarum, Lysiphlebus japonicus, Lysiphlebus testaceipes, Monoctonus nervosus, Pauesia abietis, Pauesia unilachni, Praon sp., Praon capitophori, Praon dorsale, Praon exsoletum, Praon occidentalis, Praon palitans, Praon simulans, Praon unicus, Praon volucre, Trioxys aceris, Trioxys angelicae, Trioxys communis, Trioxys complanatus, Trioxys pallidus, Trioxys utilis* (Braconidae), *Charips dolichocerus, Charips victrix* (Charipidae), *Epidinocarsis lopezi, Homalotylus terminalis, Ooencyrtus kuvanae, Psyllaephagus sp., Psyllaephagus yaseeni* (Encyrtidae), *Alloxysta sp., Alloxysta fractricornis* (Figitidae), *Telenomus dendrolimusi* (Scelionidae) (Noyes, 2019).

Distribution: Argentina, Armenia, Australia, Belgium, Brazil, Bulgaria, Canada, Canary Islands, Caucasus, Chile, Colombia, Croatia, Cuba, Czech Republic, Czechoslovakia, Egypt, El Salvador, Europe, France, Germany, Greece, Hawaii, Hungary, India, Iran, Iraq, Israel, Italy, Japan, Kazakhstan, Kirgizia, Korea, Libya, Macedonia, Mexico, Moldova, Montenegro, Morocco, Nearctic, Netherlands, New Zealand, Pakistan, People's Republic of China, Peru, Poland, Puerto Rico, Romania, Russia, Rwanda, Serbia, Slovakia, Spain, Sweden, Switzerland, Syria, Tadzhikistan, Taiwan, Trinidad & Tobago, Tselinograd Obl., Turkey, Turkmenistan, Ukraine, United Kingdom, United States of America, Uruguay, USSR, Venezuela, Yemen, Yugoslavia (pre-1991) (Noyes, 2019).

Pachyneuron erzurumicum (Doğanlar, 1986)

Publications with Turkey records: Doğanlar (1986), Noyes (2019), Öncüer (1991).

Associates: *Euphorbia sp.* (Euphorbiaceae), *Prunus mahaleb, Pyrus sp.* (Rosaceae) (Noyes, 2019).
Hosts: *Aphis sp.* (Hemiptera: Aphididae) (Noyes, 2019).
Distribution: Europe, Iran, Kazakhstan, Turkey (Noyes, 2019).

Pachyneuron formosum (Walker, 1833)
Publications with Turkey records: Avci and Özbek (1991), Noyes (2019), Öncüer (1991).
Associates: *Eryngium campestre, Pastinaca sp.* (Apiaceae), *Artemisia sp., Carduus bourgeanus* (Asteraceae), *Lepidium subulatum* (Brassicaceae), *Atriplex halimus* (Chenopodiaceae), *Juniperus thurifera* (Cupressaceae), *Medicago sativa, Retama sphaerocarpa* (Fabaceae), *Fagus sylvatica* (Fagaceae), *Ephedra nebrodensis* (Gnetaceae), *Pinus halepensis* (Pinaceae), *Triticum sp., Triticum aestivum* (Poaceae), *Rumex sp.* (Polygonaceae), *Reseda lutea* (Resedaceae), *Rhamnus lycioides* (Rhamnaceae), *Prunus avium, Prunus cerasus, Prunus persica, Pyrus communis* (Rosaceae), *Citrus sp.* (Rutaceae), *Tamarix canariensis* (Tamaricaceae) (Noyes, 2019).
Hosts: Diptera, *Betasyrphus serarius, Epistrophe balteata, Episyrphus balteatus, Metasyrphus corollae, Neocnemodon fulvimanus, Sphaerophoria scripta, Syrphus balteatus, Syrphus luniger, Syrphus ribesii, Xanthandrus comtus* (Syrphidae), *Aphis sp., Cavariella aegopodii, Hyalopterus pruni* (Hemiptera: Aphididae), *Pseudococcus sp.* (Pseudococcidae), *Dendrolimus sp.* (Lepidoptera: Lasiocampidae), *Delias sp.* (Pieridae) (Noyes, 2019).
Parasitoid hosts: *Diplazon laetatorius* (Hymenoptera: Ichneumonidae) (Noyes, 2019).
Distribution: Belgium, Bulgaria, Canary Islands, England, Europe, France, Germany, Hungary, Iran, Italy, Japan, Kazakhstan, Kirgizia, Middle East, Morocco, Netherlands, Palearctic, People's Republic of China, Romania, Spain, Sweden, Switzerland, Tadzhikistan, Tselinograd Obl., Turkey (Noyes, 2019).

Pachyneuron gibbiscuta (Thomson, 1878)
Publications with Turkey records: Noyes (2019), Öncüer (1991).
Associates: *Prunus avium* (Rosaceae) (Noyes, 2019).
Hosts: *Myzus cerasi* (Hemiptera: Aphididae) (Noyes, 2019).
Parasitoid hosts: *Aphidius sp.* (Hymenoptera: Braconidae) (Noyes, 2019).
Distribution: Austria, Czech Republic, Czechoslovakia, Germany, People's Republic of China, Sweden, Turkey (Noyes, 2019).

Pachyneuron groenlandicum (Holmgren, 1872)
Publications with Turkey records: Noyes (2019), Öncüer (1991).
Associates: *Brachycaudus sp.* (Hemiptera: Aphididae), *Pistacia terebinthus* (Anacardiaceae), *Daucus carota* (Apiaceae), *Artemisia sp., Artemisia montana, Carduus sp., Chromolaena odorata* (Asteraceae), *Betula sp.* (Betulaceae), *Euphorbia sp.* (Euphorbiaceae), *Cassia sp., Glycine max, Vicia faba* (Fabaceae*), Abies sachalinensis, Larix leptolepis, Pinus densiflora, Pinus strobus* (Pinaceae), *Phragmites communis, Triticum sp., Triticum aestivum, Zea mays* (Poaceae), *Malus pumila, Prunus persica, Pyrus communis* (Rosaceae), *Populus sp., Salix sp., Salix cinerea* (Salicaceae), *Solanum tuberosum* (Solanaceae) (Noyes, 2019).
Hosts: *Oscinella frit* (Diptera: Chloropidae), *Psila rosae* (Psilidae), *Epistrophe balteata, Epistrophe cinctipes, Episyrphus balteata, Episyrphus balteatus, Ischiodon sp., Macrosyrphus confrator, Sphaerophoria sp., Sphaerophoria scripta, Syrphus sp., Syrphus balteatus* (Syrphidae), *Callipterinella calliptera, Forda marginata* (Hemiptera: Aphididae), *Physokermes jezoensis* (Coccidae), *Dendrolimus sp.* (Lepidoptera: Lasiocampidae), *Autographa gamma* (Noctuidae), *Pieris sp.* (Pieridae), (Neuroptera: Hemerobiidae) (Noyes, 2019).
Parasitoid hosts: *Aleiodes borealis, Aphidius sp., Aphidius sicarius* (Hymenoptera: Braconidae) (Noyes, 2019).
Distribution: Belgium, Bulgaria, Czech Republic, Czechoslovakia, Europe, France, Germany, Greenland, India, Iran, Italy, Japan, Kazakhstan, Korea, Moldova, Nearctic, Netherlands, Palearctic, People's Republic of China, Poland, Romania, Slovakia, Sweden, Switzerland, Tselinograd Obl., Turkey, United Kingdom, USSR, Yemen (Noyes, 2019).

Pachyneuron leucopiscida (Mani, 1939)
Publications with Turkey records: Alaoglu (1994), Noyes (2019), Öncüer (1991).
Associates: *Dactylopius sp.* (Hemiptera: Pseudococcidae), *Pistacia atlantica* (Anacardiaceae), *Ribes aureum* (Grossulariaceae), *Gossypium sp.* (Malvaceae), *Zea mays* (Poaceae), *Prunus avium* (Rosaceae) (Noyes, 2019).
Hosts: *Cremifania nigrocellulata, Leucopis sp., Leucopis nigricornis* (Diptera: Chamaemyiidae), *Acletoxenus indicus* (Drosophilidae), *Aphis sp., Aphis craccivora, Aphis gossypii, Cryptomyzus ribis, Dysaphis reaumuri, Hyalopterus pruni, Phloeomyzus passerine, Smynthurodes betae* (Hemiptera: Aphididae), *Dactylopius sp.* (Pseudococcidae) (Noyes, 2019).
Parasitoid hosts: *Lysiphlebus fabarum* (Hymenoptera: Braconidae) (Noyes, 2019).

Parasitoids: *Marietta leopardina* (Aphelinidae) (Noyes, 2019).
Distribution: Croatia, Czech Republic, Czechoslovakia, Europe, Germany, India, Iran, Israel, Kazakhstan, Kirgizia, Moldova, Slovakia, Sweden, Switzerland, Transcaucasus, Tselinograd Obl., Turkey, United Kingdom, Yemen (Noyes, 2019).

Pachyneuron muscarum (Linnaeus, 1758)
Publications with Turkey records: Japoshvili and Karaca (2002), Noyes (2019), Öncüer (1974; 1991), Ozsemerci and Aksit (2003), Ulgenturk (2001), Ulgenturk et al. (2001), Ulgenturk and Toros (2001).
Associates: *Trichomasthus frontalis, Trichomasthus marsus* (Encyrtidae), *Formicidae, Sonchus sp.* (Asteraceae), *Symphytum officinale* (Boraginaceae), *Brassica napus, Brassica oleracea* (Brassicaceae), *Chenopodium album* (Chenopodiaceae), *Carex brizoides* (Cyperaceae), *Robinia pseudoacacia* (Fabaceae), *Ribes aureum* (Grossulariaceae), *Olea europaea* (Oleaceae), *Picea abies, Pinus sp.* (Pinaceae), *Agropyron intermedium, Brachypodium pinnatum, Brachypodium sylvaticum, Festuca sp., Zea mays* (Poaceae), *Crataegus sp., Malus domestica, Prunus sp., Prunus armeniaca, Prunus cerasifera, Prunus domestica, Prunus institia, Prunus padus, Prunus spinosa, Pyrus communis* (Rosaceae), *Citrus sp., Citrus sinensis, Murraya paniculata* (Rutaceae), *Tilia sp.* (Tiliaceae), *Ulmus sp.* (Ulmaceae), *Vitis vinifera* (Vitaceae) (Noyes, 2019).
Hosts: *Chilocorus bipustulatus* (Coleoptera: Coccinellidae), *Ceutorhynchus assimilis* (Curculionidae), *Agromyza andalusiaca, Agromyza apfelbecki, Melanagromyza symphyti, Napomyza lateralis, Ophiomyia pinguis, Phytomyza cineracea, Phytomyza lateralis, Phytomyza ramose, Phytomyza rufipes, Tylomyza pinguis* (Diptera: Agromyzidae), *Dasineura brassicae* (Cecidomyiidae), *Chlorops pumilionis, Chlorops taeniopus, Meromyza saltatrix, Oscinella frit* (Chloropidae), *Aphis fabae, Cryptomyzus ribis, Hyperomyzus lactucae, Schizolachnus pineti, Uroleucon sonchi* (Hemiptera: Aphididae), *Ceroplastes floridensis, Ceroplastes rusci, Coccus capparidis, Coccus hesperidum, Coccus pseudomagnoliarum, Didesmococcus sp., Eriopeltis araxis, Eriopeltis strelkovi, Eulecanium ciliatum, Eulecanium coryli, Eulecanium tiliae, Filippia dece, Filippia oleae, Luzulaspis frontalis, Parthenolecanium sp., Parthenolecanium corni, Parthenolecanium coryli, Parthenolecanium persicae, Parthenolecanium rufulum, Philippia oleae, Physokermes hemicryphus, Physokermes piceae, Saissetia oleae, Sphaerolecanium prunastri* (Coccidae), *Aonidiella aurantii* (Diaspididae), *Acanthococcus aceris, Gossyparia spuria, Greenisca brachypodii, Rhizococcus agropyri* (Eriococcidae), *Nidularia pulvinata* (Kermesidae),

Heliococcus bohemicus, Maconellicoccus hirsutus, Nipaecoccus viridis, Phenacoccus aceris, Planococcus citri, Pseudococcus sp., Pseudococcus citriculus (Pseudococcidae), *Cacopsylla sp., Cacopsylla crataegi, Cacopsylla melanoneura, Cacopsylla peregrina, Cacopsylla pyri, Psylla sp., Psylla mali, Psylla pyricola* (Psyllidae), *Acantholyda nemoralis* (Hymenoptera: Pamphiliidae), *Dendrolimus pini* (Lepidoptera: Lasiocampidae), *Tortrix viridana* (Tortricidae), *Yponomeuta padellus* (Yponomeutidae) (Noyes, 2019).
Parasitoid hosts: *Coccophagus lycimnia* (Hymenoptera: Aphelinidae), *Aphidius sp.* (Braconidae), *Blastothrix confuse, Diaphorencyrtus diaphorinae, Metaphycus insidiosus, Metaphycus lounsburyi, Microterys flavus, Microterys sylvius, Prionomitus mitratus, Prionomitus tiliaris* (Encyrtidae), *Tetrastichus ceroplastae* (Eulophidae), *Trichogramma embryophagum* (Trichogrammatidae) (Noyes, 2019).
Distribution: Armenia, Belgium, Bulgaria, Caucasus, Croatia, Czech Republic, Czechoslovakia, Denmark, Europe, France, Georgia, Germany, Greece, Hungary, India, Iran, Israel, Italy, Kazakhstan, Moldova, Netherlands, People's Republic of China, Poland, Romania, Russia, Saudi Arabia, Serbia, Slovakia, Spain, St Vincent & Grenadines, Sweden, Switzerland, Taiwan, Tselinograd Obl., Turkey, Ukraine, United Kingdom, Yugoslavia (Federal Republic) (Noyes, 2019).

Pachyneuron nelsoni (Girault, 1928)
Publications with Turkey records: Noyes (2019), Öncüer (1991).
Associates: *Patchiella reaumuri* (Hemiptera: Aphididae), *Medicago sativa* (Fabaceae), *Angophora costata, Eucalyptus sp., Eucalyptus camaldulensis* (Myrtaceae), *Flindersia maculosa* (Rutaceae) (Noyes, 2019).
Hosts: *Epistrophe balteata, Episyrphus balteatus* (Diptera: Syrphidae), *Dysaphis reaumuri, Patchiella reaumuri* (Hemiptera: Aphididae) (Noyes, 2019).
Distribution: Afrotropical, Australia, Azerbaijan, Canary Islands, Croatia, Europe, India, Kazakhstan, Kirgizia, Libya, Moldova, People's Republic of China, Russia, Spain, Tadzhikistan, Tselinograd Obl., Turkey, Uzbekistan, Yemen (Noyes, 2019).
Genus *Platneptis* Boucek, 1961
Platneptis laeta (Walker, 1848)
Publications with Turkey records: Noyes (2019), Öncüer (1991).
Associates: Unknown (Noyes, 2019).
Hosts: Unknown (Noyes, 2019).
Distribution: Czech Republic, Czechoslovakia, Europe, Sweden, Turkey, United Kingdom (Noyes, 2019).

Genus *Pseudocatolaccus* (Masi, 1908)

Pseudocatolaccus nitescens (Walker, 1834)

Publications with Turkey records: Noyes (2019), Öncüer (1991).

Associates: *Artemisia herba-alba* (Asteraceae), *Carpinus sp.* (Betulaceae), *Capparis spinosa* (Capparidaceae), *Salsola vermiculata* (Chenopodiaceae), *Calycotome spinosa, Calycotome villosa, Sarothamnus sp., Sarothamnus scoparius, Ulex europaeus* (Fabaceae), *Ephedra nebrodensis* (Gnetaceae), *Mentha arvensis* (Lamiaceae), *Osyris alba* (Santalaceae) (Noyes, 2019).

Hosts: *Bruchidius unicolor* (Coleoptera: Bruchidae), *Cionus thapsi* (Curculionidae), *Asphondylia sp., Asphondylia calycotomae, Asphondylia capparis, Asphondylia gennadii, Asphondylia ignorata, Asphondylia mayeri, Asphondylia mikii, Asphondylia ononidis, Asphondylia sarothamni, Asphondylia ulicis, Asphondylia verbasci, Camptomyia strobi, Cecidomyia ulicis, Eumarchalia gennadii, Ischnonyx prunorum, Kaltenbachiola strobi* (Diptera: Cecidomyiidae) (Noyes, 2019).

Distribution: Belgium, Bulgaria, Canary Islands, Croatia, Cyprus, Czech Republic, Czechoslovakia, Europe, France, Germany, Greece, Hungary, Italy, Kazakhstan, Macedonia, Madeira, Moldova, Montenegro, Morocco, Netherlands, Romania, Russia, Serbia, Slovakia, Slovenia, Spain, Sweden, Switzerland, Tselinograd Obl., Turkey, Ukraine, United Kingdom (Noyes, 2019).

Genus *Psilocera* (Walker, 1833)

Psilocera obscura (Walker, 1833)

Publications with Turkey records: Noyes (2019), Öncüer (1991).

Parasitoid hosts: *Apanteles glomeratus* (Hymenoptera: Braconidae) (Noyes, 2019).

Distribution: Bulgaria, Croatia, Czech Republic, Czechoslovakia, England, Europe, France, Germany, Italy, Kazakhstan, Moldova, Montenegro, Netherlands, Romania, Russia, Serbia, Slovakia, Slovenia, Spain, Sweden, Tselinograd Obl., Turkey, Yugoslavia (Federal Republic) (Noyes, 2019).

Genus *Psilonotus* (Walker, 1834)

Psilonotus achaeus (Walker, 1848)

Publications with Turkey records: Noyes (2019), Öncüer (1991).

Associates: *Arctium lappa* (Asteraceae), *Betula sp., Betula papyrifera* (Betulaceae) (Noyes, 2019).

Hosts: *Oligotrophus betulae, Semudobia betulae* (Diptera: Cecidomyiidae) (Noyes, 2019).

Distribution: Belgium, Bulgaria, Canada, Czech Republic, Czechoslovakia, Europe, Germany, Hungary, Kazakhstan, Nearctic, Netherlands, Sweden, Tselinograd Obl., Turkey, Ukraine, United Kingdom, United States of America (Noyes, 2019).

Psilonotus adamas (Walker, 1834)
Publications with Turkey records: Noyes (2019), Öncüer (1991).
Associates: *Betula sp., Betula pendula, Betula pubescens* (Betulaceae) (Noyes, 2019).
Hosts: *Semudobia betulae* (Diptera: Cecidomyiidae), *Kleidocerys resedae* (Hemiptera: Lygaeidae) (Noyes, 2019).
Distribution: Belgium, Bulgaria, Czech Republic, Europe, Germany, Hungary, Kazakhstan, Netherlands, Romania, Russia, Sweden, Tselinograd Obl., Turkey, Ukraine, United Kingdom, USSR (Noyes, 2019).

Psilonotus hortensia (Walker, 1846)
Publications with Turkey records: Noyes (2019), Öncüer (1991).
Associates: *Betula sp., Betula pubescens* (Betulaceae), *Molinia caerulea* (Poaceae) (Noyes, 2019).
Hosts: *Semudobia betulae* (Diptera: Cecidomyiidae) (Noyes, 2019).
Distribution: Belgium, Germany, Hungary, Netherlands, Sweden, Turkey, United Kingdom (Noyes, 2019).
Genus *Psychophagus* (Mayr, 1904)
Psychophagus omnivorus (Walker, 1835)
Publications with Turkey records: Noyes (2019), Öncüer (1991).
Associates: *Acer sp.* (Aceraceae) (Noyes, 2019).
Hosts: *Ips typographus* (Coleoptera: Scolytidae), *Diprion hercyniae, Gilpinia hercyniae, Neodiprion sp., Neodiprion lecontei* (Hymenoptera: Diprionidae), *Cephalcia marginata* (Pamphiliidae), *Pristiphora geniculata* (Tenthredinidae), *Arctia caja, Diacrisia virginica, Estigmene Acraea, Hyphantria cunea, Panaxia dominula, Tyria jacobaeae, Utetheisa bella* (Lepidoptera: Arctiidae), *Coleophora laricella* (Coleophoridae), *Malacosoma Americanum, Malacosoma disstria, Malacosoma neustria* (Lasiocampidae), *Cnidocampa flavescens* (Limacodidae), *Strymonidia w-album* (Lycaenidae), *Euproctis chrysorrhoea, Hemerocampa leucostigma, Leucoma salicis, Lymantria dispar, Lymantria monachal, Notolophus antiquus, Nygmia phaeorrhoea, Ocneria dispar, Orgyia leucostigma, Porthetria dispar, Stilpnotia salicis* (Lymantriidae), *Achatodes zeae, Acronicta aceris, Catocala sp., Chrysodeixis chalcites, Lacanobia oleracea, Macronoctua onusta, Spodoptera exigua* (Noctuidae), *Cerura*

vinula, Datana ministra, Hoplitis milhauseri, Ichthyura inclusa, Thaumetopoea pityocampa, Thaumetopoea processionea (Notodontidae), *Polygonia c-album, Vanessa atalanta, Vanessa c-album* (Nymphalidae), *Depressaria heracliana* (Oecophoridae), *Colias sp., Pieris brassicae* (Pieridae), *Dioryctria auranticella, Galleria mellonella* (Pyralidae), *Hyalophora cecropia, Platysamia cecropia, Samia cecropia, Saturnia pavonia* (Saturniidae), *Archippus packardianus, Archips fumiferana, Choristoneura fumiferana, Epinotia pygmaeana, Grapholita molesta, Grapholita packardi* (Tortricidae) (Noyes, 2019).

Parasitoid hosts: *Agria monachae, Sarcophaga sp.* (Diptera: Sarcophagidae), *Compsilura concinnata, Parasetigena silvestris* (Tachinidae), *Apanteles affinis, Apanteles congestus, Apanteles glomeratus, Apanteles liparidis, Apanteles spurius, Macrocentrus peroneae* (Hymenoptera: Braconidae), *Itoplectis conquisitor* (Ichneumonidae) (Noyes, 2019).

Parasitoids: *Dimmockia incongrua, Pediobius pyrgo* (Eulophidae), *Eurytoma sp.* (Eurytomidae) (Noyes, 2019).

Distribution: Austria, Azerbaijan, Belgium, Canada, Croatia, Czech Republic, Czechoslovakia, England, Europe, France, Germany, Hungary, Iran, Macedonia, Moldova, Nearctic, Netherlands, North Africa, Poland, Romania, Russia, Serbia, Slovakia, Spain, Sweden, Transcaucasus, Turkey, Ukraine, United States of America, USSR, Yugoslavia (pre-1991) (Noyes, 2019).

Genus *Pteromalus* (Swederus, 1795)

Pteromalus albipennis (Walker, 1835)

Publications with Turkey records: Noyes (2019), Öncüer (1991).

Associates: *Paroxyna sp., Tephritis sp., Trupanea stellata, Urophora sp.* (Diptera: Tephritidae), *Arctium sp., Arctium lappa, Arctium tomentosum, Arnica montana, Carduus sp., Carduus bourgeanus, Carduus crispus, Carduus pycnocephalus, Centaurea sp., Centaurea nemoralis, Centaurea scabiosa, Crepis biennis, Crepis blattaroides, Inula hirta, Saussurea amara, Saussurea salsa, Serratula coronate, Sonchus oleraceus* (Asteraceae), *Eruca vesicaria* (Brassicaceae), *Hordeum leporinum* (Poaceae), *Osyris alba* (Santalaceae) (Noyes, 2019).

Hosts: *Chaetorellia jaceae, Chaetostomella cylindrica, Chaetostomella onotrophes, Myopites inulaedyssentericae, Orellia colon, Orellia tussilaginis, Orellia winthemi, Tephritis sp., Tephritis arnica, Tephritis bardanae, Tephritis crepidis, Tephritis formosa, Terellia ruficauda, Terellia tussilaginis, Terellia winthemi, Trypeta sp., Trypeta tussilaginis, Trypeta winthemi, Urophora cardui* (Diptera: Tephritidae) (Noyes, 2019).

Distribution: Belgium, Canary Islands, Czech Republic, Czechoslovakia, England, Europe, Germany, Hungary, Iran, Ireland, Italy, Kazakhstan, Moldova, Netherlands, Romania, Serbia, Slovakia, Spain, Sweden, Switzerland, Tselinograd Obl., Turkey (Noyes, 2019).

Pteromalus apum (Retzius, 1783)
Publications with Turkey records: Noyes (2019), Öncüer (1991).
Associates: *Medicago sativa* (Fabaceae) (Noyes, 2019).
Hosts: *Oscinella frit* (Diptera: Chloropidae), *Apis mellifera, Megachile sp., Megachile centuncularis, Megachile pyrenaea, Megachile pyrenaica, Megachile relativa, Megachile rotundata, Megachile willughbiella* (Hymenoptera: Apidae), *Hyphantria cunea* (Lepidoptera: Arctiidae), *Eurodryas aurinia, Hypodryas maturna, Melitaea cinxia, Melitaea didyma, Mellicta athalia* (Nymphalidae), *Gonepteryx rhamni* (Pieridae) (Noyes, 2019).
Distribution: Argentina, Belgium, Bulgaria, Canada, Czech Republic, Czechoslovakia, Denmark, England, Europe, Finland, France, Montenegro, Netherlands, Russia, Slovakia, Slovenia, Spain, Sweden, Switzerland, Turkey, United States of America, USSR, Yugoslavia (pre-1991) (Noyes, 2019).

Pteromalus bedeguaris (Thomson, 1878)
Publications with Turkey records: Bayram et al. (1998), Gencer (2003), Mete and Mergen (2017), Noyes (2019), Öncüer (1991), Özbek et al. (1999).
Associates: *Wolbachia sp.* (Bacteria: Wolbachiaceae), *Sigmophora brevicornis* (Hymenoptera: Eulophidae), *Eupelmus urozonus* (Eupelmidae), *Eurytoma rosae* (Eurytomidae), *Rosa sp., Rosa canina, Rosa damascene, Rosa rubiginosa* (Rosaceae), *Sphaeropyx bicolor* (Hymenoptera: Braconidae) (Noyes, 2019).
Hosts: *Diplolepis sp., Diplolepis centifoliae, Diplolepis eglanteriae, Diplolepis fructuum, Diplolepis mayri, Diplolepis nervosa, Diplolepis rosae, Diplolepis rosarum, Diplolepis spinosissimae, Periclistus sp., Periclistus brandtii, Rhodites rosae* (Hymenoptera: Cynipidae) (Noyes, 2019).
Parasitoid hosts: *Orthopelma luteolator, Orthopelma mediator* (Ichneumonidae), *Glyphomerus stigma, Torymus bedeguaris* (Torymidae) (Noyes, 2019).
Distribution: Andorra, Austria, Belgium, Bulgaria, Canada, Czech Republic, Czechoslovakia, Denmark, England, Europe, France, Germany, Hungary, Iran, Italy, Kazakhstan, Kirgizia, Nearctic, Netherlands,

Romania, Russia, Serbia, Slovakia, Spain, Sweden, Switzerland, Tadzhikistan, Transcaucasus, Tselinograd Obl., Turkey, Ukraine, United States of America, USSR, Uzbekistan, Yugoslavia (pre-1991) (Noyes, 2019).

Pteromalus bifoveolatus (Foerster, 1861)
Publications with Turkey records: Noyes (2019), Öncüer (1991).
Hosts: *Malacosoma franconicum, Malacosoma Neustria, Philudoria potatoria* (Lepidoptera: Lasiocampidae), *Orgyia antiqua* (Lymantriidae), *Cerura vinula, Dicranura vinula* (Notodontidae), *Saturnia pyri* (Saturniidae) (Noyes, 2019).
Distribution: Bulgaria, Croatia, Czech Republic, Czechoslovakia, England, Europe, Germany, Iran, Italy, Kazakhstan, Mauritania, Moldova, Morocco, North Africa, Romania, Serbia, Slovakia, Spain, Sweden, Switzerland, Tselinograd Obl., Turkey (Noyes, 2019).

Pteromalus chrysos (Walker, 1836)
Publications with Turkey records: Noyes (2019), Öncüer (1991).
Associates: *Andricus callidoma* (Cynipidae), *Robinia sp.* (Fabaceae), *Quercus pubescens, Quercus robur* (Fagaceae), *Triticum aestivum* (Poaceae), *Malus communis, Malus domestica, Prunus sp., Prunus padus* (Rosaceae), *Ulmus glabra* (Ulmaceae) (Noyes, 2019).
Hosts: *Oulema gallaeciana, Oulema lichenis* (Coleoptera: Chrysomelidae), *Noeeta pupillata, Tephritis arnica, Tephritis lappae* (Diptera:Tephritidae), *Diprion pini* (Hymenoptera: Diprionidae), *Trichiocampus ulmi* (Tenthredinidae), *Ocnogyna baetica* (Lepidoptera: Arctiidae), *Coleotechnites piceaella, Exoteleia dodecella* (Gelechiidae), *Ectropis bistortata* (Geometridae), *Phyllonorycter robiniella* (Gracillariidae), *Euproctis chrysorrhoea, Lymantria dispar, Stilpnotia salicis* (Lymantriidae), *Thaumetopoea pityocampa* (Notodontidae), *Aporia crataegi, Pieris brassicae, Pieris rapae* (Pieridae), *Adoxophyes orana, Adoxophyes reticulana, Archips crataeganus, Archips murinana, Archips rosanus, Clysia ambiguella, Epinotia nigricana, Pandemis ribeana, Polychrosis botrana, Rhyacionia buoliana, Sparganothis pilleriana, Tortrix viridana* (Tortricidae), *Prays oleae, Yponomeuta sp., Yponomeuta evonymellus, Yponomeuta malinellus, Yponomeuta padellus* (Yponomeutidae) (Noyes, 2019).
Parasitoid hosts: *Nemorilla floralis, Phryxe caudata* (Diptera: Tachinidae), *Goniozus claripennis* (Hymenoptera: Bethylidae), *Apanteles sp., Apanteles glomeratus, Apanteles melanoscelus, Apanteles murinanae, Apanteles pieridis, Apanteles solitarius, Chelonus annulipes, Cotesia*

plutellae, Macrocentrus linearis, Meteorus pulchricornis, Meteorus versicolor, Microgaster meridiana, Petalodes unicolor, Pholetesor sp. (Braconidae), *Ageniaspis fuscicollis* (Encyrtidae), *Angitia armillata, Cremastus interruptor, Hyposoter sp.* (Ichneumonidae), *Megastigmus spermotrophus* (Torymidae) (Noyes, 2019).
Parasitoids: *Baryscapus pospelovi, Baryscapus szocsi* (Eulophidae) (Noyes, 2019).
Distribution: Algeria, Austria, Belgium, Croatia, Czech Republic, Czechoslovakia, England, Europe, Finland, France, Germany, Hungary, Italy, Kazakhstan, Macedonia, Moldova, Montenegro, Morocco, Netherlands, Poland, Romania, Russia, Serbia, Slovakia, Spain, Sweden, Switzerland, Tselinograd Obl., Turkey, Ukraine, Yugoslavia (pre-1991) (Noyes, 2019).

Pteromalus cionobius (Erdos, 1953)
Publications with Turkey records: Noyes (2019), Öncüer (1991).
Associates: *Salsola vermiculata* (Chenopodiaceae), *Linaria sp., Verbascum sp., Verbascum phlomoides* (Scrophulariaceae) (Noyes, 2019).
Hosts: *Cionus thapsi* (Coleoptera: Curculionidae) (Noyes, 2019).
Distribution: Bulgaria, Czech Republic, Czechoslovakia, Europe, France, Germany, Hungary, Moldova, Spain, Sweden, Turkey, Yugoslavia (Federal Republic) (Noyes, 2019).

Pteromalus elevatus (Walker, 1834)
Publications with Turkey records: Noyes (2019), Öncüer (1991).
Associates: *Allocalvolia habrocytus* (Acarina: Winterschmidtiidae), *Ensina sonchi, Trupanea dubautiae* (Diptera: Tephritidae), *Acroptilon repens, Carduus sp., Centaurea sp., Centaurea nemoralis, Centaurea nigra, Centaurea phrygia, Centaurea scabiosa, Centaurea squarrosa, Chartolepis intermedia, Cirsium sp., Cirsium arvense, Cirsium oleraceum, Cirsium palustre, Cirsium vulgare, Dubautia raillardioides, Sonchus oleraceus* (Asteraceae), *Cephalaria uralensis* (Dipsacaceae), *Olea europaea* (Oleaceae) (Noyes, 2019).
Hosts: *Bactrocera oleraceae, Cerajocera ceratocera, Chaetorellia loricate, Hoplochaeta pupillata, Sphenella marginata, Tephritis conura, Terellia ruficauda, Urophora sp., Urophora cardui, Urophora cuspidate, Urophora jaceana, Urophora stylata, Urophora tenuis, Urophora xanthippe* (Diptera: Tephritidae), *Isocolus sp., Isocolus chartolepii* (Hymenoptera: Cynipidae), *Eurytoma tibialis* (Eurytomidae), *Pieris brassicae* (Lepidoptera: Pieridae) (Noyes, 2019).

Parasitoids: *Baryscapus gradwelli* (Hymenoptera: Eulophidae) (Noyes, 2019).
Distribution: Austria, Belgium, Bulgaria, Canada, Czech Republic, Czechoslovakia, England, Europe, Germany, Hawaii, Hungary, Iran, Kazakhstan, Kirgizia, Macedonia, Moldova, Montenegro, Netherlands, Romania, Russia, Spain, Sweden, Switzerland, Tadzhikistan, Transcaucasus, Tselinograd Obl., Turkey, Ukraine, Uzbekistan (Noyes, 2019).

Pteromalus intermedius (Walker, 1834)
Publications with Turkey records: Noyes (2019), Öncüer (1991).
Associates: *Artemisia sp., Artemisia absinthium, Artemisia vulgaris, Carduus bourgeanus, Cirsium oleraceum, Tragopogon sp.* (Asteraceae), *Atriplex halimus, Ceratoides papposa* (Chenopodiaceae), *Marrubium vulgare* (Lamiaceae), *Tamarix canariensis* (Tamaricaceae) (Noyes, 2019).
Hosts: *Smicronyx menozzii* (Coleoptera: Curculionidae), *Oxyna nebulosa, Oxyna parietina, Tephritis conura* (Diptera: Tephritidae), *Carpochena ceratoidis* (Lepidoptera: Coleophoridae) (Noyes, 2019).
Distribution: Bulgaria, England, Europe, France, Germany, Iran, Italy, Kazakhstan, Madeira, Netherlands, Romania, Spain, Sweden, Tselinograd Obl., Turkey, USSR (Noyes, 2019).

Pteromalus platyphilus (Walker, 1874)
Publications with Turkey records: Noyes (2019), Öncüer (1991).
Associates: *Lythrum salicaria* (Lythraceae), *Malus domestica, Rosa sp.* (Rosaceae), *Tamarix canariensis* (Tamaricaceae), *Ulmus sp.* (Ulmaceae) (Noyes, 2019).
Hosts: *Dictyna sp., Dictyna arundinacea, Dictyna uncinata* (Araneae: Dictynidae), *Eriosoma sp.* (Hemiptera: Aphididae) (Noyes, 2019).
Distribution: Afghanistan, Croatia, Czech Republic, Czechoslovakia, England, Europe, Germany, Hungary, Kazakhstan, Kirgizia, Macedonia, Moldova, Netherlands, Palearctic, Russia, Serbia, Spain, Sweden, Tselinograd Obl., Turkey, Turkmenistan, USSR (Noyes, 2019).

Pteromalus puparum (Linnaeus, 1758)
Publications with Turkey records: Avci and Özbek (1990), Noyes (2019), Öncüer (1991), Uzun (1987).
Associates: *Sceliphron sp.* (Sphecidae), *Carduus pycnocephalus* (Asteraceae), *Brassica sp., Brassica oleracea* (Brassicaceae), *Atriplex halimus* (Chenopodiaceae), *Quercu coccifera* (Fagaceae), *Hordeum leporinum* (Poaceae), *Fagopyrum esculentum* (Polygonaceae), *Malus*

pumila, Prunus persica, Pyrus malus (Rosaceae), *Murraya koenigii, Citrus sp.* (Rutaceae), *Urtica dioica, Urtica ferox* (Urticaceae) (Noyes, 2019).

Hosts: *Bruchidius unicolor* (Coleoptera: Bruchidae), *Ceutorhynchus obstrictus* (Curculionidae), *Pityocteines curvidens* (Scolytidae), *Oscinella frit* (Diptera: Chloropidae), *Lepidosaphes malicola* (Hemiptera: Diaspididae), *Apanteles sp.* (Hymenoptera: Braconidae), *Cynipidae, Sceliphron sp.* (Sphecidae), *Polistes fuscatus* (Vespidae), *Phragmatobia fuliginosa* (Lepidoptera: Arctiidae), *Coleophora fuscedinella* (Coleophoridae), *Lambdina fiscellaria* (Geometridae), *Epargyreus tityrus, Thymelicus lineola* (Hesperiidae), *Malacosoma americanum, Malacosoma disstria, Malacosoma neustria* (Lasiocampidae), *Cosmolyce baeticus* (Lycaenidae), *Cynthia cardui, Cynthia carye, Euproctis chrysorrhoea, Lymantria monachal, Nygmia phaeorrhoea* (Lymantriidae), *Acronicta psi, Autographa gamma, Mamestra brassicae, Panolis flammea, Trichoplusia ni* (Noctuidae), *Thaumetopoea processionea* (Notodontidae), *Aglais urticae, Agraulis vanilla, Argynnis adippe, Argynnis cydippe, Argynnis pandora, Bassaris gonerilla, Inachis io, Ladoga Camilla, Libythea celtis, Limenitis Archippus, Limenitis Camilla, Melitaea cinxia, Nymphalis sp., Nymphalis antiopa, Nymphalis polychloros, Pandoriana pandora, Polygonia c-album, Polygonia comma, Polygonia interrogationis, Polygonia satyrus, Pyrameis cardui, Vanessa sp., Vanessa antiopa, Vanessa atalanta, Vanessa c-album, Vanessa cardui, Vanessa carye, Vanessa io, Vanessa polychloros, Vanessa urticae, Vanessa virginiensis* (Nymphalidae), *Papilio sp., Papilio ajax, Papilio cresphontes, Papilio demodocus, Papilio demoleus, Papilio machaon, Papilio Memnon, Papilio Palamedes, Papilio polytes, Papilio polyxenes, Papilio thoas, Papilio xuthus* (Papilionidae), *Aporia crataegi, Artogeia rapae, Ascia monuste, Colias electo, Colias philodice, Eurema lisa, Leucochloe daplidice, Pieris sp., Pieris brassicae, Pieris daplidice, Pieris napi, Pieris protodice, Pieris rapae, Pontia daplidice* (Pieridae), *Thyridopteryx ephemeraeformis* (Psychidae), *Philosamia cynthia* (Saturniidae), *Archips rosanus, Argyrotaenia velutinana, Lobesia botrana* (Tortricidae), *Plutella xylostella, Yponomeuta malinellus* (Yponomeutidae), *Zygaena sp., Zygaena meliloti* (Zygaenidae) (Noyes, 2019).

Parasitoid hosts: *Apanteles sp., Apanteles glomeratus, Bracon sp., Bracon gelechiae, Rogas sp., Rogas stigmator* (Hymenoptera: Braconidae), *Hyposoter fugitivus, Phobocampe clisiocampae, Polistiphaga fulva* (Ichneumonidae), *Dibrachys cavus* (Pteromalidae) (Noyes, 2019).

Parasitoids: *Eupteromalus peregrinus* (Pteromalidae) (Noyes, 2019).

Distribution: Algeria, Australia, Austria, Azores, Barbados, Belgium, Bermuda, Bolivia, Bulgaria, Canada, Canary Islands, Chile, Croatia, Czech Republic, Czechoslovakia, Egypt, El Salvador, Europe, Finland, France, Germany, Greece, Hawaii, Hungary, India, Iran, Iraq, Ireland, Israel, Italy, Japan, Kazakhstan, Kirgizia, Korea, Macedonia, Madeira, Malaysia, Moldova, Mongolia, Nearctic, Nepal, Netherlands, New Zealand, North Africa, Pakistan, Papua New Guinea, People's Republic of China, Poland, Portugal, Romania, Russia, Saint Lucia, Saudi Arabia, Serbia, Slovakia, South Africa, Spain, Sweden, Switzerland, Tadzhikistan, Taiwan, Tselinograd Obl., Turkey, Ukraine, United Kingdom, United States of America, Uruguay, USSR, Uzbekistan, Yugoslavia (pre-1991) (Noyes, 2019).

Pteromalus semotus (Walker, 1834)
Publications with Turkey records: Avci and Özbek (1990), Noyes (2019), Öncüer (1991).
Associates: *Bruchus loti* (Coleoptera: Bruchidae), *Chilaspis nitida* (Hymenoptera: Cynipidae), *Evetria buoliana* (Lepidoptera: Tortricidae), *Artemisia sp., Artemisia herba-alba, Carduus bourgeanus, Onopordum nervosum* (Asteraceae), *Brassica oleracea* (Brassicaceae), *Salsola vermiculata* (Chenopodiaceae), *Juniperus phoeniciae, Juniperus thurifera* (Cupressaceae), *Genista scorpius, Trifolium sp., Vicia sp.* (Fabaceae), *Quercu coccifera, Quercus cerris, Quercus pubescens* (Fagaceae), *Frankenia thymifolia* (Frankeniaceae), *Ephedra nebrodensis* (Gnetaceae), *Aesculus hippocastanum* (Hippocastanaceae), *Olea europaea* (Oleaceae), *Larix decidua, Pinus sp., Pinus contorta, Pinus halepensis, Pinus nigra* (Pinaceae), *Triticum aestivum* (Poaceae), *Malus sp., Malus communis, Malus domestica, Malus pumila, Prunus sp., Prunus padus, Prunus spinosa* (Rosaceae), *Populus canadensis, Populus cathayana, Populus simonii* (Salicaceae), *Tamarix canariensis* (Tamaricaceae), *Tribulus terrestris* (Zygophyllaceae) (Noyes, 2019).
Hosts: *Ernobius abietis* (Coleoptera: Anobiidae), *Apion pomonae* (Apionidae), *Oulema gallaeciana* (Chrysomelidae), *Brachonyx pineti, Ceutorhynchus assimilis, Hypera contaminate, Hypera nigrirostris, Microlarinus lypriformis, Rhynchaenus empopulifolis* (Curculionidae), *Neodiprion sertifer* (Hymenoptera: Diprionidae), *Coleophora sp., Coleophora alcyonipennella, Coleophora congeriella, Coleophora frischella, Coleophora fuscedinella, Coleophora hemerobiella, Coleophora joannisella, Coleophora laricella, Coleophora longisignella, Coleophora montegella, Coleophora saponariella, Coleophora spissicornis* (Lepidoptera: Coleophoridae), *Coleotechnites piceaella,*

Mesophleps oxycedrella (Gelechiidae), *Ectropis bistortata, Eupithecia innotata* (Geometridae), *Callisto denticulella, Caloptilia stigmatella, Cameraria ohridella, Gracillaria syringella, Phyllonorycter blancardella, Phyllonorycter stettinensis* (Gracillariidae), *Leucoma salicis, Lymantria dispar* (Lymantriidae), *Leucoptera malifoliella* (Lyonetiidae), *Nycteola pseudasiatica, Spodoptera littoralis* (Noctuidae), *Cerura vinula* (Notodontidae), *Aporia crataegi, Pieris brassicae, Pieris rapae* (Pieridae), *Tischeria heinemanni* (Tischeriidae), *Archippus abiephaga, Archippus pulchra, Archippus similis, Archips rosanus, Archips xylosteanus, Carpocapsa pomonella, Cydia pomonella, Cymolomia hartigiana, Epiphyas sp., Epiphyas postvittana, Evetria purdeyi, Rhyacionia buoliana, Rhyacionia duplana, Rhyacionia simulate, Spilonota eremitana, Tortrix viridana* (Tortricidae), *Prays oleae, Yponomeuta sp., Yponomeuta evonymellus, Yponomeuta malinellus, Yponomeuta padellus* (Yponomeutidae), *Zygaena filipendulae* (Zygaenidae) (Noyes, 2019).

Parasitoid hosts: *Apanteles sp., Apanteles affinis, Apanteles congestus, Apanteles glomeratus, Apanteles liparidis, Apanteles pieridis, Apanteles zygaenarum, Ascogaster quadridentate, Bracon osculator, Cotesia glomerata, Macrocentrus resinellae, Macrocentrus thoracicus, Meteorus ictericus, Microplitis rufiventris, Rogas sp.* (Hymenoptera: Braconidae), *Cephaloglypta murinanae, Diadegma sp., Diadegma interruptum, Enytus apostatus, Eriborus sp., Gregopimpla kuwanae, Hyposoter sp., Ichneumon trilineatus, Itoplectis maculator, Tranosema nigricans* (Ichneumonidae) (Noyes, 2019).

Parasitoids: *Baryscapus pospelovi, Baryscapus szocsi* (Eulophidae) (Noyes, 2019).

Distribution: Austria, Belgium, Bosnia Hercegovina, Bulgaria, Canary Islands, Cape Verde Islands, Czech Republic, Czechoslovakia, Denmark, Egypt, Estonia, Europe, France, Germany, Hungary, India, Italy, Japan, Kazakhstan, Luxembourg, Mexico, Moldova, Netherlands, New Zealand, Norway, Pakistan, People's Republic of China, Poland, Romania, Russia, Serbia, Slovakia, Spain, Sweden, Switzerland, Syria, Tselinograd Obl., Turkey, Turkmenistan, Ukraine, United Kingdom, USSR, Yugoslavia (pre-1991) (Noyes, 2019).

Pteromalus sequester (Walker, 1835)

Publications with Turkey records: Noyes (2019), Öncüer (1991).

Associates: *Artemisia austriaca, Scariola orientalis* (Asteraceae), *Silene odoratissima* (Caryophyllaceae), *Atriplex halimus* (Chenopodiaceae), *Convolvulus arvensis* (Convolvulaceae), *Juniperus thurifera* (Cupressaceae), *Euphorbia sp.* (Euphorbiaceae), *Alhagi graecorum,*

Alhagi pseudalhagi, Anthyllis vulneraria, Astragalus sp., Astragalus ammonodendron, Astragalus bungee, Astragalus cornutus, Astragalus glycyphyllos, Astragalus turczaninovii, Calycotome spinosa, Caragana arborescens, Cytisus sp., Cytisus scoparius, Glycyrrhiza uralensis, Hedysarum semenovii, Lathyrus niger, Lathyrus sylvestris, Lotus corniculatus, Medicago sp., Medicago sativa, Onobrychis viciifolia, Sarothamnus sp., Sarothamnus scoparius, Sesbania acutifolia, Trifolium pratense, Ulex europaeus, Vicia sp., Vicia hirsuta (Fabaceae), *Althaea nudiflora* (Malvaceae), *Plantago depressa, Plantago lanceolata* (Plantaginaceae), *Linaria pedicellate* (Scrophulariaceae), *Tamarix canariensis* (Tamaricaceae), *Tribulus terrestris* (Zygophyllaceae) (Noyes, 2019).

Hosts: *Apion sp., Apion cerdo, Apion flavofemoratum, Apion fulvirostre, Apion fuscirostre, Apion loti, Apion ulicis, Exapion fuscirostre* (Coleoptera: Apionidae), *Bruchidius sp., Bruchidius ater, Bruchidius fasciatus, Bruchidius fulvus, Bruchidius glyzyrrhizae, Bruchidius marginalis, Bruchus affinis, Bruchus atomarius, Bruchus lividimanus* (Bruchidae), *Gymnetron asellus, Hypera venusta, Lepidotychius sp., Lepidotychius morawitzi, Mecinus collaris, Microlarinus lypriformis, Pachytychius hordei* (Curculionidae), *Asphondylia sp., Asphondylia mayeri, Contarinia hedysarocarpi, Seriphidomyia rerichi* (Diptera: Cecidomyiidae), *Hypenidium oculatum* (Tephritidae), *Bruchophagus sp., Bruchophagus caraganae, Bruchophagus funebris, Bruchophagus gibbus, Bruchophagus kolobovae, Bruchophagus platypterus, Bruchophagus roddi, Eurytoma onobrychidis* (Hymenoptera: Eurytomidae), *Galleria mellonella* (Lepidoptera: Pyralidae) (Noyes, 2019).

Parasitoid hosts: *Oedaule italica* (Hymenoptera: Pergidae) (Noyes, 2019).

Distribution: Afrotropical, Australia, Belgium, Bulgaria, Canada, Canary Islands, Chile, Czech Republic, Czechoslovakia, Europe, France, Germany, Hawaii, Hungary, India, Iran, Iraq, Israel, Kazakhstan, Kirgizia, Mexico, Nearctic, Netherlands, New Zealand, Palearctic, Poland, Romania, Russia, Serbia, Slovakia, Spain, Sweden, Tselinograd Obl., Turkey, Ukraine, United Kingdom, United States of America, USSR (Noyes, 2019).

Genus *Rhaphitelus* (Walker, 1834)

Rhaphitelus maculatus (Walker, 1834)

Publications with Turkey records: Narendran et al. (1995), Noyes (2019), Öncüer (1991).

Associates: *Scolytus butovitschi* (Coleoptera: Scolytidae), *Cotinus coggygria* (Anacardiaceae), *Platycladus orientalis, Sabina chinensis* (Cupressaceae), *Cytisus scoparius, Sarothamnus sp.* (Fabaceae), *Juglans regia* (Juglandaceae), *Fraxinus pennsylvanica, Olea europaea* (Oleaceae), *Pinus sp.* (Pinaceae), *Crataegus sp., Cydonia oblonga, Malus pumila, Malus sylvestris, Prunus sp., Prunus armeniaca, Prunus cerasifera, Prunus glandulosa, Prunus persica, Pyrus bretschneideri* (Rosaceae), *Ulmus pumila* (Ulmaceae) (Noyes, 2019).
Hosts: *Magdalis sp., Magdalis armicollis, Magdalis armigera, Magdalis barbata, Magdalis barbital, Pissodes sp., Pissodes strobi* (Coleoptera: Curculionidae), *Chaetoptelius vestitus, Hylesinus sp., Hylesinus californicus, Hylesinus fraxini, Hylesinus oleiperda, Hylesinus orni, Hylesinus vestitus, Hypoborus ficus, Phloeophthorus latus, Phloeophthorus rhododactylus, Phloeosinus sp., Phloeosinus armatus, Phloeosinus aubei, Phloeosinus bicolor, Phloeosinus cupressi, Phloeosinus punctatus, Phloeosinus serrifer, Phloeotribus sp., Phloeotribus scarabaeoides, Scolytus sp., Scolytus amygdali, Scolytus butovitschi, Scolytus ensifer, Scolytus intricatus, Scolytus japonicus, Scolytus mali, Scolytus multistriatus, Scolytus nitidus, Scolytus pygmaeus, Scolytus rugulosus, Scolytus schevyrewi, Scolytus scolytus, Scolytus semonovi, Scolytus seulensis, Scolytus sulcifrons* (Scolytidae), *Zeuzera pyrina* (Lepidoptera: Cossidae) (Noyes, 2019).
Parasitoid hosts: *Coeloides filiformis, Coeloides melanotus* (Hymenoptera: Braconidae) (Noyes, 2019).
Distribution: Argentina, Australia, Belgium, Bulgaria, Canada, Chile, Croatia, Czech Republic, Czechoslovakia, Egypt, England, Europe, France, Germany, Hungary, India, Iran, Israel, Italy, Japan, Kazakhstan, Kirgizia, Moldova, Nearctic, Netherlands, New World, New Zealand, People's Republic of China, Poland, Romania, Russia, Serbia, Slovakia, Spain, Sweden, Tselinograd Obl., Tunisia, Turkey, Turkmenistan, Ukraine, United States of America, USSR, Uzbekistan, Yugoslavia (Federal Republic), Yugoslavia (pre-1991) (Noyes, 2019).
Genus *Rhopalicus* (Foerster, 1856)
Rhopalicus guttatus (Ratzeburg, 1844)
Publications with Turkey records: Noyes (2019), Öncüer (1991).
Associates: *Cryphalus tabulaeformis* (Coleoptera: Scolytidae), *Picea sp., Picea abies, Pinus sp., Pinus armandi, Pinus sylvestris, Pinus tabulaeformis, Pinus yunnanensis* (Pinaceae) (Noyes, 2019).
Hosts: *Agrilus viridis* (Coleoptera: Buprestidae), *Pissodes sp., Pissodes notatus, Pissodes piceae, Pissodes pini, Pissodes validirostris* (Curculionidae), *Blastophagus piniperda, Ips sp., Ips typographus,*

Orthotomicus laricis, Tomicus piniperda (Scolytidae), *Dioryctria mutatella* (Lepidoptera: Pyralidae) (Noyes, 2019).
Distribution: Czech Republic, Czechoslovakia, England, Europe, Finland, France, Germany, Hungary, Lithuania, Netherlands, People's Republic of China, Portugal, Russia, Spain, Sweden, Turkey, Ukraine, United Kingdom, USSR, Yugoslavia (Federal Republic) (Noyes, 2019).

Rhopalicus tutela (Walker, 1836)
Publications with Turkey records: Noyes (2019), Öncüer (1991).
Associates: *Pityogenes chalcographus, Tomicus minor, Xyloterus lineatus* (Coleoptera: Scolytidae), *Abies nephrolepis, Abies nordmanniana, Larix laricina, Picea sp., Picea abies, Picea jezoensis, Picea koraiensis, Picea wilsoni, Pinus sp., Pinus armandi, Pinus contorta, Pinus koraiensis, Pinus strobus, Pinus sylvestris, Pinus tabulaeformis, Pinus taeda, Pinus yunnanensis* (Pinaceae) (Noyes, 2019).
Hosts: *Bostrichus laricis* (Coleoptera: Bostrychidae), *Pissodes sp., Pissodes harcyniae, Pissodes notatus, Pissodes pini, Pissodes piniphilus, Pissodes strobi* (Curculionidae), *Blastophagus minor, Blastophagus piniperda, Dendroctonus armandi, Dendroctonus micans, Dendroctonus simplex, Dryocoetes autographus, Dryocoetes hectographus, Hylastes ater, Hylastes glabratus, Hylesinus fraxini, Hylurgops sp., Hylurgops palliates, Ips sp., Ips acuminatus, Ips amitinus, Ips cembrae, Ips chalcographus, Ips curvidens, Ips duplicatus, Ips grandicollis, Ips pilidens, Ips pini, Ips sexdentatus, Ips suturalis, Ips typographus, Myelophilus piniperda, Orthotomicus laricis, Orthotomicus suturalis, Phloeophthorus spinulosus, Pityocteines curvidens, Pityogenes bidentatus, Pityogenes chalcographus, Pityophthorus lichtensteini, Polygraphus poligraphus, Polygraphus subopacus, Scolytus intricatus, Scolytus multistriatus, Scolytus rugulosus, Scolytus scolytus, Tomicus piniperda* (Scolytidae), *Cecidomyia fagi* (Diptera: Cecidomyiidae) (Noyes, 2019).
Distribution: Armenia, Austria, Belarus, Belgium, Bulgaria, Canada, Croatia, Czech Republic, Czechoslovakia, Europe, Finland, France, Georgia, Germany, Hungary, Japan, Kazakhstan, Kirgizia, Lithuania, Nearctic, Netherlands, New Zealand, Norway, People's Republic of China, Poland, Romania, Russia, Serbia, Slovakia, Spain, Sweden, Switzerland, Transcaucasus, Tselinograd Obl., Turkey, Ukraine, United Kingdom, United States of America (Noyes, 2019).

Genus *Schizonotus* (Ratzeburg, 1852)
Schizonotus sieboldi (Ratzeburg, 1848)

Publications with Turkey records: Noyes (2019), Zeki and Toros (1990).
Hosts: *Chrysomela coerulans, Chrysomela lapponica, Chrysomela populi, Chrysomela tremula, Chrysomela vigintipunctata, Chrysomesa coerufans, Melasoma interrupta, Melasoma lapponica, Melasoma populi, Melasoma scripta, Melasoma tremulae, Melasoma vigintipunctata, Plagiodera versicolora* (Coleoptera: Chrysomelidae), *Calliphora azurea, Lucilia dispar* (Diptera: Calliphoridae), *Clostera anastomosis* (Lepidoptera: Notodontidae), *Abies sp.* (Pinaceae), *Populus sp., Populus nigra, Populus tremula, Salix sp., Salix fragilis, Salix rubens* (Salicaceae) (Noyes, 2019).
Distribution: Bulgaria, Caucasus, Commonwealth of Independent States, Croatia, Czech Republic, Czechoslovakia, Europe, France, Germany, Hungary, Iran, Italy, Kazakhstan, Lithuania, Macedonia, Moldova, Norway, Pakistan, Palearctic, People's Republic of China, Russia, Serbia, Slovakia, Sweden, Switzerland, Tselinograd Obl., Turkey, Turkmenistan, United Kingdom, United States of America, USSR, Uzbekistan (Noyes, 2019).

Genus *Sphegigaster* (Spinola, 1811)
Sphegigaster brevicornis (Walker, 1833)
Publications with Turkey records: Noyes (2019), Öncüer (1991).
Associates: *Beta vulgaris* (Chenopodiaceae), *Medicago sativa* (Fabaceae) (Noyes, 2019).
Hosts: Diptera: Agromyzidae (Noyes, 2019).
Distribution: England, Italy, Romania, Spain, Sweden, Turkey (Noyes, 2019).

Sphegigaster pallicornis (Spinola, 1808)
Publications with Turkey records: Noyes (2019), Öncüer (1991).
Associates: *Ilex sp., Ilex aquifolium* (Aquifoliaceae), *Crambe tataria* (Brassicaceae), *Orobanche ramose* (Orobanchaceae) (Noyes, 2019).
Hosts: *Phytomyza sp., Phytomyza ilicis, Phytomyza ramose, Phytomyza rufipes* (Diptera: Agromyzidae) (Noyes, 2019).
Parasitoids: *Pediobius metallicus* (Hymenoptera: Eulophidae) (Noyes, 2019).
Distribution: Austria, Belgium, Bulgaria, Canada, Czech Republic, France, Germany, Hungary, Ireland, Italy, Kazakhstan, Moldova, Nearctic, Netherlands, Romania, Serbia, Slovakia, Spain, Sweden, Tselinograd Obl., Turkey, United Kingdom (Noyes, 2019).

Sphegigaster pedunculiventris (Spinola, 1808)
Publications with Turkey records: Noyes (2019), Öncüer (1991).
Associates: *Angelica sylvestris* (Apiaceae), *Centaurea nigra, Senecio jacobaea* (Asteraceae) (Noyes, 2019).
Hosts: *Agromyza aeneoventris, Melanagromyza aeneoventris, Melanagromyza albocilia, Melanagromyza lappae, Napomyza lateralis* (Diptera: Agromyzidae) (Noyes, 2019).
Distribution: Belgium, Bulgaria, Czech Republic, England, Europe, Germany, Hungary, Italy, Kazakhstan, Moldova, Netherlands, Romania, Slovakia, Sweden, Tselinograd Obl., Turkey (Noyes, 2019).

Genus *Spintherus* (Thomson, 1878)
Spintherus dubius (Nees, 1834)
Publications with Turkey records: Noyes (2019), Öncüer (1991).
Associates: *Carduus bourgeanus* (Asteraceae), *Astragalus asper, Trifolium sp., Trifolium fragiferum, Trifolium medium, Trifolium pratense, Trifolium repens* (Fabaceae) (Noyes, 2019).
Hosts: *Apion sp., Apion africum, Apion apricans, Apion trifolii* (Coleoptera: Apionidae), *Ostrinia nubilalis* (Lepidoptera: Pyralidae) (Noyes, 2019).
Parasitoid hosts: *Bruchophagus astragali* (Hymenoptera: Eurytomidae), *Spintherus dubius* (Pteromalidae) (Noyes, 2019).
Distribution: Belgium, Bulgaria, Caucasus, Croatia, Czech Republic, England, Europe, France, Germany, Hungary, Kazakhstan, Kirgizia, Moldova, Montenegro, Netherlands, Palearctic, Poland, Romania, Russia, Serbia, Slovakia, Spain, Sweden, Switzerland, Tselinograd Obl., Turkey, Ukraine, USSR (Noyes, 2019).

Genus *Stenomalina* (Ghesquiere, 1946)
Stenomalina communis (Nees, 1834)
Publications with Turkey records: Noyes (2019), Öncüer (1991).
Associates: *Corylus avellana* (Betulaceae) (Noyes, 2019).
Hosts: *Chlorops laetus, Chlorops pumilionis* (Diptera: Chloropidae), *Andricus quercuscalicis* (Hymenoptera: Cynipidae), *Dioryctria abietella* (Lepidoptera: Pyralidae) (Noyes, 2019).
Distribution: Austria, Belarus, Croatia, Czech Republic, England, Europe, Germany, Hungary, Kazakhstan, Moldova, Netherlands, Romania, Russia, Spain, Sweden, Tselinograd Obl., Turkey (Noyes, 2019).

Stenomalina favorinus (Walker, 1839)
Publications with Turkey records: Noyes (2019), Öncüer (1991).
Associates: Unknown (Noyes, 2019).
Hosts: Unknown (Noyes, 2019).
Distribution: Belgium, Bulgaria, Croatia, Czech Republic, Czechoslovakia, England, Europe, Germany, Ireland, Netherlands, Romania, Sweden, Turkey (Noyes, 2019).

Stenomalina laticeps (Walker, 1850)
Publications with Turkey records: Noyes (2019), Öncüer (1991).
Associates: Unknown (Noyes, 2019).
Hosts: Unknown (Noyes, 2019).
Distribution: Czech Republic, England, Europe, Germany, Hungary, Kazakhstan, Netherlands, People's Republic of China, Spain, Sweden, Tselinograd Obl., Turkey (Noyes, 2019).

Genus *Syntomopus* (Walker, 1833)
Syntomopus incisus (Thomson, 1878)
Publications with Turkey records: Noyes (2019), Öncüer (1991).
Associates: *Arctium sp., Arctium vulgare, Carduus sp., Cirsium sp., Cirsium eriophorum, Hypochaeris radicata* (Asteraceae), *Urtica dioica* (Urticaceae) (Noyes, 2019).
Hosts: *Melanagromyza sp., Melanagromyza aenea, Melanagromyza aeneoventris, Melanagromyza dettmeri, Melanagromyza lappae* (Diptera: Agromyzidae), *Phanacis hypochoeridis* (Hymenoptera: Cynipidae) (Noyes, 2019).
Distribution: Bulgaria, Canary Islands, Czech Republic, England, Europe, Germany, Hungary, Isle of Man, Italy, Kazakhstan, Netherlands, People's Republic of China, Romania, Serbia, Slovakia, Spain, Sweden, Tselinograd Obl., Turkey, Yugoslavia (Federal Republic) (Noyes, 2019).

Syntomopus incurvus (Walker, 1833)
Publications with Turkey records: Noyes (2019), Öncüer (1991).
Associates: *Urtica dioica* (Urticaceae) (Noyes, 2019).
Hosts: *Melanagromyza sp., Melanagromyza aenea, Melanagromyza dettmeri, Melanagromyza tripolii* (Diptera: Agromyzidae), *Adaina microdactyla* (Lepidoptera: Pterophoridae) (Noyes, 2019).
Distribution: Belgium, Czech Republic, Europe, Germany, Italy, Kazakhstan, Netherlands, People's Republic of China, Serbia, Sweden, Tselinograd Obl., Turkey, United Kingdom (Noyes, 2019).

Genus *Tomicobia* (Ashmead, 1899)
Tomicobia promulus (Walker, 1840)
Publications with Turkey records: Noyes (2019), Öncüer (1991).
Hosts: *Otiorhynchus ligustici* (Coleoptera: Curculionidae) (Noyes, 2019).
Distribution: Austria, Belgium, Bulgaria, Czech Republic, Czechoslovakia, Europe, Macedonia, Moldova, Romania, Slovakia, Sweden, Turkey, United Kingdom (Noyes, 2019).

Genus *Trichomalopsis* (Crawford, 1913)
Trichomalopsis hemiptera (Walker, 1835)
Publications with Turkey records: Noyes (2019), Öncüer (1991).
Associates: *Quercus sp.* (Fagaceae) (Noyes, 2019).
Hosts: *Pegomya betae, Pegomya hyoscyami* (Diptera: Anthomyiidae), *Mayetiola destructor* (Cecidomyiidae), *Chlorops pumilionis, Chlorops taeniopus, Hippelates collusor, Oscinella frit* (Chloropidae), *Cynips quercusfolii* (Hymenoptera: Cynipidae), *Hyphantria cunea* (Lepidoptera: Arctiidae), *Euproctis chrysorrhoea, Nygmia phaeorrhoea, Porthetria dispar, Stilpnotia salicis* (Lymantriidae), *Spodoptera exigua* (Noctuidae), *Aporia crataegi, Pieris brassicae, Pieris rapae* (Pieridae), *Ostrinia nubilalis, Pyrausta aurata* (Pyralidae), *Acrolepia assectella, Yponomeuta padellus* (Yponomeutidae) (Noyes, 2019).
Parasitoid hosts: *Compsilura concinnata, Lydella stabulans, Lydella thompsoni, Paraphorocera senilis, Voria ruralis* (Diptera: Tachinidae), *Apanteles euchaetis, Apanteles fulvipes, Apanteles glomeratus, Apanteles hyphantriae, Apanteles lacteicolor, Apanteles melanoscelus, Apanteles solitarius, Hormius sp., Meteorus versicolor, Microgaster globate, Petalodes unicolor* (Hymenoptera: Braconidae), *Angitia armillata, Angitia punctoria, Diadromus collaris, Microcryptus abdominator, Phaeogenes nigridens* (Ichneumonidae) (Noyes, 2019).
Parasitoids: *Pediobius pyrgo* (Eulophidae) (Noyes, 2019).
Distribution: Austria, Belgium, Canada, Czech Republic, Czechoslovakia, Europe, Finland, France, Germany, Hungary, Ireland, Italy, Jamaica, Japan, Kazakhstan, Korea, Moldova, Morocco, Nearctic, Neotropical, Netherlands, New Zealand, People's Republic of China, Puerto Rico, Russia, Spain, Sweden, Switzerland, Tselinograd Obl., Turkey, Ukraine, United Kingdom, United States of America, Yugoslavia (pre-1991) (Noyes, 2019).

Trichomalopsis littoralis (Graham, 1969)
Publications with Turkey records: Noyes (2019), Öncüer (1991).
Associates: Unknown (Noyes, 2019).

Hosts: Unknown (Noyes, 2019).
Distribution: England, Netherlands, Spain, Sweden, Turkey (Noyes, 2019).

Trichomalopsis peregrina (Graham, 1969)
Publications with Turkey records: Noyes (2019), Öncüer (1991).
Associates: *Euproctis chrysorrhoea* (Lepidoptera: Lymantriidae), *Brassica oleracea* (Brassicaceae) (Noyes, 2019).
Hosts: *Rhagoletis sp.* (Diptera: Tephritidae), *Euproctis chrysorrhoea, Euproctis phaeorrhoea, Leucoma salicis, Nygmia phaeorrhoea, Porthetria dispar, Stilpnotia salicis* (Lepidoptera: Lymantriidae), *Pieris brassicae* (Pieridae) (Noyes, 2019).
Parasitoid hosts: *Lydella sp.* (Diptera: Tachinidae), *Apanteles lacteicolor, Apanteles melanoscelus, Meteorus versicolor, Rogas unicolor* (Hymenoptera: Braconidae), *Pteromalus puparum* (Pteromalidae) (Noyes, 2019).
Parasitoids: *Pediobius tarsalis* (Eulophidae) (Noyes, 2019).
Distribution: Austria, Bulgaria, Canada, Czech Republic, England, Europe, France, Germany, Hungary, Iran, Italy, Kazakhstan, Moldova, Nearctic, Netherlands, Romania, Russia, Slovakia, Spain, Tselinograd Obl., Turkey, Ukraine, United States of America, USSR (Noyes, 2019).

Genus *Trichomalus* (Thomson, 1878)
Trichomalus apertus (Walker, 1835)
Publications with Turkey records: Noyes (2019), Öncüer (1991).
Associates: *Carduus bourgeanus, Cirsium acanthoides* (Asteraceae) (Noyes, 2019).
Hosts: *Melanagromyza aeneoventris* (Diptera: Agromyzidae) (Noyes, 2019).
Distribution: England, Germany, Hungary, Netherlands, Spain, Sweden, Turkey (Noyes, 2019).

Trichomalus campestris (Walker, 1834)
Publications with Turkey records: Noyes (2019), Öncüer (1991).
Associates: *Turdus merula* (Birds), *Pastinaca sativa* (Apiaceae), *Carpinus sp.* (Betulaceae), *Cardaria draba, Sisymbrium irio* (Brassicaceae), *Lotus sp., Medicago sativa, Retama sphaerocarpa, Trifolium sp., Trifolium pratense* (Fabaceae), *Rumex acetosa* (Polygonaceae), *Rhamnus lycioides* (Rhamnaceae).

Hosts: *Apion sp., Apion apricans, Apion arrogans, Apion loti, Apion pisi, Apion seniculus, Apion trifolii* (Coleoptera: Apionidae), *Ceutorhynchus cardariae* (Curculionidae) (Noyes, 2019).
Distribution: Belgium, Bosnia Hercegovina, Bulgaria, Croatia, Czech Republic, Czechoslovakia, Europe, France, Germany, Hungary, Israel, Italy, Moldova, Montenegro, Netherlands, Romania, Serbia, Slovakia, Spain, Sweden, Switzerland, Turkey, Ukraine, United Kingdom (Noyes, 2019).

Trichomalus elongatus (Delucchi and Graham, 1956)
Publications with Turkey records: Noyes (2019), Öncüer (1991).
Associates: *Althaea rosea* (Malvaceae) (Noyes, 2019).
Hosts: *Apion radiolus* (Coleoptera: Apionidae) (Noyes, 2019).
Parasitoid hosts: *Eurytoma sp.* (Hymenoptera: Eurytomidae) (Noyes, 2019).
Distribution: Austria, Belgium, Canary Islands, England, Europe, Germany, Kazakhstan, Kirgizia, Spain, Sweden, Tselinograd Obl., Turkey (Noyes, 2019).

Trichomalus fulvipes (Walker, 1836)
Publications with Turkey records: Noyes (2019), Öncüer (1991).
Associates: *Trifolium pratense* (Fabaceae), *Pinus halepensis* (Pinaceae), *Pyrus communis* (Rosaceae) (Noyes, 2019).
Hosts: *Apion sp., Apion arrogans* (Coleoptera: Apionidae) (Noyes, 2019).
Distribution: Belgium, Bulgaria, Czech Republic, Europe, Germany, Hungary, Israel, Netherlands, Romania, Slovakia, Spain, Sweden, Switzerland, Turkey, United Kingdom (Noyes, 2019).

Trichomalus helvipes (Walker, 1834)
Publications with Turkey records: Noyes (2019), Öncüer (1991).
Associates: *Apion sp.* (Apionidae), *Pastinaca sativa* (Apiaceae), *Carduus bourgeanus* (Asteraceae), *Ononis tridentata, Retama sphaerocarpa, Trifolium sp., Trifolium pratense* (Fabaceae), *Ephedra nebrodensis* (Gnetaceae), *Reseda lutea* (Resedaceae) (Noyes, 2019).
Hosts: *Apion sp., Apion apricans, Apion trifolii* (Coleoptera: Apionidae), *Brachonyx pineti* (Curculionidae) (Noyes, 2019).
Distribution: Belgium, Bulgaria, Czech Republic, Europe, France, Germany, Hungary, Netherlands, Norway, Romania, Spain, Sweden, Switzerland, Turkey, United Kingdom (Noyes, 2019).

Trichomalus nanus (Walker, 1836)
Publications with Turkey records: Noyes (2019), Öncüer (1991).
Associates: *Alliaria petiolate, Brassica napus, Brassica oleracea* (Brassicaceae) (Noyes, 2019).
Hosts: *Ceutorhynchus alliariae, Ceutorhynchus obstrictus, Ceutorhynchus pallidactyllus (Coleoptera: Curculionidae), Oscinella frit* (Diptera: Chloropidae) (Noyes, 2019).
Distribution: Austria, Bulgaria, Czech Republic, Czechoslovakia, England, Europe, Germany, Hungary, Kazakhstan, Latvia, Moldova, Netherlands, Russia, Slovakia, Spain, Sweden, Switzerland, Tselinograd Obl., Turkey, Ukraine, USSR (Noyes, 2019).

Trichomalus posticus (Walker, 1834)
Publications with Turkey records: Noyes (2019), Öncüer (1991).
Associates: *Trifolium hybridum* (Fabaceae) (Noyes, 2019).
Hosts: *Apion meieri* (Coleoptera: Apionidae), *Chloropisca notata, Chlorops sp., Oscinella frit* (Diptera: Chloropidae).
Parasitoid hosts: *Pteromalus puparum* (Hymenoptera: Pteromalidae) (Noyes, 2019).
Distribution: Belgium, Czech Republic, Europe, Germany, Iran, Kazakhstan, Netherlands, Spain, Sweden, Tselinograd Obl., Turkey, United Kingdom, USSR (Noyes, 2019).

Trichomalus rufinus (Walker, 1835)
Publications with Turkey records: Noyes (2019), Öncüer (1991).
Associates: *Anagyris foetida, Lotus corniculatus, Medicago sativa, Trifolium sp.* (Fabaceae), *Ephedra nebrodensis* (Gnetaceae), *Pyrus communis* (Rosaceae) (Noyes, 2019).
Hosts: *Apion sp., Apion apricans, Apion croceifemoratum, Apion loti, Apion pisi, Apion trifolii* (Coleoptera: Apionidae) (Noyes, 2019).
Distribution: Austria, Belgium, Bosnia Hercegovina, Bulgaria, Canary Islands, Czech Republic, Czechoslovakia, Europe, France, Germany, Greece, Hungary, Kazakhstan, Lithuania, Montenegro, Netherlands, Norway, Romania, Slovakia, Spain, Sweden, Tselinograd Obl., Turkey, Ukraine, United Kingdom (Noyes, 2019).

Genus *Trjapitzinia* (Dzhanokmen, 1975)
Trjapitzinia leucomae (Dzhanokmen, 1975)
Publications with Turkey records: Kilic and Alaoglu (1996), Noyes (2019), Öncüer (1991).
Associates: *Populus alba, Salix sp.* (Salicaceae) (Noyes, 2019).

Hosts: *Leucoma salicis, Stilpnotia salicis* (Lepidoptera: Lymantriidae) (Noyes, 2019).
Distribution: Kazakhstan, Turkey (Noyes, 2019).

Subfamily Spalangiinae (Haliday, 1833)
Genus *Spalangia* (Latreille, 1805)
Spalangia fuscipes (Nees, 1834)
Publications with Turkey records: Noyes (2019), Öncüer (1991).
Associates: *Wolbachia sp.* (Bacteria: Wolbachiaceae) (Noyes, 2019).
Hosts: *Lasioptera eryngii, Thomasiella eryngii* (Diptera: Cecidomyiidae), *Oscinella sp., Oscinella frit, Oscinella pusilla* (Chloropidae) (Noyes, 2019).
Distribution: Algeria, Austria, Bulgaria, Canary Islands, Croatia, Czech Republic, Czechoslovakia, Europe, France, Germany, Hungary, India, Ireland, Italy, Malaysia, Moldova, Morocco, Nearctic, Netherlands, North Africa, Palearctic, People's Republic of China, Romania, Russia, Slovakia, Spain, Sweden, Transcaucasus, Turkey, Ukraine, United States of America, USSR (Noyes, 2019).

Spalangia kocakeri (Doğanlar, 2017)
Publications with Turkey records: Doğanlar (2017), Noyes (2019).
Associates: Unknown (Noyes, 2019).
Hosts: Unknown (Noyes, 2019).
Distribution: Turkey (Noyes, 2019).

Spalangia marashensis (Doğanlar, 2017)
Publications with Turkey records: Doğanlar (2017), Noyes (2019).
Associates: Unknown (Noyes, 2019).
Hosts: Unknown (Noyes, 2019).
Distribution: Turkey (Noyes, 2019).

Spalangia subpunctata (Foerster, 1850)
Publications with Turkey records: Noyes (2019), Öncüer (1991).
Hosts: *Hydrotaea irritans* (Diptera: Anthomyiidae), *Pycnosoma rufifacies* (Calliphoridae), *Musca domestica* (Muscidae), *Physiphora sp., Physiphora demandata* (Otitidae), *Ravinia striata* (Sarcophagidae), *Syritta sp., Syritta pipiens* (Syrphidae) (Noyes, 2019).
Distribution: Belgium, Bulgaria, Caucasus, Czech Republic, Czechoslovakia, Denmark, England, Europe, France, Germany, Hungary, Kazakhstan, Madeira, Moldova, Montenegro, Morocco, Netherlands, North Africa, People's Republic of China, Poland, Romania, Serbia,

Slovakia, Spain, Sweden, Tadzhikistan, Tselinograd Obl., Turkey, USSR, Uzbekistan (Noyes, 2019).

Subfamily Sycoryctinae (Wiebes, 1966)
Genus *Philotrypesis* (Foerster, 1878)
Philotrypesis caricae (Linnaeus, 1762)
Publications with Turkey records: Doğanlar (2012), Noyes (2019).
Associates: *Blastophaga psenes* (Hymenoptera: Agaonidae), *Crematogaster scutellaris* (Formicidae), *Schistonchus caprifici* (Nematoda: Aphelenchoididae), *Ficus carica* (Moraceae) (Noyes, 2019).
Hosts: *Blastophaga psenes* (Hymenoptera: Agaonidae), *Ficus sp., Ficus carica* (Moraceae) (Noyes, 2019).
Distribution: Caucasus, Europe, France, Israel, Italy, Turkey, Ukraine, United States of America, USSR (Noyes, 2019).

Philotrypesis emeryi (Grandi, 1926)
Publications with Turkey records: Doğanlar (2012), Noyes (2019).
Associates: *Ficus macrocarpa* (Moraceae).
Hosts: *Ficus microcarpa* (Moraceae) (Noyes, 2019).
Distribution: Brazil, Hawaii, Indonesia, Japan, People's Republic of China, Taiwan, Turkey, United States of America (Noyes, 2019).

Philotrypesis taiwanensis (Chen, 1999)
Publications with Turkey records: Doğanlar (2012), Noyes (2019).
Associates: *Ficus macrocarpa* (Moraceae) (Noyes, 2019).
Hosts: *Eupristina verticillata* (Hymenoptera: Agaonidae) (Noyes, 2019).
Distribution: Brazil, Taiwan, Turkey (Noyes, 2019).

References

Avcı, Ümit, and Özbek, Hikmet. 1991. "Natural enemies of the cabbage aphid, *Brevicoryne brassicae* (L.) (Homoptera, Aphididae) in Erzurum". *Turkish Journal of Entomology*, 15(1):37-41.

Bayram, Şerife, and Ülgentürk, Selma, and Toros, Seval. 1998. "Research on the insects causing galls on dog rose (*Rosa spp.*) and their parasitoids in Ankara province". *Turkish Journal of Entomology*, 22(4):259-268.

Bulut, Ebrucan, and Göçmen, Hüseyin. 2000. "Pests and their natural enemies on greenhouse vegetables in Antalya". *Bulletin. Section Regionale Ouest Palaearctique, Organisation Internationale de Lutte Biologique*, 23(1):33-37.

Çikman, Emine. 2012. "Parasitoids of the leafminers (Diptera: Agromyzidae) from Elazig Province, Turkey". *African Journal of Agricultural Research*, 7(12):1937-1943.

Doğanlar, Mikdat, and Bolu, Halil. 2004. "A new species of *Gugolzia* Delucchi & Steffan (Hym., Pteromalidae) from Turkey, a parasitoid of Eurytoma amygdali Enderlein (Hym., Eurytomidae)". *Zoology in the Middle East*, 32:75-78.

Doğanlar, Mikdat, and Doğanlar, Oğuzhan. 2010. "Review of the species of *Gugolzia* Delucchi and Steffan (Hymenoptera: Pteromalidae) in Europe and Turkey, with descriptions of new species". *Turkish Journal of Zoology*, 34(1):23-34.

Doğanlar, Mikdat, and Gözüaçak, Celalettin, and Subaşı, Yüksel. 2018. "A new species of *Gugolzia* Delucchi & Steffan (Hym.: Pteromalidae) from Igdir, Turkey, reared from fruit of Sophora alopecuroides L. (Fabaceae)". *Munis Entomology and Zoology*, 13(2):374-377.

Doğanlar, Mikdat, and Gözüaçık, Celalettin, and Subaşı, Yüksel. 2018. "A New Species of *Gugolzia* Delucchi & Steffan (Hym., Pteromalidae) from Iğdir, Turkey, Reared from fruit of *Sophora alopecuroides* L. (Fabaceae)". *Munis Entomology and Zoology*. Vol. 13, No: 2; 374-377.

Doğanlar, Mikdat, and Sertkaya, Erdal, and & Skuhravá, Marcela. 2011a. "Pest status of olive leaf gall midge *Dasineura oleae* (Angelini, 1831), description of *Lasioptera oleicola* Skuhravá *sp.* new (Diptera: Cecidomyiidae) and effectiveness of parasitoids on their populations in Hatay, Turkey". *Turkish Journal of Entomology*, 35(2):265-284.

Doğanlar, Mikdat, and Yiğit, Talip. 2019. "A new species of *Gastrancistrus* Westwood (Hymenoptera: Pteromalidae) reared from apricot flower midge, *Contarina pruniflorum* Coutin & Rambier (Diptera: Cecidomyiidae) in Malatya province, Turkey". *Munis Entomology and Zoology*, 14 (1): 24-27.

Doğanlar, Mikdat, and Yiğit, Talip. 2019. "A new species of *Gastrancistrus* Westwood (Hymenoptera: Pteromalidae) reared from apricot flower midge, *Contarinia pruniflorum* Coutin & Rambier (Diptera: Cecidomyiidae) in Malatya province, Turkey". *Munis Entomology and Zoology*, 14(1):24-27.

Doğanlar, Mikdat. 1986. "Morphological studies of the hypopygium and its importance to the taxonomy of the genera Pachyneuron and Euneura (Hymenoptera: Pteromalidae), with description of a new species of Pachyneuron from Turkey". *Cumhuriyet Üniversitesi, Fen Bilimleri Dergisi*, 4: 23-32.

—. 1987. "Hypopygia of most Nearctic and Palaearctic species of *Dibrachys* Foerster, key to most species of the genus and descriptions of three new species (Hymenoptera: Pteromalidae)". *Spixiana*, München 10:191-206.

—. 2006. "Systematic studies on some species of *Halticoptera* Spinola, 1811 (Hymenoptera: Chalcidoidea, Pteromalidae), with descriptions of new species". *Journal of Applied Sciences Research*, 2(3):168-183.

—. 2007. "A new species of *Muscidifurax* Girault & Sanders, 1910 (Hymenoptera: Pteromalidae) from Adana province, Turkey". *Turkish Journal of Entomology*, 31(4):243-252.

—. 2011. "Parastioid complex of the olive leaf gall midges, *Dasineura oleae* (Angelini 1831) and *Lasioptera oleicola* Skuhravá, 2011 (Diptera: Cecidomyiidae) in Hatay Turjey, with descriptions of new genus and species from Tetrastichinae (Hymenoptera: Eulophidae)". *Turkish Journal of Entomology*, 35(2): 245-264.

—. 2012. "Occurrence of fig wasps (Hymenoptera: Chalcidoidea) in *Ficus carica* [sic] and *F. microcarcapa* in Hatay, Turkey". *Turkish Journal of Zoology*, 36(5):721-724.

—. 2014. "First reports on the natural enemy fauna of the chestnut gall wasp, *Dryocosmus kuriphilus* Yasumatsu (Hymenoptera: Cynipidae) in Yalova, Turkey". *Türkiye Biyolojik Mücadele Dergisi*, 5(1):67-74.

—. 2017. "Species of *Spalangia* Latreille, 1805 (Hymenoptera: Pteromalidae, Spalangiinae) from Turkey, with descriptions of new species". *Entomofauna*, 38(21):477-492.

—. 2017. "Species of *Spalangia* Latreille, 1805 (Hymenoptera: Pteromalidae, Spalangiinae) from Turkey, with descriptions of new species". *Entomofauna*, 38(21):477-492.

—. 2018a. "Species of *Colotrechnus* Thomson, 1878 (Hymenoptera: Pteromalidae, Colotrechninae) from Turkey, with description of a new species". *Munis Entomology and Zoology*, 13(1):214-241.

—. 2018a. "Species of *Colotrechnus* Thomson, 1878 (Hymenoptera: Pteromalidae, Colotrechninae) from Turkey, with description of a new species". *Munis Entomology and Zoology*, 13(1):214-241.

—. 2018b. "Species of *Macroglenes* Westwood, 1832 (Hymenoptera: Pteromalidae, Spalangiinae) from Turkey, with description of a new species". *Entomofauna*, 39(1):317-324.

—. 2018b. "Species of *Macroglenes* Westwood, 1832 (Hymenoptera: Pteromalidae, Spalangiinae) from Turkey, with description of a new species". *Entomofauna*, 39(1):317-324.

—. 2018c. "A new species of *Lampoterma* (Hymenoptera: Pteromalidae) from Erzurum, Turkey". *Munis Entomology and Zoology*, 13(2):386-394.

—. 2018c. "A new species of *Lampoterma* (Hymenoptera: Pteromalidae) from Erzurum, Turkey". *Munis Entomology and Zoology*, 13(2):386-394.

—. 2019. "Species of *Gugolzia* Delucchi & Steffan (Hymenoptera: Pteromalidae) from Turkey, with descriptions of two new species". *Munis Entomology and Zoology*, 14(1):1-8.

—. 2019. "Species of *Gugolzia* Delucchi & Steffan (Hymenoptera: Pteromalidae) from Turkey, with descriptions of two new species". *Munis Entomology and Zoology*, 14(1):1-8.

Gençer, Lütfiye. 2003. "The chalcidoid parasitoids of *Diplolepis mayri* Schld. (Hymenoptera: Cynipidae), a pest of *Rosa canina* L. in Sivas province". *Turkish Journal of Entomology*, 27(2):107-111.

Gülel, A. 1988. "Effects of quantitative food shortage on adult size and progeny production in the parasitoid *Dibrachys boarmiae* (Hymenoptera: Pteromalidae)". Doga Türk Zooloji Dergisi, 12(1):48-54.

Hızal, Erdem, and Parlak, Nebahat Nihan. 2013. "*Bruchidius terrenus* and *Bruchidius siliquastri* (Coleoptera: Chrysomelidae: Bruchinae) - first records for Turkey". *Florida Entomologist*, 96(1):66-70.

Japoshvili, George, and Karaca, İsmail. 2002b. "Coccid (Homoptera: Coccoidea) species of Isparta Province, and their parasitoids from Turkey and Georgia". *Turkish Journal of Zoology*, 26(4):371-376.

Kılıç, Nuray, and Alaoğlu, Özdemir. 1996. "Biology and parasitoids of the satin moth *Leucoma salicis* (L.) (Lepidoptera, Lymantriidae) a pest of poplar trees in Erzurum Province (Turkey)". *Turkish Journal of Entomology*, 20(4):269-279.

Kılınçer, Neşet. 1982. "Investigations on the parasite complex of the small cabbage white butterfly (*Pieris rapae* L.) (Lepidoptera: Pieridae) in the Ankara region". *Bitki Koruma Bülteni 22*: 107-119.

Koçak, Erhan, and Özdemir, Mustafa. 2012. "Parasitic Hymenoptera reared from the insects on *Heracleum platytaenium* Boiss. (Apiaceae) with new faunistic and biological records". *Turkish Journal of Zoology*, 36(2):201-208.

Korkmaz, Ertan Mahir, and Hastaoğlu Örgen, Sevda, and Gençer, Lütfiye, and Ülgentürk, Selma, and Başıbüyük, Hasan Hüseyin. 2010. "Determination of some wheat pests and their parasitoids in wheat fields of Central Anatolia Region". *Türkiye Entomoloji Dergisi*, 34(3):361-377.

Mete, Özlem, and Mergen, Orhan. 2017. "The community components associated with two common rose gall wasps (Hymenoptera:

Cynipidae: Diplolepidini) in Turkey". *Turkish Journal of Zoology*, 41:696-701.

Narendran, Thekke Curuppathe, and Tezcan, Serdar, and Civelek, Hasan Sungur. 1995. "A new species of *Eurytoma illiger* (Hymenoptera, Eurytomidae) parasitic on Scolytus rugulosus Ratzeburg (Coleoptera, Scolytidae) in Turkey and some notes about it". *Turkish Journal of Entomology*, 19(2):81-86.

Noyes, John S. 2019. *Universal Chalcidoidea Database*. Available at: www.nhm.ac.uk/chalcidoids (accessed January 2022).

Öncüer, Cezmi. 1974. "Ege bölgesinde turunçgil bahçelerinde zararli Coccus (Homoptera: Coccidae) türlerinin taninmasi yayilisi ve dogal düsmanlari üzerinde arastirmalar". *Bitki Koruma Bülteni*, Supplement 1:1-59.

—. 1991. "A catalogue of the parasites and predators of insect pests of Turkey". *Ege Üniversitesi Ziraat Fakultesi Yayınları* [Ege University Agriculture Faculty Publ.] 505.

Özbek, Hikmet, and Güçlü, Şaban, and Tozlu, Göksel. 1999. "Biology and natural enemies of *Diplolepis mayri* Schld. (Hymenoptera: Cynipidae), a pest of *Rosa canina* L. in Erzurum Province". *Turkish Journal of Entomology*, 23(1):39-50.

Özsemerci, Fatma, and Aksit, Tülin. 2003. "Investigations on some biological characteristics and populations fluctuation of *Ceroplastes rusci* L. (Homoptera: Coccidae) harmful to fig trees in Aydin province". *Turkish Journal of Entomology*, 27(1):13-25.

Polat, Alper, and Tozlu, Göksel. 2010. "Investigations on the brief biology, host plants and parasitoids of *Archips rosana* (Linnaeus, 1758) (Lepidoptera: Tortricidae) in Erzurum". *Turkish Journal of Entomology*, 34(4):529-542.

Stringer, Danielle N., and Mantel, Sarah, and Jennings, John T., and Austin, Andrew D. 2012. "Family Pteromalidae", *Australian Centre for Evolutionary Biology and Biodiversity, and the School of Earth and Environmental Science*. The University of Adelaide, Australia (Available from: https://www.researchgate.net/publication/268079507_Famiy_PTEROMALIDAE, accessed October 2022).

Ülgentürk, Selma, and Kaydan, Mehmet Bora, and Zeki, Cevdet, and Toros, Seval. 2001. "*Sphaerolecanium prunastri* Boyer de Fonscolombe (Hemiptera: Coccidae): distribution, host plants and natural enemies in the Turkish Lake district". *Bolletino Zoologia Agraria di Bachicoltura Ser*. II, 33:3, 357-363.

Ülgentürk, Selma, and Toros, Seval. 2001. "Natural enemies of the oak scale insect, *Eulecanium ciliatum* (Douglas) (Hemiptera: Coccidae) in Turkey". *Entomologica*, Bari 33:219-224.

Ülgentürk, Selma. 2001. "Parasitoids and predators of Coccidae (Homoptera: Coccoidea) species on ornamental plants in Ankara, Turkey". *Acta Phytopathologica et Entomologica Hungarica*, 36(3/4):369-375.

Uzun, Sündüs. 1987. "Parasites of the cabbage butterfly (*Pieris brassicae* (L.)) (Lepidoptera: Pieridae) harmful on cabbage and cauliflower in the Izmir district". *Turkish Journal of Entomology*, 11(4):237-245.

Zeki, Huriye, and Toros, Seval. 1990. "Investigations on the determination of natural enemies of *Chrysomela populi* L. and *Chrysomela tremulae* F. (Col., Chrysomelidae) harmful to poplars and the efficiency of their parasitoids in central Anatolia region". *Proceedings of the Second Turkish National Congress of Biological Control*, (26-29 September 1990) pp.251-260.

CHAPTER SIXTEEN

FAMILY: SIGNIPHORIDAE HOWARD, 1894

The Signiphoridae family comprises small, thick-bodied Chalcids that attack scale insects, whiteflies, and other Homoptera or act as hyperparasitoids of the Chalcids that attack Homoptera. Signiphoridae range in size from 0.22 to 1.7 mm. Their color varies from black to salmon pink or white, but they do not exhibit metallic hues. The cuticle of Signiphoridae has less sculpturing compared to the families Chalcididae and Eurytomidae.

The main characteristics of Signiphoridae include a sessile metasoma (lacking a thread-waist), a propodeum with a medium triangular zone, and long and unsegmented club of the antennae preceded by 1-4 ring-like annelli. The wings are fringed with medium to long setae, and there are short post-marginal and stigmal veins. Setae on the wing membrane are sparser compared to other families (see www.faculty.ucr.edu).

The family currently includes four genera and 76 species, and in Turkey, there are two genera and three species (Noyes, 2019).

Genus *Chartocerus* (Motschulsky, 1859)
Chartocerus novitzkyi (Domenichini, 1955)
Publications with Turkey records: Noyes (2019).
Associates: Unknown (Noyes, 2019).
Hosts: Unknown (Noyes, 2019).
Distribution: Turkey (Noyes, 2019).

Chartocerus subaeneus (Foerster, 1878)
Publications with Turkey records: Noyes (2019), Öncüer (1991).
Associates: *Juniperus thurifera* (Cupressaceae), *Manihot esculenta* (Euphorbiaceae), *Morus sp.* (Moraceae), *Saccharum officinarum* (Poaceae), *Lantana camara* (Verbenaceae) (Noyes, 2019).

Hosts: *Sidis sp.* (Coleoptera: Coccinellidae), *Leucopis sp.* (Diptera: Chamaemyiidae), (Hemiptera: Aleyrodidae), Aphididae, *Physokermes piceae* (Coccidae), *Lepidosaphes sp., Quadraspidiotus gigas, Quadraspidiotus perniciosus* (Diaspididae), *Eriococcus insignis* (Eriococcidae), *Phenacoccus manihoti, Phenacoccus mespili, Phenacoccus solenopsis, Planococcus citri, Pseudococcus sp., Pseudococcus comstocki, Pseudococcus longispinus, Saccharicoccus sacchari, Spilococcus nanae* (Pseudococcidae) (Noyes, 2019).
Parasitoid hosts: *Anagyrus sp., Blepyrus insularis, Coccidencyrtus dissimilis, Epidinocarsis lopezi, Euaphycus duplus, Homalotylus quaylei, Hungariella peregrina, Pseudaphycus malinus, Pseudaphycus phenacocci* (Hymenoptera: Encyrtidae), *Allotropa sp., Allotropa mecrida* (Platygastridae) (Noyes, 2019).
Distribution: Canary Islands, Caucasus, Croatia, Czech Republic, Egypt, Europe, Gambia, Georgia, Germany, Hungary, Israel, Italy, Moldova, Nigeria, Palearctic, People's Republic of China, Poland, Serbia, Slovakia, Spain, Sweden, Turkey, Turkmenistan, Ukraine, United Kingdom, USSR (Noyes, 2019).

Genus *Thysanus* (Walker, 1840)
Thysanus ater (Walker, 1840)
Catalagues (Journals) with Turkiye records: Japoshvili and Karaca (2002), Noyes (2019), Öncüer (1991).
Associates: *Corylus avellana* (Betulaceae), *Elaeagnus angustifolia* (Elaeagnaceae), *Pinus halepensis* (Pinaceae), *Malus pumila* (Rosaceae), *Populus hybrida, Populus transcaucasica* (Salicaceae) (Noyes, 2019).
Hosts: *Eulecanium corni, Parthenolecanium corni, Pulvinaria maxima, Sphaerolecanium prunastri* (Hemiptera: Coccidae), *Aspidiotus ostreaeformis, Chionaspis lepineyi, Chionaspis salicis, Diaspidiotus caucasicus, Diaspidiotus elaeagni, Diaspidiotus spurcatus, Lepidosaphes ulmi, Leucaspis pini, Melanaspis obscura, Parlatoria oleae, Pseudaulacaspis pentagona, Quadraspidiotus gigas, Quadraspidiotus ostreaeformis, Quadraspidiotus perniciosus* (Diaspididae) (Noyes, 2019).
Parasitoid hosts: *Aphytis mytilaspidis* (Hymenoptera: Aphelinidae), *Coccidencyrtus dissimilis, Euaphycus duplus* (Encyrtidae) (Noyes, 2019).
Distribution: Argentina, Bulgaria, Caucasus, Croatia, Czech Republic, Europe, Georgia, Germany, Hungary, India, Italy, Moldova, Norway, Palearctic, Serbia, Spain, Sweden, Turkey, Ukraine, United Kingdom, United States of America, Uruguay, USSR, Yugoslavia (pre-1991) (Noyes, 2019).

References

Japoshvili, George, and Karaca, İsmail. 2002b. "Coccid (Homoptera: Coccoidea) species of Isparta Province, and their parasitoids from Turkey and Georgia". *Turkish Journal of Zoology*, 26(4):371-376.

Noyes, John S. 2019. *Universal Chalcidoidea Database*. Available at: www.nhm.ac.uk/chalcidoids (accessed January 2022).

Öncüer, Cezmi. 1991. "A catalogue of the parasites and predators of insect pests of Turkey". *Ege Üniversitesi Ziraat Fakultesi Yayınları* [Ege University Agriculture Faculty Publ.] 505.

Signiphoridae, http://www.faculty.ucr.edu/~legneref/identify/signipho.htm (accessed October 2022).

CHAPTER SEVENTEEN

FAMILY: TETRACAMPIDAE FOERSTER, 1856

The Tetracampidae is one of the smallest families within the Chalcidoidea (Hymenoptera). Currently, the family includes 15 genera and 50 species placed in six subfamilies. In Turkey, there are three subfamilies, four genera and eight species identified.

The main characteristics of Tetracampidae include forewings lacking a speculum, 5-segmented tarsi (4-segmented in males of Tetracampinae and Baeomorphinae), a shining greenish body, scutellum with four long bristles, an unusual propodeum often bearing several setae medially or submedially, a weak and nearly straight spur on the fore tibia, a gaster with a short petiole, and a strongly invaginated hind edge of the basal tergite, creating the appearance of a deep groove between it and second tergite (Noyes, 2019).

Subfamily Mongolocampinae (Sugonjaev, 1971)
Genus *Eremocampe* (Sugonjaev, 1971)
Eremocampe choresmiana (Sugonjaev, 1971)
Publications with Turkey records: Noyes (2019), Sugonjaev, 1986.
Hosts: *Nitraria schoberi* (Zygophyllaceae) (Noyes, 2019).
Distribution: Turkey, Turkmenistan (Noyes, 2019).

Subfamily Platynocheilinae (Boucek, 1958)
Genus *Platynocheilus* (Westwood, 1837)
Platynocheilus cuprifrons (Nees, 1834)
Publications with Turkey records: Doğanlar (2003), Noyes (2019), Öncüer (1991).
Associates: *Centaurea aspera, Centaurea nicaeensis, Centaurea sphaerocephala* (Asteraceae), *Eruca vesicaria, Sisymbrium irio* (Brassicaceae) (Noyes, 2019).

Hosts: Diptera: Agromyzidae, *Cecidomyia fagi, Mikiola fagi* (Cecidomyiidae), *Andricus curvator, Andricus terminalis, Biorhiza pallida, Isocolus lichtensteini* (Hymenoptera: Cynipidae), *Mesopolobus sp.* (Pteromalidae) (Noyes, 2019).
Distribution: Algeria, Austria, Canary Islands, Caucasus, Croatia, Czech Republic, Czechoslovakia, Denmark, Egypt, Europe, France, Georgia, Germany, Hungary, Ireland, Italy, Moldova, Montenegro, Netherlands, Romania, Russia, Slovakia, Slovenia, Spain, Sweden, Transcaucasus, Turkey, United Kingdom, USSR (Noyes, 2019).

Platynocheilus karsensis (Doğanlar, 2003)
Publications with Turkey records: Doğanlar (2003), Noyes (2019).
Associates: Unknown (Noyes, 2019).
Hosts: Unknown (Noyes, 2019).
Distribution: Turkey (Noyes, 2019).

Subfamily Tetracampinae (Foerster, 1856)
Genus *Epiclerus* (Haliday, 1844)
Epiclerus alahanensis (Doğanlar, 2003)
Publications with Turkey records: Doğanlar (2003), Noyes (2019).
Associates: *Sonchus oleraceus* (Asteraceae) (Noyes, 2019).
Hosts: *Phytomyza horticola* (Diptera: Agromyzidae) (Noyes, 2019).
Distribution: Turkey (Noyes, 2019).

Epiclerus guemenekus (Doğanlar, 2003)
Publications with Turkey records: Doğanlar (2003), Noyes (2019).
Associates: Unknown (Noyes, 2019).
Hosts: Unknown (Noyes, 2019).
Distribution: Turkey (Noyes, 2019).

Epiclerus panyas (Walker, 1839)
Publications with Turkey records: Doğanlar (2003), Gencer (2009), Noyes (2019).
Associates: *Centaurea scabiosa, Cirsium arvense, Cirsium palustre, Eupatorium cannabinum, Inula conyza, Senecio sp., Sonchus sp., Sonchus asper, Tanacetum sp., Taraxacum sp.* (Asteraceae), *Atriplex halimus* (Chenopodiaceae), *Cucumis sativus* (Cucurbitaceae), *Plantago lanceolata* (Plantaginaceae), *Holcus sp., Holcus lanatus, Holcus mollis, Poa sp.* (Poaceae), *Tamarix canariensis* (Tamaricaceae) (Noyes, 2019).
Hosts: *Chromatomyia sp., Chromatomyia horticola, Chromatomyia syngenesiae, Melanagromyza beckeri, Ophiomyia beckeri, Phytomyza*

conyzae, Phytomyza nigra, Phytomyza plantaginis, Phytomyza primulae, Phytomyza ranunculi, Phytomyza spinaciae, Phytomyza syngenesiae (Diptera: Agromyzidae) (Noyes, 2019).
Distribution: Canary Islands, Czech Republic, Czechoslovakia, Europe, Germany, Hungary, Italy, Macedonia, Slovenia, Spain, Sweden, Turkey, United Kingdom (Noyes, 2019).

Epiclerus tokatus (Doğanlar, 2003)
Publications with Turkey records: Doğanlar (2003), Noyes (2019).
Associates: Unknown (Noyes, 2019).
Hosts: Unknown (Noyes, 2019).
Distribution: Turkey (Noyes, 2019).

Genus *Foersterella* (Dalla Torre, 1897)
Foersterella seyhanensis (Doğanlar, 2003)
Publications with Turkey records: Doğanlar (2003), Noyes (2019).
Associates: *Cynara scolymus* (Asteraceae) (Noyes, 2019).
Hosts: *Cassida rubiginosa* (Coleoptera: Chrysomelidae) (Noyes, 2019).
Distribution: Turkey (Noyes, 2019).

References

Doğanlar, Mikdat. 2003. "Notes on the species of Tetracampidae, with descriptions of some new species from Turkey". *Entomofauna*, 24(27):381-396.

Gençer, Lütfiye. 2009. "Contribution to the knowledge of the chalcid parasitoid complex (Hymenoptera: Chalcidoidea) of agromyzid leafminers (Diptera: Agromizidae) from Turkey, with new hosts and records". *Journal of Plant Protection Research*, 49(2):158-161.

Noyes, John S. 2019. *Universal Chalcidoidea Database*. Available at: www.nhm.ac.uk/chalcidoids (accessed January 2022).

Öncüer, Cezmi. 1991. "A catalogue of the parasites and predators of insect pests of Turkey". *Ege Üniversitesi Ziraat Fakultesi Yayınları* [Ege University Agriculture Faculty Publ.] 505.

CHAPTER EIGHTEEN

FAMILY: TORYMIDAE WALKER, 1833

Many Torymids oviposit through plant tissue, usually targeting galls or developing seeds. However, some species also oviposit into pupae concealed in silk or plant tissue. Torymid larvae can exhibit entomophagous behavior, feeding on other insects, or, less commonly, phytophagous. In some cases, they may even display both feeding strategies, consuming both the gall maker and the gall tissue as observed in some Eurytomidae species. The subfamily Toryminae, many Monodontomerinae and a few species of Megastigminae mainly act as idiobiont ectoparasitoids of the inhabitants of plant galls.

The main diagnostic characteristics of Torymidae are as follows: the female has an ovipositor clearly exserted, sometimes much longer than the abdomen; the stigmal vein is very short, although sometimes with a greatly enlarged stigma, so that the apex of the uncus almost touches the anterior margin of the forewing; the petiole is strongly transverse, and not distinct; the body is generally elongate, measuring about 1.1-7.5 mm in length, excluding the ovipositor; with the ovipositor included, the length can reach up to 16.0 mm; and the cercal plates are not flat, but slightly raised and papillar in form.

The family currently includes 68 genera and 986 species classified in two subfamilies, and in Turkey, there are seven subfamilies, 19 genera and 111 species (Noyes, 2019).

Subfamily Chalcimerinae
Genus *Exopristoides* (Boucek, 1982)
Exopristoides dentatus (Boucek, 1982)
Publications with Turkey records: Noyes (2019).
Associates: *Phlomis tuberosa* (Lamiaceae) (Noyes, 2019).

Hosts: *Panteliella fedtschenkoi, Vetustia investigata* (Hymenoptera: Cynipidae) (Noyes, 2019).
Distribution: Tunisia, Turkey, Ukraine (Noyes, 2019).

Subfamily Erimerinae (Crawford, 1914)
Genus *Adontomerus* (Nikolskaya, 1955)
Adontomerus amygdali (Boucek, 1958)
Publications with Turkey records: Noyes (2019), Stojanova et al. (2012).
Hosts: *Eurytoma amygdali* (Hymenoptera: Eurytomidae) (Noyes, 2019).
Distribution: Jordan, Russia, Turkey (Noyes, 2019).

Adontomerus impolitus (Askew and Nieves Aldrey, 1988)
Publications with Turkey records: Noyes (2019), Stojanova et al. (2012).
Associates: *Acroptilon repens, Tragopogon sp.* (Asteraceae) (Noyes, 2019).
Hosts: *Aulacidea acroptilonica, Aulacidea tragopogonis* (Hymenoptera: Cynipidae) (Noyes, 2019).
Distribution: Iran, Spain, Turkey, Uzbekistan (Noyes, 2019).

Genus *Eridontomerus* (Crawford, 1907)
Eridontomerus fulviventris (Erdos, 1954)
Publications with Turkey records: Noyes (2019), Stojanova et al. (2012).
Associates: *Centaurea sp.* (Asteraceae) (Noyes, 2019).
Distribution: Europe, Hungary, Romania, Turkey, Ukraine (Noyes, 2019).

Eridontomerus golbasinensis (Doğanlar, 2016)
Publications with Turkey records: Doğanlar (2016), Noyes (2019).
Associates: Unknown (Noyes, 2019).
Hosts: Unknown (Noyes, 2019).
Distribution: Turkey (Noyes, 2019).

Eridontomerus karatayensis (Doğanlar, 2016)
Publications with Turkey records: Doğanlar (2016), Noyes (2019).
Associates: Unknown (Noyes, 2019).
Hosts: Unknown (Noyes, 2019).
Distribution: Turkey (Noyes, 2019).

Genus *Erimerus* (Crawford, 1914)
Erimerus adananensis (Doğanlar, 2016)
Publications with Turkey records: Doğanlar (2016), Noyes (2019).

Associates: Unknown (Noyes, 2019).
Hosts: Unknown (Noyes, 2019).
Distribution: Turkey (Noyes, 2019).

Erimerus sanliurfanensis (Doğanlar, 2016)
Publications with Turkey records: Doğanlar (2016), Noyes (2019).
Associates: Unknown (Noyes, 2019).
Hosts: Unknown (Noyes, 2019).
Distribution: Turkey (Noyes, 2019).

Genus *Exopristus* (Ruschka, 1923)
Exopristus trigonomerus (Masi, 1916)
Publications with Turkey records: Doğanlar (2016), Noyes (2019), Öncüer (1991), Stojanova et al. (2012).
Associates: Coleoptera: Curculionidae, *Centaurea sp., Cirsium vulgare* (Asteraceae), *Cistus incanus* (Cistaceae), *Retama sphaerocarpa* (Fabaceae), *Ephedra fragilis* (Gnetaceae), *Papaver sp.* (Papaveraceae), *Linaria dalmatica, Scrophularia sp., Verbascum sp.* (Scrophulariaceae) (Noyes, 2019).
Hosts: *Gymnetron bipunctulatum, Pachytychius hordei* (Coleoptera: Curculionidae), *Lecanodiaspis sardoa* (Hemiptera: Lecanodiaspididae), *Bracon sp.* (Hymenoptera: Braconidae), *Etiella zinckenella* (Lepidoptera: Pyralidae) (Noyes, 2019).
Parasitoid hosts: *Bracon sp., Bracon pectoralis, Bracon piger* (Hymenoptera: Braconidae) (Noyes, 2019).
Distribution: Bosnia Hercegovina, Bulgaria, Croatia, Czech Republic, Czechoslovakia, Europe, France, Greece, Hungary, Iran, Italy, Macedonia, Moldova, Mongolia, North Africa, Palearctic, Romania, Slovakia, Spain, Transcaucasus, Turkey, Ukraine, USSR (Noyes, 2019).

Genus *Idarnotorymus* (Masi, 1916)
Idarnotorymus burdurensis (Doğanlar and Doğanlar, 2008)
Publications with Turkey records: Doğanlar and Doğanlar (2008), Noyes (2019), Stojanova *et al.* (2012).
Associates: Unknown (Noyes, 2019).
Hosts: Unknown (Noyes, 2019).
Distribution: Turkey (Noyes, 2019).

Idarnotorymus pulcher (Masi, 1916)
Publications with Turkey records: Noyes (2019), Stojanova et al. (2012).
Associates: *Bromus sp.* (Poaceae) (Noyes, 2019).

Hosts: *Tetramesa sp.* (Hymenoptera: Eurytomidae) (Noyes, 2019).
Distribution: Bulgaria, Croatia, Czechoslovakia, Europe, Hungary, Italy, Slovakia, Turkey, Ukraine, USSR, Uzbekistan (Noyes, 2019).

Genus *Idiomacromerus* (Crawford, 1914)
Idiomacromerus akdenizeus (Doğanlar, 1989)
Publications with Turkey records: Doğanlar (1989; 2016), Noyes (2019).
Associates: Unknown (Noyes, 2019).
Hosts: Unknown (Noyes, 2019).
Distribution: Turkey (Noyes, 2019).

Idiomacromerus aladagensis (Doğanlar, 2016)
Publications with Turkey records: Doğanlar (2016), Noyes (2019).
Associates: Unknown (Noyes, 2019).
Hosts: Unknown (Noyes, 2019).
Distribution: Turkey (Noyes, 2019).

Idiomacromerus aslihanae (Doğanlar, 2016)
Publications with Turkey records: Doğanlar (2016), Noyes (2019).
Associates: Unknown (Noyes, 2019).
Hosts: Unknown (Noyes, 2019).
Distribution: Turkey (Noyes, 2019).

Idiomacromerus curticaudatus (Szelenyi, 1981)
Publications with Turkey records: Doğanlar (2016), Noyes (2019).
Associates: *Medicago sativa* (Fabaceae) (Noyes, 2019).
Distribution: Hungary, Turkey (Noyes, 2019).

Idiomacromerus erzurumensis (Doğanlar, 2016)
Publications with Turkey records: Doğanlar (2016), Noyes (2019).
Hosts: Diptera: Tephritidae, *Carduus sp.* (Asteraceae), *Medicago sativa* (Fabaceae) (Noyes, 2019).
Distribution: Turkey (Noyes, 2019).

Idiomacromerus fursovi (Doğanlar, 2016)
Publications with Turkey records: Doğanlar (2016), Noyes (2019).
Associates: Unknown (Noyes, 2019).
Hosts: Unknown (Noyes, 2019).
Distribution: Turkey (Noyes, 2019).

Idiomacromerus golbasinensis (Doğanlar, 2016)
Publications with Turkey records: Doğanlar (2016), Noyes (2019).
Associates: Unknown (Noyes, 2019).
Hosts: Unknown (Noyes, 2019).
Distribution: Turkey (Noyes, 2019).

Idiomacromerus gozuaciki (Doğanlar, 2016)
Publications with Turkey records: Doğanlar (2016), Noyes (2019).
Associates: Unknown (Noyes, 2019).
Hosts: Unknown (Noyes, 2019).
Distribution: Turkey (Noyes, 2019).

Idiomacromerus gumusensis (Doğanlar, 2016)
Publications with Turkey records: Doğanlar (2016), Noyes (2019).
Associates: Unknown (Noyes, 2019).
Hosts: Unknown (Noyes, 2019).
Distribution: Turkey (Noyes, 2019).

Idiomacromerus haliti (Doğanlar, 2016)
Publications with Turkey records: Doğanlar (2016), Noyes (2019).
Associates: Unknown (Noyes, 2019).
Hosts: Unknown (Noyes, 2019).
Distribution: Turkey (Noyes, 2019).

Idiomacromerus hasandagus (Doğanlar, 2016)
Publications with Turkey records: Doğanlar (2016), Noyes (2019).
Associates: Unknown (Noyes, 2019).
Hosts: Unknown (Noyes, 2019).
Distribution: Turkey (Noyes, 2019).

Idiomacromerus karakurtensis (Doğanlar, 2016)
Publications with Turkey records: Doğanlar (2016), Noyes (2019).
Associates: Unknown (Noyes, 2019).
Hosts: Unknown (Noyes, 2019).
Distribution: Turkey (Noyes, 2019).

Idiomacromerus lutfiyeae (Doğanlar, 2016)
Publications with Turkey records: Doğanlar (2016), Noyes (2019).
Associates: Unknown (Noyes, 2019).
Hosts: Unknown (Noyes, 2019).
Distribution: Turkey (Noyes, 2019).

Idiomacromerus mbahadiri (Doğanlar, 2016)
Publications with Turkey records: Doğanlar (2016), Noyes (2019).
Associates: Unknown (Noyes, 2019).
Hosts: Unknown (Noyes, 2019).
Distribution: Turkey (Noyes, 2019).

Idiomacromerus neslihanae (Doğanlar, 2016)
Publications with Turkey records: Doğanlar (2016), Noyes (2019).
Associates: Unknown (Noyes, 2019).
Hosts: Unknown (Noyes, 2019).
Distribution: Turkey (Noyes, 2019).

Idiomacromerus nigdenensis (Doğanlar, 2016)
Publications with Turkey records: Doğanlar (2016), Noyes (2019).
Associates: *Onobrychis sativa* (Fabaceae) (Noyes, 2019).
Distribution: Turkey (Noyes, 2019).

Idiomacromerus oguzhani (Doğanlar, 2016)
Publications with Turkey records: Doğanlar (2016), Noyes (2019).
Associates: Unknown (Noyes, 2019).
Hosts: Unknown (Noyes, 2019).
Distribution: Turkey (Noyes, 2019).

Idiomacromerus pannonicus (Ruschka, 1923)
Publications with Turkey records: Noyes (2019), Öncüer (1991), Stojanova et al. (2012).
Associates: *Carduus sp.* (Asteraceae), *Medicago sativa* (Fabaceae) (Noyes, 2019).
Hosts: *Bruchophagus roddi, Eurytoma roddi* (Hymenoptera: Eurytomidae) (Noyes, 2019).
Parasitoid hosts: *Bruchophagus roddi* (Hymenoptera: Eurytomidae) (Noyes, 2019).
Distribution: Austria, Bulgaria, Europe, France, Hungary, Italy, Palearctic, Romania, Slovakia, Turkey (Noyes, 2019).

Idiomacromerus papaveris (Foerster, 1856)
Publications with Turkey records: Doğanlar (2016), Noyes (2019), Stojanova et al. (2012).
Associates: *Centaurea jacea, Centaurea nigra, Hieracium sp., Lactuca viminea* (Asteraceae), *Gypsophila struthium* (Caryophyllaceae), *Phlomis tuberosa, Salvia fruticose, Salvia triloba* (Lamiaceae), *Papaver sp.,*

Papaver dubium, Papaver rhoeas, Papaver somniferum (Papaveraceae), *Potentilla sp., Potentilla recta* (Rosaceae), *Aylax oraniensis* (Hymenoptera: Cynipidae) (Noyes, 2019).
Hosts: *Aulacidea sp., Aylax sp., Aylax jaceae, Aylax minor, Aylax oraniensis, Aylax papaveris, Aylax salvia, Barbotinia oraniensis, Diastrophus sp., Isocolus jaceae, Rhodus cyprius, Salviela kezivi, Timaspis phoenixopodos, Xestophanes sp., Xestophanes szepligetii* (Hymenoptera: Cynipidae) (Noyes, 2019).
Distribution: Andorra, Austria, Bulgaria, Croatia, Czech Republic, England, Europe, France, Germany, Hungary, Iran, Israel, Italy, Macedonia, Moldova, Spain, Turkey, Ukraine, Yugoslavia (pre-1991) (Noyes, 2019).

Idiomacromerus sebnemae (Doğanlar, 2016)
Publications with Turkey records: Doğanlar (2016), Noyes (2019).
Associates: Unknown (Noyes, 2019).
Hosts: Unknown (Noyes, 2019).
Distribution: Turkey (Noyes, 2019).

Idiomacromerus selimensis (Doğanlar, 2016)
Publications with Turkey records: Doğanlar (2016a), Noyes (2019).
Associates: Unknown (Noyes, 2019).
Hosts: Unknown (Noyes, 2019).
Distribution: Turkey (Noyes, 2019).

Idiomacromerus turhalensis (Doğanlar, 2016)
Publications with Turkey records: Doğanlar (2016a), Noyes (2019).
Associates: Unknown (Noyes, 2019).
Hosts: Unknown (Noyes, 2019).
Distribution: Turkey (Noyes, 2019).
Idiomacromerus uguranlari (Doğanlar, 2016)
Publications with Turkey records: Doğanlar (2016a), Noyes (2019).
Associates: *Olea europaea* (Oleaceae) (Noyes, 2019).
Distribution: Turkey (Noyes, 2019).

Idiomacromerus yunusi (Doğanlar, 2016)
Publications with Turkey records: Doğanlar (2016a), Noyes (2019).
Associates: *Echinops sp.* (Asteraceae) (Noyes, 2019).
Distribution: Turkey (Noyes, 2019).

Idiomacromerus zerovaae (Doğanlar, 2016)
Publications with Turkey records: Doğanlar (2016a), Noyes (2019).
Associates: Unknown (Noyes, 2019).
Hosts: Unknown (Noyes, 2019).
Distribution: Turkey (Noyes, 2019).

Idiomacromerus zeynepbanuae (Doğanlar, 2016)
Publications with Turkey records: Doğanlar (2016a), Noyes (2019).
Associates: Unknown (Noyes, 2019).
Hosts: Unknown (Noyes, 2019).
Distribution: Turkey (Noyes, 2019).

Genus *Microdontomerus* (Crawford, 1907)
Microdontomerus altinekinesis (Doğanlar, 2016)
Publications with Turkey records: Doğanlar (2016a; 2016b), Noyes (2019).
Associates: Unknown (Noyes, 2019).
Hosts: Unknown (Noyes, 2019).
Distribution: Turkey (Noyes, 2019).

Microdontomerus annulatus (Spinola, 1808)
Publications with Turkey records: Doğanlar (2016a; 2016b), Noyes (2019), Öncüer (1991), Stojanova et al. (2012).
Associates: *Aylax sp., Phanacis sp.* (Hymenoptera: Cynipidae), *Olethreutes sororiana* (Lepidoptera: Tortricidae), *Eryngium campestre, Ferula communis* (Apiaceae), *Carduus edelbergii, Carthamus sp., Carthamus tinctorius, Centaurea sp., Centaurea adpressa, Centaurea scabiosa, Centaurea solstitialis, Centaurea Taurica, Cousinia refracta, Jurinea arachnoidea, Onopordum nervosum, Senecio borisphonicum* (Asteraceae), *Atriplex halimus* (Chenopodiaceae), *Phlomis sp., Phlomis fruticosa* (Lamiaceae), *Triticum sp.* (Poaceae) (Noyes, 2019).
Hosts: *Asphondylia punica* (Diptera: Cecidomyiidae), *Acanthiophilus helianthin, Chaetorellia carthami, Orellia colon, Terellia serratulae, Terellia virens* (Tephritidae), *Aylax sp., Phanacis sp.* (Hymenoptera: Cynipidae), *Aphania sauciana, Argyroploce salicella, Endothenia soreriana, Olethreutes sp., Olethreutes sororiana* (Lepidoptera: Tortricidae) (Noyes, 2019).
Distribution: Austria, Croatia, Czech Republic, Czechoslovakia, England, Egypt, Europe, France, Germany, Hungary, Iran, Italy, Libya, Moldova, Montenegro, Nearctic, Pakistan, Romania, Serbia, Slovakia, Spain, Sweden, Transcaucasus, Turkey, Ukraine (Noyes, 2019).

Microdontomerus balikesirensis (Doğanlar, 2016)
Publications with Turkey records: Doğanlar (2016a), Noyes (2019).
Associates: *Olea europaea* (Oleaceae) (Noyes, 2019).
Distribution: Turkey (Noyes, 2019).

Microdontomerus direklinensis (Doğanlar, 2016)
Publications with Turkey records: Doğanlar (2016a; 2016b), Noyes (2019).
Associates: Unknown (Noyes, 2019).
Hosts: Unknown (Noyes, 2019).
Distribution: Turkey (Noyes, 2019).

Microdontomerus erzurumensis (Doğanlar, 2016)
Publications with Turkey records: Doğanlar (2016a), Noyes (2019).
Associates: *Medicago sativa* (Fabaceae) (Noyes, 2019).
Distribution: Turkey (Noyes, 2019).

Microdontomerus gurcukoyensis (Doğanlar, 2016)
Publications with Turkey records: Doğanlar (2016a; 2016b), Noyes (2019).
Associates: Unknown (Noyes, 2019).
Hosts: Unknown (Noyes, 2019).
Distribution: Turkey (Noyes, 2019).

Microdontomerus kahramanmarasensis (Doğanlar and Aslan, 2018)
Publications with Turkey records: Doğanlar and Aslan (2018), Noyes (2019).
Associates: *Echinops orientalis* (Asteraceae) (Noyes, 2019).
Distribution: Turkey (Noyes, 2019).

Microdontomerus kayserinensis (Doğanlar, 2016)
Publications with Turkey records: Doğanlar (2016a), Noyes (2019).
Associates: *Sinapis sp.* (Brassicaceae) (Noyes, 2019).
Distribution: Turkey (Noyes, 2019).
Microdontomerus kocakeri (Doğanlar, 2016)
Publications with Turkey records: Doğanlar (2016a), Noyes (2019).
Associates: *Prunus cerasi* (Rosaceae) (Noyes, 2019).
Hosts: *Centaurea sibirica* (Asteraceae) (Noyes, 2019).
Distribution: Turkey (Noyes, 2019).

Microdontomerus maculosacus (Doğanlar, 2016)
Publications with Turkey records: Doğanlar (2016a), Noyes (2019).
Associates: *Centaurea maculosa* (Asteraceae) (Noyes, 2019).
Distribution: Turkey (Noyes, 2019).

Microdontomerus sivasensis (Doğanlar, 2016)
Publications with Turkey records: Doğanlar (2016a), Noyes (2019).
Associates: Unknown (Noyes, 2019).
Hosts: Unknown (Noyes, 2019).
Distribution: Turkey (Noyes, 2019).

Microdontomerus ulukislanensis (Doğanlar, 2016)
Publications with Turkey records: Doğanlar (2016a), Noyes (2019).
Associates: Unknown (Noyes, 2019).
Hosts: Unknown (Noyes, 2019).
Distribution: Turkey (Noyes, 2019).

Microdontomerus yigiti (Doğanlar, 2016)
Publications with Turkey records: Doğanlar (2016a), Noyes (2019).
Associates: Unknown (Noyes, 2019).
Hosts: Unknown (Noyes, 2019).
Distribution: Turkey (Noyes, 2019).

Subfamily Glyphomerinae (Jansta, Cruaud, Delvare, Genson, Heraty, Krizkova and Rasplus, 2018)
Genus *Glyphomerus* (Foerster, 1856)
Glyphomerus stigma (Fabricius, 1793)
Publications with Turkey records: Bayram et al. (1998), Gencer (2003), Noyes (2019), Mete and Mergen (2017), Öncüer (1991).
Hosts: *Melitoma taurea* (Hymenoptera: Apidae), *Diplolepis sp., Diplolepis centifoliae, Diplolepis eglanteriae, Diplolepis fructuum, Diplolepis fukudae, Diplolepis mayri, Diplolepis multispinosa, Diplolepis nervosa, Diplolepis opaca, Diplolepis polita, Diplolepis rosae, Diplolepis spinosissimae, Periclistus brandtii, Periclistus pirate, Rhodites opaca, Rhodites polita, Rhodites rosae, Rhodites spinosissima, Xestophanes potentillae* (Cynipidae) (Noyes, 2019).
Parasitoid hosts: *Periclistus brandtii* (Cynipidae) (Noyes, 2019).
Parasitoids: *Pteromalus bedeguaris* (Pteromalidae), *Rosa sp., Rosa canina, Rosa pimpinellifolia, Rosa rubiginosa, Rosa rugosa* (Rosaceae) (Noyes, 2019).

Distribution: Andorra, Austria, Bulgaria, Canada, Croatia, Czech Republic, Czechoslovakia, England, Europe, France, Germany, Greece, Hungary, Iran, Italy, Japan, Kirgizia, Nearctic, Netherlands, Romania, Russia, Serbia, Slovakia, Spain, Sweden, Switzerland, Turkey, Ukraine, United States of America, USSR, Yugoslavia (Federal Republic), Yugoslavia (pre-1991) (Noyes, 2019).

Glyphomerus tibialis (Foerster, 1856)
Publications with Turkey records: Noyes (2019), Stojanova et al. (2012).
Associates: *Hieracium pilosella, Hypochaeris radicata* (Asteraceae), *Glechoma hederacea, Salvia lavandulifolia, Salvia nemorosa, Salvia officinalis, Salvia pratensis, Salvia sclarea, Salvia triloba* (Lamiaceae), *Papaver sp., Papaver dubium, Papaver rhoeas* (Papaveraceae), *Festuca ovina* (Poaceae), *Potentilla argentea, Potentilla recta, Rosa sp., Rosa pimpinellifolia* (Rosaceae) (Noyes, 2019).
Hosts: *Aulacidea sp., Aulacidea subterminalis, Aylax sp., Aylax glechomae, Barbotinia oraniensis, Diastrophus sp., Diastrophus mayri, Diplolepis centifoliae, Diplolepis nervosa, Diplolepis spinosissimae, Hedickiana levantina, Liposthenes glechomae, Neaylax sp., Phanacis hypochoeridis, Rhodus cyprius* (Hymenoptera: Cynipidae), *Harmolita sp., Tetramesa sp., Tetramesa brevicollis, Tetramesa brevicornis* (Eurytomidae) (Noyes, 2019).
Distribution: Austria, Croatia, Czech Republic, Czechoslovakia, England, Europe, Finland, France, Germany, Greece, Jordan, Macedonia, Moldova, Mongolia, Montenegro, Netherlands, Slovakia, Spain, Sweden, Transcaucasus, Turkey, Ukraine, USSR, Yugoslavia (pre-1991) (Noyes, 2019).
Subfamily Monodontomerinae (Ashmead, 1899)
Genus *Monodontomerus* (Westwood, 1833)
Monodontomerus aereus (Walker, 1834)
Publications with Turkey records: Avci (2009), Kilic and Alaoglu (1996), Noyes (2019), Öncüer (1991), Öncüer et al. (1978), Polat and Tozlu (2010), Stojanova et al. (2012).
Associates: *Arbutus sp.* (Ericaceae), *Quercus sp., Quercus suber* (Fagaceae), *Juglans regia* (Juglandaceae), *Abies alba, Larix decidua* (Pinaceae), *Malus pumila, Prunus sp., Prunus spinosa* (Rosaceae), *Citrus sp.* (Rutaceae), *Populus sp., Populus alba* (Salicaceae), *Ulmus glabra* (Ulmaceae) (Noyes, 2019).
Hosts: *Megachile muraria* (Hymenoptera: Apidae), *Diplolepis rosae* (Cynipidae), *Diprion pini* (Diprionidae), *Trichiocampus ulmi*

(Tenthredinidae), *Hyphantria cunea* (Lepidoptera: Arctiidae), *Dendrolimus pini, Malacosoma Americanum, Malacosoma franconicum, Malacosoma neustria* (Lasiocampidae), *Euproctis chrysorrhoea, Euproctis phaeorrhoea, Euproctis similis, Hemerocampa leucostigma, Leucoma salicis, Liparis chrysorrhoea, Lymantria dispar, Lymantria monachal, Lymantria obfuscate, Nygmia phaeorrhoea, Orgyia leucostigma, Porthetria sp., Porthetria dispar, Stilpnotia salicis* (Lymantriidae), *Thaumetopoea processionea* (Notodontidae), *Nymphalis polychloros* (Nymphalidae), *Aporia crataegi, Pieris brassicae, Pieris rapae* (Pieridae), *Archips argyrospilus, Archips murinana, Archips rosanus, Carpocapsa pomonella, Choristoneura murinana, Choristoneura rosaceana, Clysia ambiguella, Cydia pomonella, Epinotia pygmaeana, Pandemis ribeana, Petrova resinella, Polychrosis botrana, Ptycholomoides aeriferanus, Sparganothis pilleriana, Spilonota lariciana, Tortrix viridana* (Tortricidae), *Scythropia crataegella, Yponomeuta malinellus, Yponomeuta rorellus* (Yponomeutidae) (Noyes, 2019).

Parasitoid hosts: *Sarcophaga sp.* (Diptera: Sarcophagidae), *Alsomyia nidicola, Blepharipa pratensis, Blepharipa scutellate, Carcelia laxifrons, Compsilura concinnata, Echinomyia praeceps, Exorista larvarum, Exorista segregate, Lydella nigripes, Opsosturmia nidicola, Sturmia nidicola, Sturmia scutellate, Tachina sp., Tachina larvarum, Tachina mella, Townsendiellomyia nidicola, Tricholyga segregate, Zenillia laxifrons* (Tachinidae), *Apanteles sp., Apanteles glomeratus, Apanteles lacteicolor, Apanteles liparidis, Apanteles melanoscelus, Apanteles pieridis, Apanteles solitarius, Cotesia melanoscela, Dolichogenidea lacteicolor, Meteorus versicolor* (Hymenoptera: Braconidae), *Brachymeria compsilurae, Brachymeria intermedia, Brachymeria tibialis* (Chalcididae), *Colpoclypeus florus* (Eulophidae), *Apechthis compunctor, Coccygomimus sp., Gambrus nuncius, Glypta murinanae, Hyposoter disparis, Iseropus coelebs, Itoplectis maculator, Phaeogenes maculicornis, Phobocampe disparis, Phobocampe unicincta, Pimpla sp., Pimpla coelebs, Pimpla instigator, Spilocryptus nuncius, Theronia atalantae* (Ichneumonidae) (Noyes, 2019).

Distribution: Armenia, Austria, Bosnia Hercegovina, Bulgaria, Canada, Croatia, Czech Republic, Czechoslovakia, Europe, France, Georgia, Germany, Greece, Hungary, India, Iran, Italy, Japan, Kazakhstan, Kirgizia, Macedonia, Moldova, Nearctic, Netherlands, North Africa, Palearctic, Poland, Romania, Russia, Serbia, Slovakia, Spain, Switzerland, Tselinograd Obl., Turkey, Ukraine, United Kingdom, United States of America, USSR, Yugoslavia (pre-1991) (Noyes, 2019).

Monodontomerus minor (Ratzeburg, 1848)
Publications with Turkey records: Polat and Tozlu (2010), Noyes (2019), Stojanova et al. (2012).
Associates: *Robinia pseudoacacia* (Fabaceae), *Quercus palustris, Quercus rubra* (Fagaceae), *Pinus massoniana* (Pinaceae), *Malus pumila, Prunus serotina* (Rosaceae), *Populus sp.* (Salicaceae) (Noyes, 2019).
Hosts: *Megachile centuncularis, Osmia coerulescens* (Hymenoptera: Apidae), *Trichiosoma sp., Trichiosoma lucorum, Trichiosoma tibialae, Trichiosoma tibiale* (Cimbicidae), *Diprion pini, Gilpinia sp., Gilpinia pallida, Neodiprion sertifer* (Diprionidae), *Hyphantria cunea* (Lepidoptera: Arctiidae), *Cystidia couggaria* (Geometridae), *Dendrolimus sp., Dendrolimus pini, Dendrolimus sibiricus, Dendrolimus spectabilis, Dendrolimus superans, Dendrolimus tabulaeformis, Gastropacha populifolia, Malacosoma Americanum, Malacosoma neustria* (Lasiocampidae), *Cnidocampa flavescens, Narosoideus flavidorsalis* (Limacodidae), *Euproctis phaeorrhoea, Euproctis similis, Lymantria dispar, Lymantria monachal, Orgyia antiqua, Orgyia pseudotsugata, Porthesia chrysorrhoea, Stilpnotia salicis* (Lymantriidae), *Chloephora bicolorana* (Noctuidae), *Cerura menciana, Clostera anachoreta, Melalopha anachoreta, Phryganidia californica, Pygaera anastomosis, Thaumetopoea processionea* (Notodontidae), Nymphalidae, *Aporia crataegi, Pieris brassicae, Pieris rapae* (Pieridae), *Cryptothelea formosicola, Cryptothelea variegate, Eumeta japonica, Eumeta minuscula, Thyridopteryx ephemeraeformis* (Psychidae), *Hyalophora cecropia, Samia cynthia* (Saturniidae), *Dilina tiliae, Mimas tiliae* (Sphingidae), *Archips sp., Archips argyrospilus, Archips fumiferana, Archips rosanus, Choristoneura fumiferana, Grapholita molesta, Laspeyresia pomonella, Pandemis heparana, Pandemis ribeana, Rhyacionia buoliana, Sparganothis sp., Tortrix viridana* (Tortricidae), *Yponomeuta evonymellus* (Yponomeutidae), *Illiberis pruni* (Zygaenidae) (Noyes, 2019).
Parasitoid hosts: *Exorista mella, Masicera zimini, Tachina sp., Zenillia libatrix* (Diptera: Tachinidae), *Aleiodes sp., Apanteles sp., Apanteles glomeratus, Cotesia glomerata, Glyptapanteles paridis, Rhogas sp., Rhogas dendrolimi, Rogas sp.* (Hymenoptera: Braconidae), *Eurytoma verticillata* (Eurytomidae), *Agrothereutes migrator, Apechthis compunctor, Apechthis rufata, Cryptus quadratus, Dicamptus nigropictus, Gambrus extrematis, Hyposoter takagii, Ichneumon garugawensis, Iseropus himalayensis, Iseropus orientalis, Itoplectis conquisitor, Phytodietus fumiferanae, Pimpla turionellae, Pimpla varicornis, Spilocryptus migrator, Therion giganteum, Theronia atalantae*

(Ichneumonidae), *Hyalophora cecropia* (Lepidoptera: Saturniidae) (Noyes, 2019).
Distribution: Austria, Canada, Croatia, Czech Republic, Czechoslovakia, Europe, Finland, France, Georgia, Germany, Hungary, India, Italy, Japan, Kazakhstan, Korea, Moldova, Nearctic, Netherlands, Pakistan, People's Republic of China, Poland, Russia, Serbia, Slovakia, Sweden, Tadzhikistan, Tselinograd Obl., Turkey, Ukraine, United States of America, USSR (Noyes, 2019).

Monodontomerus obscurus (Westwood, 1833)
Publications with Turkey records: Noyes (2019), Öncüer (1991), Stojanova et al. (2012).
Associates: *Chalicodoma sp., Xylocopa sp.* (Hymenoptera: Apidae), *Sceliphron sp., Sceliphron destillatorium* (Sphecidae), *Eumenes sp.* (Vespidae), *Shorea robusta* (Dipterocarpaceae), *Euphorbia sp.* (Euphorbiaceae), *Medicago sativa* (Fabaceae), *Quercus sp.* (Fagaceae), *Phragmites sp.* (Poaceae), *Populus ciliate* (Salicaceae), *Heritiera fomes* (Sterculiaceae) (Noyes, 2019).
Hosts: *Pachytychius hordei* (Coleoptera: Curculionidae), *Eulalia viridula, Hoplodonta viridula, Odontomyia viridula* (Diptera: Stratiomyiidae), *Anthophora abrupta, Anthophora pilipes, Chalicodoma sp., Chalicodoma muraria, Hoplitis adunca, Megachile sp., Megachile cephalotes, Megachile flavipes, Megachile lanata, Megachile pacifica, Megachile rotundata, Megachile willughbiella, Osmia sp., Osmia adunca, Osmia cordata, Osmia cornifrons, Osmia cornuta, Osmia latreillei, Osmia lignaria, Osmia ribifloris, Osmia rufa, Osmia sanrafaelae, Xylocopa fenestrate* (Hymenoptera: Apidae), *Chrysis sp.* (Chrysididae), *Diprion pini* (Diprionidae), *Trypoxylon sp.* (Sphecidae), *Eumenes pomiformis, Eumenes unguiculata* (Vespidae), *Coleotechnites piceaella* (Lepidoptera: Gelechiidae), *Lymantria obfuscata* (Lymantriidae), *Carpocapsa pomonella* (Tortricidae) (Noyes, 2019).
Distribution: Azores, Canada, Chile, Croatia, Czech Republic, Czechoslovakia, Denmark, Egypt, England, Europe, France, Germany, Hungary, India, Iran, Israel, Italy, Kazakhstan, Lebanon, Macedonia, Moldova, Nearctic, Netherlands, Pakistan, Romania, Russia, Slovakia, Spain, Sweden, Switzerland, Transcaucasus, Tselinograd Obl., Turkey, Turkmenistan, Ukraine, United States of America, USSR, Yugoslavia (pre-1991) (Noyes, 2019).

Genus *Oopristus* (Steffan, 1968)
Oopristus erganicus (Tarla and Doğanlar, 2010)

Publications with Turkey records: Noyes (2019), Stojanova et al. (2012), Tarla et al. (2010).
Associates: *Punica granatum* (Lythraceae), *Platanus orientalis* (Platanaceae), *Amygdalus sp.* (Rosaceae) (Noyes, 2019).
Hosts: *Apodiphus amygdali* (Hemiptera: Pentatomidae) (Noyes, 2019).
Distribution: Turkey (Noyes, 2019).

Oopristus tayfursokmeni (Tarla and Doğanlar, 2010)
Publications with Turkey records: Noyes (2019), Stojanova et al. (2012), Tarla et al. (2010).
Associates: *Acacia sp.* (Fabaceae), *Platanus orientalis* (Platanaceae) (Noyes, 2019).
Hosts: *Apodiphus amygdali* (Hemiptera: Pentatomidae) (Noyes, 2019).
Distribution: Turkey (Noyes, 2019).

Oopristus turkestanicus (Skriptshinsky, 1929)
Publications with Turkey records: Noyes (2019), Stojanova et al. (2012), Tarla et al. (2010).
Associates: *Acer sp.* (Aceraceae), *Acacia sp., Indigofera articulata* (Fabaceae), *Platanus orientalis* (Platanaceae), *Malus pumila* (Rosaceae), *Populus alba* (Salicaceae) (Noyes, 2019).
Hosts: *Apodiphus sp., Apodiphus amygdali, Sarju eremica* (Hemiptera: Pentatomidae) (Noyes, 2019).
Distribution: Iran, Iraq, Pakistan, Turkey, Uzbekistan (Noyes, 2019).

Subfamily Podagrioninae: Podagrionini (Ashmead, 1904)
Genus *Podagrion* (Spinola, 1811)
Podagrion splendens (Spinola, 1811)
Publications with Turkey records: Noyes (2019), Stojanova et al. (2012).
Associates: *Eryngium campestre* (Apiaceae), *Onopordum nervosum* (Asteraceae), *Suaeda vera* (Chenopodiaceae), *Ephedra nebrodensis* (Gnetaceae), *Marrubium vulgare, Teucrium capitatum* (Lamiaceae), *Reseda lutea* (Resedaceae), *Tamarix canariensis* (Tamaricaceae) (Noyes, 2019).
Hosts: *Empusa fasciata, Iris oratoria, Mantis religiosa* (Orthoptera: Mantidae) (Noyes, 2019).
Distribution: Bosnia Hercegovina, Bulgaria, Caucasus, Croatia, Czechoslovakia, Europe, France, Greece, Hungary, Italy, Morocco, Russia, Slovakia, Spain, Transcaucasus, Turkey (Noyes, 2019).

Genus *Podagrionella* (Girault, 1914)
Podagrionella eremiaphilae (Doğanlar and Doğanlar, 2009)
Publications with Turkey records: Doğanlar and Doğanlar (2009), Noyes (2019), Stojanova et al. (2012).
Hosts: *Eremiaphila turcica* (Orthoptera: Eremiaphilidae), Mantidae (Noyes, 2019).
Distribution: Turkey (Noyes, 2019).

Podagrionella konyaensis (Doğanlar and Doğanlar, 2009)
Publications with Turkey records: Doğanlar and Doğanlar (2009), Noyes (2019), Stojanova et al. (2012).
Associates: Unknown (Noyes, 2019).
Hosts: Unknown (Noyes, 2019).
Distribution: Turkey (Noyes, 2019).

Subfamily Toryminae: Torymini (Walker, 1833)
Genus *Diomorus* (Walker, 1834)
Diomorus calcaratus (Nees, 1834)
Publications with Turkey records: Noyes (2019).
Associates: *Omalus auratus* (Coleoptera: Chrysomelidae), *Andricus quercustozae* (Hymenoptera: Cynipidae), *Pemphredon rufiger, Stigmus pendulus* (Sphecidae), *Quercus sp., Quercus petraea* (Fagaceae), *Rubus sp.* (Rubiaceae) (Noyes, 2019).
Hosts: *Adleria quercustozae, Andricus kollari, Andricus quercustozae, Cynips kollari* (Hymenoptera: Cynipidae), *Cemonus unicolor, Ectemnius rubicola, Pemphredon sp., Pemphredon lethifer, Pemphredon luctuosus, Pemphredon unicolor, Rhopalum coarctatum, Stigmus sp., Stigmus pendulus, Stigmus solskyi, Trypoxylon figulus* (Sphecidae) (Noyes, 2019).
Distribution: Armenia, Austria, Belgium, Bosnia Hercegovina, Bulgaria, Caucasus, Croatia, Czech Republic, Czechoslovakia, Europe, France, Hungary, India, Iran, Italy, Macedonia, Moldova, Netherlands, People's Republic of China, Romania, Russia, Serbia, Slovakia, Spain, Turkey, Ukraine, USSR (Noyes, 2019).
Genus *Torymus* (Dalman, 1820)
Torymus apiomyiae (Boucek and Mihajlovic, 1986)
Publications with Turkey records: Noyes (2019), Stojanova et al. (2012).
Associates: *Pyrus sp., Pyrus communis* (Rosaceae) (Noyes, 2019).
Hosts: *Apiomyia bergenstammi* (Diptera: Cecidomyiidae) (Noyes, 2019).
Distribution: Macedonia, Turkey, Yugoslavia (pre-1991) (Noyes, 2019).

Torymus arcella (Graham and Gijswijt, 1998)
Publications with Turkey records: Noyes (2019), Stojanova et al. (2012).
Associates: Unknown (Noyes, 2019).
Hosts: Unknown (Noyes, 2019).
Distribution: Kazakhstan, Turkey, Ukraine (Noyes, 2019).

Torymus auratus (Mueller, 1764)
Publications with Turkey records: Noyes (2019), Stojanova et al. (2012).
Associates: *Andricus burgundus, Andricus conglomeratus, Andricus coronatus, Andricus curvator, Andricus dentimitratus, Andricus fecundatrix, Andricus glutinosus, Andricus grossulariae, Andricus hispanicus, Andricus hungaricus, Andricus infectorius, Andricus inflator, Andricus kollari, Andricus lignicolus, Andricus lucidus, Andricus multiplicatus, Andricus pictus, Andricus polycerus, Andricus pseudoinflator, Andricus quercuscalicis, Andricus quercusramuli, Andricus quercustozae, Aphelonyx cerricola, Biorhiza pallida, Cerroneuroterus lanuginosus, Chilaspis nitida, Cynips agama, Cynips disticha, Cynips divisa, Cynips longiventris, Cynips quercus, Cynips quercusfolii, Dryocosmus cerriphilus, Dryocosmus kuriphilus, Neuroterus albipes, Neuroterus numismalis, Neuroterus quercusbaccarum, Neuroterus tricolor, Plagiotrochus amenti, Pseudoneuroterus saliens, Synophrus politus, Trigonaspis Megaptera, Trigonaspis synaspis* (Cynipidae), *Quercus sp., Quercus cerris, Quercus faginea, Quercus frainetto, Quercus infectoria, Quercus lusitanica, Quercus petraea, Quercus pubescens, Quercus pyrenaica, Quercus robur, Quercus suber* (Fagaceae), *Rosa canina* (Rosaceae), *Rubus sp.* (Rubiaceae) (Noyes, 2019).
Hosts: *Andricus aestivalis, Andricus conglomeratus, Andricus fecundatrix, Andricus hispanicus, Andricus kollari, Andricus lignicolus, Andricus mayri, Andricus quercuscalicis, Andricus quercustozae, Aphelonyx cerricola, Biorhiza pallida, Chilaspis nitida, Cynips sp., Cynips divisa, Cynips longiventris, Cynips quercus, Cynips quercusfolii, Diastrophus rubi, Diplolepis fructuum, Diplolepis quercusfolii, Dryocosmus kuriphilus, Neuroterus quercusbaccarum, Synergus gallaepomiformis* (Hymenoptera: Cynipidae) (Noyes, 2019).
Parasitoid hosts: *Synergus pallicornis, Synergus reinhardi, Synergus umbraculus* (Hymenoptera: Cynipidae), *Eudecatoma biguttata, Eurytoma brunniventris* (Eurytomidae) (Noyes, 2019).

Distribution: Andorra, Austria, Belgium, Bulgaria, Croatia, Czech Republic, Europe, France, Georgia, Germany, Greece, Hungary, Iran, Italy, Moldova, Morocco, Netherlands, Poland, Romania, Slovakia, Spain, Sweden, Turkey, Ukraine, United Kingdom, United States of America (Noyes, 2019).

Torymus basarani (Doğanlar, 2017)
Publications with Turkey records: Doğanlar (2017), Noyes (2019).
Associates: Unknown (Noyes, 2019).
Hosts: Unknown (Noyes, 2019).
Distribution: Turkey (Noyes, 2019).

Torymus bedeguaris (Linnaeus, 1758)
Publications with Turkey records: Bayram et al. (1998), Gencer (2003), Mete and Mergen (2017), Noyes (2019), Ozbek et al. (1999), Stojanova et al. (2012).
Associates: *Wolbachia sp.* (Bacteria: Wolbachiaceae), *Asteraceae, Quercus sp.* (Fagaceae), *Hypecoum imberbe* (Papaveraceae), *Rosa sp., Rosa beggeriana, Rosa canina, Rosa rubiginosa, Rosa tshatyrdagi* (Rosaceae), *Rubus sp.* (Rubiaceae) (Noyes, 2019).
Hosts: *Dasineura rosarum, Wachtliella rosarum* (Diptera: Cecidomyiidae), *Aulacidea rubi, Aylax hypecoi, Cynips quercus, Cynips rosae, Diplolepis sp., Diplolepis bicolor, Diplolepis centifoliae, Diplolepis eglanteriae, Diplolepis fructuum, Diplolepis ignota, Diplolepis mayri, Diplolepis multispinosa, Diplolepis polita, Diplolepis radicum, Diplolepis rosae, Diplolepis rosarum, Diplolepis spinosa, Diplolepis spinosissimae, Diplolepis tuberculatrix, Diplolepis variabilis, Nipporhodites magna, Periclistus pirate, Rhodites eglanteriae, Rhodites mayri, Rhodites rosae* (Hymenoptera: Cynipidae) (Noyes, 2019).
Parasitoid hosts: *Orthopelma mediator* (Ichneumonidae) (Noyes, 2019)
Parasitoids: *Quadrastichus rosarum* (Eulophidae) (Noyes, 2019).
Distribution: Andorra, Armenia, Austria, Azerbaijan, Belgium, Bulgaria, Canada, Caucasus, Croatia, Czech Republic, Czechoslovakia, Denmark, England, Europe, France, Georgia, Germany, Greece, Hungary, Iran, Italy, Kazakhstan, Kirgizia, Moldova, Nearctic, Netherlands, People's Republic of China, Romania, Russia, Serbia, Slovakia, Spain, Sweden, Switzerland, Tadzhikistan, Tselinograd Obl., Turkey, Turkmenistan, Ukraine, United States of America, USSR, Yugoslavia (Federal Republic), Yugoslavia (pre-1991) (Noyes, 2019).

Torymus bingoelensis (Doğanlar, 2017)
Publications with Turkey records: Doğanlar (2017), Noyes (2019).
Associates: *Quercus sp.* (Fagaceae) (Noyes, 2019).
Hosts: Hymenoptera: Cynipidae (Noyes, 2019).
Distribution: Turkey (Noyes, 2019).

Torymus cultratus (Graham and Gijswijt, 1998)
Publications with Turkey records: Noyes (2019), Stojanova et al. (2012).
Associates: Unknown (Noyes, 2019).
Hosts: Unknown (Noyes, 2019).
Distribution: Turkey, Ukraine (Noyes, 2019).

Torymus cultriventris (Ratzeburg, 1844)
Publications with Turkey records: Öncüer (1991), Noyes (2019), Stojanova et al. (2012), Ural and Kurt (1973).
Associates: *Corylus avellana* (Betulaceae), *Fagus sp., Fagus sylvatica* (Fagaceae) (Noyes, 2019).
Hosts: *Hartigiola annulipes, Mikiola fagi, Mikomyia coryli* (Diptera: Cecidomyiidae), *Andricus curvator* (Hymenoptera: Cynipidae) (Noyes, 2019).
Parasitoid hosts: *Aprostocetus elongatus, Hyperteles elongatus* (Eulophidae) (Noyes, 2019).
Distribution: Belgium, Croatia, Czech Republic, Czechoslovakia, Europe, Germany, Hungary, Montenegro, Netherlands, Poland, Slovakia, Turkey, Ukraine, United Kingdom, Yugoslavia (Federal Republic) (Noyes, 2019).

Torymus cyaneus (Walker, 1847)
Publications with Turkey records: Noyes (2019), Stojanova et al. (2012).
Associates: *Andricus coriarius, Andricus coronatus, Andricus infectorius, Andricus lucidus, Cynips agama, Cynips cornifex, Cynips disticha, Cynips divisa, Cynips longiventris, Cynips quercus, Cynips quercusfolii, Neuroterus anthracinus, Pseudoneuroterus saliens, Synophrus politus, Trigonaspis megaptera, Quercus sp., Quercus cerris, Quercus faginea, Quercus frainetto, Quercus infectoria, Quercus lusitanica, Quercus petraea, Quercus pubescens, Quercus pyrenaica, Quercus robur* (Fagaceae) (Noyes, 2019).
Hosts: *Adleria sp., Andricus sp., Biorhiza pallida, Cynips sp., Cynips agama, Cynips cortices, Cynips disticha, Cynips divisa, Cynips longiventris, Cynips quercus, Cynips quercusfolii, Diplolepis sp.,*

Diplolepis divisa, Neuroterus tricolor (Hymenoptera: Cynipidae) (Noyes, 2019).

Parasitoid hosts: *Synergus gallaepomiformis, Synergus vulgaris* (Hymenoptera: Cynipidae) (Noyes, 2019).

Distribution: Austria, Azerbaijan, Belgium, Bulgaria, Croatia, Czech Republic, Czechoslovakia, Europe, Finland, France, Germany, Greece, Hungary, Moldova, Netherlands, Romania, Russia, Slovakia, Spain, Sweden, Turkey, Ukraine, United Kingdom (Noyes, 2019).

Torymus erucarum (Schrank, 1781)

Publications with Turkey records: Noyes (2019), Stojanova et al. (2012).

Associates: *Andricus hungaricus, Andricus infectorius, Andricus kollari, Andricus quercusradicis, Andricus sieboldin, Aphelonyx cerricola, Cynips longiventris, Cynips quercusfolii* ((Hymenoptera: Cynipidae), *Fagus sp., Fagus sylvatica, Quercus sp., Quercus faginea, Quercus petraea, Quercus pyrenaica, Quercus robur* (Fagaceae), *Rhamnus alaternus* (Rhamnaceae) (Noyes, 2019).

Hosts: *Mikiola fagi* (Diptera: Cecidomyiidae), *Andricus sp., Andricus inflator, Andricus kollari, Andricus quercusradicis, Andricus testaceipes, Aphelonyx cerricola, Biorhiza pallida, Diastrophus rubi* (Hymenoptera: Cynipidae), *Diprion pini* (Diprionidae), *Carpocapsa pomonella* (Lepidoptera: Tortricidae) (Noyes, 2019).

Parasitoid hosts: *Synergus incrassatus* (Hymenoptera: Cynipidae) (Noyes, 2019).

Distribution: Andorra, Austria, Croatia, Czech Republic, Czechoslovakia, Denmark, England, Europe, France, Germany, Hungary, Iran, Italy, Macedonia, Moldova, Netherlands, Poland, Portugal, Romania, Russia, Serbia, Slovakia, Spain, Sweden, Turkey, Ukraine, Yugoslavia (Federal Republic) (Noyes, 2019).

Torymus fastuosus (Boheman, 1834)

Publications with Turkey records: Noyes (2019), Stojanova et al. (2012).

Associates: *Trigonaspis Megaptera, Trigonaspis synaspis* (Hymenoptera: Cynipidae), *Quercus sp., Quercus faginea, Quercus infectoria, Quercus petraea, Quercus pyrenaica, Quercus robur* (Fagaceae) (Noyes, 2019).

Hosts: *Trigonaspis Megaptera, Trigonaspis synaspis* (Hymenoptera: Cynipidae) (Noyes, 2019).

Distribution: Austria, Croatia, Czech Republic, England, Europe, France, Georgia, Germany, Moldova, Netherlands, Poland, Russia, Spain, Sweden, Turkey, Ukraine (Noyes, 2019).

Torymus flavipes (Walker, 1833)
Publications with Turkey records: Doğanlar (2014), Noyes (2019), Stojanova et al. (2012).
Associates: *Andricus callidoma, Andricus conglomeratus, Andricus conificus, Andricus corruptrix, Andricus curvator, Andricus glandulae, Andricus grossulariae, Andricus infectorius, Andricus inflator, Andricus multiplicatus, Andricus pseudoinflator, Andricus quadrilineatus, Andricus quercuscalicis, Andricus quercusradicis, Andricus quercusramuli, Andricus schroeckingeri, Andricus seminationis, Andricus singularis, Andricus solitarius, Aphelonyx cerricola, Biorhiza pallida, Chilaspis nitida, Cynips disticha, Cynips divisa, Cynips longiventris, Cynips quercusfolii, Dryocosmus kuriphilus, Neuroterus albipes, Neuroterus anthracinus, Neuroterus numismalis, Neuroterus quercusbaccarum, Neuroterus saliens, Neuroterus tricolor, Plagiotrochus australis, Plagiotrochus quercusilicis, Pseudoneuroterus saliens, Trigonaspis megaptera* (Cynipidae), *Daucus carota, Pastinaca sp., Pimpinella sp., Seseli sp., Torilis sp.* (Apiaceae), *Castanea sativa, Fagus sylvatica, Quercu coccifera, Quercus sp., Quercus cerris, Quercus dalechampi, Quercus faginea, Quercus ilex, Quercus lusitanica, Quercus pedunculiflora, Quercus petraea, Quercus pubescens, Quercus pyrenaica, Quercus robur* (Fagaceae), *Rosa sp.* (Rosaceae) (Noyes, 2019).
Hosts: *Cecidomyia baccharicola, Dasineura asperulae, Kiefferia pimpinellae, Putoniella marsupialis, Rabdophaga dubia* (Diptera: Cecidomyiidae), *Chaetostomella onotrophes, Urophora jaceana, Urophora quadrifasciata* (Tephritidae), *Andricus sp., Andricus conglomeratus, Andricus curvator, Andricus inflator, Andricus multiplicatus, Andricus quadrilineatus, Andricus quercusramuli, Andricus seminationis, Andricus singularis, Aphelonyx cerricola, Biorhiza sp., Biorhiza pallida, Chilaspis nitida, Cynips disticha, Cynips divisa, Cynips longiventris, Cynips quercusfolii, Diplolepis eglanteriae, Diplolepis mayri, Diplolepis nervosa, Diplolepis rosae, Diplolepis spinosa, Dryocosmus kuriphilus, Neuroterus sp., Neuroterus albipes, Neuroterus numismalis, Neuroterus quercusbaccarum, Neuroterus tricolor, Plagiotrochus ilicis, Plagiotrochus quercusilicis* (Hymenoptera: Cynipidae) (Noyes, 2019).
Parasitoid hosts: *Periclistus brandtii, Synergus sp., Synergus albipes, Synergus gallaepomiformis, Synergus nervosus* (Hymenoptera: Cynipidae), *Olynx arsames, Pediobius lysis* (Eulophidae), *Eurytoma brunniventris* (Eurytomidae), *Mesopolobus sp., Mesopolobus tibialis* (Pteromalidae), *Syntomaspis notata, Torymus notatus* (Torymidae) (Noyes, 2019).

Distribution: Andorra, Austria, Azerbaijan, Belgium, Bulgaria, Caucasus, Croatia, Czech Republic, Czechoslovakia, Europe, Finland, France, Germany, Hungary, Israel, Italy, Netherlands, Palearctic, People's Republic of China, Poland, Romania, Russia, Slovakia, Slovenia, Spain, Sweden, Switzerland, Turkey, Ukraine, United Kingdom, United States of America (Noyes, 2019).

Torymus geranii (Walker, 1833)
Publications with Turkey records: Noyes (2019).
Associates: *Andricus curvator, Andricus fecundatrix, Andricus glutinosus, Andricus grossulariae, Andricus hartigi, Andricus hungaricus, Andricus kollari, Andricus quercuscalicis, Andricus quercusradicis, Andricus quercusramuli, Aphelonyx cerricola, Biorhiza pallida, Cerroneuroterus lanuginosus, Chilaspis nitida, Cynips agama, Cynips disticha, Cynips divisa, Cynips longiventris, Cynips quercus, Cynips quercusfolii, Dryocosmus cerriphilus, Neuroterus anthracinus, Neuroterus quercusbaccarum, Trigonaspis synaspis (Hymenoptera: Cynipidae), Castanea sp., Castanea sativa, Quercus sp., Quercus cerris, Quercus infectoria, Quercus ithaburensis, Quercus lusitanica, Quercus petraea, Quercus pubescens, Quercus pyrenaica, Quercus robur* (Fagaceae), *Rosa sp.* (Rosaceae) (Noyes, 2019).
Hosts: *Helicomyia saliciperda* (Diptera: Cecidomyiidae), *Andricus curvator, Andricus kollari, Andricus quercuscalicis, Biorhiza pallida, Cynips divisa, Cynips longiventris, Diplolepis nervosa, Dryocosmus kuriphilus, Trigonaspis synaspis* (Hymenoptera: Cynipidae) (Noyes, 2019).
Distribution: Andorra, Austria, Azerbaijan, Belgium, Bulgaria, Croatia, England, Finland, France, Germany, Hungary, Iran, Italy, Japan, Jordan, Korea, Netherlands, People's Republic of China, Poland, Portugal, Romania, Russia, Serbia, Slovakia, Spain, Sweden, Turkey, Ukraine (Noyes, 2019).

Torymus igniceps (Mayr, 1874)
Publications with Turkey records: Noyes (2019), Stojanova et al. (2012).
Hosts: *Rhodites rosae* (Hymenoptera: Cynipidae) (Noyes, 2019).
Distribution: Austria, Czech Republic, Czechoslovakia, Europe, Germany, Hungary, Italy, Netherlands, Sweden, Turkey, Ukraine, United Kingdom (Noyes, 2019).

Torymus longicalcar (Graham, 1994)
Publications with Turkey records: Noyes (2019), Stojanova et al. (2012).
Associates: *Dryocosmus cerriphilus, Neuroterus quercusbaccarum* (Hymenoptera: Cynipidae), *Acer sp., Acer pseudoplatanus* (Aceraceae), *Alnus glutinosa* (Betulaceae), *Cornus mas, Cornus sanguinea* (Cornaceae), *Quercus sp.* (Fagaceae), *Salix caprea* (Salicaceae) (Noyes, 2019).
Hosts: *Dryomyia circinnans, Dryomyia concinna* (Diptera: Cecidomyiidae), *Dryocosmus nervosus, Pediaspis aceris* (Hymenoptera: Cynipidae) (Noyes, 2019).
Distribution: Austria, Denmark, Germany, Greece, Hungary, Italy, Slovakia, Sweden, Switzerland, Turkey (Noyes, 2019).

Torymus nitidulus (Walker, 1833)
Publications with Turkey records: Noyes (2019), Öncüer (1991), Stojanova et al. (2012).
Associates: *Semudobia sp.* (Diptera: Cecidomyiidae), *Betula sp., Betula exilis, Betula pendula* (Betulaceae) (Noyes, 2019).
Hosts: *Oligotrophus betulae, Semudobia sp., Semudobia betulae* (Diptera: Cecidomyiidae) (Noyes, 2019).
Distribution: Czech Republic, Czechoslovakia, England, Europe, Germany, Hungary, Mongolia, Nearctic, Netherlands, People's Republic of China, Romania, Russia, Sweden, Turkey, Ukraine, United States of America, USSR (Noyes, 2019).

Torymus phillyreae (Ruschka, 1921)
Publications with Turkey records: Doğanlar (2011), Doğanlar et al. (2011), Noyes (2019).
Associates: *Dasineura oleae, Lasioptera oleicola* (Diptera: Cecidomyiidae), *Artemisia sp.* (Asteraceae), *Atriplex halimus* (Chenopodiaceae), *Calicotome spinosa, Cytisus sp., Cytisus scoparius, Genista sp., Genista florida, Ulex europaeus* (Fabaceae), *Fagus sylvatica* (Fagaceae), *Phillyrea sp.* (Oleaceae), *Scrophularia sp., Scrophularia peregrine* (Scrophulariaceae) (Noyes, 2019).
Hosts: *Asphondylia sp., Asphondylia conglomerate, Asphondylia punica, Asphondylia sarothamni, Asphondylia ulicis, Braueriella sp., Braueriella phillyreae, Dasineura oleae, Stictodiplosis sp., Stictodiplosis scrophulariae* (Diptera: Cecidomyiidae), *Dryocosmus australis* (Hymenoptera: Cynipidae) (Noyes, 2019).

Distribution: Armenia, Egypt, England, Europe, France, Germany, Greece, Hungary, Italy, Romania, Russia, Spain, Sweden, Turkey, Ukraine (Noyes, 2019).

Torymus pygmaeus (Mayr, 1874)
Publications with Turkey records: Noyes (2019), Stojanova et al. (2012).
Associates: *Quercus sp., Quercus cerris* (Fagaceae) (Noyes, 2019).
Hosts: *Contarinia subulifex* (Diptera: Cecidomyiidae), *Biorhiza pallida* (Hymenoptera: Cynipidae) (Noyes, 2019).
Distribution: Austria, Europe, Sweden, Turkey, Ukraine (Noyes, 2019).

Torymus ramicola (Ruschka, 1921)
Publications with Turkey records: Doğanlar (2014), Noyes (2019).
Associates: *Artemisia sp.* (Asteraceae) (Noyes, 2019).
Hosts: *Diplosis ramicola* (Diptera: Cecidomyiidae), *Dryocosmus kuriphilus* (Hymenoptera: Cynipidae) (Noyes, 2019).
Distribution: Austria, Europe, Hungary, Sweden, Turkey, Ukraine (Noyes, 2019).

Torymus rubi (Schrank, 1781)
Publications with Turkey records: Gencer (2003), Noyes (2019), Stojanova *et al.* (2012).
Associates: *Carduus bourgeanus, Hieracium virosum* (Asteraceae), *Pteridium aquilinum* (Dennstaedtiaceae), G*lechoma hederacea* (Lamiaceae), *Fraxinus sp.* (Oleaceae), *Rosa sp., Rosa canina, Rosa rubiginosa* (Rosaceae), *Rubus sp., Rubus caesius, Rubus fruticosus, Rubus idaeus* (Rubiaceae), *Sphaeropyx bicolor* (Hymenoptera: Braconidae) (Noyes, 2019).
Hosts: *Pteridium aquilinum* (Dennstaedtiaceae), *Stereonychus fraxini* (Coleoptera: Curculionidae), *Lasioptera rubi, Perrisia acrophilae* (Diptera: Cecidomyiidae), *Aulacidea rubi, Cynips longiventris, Diastrophus rubi, Diplolepis sp., Diplolepis centifoliae, Diplolepis mayri, Diplolepis nervosa, Diplolepis rosae, Diplolepis spinosa, Diplolepis spinosissimae, Liposthenes glechomae, Periclistus brandtii, Rhodites mayri, Rhodites rosae* (Hymenoptera: Cynipidae) (Noyes, 2019).
Parasitoid hosts: *Orthopelma luteolator* (Ichneumonidae), *Glyphomerus stigma* (Torymidae) (Noyes, 2019).
Distribution: Austria, Azerbaijan, Belgium, Bulgaria, Caucasus, Croatia, Czech Republic, Czechoslovakia, Denmark, England, Europe, France, Germany, Hungary, Italy, Kazakhstan, Kirgizia, Moldova, Netherlands,

Poland, Romania, Russia, Serbia, Slovakia, Spain, Sweden, Switzerland, Tselinograd Obl., Turkey, Ukraine, USSR, Yugoslavia (pre-1991) (Noyes, 2019).

Torymus triangularis (Thomson, 1876)
Publications with Turkey records: Noyes (2019), Stojanova et al. (2012).
Associates: Unknown (Noyes, 2019).
Hosts: Unknown (Noyes, 2019).
Distribution: Europe, Hungary, Sweden, Turkey (Noyes, 2019).

Subfamily Toryminae: Torymoidini (Grissell, 1995)
Genus: ***Ameromicrus*** (Nikolskaya, 1954)
Ameromicrus yumurtalikiensis (Doğanlar, 1989)
Publications with Turkey records: Doğanlar (1989), Noyes (2019), Stojanova et al. (2012).
Associates: Unknown (Noyes, 2019).
Hosts: Unknown (Noyes, 2019).
Distribution: Turkey (Noyes, 2019).

Genus *Didactyliocerus* (Masi, 1916)
Didactyliocerus dispar (Masi, 1916)
Publications with Turkey records: Noyes (2019), Öncüer (1991), Stojanova et al. (2012).
Associates: Unknown (Noyes, 2019).
Hosts: Unknown (Noyes, 2019).
Distribution: Albania, Bulgaria, Croatia, Czechoslovakia, Hungary, Italy, Macedonia, Moldova, Montenegro, Slovakia, Syria, Transcaucasus, Turkey (Noyes, 2019).

Genus *Pseudotorymus* (Masi, 1921)
Pseudotorymus adananensis (Doğanlar, 2016)
Publications with Turkey records: Doğanlar (2016), Noyes (2019).
Associates: *Diplotaxis muralis* (Brassicaceae) (Noyes, 2019).
Hosts: Diptera: Cecidomyiidae (Noyes, 2019).
Distribution: Turkey (Noyes, 2019).

Pseudotorymus aladagensis (Doğanlar, 2016)
Publications with Turkey records: Doğanlar (2016), Noyes (2019).
Associates: Unknown (Noyes, 2019).
Hosts: Unknown (Noyes, 2019).

Distribution: Turkey (Noyes, 2019).

Pseudotorymus arvernicus (Walker, 1833)
Publications with Turkey records: Noyes (2019), Stojanova et al. (2012).
Associates: *Apion apricans, Apion craccae* (Coleoptera: Apionidae), *Dasineura sp., Lestodiplosis sp.* (Diptera: Cecidomyiidae), *Tragopogon dubius* (Asteraceae), *Astragalus cicer, Lathyrus sylvestris, Medicago sativa, Onobrychis sativa, Trifolium sp., Trifolium pratense* (Fabaceae) (Noyes, 2019).
Hosts: *Apion sp., Apion apricans, Apion craccae, Apion holosericeum, Apion trifolii* (Coleoptera: Apionidae), *Bruchidius sp., Bruchidius unicolor, Bruchidius virescens* (Bruchidae), *Aulacidea tragopogonis* (Hymenoptera: Cynipidae) (Noyes, 2019).
Distribution: Austria, Croatia, Czech Republic, Czechoslovakia, England, Europe, France, Germany, Hungary, Italy, Mongolia, Montenegro, Netherlands, Palearctic, Romania, Serbia, Slovakia, Sweden, Turkey, Ukraine (Noyes, 2019).

Pseudotorymus bedirlinensis (Doğanlar, 2016)
Publications with Turkey records: Doğanlar (2016), Noyes (2019).
Associates: Unknown (Noyes, 2019).
Hosts: Unknown (Noyes, 2019).
Distribution: Turkey (Noyes, 2019).

Pseudotorymus bollinensis (Askew, 2002)
Publications with Turkey records: Doğanlar (2016), Noyes (2019).
Associates: *Geranium pratense* (Geraniaceae) (Noyes, 2019).
Hosts: Diptera: Cecidomyiidae (Noyes, 2019).
Distribution: England, Turkey (Noyes, 2019).

Pseudotorymus celikhanensis (Doğanlar, 2016)
Publications with Turkey records: Doğanlar (2016), Noyes (2019).
Associates: Unknown (Noyes, 2019).
Hosts: Unknown (Noyes, 2019).
Distribution: Turkey (Noyes, 2019).

Pseudotorymus dogubayezitensis (Doğanlar, 2016)
Publications with Turkey records: Doğanlar (2016), Noyes (2019).
Associates: Unknown (Noyes, 2019).
Hosts: Unknown (Noyes, 2019).

Distribution: Turkey (Noyes, 2019).

Pseudotorymus haliti (Doğanlar, 2016)
Publications with Turkey records: Doğanlar (2016), Noyes (2019).
Associates: Unknown (Noyes, 2019).
Hosts: Unknown (Noyes, 2019).
Distribution: Turkey (Noyes, 2019).

Pseudotorymus hasanberkayi (Doğanlar and Aslan, 2018)
Publications with Turkey records: Doğanlar and Aslan (2018), Noyes (2019).
Associates: *Echinops orientalis* (Asteraceae) (Noyes, 2019).
Distribution: Turkey (Noyes, 2019).

Pseudotorymus horasanensis (Doğanlar, 2016)
Publications with Turkey records: Doğanlar (2016), Noyes (2019).
Associates: Unknown (Noyes, 2019).
Hosts: Unknown (Noyes, 2019).
Distribution: Turkey (Noyes, 2019).

Pseudotorymus ispirlii (Doğanlar, 2016)
Publications with Turkey records: Doğanlar (2016), Noyes (2019).
Associates: Unknown (Noyes, 2019).
Hosts: Unknown (Noyes, 2019).
Distribution: Turkey (Noyes, 2019).
Pseudotorymus kecikuyusunensis (Doğanlar, 2016)
Publications with Turkey records: Doğanlar (2016), Noyes (2019).
Associates: Unknown (Noyes, 2019).
Hosts: Unknown (Noyes, 2019).
Distribution: Turkey (Noyes, 2019).

Pseudotorymus leguminus (Ruschka, 1923)
Publications with Turkey records: Doğanlar (2016), Noyes (2019).
Associates: *Lathyrus sylvestris, Lotus pedunculatus, Onobrychis sativa, Onobrychis viciifolia* (Fabaceae) Noyes, 2019).
Hosts: *Contarinia loti, Contarinia medicaginis, Contarinia onobrychidis, Contarinia silvestris, Contarinia sylvestris, Dasineura silvestris* (Diptera: Cecidomyiidae) (Noyes, 2019).
Distribution: Austria, Czech Republic, Czechoslovakia, England, Europe, Germany, Turkey, Ukraine (Noyes, 2019).

Pseudotorymus lutfiyeae (Doğanlar, 2016)
Publications with Turkey records: Doğanlar (2016), Noyes (2019).
Associates: Unknown (Noyes, 2019).
Hosts: Unknown (Noyes, 2019).
Distribution: Turkey (Noyes, 2019).

Pseudotorymus militaris (Boheman, 1834)
Publications with Turkey records: Noyes (2019), Öncüer (1991), Stojanova et al. (2012).
Associates: *Medicago sativa* (Fabaceae), *Papaver papaveris* (Papaveraceae), *Salix sp.* (Salicaceae) (Noyes, 2019).
Distribution: Czech Republic, England, Europe, Germany, Iran, Sweden, Turkey, Ukraine (Noyes, 2019).

Pseudotorymus nigdenensis (Doğanlar, 2016)
Publications with Turkey records: Doğanlar (2016), Noyes (2019).
Associates: Unknown (Noyes, 2019).
Hosts: Unknown (Noyes, 2019).
Distribution: Turkey (Noyes, 2019).

Pseudotorymus ocaklii (Doğanlar, 2016)
Publications with Turkey records: Doğanlar (2016), Noyes (2019).
Associates: Unknown (Noyes, 2019).
Hosts: Unknown (Noyes, 2019).
Distribution: Turkey (Noyes, 2019).

Pseudotorymus pannonicus (Mayr, 1874)
Publications with Turkey records: Noyes (2019), Öncüer (1991).
Associates: *Medicago sativa* (Fabaceae) (Noyes, 2019).
Distribution: Europe, Hungary, Romania, Turkey, Ukraine (Noyes, 2019).

Pseudotorymus papaveris (Thomson, 1876)
Publications with Turkey records: Doğanlar (2016), Noyes (2019), Stojanova et al. (2012).
Associates: *Papaver sp., Papaver dubium, Papaver rhoeas, Papaver somniferum* (Papaveraceae) (Noyes, 2019).
Hosts: *Dasineura papaveris, Lestodiplosis callida* (Diptera: Cecidomyiidae), *Aylax minor, Aylax papaveris, Barbotinia oraniensis, Iraella luteipes* (Hymenoptera: Cynipidae) (Noyes, 2019).

Distribution: Austria, Bulgaria, Czech Republic, Europe, France, Germany, Hungary, Moldova, Mongolia, Netherlands, Palearctic, Romania, Serbia, Spain, Sweden, Turkey, Ukraine (Noyes, 2019).

Pseudotorymus pazarcikensis (Doğanlar, 2016)
Publications with Turkey records: Doğanlar (2016), Noyes (2019).
Associates: Unknown (Noyes, 2019).
Hosts: Unknown (Noyes, 2019).
Distribution: Turkey (Noyes, 2019).

Pseudotorymus rosarum (Zerova and Seregina, 1992)
Publications with Turkey records: Noyes (2019), Stojanova et al. (2012).
Associates: *Rosa sp., Rosa canina* (Rosaceae) (Noyes, 2019).
Hosts: *Diplolepis fructuum* (Hymenoptera: Cynipidae) (Noyes, 2019).
Distribution: Tadzhikistan, Turkey, Ukraine (Noyes, 2019).

Pseudotorymus samsatensis (Doğanlar, 2016)
Publications with Turkey records: Doğanlar (2016), Gencer (2003), Noyes (2019), Stojanova *et al.* (2012).
Associates: Unknown (Noyes, 2019).
Hosts: Unknown (Noyes, 2019).
Distribution: Turkey (Noyes, 2019).

Pseudotorymus sapphyrinus (Fonscolombe, 1832)
Publications with Turkey records: Doğanlar (2016), Gencer (2003), Noyes (2019), Stojanova et al. (2012).
Associates: *Eryngium campestre* (Apiaceae), *Centaurea diffusa, Centaurea substitute* (Asteraceae), *Rosa canina, Rosa eglanteria* (Rosaceae) (Noyes, 2019).
Hosts: *Lasioptera eryngii, Thomasiella eryngii* (Diptera: Cecidomyiidae), *Diplolepis mayri, Diplolepis rosae, Isocolus centaureae* (Hymenoptera: Cynipidae) (Noyes, 2019).
Distribution: Azerbaijan, Bulgaria, Croatia, Czech Republic, Czechoslovakia, Europe, France, Germany, Hungary, Moldova, Mongolia, Romania, Serbia, Slovakia, Sweden, Turkey, Ukraine (Noyes, 2019).

Pseudotorymus sivasensis (Doğanlar, 2016)
Publications with Turkey records: Doğanlar (2016), Noyes (2019).
Associates: Unknown (Noyes, 2019).
Hosts: Unknown (Noyes, 2019).

Distribution: Turkey (Noyes, 2019).

Pseudotorymus verbasci (Erdos, 1955)
Publications with Turkey records: Doğanlar (2016), Noyes (2019).
Associates: *Verbascum sp., Verbascum austriacum* (Scrophulariaceae) (Noyes, 2019).
Hosts: *Asphondylia verbasci* (Diptera: Cecidomyiidae) (Noyes, 2019).
Distribution: Bulgaria, Europe, Hungary, Turkey (Noyes, 2019).

Genus *Torymoides* (Walker, 1871)
Torymoides kiesenwetteri (Mayr, 1874)
Publications with Turkey records: Noyes (2019), Öncüer (1991), Stojanova et al. (2012).
Associates: *Myopites inulaedyssentericae* (Tephritidae), *Eryngium maritimum* (Apiaceae), *Carduus pycnocephalus, Centaurea sp., Centaurea cyanus, Eupatorium sp., Eupatorium adenophorum, Eupatorium glandulosum, Inula sp., Inula crithmoides, Inula hirta, Inula viscosa, Otanthus maritimus, Pulicaria sp., Pulicaria dysenterica* (Asteraceae), *Gypsophila struthium* (Caryophyllaceae), *Medicago sativa* (Fabaceae), *Marrubium vulgare* (Lamiaceae), *Hordeum leporinum* (Poaceae), *Reseda lutea* (Resedaceae), *Tamarix canariensis* (Tamaricaceae) (Noyes, 2019).
Hosts: *Lasioptera eryngii* (Diptera: Cecidomyiidae), *Acanthiophilus helianthin, Acanthophilus helianthin, Aciura coryli, Chaetostoma completum, Chetostoma completum, Myopites sp., Myopites blotii, Myopites eximia, Myopites frauenfeldi, Myopites inulae, Myopites inulaedyssentericae, Myopites longirostris, Myopites stylata, Procecidochares utilis, Tephritis stictica, Urophora stylata* (Tephritidae) (Noyes, 2019).
Distribution: Andorra, Bulgaria, Canary Islands, Croatia, Czech Republic, Czechoslovakia, Egypt, England, Europe, France, Germany, Greece, Hungary, India, Iran, Italy, Macedonia, Madeira, Moldova, Nepal, Palearctic, People's Republic of China, Poland, Romania, Serbia, Slovakia, Spain, Switzerland, Turkey, Yemen, Yugoslavia (pre-1991) (Noyes, 2019).

References

Avci, Mustafa. 2009. "Parasitoid complex and new host plants of the gypsy moth, *Lymantria dispar* L. in the Lakes District, Turkey". *Journal of Animal and Veterinary Advances*, 8(7):1402-1405.

Bayram, Şerife, and Ülgentürk, Selma, and Toros, Seval. 1998. "Research on the insects causing galls on dog rose (*Rosa spp.*) and their parasitoids in Ankara province". *Turkish Journal of Entomology*, 22(4):259-268.

Doğanlar, Mikdat, and Aslan, M. Murat. 2018. "Two new species of Torymidae (Hymenoptera: Chalcidoidea) reared from flower heads of *Echinops orientalis* Trautvetter in south-eastern Turkey". *Zoology in the Middle East*, 64(3):238.243.

Doğanlar, Mikdat, and Sertkaya, Erdal, and & Skuhravá, Marcela. 2011a. "Pest status of olive leaf gall midge *Dasineura oleae* (Angelini, 1831), description of *Lasioptera oleicola* Skuhravá sp. new (Diptera: Cecidomyiidae) and effectiveness of parasitoids on their populations in Hatay, Turkey". *Turkish Journal of Entomology*, 35(2):265-284.

Doğanlar, Mikdat. 1989. "Two new species of Ameromicrus Nikolskaja (Hymenoptera: Toryimidae) from Adana district, Turkey". *Turkish Journal of Entomology*, 13 (3):141-146.

—. 2011. "Parastioid complex of the olive leaf gall midges, *Dasineura oleae* (Angelini 1831) and *Lasioptera oleicola* Skuhravá, 2011 (Diptera: Cecidomyiidae) in Hatay Turkey, with descriptions of new genus and species from Tetrastichinae (Hymenoptera: Eulophidae)". *Turkish Journal of Entomology*, 35(2):245-264.

—. 2014. "First reports on the natural enemy fauna of the chestnut gall wasp, *Dryocosmus kuriphilus* Yasumatsu (Hymenoptera: Cynipidae) in Yalova, Turkey". *Türkiye Biyolojik Mücadele Dergisi*, 5(1):67-74.

—. 2016. "*Ditropinotus* Crawford, 1907 (Hymenoptera: Torymidae: Microdontomerini) species from Turkey, with descriptions of new species". *Munis Entomology and Zoology*, 11(2):269-277.

—. 2016a. "Species of *Idiomacromerus* Crawford in Turkey (Hymenoptera: Torymidae: Microdontomerini)". *Munis Entomology and Zoology*, 11(2):312-354.

—. 2016b. "Species of *Pseudotorymus* Masi in Turkey (Hymenoptera: Torymidae: Microdontomerini)". *Munis Entomology and Zoology*, 11(2):393- 432.

—. 2016c. "Turkish species of *Microdontomerus* Crawford, 1907 with descriptions of new species (Hymenoptera: Toryidae: Microdontomerini)", *Munis Entomology and Zoology*, 11(2):363-381.

—. 2016d. "Species of *Microdontomerus* (Crawford, 1907) and *Eridontomerus* (Crawford, 1907) (Hymenoptera: Torymidae: Microdontomerini) from Turkey, with descriptions of a new species". *Entomofauna*, 37(33):505-520.

—. 2017. "Turkish species of *Torymus* Dalman, 1820 (Hymenoptera: Toryidae [sic]: Torymini), with descriptions of new species". *Munis Entomology and Zoology*, 12(1):5-22.

Doğanlar, Oğuzhan, and Doğanlar, Mikdat. 2008. "A new species of *Idarnotorymus* Masi, 1916 (Hymenoptera: Torymidae) from Turkey". *Turkish Journal of Zoology*, 32:85-90.

Doğanlar, Oğuzhan, and Doğanlar, Mikdat. 2009. "*Podagrionella* Girault, 1913 (Hymenoptera: Torymidae: Podagrionidae) species from Turkey with descriptions of new species". *Turkish Journal of Zoology*, 33(2):181-186.

Gençer, Lütfiye. 2003. "The chalcidoid parasitoids of *Diplolepis mayri* Schld. (Hymenoptera: Cynipidae), a pest of *Rosa canina* L. in Sivas province". *Turkish Journal of Entomology*, 27(2):107-111.

Kılıç, Nuray, and Alaoğlu, Özdemir. 1996. "Biology and parasitoids of the satin moth *Leucoma salicis* (L.) (Lepidoptera, Lymantriidae) a pest of poplar trees in Erzurum Province (Turkey)". *Turkish Journal of Entomology*, 20(4):269-279.

Mete, Özlem, and Mergen, Orhan. 2017. "The community components associated with two common rose gall wasps (Hymenoptera: Cynipidae: Diplolepidini) in Turkey". *Turkish Journal of Zoology*, 41:696-701.

Noyes, John S. 2019. *Universal Chalcidoidea Database*. Available at: www.nhm.ac.uk/chalcidoids (accessed January 2022).

Öncüer Cezmi. 1991. "A catalogue of the parasites and predators of insect pests of Turkey". – *E.Ü. Ziraat Fakültesi Yayınları No. 505*, Bornova-İzmir, pp. 354.

Öncüer, Cezmi, and Yalçın, E. & Erkin, E. 1978. "The natural enemies of the pupae of *Euproctis chrysorrhoea* L. (Lepidoptera: Lymantriidae), which damages fruit trees in the Aegean region, and their effectiveness". *Türkiye Bitki Koruma Dergisi*, 2(1):31-36.

Özbek, Hikmet, and Güçlü, Şaban, and Tozlu, Göksel. 1999. "Biology and natural enemies of *Diplolepis mayri* Schld. (Hymenoptera: Cynipidae), a pest of Rosa canina L. in Erzurum Province". *Turkish Journal of Entomology*, 23(1):39-50.

Polat, Alper, and Tozlu, Göksel. 2010. "Investigations on the brief biology, host plants and parasitoids of *Archips rosana* (Linnaeus, 1758) (Lepidoptera: Tortricidae) in Erzurum". *Turkish Journal of Entomology*, 34(4):529-542.

Stojanova, Anelia, and Civelek, Hasan Sungur, and Yörük, Berna, and Sari, Sabahat, and Atahan, Tuğçe. 2012. "Checklists of Turkish Eurytomidae Walker, 1832 and Torymidae Walker, 1833

(Hymenoptera: Chalcidoidea)". *Turkish Journal of Entomology*, 36(1):69-82.

Tarla, Şener, and Doğanlar, Mikdat, and Gözüaçık, Celalettin. 2010. "The species of *Oopristus* Steffan, 1968 (Hymenoptera: Torymidae: Toryminae: Monodontomerini) of Turkey with descriptions of two new species". *Turkish Journal of Zoology*, 34(4):487-495.

Ural, İhsan, and Kurt, Ayhan. 1973. "Doğu Karadeniz Bölgesi Fındık bahçelerinde zarar yapan Fındık Gal Sineği (*Mikomyia coryli* Kieffer)'nin biyolojisi ve mücadelesi üzerinde araştırmalar". *Bitki Koruma Bülteni*, 13(1):1-18.

CHAPTER NINETEEN

FAMILY: TRICHOGRAMMATIDAE HALIDAY, 1851

Trichogrammatidae are parasitic on the eggs of other insects. To complete their development, their adult size cannot exceed that of a single insect egg. Often, multiple individuals will develop within a single egg, resulting in emerging adults that are only a fraction of the size of their host. Therefore, Trichogrammatidae are among the smallest known insects, ranging in size from 0.2 to 1.5 mm.

Certain genera, such as Trichogramma, are known to parasitize eggs of various insect orders, while other genera appear to be restricted to a single host order. The eggs of Hemiptera (true bugs, leafhoppers, etc.) are parasitized by the largest number of genera including, Aphelinoidea, Paracentrobia, and Ufens. However, several genera also utilize eggs of Coleoptera (beetles) and Lepidoptera (moths and butterflies). Most trichogrammatids parasitize eggs that are laid in or on plant tissues.

Several genera, such as *Hydrophylita*, Lathromeroidea and *Prestwichia*, specialize in parasitizing eggs of aquatic insects and have been reported to swim underwater in search of hosts. A few genera of Trichogrammatidae are of interest for use in biological control. Trichogramma has received significant attention in applied entomology due to its ability to parasitize numerous pest Lepidoptera. It can be mass propagated and released with relative ease, and thus, Trichogramma has been the most widely used arthropod for augmentative biological control programs worldwide (see hymenoptera.ucr.edu).

The main diagnostic characteristics of Trichogrammatidae are that they are minute to small insects range from 0.3 to 1.2 mm in length, excluding the ovipositor, or up to 1.8 mm in length including the ovipositor. They are not metallic in appearence, and their funicle typically has no more than two segments. The tarsi are 3-segmented.

The family currently includes 83 genera and 839 species. In Turkey, there are three subfamilies, nine genera and 23 species (Noyes, 2019).

Subfamily Oligositinae: Chaetostrichini (Girault, 1912)
Genus *Aphelinoidea* (Girault, 1911)
Aphelinoidea anatolica (Novicky, 1936)
Publications with Turkey records: Noyes (2019).
Associates: *Chenopodium sp.* (Chenopodiaceae) (Noyes, 2019).
Hosts: *Circulifer tenellus* (Hemiptera: Cicadellidae) (Noyes, 2019).
Distribution: Algeria, Iran, Libya, People's Republic of China, Syria, Turkey, United States of America (Noyes, 2019).

Genus *Bloodiella* (Nowicki, 1935)
Bloodiella andalusica (Nowicki, 1935)
Publications with Turkey records: Noyes (2019), Öncüer (1991).
Associates: Unknown (Noyes, 2019).
Hosts: Unknown (Noyes, 2019).
Distribution: Europe, France, Hungary, Poland, Spain, Turkey, Yugoslavia (pre-1991) (Noyes, 2019).

Genus *Chaetostricha* (Walker, 1851)
Chaetostricha steineri (Nowicki, 1935)
Publications with Turkey records: Noyes (2019).
Associates: Unknown (Noyes, 2019).
Hosts: Unknown (Noyes, 2019).
Distribution: Turkey (Noyes, 2019).

Genus *Monorthochaeta* (Blood, 1923)
Monorthochaeta adanaensis (Doğanlar, 2002)
Publications with Turkey records: Doğanlar (2002), Noyes (2019).
Associates: *Cynara scolymus* (Asteraceae) (Noyes, 2019).
Hosts: *Cassida rubiginosa* (Coleoptera: Chrysomelidae) (Noyes, 2019).
Distribution: Turkey (Noyes, 2019).

Monorthochaeta galatica (Nowicki, 1935)
Publications with Turkey records: Noyes (2019), Öncüer (1991).
Associates: Unknown (Noyes, 2019).
Hosts: Unknown (Noyes, 2019).
Distribution: Hungary, Poland, Turkey (Noyes, 2019).

Genus *Ufens* (Girault, 1911)
Ufens foersteri (Kryger, 1919)
Publications with Turkey records: Noyes (2019).
Associates: *Lepidosaphes ulmi* (Hemiptera: Diaspididae), *Asphodelus sp.* (Asphodelaceae), *Heliotropium sp.* (Boraginaceae), *Atriplex sp., Chenopodium sp., Chenopodium hybridum, Salsola sp.* (Chenopodiaceae), *Cistus sp.* (Cistaceae), *Quercus sp., Quercus ilex* (Fagaceae), *Prunus sp.* (Rosaceae), *Populus italicus* (Salicaceae) (Noyes, 2019).
Hosts: *Circulifer sp.* (Hemiptera: Cicadellidae), *Aspidiotus perniciosus* (Diaspididae), *Reuteria marqueti* (Miridae), *Uromenus confusus* (Orthoptera: Tettigoniidae) (Noyes, 2019).
Distribution: Australia, Czech Republic, Denmark, Egypt, England, Europe, France, Germany, Ghana, Greece, Guinea, Hungary, Iran, Israel, Italy, Kenya, Kirgizia, Madagascar, Morocco, Netherlands, Nigeria, Pakistan, People's Republic of China, Poland, Romania, Russia, South Africa, Spain, Tanzania, Thailand, Turkey, Turkmenistan, Uganda, Ukraine (Noyes, 2019).

Ufens similis (Kryger, 1932)
Publications with Turkey records: Noyes (2019), Öncüer (1991).
Associates: *Convolvulus arvensis* (Convolvulaceae) (Noyes, 2019).
Distribution: Denmark, Egypt, France, Gambia, Greece, India, Italy, Kenya, Kirgizia, Oman, People's Republic of China, Poland, South Africa, Turkey (Noyes, 2019).

Genus *Uscana* (Girault, 1911)
Uscana johnstoni (Waterston, 1926)
Publications with Turkey records: Noyes (2019).
Hosts: *Sphenoptera gossypii* (Coleoptera: Buprestidae) (Noyes, 2019).
Distribution: North Africa, Romania, Sudan, Turkey (Noyes, 2019).

Subfamily Oligositinae: Oligositini (Ashmead, 1904)
Genus *Oligosita* (Walker, 1851)
Oligosita collina (Walker, 1851)
Publications with Turkey records: Noyes (2019), Öncüer (1991).
Associates: Unknown (Noyes, 2019).
Hosts: Unknown (Noyes, 2019).
Distribution: Czech Republic, France, Ireland, Thailand, Turkey, United Kingdom (Noyes, 2019).

Oligosita foersteri (Girault, 1914)
Publications with Turkey records: Noyes (2019).
Associates: *Andropogon ischaemum* (Poaceae) (Noyes, 2019).
Distribution: Europe, Germany, Hungary, Spain, Turkey (Noyes, 2019).

Oligosita pallida (Kryger, 1919)
Publications with Turkey records: Noyes (2019), Öncüer (1991), Yigit and Erkiliç (1987; 1992).
Hosts: *Arboridia adanae, Arboridia Kermanshah, Edwardsiana rosae, Erythroneura eburnean, Zygina rhamni* (Hemiptera: Cicadellidae) (Noyes, 2019).
Associates: *Corylus avellana* (Betulaceae), *Rosa sp.* (Rosaceae), *Rubus fruticosus* (Rubiaceae), *Vitis vinifera* (Vitaceae) (Noyes, 2019).
Distribution: Croatia, Czech Republic, Denmark, Hungary, Iran, Italy, Netherlands, Turkey, United Kingdom, Yugoslavia (pre-1991) (Noyes, 2019).

Genus *Pseudoligosita* (Girault, 1913)
Pseudoligosita gracilior (Nowicki, 1935)
Publications with Turkey records: Noyes (2019).
Hosts: *Eriococcus buxi* (Hemiptera: Eriococcidae) (Noyes, 2019).
Distribution: Czech Republic, Czechoslovakia, Europe, France, Romania, Turkey, USSR (Noyes, 2019).

Pseudoligosita paphlagonica (Nowicki, 1935)
Publications with Turkey records: Noyes (2019).
Associates: Unknown (Noyes, 2019).
Hosts: Unknown (Noyes, 2019).
Distribution: Turkey (Noyes, 2019).

Subfamily Trichogrammatinae: Trichogrammatini
Genus *Pterandrophysalis* (Nowicki, 1935)
Pterandrophysalis levantina (Nowicki, 1935)
Publications with Turkey records: Noyes (2019), Öncüer (1991).
Hosts: *Bangasternus orientalis* (Coleoptera: Curculionidae) (Noyes, 2019).
Associates: *Cirsium sp., Cirsium vulgare* (Asteraceae) (Noyes, 2019).
Distribution: England, France, Greece, Italy, Syria, Tunisia, Turkey, Ukraine, United States of America (Noyes, 2019).

Genus *Trichogramma* (Westwood, 1833)
Trichogramma brassicae (Bezdenko, 1968)

Publications with Turkey records: Avci (2003), Caner and Tuncer (2001), Noyes (2019), Özpinar et al. (1999).
Associates: *Lydella thompsoni* (Diptera: Tachinidae), *Trichogramma sp., Trichogramma cacaeciae* (Hymenoptera: Trichogrammatidae), *Brassica sp., Brassica oleracea* (Brassicaceae), *Beta vulgaris* (Chenopodiaceae), *Carya illinoinensis, Juglans sp.* (Juglandaceae), *Punica granatum* (Lythraceae), *Cedrus libani* (Pinaceae), *Lolium multiflorum, Triticum aestivum, Zea mays* (Poaceae), *Fragaria sp.* (Rosaceae), *Capsicum sp., Capsicum annuum, Lycopersicon esculentum* (Solanaceae) (Noyes, 2019).
Hosts: Diptera: Tachinidae, *Lydella thompsoni, Keiferia lycopersicella, Pectinophora gossypiella, Sitotroga cerealella* (Lepidoptera: Gelechiidae), *Hesperia comma* (Hesperiidae), *Cyaniris semiargus, Plebejus idas, Polyommatus icarus* (Lycaenidae), *Laelia salicis* (Lymantriidae), *Actebia fennica, Agrotis exclamationis, Agrotis ipsilon, Agrotis segetum, Archanara geminipunctata, Autographa gamma, Chrysodeixis chalcites, Helicoverpa armigera, Helicoverpa zea, Heliothis armigera, Heliothis zea, Mamestra brassicae, Mamestra trifolii, Scotia ipsilon, Sesamia cretica, Sesamia nonagrioides, Trichoplusia ni* (Noctuidae), *Traumatocampa ispartaensis* (Notodontidae), *Aphantopus hyperanthus, Argynnis adippe, Argynnis Niobe, Clossiana selene, Clossiana titania, Coenonympha pamphilus, Erebia ligea, Hipparchia Alcyone, Maniola jurtina, Melanargia galathea, Mellicta athalia, Vanessa atalanta* (Nymphalidae), *Maroga melanostigma* (Oecophoridae), *Papilio demoleus, Papilio machaon* (Papilionidae), *Artogeia napi, Pieris sp., Pieris brassicae, Pieris rapae* (Pieridae), *Anagasta kuehniella, Chilo phragmitellus, Chilo suppressalis, Ephestia cautella, Ephestia kuehniella, Evergestis forficalis, Galleria mellonella, Ostrinia nubilalis, Palpita unionalis, Plodia interpunctella, Spectrobates ceratoniae* (Pyralidae), *Samia cynthia* (Saturniidae), *Deilephila elpenor, Sphinx ligustri* (Sphingidae), *Cydia pomonella, Eupoecilia ambiguella, Lobesia botrana* (Tortricidae), *Plutella xylostella, Prays oleae* (Yponomeutidae), *Zygaena filipendulae* (Zygaenidae) (Noyes, 2019).
Distribution: Argentina, Australia, Austria, Belgium, Bulgaria, Canada, Europe, France, Germany, Iran, Italy, Japan, Moldova, Netherlands, Palearctic, People's Republic of China, Romania, Russia, Slovenia, Spain, Switzerland, Turkey, Ukraine, United States of America, USSR, Yugoslavia (pre-1991) (Noyes, 2019).

Trichogramma buluti (Bulut and Kilincer, 1991)
Publications with Turkey records: Bulut and Kilinçer (1991), Noyes (2019), Özbek and Coruh (2010).

Hosts: *Malacosoma neustria* (Lepidoptera: Lasiocampidae) (Noyes, 2019).
Distribution: Turkey (Noyes, 2019).

Trichogramma cacaeciae (Marchal, 1927)
Publications with Turkey records: Noyes (2019), Öncuer and Uzun (1991), Uzun and Öncüer (1990).
Associates: *Bacillus thuringiensis* (Bacteria: Bacillaceae), *Wolbachia sp.* (Wolbachiaceae), *Trichogramma brassicae* (Hymenoptera: Trichogrammatidae), *Brassica oleracea* (Brassicaceae), *Dianthus caryophyllus* (Caryophyllaceae), *Spinacia oleracea* (Chenopodiaceae), *Quercus sp.* (Fagaceae), *Allium porrum* (Liliaceae), *Punica granatum* (Lythraceae), *Pinus sylvestris* (Pinaceae), *Zea mays* (Poaceae), *Malus sp., Malus pumila, Prunus sp.* (Rosaceae), *Salix sp.* (Salicaceae), *Vitis vinifera* (Vitaceae) (Noyes, 2019).
Hosts: *Strobilomyia anthracina* (Diptera: Tachinidae), *Cimex lectularius* (Hemiptera: Cimicidae), *Caliroa cerasi* (Hymenoptera: Tenthredinidae), *Arctia caja, Hyphantria cunea* (Lepidoptera: Arctiidae), *Sitotroga cerealella* (Gelechiidae), *Bupalus piniarius* (Geometridae), *Dendrolimus pini, Malacosoma neustria* (Lasiocampidae), *Euproctis phaeorrhoea, Leucoma salicis, Orgyia antiqua, Stilpnotia salicis* (Lymantriidae), *Acronicta aceris, Agrochola circellaris, Autographa gamma, Mamestra brassicae, Noctua pronuba, Panolis flammea* (Noctuidae), *Pieris brassicae* (Pieridae), *Anagasta kuehniella, Chilo sacchariphagus, Ectomyelois ceratoniae, Ephestia sp., Ephestia elutella, Ephestia kuehniella, Galleria mellonella, Ostrinia nubilalis, Plodia interpunctella, Spectrobates ceratoniae* (Pyralidae), *Adoxophyes orana, Archips crataeganus, Archips murinana, Archips rosanus, Argyrotaenia velutinana, Cacoecia rosanus, Cacoecimorpha pronubana, Carpocapsa pomonella, Choristoneura murinana, Cydia funebrana, Cydia molesta, Cydia pomonella, Cydia strobilella, Enarmonia formosana, Epichoristodes acerbella, Eupoecilia ambiguella, Grapholita funebrana, Laspeyresia funebrana, Laspeyresia pomonella, Lobesia botrana, Pandemis heparana, Pandemis ribeana, Tetramoera schistaceana, Tortrix viridana, Zeiraphera sp., Zeiraphera canadensis* (Tortricidae), *Acrolepiopsis assectella, Plutella xylostella* (Yponomeutidae) (Noyes, 2019).
Distribution: Argentina, Austria, Belarus, Bulgaria, Cuba, Czechoslovakia, Denmark, England, Estonia, Europe, France, Germany, Greece, Iran, Italy, Kazakhstan, Kirgizia, Latvia, Lithuania, Moldova, Morocco, Netherlands, People's Republic of China, Peru, Poland,

Portugal, Russia, Spain, Switzerland, Syria, Tselinograd Obl., Tunisia, Turkey, Ukraine, United States of America, USSR, Uzbekistan, Yugoslavia (pre-1991) (Noyes, 2019).

Trichogramma dendrolimi (Matsumura, 1926)
Publications with Turkey records: Bulut and Kilincer (1991; 1993), Noyes (2019), Öncüer (1991).
Associates: *Bacillus thuringiensis* (Bacteria: Bacillaceae), *Wolbachia sp.* (Wolbachiaceae), *Anastatus japonicus* (Hymenoptera: Eupelmidae), *Trichogramma sp., Trichogramma embryophagum* (Trichogrammatidae), *Colocasia esculentum* (Araceae), *Brassica sp., Brassica oleracea* (Brassicaceae), *Humulus lupulus* (Cannabidaceae), *Dianthus caryophyllus* (Caryophyllaceae), *Beta vulgaris* (Chenopodiaceae), *Aleurites fordii* (Euphorbiaceae), *Glycine max* (Fabaceae), *Quercus sp., Quercus robur* (Fagaceae), *Allium porrum* (Liliaceae), *Gossypium sp.* (Malvaceae), *Ficus sp.* (Moraceae), *Olea europaea* (Oleaceae), *Cedrus deodara, Larix sp., Picea abies, Pinus sp., Pinus koraiensis, Pinus malus, Pinus massoniana, Pinus thunbergii* (Pinaceae), *Andropogon ischaemum, Miscanthus sacchariflorus, Oryza sativa, Phragmites communis, Phyllostachys pubescens, Triticum aestivum, Zea mays* (Poaceae), *Malus communis, Malus pumila, Prunus sp., Prunus persica, Pyrus communis* (Rosaceae), *Citrus sp.* (Rutaceae), *Populus sp., Salix sp.* (Salicaceae), *Capsicum sp., Lycopersicon esculentum* (Solanaceae), *Camellia sp.* (Theaceae), *Vitis vinifera* (Vitacea) (Noyes, 2019).
Hosts: *Zanthoxylum ailanthoides* (Rutaceae), *Deporaus betulae* (Coleoptera: Rhynchitidae), *Acantholyda posticalis* (Hymenoptera: Pamphiliidae), *Hyphantria cunea* (Lepidoptera: Arctiidae), *Plusiodonta coelonota* (Erebidae), *Sitotroga cerealella* (Gelechiidae), *Buzura suppressaria* (Geometridae), *Parnara guttata* (Hesperiidae), *Hyblaea puera* (Hyblaeidae), *Dendrolimus sp., Dendrolimus albolineatus, Dendrolimus latipennis, Dendrolimus pini, Dendrolimus punctatus, Dendrolimus sibiricus, Dendrolimus spectabilis, Dendrolimus superans, Lebeda nobilis, Malacosoma neustria* (Lasiocampidae), *Cnidocampa flavescens, Monema flavescens, Scopelodes contractus* (Limacodidae), *Dasychira abietis, Euproctis chrysorrhoea, Euproctis pseudoconspersa, Lymantria dispar, Orgyia antiqua* (Lymantriidae), *Agrotis exclamationis, Agrotis ipsilon, Agrotis segetum, Agrotis ypsilon, Anomis flava, Asota ficus, Autographa gamma, Barathra brassicae, Cocytodes coerulea, Discestra trifolii, Helicoverpa armigera, Helicoverpa assulta, Heliothis armigera, Heliothis virescens, Leucania unipuncta, Mamestra brassicae, Mamestra trifolii, Mythimna separata, Phytometra aganata, Spodoptera litura* (Noctuidae), *Clostera anastomosis, Fentonia ocypete,*

Lampronadata cristata, Micromelalopha troglodyte, Phalera assimilis (Notodontidae), *Inacbis io, Inachis io* (Nymphalidae), *Opisina arenosella* (Oecophoridae), *Papilio protenor, Papilio xuthus* (Papilionidae), *Pieris brassicae, Pieris rapae* (Pieridae), *Algedonia coclesalis, Anagasta kuehniella, Chilo partellus, Chilo sacchariphagus, Chilo suppressalis, Cnaphalocrocis medinalis, Corcyra sp., Corcyra cephalonica, Ephestia cautella, Ephestia kuehniella, Galleria mellonella, Loxostege sticticalis, Notarcha derogate, Ostrinia sp., Ostrinia furnacalis, Ostrinia nubilalis* (Pyralidae), *Actias artemis, Actias selene, Antheraea pernyi, Caligula japonica, Dictyoploca japonica, Eriogyna pyretorum, Philosamia Cynthia, Philosamia ricini* (Saturniidae), *Clanis bilineata* (Sphingidae), *Adoxophyes sp., Adoxophyes orana, Adoxophyes privatana, Archips sp., Archips rosanus, Cacoecia crataeganus, Cacoecimorpha pronubana, Capua reticulana, Carpocapsa pomonella, Cydia funebrana, Cydia pomonella, Epichoristodes acerbella, Eupoecilia ambiguella, Grapholita funebrana, Grapholita molesta, Homona coffearia, Homona magnanima, Laspeyresia pomonella, Laspeyresia strobilella, Leguminivora glycinivorella, Lobesia botrana, Pandemis heparana, Rhyacionia buoliana, Spilonota lechriaspis, Tetramoera schistaceana* (Tortricidae), *Acrolepiopsis assectella, Plutella sp., Plutella xylostella, Prays oleae* (Yponomeutidae) (Noyes, 2019).

Parasitoids: *Pachyneuron nawai, Pachyneuron solitarium* (Hymenoptera: Pteromalidae) (Noyes, 2019).

Distribution: Austria, Belarus, Bulgaria, Chile, Europe, France, Germany, Greece, Hungary, India, Iran, Italy, Japan, Kazakhstan, Korea, Latvia, Lithuania, Moldova, Netherlands, Pakistan, People's Republic of China, Poland, Romania, Russia, Taiwan, Tselinograd Obl., Turkey, Ukraine, USSR, Vietnam (Noyes, 2019).

Trichogramma embryophagum (Hartig, 1838)

Publications with Turkey records: Avci (2000), Bulut (1990), Kilinçer (1991), Bulut (1992), Ciftci et al. (1995), Noyes (2019), Öncüer (1991), Özder and Kilincer (1996), Üncal et al. (1994), Zeki et al. (1995).

Associates: *Wolbachia sp.* (Bacteria: Wolbachiaceae), *Trichogramma sp., Trichogramma dendrolimi* (Hymenoptera: Trichogrammatidae), *Chrysoperla carnea* (Neuroptera: Chrysopidae), *Phoenix dactylifera* (Arecaceae), *Brassica oleracea* (Brassicaceae), *Fagus sylvatica* (Fagaceae), *Punica granatum* (Lythraceae), *Picea sp., Picea abies, Pinus sp., Pinus brutia, Pinus halepensis, Pinus insignis, Pinus nigra, Pinus pinaster, Pinus radiata, Pinus sylvestris* (Pinaceae), *Triticum aestivum, Zea mays* (Poaceae), *Crataegus oxyacantha, Malus pumila, Prunus sp.*

(Rosaceae), *Populus sp., Salix sp., Salix caprea* (Salicaceae), *Vitis vinifera* (Vitaceae) (Noyes, 2019).

Hosts: *Acantholyda erythrocephala, Acantholyda nemoralis, Acantholyda posticalis, Cephalcia sp., Cephalcia abietis* (Hymenoptera: Pamphiliidae), *Nematus melanaspis* (Tenthredinidae), *Pectinophora gossypiella, Sitotroga cerealella* (Lepidoptera: Gelechiidae), *Bupalus piniarius, Ellopia prospiaria, Eupithecia sp., Gonodontis bidentata, Semiothisa liturata, Thera fermata, Thera obeliscata* (Geometridae), *Hyblaea puera* (Hyblaeidae), *Dendrolimus pini, Gastropacha quercifolia, Malacosoma neustria, Selenephera lunigera* (Lasiocampidae), *Dasychira pudibunda, Euproctis similis, Leucoma salicis, Orgyia antiqua* (Lymantriidae), *Acronicta tridens, Agrotis segetum, Cirrhia togata, Heliothis zea, Lithophane ornitopus, Mamestra brassicae, Panolis flammea, Sarrothripus asiatica, Sesamia cretica, Sesamia nonagrioides* (Noctuidae), *Clostera anachoreta, Clostera anastomosis, Thaumetopoea pityocampa, Thaumetopoea wilkinsoni, Traumatocampa pityocampa* (Notodontidae), *Opisina arenosella* (Oecophoridae), *Pieris brassicae, Pieris rapae* (Pieridae), *Anagasta kuehniella, Corcyra cephalonica, Ectomyelois ceratoniae, Ephestia elutella, Ephestia kuehniella, Ostrinia furnacalis, Ostrinia nubilalis, Pyrausta nubilalis, Spectrobates ceratoniae* (Pyralidae), *Antheraea pernyi* (Saturniidae), *Macroglossum stellatarum* (Sphingidae), *Adoxophyes orana, Adoxophyes reticulana, Ancylis achatina, Archips rosanus, Argyroploce nubiferana, Carpocapsa pomonella, Clysia ambiguella, Cydia pomonella, Eupoecilia ambiguella, Grapholita funebrana, Grapholita molesta, Grapholita prunifoliae, Hedya nubiferana, Laspeyresia pomonella, Lobesia botrana, Pandemis heparana, Pandemis ribeana, Ptycholoma lecheana, Rhyacionia buoliana, Spilonota ocellana, Tmetocera ocellana, Tortrix viridana, Zeiraphera diniana, Zeiraphera griseana* (Tortricidae) (Noyes, 2019).

Distribution: Albania, Algeria, Armenia, Austria, Belarus, Bulgaria, Chile, Czech Republic, Czechoslovakia, Europe, France, Georgia, Germany, Greece, Hungary, India, Iran, Israel, Italy, Kazakhstan, Kirgizia, Latvia, Moldova, Morocco, Netherlands, Norway, People's Republic of China, Poland, Portugal, Romania, Russia, Slovakia, Spain, Taiwan, Tselinograd Obl., Turkey, Turkmenistan, Ukraine, United States of America, USSR, Vietnam, Yugoslavia (pre-1991) (Noyes, 2019).

Trichogramma evanescens (Westwood, 1833)

Publications with Turkey records: Koclu and Karsavuran (1999), Noyes (2019), Öncüer (1991), Özpinar (1996), Özpinar and Kornosor (1994; 1995), Özpinar et al. (1999).

Associates: *Bacillus thuringiensis* (Bacteria: Bacillaceae), *Wolbachia sp.* (Wolbachiaceae), *Coleomegilla maculata* (Coleoptera: Coccinellidae), *Cales noacki* (Hymenoptera: Aphelinidae), *Chelonus texanus* (Braconidae), *Rhus continus, Rhus cotinus* (Anacardiaceae), *Lactuca sativa* (Asteraceae), *Alnus sp., Alnus glutinosa* (Betulaceae), *Brassica sp., Brassica oleracea* (Brassicaceae), *Cannabis sativa* (Cannabidaceae), *Beta vulgaris* (Chenopodiaceae), *Ipomoea triloba* (Convolvulaceae), *Astragalus sp., Pisum sativum, Vigna sesquipedalis* (Fabaceae), *Ribes nigrum* (Grossulariaceae), *Gladiolus sp.* (Iridaceae), *Allium porrum* (Liliaceae), *Punica granatum* (Lythraceae), *Gossypium sp.* (Malvaceae), *Olea europaea* (Oleaceae), *Larix decidua, Pinus sp.* (Pinaceae), *Andropogon ischaemum, Miscanthus sacchariflorus, Phragmites communis, Saccharum officinarum, Secale cereale, Sorghum sp., Triticum aestivum, Zea mays* (Poaceae), *Malus communis, Malus pumila, Prunus sp., Prunus persica* (Rosaceae), *Citrus limon* (Rutaceae), *Populus sp., Salix sp.* (Salicaceae), *Lycopersicon esculentum* (Solanaceae), *Tectona grandis* (Verbenaceae), *Vitis vinifera* (Vitaceae) (Noyes, 2019).
Hosts: *Brassica oleracea* (Brassicaceae), *Bruchus obtectus* (Coleoptera: Bruchidae), *Cassida deflorate,Cassida nebulosa, Cassida nobilis, Cassida vittate, Donacia simplex* (Chrysomelidae), *Rhynchaenus testaceus* (Curculionidae), *Dermestes maculatus* (Dermestidae), *Rhynchites auratus, Rhynchites betulae* (Rhynchitidae), *Tribolium castaneum* (Tenebrionidae), *Atherigona soccata, Pegomya betae, Pegomya hyoscyami* (Diptera: Anthomyiidae), *Oxycera sp., Stratiomys sp.* (Stratiomyiidae), *Melanostoma mellinum, Paragus quadrifasciatus, Syrphus sp., Syrphus balteatus, Syrphus pyrastri, Syrphus vitripennis* (Syrphidae), *Chrysops sp., Chrysops caecutiens, Chrysops relictus, Tabanus sp.* (Tabanidae), *Cimex lectularius* (Hemiptera: Cimicidae), *Acantholyda erythrocephala, Acantholyda posticalis, Acantholyda stellata, Cephalcia abietis, Cephalcia arvensis, Cephalcia signata* (Hymenoptera: Pamphiliidae), *Caliroa cerasi, Croesus septentrionalis, Emphytus tener, Pteronidea ferruginea, Pteronidea ribesii* (Tenthredinidae), *Arctia caja, Diacrisia obliqua, Eilema sp., Hyphantria cunea, Spilosoma sp. (Lepidoptera: Arctiidae), Rondotia menciana* (Bombycidae), *Danaus chrysippus* (Danaidae), *Pectinophora sp., Pectinophora gossypiella, Phthorimaea operculella, Sitotroga cerealella* (Gelechiidae), *Boarmia grisescens, Bupalus piniarius, Cidaria bilineata, Cidaria didymata, Crocallis elinguaria, Erannis defoliaria, Lambdina fiscellaria, Operophtera brumata* (Geometridae), *Anthophila atrilineata* (Glyphipterygidae), *Cosmotriche potatoria, Dendrolimus pini, Dendrolimus punctatus, Dendrolimus segregatus, Dendrolimus spectabilis, Malacosoma disstria, Malacosoma neustria*

(Lasiocampidae), *Atherix sp.* (Leptidae), *Plebejus hespericus, Thecla betulae, Virachola livia* (Lycaenidae), *Arctornis chrysorrhoea, Euproctis lunata, Euproctis phaeorrhoea, Laelia salicis, Lymantria dispar, Lymantria monachal, Notolophus antiquus, Nygmia phaeorrhoea, Orgyia antiqua, Orgyia gonostigma, Porthetria dispar, Stilpnotia salicis* (Lymantriidae), *Acontia luctuosa, Acronicta aceris, Acronicta major, Acronicta rumicis, Acronicta tridens, Agrotis exclamationis, Agrotis ipsilon, Agrotis segetum, Amathes c-nigrum, Antitype flavicincta, Apopestes spectrum, Autographa sp., Autographa gamma, Barathra brassicae, Catocala elocata, Chrysodeixis chalcites, Cirrhia gilvago, Discestra trifolii, Earias sp., Earias cupreoviridis, Earias fabia, Earias insulana, Euxoa obelisca, Euxoa segetum, Gonospileia glyphica, Helicoverpa armigera, Heliothis armigera, Heliothis virescens, Heliothis zea, Mamestra sp., Mamestra brassicae, Mamestra oleracea, Mamestra trifolii, Naranga aenescens, Oria musculosa, Panolis flammea, Parallelia algira, Phalaena typica, Phlogophora meticulosa, Phytometra gamma, Plusia gamma, Polia oleracea, Polia pisi, Polia suasa, Prodenia litura, Pyrrhia umbra, Rivula atimeta, Sarrothripus musculana, Scotia ipsilon, Sesamia cretica, Sesamia nonagrioides, Spaelotis pronubana, Spodoptera sp., Spodoptera littoralis, Spodoptera litura, Tholera popularis, Trachea triplicis, Trichoplusia ni, Triphaena pronuba* (Noctuidae), *Lampronadata cristata, Phalera bucephala, Phalera bucephaloides, Thaumetopoea pityocampa* (Notodontidae), *Nymphalis polychloros* (Nymphalidae), *Depressaria nervosa, Endrosis lactella, Opisina arenosella* (Oecophoridae), *Iphiclides podalirius, Papilio polytes* (Papilionidae), *Aporia crataegi, Artogeia rapae, Leptidea sinapis, Pieris sp., Pieris brassicae, Pieris daplidice, Pieris napi, Pieris rapae* (Pieridae), *Achroia grisella, Anagasta sp., Anagasta kuehniella, Cadra cautella, Chilo sp., Chilo Agamemnon, Chilo Indicus, Chilo infuscatellus, Chilo partellus, Chilo sacchariphagus, Chilo simplex, Chilo suppressalis, Cnaphalocrocis medinalis, Corcyra sp., Corcyra cephalonica, Crambus geniculeus, Diatraea sp., Ectomyelois ceratoniae, Ephestia sp., Ephestia calidella, Ephestia cautella, Ephestia elutella, Ephestia kuehniella, Etiella zinckenella, Evergestis forficalis, Galleria sp., Galleria mellonella, Glyphodes pyloalis, Hymenia recurvalis, Loxostege sticticalis, Marasmia patnalis, Maruca vitrata, Ostrinia furnacalis, Ostrinia nubilalis, Palpita unionalis, Phlyctaenia forficalis, Plodia interpunctella, Pyrausta machaeralis, Pyrausta nubilalis, Salebria semirubella, Spectrobates ceratoniae, Susumia exigua* (Pyralidae), *Attacus Cynthia, Attacus ricini, Philosamia cynthia* (Saturniidae), *Celerio lineata, Manduca sexta, Smerinthus populi, Sphinx pinastri* (Sphingidae), *Adoxophyes orana,*

Adoxophyes reticulana, Archips crataeganus, Archips pronubana, Archips rosanus, Cacoecia rosanus, Cacoecimorpha pronubana, Carpocapsa pomonella, Choristoneura fumiferana, Clysia ambiguella, Cnephasia longana, Cnephasia pumicana, Cydia funebrana, Cydia pomonella, Epinotia pygmaeana, Epinotia tedella, Eupoecilia ambiguella, Grapholita delineana, Grapholita funebrana, Grapholita molesta, Gypsonoma aceriana, Homona coffearia, Laspeyresia microgrammana, Laspeyresia molesta, Laspeyresia nigricana, Laspeyresia pomonella, Lobesia botrana, Pandemis chondrillana, Pandemis heparana, Petrova resinella, Polychrosis botrana, Rhopobota naevana, Rhyacionia buoliana, Sparganothis pilleriana, Tetramoera schistaceana, Zeiraphera diniana (Tortricidae), *Acrolepiopsis assectella, Argyresthia conjugella, Plutella sp., Plutella maculipennis, Plutella xylostella, Prays citri, Prays oleae* (Yponomeutidae), *Theresimima ampelophaga, Zygaena sp.* (Zygaenidae), *Sialis lutaria* (Megaloptera: Sialidae), *Chrysopa carnea, Chrysopa ventralis, Chrysoperla sp., Nothochrysa italica* (Neuroptera: Chrysopidae) (Noyes, 2019).
Parasitoid hosts: *Euphorocera edwardsii* (Diptera: Tachinidae) (Noyes, 2019).
Distribution: Armenia, Austria, Azerbaijan, Belgium, Bulgaria, Canada, Canary Islands, Chile, Comores, Cuba, Czech Republic, Czechoslovakia, Denmark, Egypt, England, Europe, France, Georgia, Germany, Hungary, India, Iran, Italy, Kazakhstan, Libya, Lithuania, Macedonia, Madagascar, Madeira, Mauritius, Moldova, Morocco, Netherlands, People's Republic of China, Philippines, Poland, Portugal, Romania, Russia, Serbia, Slovakia, Spain, Sri Lanka, Sweden, Switzerland, Tselinograd Obl., Tunisia, Turkey, Turkmenistan, Ukraine, United States of America, USSR, Uzbekistan, Vietnam, Yugoslavia (pre-1991) (Noyes, 2019).

Trichogramma kilinceri (Bulut and Kilincer, 1991)
Publications with Turkey records: Bulut and Kilinçer (1991), Bulut (1992), Noyes (2019).
Hosts: *Cydia pomonella* (Lepidoptera: Tortricidae) (Noyes, 2019).
Distribution: Turkey (Noyes, 2019).
Trichogramma pintoi Voegele, 1982
Publications with Turkey records: Koclu and Karsavuran (1999), Noyes (2019).
Associates: *Wolbachia sp.* (Bacteria: Wolbachiaceae), *Schinus molle* (Anacardiaceae), *Baccharis sp., Chrysothamnus sp., Cirsium vulgare* (Asteraceae), *Astragalus sp., Glycyrrhiza lepidota* (Fabaceae), *Gossypium sp., Malva sp.* (Malvaceae), *Azadirachta indica* (Meliaceae), *Eriogonum*

effusum (Polygonaceae), *Salix sp.* (Salicaceae), *Lycopersicon esculentum* (Solanaceae), *Lantana camara* (Verbenaceae) (Noyes, 2019).
Hosts: *Sitotroga cerealella* (Lepidoptera: Gelechiidae), *Icaricia acmon, Lycaeides melissa, Plebejus hespericus* (Lycaenidae), *Leucoma salicis* (Lymantriidae), *Actebia fennica, Agrotis segetum, Barathra brassicae, Helicoverpa armigera, Helicoverpa zea, Heliothis armigera, Plusia sp.* (Noctuidae), *Vanessa sp.* (Nymphalidae), *Platyptilia carduidactyla* (Pterophoridae), *Corcyra cephalonica, Ephestia kuehniella, Ostrinia sp., Ostrinia nubilalis* (Pyralidae), *Antheraea pernyi, Philosamia cynthia* (Saturniidae), *Carpocapsa pomonella, Cydia nigricana, Cydia pomonella* (Tortricidae), *Plutella xylostella* (Yponomeutidae), *Chrysoperla carnea* (Neuroptera: Chrysopidae) (Noyes, 2019).
Distribution: Argentina, Bulgaria, Canada, Cuba, France, India, Iran, Mexico, People's Republic of China, Peru, Portugal, Russia, Spain, Turkey, Ukraine, United States of America, USSR, Uzbekistan (Noyes, 2019).

Trichogramma turkeiense (Bulut and Kilincer, 1991)
Publications with Turkey records: Bulut and Kilincer (1991), Noyes (2019), Öncüer (1991), Özder and Kilincer (1996).
Associates: *Malus communis* (Rosaceae) (Noyes, 2019).
Hosts: *Sitotroga cerealella* (Lepidoptera: Gelechiidae), *Euproctis chrysorrhoea* (Lymantriidae), *Agrotis segetum* (Noctuidae), *Ephestia kuehniella* (Pyralidae), *Cydia pomonella* (Tortricidae) (Noyes, 2019).
Distribution: Turkey (Noyes, 2019).

Trichogramma turkestanicum (Meyer, 1940)
Publications with Turkey records: Koclu and Karsavuran (1999), Noyes (2019).
Associates: (Neuroptera: Chrysopidae), *Wolbachia sp.* (Bacteria: Wolbachiaceae), *Gossypium sp.* (Malvaceae), *Lycopersicon esculentum* (Solanaceae) (Noyes, 2019).
Hosts: *Helicoverpa armigera* (Lepidoptera: Noctuidae), *Ephestia kuehniella* (Pyralidae), *Chrysoperla carnea* (Noyes, 2019).
Distribution: Denmark, Kazakhstan, Portugal, Tselinograd Obl., Turkey (Noyes, 2019).

References

Avcı, Mustafa. 2000. "Investigations on structure of egg-batches, parasitism and egg layinh habits of *Thaumetopoea pityocampa* (Den. &

Schiff.) (Lepidoptera: Thaumetopoeidae) in various regions of Turkey". *Turkish Journal of Entomology*, 24(3):167-178.

—. 2003. "Parasitism of egg-batches of the cedar processionary moth *Traumatocampa ispartaensis* in Turkey". *Phytoparasitica*, 31(2):118-123.

Bulut, Hüseyin, and Kılınçer, Neşet. 1989. "Investigations on the species of *Trichogramma spp*. (Hym.: Trichogrammatidae), egg parasites of important Lepidoptera pests of fruit trees and their distribution in Ankara". *Bitki Koruma Bülteni*, 29(1-2):19-46.

—. 1990. "Investigations on the biology and natural effectivenes of the egg parasite, *Trichogramma dendrolimi* Matsumura (Hym.: Trichogrammatidae) of the leaf roller species (Archips spp.) in Ankara". *Bitki Koruma Bülteni*, 30(1-4):45-69.

Bulut, Hüseyin. 1990. "Investigations on the distribution and natural effectiveness of *Trichogramma embryophagum* (Hartig) found in the eggs of *Hedya nubifera* Haw. *Spilonota ocellana* F". *Proceedings of the Second Turkish National Congress of Biological Control*, pp.25-35.

—. 1992. "Investigations on the natural effeciveness of egg parasites *Trichogramma embryophagum* (Hartig) and *T. kilinceri* Kostadinov of *Cydia pomonella* L. in Ankara". *Bitki Koruma Bülteni*, 29(3-4):165-194.

Caner, Yasemin Kahveci, and Tuncer, Celal. 2001. "Effects of intercropping of corn with bean, soybean and squash on the damage of the European corn borer (*Ostrinia nubilalis* Hbn. Lepidoptera: Pyralidae) and parasitism of egg masses". *Ondokuz Mayis Universitesi, Ziraat Fakultesi Dergisi*, 16(3):55-62.

Çiftçi, Kemal, and Özkan, Ahmet, and Türkyılmaz, Naci. 1995. "Research on possibilities of biological control of apple pests in Antalya province". *Bitki Koruma Bülteni*, 35(1-2):45-61.

Doğanlar, Mikdat. 2002. "A new species of *Monorthocaeta* [sic] Blood (Hymenoptera: Chalcidoidea, Trichogrammatidae) from Adana, Türkiye". *Turkish Journal of Entomology*, 26(4):251-255.

Koçlu, Türkan, and Karsavuran, Yusuf. 1999. "Natural effectiveness of the egg parasitoids of *Helicoverpa armigera* (Hubner) (Lepidoptera: Noctuidae)". *Turkish Journal of Entomology*, 22(4):269-283.

Noyes, John S. 2019. *Universal Chalcidoidea Database*. Available at: www.nhm.ac.uk/chalcidoids (accessed January 2022).

Öncüer, Cezmi, and Uzun, Sündüs. 1991. "Investigations on the possible usefulness of *Trichogramma cacaeciae* Marchal (Hymenoptera: Trichogrammatidae) for the biological control against *Archips spp*. (Lepidoptera: Tortricidae) which are harmful on cherry trees in Izmir and Manisa Provinces". *Doğa, Türk Tarim ve Ormancilik Dergisi*, 15(2):432-444.

Öncüer, Cezmi. 1991. "A catalogue of the parasites and predators of insect pests of Turkey". *Ege Üniversitesi Ziraat Fakültesi Yayınları* [Ege University Agriculture Faculty Publ.] 505.

Özbek, Hikmet, and Çoruh, Saliha. 2010. "Egg parasitoids of *Malacosoma neustria* (Linnaeus, 1758) (Lepidoptera: Lasiocampidae) in Erzurum province of Turkey". *Turkish Journal of Entomology*, 34(4):551-560.

Özder, Nihal Aydın, and Kılınçer, Neşet. 1996. "Investigations on some biological relations between *Agrotis segetum* (Denis and Schiff.) (Lepidoptera, Noctuidae) and *Trichogramma embryophagum* (Hartig) and *T. turkeiensis* Kostadinov (Hymenoptera; Trichogrammatidae)". *Turkish Journal of Entomology*, 20(1):35-49.

Özpınar, Ali, and Kornoşor, Serpil. 1994. "The effectivenes of the released *Trichogramma evanescens* Westwood (Hymenoptera: Trichogrammatidae) on the egg population of *Ostrinia nubilalis* Hübner (Lepidoptera: Pyralidae)". *Proceedings of the Third Turkish National Congress of Biological Control*, Izmir, 25-28 January, 1994 pp.421-430 ISBN 975-483-267-6.

—. 1995. "Population development of the egg parasitoid, *Trichogramma evanescens* Westwood (Hymenoptera: Trichogrammatidae) on the eggs of *Ostrinia nubilalis* Hubner (Lepidoptera: Pyralidae) damaged". *Turkish Journal of Entomology*, 19(1):5-16.

Özpınar, Ali, and Uzun, Sündüs, and Hassan, Serif Ali. 1999. "A study on selection of the most effective of species or strains of Trichogramma for biological control of *Ostrinia nubilalis* Hubner". *Turkish Journal of Agriculture and Forestry*, 23(1):83-86.

Özpınar, Ali. 1996. "The effectiveness of *Trichogramma evanescens* Westwood (Hymenoptera, Trichogrammatidae) relesed against *Ostrinia nubilalis* Hubner (Lepidoptera, Pyralidae)". *Turkish Journal of Entomology*, 20(2):97-104. Trichogrammatidae; https://hymenoptera.ucr.edu/Trichogrammatidae.html (accessed November 2022).

Üncal, G., Kedici, Ramazan, and Melan, Kadir. 1994. "The side effects of some pesticides on *Trichogramma embryophagum* (Hartig) in the field conditions". *Proceedings of the Third Turkish National Congress of Biological Control*, Izmir, pp.441-448.

Uzun, Sündüs, and Öncüer, Cezmi. 1990. "Evaluation in respect of integrated control program in cherry orchards between the egg parasitoid *Trichogramma cacoeciae* March (Hymenoptera: Trichogrammatidae) and *Archips rosanus* L. (Lepidoptera:

Tortricidae)". *Proceedings of the Second Turkish National Congress of Biological Control*, pp.1-13.

Yiğit, Abdurrahman, and Erkılıç, Lerzan. 1987. "Studies on egg parasitoids of grape leafhopper, Arboridia adanae Dlab. (Homoptera: Cicadellidae) and their effects in the region of South Anotolia". Türkiye I. Entomoloji Kongresi Bildirileri, Ege Universitesi Atatürk Kültür Merkezi, Bornova/Izmir, Turkey, pp.35-42.

Yiğit, Abdurrahman, and Erkılıç, Lerzan. 1992. Studies on bio-ecology and control of grape leaf-hopper (*Arboridia* (=Erythroneura) *adanae* Dlab.) (Homoptera: Cicadellidae) in southern Anatolia region. Zirai Mücadele Arastirma Yilligi, Turkey (Plant Protection Research Annual), No: 22-28.

Zeki, Cevdet, and Kedici, Ramazan, and Çevik, Tuncer, and Halıcı, Semet, and Er, Hikmet. 1995. "Investigations on the effectiveness of IGR (fenoxycarb) and egg parasitoid (*Trichogramma embryophagum* Hartig) against codling moth (*Cydia pomonella* L.)". *Zirai Mücadele Araştırma Yıllığı*, No 30:19-20.

APPENDIX

THE SPECIES NAMES OF CHALCIDOIDEA FOUND IN TURKEY

Family: Agaonidae (Walker, 1846)
Blastophaga psenes (Linnaeus, 1758)
Eupristina verticillata (Waterston, 1921)

Family: Aphelinidae (Thomson, 1876)
Aphelinus asychis (Walker, 1839)
Aphelinus chaonia (Walker, 1839)
Aphelinus mali (Haldeman, 1851)
Aphelinus semiflavus (Howard, 1908)
Aphelinus varipes (Foerster, 1841)
Aphytis abnormis (Howard, 1881)
Aphytis aonidiae (Mercet, 1911)
Aphytis chilensis (Howard, 1900)
Aphytis chrysomphali (Mercet, 1912)
Aphytis diaspidis (Howard, 1881)
Aphytis hispanicus (Mercet, 1912)
Aphytis lepidosaphes (Compere, 1955)
Aphytis libanicus (Traboulsi, 1969)
Aphytis lingnanensis (Compere, 1955)
Aphytis maculicornis (Masi, 1911)
Aphytis melinus (DeBach, 1959)
Aphytis mytilaspidis (Le Baron, 1870)
Aphytis neuter (Yasnosh and Myartseva, 1971)
Aphytis proclia (Walker, 1839)
Cales noacki (Howard, 1907)
Coccobius annulicornis (Ratzeburg, 1852)
Coccophagus differens (Yasnosh, 1966)
Coccophagus insidiator (Dalman, 1826)
Coccophagus kvavadze (Japoshvili and Karaca, 2002)
Coccophagus lycimnia (Walker, 1839)

Coccophagus palaeolecanii (Yasnosh, 1957)
Coccophagus piceae (Erdos, 1956)
Coccophagus proximus (Yasnosh, 1966)
Coccophagus scutellaris (Dalman, 1826)
Coccophagus spartanus (Japoshvili and Karaca, 2002)
Diaspiniphagus moeris (Walker, 1839)
Encarsia aspidioticola (Mercet, 1929)
Encarsia berlesei (Howard, 1906)
Encarsia citrina (Craw, 1891)
Encarsia formosa (Gahan, 1924)
Encarsia inaron (Walker, 1839)
Encarsia lahorensis (Howard, 1911)
Encarsia leucaspidis (Mercet, 1912)
Encarsia lounsburyi (Berlese and Paoli, 1916)
Encarsia perniciosi (Tower, 1913)
Encarsia sophia (Girault and Dodd, 1915)
Encarsia tricolor (Foerster, 1878)
Eretmocerus debachi (Rose and Rosen, 1992)
Eretmocerus diversiciliatus (Silvestr, 1915)
Eretmocerus mundus (Mercet, 1931)
Marietta leopardina (Motschulsky, 1863)
Marietta picta (Andre, 1878)
Pteroptrix bicolor (Howard, 1898)
Pteroptrix lauri (Mercet, 1911)

Family: Azotidae (Nikolskaya and Yasnosh, 1966)
Ablerus atomon (Walker, 1847)
Ablerus celsus (Walker, 1839)
Ablerus perspeciosus (Girault, 1916)

Family: Chalcididae (Latreille, 1817)
Anachalcis major (Boucek, 1952)
Anachalcis rubra (Steffan, 1951)
Aphasganophora armeniaca (Boucek, 1952)
Belaspidia nigra (Siebold, 1856)
Belaspidia obscura (Masi, 1916)
Brachymeria femorata (Panzer, 1801)
Brachymeria inermis (Fonscolombe, 1840)
Brachymeria minuta (Linnaeus, 1767)
Brachymeria moerens (Ruschka, 1922)
Brachymeria obtusata (Foerster, 1859)

Brachymeria pseudorugosa Masi, 1951
Brachymeria rugulosa (Foerster, 1859)
Brachymeria secundaria (Ruschka, 1922)
Brachymeria tibialis (Walker, 1834)
Brachymeria vitripennis (Foerster, 1859)
Dirhinus hesperidum (Rossi, 1790)
Euchalcis magna (Boucek, 1952)
Euchalcis miegii (Dufour, 1861)
Euchalcis unicolor (Lucas, 1871)
Haltichella rufipes (Olivier, 1791)
Hockeria apani (Doğanlar, 1990)
Hockeria bifasciata (Walker, 1834)
Hockeria brachygaster (Boucek, 1956)
Hockeria karatasensis (Boucek, 1952)
Hockeria mengenillarum (Silvestri, 1943)
Hockeria unicolor (Walker, 1834)
Kriechbaumerella mansues (Nikolskaya, 1952)
Lasiochalcidia dargelasii (Latreille, 1805)
Neochalcis fertoni (Kieffer, 1899)
Neochalcis hippotoides (Masi, 1916)
Proconura nigripes (Fonscolombe, 1832)
Psilochalcis benoisti (Steffan, 1948)
Psilochalcis immaculata (Rossi, 1792)
Psilochalcis ligustica (Masi, 1929)
Psilochalcis mirabilis (Boucek, 1952)
Psilochalcis nitida (Boucek, 1952)
Psilochalcis rufitarsis (Illiger, 1807)
Psilochalcis subaenea (Masi, 1929)
Psilochalcis subarmata (Foerster, 1855)

Family: Encyrtidae (Walker, 1837)
Acerophagus malinus (Gahan, 1946)
Achalcerinys lindus (Mercet, 1921)
Adelencyrtus aulacaspidis (Brethes, 1914)
Adelencyrtus intersectus (Fonscolombe, 1832)
Aenasius arizonensis (Girault, 1915)
Ageniaspis fuscicollis (Dalman, 1820)
Anagyrus alienus (Japoshvili, 2012)
Anagyrus aligarhensis (Agarwal and Alam, 1959)
Anagyrus dactylopii (Howard, 1898)
Anagyrus descriptus (Japoshvili, 2012)

Anagyrus kilinceri (Japoshvili, 2010)
Anagyrus matritensis (Mercet, 1921)
Anagyrus pseudococci (Girault, 1915)
Anagyrus schmuttereri (Ferriere, 1955)
Anagyrus schoenherri (Westwood, 1837)
Anagyrus securicornis (Domenichini, 1953)
Anagyrus tamaricicola (Trjapitzin, 1968)
Anicetus italicus (Masi, 1917)
Aphycoides clavellatus (Dalman, 1820)
Aphycus moravicus (Hoffer, 1952)
Aphycus primus (Mercet, 1917)
Blastothrix erythrostetha (Walker, 1847)
Blastothrix gurselae (Japoshvili and Karaca, 2004)
Blastothrix hungarica (Erdos, 1959)
Blastothrix longipennis (Howard, 1881)
Blastothrix sericea (Dalman, 1820)
Blastothrix turanica (Sugonjaev, 1964)
Bothriothorax clavicornis (Dalman, 1820)
Bothriothorax serratellus (Dalman, 1820)
Callipteroma sexguttata Motschulsky, 1863
Cerapterocerus mirabilis Westwood, 1833
Cerchysiella planiscutellum (Mercet, 1921)
Cerchysius subplanus (Dalman, 1820)
Charitopus bulentyasari (Japoshvili, 2012)
Charitopus desertus (Myartseva, 1981)
Charitopus eristoi (Japoshvili, 2010)
Charitopus fulviventris (Foerster, 1860)
Charitopus ismailkaracai (Japoshvili, 2012)
Cheiloneurus claviger (Thomson, 1876)
Cheiloneurus elegans (Dalman, 1820)
Cheiloneurus paralia (Walker, 1837)
Cheiloneurus quercus (Mayr, 1876)
Choreia maculata (Hoffer, 1954)
Coelopencyrtus arenarius (Erdos, 1957)
Coelopencyrtus callidii (Jansson, 1957)
Comperiella bifasciata (Howard, 1906)
Copidosoma agrotis (Fonscolombe, 1832)
Copidosoma aithyia (Walker, 1837)
Copidosoma boucheanum (Ratzeburg, 1844)
Copidosoma breviclava (Hoffer, 1970)
Copidosoma floridanum (Ashmead, 1900)

Copidosoma hispanicum (Mercet, 1921)
Copidosoma primulum (Mercet, 1921)
Copidosoma truncatellum (Dalman, 1820)
Copidosoma varicorne (Nees, 1834)
Cowperia areolata (Walker, 1872)
Dicarnosis sugonjaevi (Myartseva, 1977)
Dicarnosis superba (Mercet, 1921)
Dinocarsis hemiptera (Dalman, 1820)
Discodes aeneus (Dalman, 1820)
Discodes coccophagus (Ratzeburg, 1848)
Discodes obscuriclavus (Myartseva, 1981)
Discodes trjapitzini (Herthevtzian, 1979)
Dusmetia ceballosi (Mercet, 1921)
Echthroplexiella aeneiventris (Mercet, 1921)
Encyrtus aurantii (Geoffroy, 1785)
Encyrtus infelix (Embleton, 1902)
Encyrtus infidus (Rossi, 1790)
Ericydnus aeneus (Nikolskaya, 1952)
Ericydnus apterogenes (Mayr, 1876)
Ericydnus caudatus (Erdos, 1957)
Ericydnus robustior (Mercet, 1921)
Ericydnus sipylus (Walker, 1837)
Eugahania fumipennis (Ratzeburg, 1852)
Ginsiana carpetana (Mercet, 1921)
Habrolepis dalmanni (Westwood, 1837)
Habrolepis italicus (Delucchi, 1965)
Homalotylus flaminius (Dalman, 1820)
Homalotylus nigricornis (Mercet, 1921)
Homalotylus quaylei (Timberlake, 1919)
Hoplopsis minuta (Fabricius, 1793)
Isodromus flaviscutum (Hoffer and Trjapitzin, 1978)
Isodromus niger (Ashmead, 1900)
Lamennaisia ambigua (Nees, 1834)
Leptomastidea abnormis (Girault, 1915)
Leptomastix algirica (Trjapitzin, 1989)
Leptomastix dactylopii (Howard, 1885)
Leptomastix ephyra (Noyes and Hayat, 1994)
Leptomastix flava (Mercet, 1921)
Leptomastix tanasijtshuki (Sharkov, 1983)
Mahencyrtus comara (Walker, 1837)
Mayrencyrtus imandes (Walker, 1837)

Mayridia formosula (Mercet, 1921)
Mayridia merceti (Trjapitzin, 1969)
Mayridia procera (Mercet, 1921)
Mayridia pulchra (Mercet, 1921)
Metanotalia maderensis (Walker, 1872)
Metaphycus asterolecanii (Mercet, 1923)
Metaphycus delos (Guerrieri and Noyes, 2000)
Metaphycus discolor (De Santis, 1970)
Metaphycus dispar (Mercet, 1925)
Metaphycus flavus (Howard, 1881)
Metaphycus garmon (Guerrieri and Noyes, 2000)
Metaphycus gennaroi (Guerrieri and Noyes, 2000)
Metaphycus helvolus (Compere, 1926)
Metaphycus ibericus (Mercet, 1921)
Metaphycus insidiosus (Mercet, 1921)
Metaphycus lounsburyi (Howard, 1898)
Metaphycus melanostomatus (Timberlake, 1916)
Metaphycus petitus (Walker, 1851)
Metaphycus piceus (Hoffer, 1954)
Metaphycus silvestrii (Sugonjaev, 1970)
Metaphycus stagnarum (Hoffer, 1954)
Metaphycus stanleyi (Compere, 1940)
Metaphycus swirskii (Annecke and Mynhardt, 1979)
Metaphycus zebratus (Mercet, 1917)
Microterys annulatus (Erdos, 1957)
Microterys bellae (Trjapitzin, 1968)
Microterys cneus (Trjapitzin and Sugonjaev, 1976)
Microterys darevskii (Trjapitzin, 1968)
Microterys golcukus (Japoshvili, 2012)
Microterys hortulanus (Erdos, 1956)
Microterys lunatus (Dalman, 1820)
Microterys masii (Silvestri, 1919)
Microterys nietneri (Motschulsky, 1859)
Microterys sylvius (Dalman, 1820)
Microterys tricoloricornis (De Stefani, 1886)
Monodiscodes dimorphus (Mercet, 1921)
Monodiscodes intermedius (Mayr, 1876)
Ooencyrtus kuvanae (Howard, 1910)
Ooencyrtus masii (Mercet, 1921)
Ooencyrtus neustriae (Mercet, 1925)
Ooencyrtus pityocampae (Mercet, 1921)

Ooencyrtus tardus (Ratzeburg, 1844)
Ooencyrtus telenomicida (Vassiliev, 1904)
Ooencyrtus zoeae (Guerrieri and Samra, 2018)
Parasauleia trjapitzini (Hoffer, 1968)
Paratetracnemoidea malenotti (Mercet, 1918)
Pentacladocerus araraticus (Trjapitzin, 1962)
Prochiloneurus bolivari (Mercet, 1919)
Prochiloneurus uyguni (Hayat, 2016)
Pseudococcobius obenbergeri (Novickij, 1926)
Psilophrys bella (Japoshvili and Noyes, 2006)
Psilophrys ghilarovi (Japoshvili, 2005)
Psilophrys parvulus (Guerrieri and Viggiani, 1991)
Psilophrys tenuicornis (Graham, 1969)
Psyllaephagus caillardiae (Sugonjaev, 1968)
Psyllaephagus elaeagni (Trjapitzin, 1967)
Psyllaephagus euphyllurae (Masi, 1911)
Psyllaephagus pistaciae (Ferriere, 1961)
Psyllaephagus procerus (Mercet, 1921)
Psyllaephagus pulchellus (Mercet, 1921)
Psyllaephagus tokgaevi (Myartseva, 1979)
Rhopus flavidus (Mercet, 1921)
Rhopus trjapitzini (Myartseva, 1982)
Tetracnemus heydeni (Mayr, 1876)
Tetracnemus terminassianae (Herthevtzian, 1978)
Trechnites flavipes (Mercet, 1921)
Trechnites insidiosus (Crawford, 1910)
Trichomasthus albimanus (Thomson, 1876)
Trichomasthus cyaneus (Dalman, 1820)
Syrphophagus aeruginosus (Dalman, 1820)
Syrphophagus aphidivorus (Mayr, 1876)
Zaomma lambinus (Walker, 1838)

Family: Eucharitidae (Walker, 1846)
Eucharis adscendens (Fabricius, 1787)
Eucharis anatolica (Boucek, 1952)
Eucharis punctata (Foerster, 1859)
Eucharis turca (Nikolskaya, 1952)
Stilbula cyniformis (Rossi, 1792)
Stilbula vitripennis (Masi, 1934)

Family: Eulophidae (Westwood, 1829)
Anaprostocetus acuminatus (Ratzeburg, 1848)
Anaprostocetus ankarensis (Gencer, 2010)
Aprostocetus agevilleae (Graham, 1987)
Aprostocetus agrus (Walker, 1839)
Aprostocetus bucculentus (Kostjukov, 1978)
Aprostocetus catius (Walker, 1839)
Aprostocetus caudatus (Westwood, 1833)
Aprostocetus ceroplastae (Girault, 1916)
Aprostocetus cracens (Graham, 1987)
Aprostocetus crino (Walker, 1838)
Aprostocetus eriophyes (Taylor, 1909)
Aprostocetus forsteri (Walker, 1847)
Aprostocetus holoxanthus (Graham, 1991)
Aprostocetus lasiopterus (Doğanlar, 2011)
Aprostocetus levadiensis (Graham, 1987)
Aprostocetus longicauda (Thomson, 1878)
Aprostocetus lycidas (Walker, 1839)
Aprostocetus neglectus (Domenichini, 1957)
Aprostocetus pausiris (Walker, 1839)
Aprostocetus productus (Graham, 1987)
Aprostocetus pygmaeus (Zetterstedt, 1838)
Aprostocetus samandagus (Doğanlar, 2011)
Aprostocetus terebrans (Erdos, 1954)
Aprostocetus toddaliae (Risbec, 1958)
Aprostocetus trjapitzini (Kostjukov, 1976)
Aprostocetus tymber (Walker, 1839)
Aulogymnus arsames (Walker, 1838)
Aulogymnus gallarum (Linnaeus, 1761)
Aulogymnus skianeuros (Ratzeburg, 1844)
Aulogymnus trilineatus (Mayr, 1877)
Baryscapus adalia (Walker, 1839)
Baryscapus babiyi (Doğanlar, 1993)
Baryscapus berhidanus (Erdos, 1954)
Baryscapus bruchidii (Erdos, 1951)
Baryscapus bruchivorus (Gahan, 1942)
Baryscapus bruchophagi (Gahan, 1913)
Baryscapus ceroplastophilus (Domenichini, 1965)
Baryscapus cirsiicola (Graham, 1991)
Baryscapus crassicornis (Erdos, 1954)
Baryscapus daira (Walker, 1839)

Baryscapus endemus (Walker, 1839)
Baryscapus euphorbiae Graham, 1991
Baryscapus evonymellae (Bouche, 1834)
Baryscapus fossarum (Graham, 1991)
Baryscapus galactopus (Ratzeburg, 1844)
Baryscapus garganus (Domenichini, 1958)
Baryscapus gaziantepensis (Doğanlar, 1993)
Baryscapus impeditus (Nees, 1834)
Baryscapus nigroviolaceus (Nees, 1834)
Baryscapus oophagus (Otten, 1942)
Baryscapus pallidae (Graham, 1991)
Baryscapus protasis (Graham, 1991)
Baryscapus servadeii (Domenichini, 1965)
Baryscapus spartifoliellae (Graham, 1991)
Baryscapus transversalis (Graham, 1991)
Baryscapus turionum (Hartig, 1838)
Ceranisus adiyamanensis (Doğanlar and Doğanlar, 2013)
Ceranisus amanosus (Doğanlar, Gumovsky and Doğanlar, 2009)
Ceranisus antalyacus (Triapitsyn, 2004)
Ceranisus atcanacus (Doğanlar and Doğanlar, 2013)
Ceranisus bozovaensis (Doğanlar, 2003)
Ceranisus filizinancae (Doğanlar and Doğanlar, 2014)
Ceranisus hirsutus (Doğanlar and Triapitsyn, 2007)
Ceranisus karacadagi (Doğanlar and Doğanlar, 2013)
Ceranisus karkamisus (Doğanlar and Doğanlar, 2013)
Ceranisus kocaki (Doğanlar and Doğanlar, 2014)
Ceranisus komurcukurus (Doğanlar and Doğanlar, 2013)
Ceranisus menes (Walker, 1839)
Ceranisus nizipus (Doğanlar and Doğanlar, 2013)
Ceranisus oezdikmeni (Doğanlar and Doğanlar, 2014)
Ceranisus oguzeliensis (Doğanlar and Doğanlar, 2013)
Ceranisus onuri (Doğanlar, 2010)
Ceranisus pacuvius (Walker, 1838)
Ceranisus planitianus (Erdos, 1966)
Ceranisus sekilinensis (Doğanlar and Doğanlar, 2013)
Ceranisus sincanus (Doğanlar and Doğanlar, 2013)
Cirrospilus elegantissimus (Westwood, 1832)
Cirrospilus ingenuus (Gahan, 1932)
Cirrospilus lyncus (Walker, 1838)
Cirrospilus pictus (Nees, 1834)
Cirrospilus staryi (Boucek, 1959)

Cirrospilus viticola (Rondani, 1877)
Cirrospilus vittatus (Walker, 1838)
Chrysocharis amanus (Walker, 1839)
Chrysocharis debussyi (Hansson, 1985)
Chrysocharis gemma (Walker, 1839)
Chrysocharis idyia (Walker, 1839)
Chrysocharis illustris (Graham, 1963)
Chrysocharis laomedon (Walker, 1839)
Chrysocharis latifossa (Hansson, 1985)
Chrysocharis liriomyzae (Delucchi, 1954)
Chrysocharis mediana (Foerster, 1861)
Chrysocharis nephereus (Walker, 1839)
Chrysocharis pentheus (Walker, 1839)
Chrysocharis phryne (Walker, 1839)
Chrysocharis polyzo (Walker, 1839)
Chrysocharis pubens (Delucchi, 1954)
Chrysocharis pubicornis (Zetterstedt, 1838)
Chrysocharis viridis (Nees, 1834)
Closterocerus chamaeleon (Girault, 1922)
Closterocerus lyonetiae (Ferriere, 1952)
Closterocerus trifasciatus (Westwood, 1833)
Crataepus marbis (Walker, 1839)
Colpoclypeus florus (Walker, 1839)
Dahlbominus fuscipennis (Zetterstedt, 1838)
Dicladocerus westwoodii (Westwood, 1832)
Diglyphus anadolucus (Doğanlar, 1982)
Diglyphus begini (Ashmead, 1904)
Diglyphus chabrias (Walker, 1838)
Diglyphus crassinervis (Erdos, 1958)
Diglyphus isaea (Walker, 1838)
Diglyphus minoeus (Walker, 1838)
Diglyphus pachyneurus (Graham, 1963)
Diglyphus poppoea (Walker, 1848)
Diglyphus pusztensis (Erdos and Novicky, 1951)
Diglyphus sabulosus (Erdos, 1951)
Diglyphus sensilis (Yefremova, 2011)
Elachertus fenestratus (Nees, 1834)
Elachertus inunctus (Nees, 1834)
Elachertus pulcher (Erdos, 1961)
Elasmus cyprianus (Ferriere, 1947)
Elasmus flabellatus (Fonscolombe, 1832)

Elasmus nudus (Nees, 1834)
Elasmus phthorimaeae (Ferriere, 1947)
Elasmus platyedrae (Ferriere, 1935)
Elasmus steffani (Viggiani, 1967)
Elasmus unicolor (Rondani, 1877)
Elasmus westwoodi (Giraud, 1856)
Entedon abdera (Walker, 1839)
Entedon adiyamanicus (Doğanlar, Doğanlar and Boyadzhiev, 2013)
Entedon akdagicus (Doğanlar and Doğanlar, 2013)
Entedon angorensis (Gumovsky, 1999)
Entedon arabanensis (Doğanlar and Doğanlar, 2013)
Entedon bakacakicus (Doğanlar, 2013)
Entedon biroi (Erdos, 1944)
Entedon calcicola (Graham, 1971)
Entedon cionobius (Thomson, 1878)
Entedon fufius (Walker, 1846)
Entedon gumushanensis (Doğanlar and Doğanlar, 2013)
Entedon hercyna (Walker, 1839)
Entedon insignis (Erdos, 1944)
Entedon lixi (Erdos, 1951)
Entedon longus (Boucek, 1968)
Entedon nevsehiricus (Doğanlar and Doğanlar, 2013)
Entedon nizipicus (Doğanlar, Doğanlar and Boyadzhiev, 2013)
Entedon pallicrus (Erdos, 1944)
Entedon parvicalcar (Thomson, 1878)
Entedon pinetorum (Ratzeburg, 1852)
Entedon procioni (Erdos, 1944)
Entedon sarkislaensis (Doğanlar and Doğanlar, 2013)
Entedon susuzensis (Doğanlar and Doğanlar, 2013)
Entedon thomsonianus (Erdos, 1944)
Euderus agrili (Boucek, 1963)
Euderus brevicornis (Boucek, 1963)
Euderus palustris (Erdos, 1951)
Eulophus larvarum (Linnaeus, 1758)
Euplectrus bicolor (Swederus, 1795)
Euplectrus ceresensis (Ferriere, 1941)
Euplectrus flavipes (Fonscolombe, 1832)
Euplectrus phthorimaeae (Ferriere, 1941)
Hemiptarsenus ornatus (Nees, 1834)
Hemiptarsenus unguicellus (Zetterstedt, 1838)
Hemiptarsenus varicornis (Girault, 1913)

Hemiptarsenus zilahisebessi (Erdos, 1951)
Hyssopus geniculatus (Hartig, 1838)
Hyssopus nigritulus (Zetterstedt, 1838)
Leptocybe invasa (Fisher and La Salle, 2004)
Melittobia acasta (Walker, 1839)
Microlycus erdoesi (Boucek, 1959)
Minotetrastichus frontalis (Nees, 1834)
Minotetrastichus platanellus (Mercet, 1922)
Miotropis unipuncta (Nees, 1834)
Necremnus artynes (Walker, 1839)
Necremnus croton (Walker, 1839)
Necremnus leucarthros (Nees, 1834)
Necremnus tidius (Walker, 1839)
Neochrysocharis aratus (Walker, 1838)
Neochrysocharis arvensis (Graham, 1963)
Neochrysocharis chlorogaster (Erdos, 1966)
Neochrysocharis clarus (Szelenyi, 1977)
Neochrysocharis cuprifrons (Erdos, 1954)
Neochrysocharis formosus (Westwood, 1833)
Neotrichoporoides beyarslani (Doğanlar, 1993)
Neotrichoporoides biogradensis (Graham, 1987)
Neotrichoporoides cavigena (Graham, 1987)
Neotrichoporoides cynodontis (Domenichini, 1967)
Neotrichoporoides mediterraneus (Graham, 1986)
Neotrichoporoides szelenyii (Erdos, 1951)
Neotrichoporoides tokatensis (Doğanlar, 1993)
Neotrichoporoides viridimaculatus (Fullaway, 1955)
Oomyzus scaposus (Thomson, 1878)
Oomyzus sempronius (Erdos, 1954)
Ophelimus maskelli (Ashmead, 1900)
Pediobius brachycerus (Thomson, 1878)
Pediobius bruchicida (Rondani, 1872)
Pediobius cassidae (Erdos, 1958)
Pediobius epigonus (Walker, 1839)
Pediobius eubius (Walker, 1839)
Pediobius flaviscapus (Thomson, 1878)
Pediobius lysis (Walker, 1839)
Pediobius metallicus (Nees, 1834)
Pediobius nigritarsis (Thomson, 1878)
Pediobius pyrgo (Walker, 1839)
Pediobius rotundatus (Fonscolombe, 1832)

Pediobius saulius (Walker, 1839)
Platyplectrus bouceki (Erdos, 1966)
Platyplectrus chlorocephalus (Nees, 1834)
Platyplectrus pannonica (Erdos, 1966)
Pnigalio agraules (Walker, 1839)
Pnigalio cristatus (Ratzeburg, 1848)
Pnigalio incompletus (Boucek, 1971)
Pnigalio longulus (Zetterstedt, 1838)
Pnigalio mediterraneus (Ferriere and Delucchi, 1957)
Pnigalio nemati (Westwood, 1838)
Pnigalio pectinicornis (Linnaeus, 1758)
Pnigalio soemius (Walker, 1839)
Puklina gelincika (Doğanlar, 1993)
Pronotalia carlinarum (Szelenyi and Erdos, 1951)
Pronotalia erzurumica (Doğanlar, 1993)
Pronotalia fiorii (Domenichini, 1958)
Pronotalia hungarica (Erdos, 1955)
Pronotalia tortumensis (Doğanlar, 1993)
Quadrastichodella nova (Girault, 1922)
Quadrastichus dasineurae (Doğanlar, LaSalle, Sertkaya and Doğanlar, 2009)
Rhicnopelte crassicornis (Nees, 1834)
Semielacher petiolata (Girault, 1915)
Sigmophora brevicornis (Panzer, 1804)
Stenomesius rufescens (Retzius, 1783)
Stepanovia eurytomae (Nees, 1834)
Stepanovia fructirosae (Boyadzhiev, Yefremova and Tozlu, 2017)
Sympiesis acalle (Walker, 1848)
Sympiesis albiflava (Fahringer, 1944)
Sympiesis euspilapterygis (Erdos, 1958)
Sympiesis flavopicta (Boucek, 1959)
Sympiesis sericeicornis (Nees, 1834)
Sympiesis turcica (Fahringer, 1944)
Sympiesis viridula (Thomson, 1878)
Tetrastichus coelarchus (Graham, 1991)
Tetrastichus decrescens (Graham, 1991)
Tetrastichus melasomae (Graham, 1991)
Tetrastichus sinope (Walker, 1839)
Wichmannia decorata (Ruschka, 1916)
Zagrammosoma talitzkii (Boucek, 1961)
Zagrammosoma variegatum (Masi, 1907)
Zeytinus hatayensis (Doğanlar, 2011)

Family: Eupelmidae (Walker, 1833)
Anastatus bifasciatus (Geoffroy, 1785)
Anastatus japonicus (Ashmead, 1904)
Calosota aestivalis (Curtis, 1836)
Calosota metallica (Gahan, 1922)
Eupelmus annulatus (Nees, 1834)
Eupelmus aseculatus (Kalina, 1981)
Eupelmus azureus (Ratzeburg, 1844)
Eupelmus impennis (Nikolskaya, 1952)
Eupelmus lanceolatus (Gibson and Fusu, 2016)
Eupelmus messene (Walker, 1839)
Eupelmus microzonus (Foerster, 1860)
Eupelmus pistaciae (Al khatib, 2015)
Eupelmus seculatus (Ferriere, 1954)
Eupelmus urozonus (Dalman, 1820)
Eupelmus vesicularis (Retzius, 1783)
Eupelmus vladimiri (Fusu, 2017)
Pentacladia hatayensis (Doğanlar, 2003)

Family: Eurytomidae (Walker, 1832)
Aximopsis augasmae (Zerova, 1977)
Aximopsis collina (Zerova, 1984)
Aximopsis nodularis (Boheman, 1836)
Bruchophagus astragali (Fedoseeva, 1954)
Bruchophagus bajariae (Erdos, 1957)
Bruchophagus coluteae (Boucek, 1954)
Bruchophagus coronillae (Erdelyi and Szelenyi, 1975)
Bruchophagus gibbus (Boheman, 1836)
Bruchophagus mutabilis (Nikolskaya, 1952)
Bruchophagus platypterus (Walker, 1834)
Bruchophagus robiniae (Zerova, 1970)
Bruchophagus roddi (Gussakovskiy, 1933)
Eurytoma adleriae (Zerova, 1995)
Eurytoma aemula (Szelenyi, 1974)
Eurytoma amygdali (Enderlein, 1907)
Eurytoma appendigaster (Swederus, 1795)
Eurytoma aspila (Walker, 1836)
Eurytoma brevicoxa (Zerova and Cam, 2003)
Eurytoma brunniventris (Ratzeburg, 1852)
Eurytoma centaureae (Claridge, 1960)
Eurytoma collaris (Walker, 1832)

Eurytoma compressa (Fabricius, 1794)
Eurytoma crambeae (Zerova, 1978)
Eurytoma curculionum (Mayr, 1878)
Eurytoma curta (Walker, 1832)
Eurytoma cypriaca (Masi, 1934)
Eurytoma dentata (Mayr, 1878)
Eurytoma erdoesi (Szelenyi, 1974)
Eurytoma flavimana (Boheman, 1836)
Eurytoma goidanichi (Boucek, 1970)
Eurytoma harmolitarum (Erdos, 1957)
Eurytoma herbaria (Zerova, 1994)
Eurytoma infracta (Mayr, 1904)
Eurytoma koeleriae (Erdos, 1969)
Eurytoma laserpitii (Mayr, 1878)
Eurytoma lobopterae (Erdos, 1964)
Eurytoma martellii (Domenichini, 1960)
Eurytoma mayri (Ashmead, 1887)
Eurytoma morio (Boheman, 1836)
Eurytoma ochraceipes (Kalina, 1970)
Eurytoma onobrychidis (Nikolskaya, 1933)
Eurytoma oophaga (Silvestri, 1920)
Eurytoma phalaridis (Graham, 1974)
Eurytoma pistaciae (Rondani, 1877)
Eurytoma plotnikovi (Nikolskaya, 1934)
Eurytoma pollux (Claridge, 1959)
Eurytoma robusta (Mayr, 1878)
Eurytoma rosae (Nees, 1834)
Eurytoma rufipes (Walker, 1832)
Eurytoma schreineri (Schreiner, 1908)
Eurytoma serratulae (Fabricius, 1798)
Eurytoma setigera (Mayr, 1878)
Eurytoma strigifrons (Thomson, 1876)
Eurytoma striolata (Ratzeburg, 1848)
Eurytoma terebinthi (Rondani, 1877)
Eurytoma tokatensis (Doğanlar and Çam, 1991)
Eurytoma turkezia (Zerova and Cam, 2003)
Eurytoma verticillata (Fabricius, 1798)
Exeurytoma anatolica (Cam, 1998)
Exeurytoma kebanensis (Doğanlar, 2005)
Macrorileya femorata (Boucek, 1952)
Macrorileya inopinata (Silvestri, 1920)

Sycophila biguttata (Swederus, 1795)
Sycophila flavicollis (Walker, 1834)
Sycophila iracemae (Nieves Aldrey, 1984)
Sycophila mayri (Erdos, 1959)
Sycophila mellea (Curtis, 1831)
Sycophila submutica (Thomson, 1876)
Sycophila variegata (Curtis, 1831)
Systole albipennis (Walker, 1832)
Systole bipunctata (Erdos, 1952)
Systole brunnicornis (Zerova and Cam, 2003)
Systole conspicua (Erdos, 1951)
Systole coriandri (Gussakovsky, 1933)
Systole minima (Zerova and Cam, 2003)
Systole plana (Zerova and Cam, 2004)
Systole salviae (Zerova, 1968)
Systole tokata (Zerova and Cam, 2004)
Tetramesa agrostidis (Howard, 1896)
Tetramesa anatolica (Zerova and Cam, 2003)
Tetramesa angustipenne (Walker, 1832)
Tetramesa brevicornis (Walker, 1832)
Tetramesa brischkei (Schlechtendal, 1891)
Tetramesa cereipes (Erdos, 1955)
Tetramesa cornuta (Walker, 1832)
Tetramesa crassicornis (Walker, 1832)
Tetramesa elongia (Zerova, 2007)
Tetramesa eximia (Giraud, 1863)
Tetramesa fulvicollis (Walker, 1832)
Tetramesa fumipennis (Walker, 1832)
Tetramesa giraudi (Schlechtendal, 1891)
Tetramesa linearis (Walker, 1832)
Tetramesa longula (Dalman, 1820)
Tetramesa maderae (Walker, 1849)
Tetramesa romana (Walker, 1873)
Tetramesa tritici (Fitch, 1859)
Rileya asiatica (Zerova, 1976)
Rileya Doğanlari (Cam, 2003)

Family Leucospidae (Walker, 1834)
Leucospis bifasciata (Klug, 1812)
Leucospis biguetina (Jurine, 1807)
Leucospis brevicauda (Fabricius, 1804)

Leucospis dorsigera (Fabricius, 1775)
Leucospis gigas (Fabricius, 1793)
Leucospis intermedia (Illiger, 1807)

Family Megastigmidae Thomson, 1876)
Bootanomyia almusiensis (Doğanlar, 1989)
Bootanomyia balikesirensis (Doğanlar, 2011)
Bootanomyia dorsalis (Fabricius, 1798)
Bootanomyia emrezaferi (Doğanlar, 2011)
Bootanomyia hepdurgunae (Doğanlar, 2011)
Bootanomyia mehmeti (Doğanlar, 2011)
Bootanomyia onuri (Doğanlar, 2011)
Bootanomyia shebnemae (Doğanlar, 2011)
Bootanomyia stigmatizans (Fabricius, 1798)
Bootanomyia synophri (Mayr, 1874)
Megastigmus juniperi (Nikolskaya, 1952)
Megastigmus leptocybus (Doğanlar and Hassan, 2013)
Megastigmus pistaciae (Walker, 1871)
Megastigmus rosae (Boucek, 1971)
Megastigmus schimitscheki (Novicky, 1954)
Megastigmus suspectus (Borries, 1895)
Megastigmus usakensis (Doğanlar and Zengin, 2018)
Megastigmus wachtli (Seitner, 1916)
Megastigmus zvimendeli (Doğanlar and Hassan, 2010)
Westralianus altinoezus (Doğanlar, 2011)

Family Mymaridae (Haliday, 1833)
Anagrus atomus (Linnaeus, 1767)
Anaphes diana (Girault, 1911)
Cosmocomoidea tremulae (Bakkendorf, 1934)
Gonatocerus fuscicornis (Walker, 1846)
Gonatocerus longicornis (Nees, 1834)
Litus cynipseus (Haliday, 1833)
Lymaenon aureus (Girault, 1911)
Lymaenon litoralis (Haliday, 1833)
Lymaenon longior (Soyka, 1946)
Lymaenon thyrides (Debauche, 1948)
Ooctonus hemipterus (Haliday, 1833)
Polynema fennicum (Soyka, 1946)
Polynema fuscipes (Haliday, 1833)
Polynema laetum (Foerster, 1847)

Polynema ovulorum (Linnaeus, 1758)
Polynema pusillum (Haliday, 1833)
Polynema tenue (Soyka, 1956)
Polynema valkenburgense (Soyka, 1931)
Stephanodes similis (Foerster, 1847)

Family: Ormyridae (Foerster, 1856)
Ormyrus ardahanensis (Doğanlar, 1991)
Ormyrus bingoeliensis (Doğanlar, 1991)
Ormyrus diffinis (Fonscolombe, 1832)
Ormyrus gratiosus (Foerster, 1860)
Ormyrus kazovaensis (Doğanlar, 1991)
Ormyrus longicornis (Boucek, 1969)
Ormyrus nitidulus (Fabricius, 1804)
Ormyrus orientalis (Walker, 1871)
Ormyrus pomaceus (Geoffroy, 1785)
Ormyrus tschami (Doğanlar, 1991)
Ormyrus wachtli (Mayr, 1904)
Ormyrus yeschilirmaka (Doğanlar, 1991)

Family: Perilampidae (Förster, 1856)
Brachyelatus viridis (Hoffer and Novicky, 1954)
Chrysolampus adiyamanus (Doğanlar and Doğanlar, 2014)
Chrysolampus akcakalenus (Doğanlar and Doğanlar, 2014)
Chrysolampus besnilus (Doğanlar and Doğanlar, 2014)
Chrysolampus dortyolus (Doğanlar and Doğanlar, 2014)
Chrysolampus gencerae (Doğanlar and Doğanlar, 2014)
Chrysolampus pazarcikus (Doğanlar and Doğanlar, 2014)
Chrysolampus splendidulus (Spinola, 1808)
Chrysolampus thenae (Walker, 1848)
Chrysolampus yayladagus (Doğanlar and Doğanlar, 2014)
Chrysomalla gulsehirica (Doğanlar and Doğanlar, 2012)
Chrysomalla nevsehirica (Doğanlar and Doğanlar, 2012)
Chrysomalla roseri (Foerster, 1859)
Chrysomalla sekilica (Doğanlar and Doğanlar, 2012)
Chrysomalla stigmatica (Boucek, 1972)
Chrysomalla turcica (Boucek, 1972)
Elatomorpha nurdagica (Doğanlar and Doğanlar, 2014)
Perilampus aeneus (Rossius, 1790)
Perilampus catilus (Argaman, 1990)

Family: Pteromalidae (Dalman, 1820)
Anisopteromalus calandrae (Howard, 1881)
Arthrolytus discoideus (Nees, 1834)
Arthrolytus geyrazensis (Doğanlar, 2018)
Arthrolytus gurderei (Doğanlar, 2018)
Arthrolytus kahvecii (Doğanlar, 2018)
Arthrolytus magarsosensis (Doğanlar, 2018)
Arthrolytus nenehatunae (Doğanlar, 2018)
Arthrolytus nuridemiragi (Doğanlar, 2018)
Arthrolytus yazicioglui (Doğanlar, 2018)
Asaphes suspensus (Nees, 1834)
Asaphes vulgaris (Walker, 1834)
Caenocrepis arenicola (Thomson, 1878)
Caenocrepis bothynoderi (Gromakov, 1940)
Catolaccus ater (Ratzeburg, 1852)
Cea pulicaris (Walker, 1837)
Cecidostiba fungosa (Geoffroy, 1785)
Cheiropachus quadrum (Fabricius, 1787)
Chlorocytus breviscapus Graham, 1965
Chlorocytus diversus (Walker, 1836)
Chlorocytus phalaridis Graham, 1965)
Chlorocytus ultonicus Graham, 1965)
Colotrechnus akifkansui (Doğanlar, 2018)
Colotrechnus bekiralkani (Doğanlar, 2018)
Colotrechnus birecikensis (Doğanlar, 2018)
Colotrechnus hasangirayi (Doğanlar, 2018)
Colotrechnus karatasensis (Doğanlar, 2018)
Colotrechnus kemaliyenensis (Doğanlar, 2018)
Colotrechnus mustafaoezeri (Doğanlar, 2018)
Colotrechnus viridis (Masi, 1921)
Colotrechnus zduzguenesae (Doğanlar, 2018)
Coelopisthia caledonica (Askew, 1980)
Conomorium patulum (Walker, 1835)
Conomorium pityocampae (Graham, 1992)
Cyrtogaster vulgaris (Walker, 1833)
Cyrtoptyx latipes (Rondani, 1874)
Cyrtoptyx robustus (Masi, 1907)
Dibrachoides dynastes (Foerster, 1841)
Dibrachys affinis (Masi, 1907)
Dibrachys microgastri (Bouche, 1834)
Dibrachys palandoekenus (Doğanlar, 1987)

Dinarmus acutus (Thomson, 1878)
Dinarmus italicus (Masi, 1922)
Dinotiscus eupterus (Walker, 1836)
Dipara claviger (Kieffer, 1906)
Erdoesina alboannulata (Ratzeburg, 1852)
Eunotus areolatus (Ratzeburg, 1852)
Gastrancistrus glabellus (Nees, 1834)
Gastrancistrus hirtulus Graham, 1969
Gastrancistrus latifrons (Thomson, 1876)
Gastrancistrus pruniflorumus (Doğanlar and Yigit, 2019)
Goidanichium atrum (Boucek, 1970)
Gugolzia bademia (Doğanlar, 2004)
Gugolzia bandirmae (Doğanlar, 2019)
Gugolzia harmolitae (Delucchi and Steffan, 1956)
Gugolzia karadagae (Doğanlar and Doğanlar, 2010)
Gugolzia melengicia (Doğanlar, 2010)
Gugolzia oezdenoerneki (Doğanlar, 2019)
Gugolzia tuzlucanensis (Doğanlar, Gozuacik and Subasi, 2018)
Halticoptera adanacus Doğanlar, 2006)
Halticoptera aenea (Walker, 1833)
Halticoptera circulus (Walker, 1833)
Halticoptera crius (Walker, 1839)
Halticoptera dimidiata (Foerster, 1841)
Halticoptera kemaliyensis (Doğanlar, 2006)
Halticoptera laevigata (Thomson, 1876)
Halticoptera longipetiolus (Hedqvist, 1975)
Halticoptera moczari (Erdos, 1954)
Halticoptera patellana (Dalman, 1818)
Halticoptera polita (Walker, 1834)
Halticoptera poreia (Walker, 1848)
Halticoptera sakipagai (Doğanlar, 2006)
Halticoptera triannulata (Erdos, 1946)
Halticoptera yoncacus (Doğanlar, 2006)
Hobbya stenonota (Ratzeburg, 1848)
Homoporus arestor (Walker, 1848)
Homoporus febriculosus (Girault, 1917)
Homoporus fulviventris (Walker, 1835)
Homoporus subniger (Walker, 1835)
Ksenoplata quadrata Boucek, 1965
Lampoterma ceutorhynchae (Doğanlar, 2018)
Lampoterma viride (Thomson, 1876)

Lampoterma yoncae (Doğanlar, 2018)
Lamprotatus truncatus (Fonscolombe, 1832)
Lariophagus distinguendus (Foerster, 1841)
Macroglenes fidanlinensis (Doğanlar, 2018)
Macroglenes varicornis (Haliday, 1833)
Merisus splendidus (Walker, 1834)
Mesopolobus aequus (Walker, 1834)
Mesopolobus amaenus (Walker, 1834)
Mesopolobus aspilus (Walker, 1835)
Mesopolobus diffinis (Walker, 1834)
Mesopolobus dubius (Walker, 1834)
Mesopolobus graminum (Hardh, 1950)
Mesopolobus incultus (Walker, 1834)
Mesopolobus laticornis (Walker, 1834)
Mesopolobus mediterraneus (Mayr, 1903)
Mesopolobus morys (Walker, 1848)
Mesopolobus nobilis (Walker, 1834)
Mesopolobus sericeus (Forster, 1770)
Mesopolobus teliformis (Walker, 1834)
Mesopolobus tibialis (Westwood, 1833)
Miscogaster elegans (Walker, 1833)
Miscogaster hortensis (Walker, 1833)
Miscogaster rufipes (Walker, 1833)
Muscidifurax adanacus (Doğanlar, 2007)
Nasonia vitripennis (Walker, 1836)
Norbanus calabrus (Masi, 1942)
Norbanus cerasiops (Masi, 1922)
Norbanus obscurus (Masi, 1922)
Notanisus oulmesiensis (Delucchi, 1962)
Odontofroggatia ishii (Wiebes, 1980)
Pachycrepoideus vindemmiae (Rondani, 1875)
Pachyneuron ahlaense (Mani and Saraswat, 1974)
Pachyneuron aphidis (Bouche, 1834)
Pachyneuron erzurumicum (Doğanlar, 1986)
Pachyneuron formosum (Walker, 1833)
Pachyneuron gibbiscuta (Thomson, 1878)
Pachyneuron groenlandicum (Holmgren, 1872)
Pachyneuron leucopiscida (Mani, 1939)
Pachyneuron muscarum (Linnaeus, 1758)
Pachyneuron nelsoni (Girault, 1928)
Philotrypesis caricae (Linnaeus, 1762)

Philotrypesis emeryi (Grandi, 1926)
Philotrypesis taiwanensis (Chen, 1999)
Platneptis laeta (Walker, 1848)
Pseudocatolaccus nitescens (Walker, 1834)
Psilocera obscura (Walker, 1833)
Psilonotus achaeus (Walker, 1848)
Psilonotus adamas (Walker, 1834)
Psilonotus hortensia (Walker, 1846)
Psychophagus omnivorus (Walker, 1835)
Pteromalus albipennis (Walker, 1835)
Pteromalus apum (Retzius, 1783)
Pteromalus bedeguaris (Thomson, 1878)
Pteromalus bifoveolatus (Foerster, 1861)
Pteromalus chrysos (Walker, 1836)
Pteromalus cionobius (Erdos, 1953)
Pteromalus elevatus (Walker, 1834)
Pteromalus intermedius (Walker, 1834)
Pteromalus platyphilus (Walker, 1874)
Pteromalus puparum (Linnaeus, 1758)
Pteromalus semotus (Walker, 1834)
Pteromalus sequester (Walker, 1835)
Oxyglypta rugosa (Ruschka, 1912)
Rhaphitelus maculatus (Walker, 1834)
Rhopalicus guttatus (Ratzeburg, 1844)
Rhopalicus tutela (Walker, 1836)
Schizonotus sieboldi (Ratzeburg, 1848)
Scutellista caerulea (Fonscolombe, 1832)
Spalangia fuscipes (Nees, 1834)
Spalangia kocakeri (Doğanlar, 2017)
Spalangia marashensis (Doğanlar, 2017)
Spalangia subpunctata (Foerster, 1850)
Sphegigaster brevicornis (Walker, 1833)
Sphegigaster pallicornis (Spinola, 1808)
Sphegigaster peduncuIiventris (Spinola, 1808)
Spintherus dubius (Nees, 1834)
Stenomalina communis (Nees, 1834)
Stenomalina favorinus (Walker, 1839)
Stenomalina laticeps (Walker, 1850)
Syntomopus incisus (Thomson, 1878)
Syntomopus incurvus (Walker, 1833)
Systasis encyrtoides (Walker, 1834)

Theocolax elegans (Westwood, 1874)
Thinodytes cyzicus (Walker, 1839)
Trichomalopsis hemiptera (Walker, 1835)
Trichomalopsis littoralis (Graham, 1969)
Trichomalopsis peregrina (Graham, 1969)
Trichomalus apertus (Walker, 1835)
Trichomalus campestris (Walker, 1834)
Trichomalus elongatus (Delucchi and Graham, 1956)
Trichomalus fulvipes (Walker, 1836)
Trichomalus helvipes (Walker, 1834)
Trichomalus nanus (Walker, 1836)
Trichomalus posticus (Walker, 1834)
Trichomalus rufinus (Walker, 1835)
Trjapitzinia leucomae (Dzhanokmen, 1975)
Tomicobia promulus (Walker, 1840)
Walkerella microcarpae (Boucek, 1993)

Family: Signiphoridae (Howard, 1894)
Chartocerus novitzkyi (Domenichini, 1955)
Chartocerus subaeneus (Foerster, 1878)
Thysanus ater (Walker, 1840)

Family: Tetracampidae (Foerster, 1856)
Epiclerus alahanensis (Doğanlar, 2003)
Epiclerus guemenekus (Doğanlar, 2003)
Epiclerus panyas (Walker, 1839)
Epiclerus tokatus (Doğanlar, 2003)
Eromocampe choresmiana (Sugonjaev, 1986)
Foersterella seyhanensis (Doğanlar, 2003)
Platynocheilus cuprifrons (Nees, 1834)
Platynocheilus karsensis Doğanlar, 2003

Family: Torymidae (Walker, 1833)
Adontomerus amygdali (Boucek, 1958)
Adontomerus impolitus (Askew and Nieves Aldrey, 1988)
Ameromicrus yumurtalikiensis (Doğanlar, 1989)
Didactyliocerus dispar (Masi, 1916)
Diomorus calcaratus (Nees, 1834)
Eridontomerus fulviventris (Erdos, 1954)
Eridontomerus golbasinensis (Doğanlar, 2016)
Eridontomerus karatayensis (Doğanlar, 2016)

Erimerus adananensis (Doğanlar, 2016)
Erimerus sanliurfanensis (Doğanlar, 2016)
Exopristoides dentatus (Boucek, 1982)
Exopristus trigonomerus (Masi, 1916)
Glyphomerus stigma (Fabricius, 1793)
Glyphomerus tibialis (Foerster, 1856)
Idarnotorymus burdurensis (O. Doğanlar and M. Doğanlar, 2008)
Idarnotorymus pulcher (Masi, 1916)
Idiomacromerus akdenizeus (Doğanlar, 1989)
Idiomacromerus aladagensis (Doğanlar, 2016)
Idiomacromerus aslihanae (Doğanlar, 2016)
Idiomacromerus curticaudatus (Szelenyi, 1981)
Idiomacromerus erzurumensis (Doğanlar, 2016)
Idiomacromerus fursovi (Doğanlar, 2016)
Idiomacromerus golbasinensis (Doğanlar, 2016)
Idiomacromerus gozuaciki Doğanlar, 2016)
Idiomacromerus gumusensis (Doğanlar, 2016)
Idiomacromerus haliti (Doğanlar, 2016)
Idiomacromerus hasandagus (Doğanlar, 2016)
Idiomacromerus karakurtensis (Doğanlar, 2016)
Idiomacromerus lutfiyeae (Doğanlar, 2016)
Idiomacromerus mbahadiri (Doğanlar, 2016)
Idiomacromerus neslihanae (Doğanlar, 2016)
Idiomacromerus nigdenensis (Doğanlar, 2016)
Idiomacromerus oguzhani (Doğanlar, 2016)
Idiomacromerus pannonicus (Ruschka, 1923)
Idiomacromerus papaveris (Foerster, 1856)
Idiomacromerus sebnemae (Doğanlar, 2016)
Idiomacromerus selimensis (Doğanlar, 2016)
Idiomacromerus turhalensis (Doğanlar, 2016)
Idiomacromerus uguranlari (Doğanlar, 2016)
Idiomacromerus yunusi (Doğanlar, 2016)
Idiomacromerus zerovaae (Doğanlar, 2016)
Idiomacromerus zeynepbanuae (Doğanlar, 2016)
Microdontomerus altinekinesis (Doğanlar, 2016)
Microdontomerus annulatus (Spinola, 1808)
Microdontomerus balikesirensis (Doğanlar, 2016)
Microdontomerus direklinensis (Doğanlar, 2016)
Microdontomerus erzurumensis (Doğanlar, 2016)
Microdontomerus gurcukoyensis (Doğanlar, 2016)
Microdontomerus kahramanmarasensis (Doğanlar and Aslan, 2018)

Microdontomerus kayserinensis (Doğanlar, 2016)
Microdontomerus kocakeri (Doğanlar, 2016)
Microdontomerus maculosacus (Doğanlar, 2016)
Microdontomerus sivasensis (Doğanlar, 2016)
Microdontomerus ulukislanensis (Doğanlar, 2016)
Microdontomerus yigiti (Doğanlar, 2016)
Monodontomerus aereus (Walker, 1834)
Monodontomerus minor (Ratzeburg, 1848)
Monodontomerus obscurus (Westwood, 1833)
Oopristus erganicus (Tarla and Doğanlar, 2010)
Oopristus tayfursokmeni (Tarla and Doğanlar, 2010)
Oopristus turkestanicus (Skriptshinsky, 1929)
Podagrion splendens (Spinola, 1811)
Podagrionella eremiaphilae (Doğanlar and Doğanlar, 2009)
Podagrionella konyaensis (Doğanlar and Doğanlar, 2009)
Pseudotorymus adananensis (Doğanlar, 2016)
Pseudotorymus aladagensis (Doğanlar, 2016)
Pseudotorymus arvernicus (Walker, 1833)
Pseudotorymus bedirlinensis (Doğanlar, 2016)
Pseudotorymus bollinensis (Askew, 2002)
Pseudotorymus celikhanensis (Doğanlar, 2016)
Pseudotorymus dogubayezitensis (Doğanlar, 2016)
Pseudotorymus haliti (Doğanlar, 2016)
Pseudotorymus hasanberkayi (Doğanlar and Aslan, 2018)
Pseudotorymus horasanensis (Doğanlar, 2016)
Pseudotorymus ispirlii (Doğanlar, 2016)
Pseudotorymus kecikuyusunensis (Doğanlar, 2016)
Pseudotorymus leguminus (Ruschka, 1923)
Pseudotorymus lutfiyeae (Doğanlar, 2016)
Pseudotorymus militaris (Boheman, 1834)
Pseudotorymus nigdenensis (Doğanlar, 2016)
Pseudotorymus ocaklii (Doğanlar, 2016)
Pseudotorymus pannonicus (Mayr, 1874)
Pseudotorymus papaveris (Thomson, 1876)
Pseudotorymus pazarcikensis (Doğanlar, 2016)
Pseudotorymus rosarum (Zerova and Seregina, 1992)
Pseudotorymus samsatensis (Doğanlar, 2016)
Pseudotorymus sapphyrinus (Fonscolombe, 1832)
Pseudotorymus sivasensis (Doğanlar, 2016)
Pseudotorymus verbasci (Erdos, 1955)
Torymus apiomyiae (Boucek and Mihajlovic, 1986)

Torymus arcella (Graham and Gijswijt, 1998)
Torymus auratus (Mueller, 1764)
Torymus basarani (Doğanlar, 2017)
Torymus bedeguaris (Linnaeus, 1758)
Torymus bingoelensis (Doğanlar, 2017)
Torymus cultratus (Graham and Gijswijt, 1998)
Torymus cultriventris (Ratzeburg, 1844)
Torymus cyaneus (Walker, 1847)
Torymus erucarum (Schrank, 1781)
Torymus fastuosus (Boheman, 1834)
Torymus flavipes (Walker, 1833)
Torymus geranii (Walker, 1833)
Torymus igniceps (Mayr, 1874)
Torymus longicalcar (Graham, 1994)
Torymus nitidulus (Walker, 1833)
Torymus phillyreae (Ruschka, 1921)
Torymus pygmaeus (Mayr, 1874)
Torymus ramicola (Ruschka, 1921)
Torymus rubi (Schrank, 1781)
Torymus triangularis (Thomson, 1876)
Torymoides kiesenwetteri (Mayr, 1874)

Family: Trichogrammatidae (Haliday, 1851)
Aphelinoidea anatolica (Novicky, 1936)
Bloodiella andalusica (Nowicki, 1935)
Chaetostricha steineri (Nowicki, 1935)
Monorthochaeta adanaensis (Doğanlar, 2002)
Monorthochaeta galatica (Nowicki, 1935)
Oligosita collina (Walker, 1851)
Oligosita foersteri (Girault, 1914)
Oligosita pallida (Kryger, 1919)
Pseudoligosita gracilior (Nowicki, 1935)
Pseudoligosita paphlagonica (Nowicki, 1935)
Pterandrophysalis levantina (Nowicki, 1935)
Trichogramma brassicae (Bezdenko, 1968)
Trichogramma buluti (Bulut and Kilincer, 1991)
Trichogramma cacaeciae (Marchal, 1927)
Trichogramma dendrolimi (Matsumura, 1926)
Trichogramma embryophagum (Hartig, 1838)
Trichogramma evanescens (Westwood, 1833)
Trichogramma kilinceri (Bulut and Kilincer, 1991)

Trichogramma pintoi (Voegele, 1982)
Trichogramma turkeiense (Bulut and Kilincer, 1991)
Trichogramma turkestanicum (Meyer, 1940)
Ufens foersteri (Kryger, 1919)
Ufens similis (Kryger, 1932)
Uscana johnstoni (Waterston, 1926)